T0211687

Lecture Notes in Computer Science 11904

More information about this series at http://www.springer.com/series/7407

Daniel Archambault · Csaba D. Tóth (Eds.)

Graph Drawing and Network Visualization

27th International Symposium, GD 2019
Prague, Czech Republic, September 17–20, 2019
Proceedings

 Springer

Editors
Daniel Archambault 🆔
Swansea University
Swansea, UK

Csaba D. Tóth 🆔
California State University, Northridge
Los Angeles, CA, USA

ISSN 0302-9743   ISSN 1611-3349 (electronic)
Lecture Notes in Computer Science
ISBN 978-3-030-35801-3   ISBN 978-3-030-35802-0 (eBook)
https://doi.org/10.1007/978-3-030-35802-0

LNCS Sublibrary: SL1 – Theoretical Computer Science and General Issues

This Springer imprint is published by the registered company Springer Nature Switzerland AG
The registered company address is: Gewerbestrasse 11, 6330 Cham, Switzerland

Preface

This volume contains the papers presented at GD 2019, the 27th International Symposium on Graph Drawing and Network Visualization, held during September 17–20, 2019, in Průhonice, near Prague. Graph drawing is concerned with the geometric representation of graphs and constitutes the algorithmic core of network visualization. Graph drawing and network visualization are motivated by applications where it is crucial to visually analyze and interact with relational datasets. Information about the conference series and past symposia is maintained at http://www.graphdrawing.org. The 2019 edition of the conference was hosted by the Charles University, with Jiří Fiala and Pavel Valtr as co-chairs of the Organizing Committee. A total of 98 participants attended the conference.

Regular papers could be submitted to one of two distinct tracks: Track 1 for papers on combinatorial and algorithmic aspects of graph drawing or Track 2 for papers on experimental, applied, and network visualization aspects. Short papers were given a separate category, which welcomed both theoretical and applied contributions. An additional track was devoted to poster submissions. All tracks were handled by a single Program Committee. In response to the call for papers, the Program Committee received a total of 113 submissions, consisting of 100 papers (56 in Track 1, 25 in Track 2, and 19 in the short paper category) and 13 posters. More than 350 single-blind reviews were provided, more than a third of which were contributed by external sub-reviewers. After extensive electronic discussions via EasyChair, the Program Committee selected 42 papers and 12 posters for inclusion in the scientific program of GD 2019. This resulted in an overall paper acceptance rate of 42% (45% in Track 1, 36% in Track 2, and 42% in the short paper category). Authors published an electronic version of their accepted papers on the arXiv e-print repository; a conference index with links to these contributions was made available before the conference.

There were three invited lectures at GD 2019, one on each day of the scientific program. John T. Stasko, from the Georgia Institute of Technology, USA, presented "Pushing the Boundaries of Interaction in Data Visualization," Bartosz Walczak, from the Jagiellonian University, Poland, talked about "Old and New Challenges in Coloring Graphs with Geometric Representations," and Giuseppe Di Battista, from the Università Roma Tre, Italy, made the case that "This is Time in/for Graph Drawing." Abstracts of all three invited lectures are included in these proceedings.

The conference gave out best paper awards in Track 1 and Track 2, as well as a best presentation award and a best poster award. As decided by a majority vote of the Program Committee, the award for the best paper in Track 1 was assigned to "Exact Crossing Number Parameterized by Vertex Cover" by Petr Hliněný and Abhisekh Sankaran, and the award for the best paper in Track 2 was assigned to "Symmetry Detection and Classification in Drawings of Graphs" by Felice De Luca, Md. Iqbal Hossain, and Stephen Kobourov. Based on a majority vote of conference participants, the best presentation award was given to Arthur van Goethem for his presentation of the paper "Optimal Morphs of Planar Orthogonal Drawings II," and the best poster award was given to "Packing Trees into 1-planar Graphs" by Felice De Luca, Emilio Di Giacomo, Seok-Hee Hong, Stephen Kobourov, William Lenhart, Giuseppe Liotta, Henk Meijer, Alessandra Tappini, and Stephen Wismath. Congratulations to all the award winners for their excellent contributions, and many thanks to Springer whose sponsorship funded the prize money for these awards.

Following the tradition, the 26th Annual Graph Drawing Contest was held during the conference. The contest was divided into two parts, creative topics and the live challenge. The creative topics featured two graphs, the Marvel Cinematic Universe graph and the Meal Ingredients graph. The live challenge focused on minimizing the number of crossings in an upward drawing on a fixed grid, and had two categories: manual and automatic. Awards were given in each of the four categories. We thank the Contest Committee, chaired by Philipp Kindermann, for preparing interesting and challenging contest problems. A report about the contest is included in these proceedings.

Many people and organizations contributed to the success of GD 2019. We would like to thank all members of the Program Committee and the external reviewers for carefully reviewing and discussing the submitted papers and posters; this was crucial for putting together a strong and interesting program. Thanks to all authors who chose GD 2019 as the publication venue for their research. We are grateful for the support of the "gold" sponsors Avast, RSJ, Tom Sawyer Software, Unicorn, and yWorks, the "bronze" sponsor Springer, and contributor Znovín Znojmo. Their generosity helped make this symposium a memorable event for all participants. Last but not least, we would like to express our appreciation of the organizing team: all members of the Organizing Committee, Martin Balko, Jiří Fiala, Anna Kotěšovcová, and Pavel Valtr, as well as all student volunteers, Jaroslav Hančl, Radek Hušek, Tomáš Masařík, Jana Novotná, Michael Skotnica, Jana Syrovátková, Aneta Šťastná, and Peter Zeman.

The 28th International Symposium on Graph Drawing and Network Visualization (GD 2020) will take place during September 16–18, 2020, in Vancouver, BC, Canada. David Auber and Pavel Valtr will co-chair the Program Committee, and Will Evans will chair the Organizing Committee.

October 2019

Daniel Archambault
Csaba D. Tóth

Organization

Steering Committee

Daniel Archambault	Swansea University, UK
David Auber	LaBRI, Université Bordeaux I, France
Therese Biedl	University of Waterloo, Canada
Giuseppe Di Battista	Università Roma Tre, Italy
Walter Didimo	University of Perugia, Italy
Andreas Kerren	Linnaeus University, Sweden
Stephen G. Kobourov (Chair)	University of Arizona, USA
Martin Nöllenburg	Technische Universität Wien, Austria
Roberto Tamassia	Brown University, USA
Ioannis G. Tollis	University of Crete, Greece, and Tom Sawyer Software, USA
Csaba D. Tóth	California State University, Northridge, USA
Pavel Valtr	Charles University, Czech Republic

Program Committee

Daniel Archambault (Co-chair)	Swansea University, UK
David Auber	LaBRI, Université Bordeaux 1, France
Benjamin Bach	The University of Edinburgh, UK
Fabian Beck	University of Duisburg-Essen, Germany
Michael Bekos	University of Tübingen, Germany
Prosenjit Bose	Carleton University, Canada
Maike Buchin	Ruhr University Bochum, Germany
Nan Cao	Tongji University, China
Giordano Da Lozzo	Università Roma Tre, Italy
Emilio Di Giacomo	University of Perugia, Italy
Tim Dwyer	Monash University, Australia
David Eppstein	University of California, Irvine, USA
Yifan Hu	Yahoo! Research, USA
Irina Kostitsyna	TU Eindhoven, The Netherlands
Jan Kynčl	Charles University, Czech Republic
Anna Lubiw	University of Waterloo, Canada
Maarten Löffler	Utrecht University, The Netherlands
Kwan-Liu Ma	University of California, Davis, USA
Silvia Miksch	Technische Universität Wien, Austria
Kazuo Misue	University of Tsukuba, Japan
Helen Purchase	University of Glasgow, UK

Ignaz Rutter   Universität Passau, Germany
Alexandru Telea   University of Groningen, The Netherlands
Csaba D. Tóth (Co-chair)   California State University, Northridge, USA
Torsten Ueckerdt   Karlsruhe Institute of Technology, Germany
Birgit Vogtenhuber   Graz University of Technology, Austria
Hsiang-Yun Wu   Technische Universität Wien, Austria

Organizing Committee

Martin Balko   Charles University, Czech Republic
Jiří Fiala (Co-chair)   Charles University, Czech Republic
Anna Kotěšovcová   Conforg Ltd, Czech Republic
Pavel Valtr (Co-chair)   Charles University, Czech Republic

Contest Committee

Philipp Kindermann (Chair)   Universität Würzburg, Germany
Tamara Mchedlidze   Karlsruhe Institute of Technology, Germany
Ignaz Rutter   Universität Passau, Germany

External Reviewers

Ábrego, Bernardo   Crnovrsanin, Tarik
Ackerman, Eyal   D'Angelo, Anthony
Aerts, Nieke   De Luca, Felice
Aichholzer, Oswin   Didimo, Walter
Akopyan, Arseniy   Dujmović, Vida
Almeida Leite, Roger   Felsner, Stefan
Angelini, Patrizio   Frati, Fabrizio
Arroyo, Alan   Fulek, Radoslav
Balko, Martin   van Garderen, Mereke
Bhore, Sujoy   Gonçalves, Daniel
Biniaz, Ahmad   Goodwin, Sarah
Binucci, Carla   Grastien, Ban
Bläsius, Thomas   Grelier, Nicolas
Borrazzo, Manuel   Grilli, Luca
Brückner, Guido   Gronemann, Martin
Buchin, Kevin   Gschwandtner, Theresia
Bärtschi, Andreas   Gupta, Siddharth
Cano, Pilar   Hidalgo-Toscano, Carlos
Cardinal, Jean   Hill, Darryl
Carrière, Mathieu   Hoffmann, Michael
Chaplick, Steven   van der Hoog, Ivor
Chimani, Markus   Isaacs, Kate
Cibulka, Josef   Itoh, Masahiko
Cornelsen, Sabine   Kaaser, Dominik

Kaufmann, Michael
Keszegh, Balázs
Kilgus, Bernhard
Kindermann, Philipp
Kleist, Linda
Klemz, Boris
Klute, Fabian
van Kreveld, Marc
Kriegel, Klaus
Kryven, Myroslav
Kwon, Oh-Hyun
Langerman, Stefan
Lhuillier, Antoine
Li, Guangping
Maheshwari, Anil
McGee, Fintan
Micek, Piotr
Miltzow, Till
Mondal, Debajyoti
Montecchiani, Fabrizio
Morin, Pat
Mulzer, Wolfgang
Mütze, Torsten
Niedermann, Benjamin
Okamoto, Yoshio
Onoue, Yosuke

Ortali, Giacomo
Palfrader, Peter
Parada, Irene
Patrignani, Maurizio
Pergel, Martin
Pupyrev, Sergey
Radermacher, Marcel
Ravsky, Alexander
van Renssen, André
Roy, Sasanka
Ryvkin, Leonie
Sallaberry, Arnaud
Schetinger, Victor
Schnider, Patrick
Schröder, Felix
Schulz, André
Selbach, Leonie
Stumpf, Peter
T. P., Sandhya
Tappini, Alessandra
Tóth, Géza
Verbeek, Kevin
Wakita, Ken
Wood, David R.
Wybrow, Michael

Sponsors

Gold Sponsors

Bronze Sponsor

Contributors

Invited Lectures

Pushing the Boundaries of Interaction in Data Visualization

John T. Stasko

Georgia Institute of Technology, Atlanta, GA, USA
stasko@cc.gatech.edu

Abstract. People use data visualization for two main purposes: communication and analysis. On the analysis side, when the data being examined is of modest size or larger, it is difficult to imagine an effective visualization system without interaction. In this talk, I'll outline the value and uses of interaction for visualization, focusing on recent challenges and opportunities that have arisen. For example, what are good ways to interact with a visualization on a small screen without a mouse and keyboard present? And how can multimodal input, including speech and touch, assist people's interactions with visualizations? To answer these questions, I'll show examples of recent visualization projects, with a specific emphasis on graph and network visualizations.

Old and New Challenges in Coloring Graphs with Geometric Representations

Bartosz Walczak

Jagiellonian University, Kraków, Poland
walczak@tcs.uj.edu.pl

Abstract. A central problem in graph theory is to compute or estimate the *chromatic number* of a graph, i.e., the minimum number of colors to be put on the vertices so that no two neighbors receive the same color. Being very hard in general, it has been considered for various restricted classes of graphs, in which the chromatic number remains in a tighter connection to the structure of the graph. This includes, in particular, classes of graphs defined on families of geometric objects: intersection graphs, disjointness graphs, visibility graphs, etc., motivated by practical applications in resource allocation, map labeling, and VLSI design. This area of research has seen remarkable progress in recent years. In particular, we have already quite a good understanding of which classes of graphs (with geometric representations) allow the chromatic number to be bounded by a function of the maximum size of a clique and which do not. Much less is known about the growth of these bounding functions, for instance, whether the chromatic number can be bounded by a *polynomial* of the size of the maximum clique.

The goal of this talk is to familiarize the audience with classical and new problems in coloring graphs with geometric representations, and to present some of the most recent developments, including a quadratic bound on the chromatic number in terms of the maximum clique size for circle graphs (intersection graphs of chords of a circle), due to Davies and McCarty.

This Is Time in/for Graph Drawing

Giuseppe Di Battista

Università Roma Tre, Rome, Italy
`giuseppe.dibattista@uniroma3.it`

Abstract. In all fields of science and technology graph-inspired models are used to represent and understand reality, and the effectiveness of such models is often related to their graphical representation. This motivates the birth and the development of Graph Drawing as a self-standing scientific discipline.

During its evolution, lasting about half a century, Graph Drawing successfully faced several challenges, in some cases originated by the requirements of the reality to be represented, and in some cases, motivated by deep theoretical questions. This happened at the meeting point of the fields whose combination is the core of Graph Drawing, namely, Algorithmics, Computational Geometry, Graph Theory and Combinatorics, and Information Visualization (in alphabetical order).

One of the main challenges for Graph Drawing is the relationship between drawings and time (i.e., the temporal evolution of the visualized graphs). This relationship has been the subject of studies throughout the entire history of the discipline, as it is witnessed by the presence of about 40 papers on this topic in the Graph Drawing Conference Proceedings. On the other hand, this challenge inspired an even larger body of literature in the Information Visualization field. For several reasons, this literature has grown in a way that is largely independent from the Graph Drawing one.

We will discuss the main methods and techniques that the Graph Drawing community devised to deal with time, emphasizing their algorithmic, combinatorial, and geometric aspects, and considering their practical applicability to Information Visualization. We will focus on dynamic algorithms, streaming, animation, and morphing.

This research was supported in part by MIUR Project "AHeAD" under PRIN 20174LF3T8.

Contents

Clustering

Quality Metrics

Arrangements

A Low Number of Crossings

Best Paper in Track 1

Morphing and Planarity

Parameterized Complexity

Collinearities

Topological Graph Theory

Best Paper in Track 2

Level Planarity

Graph Drawing Contest Report

Poster Abstracts

Cartograms and Intersection Graphs

Stick Graphs with Length Constraints

Steven Chaplickⓘ, Philipp Kindermannⓘ, Andre Löffler, Florian Thiele,
Alexander Wolffⓘ, Alexander Zaft, and Johannes Zink(✉)ⓘ

Institut für Informatik, Universität Würzburg, Würzburg, Germany
zink@informatik.uni-wuerzburg.de

Abstract. Stick graphs are intersection graphs of horizontal and vertical line segments that all touch a line of slope -1 and lie above this line. De Luca et al. [GD'18] considered the recognition problem of stick graphs when no order is given (STICK), when the order of either one of the two sets is given (STICK$_A$), and when the order of both sets is given (STICK$_{AB}$). They showed how to solve STICK$_{AB}$ efficiently.

In this paper, we improve the running time of their algorithm, and we solve STICK$_A$ efficiently. Further, we consider variants of these problems where the lengths of the sticks are given as input. We show that these variants of STICK, STICK$_A$, and STICK$_{AB}$ are all NP-complete. On the positive side, we give an efficient solution for STICK$_{AB}$ with fixed stick lengths if there are no isolated vertices.

1 Introduction

For a given collection $\mathcal{S}$ of geometric objects, the *intersection graph of $\mathcal{S}$* has $\mathcal{S}$ as its vertex set and an edge whenever $S \cap S' \neq \emptyset$, for $S, S' \in \mathcal{S}$. This paper concerns *recognition* problems for classes of intersection graphs of restricted geometric objects, i.e., determining whether a given graph is an intersection graph of a family of restricted sets of geometric objects. A classic (general) class of intersection graphs is that of *segment graphs*, the intersection graphs of line segments in the plane[1]. For example, segment graphs are known to include planar graphs [4]. The recognition problem for segment graphs is $\exists\mathbb{R}$-complete[2] [18,22]. On the other hand, one of the simplest natural subclasses of segment graphs is that of the *permutation* graphs, the intersection graphs of line segments where there are two parallel lines such that each line segment has its two end points on these parallel lines[3], we say that the segments are *grounded* on these two

---

[1] We follow the common convention that parallel segments do not intersect and each point in the plane belongs to at most two segments.

[2] Note that $\exists\mathbb{R}$ includes NP, see [22,24] for background on the complexity class $\exists\mathbb{R}$.

[3] i.e., we think of the sequence of end points on the "bottom" line as one permutation π on the vertices and the sequence on the top line as another permutation π', where uv is an edge if and only if the order of u and v differs in π and π'.

The full version of this article is available at ArXiv [8]. S.C. and A.W. acknowledge support from DFG grants WO 758/11-1 and WO 758/9-1, respectively.

D. Archambault and C. D. Tóth (Eds.): GD 2019, LNCS 11904, pp. 3–17, 2019.
https://doi.org/10.1007/978-3-030-35802-0_1

lines. The recognition problem for permutation graphs can be solved in linear time [19]. *Bipartite* permutation graphs have an even simpler intersection representation [25]: they are the intersection graphs of unit-length vertical and horizontal line segments which are again double-grounded (without loss of generality both lines have a slope of -1). The simplicity of bipartite permutation graphs leads to a simpler linear-time recognition algorithm [27] than that of permutation graphs.

Several recent articles [1,2,6,7] compare and study the geometric intersection graph classes occurring between the simple classes, such as bipartite permutation graphs, and the general classes, such a segment graphs. Cabello and Jejčič [1] mention that studying such classes with constraints on the sizes or lengths of the objects is an interesting direction for future work (and such constraints are the focus of our work). Note that similar length restrictions have been considered for other geometric intersection graphs such as interval graphs [15,16,23].

When the segments are not grounded, but still are only horizontal and vertical, the class is referred to as *grid intersection graphs* and it also has a rich history, see, e.g., [6,7,13,17]. In particular, note that the recognition problem is NP-complete for grid intersection graphs [17]. But, if both the permutation of the vertical segments and the permutation of the horizontal segments are given, then the problem becomes a trivial check on the bipartite adjacency matrix [17]. However, for the variant where only one such permutation, e.g., the order of the horizontal segments, is given, the complexity remains open. A few special cases of this problem have been solved efficiently [5,9,10], e.g., one such case [5] is equivalent to the problem of *level planarity testing* which can be solved in linear time [14].

In this paper we study recognition problems concerning so-called *stick* graphs, the intersection graphs of grounded vertical and horizontal line segments (i.e., grounded grid intersection graphs). Classes closely related to stick graphs appear in several application contexts, e.g., in *nano PLA-design* [26] and detecting *loss of heterozygosity events in the human genome* [3,12]. Note that, similar to the general case of segment graphs, it was recently shown that the recognition problem for grounded segments (where arbitrary slopes are allowed) is $\exists\mathbb{R}$-complete [2]. So, it seems likely that the recognition problem for stick graphs is NP-complete (similar to grid intersection graphs), but thus far it remains open. The primary prior work on recognizing stick graphs is due to De Luca et al. [9]. Similarly to Kratochvíl's approach to grid intersection graphs [17], De Luca et al. characterized stick graphs through their bipartite adjacency matrix and used this result as a basis to develop polynomial-time algorithms to solve two constrained cases of the stick graph recognition problem called STICK$_A$ and STICK$_{AB}$, defined next. However, their algorithm for STICK$_A$ is incorrect [21], leaving STICK$_A$ open.

Definition 1 (STICK). *Let G be a bipartite graph with vertex set $A \dot\cup B$, and let ℓ be a line with slope -1. Decide whether G has an intersection representation where the vertices in A are vertical line segments[4] whose bottom end-points*

---

[4] Note that De Luca et al. [9] regarded A as horizontal segments.

Table 1. Previously known and new results for deciding whether a given bipartite graph $G = (A \dot\cup B, E)$ is a stick graph. In $O(\cdot)$, we dropped $|\cdot|$. NPC means NP-complete.

Given order	Variable length		Fixed length			
	Old results	New results	Isolated vertices		No isolated vertices	
$\emptyset$	Unknown	Unknown	NPC	[Theorem 3]	NPC	[Theorem 3]
A	Unknown	$O(AB)$ [Theorem 2]	NPC	[Theorem 4]	NPC	[Theorem 4]
A,B	$O(AB)$ [9]	$O(E)$ [Theorem 1]	NPC	[Corollary 2]	$O((A+B)^2)$	[Corollary 3]

lie on ℓ and the vertices in B are horizontal line segments whose left end-points lie on ℓ. Such a representation is a stick representation of G, the line ℓ is the ground line, the segments are called sticks, and the point where a stick meets ℓ is its foot point.

Definition 2 (STICK$_A$/STICK$_{AB}$). *In the problem STICK$_A$ (STICK$_{AB}$) we are given an instance of the STICK problem and additionally an order σ_A (orders σ_A, σ_B) of the vertices in A (in A and B). The task is to decide whether there is a stick representation that respects σ_A (σ_A and σ_B).*

Our Contribution. We first revisit the problems STICK$_A$ and STICK$_{AB}$ defined by De Luca et al. [9]. We provide the first efficient algorithm for STICK$_A$ and a faster algorithm for STICK$_{AB}$; see Sect. 2. Then we investigate the direction suggested by Cabello and Jejčič [1] where specific lengths are given for the segments of each vertex. In particular, this can be thought of as generalizing from unit stick graphs (i.e., bipartite permutation graphs), where every segment has the same length. While bipartite permutation graphs can be recognized in linear time [27], it turns out that all of the new problem variants (which we call STICK^{fix}, STICK$_A^{fix}$, and STICK$_{AB}^{fix}$) are NP-complete; see Sect. 3. Finally, we give an efficient solution for STICK$_{AB}^{fix}$ (that is, STICK$_{AB}$ with fixed stick lengths) for the special case that there are no isolated vertices (see Sect. 3.3). We conclude and state some open problems in Sect. 4. Our results are summarized in Table 1.

2   Sticks of Variable Lengths

In this section, we provide algorithms for the STICK$_A$ problem in $O(|A||B|)$ time (Theorem 2) and the STICK$_{AB}$ problem in $O(|A| + |B| + |E|)$ time (Theorem 1). Both algorithms apply a sweep-line approach (with a vertical sweep-line moving rightwards) where each vertical stick $a_i \in A$ corresponds to two events: the *enter event* of a_i (abbreviated by i) and the *exit event* of a_i (abbreviated by $i{\rightarrow}$).

Theorem 1. *STICK$_{AB}$ can be solved in $O(|A| + |B| + |E|)$ time.*

Proof. Let $\sigma_A = (a_1, \ldots, a_{|A|})$ and $\sigma_B = (b_1, \ldots, b_{|B|})$. Let β_i denote the largest index such that b_{β_i} has a neighbor in $a_1, \ldots, a_i$. Let $\hat{B}^i$ be the elements of $(b_1, \ldots, b_{\beta_i})$ that have a neighbor in $a_i, \ldots, a_{|A|}$ ordered by σ_B, and let $\hat{B}^{i{\rightarrow}}$ be the elements of $(b_1, \ldots, b_{\beta_i})$ that have a neighbor in $a_{i+1}, \ldots, a_{|A|}$. At every event

$p \in \{i, i{\rightarrow}\}$, we maintain the invariants that (i) we have a valid representation of the subgraph of G induced by $b_1, \ldots, b_{\beta_i}, a_1, \ldots, a_i$; (ii) for all these vertices, their foot points are set as consecutive integers from 1 to $\beta_i + i$; and (iii) for those not in $\hat{B}^p$, their lengths are set.

Consider the enter event of a_i. We place a_i at position $\beta_i + i$. We place the vertices $b_{\beta_{i-1}+1}, \ldots b_{\beta_i}$ (if they exist) between a_{i-1} and a_i in this order and create $\hat{B}^i$ by appending them to $\hat{B}^{(i-1){\rightarrow}}$ in this order. All neighbors of a_i have to be before a_i, and they have to be a suffix of $\hat{B}^i$. This is easily checked in $\deg(a_i)$ time. The end point of a_i is placed directly above the foot point of its first neighbor in this suffix. As such, the invariants (i)–(iii) are maintained.

Consider the exit event of a_i and each neighbor b_j of a_i. If a_i is the last neighbor of b_j in σ_A, then we end b_j and set its endpoint at $\beta_i + i + 1/2$. We create $\hat{B}^{i{\rightarrow}}$ by removing each such b_j from $\hat{B}^i$. This again maintains invariants (i)–(iii). Hence, if we complete the exit event of $a_{|A|}$, we obtain a $\mathsf{STICK_{AB}}$ representation of G. Otherwise, G has no such representation. Clearly, the whole algorithm works in $O(|A| + |B| + |E|)$ time. Note that, even though we have not explicitly discussed isolated vertices, these are easily handled with length 0.   □

Theorem 2. $\mathsf{STICK_A}$ *can be solved in* $O(|A| \cdot |B|)$ *time.*

Proof. We assume that G is connected and discuss otherwise in the full version [8].

Overview. For each event $p \in \{i, i{\rightarrow}\}$, we maintain a data structure T^p that compactly encodes all *realizable* permutations of certain horizontal sticks $B^p \subseteq B$. Namely, each B^i (resp. $B^{i{\rightarrow}}$) consists of all sticks of B with a neighbor in $a_1, \ldots, a_i$ and a neighbor in $a_i, \ldots, a_{|A|}$ (resp. $a_{i+1}, \ldots, a_{|A|}$). We denote by G^p the induced subgraph of G containing $a_1, \ldots, a_i$ and their neighbors. A permutation π of B^p is realizable if there is a stick representation of the graph obtained from G^p by adding a vertical stick to the right of a_i neighboring all horizontal sticks in B^p where B^p is drawn top-to-bottom in order π. In the enter event of a_i, we add to the data structure all the vertices of B that neighbor a_i and aren't in the data structure yet (we call these *entering vertices*), and constrain the data structure so that all the neighbors of a_i must occur after (below) the non-neighbors of a_i. In the exit event of a_i, we remove all sticks of B that do not have any neighbor a_j with $j > i$, i.e., they have a_i as their last neighbor (we call these *leaving vertices*).

Data Structure. See Fig. 1 for an example. Consider any event p. Observe that G^p may consist of several connected components $G_1^p, \ldots, G_{k_p}^p$. Since G is connected, the components are naturally ordered from left to right by σ_A. Let B_j^p denote the vertices of B^p in G_j^p. In this case, in every realizable permutation of B^p, the vertices of B_j^p must come before the vertices of B_{j+1}^p. Furthermore, the vertices that will be introduced any time later can only be placed at the beginning, end, or between the components. Hence, to compactly encode the realizable permutations, it suffices to do so for each component G_j^p individually via a data structure T_j^p. Namely, our data structure will be $T^p = (T_1^p, \ldots, T_{k_p}^p)$.

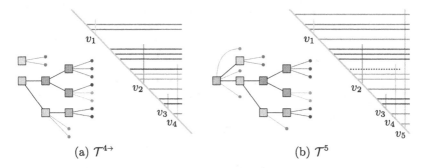

(a) $\mathcal{T}^{4\rightarrow}$ (b) $\mathcal{T}^5$

Fig. 1. An example for the data structure. In (b), the dotted stick has left the data structure and the leaves are permuted among the children to match the representation.

Each data structure T_j^p is a rooted tree. At each node, its children consist of two types: the leaves (which correspond to the vertices of B_j^p) and the non-leaves. The non-leaves are ordered, while the leaves are unordered and can be placed anywhere before, after, or between the non-leaves with the same parent. A *valid* traversal of T_j^p is a pre-order traversal where, for each node, the non-leaf children are visited in the specified order and the leaves are permuted among the non-leaf children. Each permutation *expressed* by T_j^p corresponds to a valid traversal. Note that the non-leaves are visited in the same order in every valid traversal.

Correctness and Event Processing. We will argue that this data structure is sufficient to express the realizable permutations of B^p by induction. In the base case, consider the enter event of a_1. Our data structure consists of a single component G_1^1 and clearly a single node with a leaf-child for every neighbor of a_1 captures all possible permutations.

Consider the exit event of a_i and assume that we have the data structure $\mathcal{T}^i = (T_1^i, \ldots, T_{k_i}^i)$. If there are no leaving vertices, we just keep the data structures and are done. Otherwise, $B^{i\rightarrow}$ is a strict subset of B^i. We delete all leaves from $\mathcal{T}^i$ corresponding to leaving vertices. If this results in any non-leaf node having only one child and that child is not a leaf, we merge it with its parent. If all children of an internal node get removed, we also remove the node. Obviously, this procedure maintains all realizable permutations of $B^{i\rightarrow}$ due to $G^{i\rightarrow}$.

Now consider the enter event of a_i and assume that we have the data structure $\mathcal{T}^{(i-1)\rightarrow} = (T_1^{(i-1)\rightarrow}, \ldots, T_{k_{i-1}}^{(i-1)\rightarrow})$. The essential observation is that the neighbors of a_i must form a suffix of $B^{(i-1)\rightarrow}$ in every realizable permutation after the enter event, which we will enforce in the following. Namely, either

- all vertices in $B^{(i-1)\rightarrow}$ are adjacent to a_i,
- none of them are adjacent to a_i, or
- there is an s such that (i) $B_s^{(i-1)\rightarrow}$ contains at least one neighbor of a_i; (ii) all vertices in $B_{s+1}^{(i-1)\rightarrow}, \ldots, B_{k_{i-1}}^{(i-1)\rightarrow}$ are neighbors of a_i; and (iii) no vertices in $B_1^{(i-1)\rightarrow}, \ldots, B_{s-1}^{(i-1)\rightarrow}$ are adjacent to a_i; see Fig. 2a.

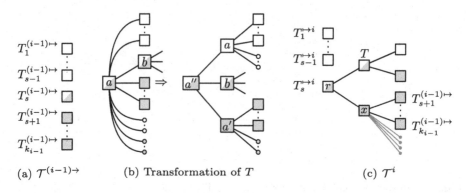

(a) $\mathcal{T}^{(i-1)\rightarrow}$ (b) **Transformation of T**   (c) $\mathcal{T}^i$

Fig. 2. Construction of $\mathcal{T}^i$. The leaves at the new node x are the entering vertices.

Otherwise, there is no realizable permutation for this event and consequently for G. The first two cases can be seen as degenerate cases (with $s = 0$ or $s = k_{i-1} + 1$) of the general case below.

We first show how to process $T_s^{(i-1)\rightarrow}$; see Fig. 2b. After that we will create the data structure $\mathcal{T}^i$. We create a tree T whose realizable permutations are precisely the subset of those of $T_s^{(i-1)\rightarrow}$ where all leaves that are neighbors of a_i occur as a suffix. We initialize $T = T_s^{(i-1)\rightarrow}$. If all vertices in $B_s^{(i-1)\rightarrow}$ are neighbors of a_i, then we are already done.

Otherwise, we define a *marked* node as one where all leaves in its subtree are neighbors of a_i; an *unmarked* node as one where no leaf in its induced subtree is a neighbor of a_i; and a node is *half-marked* otherwise. Note that the root is half-marked. Since the neighbors of a_i must form a suffix, the marked non-leaf children of a half-marked node form a suffix, the unmarked non-leaf children form a prefix, and there is at most one half-mark child. Hence, the half-marked nodes form a path in T that starts in the root; otherwise, there are no realizable permutations for this event and subsequently for G.

We traverse the path leaf-to-root. Let a be a half-marked node, and let b be its half-marked child (if it exists). We have to enforce that in any valid traversal of T the unmarked children of a are visited before b and the marked children of a are visited after b. We create a new (marked) vertex a' and move all marked children of a to a', preserving the order among the non-leaf children. Then we create a new (half-marked) node a'' and we hang a, b, and a' from a'' in this order. Finally, we put a'' into the former position of a in T. If this results in any internal node z with no leaf-children and only one child, we merge z with its parent. This ensures that all permutations realized by T have the neighbors of a_i as a suffix. Further, observe that the non-leaves of $T_s^{(i-1)\rightarrow}$ are visited in the same order in any valid traversal of T as in a valid traversal of $T_s^{(i-1)\rightarrow}$. The marked (unmarked) leaf-children of any half-marked node a of $T_s^{(i-1)\rightarrow}$ can be placed anywhere before, between, or after its marked (unmarked) children, but not before (after) b, since b has both marked and unmarked children. Hence, the permutations realized by T are exactly those realized by $T_s^{(i-1)\rightarrow}$ that have the neighbors of a_i as a suffix.

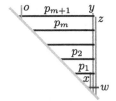

(a) frame providing the pockets

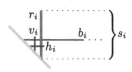

(b) number gadget for the number s_i

Fig. 3. Gadgets of our reduction from 3-PARTITION to STICK^{fix}. (Color figure online)

Now, we create the data structure T^i; see Fig. 2c. We set $T^i_1 = T^{(i-1)\rightarrow}_1, \ldots, T^i_{s-1} = T^{(i-1)\rightarrow}_{s-1}$. We additionally create T^i_s as follows. We hang $T^{(i-1)\rightarrow}_{s+1}, \ldots, T^{(i-1)\rightarrow}_{k_{i-1}}$ from a new node x in this order. We further insert the entering vertices as leaf-children of x (note that this allows them to mix freely before, after, or between the components $G^{(i-1)\rightarrow}_{s+1}, \ldots, G^{(i-1)\rightarrow}_{k_{i-1}}$. Then, we hang T followed by x off another node r, and make r the root of T^i_s. Finally, we set $T^i = (T^i_1, \ldots, T^i_s)$. This way, the order of the components $G^{(i-1)\rightarrow}_1, \ldots, G^{(i-1)\rightarrow}_{k_{i-1}}$ of $G^{(i-1)\rightarrow}$ is maintained in the data structures for G^i. Furthermore, we ensure that the entering vertices can be placed exactly before, after, or between the components of $G^{(i-1)\rightarrow}$ that are completely adjacent to a_i. Hence, this data structure captures all realizable permutations of B^i due to G^i.

The decision problem of STICK$_A$ can easily be solved by this algorithm. To find a stick representation, however, one has to backtrack through the data structures to find a valid permutation for the input problem. In the full version [8], we show how to do the backtracking and that the whole algorithm takes $O(|A||B|)$ time. □

3   Sticks of Fixed Lengths

In this section, we consider the case that, for each vertex of the input graph, its stick length is part of the input and fixed. We denote the variants of this problem by STICK^{fix}, by STICK$^{\text{fix}}_A$ if additionally σ_A is given, and by STICK$^{\text{fix}}_{AB}$ if σ_A and σ_B given. Unlike the case with variable stick length, all three variants are NP-hard; see Sects. 3.1 and 3.2. Surprisingly, STICK$^{\text{fix}}_{AB}$ can be solved efficiently by a simple linear program if the input graph contains no isolated vertices (i.e., vertices of degree 0); see Sect. 3.3. With our linear program, we can check the feasibility of any instance of STICK^{fix} if we are given a total order of the sticks on the ground line. With our NP-hardness results, this implies NP-completeness.

3.1   STICK^{fix}

We show that STICK^{fix} is NP-hard by reduction from 3-PARTITION, which is strongly NP-complete [11]. In 3-PARTITION, one is given a multiset S of $3m$

integers $s_1, \ldots, s_{3m}$ such that, for $i \in \{1, \ldots, 3m\}$, $C/4 < s_i < C/2$, where $C = (\sum_{i=1}^{3m} s_i)/m$, and the task is to decide whether S can be split into m sets of three integers, each summing up to C.

Theorem 3. *STICK^{fix} is NP-complete.*

Proof. We describe a polynomial-time reduction from 3-PARTITION. Given a 3-PARTITION instance $I = (S, C, m)$, we construct a fixed cage-like frame structure and introduce a number gadget for each number of S. A sketch of the frame is given in Fig. 3a. The purpose of the frame is to provide pockets, which will host our number gadgets (defined below). We add two long vertical (green) sticks y and z of length $mC+1+2\varepsilon$ and a shorter vertical (green) stick x of length 1 that are all kept together by a short horizontal (violet) stick w of some length $\varepsilon \ll 1$. We use $m + 1$ horizontal (black) sticks $p_1, p_2, \ldots, p_{m+1}$ to separate the pockets. Each of them intersects y but not z and has a specific length such that the distance between two of these sticks is $C \pm \varepsilon$. Additionally, p_1 intersects x and p_{m+1} intersects a vertical (orange) stick o of length $2C$. We use x and o to prevent the number gadgets from being placed below the bottommost and above the topmost pocket, respectively. It does not matter on which side of y the stick x ends up since each b_i of a number gadget intersects y but neither x nor z.

For each number s_i in S, we construct a number gadget; see Fig. 3b. We introduce a vertical (red) stick r_i of length s_i. Intersecting r_i, we add a horizontal (blue) stick b_i of length at least $mC+2$. The stick b_i intersects y and z, but neither x nor o. Due to these adjacencies, every number gadget can only be placed in one of the m pockets defined by $p_1, \ldots, p_{m+1}$. It cannot span multiple pockets. We require that r_i and b_i intersect each other close to their foot points, so we introduce two short (violet) sticks h_i and v_i—one horizontal, the other vertical—of lengths ε; they intersect each other, h_i intersects r_i, and v_i intersects b_i.

Given a yes-instance $I = (S, C, m)$ and a valid 3-partition P of S, the graph obtained by our reduction is realizable. Construct the frame as described before and place the number gadgets into the pockets according to P. Since the lengths of the three number gadgets' r_i sum up to $C \pm 3\varepsilon$, all three can be placed into one pocket. After distributing all number gadgets, we have a stick representation.

Given a stick representation of a graph G obtained from our reduction, we can obtain a valid solution of the corresponding 3-PARTITION instance $I = (S, C, m)$ as follows. Clearly, the shape of the frame is fixed, creating m pockets. Since the sticks $b_1, \ldots, b_{3m}$ are incident to y and z but neither to x nor to o, they can end up inside any of the pockets. In the y-dimension, each two number gadgets of numbers s_k and s_ℓ overlap at most on a section of length ε; otherwise r_k and b_ℓ or r_ℓ and b_k would intersect. Each pocket hosts precisely three number gadgets: we have $3m$ number gadgets, m pockets, and no pocket can contain four (or more) number gadgets; otherwise, there would be a number gadget of height at most $(C + \varepsilon)/4 + 2\varepsilon$, contradicting the fact that s_i is an integer with $s_i > C/4$. In each pocket, the height of the number gadgets would be too large if the three corresponding numbers of S would sum up to $C + 1$ or more. Thus, the assignment of number gadgets to pockets defines a valid 3-partition of S. $\square$

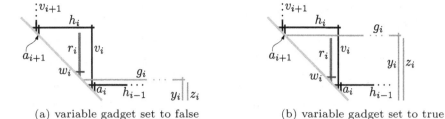

(a) variable gadget set to false   (b) variable gadget set to true

Fig. 4. Variable gadget in our reduction from MONOTONE-3-SAT to $\text{STICK}_A^{\text{fix}}$. (Color figure online)

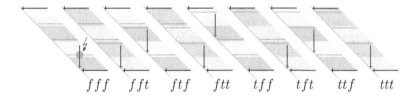

$$fff \quad fft \quad ftf \quad ftt \quad tff \quad tft \quad ttf \quad ttt$$

Fig. 5. Positive clause gadget (empty sub-stripe at the bottom). Here, a clause gadget for each of the eight possible truth assignments of a MONOTONE-3-SAT clause is depicted. E.g., tft means that the first variable is set to true, the second to false, and the third to true. Similarly, a negative clause gadget has an empty sub-stripe at the top.

The sticks of lengths $s_1, \ldots, s_{3m}$ can be simulated by paths of sticks with length ϵ each. Exploiting this, we can modify our reduction to use only three distinct stick lengths. We prove the following corollary in the full version [8].

Corollary 1. $\text{STICK}^{\text{fix}}$ *with only three different stick lengths is NP-complete.*

3.2 $\text{STICK}_A^{\text{fix}}$ and $\text{STICK}_{AB}^{\text{fix}}$

We show that $\text{STICK}_A^{\text{fix}}$ and $\text{STICK}_{AB}^{\text{fix}}$ are NP-hard by reduction from MONOTONE-3-SAT, which is NP-complete [20]. In MONOTONE-3-SAT, one is given a Boolean formula Φ in conjunctive normal form where each clause contains three distinct literals that are all positive or all negative. The task is to decide if Φ is satisfiable.

Theorem 4. $\text{STICK}_A^{\text{fix}}$ *is NP-complete.*

Proof. We describe a polynomial-time reduction from MONOTONE-3-SAT. A schematization of our reduction is depicted in Figs. 4, 5 and 6. Given a MONOTONE-3-SAT instance Φ over variables $x_1, \ldots, x_n$, we construct for each variable x_i (with $i \in \{1, \ldots, n\}$) a *variable gadget* as depicted in Fig. 4. Inside a (black) *cage*, there is a vertical (red) stick r_i with length 1 and from inside, a long horizontal (green) stick g_i leaves this cage. We can enforce the structure to look like in Fig. 4 as follows. We prescribe the order σ_A of the vertical sticks as in Fig. 4. Since a_{i+1} has length $\varepsilon \ll 1$, the horizontal (black) stick h_i intersects the two vertical (black) sticks v_{i+1} and a_{i+1} close to its foot point. We have $\sigma_A(a_{i+1}) < \sigma_A(r_i) < \sigma_A(v_i)$, so r_i is inside the cage bounded by h_i and v_i and fixed its height—as it does not intersect h_i— making sticks h_i and v_i intersect close to their end points (both have length $1 + 2\varepsilon$). Moreover, r_i cannot be below h_{i-1} because a_i is shorter than r_i and intersects h_{i-1} to the right of r_i. The stick w_i intersects r_i close to r_i's foot point because w_i has length ε. This leaves the freedom of placing g_i above or below r_i (as g_i does not intersect r_i) but still with its foot point inside the cage formed by h_i and v_i because it intersects v_i, but neither v_{i-1} nor v_{i+1}.

We say that the variable x_i is set to false if the foot point of g_i is below the foot point of r_i, and true otherwise. For each x_i, we add two long vertical (green) sticks y_i and z_i that we keep close together by a short horizontal (violet) stick of length ε (see Fig. 6 on the bottom right). We make g_i intersect y_i but not z_i. The three sticks g_i, y_i, and z_i get the same length ℓ_i. Hence, y_i and g_i intersect each other close to their end points as otherwise g_i would intersect z_i. We choose ℓ_1 sufficiently large such that the foot point of y_1 is to the right of the clause gadgets (see Fig. 6) and for each ℓ_i with $i \geq 2$, we set $\ell_i = \ell_{i-1} + 1 + 3\varepsilon$. Now compare the end points of g_i when x_i is set to false and when x_i is set to true relative to the (black) cage structure. When x_i is set to true, the end point of g_i is $1 \pm 2\varepsilon$ above and $1 \pm 2\varepsilon$ to the left compared to the case when x_i is set to false. Observe that, since g_i and y_i intersect each other close to their end points, this offset is also pushed to y_i and z_i and their foot points. Consequently, the position of the foot point of y_i (and z_i) differs by $1 \pm 2\varepsilon$ relative to the (black) frame structure depending on whether x_i is set to true or false. Our choice of ℓ_i allows this movement. In other words, no matter which truth value we assign to each x_i, there is a stick representation of the variable gadgets respecting σ_A.

For each clause, we add a *clause gadget* (see Fig. 5) as shown in Fig. 6. It is a stripe that is bounded by horizontal (black) sticks on its top and bottom. To fix the height of each stripe, we introduce two vertical (black) sticks that we keep close together by a short horizontal (black) stick of length ε. We make each horizontal (black) stick intersect only the first of these vertical (black) sticks to obtain clause gadgets of height of $4 + 2\varepsilon \pm \varepsilon$. Moreover, we make the topmost horizontal (black) stick intersect a_1 and v_1 to keep them connected to the variable gadgets. We (virtually) divide each clause gadget into four horizontal sub-stripes of height ≥ 1. For *positive clause gadgets* corresponding to all-positive clauses, we leave the bottommost sub-stripe empty; for *negative clause gadgets* corresponding to all-negative clauses, we leave the topmost sub-stripe empty. We add three horizontal (orange) sticks—one per remaining horizontal sub-stripe—and assign them bijectively to the variables of the clause. We make each horizontal (orange) stick o that is assigned to x_i intersect y_i and all y_j and z_j for $j < i$, but not z_i or y_k or z_k for any $k > i$. Thus, o intersects y_i close to o_i's end points. We choose the length of each such o so that its foot point is at the bottom of its sub-stripe if x_i is set to false or is at the top of its sub-stripe if x_i is set to true. Within the

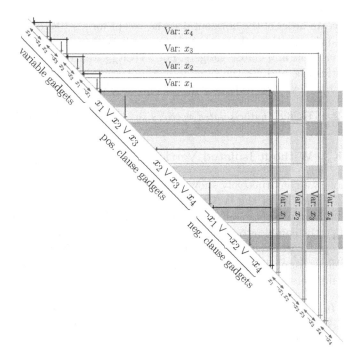

Fig. 6. Illustration of our reduction from MONOTONE-3-SAT to STICK$_A^{\text{fix}}$ (Color figure online)

positive and the negative clause gadgets, this gives us two times eight possible configurations of the orange sticks depending on the truth assignment of the three variables of the clause (see Fig. 5). Within each clause gadget, we have a vertical (blue) stick b of length 2. Each horizontal (black) stick that bounds a clause gadget intersects a short vertical (black) stick of length ε to force b into its designated clause gadget. Moreover, b is not isolated because it intersects a short (violet) stick of length ε.

Clearly, if Φ is satisfiable, there is a stick representation of the STICK$_A^{\text{fix}}$ instance obtained from Φ by our reduction by placing the sticks as described before (see also Fig. 6). In particular, the blue sticks can be placed as depicted in Fig. 5.

On the other hand, if there is a stick representation of the STICK$_A^{\text{fix}}$ instance obtained by our reduction, Φ is satisfiable. As argued before, the shape of the (black) frame structure of all gadgets is fixed by the choice of the adjacencies and lengths in the graph and σ_A. The only flexibility is, for each $i \in \{1, \ldots n\}$, whether g_i has its foot point above or below r_i. This enforces one of eight distinct configurations per clause gadget. As depicted in Fig. 5, precisely the configurations that correspond to satisfying truth assignments are realizable. Thus, we can read a satisfying truth assignment of Φ from the variable gadgets.   □

We enforce an order of the horizontal sticks except for a set W of sticks, which are the short (violet) sticks of length ε that are adjacent to the red and the blue sticks in the variable and clause gadgets. For STICK$_{AB}$ we can prescribe σ_B if we remove the sticks W and use the same reduction to obtain Corollary 2. Observe that we now have isolated vertices (the red and blue vertical sticks).

Corollary 2. *STICK*$_{AB}^{fix}$ *with isolated vertices in A or B is NP-complete.*

3.3 STICK$_{AB}^{fix}$ Without Isolated Vertices

In this section, we constructively show that STICK^{fix} is efficiently solvable if we are given a total order of the vertices in $A \cup B$ on the ground line. Note that if there is a stick representation for an instance of STICK$_{AB}$ (and consequently also STICK$_{AB}^{fix}$), the combinatorial order of the sticks on the ground line is always the same except for isolated vertices, which we formalize in the following lemma. The proof follows implicitly from the proof of Theorem 1. An explicit proof is given in the full version [8].

Lemma 1. *In all stick representations of an instance of STICK$_{AB}$, the order of the vertices $A \cup B$ on the ground line is the same after removing all isolated vertices. This order can be found in time $O(|E|)$.*

We are given an instance of STICK^{fix} and a total order $v_1, \ldots, v_n$ of the vertices ($n = |A| + |B|$) with stick lengths $\ell_1, \ldots, \ell_n$. We create a system of difference constraints, that is, a linear program $Ax \leq b$ where each constraint is a simple linear inequality of the form $x_j - x_i \leq b_k$, with n variables and $m \leq 3n - 1$ constraints. Such a system can be modeled as a weighted graph with a vertex per variable x_i and a directed edge (x_i, x_j) with weight b_k per constraint. The system is solvable if and only if there is no directed cycle of negative weights, and a solution can be found in $O(nm)$ time with the Bellman–Ford algorithm.

For each stick v_i, we create a variable x_i that corresponds to the x-coordinate of v_i's foot point on the ground line, with $x_1 = 0$. To ensure the unique order, we add $n-1$ constraints $x_{i+1} - x_i \leq -\varepsilon$ for some suitably small ε and $i = 1, \ldots, n-1$.

Let $v_i \in A$ and $v_j \in B$. If $(v_i, v_j) \in E$, then the corresponding sticks have to intersect, which they do if and only if $x_j - x_i \leq \min\{\ell_i, \ell_j\}$. If $i < j$ and $(v_i, v_j) \notin E$, then the corresponding sticks must not intersect, so we require $x_j - x_i > \min\{\ell_i, \ell_j\} \geq \min\{\ell_i, \ell_j\} + \varepsilon$. This easily gives a system of difference constraints with $O(n^2)$ constraints. We argue that a linear number suffices.

Let $v_i \in A$. Let j be the largest j such that $(v_i, v_j) \in E$ and $\ell_j \geq \ell_i$. We add a constraint $x_j - x_i \leq \ell_i$. Further, let k be the smallest k such that $(v_i, v_k) \notin E$ and $\ell_k \geq \ell_i$. We add a constraint $x_k - x_i > \ell_i \Leftrightarrow x_i - x_k \leq -\ell_i - \varepsilon$. Symmetrically, let $v_i \in B$. Let j be the smallest j such that $(v_j, v_i) \in E$ and $\ell_j > \ell_i$. We add a constraint $x_i - x_j \leq \ell_i$. Further, let k be the largest k such that $(v_k, v_i) \notin E$ and $\ell_k > \ell_i$. We add a constraint $x_i - x_k > \ell_i \Leftrightarrow x_k - x_i \leq -\ell_i - \varepsilon$.

We now argue that these constraints are sufficient to ensure that G is represented by a solution of the system. Let $v_i \in A$ and $v_j \in B$. If $i > j$, then the corresponding sticks cannot intersect, which is ensured by the fixed order. So assume that $i < j$. If $\ell_j \geq \ell_i$ and $(v_i, v_j) \in E$, then we either have the constraint $x_j - x_i \leq \ell_i$, or we have a constraint $x_k - x_i \leq \ell_i$ with $i < j < k$; together with the order constraints, this ensure that $x_j - x_i \leq x_k - x_i \leq \ell_i$. If $\ell_j \geq \ell_i$ and $(v_i, v_j) \notin E$, then we either have the constraint $x_i - x_j \leq -\ell_i - \varepsilon$, or we have a constraint $x_i - x_k \leq -\ell_i - \varepsilon$ with $i < k < j$; together with the order constraints, this ensure that $x_i - x_j \leq x_i - x_k \leq -\ell_i - \varepsilon$. Symmetrically, the constraints are also sufficient for $\ell_j < \ell_i$. We obtain a system of difference constraints with n variables and at most $3n - 1$ constraints proving Theorem 5. By Lemma 1, there is at most one realizable order of vertices for a $\mathsf{STICK}_{\mathsf{AB}}^{\mathsf{fix}}$ instance without isolated vertices, which can be found in linear time and proves Corollary 3.

Theorem 5. $\mathsf{STICK}^{\mathsf{fix}}$ *can be solved in* $O((|A| + |B|)^2)$ *time if we are given a total order of the vertices.*

Corollary 3. $\mathsf{STICK}_{\mathsf{AB}}^{\mathsf{fix}}$ *without isolated vertices is solvable in* $O((|A| + |B|)^2)$ *time.*

4   Open Problems

We have shown that $\mathsf{STICK}^{\mathsf{fix}}$ is NP-complete even if the sticks have only three different lengths, while $\mathsf{STICK}^{\mathsf{fix}}$ for unit-length sticks is solvable in linear time. But what is the computational complexity of $\mathsf{STICK}^{\mathsf{fix}}$ for sticks with one of *two* lengths? Also, the three different lengths in our proof depend on the number of sticks. Is $\mathsf{STICK}^{\mathsf{fix}}$ still NP-complete if the fixed lengths are bounded? Beside this, the complexity of the original problem STICK is still open.

References

1. Cabello, S., Jejčič, M.: Refining the hierarchies of classes of geometric intersection graphs. Electr. J. Comb. **24**(1), P1.33 (2017). http://www.combinatorics.org/ojs/index.php/eljc/article/view/v24i1p33
2. Cardinal, J., Felsner, S., Miltzow, T., Tompkins, C., Vogtenhuber, B.: Intersection graphs of rays and grounded segments. J. Graph Algorithms Appl. **22**(2), 273–295 (2018). https://doi.org/10.7155/jgaa.00470
3. Catanzaro, D., Chaplick, S., Felsner, S., Halldórsson, B.V., Halldórsson, M.M., Hixon, T., Stacho, J.: Max point-tolerance graphs. Discrete Appl. Math. **216**, 84–97 (2017). https://doi.org/10.1016/j.dam.2015.08.019
4. Chalopin, J., Gonçalves, D.: Every planar graph is the intersection graph of segments in the plane: extended abstract. In: STOC 2009, pp. 631–638. ACM (2009). https://doi.org/10.1145/1536414.1536500
5. Chaplick, S., Dorbec, P., Kratochvíl, J., Montassier, M., Stacho, J.: Contact representations of planar graphs: extending a partial representation is hard. In: Kratsch, D., Todinca, I. (eds.) WG 2014. LNCS, vol. 8747, pp. 139–151. Springer, Heidelberg (2014). https://doi.org/10.1007/978-3-319-12340-0_12
6. Chaplick, S., Felsner, S., Hoffmann, U., Wiechert, V.: Grid intersection graphs and order dimension. Order **35**(2), 363–391 (2018). https://doi.org/10.1007/s11083-017-9437-0

7. Chaplick, S., Hell, P., Otachi, Y., Saitoh, T., Uehara, R.: Ferrers dimension of grid intersection graphs. Discrete Appl. Math. **216**, 130–135 (2017). https://doi.org/10.1016/j.dam.2015.05.035

8. Chaplick, S., Kindermann, P., Löffler, A., Thiele, F., Wolff, A., Zaft, A., Zink, J.: Stick graphs with length constraints. Arxiv report (2019). http://arxiv.org/abs/1907.05257

9. De Luca, F., Hossain, M.I., Kobourov, S.G., Lubiw, A., Mondal, D.: Recognition and drawing of stick graphs. In: Biedl, T.C., Kerren, A. (eds.) GD 2018. LNCS, vol. 11282, pp. 303–316. Springer, Heidelberg (2018). https://doi.org/10.1007/978-3-030-04414-5_21

10. Felsner, S., Knauer, K., Mertzios, G.B., Ueckerdt, T.: Intersection graphs of L-shapes and segments in the plane. In: Csuhaj-Varjú, E., Dietzfelbinger, M., Ésik, Z. (eds.) MFCS 2014. LNCS, vol. 8635, pp. 299–310. Springer, Heidelberg (2014). https://doi.org/10.1007/978-3-662-44465-8_26

11. Garey, M.R., Johnson, D.S.: Computers and Intractability: A Guide to the Theory of NP-Completeness. W. H. Freeman, New York (1979)

12. Halldórsson, B.V., Aguiar, D., Tarpine, R., Istrail, S.: The Clark phaseable sample size problem: long-range phasing and loss of heterozygosity in GWAS. J. Comput. Biol. **18**(3), 323–333 (2011). https://doi.org/10.1089/cmb.2010.0288

13. Hartman, I.B., Newman, I., Ziv, R.: On grid intersection graphs. Discrete Math. **87**(1), 41–52 (1991). https://doi.org/10.1016/0012-365X(91)90069-E

14. Jünger, M., Leipert, S., Mutzel, P.: Level planarity testing in linear time. In: Whitesides, S.H. (ed.) GD 1998. LNCS, vol. 1547, pp. 224–237. Springer, Heidelberg (1998). https://doi.org/10.1007/3-540-37623-2_17

15. Klavík, P., Otachi, Y., Sejnoha, J.: On the classes of interval graphs of limited nesting and count of lengths. Algorithmica **81**(4), 1490–1511 (2019). https://doi.org/10.1007/s00453-018-0481-y

16. Köbler, J., Kuhnert, S., Watanabe, O.: Interval graph representation with given interval and intersection lengths. J. Discrete Algorithms **34**, 108–117 (2015). https://doi.org/10.1016/j.jda.2015.05.011

17. Kratochvíl, J.: A special planar satisfiability problem and a consequence of its NP-completeness. Discrete Appl. Math. **52**(3), 233–252 (1994). https://doi.org/10.1016/0166-218X(94)90143-0

18. Kratochvíl, J., Matoušek, J.: Intersection graphs of segments. J. Comb. Theory, Series B **62**(2), 289–315 (1994). https://doi.org/10.1006/jctb.1994.1071

19. Kratsch, D., McConnell, R.M., Mehlhorn, K., Spinrad, J.P.: Certifying algorithms for recognizing interval graphs and permutation graphs. SIAM J. Comput. **36**(2), 326–353 (2006). https://doi.org/10.1137/S0097539703437855

20. Li, W.N.: Two-segmented channel routing is strong NP-complete. Discrete Appl. Math. **78**(1–3), 291–298 (1997). https://doi.org/10.1016/S0166-218X(97)00020-6

21. Lubiw, A.: Private communication (2019)

22. Matoušek, J.: Intersection graphs of segments and $\exists \mathbb{R}$. ArXiv. https://arxiv.org/abs/1406.2636 (2014)

23. Pe'er, I., Shamir, R.: Realizing interval graphs with size and distance constraints. SIAM J. Discrete Math. **10**(4), 662–687 (1997). https://doi.org/10.1137/S0895480196306373

24. Schaefer, M.: Complexity of some geometric and topological problems. In: GD 2009. LNCS, vol. 5849, pp. 334–344. Springer, Heidelberg (2009). https://doi.org/10.1007/978-3-642-11805-0_32

25. Sen, M.K., Sanyal, B.K.: Indifference digraphs: a generalization of indifference graphs and semiorders. SIAM J. Discrete Math. **7**(2), 157–165 (1994). https://doi.org/10.1137/S0895480190177145
26. Shrestha, A.M.S., Takaoka, A., Tayu, S., Ueno, S.: On two problems of nano-PLA design. IEICE Trans. **94-D**(1), 35–41 (2011). https://doi.org/10.1587/transinf.E94.D.35
27. Spinrad, J., Brandstädt, A., Stewart, L.: Bipartite permutation graphs. Discrete Appl. Math. **18**(3), 279–292 (1987). https://doi.org/10.1016/S0166-218X(87)80003-3

Representing Graphs and Hypergraphs by Touching Polygons in 3D

William Evans[1], Paweł Rzążewski[2(✉)] ⓘ, Noushin Saeedi[1], Chan-Su Shin[3] ⓘ, and Alexander Wolff[4] ⓘ

[1] University of British Columbia, Vancouver, Canada
[2] Faculty of Mathematics and Information Science,
Warsaw University of Technology, Warszawa, Poland
p.rzazewski@mini.pw.edu.pl
[3] Hankuk University of Foreign Studies, Yongin, Republic of Korea
[4] Universität Würzburg, Würzburg, Germany

Dedicated to Honza Kratochvíl on his 60th birthday.

Abstract. Contact representations of graphs have a long history. Most research has focused on problems in 2d, but 3d contact representations have also been investigated, mostly concerning fully-dimensional geometric objects such as spheres or cubes. In this paper we study contact representations with convex polygons in 3d. We show that every graph admits such a representation. Since our representations use super-polynomial coordinates, we also construct representations on grids of polynomial size for specific graph classes (bipartite, subcubic). For hypergraphs, we represent their duals, that is, each vertex is represented by a point and each edge by a polygon. We show that even regular and quite small hypergraphs do not admit such representations. On the other hand, the two smallest Steiner triple systems can be represented.

1 Introduction

Representing graphs as the contact of geometric objects has been an area of active research for many years (see Hliněný and Kratochvíl's survey [15] and Alam's thesis [1]). Most of this work concerns representation in two dimensions, though there has been some interest in three-dimensional representation as well [2,3,5,13,25]. Representations in 3d typically use 3d geometric objects

The full version of this article is available at ArXiv [12]. W.E. and N.S. were funded by an NSERC Discovery grant and in part by the Institute for Computing, Information and Cognitive Systems (ICICS) at UBC. P.Rz. was partially supported by the ERC grant CUTACOMBS (no. 714704). A.W. was funded by the German Research Foundation (DFG) under grant 406987503 (WO 758/10-1). C.-S.Sh. was supported by the National Research Foundation of Korea (NRF) grant funded by the Korea government (MSIT) (no. 2019R1F1A1058963).

ⓒ Springer Nature Switzerland AG 2019
D. Archambault and C. D. Tóth (Eds.): GD 2019, LNCS 11904, pp. 18–32, 2019.
https://doi.org/10.1007/978-3-030-35802-0_2

that touch properly i.e., their intersection is a positive area 2d face. In contrast, our main focus is on contact representation of graphs and hypergraphs using non-intersecting (open, "filled") planar polygons in 3d. Two polygons are in *contact* if they share a corner vertex. Note that two triangles that share two corner vertices do not intersect and a triangle and rectangle that share two corners, even diagonally opposite ones, also do not intersect. However, no polygon contains a corner of another except at its own corner. A *contact representation of a graph in 3d* is a set of non-intersecting polygons in 3d that represent vertices. Two polygons share a corner point if and only if they represent adjacent vertices and each corner point corresponds to a distinct edge. We can see a contact representation of a graph $G = (V, E)$ as a certain drawing of its *dual hypergraph* $H_G = (E, \{E(v) \mid v \in V\})$ which has a vertex for every edge of G, and a hyperedge for every vertex v of G, namely the set $E(v)$ of edges incident to v. We extend this idea to arbitrary hypergraphs: A *non-crossing drawing of a hypergraph in 3d* is a set of non-intersecting polygons in 3d that represent *edges*. Two polygons share a corner point if and only if they represent edges that contain the same vertex and each corner point corresponds to a distinct vertex. It is straightforward to observe that the set of contact representations of a graph G is the same as the set of non-crossing drawings of H_G.

Many people have studied ways to represent hypergraphs geometrically [4, 6, 16], perhaps starting with Zykov [29]. A natural motivation of this line of research was to find a nice way to represent combinatorial configurations [14] such as Steiner systems (for an example, see Fig. 7). The main focus in representing hypergraphs, however, was on drawings in the plane. By using polygons to represent hyperedges in 3d, we gain some additional flexibility though still not all hypergraphs can be realized. Our work is related to Carmensin's work [8] on a Kuratowski-type characterization of 2d simplicial complexes (sets composed of points, line segments, and triangles) that have an embedding in 3-space. Our representations are sets of planar polygons (not just triangles) that arise from hypergraphs. Thus they are less expressive than Carmensin's topological 2d simplicial complexes and are more restricted. In particular, if two hyperedges share three vertices, the hyperedges must be coplanar in our representation.

Our work is also related to that of Ossona de Mendez [21]. He showed that a hypergraph whose vertex–hyperedge inclusion order has poset dimension d can be embedded into $\mathbb{R}^{d-1}$ such that every vertex corresponds to a unique point in $\mathbb{R}^{d-1}$ and every hyperedge corresponds to the convex hull of its vertices. The embedding ensures that the image of a hyperedge does not contain the image of a vertex and, for any two hyperedges e and e', the convex hulls of $e \setminus e'$ and of $e' \setminus e$ don't intersect. In particular, the images of disjoint hyperedges are disjoint. Note that both Ossona de Mendez and we use triangles to represent hyperedges of size 3, but for larger hyperedges, he uses higher-dimensional convex subspaces.

Our Contribution. All of our representations in this paper use convex polygons while our proofs of non-representability hold even permitting non-convex polygons. We first show that recognizing segment graphs in 3d is $\exists\mathbb{R}$-complete.

We show that every graph on n vertices with minimum vertex-degree 3 has a contact representation by convex polygons in 3d, though the volume of the drawing using integer coordinates is at least exponential in n; see Sect. 2.

Table 1. Required volume and running times of our algorithms for drawing n-vertex graphs of certain graph classes in 3d

Graph class	General	Bipartite	1-plane cubic	2-edge-conn. cubic	Subcubic
Grid volume	super-poly	$O(n^4)$	$O(n^2)$	$O(n^2)$	$O(n^3)$
Running time	$O(n^2)$	linear	linear	$O(n \log^2 n)$	$O(n \log^2 n)$
Reference	Theorem 2	Theorem 3	Theorem 4	Lemma 2	Theorem 5

For some graph classes, we give 3d drawing algorithms which require polynomial volume. Table 1 summarizes our results. When we specify the volume of the drawing, we take the product of the number of grid lines in each dimension (rather than the volume of a bounding box), so that a drawing in the xy-plane has non-zero volume. Some graphs, such as the squares of even cycles, have particularly nice representations using only unit squares; see the full version of this paper [12].

For hypergraphs our results are more preliminary. There are examples as simple as the hypergraph on six vertices with all triples of vertices as hyperedges that cannot be drawn using non-intersecting triangles; see Sect. 3. Similarly, hypergraphs with too many edges of cardinality 4 such as Steiner quadruple systems do not admit 3d drawings using convex quadrilaterals. On the other hand, we show that the two smallest Steiner triple systems can be drawn using triangles. (We define these two classes of hypergraphs in Sect. 3.)

2   Graphs

It is easy to draw graphs in 3d using points as vertices and non-crossing line segments as edges – any set of points in general position (no three colinear and no four coplanar) will support any set of edge segments without crossings. A more difficult problem is to represent a graph in 3d using polygons as vertices where two polygons intersect to indicate an edge (note that here we do not insist on a *contact* representation, i.e., polygons are allowed to intersect arbitrarily). Intersection graphs of convex polygons in 2d have been studied extensively [19]. Recognition is $\exists \mathbb{R}$-complete [23] (and thus in PSPACE since $\exists \mathbb{R} \subseteq$ PSPACE [7]) even for segments (polygons with only two vertices).

Every complete graph trivially admits an intersection representation by line segments in 2d. Not every graph, however, can be represented in this way, see e.g., Kratochvíl and Matoušek [18]. Moreover, they show that recognizing intersection graphs of line segments in the plane, called *segment graphs*, is $\exists \mathbb{R}$-complete.

It turns out that a similar hardness result holds for recognizing intersection graphs of straight-line segments in 3d (and actually in any dimension). The proof modifies the corresponding proof for 2d by Schaefer [23]. See also the excellent exposition of the proof by Matoušek [20]. For the proof, as well as the proofs of other theorems marked with ♠, see the full version of this paper [12].

Theorem 1 (♠). *Recognizing segment graphs in 3d is $\exists \mathbb{R}$-complete.*

We consider *contact representation* of graphs in 3d where no polygons are allowed to intersect except at their corners, and two polygons share a corner if and only if they represent adjacent vertices. We start by describing how to construct a contact representation for any graph using convex polygons, which requires at least exponential volume, and then describe contructions for graph families that use only polynomial volume.

2.1 General Graphs

Lemma 1. *For every positive integer $n \geq 3$, there exists an arrangement of n lines $\ell_1, \ell_2, \ldots, \ell_n$ with the following two properties:*

(A1) line ℓ_i intersects lines $\ell_1, \ell_2, \ldots, \ell_{i-1}, \ell_{i+1}, \ldots \ell_n$ in this order, and
(A2) distances between the intersection points on line ℓ_i decrease exponentially, i.e., for every i it holds that

$$d_i(j+2, j+1) \leq d_i(j+1, j)/2 \qquad for\ j \in \{1, \ldots, i-3\} \tag{1}$$
$$d_i(i+1, i-1) \leq d_i(i-1, i-2)/2 \tag{2}$$
$$d_i(i+2, i+1) \leq d_i(i+1, i-1)/2 \tag{3}$$
$$d_i(j+2, j+1) \leq d_i(j+1, j)/2 \qquad for\ j \in \{i+1, \ldots, n-2\}, \tag{4}$$

where $d_i(j, k)$ is the xy-plane distance between $p_{i,j}$ and $p_{i,k}$ and $p_{i,j} = p_{j,i}$ is the intersection point of ℓ_i and ℓ_j.

Proof. We construct the grid incrementally. We start with the x-axis as ℓ_1, the y-axis as ℓ_2, and the line through $(1, 0)$ and $(0, -1)$ as ℓ_3; see Fig. 1. Now suppose that $i > 3$, we have constructed lines $\ell_1, \ell_2, \ldots, \ell_{i-1}$, and we want to construct ℓ_i. We fix $p_{i-1,i}$ to satisfy $d_{i-1}(i, i-2) = d_{i-1}(i-2, i-3)/2$ then rotate a copy of line ℓ_{i-1} clockwise around $p_{i-1,i}$ until it (as ℓ_i) satisfies another of the inequalities in (1) with equality. Note that during this rotation, all inequalities in (A2) are satisfied and we do not move any previously constructed lines, so the claim of the lemma follows.   □

Theorem 2. *For every $n \geq 3$, the complete graph K_n admits a contact representation by non-degenerate convex polygons in 3d, each with at most $n - 1$ vertices. Such a representation can be computed in $O(n^2)$ time (assuming unit cost for arithmetic operations on coordinates).*

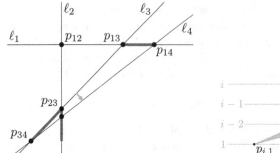

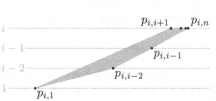

Fig. 1. Construction of ℓ_4 in the proof of Lemma 1.

Fig. 2. The polygon P_i that represents vertex i of K_n.

Proof. Take a grid according to Lemma 1. Set the z-coordinate of point $p_{i,j}$ to $\min\{i,j\}$ and represent vertex i by the polygon P_i, which we define to be the convex hull of $\{p_{i,1}, p_{i,2}, \ldots, p_{i,i-1}, p_{i,i+1}, \ldots p_{i,n}\}$. Note that P_i is contained in the vertical plane that contains line ℓ_i; see Fig. 2. To avoid that P_1 is degenerate, we reduce the z-coordinate of $p_{1,2}$ slightly.

Note that, for $i = 2, \ldots, n-1$, the counterclockwise order of the vertices around P_i is

$$p_{i,1}, p_{i,2}, \ldots, p_{i,i-1}, p_{i,n}, p_{i,n-1}, \ldots, p_{i,i+1}, p_{i,1}.$$

We show that all these points are on the boundary of P_i by ensuring that the angles formed by three consecutive points are bounded by π. Clearly the angles $\angle p_{i,i+1} p_{i,1} p_{i,2}$ and $\angle p_{i,i-1} p_{i,n} p_{i,n-1}$ are at most π. For $j = 2, \ldots, i-2$, we have that $\angle p_{i,j-1} p_{i,j} p_{i,j+1} < \pi$, which is due to the fact that the z-coordinates increase in each step by 1, while the distances decrease (property (A2)). Note that $\angle p_{i,i+1}, p_{i,i+2}, p_{i,i+3} = \cdots = \angle p_{i,n-2}, p_{i,n-1}, p_{i,n} = \pi$. Finally, we claim that $\angle p_{i,i-2}, p_{i,i-1}, p_{i,n} < \pi$. Clearly, $z(p_{i,i-1}) - z(p_{i,i-2}) = 1 = z(p_{i,n}) - z(p_{i,i-1})$. The claim follows by observing that, due to property (A2) and the geometric series formed by the distances,

$$d_i(i-1,n) = d_i(i-1,i+1) + \sum_{k=i+1}^{n-1} d_i(k,k+1) < 2d_i(i-1,i+1) \le d_i(i-2,i-1).$$

It remains to show that, for $1 \le i < j \le n$, polygons P_i and P_j do not intersect other than in $p_{i,j}$. This is simply due to the fact that P_j is above P_i in $p_{i,j}$, and lines ℓ_i and ℓ_j only intersect in (the projection of) this point. □

Corollary 1. *Every graph with minimum vertex-degree 3 admits a contact representation by convex polygons in 3d.*

Proof. Let n be the number of vertices of the given graph $G = (V, E)$. We use the contact representation of K_n and modify it as follows. For every pair $\{i,j\} \notin E$, just remove the point $p_{i,j}$ before defining the convex hulls. □

We can make the convex polygons of our construction strictly convex if we slightly change the z-coordinates. For example, decrease the z-coordinate of $p_{i,j}$ by $\delta/d_{\min\{i,j\}}(1, \max\{i,j\})$, where δ is such that moving every point by at most δ doesn't change the orientation of any four non-coplanar points.

Let us point out that Erickson and Kim [11] describe a construction of pairwise face-touching 3-polytopes in 3d that may provide the basis for a different representation in our model of a complete graph.

While we have shown that all graphs admit a 3d contact representation, these representations may be very non-symmetric and can have very large coordinates. This motivates the following question and specialized 3d drawing algorithms for certain classes of (non-planar) graphs; see the following subsections.

Open Problem 1. *Is there a polynomial p such that any n-vertex graph has a 3d contact representation with convex polygons on a grid of size $p(n)$?*

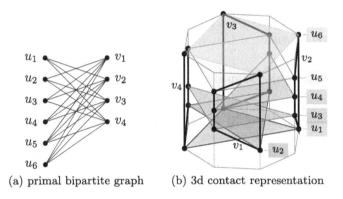

(a) primal bipartite graph   (b) 3d contact representation

Fig. 3. A 3d contact representation of a bipartite graph.

2.2 Bipartite Graphs

Theorem 3. *Every bipartite graph $G = (A \cup B, E)$ admits a contact representation by convex polygons whose vertices are restricted to a cylindrical grid of size $|A| \times 2|B|$ or to a 3d integer grid of size $|A| \times 2|B| \times 4|B|^2$. Such a representation can be computed in $O(|E|)$ time.*

Proof. Let G be the given bipartite graph with bipartition (A, B). We place the vertices of the A-polygons vertically above the corners of a regular $2|B|$-gon in the xy-plane. Each A-polygon goes to its own horizontal plane; the planes are one unit apart. For an example, see Fig. 3. For each $v \in B$, the polygon p_v that represents v has a vertical edge above a unique even corner of the $2|B|$-gon. This vertical edge connects the bottommost A-polygon incident to p_v to the topmost A-polygon incident to p_v. All the intermediate vertices of p_v are placed on the vertical line through the clockwise next corner of the $2|B|$-gon. This makes sure that all vertices of p_v lie in one plane, and p_v does not intersect any other B-polygon.

Due to convexity, the interiors of the A-polygons project to the interior of the $2|B|$-gon. Each B-polygon projects to an edge of the $2|B|$-gon. Hence, the A- and B-polygons are interior-disjoint.

Note that the polygons constructed by the argument above are not *strictly convex*. We can obtain a representation with strictly convex polygons by using a finer grid ($|A| \times |E|/2$) on the cylinder. If we insist on representations on the integer grid, we can replace the regular $2|B|$-gonal base of the cylinder by a strictly convex drawing of the $2|B|$-cycle. Using grid points on the 2d unit parabola, we obtain a 3d representation of size $|A| \times 2|B| \times 4|B|^2$.   □

If we apply Theorem 3 to $K_{3,3}$, we obtain a representation with three horizontal equilateral triangles and three vertical isosceles triangles, but with a small twist we can make all triangles equilateral. For the proof, see the full version.

Proposition 1 (♠). *The graph $K_{3,3}$ admits a contact representation in 3d using unit equilateral triangles.*

2.3   1-Planar Cubic Graphs

A simple consequence of the circle-packing theorem [17] is that every planar graph (of minimum degree 3) is the contact graph of convex polygons in the plane. In this section, we consider a generalization of planar graphs called *1-planar graphs* that have a drawing in 2d in which every edge (Jordan curve) is crossed at most once.

Our approach to realizing these graphs will use the *medial graph* G_{med} associated with a plane graph G (or, to be more general, with any graph that has an edge ordering). The vertices of G_{med} are the edges of G, and two vertices of G_{med} are adjacent if the corresponding edges of G are incident to the same vertex of G and consecutive in the circular ordering around that vertex. The medial graph is always 4-regular. If G has no degree-1 vertices, G_{med} has no loops. If G has minimum degree 3, G_{med} is simple. Also note that G_{med} is connected if and only if G is connected.

Theorem 4. *Every 1-plane cubic graph with n vertices can be realized as a contact graph of triangles with vertices on a grid of size $(3n/2-1) \times (3n/2-1) \times 3$. Given a 1-planar embedding of the graph, it takes linear time to construct such a realization.*

Proof. Let G be the given 1-plane graph. Let G'_{med} be the medial graph of G with the slight modification that, for each pair $\{e, f\}$ of crossing edges, G'_{med} has only one vertex v_{ef}, which is incident to all (up to eight) edges that immediately precede or succeed e and f in the circular order around their endpoints;

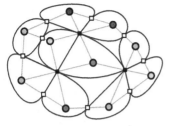 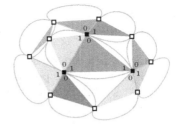

(a) a 1-plane cubic graph G and its (modified) medial graph G'_{med}

(b) representation of G with triangles; the numbers indicate the z-coordinates of the triangle corners

Fig. 4. 1-plane cubic graphs admit compact triangle contact representations.

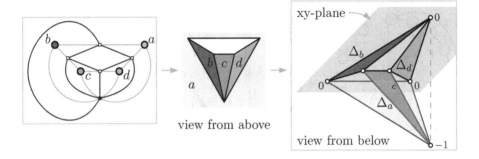

view from above

view from below

Fig. 5. left: graphs G (here a B-configuration, gray) and G'_{med}; center: straight-line drawing of G'_{med}; right: resulting 3d representation of G (numbers are z-coordinates).

see Fig. 4a. The order of the edges around v_{ef} is the obvious one. Using Schnyder's linear-time algorithm [24] for drawing 3-connected graphs[1] straight-line, we draw G'_{med} on a planar grid of size $(3n/2 - 1) \times (3n/2 - 1)$. Note that this is nearly a contact representation of G except that, in each crossing point, *all* triangles of the respective four vertices touch. Figure 4b is a sketch of the resulting drawing (without using Schnyder's algorithm) for the graph in Fig. 4a.

We add, for each crossing $\{e, f\}$, a copy v'_{ef} of the crossing point v_{ef} one unit above. Then we select an arbitrary one of the two edges, say $e = uv$. Finally we make the two triangles corresponding to u and v incident to v'_{ef} without modifying the coordinates of their other vertices. The labels in Fig. 4b are the resulting z-coordinates for our example; all unlabeled triangle vertices lie in the xy-plane.

If a crossing is on the outer face of G, it can happen that a vertex of G incident to the crossing becomes the outer face of G'_{med}; see Fig. 5 where this vertex is called a and the crossing edges are ac and bd. Consider the triangle Δ_a that represents a in G'_{med}. It covers the whole drawing of G'_{med}. To avoid intersections

---

[1] If G'_{med} is not 3-connected, we add dummy edges to fully triangulate it and then remove these edges to obtain a drawing of G'_{med}.

with triangles that participate in other crossings, we put the vertex of Δ_a that represents the crossing to $z = -1$, together with the vertex of the triangle Δ_c that represents c.

Our 3d drawing projects vertically back to the planar drawing, so all triangles are interior disjoint (with the possible exception of a triangle that represents the outer face of G'_{med}). Triangles that share an edge in the projection are incident to the same crossing – but this means that at least one of the endpoints of the shared edge has a different z-coordinate. Hence, all triangle contacts are vertex–vertex contacts. Note that some triangles may touch each other at $z = 1/2$ (as the two central triangles in Fig. 4b), but our contact model tolerates this. ☐

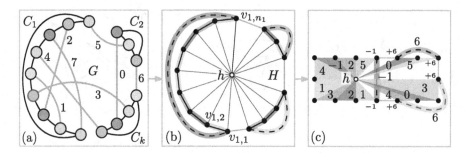

Fig. 6. Representing a 2-edge-connected cubic graph G by touching triangles in 3d: (a) partition of the edge set into disjoint cycles and a perfect matching (the numbers denote a permutation of the matching edges); (b) the graph H; (c) 3d contact representation of G; the numbers inside the triangles indicate the z-coordinates of the triangle apexes (above h), the small numbers denote the non-zero z-coordinates of the vertices.

2.4 Cubic Graphs

We first solve a restricted case and then show how this helps us to solve the general case of cubic graphs.

Lemma 2. *Every 2-edge-connected cubic graph with n vertices can be realized as a contact graph of triangles with vertices on a grid of size $3 \times n/2 \times n/2$. It takes $O(n \log^2 n)$ time to construct such a realization.*

Proof. By Petersen's theorem [22], any given 2-edge-connected cubic graph G has a perfect matching. Note that removing this matching leaves a 2-regular graph, i.e., a set of vertex-disjoint cycles $C_1, \ldots, C_k$; see Fig. 6(a). Such a partition can be computed in $O(n \log^2 n)$ time [10]. Let $n = |V(G)|$ and $n_1 = |V(C_1)|, \ldots, n_k = |V(C_k)|$. Note that $n = n_1 + \cdots + n_k$. We now construct a planar graph $H = (V, E)$ with $n+1$ vertices that will be the "floorplan" for our drawing of G. The graph H consists of an n-wheel with outer cycle $v_{1,1}, \ldots, v_{1,n_1}, \ldots, v_{k,1}, \ldots, v_{k,n_k}$, n spokes and a hub h, with additional *chords* $v_{1,1}v_{1,n_1}, v_{2,1}v_{2,n_2}, \ldots, v_{k,1}v_{k,n_k}$. We call the edges $v_{1,n_1}v_{2,1}, \ldots, v_{k,n_k}v_{1,1}$ *dummy edges* (thin gray in Fig. 6(b) and (c)) and the other edges on the outer face of the wheel *cycle edges*.

The chords and cycle edges form triangles with apex h. More precisely, for every $i \in \{1, \ldots, k\}$, the chord-based triangle $\Delta v_{i,1} v_{i,n_i} h$ and the $n_i - 1$ cycle-based triangles $\Delta v_{i,1} v_{i,2} h, \ldots, \Delta v_{i,n_i-1} v_{i,n_i} h$ together represent the n_i vertices in the cycle C_i of G. For each C_i, we still have the freedom to choose which vertex of G will be mapped to the chord-based triangle of H. This will depend on the perfect matching in G. The cycle edges will be drawn in the xy-plane (except for those incident to a chord edge); their apexes will be placed at various grid points above h such that matching triangles touch each other. The chord-based triangles will be drawn horizontally, but not in the xy-plane.

In order to determine the height of the triangle apexes, we go through the edges of the perfect matching in an arbitrary order; see the numbers in Fig. 6(a). Whenever an endpoint v of the current edge e is the *last* vertex of a cycle, we represent v by a triangle with chord base. We place the apexes of the two triangles that represent e at the lowest free grid point above h; see the numbers in Fig. 6(c). Our placement ensures that, in every cycle (except possibly one, to be determined later), the chord-based triangle is the topmost triangle. This guarantees that the interiors of no two triangles intersect (and the triangles of adjacent vertices touch).

Now we remove the chords from H. The resulting graph is a wheel; we can simply draw the outer cycle using grid points on the boundary of a $(3 \times n/2)$-rectangle and the hub on any grid point in the interior. (For the smallest cubic graph, K_4, we would actually need a (3×3)-rectangle, counting grid lines, in order to have a grid point in the interior, but it's not hard to see that K_4 can be realized on a grid of size $3 \times 2 \times 2$.) If one of the k cycles encloses h in the drawing (as C_1 in Fig. 6(c)), we move its chord-based triangle from $z = z^\star > 0$ to the plane $z = -1$, that is, below all other triangles. Let $i^\star$ be the index of this cycle (if it exists). Note that this also moves the apex of the triangle that is matched to the chord-based triangle from $z = z^\star$ to $z = -1$. In order to keep the drawing compact, we move each apex with z-coordinate $z' > z^\star$ to $z' - 1$. Then the height of our drawing equals exactly the number of edges in the perfect matching, that is, $n/2$.

The correctness of our representation follows from the fact that, in the orthogonal projection onto the xy-plane, the only pairs of triangles that overlap are the pairs formed by a chord-based triangle with each of the triangles in its cycle and, if it exists, the chord-based triangle of $C_{i^\star}$ with all triangles of the other cycles. Also note that two triangles $\Delta v_{i,j-1} v_{i,j} h$ and $\Delta v_{i,j} v_{i,j+1} h$ (the second indices are modulo n_i) that represent consecutive vertices in C_i (for some $i \in \{1, \ldots, k\}$ and $j \in \{1, \ldots, n_i\}$) touch only in a single point, namely in the image of $v_{i,j}$. This is due to the fact that vertices of G that are adjacent on C_i are not adjacent in the matching, and for each matched pair its two triangle apexes receive the same, unique z-coordinate.

We do not use all edges of H for our 3d contact representation of G. The spokes of the wheel are the projections of the triangle edges incident to h. The

k dummy edges don't appear in the representation (but play a role in the proof of Theorem 5 ahead). □

In order to generalize Lemma 2 to any cubic graph G, we use the *bridge-block tree* of G. This tree has a vertex for each 2-edge-connected component and an edge for each bridge of G. The bridge-block tree of a graph can be computed in time linear in the size of the graph [28]. The general idea of the construction is the following. First, remove all bridges from G and, using some local replacements, transform each connected component of the obtained graph into a 2-edge-connected cubic graph. Then, use Lemma 2 to construct a representation of each of these graphs. Finally, modify the obtained representations to undo the local replacements and use the bridge-block tree structure to connect the constructed subgraphs, restoring the bridges of G. The proof is in the full version.

Theorem 5 (♠). *Every cubic graph with n vertices can be realized as a contact graph of triangles with vertices on a grid of size $3n/2 \times 3n/2 \times n/2$. It takes $O(n \log^2 n)$ time to construct such a realization.*

Corollary 2. *Every graph with n vertices and maximum degree 3 can be realized as a contact graph of triangles, line segments, and points whose vertices lie on a grid of size $3\lceil n/2 \rceil \times 3\lceil n/2 \rceil \times \lceil n/2 \rceil$. It takes $O(n \log^2 n)$ time to construct such a realization.*

Proof. If n is odd, add a dummy vertex to the given graph. Then add dummy edges until the graph is cubic. Apply Theorem 5. From the resulting representation, remove the triangle that corresponds to the dummy vertex, if any. Disconnect the pairs of triangles that correspond to dummy edges. □

3 Hypergraphs

We start with a negative result. Hypergraphs that give rise to simplicial 2-complexes that are not embeddable in 3-space also do not have a realization using touching polygons. Carmesin's example of the cone over the complete graph K_5 is such a 2-complex[2], which arises from the 3-uniform hypergraph on six vertices whose edges are $\{\{i, j, 6\} : \{i, j\} \in [5]^2\}$. Recall that d-uniform means that all hyperedges have cardinality d. Any 3-uniform hypergraph that contains these edges also cannot be drawn. For example, $\mathcal{K}_n^d$, the complete d-uniform hypergraph on $n \geq 6$ vertices for $d = 3$ does not have a non-crossing drawing in 3d. For an elementary proof of this fact, see the full version.

Note that many pairs of hyperedges share two vertices in these graphs. This motivates us to consider 3-uniform *linear* hypergraphs, i.e., hypergraphs where pairs of edges intersect in at most one vertex. Very symmetric examples of such

---

[2] Carmesin [8] credits John Pardon with the observation that the *link graph* at a vertex v, which contains a node for every edge at v and an arc connecting two such nodes if they share a face at v, must be planar for the 2-complex to be embeddable.

hypergraphs are *Steiner systems*. Recall that a Steiner system $S(t, k, n)$ is an n-element set S together with a set of k-element subsets of S (called *blocks*) such that each t-element subset of S is contained in exactly one block. In particular, examples of 3-uniform hypergraphs are Steiner triple systems $S(2, 3, n)$ [27]. They exist for any vertex number in $\{6k + 1, 6k + 3 : k \in \mathbb{N}\}$. For $n = 7, 9, 13, \ldots$, the corresponding 3-uniform hypergraph has $n(n-1)/6$ hyperedges and is $((n-1)/2)$-regular.

First we show that the two smallest triple systems, i.e., $S(2, 3, 7)$ (also called the *Fano plane*) and $S(2, 3, 9)$, admit non-crossing drawings in 3d. See Fig. 7 for the picture of the representation of the Fano plane. The proofs of the results stated in this section can be found in the full version. Actually, the existence of such representations follows from Ossona de Mendez' work [21] (see introduction) since both hypergraphs have incidence orders of dimension 4 (which can be checked by using an integer linear program). His approach, however, yields coordinates that are exponential in the number of vertices.

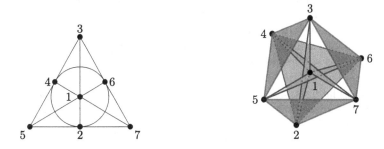

Fig. 7. The Fano plane and a drawing using touching triangles in 3d

Proposition 2 (♠). *The Fano plane $S(2, 3, 7)$ and the Steiner triple system $S(2, 3, 9)$ admit non-crossing drawings using triangles in 3d.*

Now we turn to a special class of 4-uniform hypergraphs; Steiner quadruple systems $S(3, 4, n)$ [26]. They exist for any vertex number in $\{6k + 2, 6k + 4 : k \in \mathbb{N}\}$. For $n = 8, 10, 14, \ldots$, the corresponding 4-uniform hypergraph has $m = \binom{n}{3}/4$ hyperedges and is $4m/n = (n-1)(n-2)/6$-regular. We now show that no Steiner quadruple system admits a drawing using convex quadrilaterals in 3d.

Observation 1. *In a non-crossing drawing of a Steiner quadruple system using quadrilaterals in 3d, every plane contains at most four vertices.*

Proof. Suppose that there is a drawing R and a plane Π that contains at least five vertices. Let ab be a maximum length edge of the convex hull of the points in the plane Π. No four, say $wxyz$ in that order, can be collinear or the quadrilateral containing wyz is either $wxyz$, which is degenerate (a line segment), or it contains x on its perimeter but x is not a corner, a contradiction. Thus the set S of vertices on Π that are not on the edge ab has size at least two. If there exist

$u, v \in S$ such that abu and abv form[3] two distinct quadrilaterals with ab then these quadrilaterals intersect in the plane (they are both on the same side of ab), a contradiction. If no such pair exists then S contains exactly two points and they form one quadrilateral with ab, which must contain the other vertex in Π (on the edge ab) that is not a corner, a contradiction.   □

Observation 1 is the starting point for the following result.

Proposition 3 (♠). *The Steiner quadruple system $S(3,4,8)$ does not admit a non-crossing drawing using (convex or non-convex) quadrilaterals in 3d.*

Theorem 6. *No Steiner quadruple system admits a non-crossing drawing using convex quadrilaterals in 3d.*

Proof. Day and Edelsbrunner [9, Lemma 2.3] used an approach similiar to that of Carmesin (mentioned in footnote 2) to show that the number of triangles spanned by n points in 3d is less than n^2 if no two triangles have a non-trivial intersection. (A trivial intersection is a common point or edge.) We need to redo their proof taking lower-order terms into account. If a Steiner quadruple system $S(3,4,n)$ can be drawn using quadrilaterals in 3d, the intersection of these quadrilaterals with a small sphere around a vertex is a planar graph. Recall that any $S(3,4,n)$ has n vertices and $m = \binom{n}{3}/4$ quadruples. Let v be any vertex. Then v is incident to $4m/n = (n-1)(n-2)/6$ quadrilaterals. Breaking these (convex) quadrilaterals into $(n-1)(n-2)/3$ triangles yields a graph on $n-1$ vertices (that is, on all but v) with $(n-1)(n-2)/3$ edges. For $n > 9$, this graph cannot be planar. The only Steiner quadruple system with at most nine vertices is $S(3,4,8)$, hence Proposition 3 yields our claim.   □

4   Conclusion and Open Problems

In Sect. 3 we discussed the Fano plane and other Steiner systems. The Fano plane is the smallest projective plane. Can the second smallest projective plane, $PG(3)$, which is the Steiner quadruple system $S(2,4,13)$, be drawn in 3d, such that each edge is a (convex) quadrilateral? To this end, we make the following observation (proved in the full version): If there is a drawing of $PG(3)$ in which every edge is a convex quadrilateral, then no two quadrilaterals are coplanar.

Acknowledgments. We are grateful to the organizers of the workshop Homonolo 2017, where the project originates. We thank Günter Rote for advice regarding strictly convex drawings of polygons on the grid, and we thank Torsten Ueckerdt for bringing Ossona de Mendez' work [21] to our attention. We are indebted to Arnaud de Mesmay and Eric Sedgwick for pointing us to the lemma of Dey and Edelsbrunner [9], which yielded Theorem 6.

---

[3] In a Steiner quadruple system, every triple of vertices appears in a unique quadruple.

References

1. Alam, M.J.: Contact representations of graphs in 2D and 3D. Ph.D. thesis, The University of Arizona (2015)
2. Alam, J., Evans, W., Kobourov, S., Pupyrev, S., Toeniskoetter, J., Ueckerdt, T.: Contact representations of graphs in 3D. In: Dehne, F., Sack, J.-R., Stege, U. (eds.) WADS 2015. LNCS, vol. 9214, pp. 14–27. Springer, Cham (2015). https://doi.org/10.1007/978-3-319-21840-3_2
3. Alam, M.J., Kaufmann, M., Kobourov, S.G.: On contact graphs with cubes and proportional boxes. In: Freivalds, R.M., Engels, G., Catania, B. (eds.) SOFSEM 2016. LNCS, vol. 9587, pp. 107–120. Springer, Heidelberg (2016). https://doi.org/10.1007/978-3-662-49192-8_9
4. Brandes, U., Cornelsen, S., Pampel, B., Sallaberry, A.: Path-based supports for hypergraphs. J. Discret. Algorithms **14**, 248–261 (2012). https://doi.org/10.1016/j.jda.2011.12.009
5. Bremner, D., Evans, W., Frati, F., Heyer, L., Kobourov, S.G., Lenhart, W.J., Liotta, G., Rappaport, D., Whitesides, S.H.: On representing graphs by touching cuboids. In: Didimo, W., Patrignani, M. (eds.) GD 2012. LNCS, vol. 7704, pp. 187–198. Springer, Heidelberg (2013). https://doi.org/10.1007/978-3-642-36763-2_17
6. Buchin, K., van Kreveld, M.J., Meijer, H., Speckmann, B., Verbeek, K.: On planar supports for hypergraphs. J. Graph Algorithms Appl. **15**(4), 533–549 (2011). https://doi.org/10.7155/jgaa.00237
7. Canny, J.F.: Some algebraic and geometric computations in PSPACE. In: Simon, J. (ed.) Proceedings of the 20th Annual ACM Symposium on Theory of Computing (STOC 1988), pp. 460–467 (1988). https://doi.org/10.1145/62212.62257
8. Carmesin, J.: Embedding simply connected 2-complexes in 3-space - I. A Kuratowski-type characterisation. ArXiv report (2019). http://arxiv.org/abs/1709.04642
9. Dey, T.K., Edelsbrunner, H.: Counting triangle crossings and halving planes. Discrete Comput. Geom. **12**(3), 281–289 (1994). https://doi.org/10.1007/BF02574381
10. Diks, K., Stańczyk, P.: Perfect matching for biconnected cubic graphs in $O(n \log^2 n)$ time. In: van Leeuwen, J., Muscholl, A., Peleg, D., Pokorný, J., Rumpe, B. (eds.) SOFSEM 2010. LNCS, vol. 5901, pp. 321–333. Springer, Heidelberg (2010). https://doi.org/10.1007/978-3-642-11266-9_27
11. Erickson, J., Kim, S.: Arbitrarily large neighborly families of congruent symmetric convex 3-polytopes. In: Bezdek, A. (ed.) Discrete Geometry, Pure and Applied Mathematics, vol. 253, pp. 267–278. Marcel Dekker, New York (2003). In Honor of W. Kuperberg's 60th Birthday
12. Evans, W., Rzążewski, P., Saeedi, N., Shin, C.S., Wolff, A.: Representing graphs and hypergraphs by touching polygons in 3D. ArXiv report (2019). http://arxiv.org/abs/1908.08273
13. Felsner, S., Francis, M.C.: Contact representations of planar graphs with cubes. In: Hurtado, F., van Kreveld, M.J. (eds.) Proceedings of the 27th Annual Symposium on Computational Geometry (SoCG 2011), pp. 315–320. ACM (2011). https://doi.org/10.1145/1998196.1998250
14. Gropp, H.: The drawing of configurations. In: Brandenburg, F.J. (ed.) GD 1995. LNCS, vol. 1027, pp. 267–276. Springer, Heidelberg (1996). https://doi.org/10.1007/BFb0021810

15. Hliněný, P., Kratochvíl, J.: Representing graphs by disks and balls (a survey of recognition-complexity results). Discret. Math. **229**(1–3), 101–124 (2001). https://doi.org/10.1016/S0012-365X(00)00204-1

16. Johnson, D.S., Pollak, H.O.: Hypergraph planarity and the complexity of drawing Venn diagrams. J. Graph Theory **11**(3), 309–325 (1987). https://doi.org/10.1002/jgt.3190110306

17. Koebe, P.: Kontaktprobleme der konformen Abbildung. Berichte über die Verhandlungen der Sächsischen Akad. der Wissen. zu Leipzig. Math.-Phys. Klasse **88**, 141–164 (1936). https://doi.org/10.1007/BF02418546

18. Kratochvíl, J., Matoušek, J.: Intersection graphs of segments. J. Comb. Theory Ser. B **62**(2), 289–315 (1994). https://doi.org/10.1006/jctb.1994.1071

19. van Leeuwen, E.J., van Leeuwen, J.: Convex polygon intersection graphs. In: Brandes, U., Cornelsen, S. (eds.) GD 2010. LNCS, vol. 6502, pp. 377–388. Springer, Heidelberg (2011). https://doi.org/10.1007/978-3-642-18469-7_35

20. Matoušek, J.: Intersection graphs of segments and $\exists\mathbb{R}$. ArXiv report (2014). http://arxiv.org/abs/1406.2636

21. de Mendez, P.O.: Realization of posets. J. Graph Algorithms Appl. **6**(1), 149–153 (2002). https://doi.org/10.7155/jgaa.00048

22. Petersen, J.: Die Theorie der regulären graphs. Acta Math. **15**, 193–220 (1891). https://doi.org/10.1007/BF02392606

23. Schaefer, M.: Complexity of some geometric and topological problems. In: Eppstein, D., Gansner, E.R. (eds.) GD 2009. LNCS, vol. 5849, pp. 334–344. Springer, Heidelberg (2010). https://doi.org/10.1007/978-3-642-11805-0_32

24. Schnyder, W.: Embedding planar graphs on the grid. In: Proceedings of the 1st ACM-SIAM Symposium on Discrete Algorithms (SODA 1990), pp. 138–148 (1990). https://dl.acm.org/citation.cfm?id=320176.320191

25. Thomassen, C.: Interval representations of planar graphs. J. Comb. Theory Ser. B **40**(1), 9–20 (1986). https://doi.org/10.1016/0095-8956(86)90061-4

26. Weisstein, E.W.: Steiner quadruple system. From MathWorld – A Wolfram Web Resource. http://mathworld.wolfram.com/SteinerQuadrupleSystem.html. Accessed 20 Aug 2019

27. Weisstein, E.W.: Steiner triple system. From MathWorld – A Wolfram Web Resource. http://mathworld.wolfram.com/SteinerTripleSystem.html. Accessed 20 Aug 2019

28. Westbrook, J., Tarjan, R.E.: Maintaining bridge-connected and biconnected components on-line. Algorithmica **7**(1), 433–464 (1992). https://doi.org/10.1007/BF01758773

29. Zykov, A.A.: Hypergraphs. Uspekhi Mat. Nauk **29**(6), 89–154 (1974). https://doi.org/10.1070/RM1974v029n06ABEH001303

Optimal Morphs of Planar Orthogonal Drawings II

Arthur van Goethem[✉], Bettina Speckmann, and Kevin Verbeek

Department of Mathematics and Computer Science, TU Eindhoven,
Eindhoven, The Netherlands
{a.i.v.goethem,b.speckmann,k.a.b.verbeek}@tue.nl

Abstract. Van Goethem and Verbeek [12] recently showed how to morph between two planar orthogonal drawings Γ_I and Γ_O of a connected graph G while preserving planarity, orthogonality, and the complexity of the drawing during the morph. Necessarily drawings Γ_I and Γ_O must be equivalent, that is, there exists a homeomorphism of the plane that transforms Γ_I into Γ_O. Van Goethem and Verbeek use $O(n)$ linear morphs, where n is the maximum complexity of the input drawings. However, if the graph is disconnected their method requires $O(n^{1.5})$ linear morphs. In this paper we present a refined version of their approach that allows us to also morph between two planar orthogonal drawings of a disconnected graph with $O(n)$ linear morphs while preserving planarity, orthogonality, and linear complexity of the intermediate drawings.

Van Goethem and Verbeek measure the structural difference between the two drawings in terms of the so-called *spirality* $s = O(n)$ of Γ_I relative to Γ_O and describe a morph from Γ_I to Γ_O using $O(s)$ linear morphs. We prove that $s+1$ linear morphs are always sufficient to morph between two planar orthogonal drawings, even for disconnected graphs. The resulting morphs are quite natural and visually pleasing.

1 Introduction

Continuous morphs of planar drawings have been studied for many years, starting as early as 1944, when Cairns [7] showed that there exists a planarity-preserving continuous morph between any two (compatible) triangulations that have the same outer triangle. These results were extended by Thomassen [10] in 1983, who gave a constructive proof of the fact that two compatible straight-line drawings can be morphed into each other while maintaining planarity. The resulting algorithm to compute such a morph takes exponential time (just as Cairns' result). Thomassen also considered the orthogonal setting and showed how to morph between two rectilinear polygons with the same turn sequence. For planar straight-line drawings the question was settled by Alamdari et al. [1], following work by Angelini et al. [3]. They showed that $O(n)$ uni-directional linear morphs are sufficient to morph between any compatible pair of planar straight-line drawings of a graph with n vertices while preserving planarity. The corresponding morph can be computed in $O(n^3)$ time.

© Springer Nature Switzerland AG 2019
D. Archambault and C. D. Tóth (Eds.): GD 2019, LNCS 11904, pp. 33–45, 2019.
https://doi.org/10.1007/978-3-030-35802-0_3

In this paper we consider the orthogonal setting, that is, we study planarity-preserving morphs between two planar orthogonal drawings Γ_I and Γ_O with maximum complexity n, of a graph G. Here the complexity of an orthogonal drawing is defined as the number of vertices and bends. All intermediate drawings must remain orthogonal, as to not disrupt the mental map of the reader. This immediately implies that the results of Alamdari et al. [1] do not apply, since they do not preserve orthogonality. Biedl et al. [5] described the first results in this setting, for so-called *parallel* drawings, where every edge has the same orientation in both drawings. They showed how to morph between two parallel drawings using $O(n)$ linear morphs while maintaining parallelity and planarity. More recently, Biedl et al. [4] showed how to morph between two planar orthogonal drawings using $O(n^2)$ linear morphs, while preserving planarity, orthogonality, and linear complexity. Van Goethem and Verbeek [12] improved this bound further to $O(n)$ linear morphs for a connected graph G. This bound is tight, based on the lower bound for straight-line graphs proven by Alamdari et al. [1].

If the graph G is disconnected, then Aloupis et al. [2] show how to connect G in a way that is compatible with both Γ_I and Γ_O while increasing the complexity of the drawings to at most $O(n^{1.5})$. They also prove a matching lower bound if G has at most $\frac{n}{4}$ connected components. This directly implies that Van Goethem and Verbeek require $O(n^{1.5})$ linear morphs for a disconnected graph G.

Paper Outline. We show how to refine the approach by Van Goethem and Verbeek [12] to also morph between two planar orthogonal drawings of a disconnected graph G using $O(n)$ linear morphs while preserving planarity, orthogonality, and linear complexity. In Sect. 2 we describe the necessary background. In particular, we discuss *wires*: equivalent sets of horizontal and vertical polylines that capture the x- and y-order of the vertices in Γ_I and Γ_O. The *spirality* of these wires guides the morph. In Sect. 3 we show how to find sets of wires with linear spirality for equivalent orthogonal planar drawings Γ_I and Γ_O of a disconnected planar graph G. Van Goethem and Verbeek are agnostic of the connectivity of the graph once they create the wires. Hence, using the wires constructed in Sect. 3, we can directly apply their approach to disconnected graphs.

In the remainder of the paper we show how to "batch" intermediate morphs. We argue solely based on sets of wires, hence the results apply to both connected and disconnected graphs. In particular, in Sect. 4 we show how to combine all intermediate morphs that act on segments of spirality s into one single linear morph. Hence we need only s linear morphs to morph from Γ_I to Γ_O. However, the rerouting and simplification operations introduced by van Goethem and Verbeek to lower the intermediate complexity are not compatible with batched linear morphs and hence intermediate drawings have complexity of $O(n^3)$. In Sect. 5 we present refined versions of both operations which allow us to maintain linear complexity through the s linear morphs. The initial setup for these operations costs one additional morph, for a total of $s + 1$ linear morphs that preserve planarity, orthogonality, and linear complexity. We implemented our algorithm and believe that the resulting morphs are natural and visually pleasing[1]. We restrict

---

[1] See https://youtu.be/n0ZaPtfg9TM for a short movie.

our arguments to proof sketches, all omitted details can be found in the full version [11].

2   Preliminaries

Orthogonal Drawings. A *drawing* Γ of a graph $G = (V, E)$ is a mapping from every vertex $v \in V$ to a unique point $\Gamma(v)$ in the Euclidean plane and from each edge (u, v) to a simple curve in the plane starting at $\Gamma(u)$ and ending at $\Gamma(v)$. A drawing is *planar* if no two curves intersect in an internal point, and no vertices intersect a curve in an internal point. A drawing is *orthogonal* if each edge is mapped to an orthogonal polyline consisting of horizontal and vertical segments meeting at *bends*. In a *straight-line drawing* every edge is represented by a single line-segment. Two planar drawings Γ and Γ' are *equivalent* if there exists a homeomorphism of the plane that transforms Γ into Γ'.

We consider morphs between two equivalent drawings of a graph G. To simplify the presentation, we assume that both drawings are straight-line drawings with n vertices. If this is not the case then we first *unify* Γ and Γ'. We subdivide segments, creating additional virtual bends, to ensure that every edge is represented by the same number of segments in Γ and Γ'. Next, we replace all bends with vertices. All edges of the resulting graph G^* are now represented by straight segments (horizontal or vertical) in both Γ and Γ'.

A *linear morph* of two drawings Γ and Γ' can be described by a continuous linear interpolation of all vertices and bends, which are connected by straight segments. For each $0 \leq t \leq 1$ there exists an intermediate drawing Γ_t where each vertex v is drawn at $\Gamma_t(v) = (1 - t)\Gamma_v + t\Gamma'_v$ ($\Gamma_0 = \Gamma$ and $\Gamma_1 = \Gamma'$). A linear morph *maintains planarity* (orthogonality, linear complexity, resp.), if every intermediate drawing Γ_t is planar (orthogonal, of linear complexity, resp.).

Wires. Following van Goethem and Verbeek [12] we use orthogonal polylines called *wires* as the main tool to determine the morph. Wires consist of horizontal or vertical segments called *links*. We use two sets of wires to capture the horizontal and vertical order of the vertices in Γ_I and Γ_O. The *lr-wires* $W_\rightarrow$ traverse the drawings from left to right, and the *tb-wires* $W_\downarrow$ traverse the drawings from top to bottom. Since the horizontal and vertical order of the vertices in Γ_O are

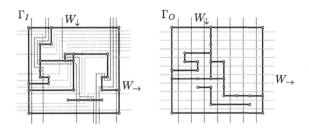

Fig. 1. Two unified drawings Γ_I and Γ_O of G (black) plus equivalent wires (red/blue). (Color figure online)

guiding our morph, the wires $W_\rightarrow$ and $W_\downarrow$ are simply horizontal and vertical lines in Γ_O separating consecutive vertices in the x- and y-order (only if their x- or y-coordinates are distinct). Γ_O and Γ_I are equivalent, hence there exist wires in Γ_I that are equivalent to the wires in Γ_O: there is a one-to-one matching between the wires of Γ_O and Γ_I such that matching wires partition the vertices identically, and cross both the segments of the drawings and the links of the other wires in the same order (see Fig. 1). Any such two wires in Γ_I do not cross if they are from the same set and cross exactly once otherwise.

Van Goethem and Verbeek use the *spirality* of wires as a measure for the distance to Γ_O (where all wires are straight lines of spirality zero). Spirality is a well-established measure in the context of orthogonal drawings and is frequently used for bend-minimization [6,8,9]. Specifically, let $w \in W_\rightarrow$ be a lr-wire, and $\ell_1, \ldots, \ell_k$ be the links ordered along w. Let b_i be the orientation of the bend from ℓ_i to ℓ_{i+1}, where $b_i = 1$ for a left turn, $b_i = -1$ for a right turn, and $b_i = 0$ otherwise. The *spirality* of a link ℓ_i is defined as $s(\ell_i) = \sum_{j=1}^{i-1} b_j$. A *maximum-spirality link* is any link with the largest absolute spirality. The spirality of a wire is the maximum absolute spirality of any link in the wire, the spirality of a set of wires is the maximum spirality of any wire in the set.

The spirality of a drawing Γ is not well defined: it is always relative to another drawing Γ' and the straight-line wires induced by Γ'. Furthermore, there are possibly multiple sets of matching wires in Γ for the straight-line wires in Γ'. Still, whenever the drawing Γ' and the matching set of wires in Γ are clear from the context, then by abuse of notation we will speak of the spirality of Γ. Unless stated otherwise, we always consider spirality relative to Γ_O.

Slides. Biedl et al. [4] introduced *slides* as a particular type of linear morph that operates on the segments of the drawing. Van Goethem and Verbeek [12] extended this concept to wires. Slides on wires may be accompanied by the insertion or deletion of bends in the drawing. In the following we exclusively consider slides on wires. A *zigzag* consists of three consecutive links of a wire and two bends β and γ that form a left turn followed by a right turn or vice versa. Consider the horizontal zigzag with bends β and γ in Fig. 2(a). Let $\mathcal{V}$ be the set of vertices and bends of both the drawing and the wires that are (1) above or at the same height as β and strictly to the left of β, (2) that are strictly above γ, and (3) β. The corresponding region is shaded in Fig. 2. A *zigzag-eliminating*

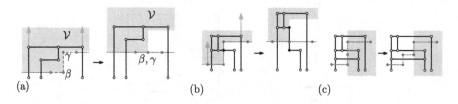

Fig. 2. A drawing (black) with vertices (open marks) and bends (closed marks). **(a)** A *zigzag-eliminating slide* with center link $\overline{\beta\gamma}$. **(b)** Introducing two additional bends in a crossing segment ensures orthogonality. **(c)** A *bend-introducing slide*.

slide is a linear morph that straightens a zigzag on a wire by moving all vertices and bends in $\mathcal{V}$ up by the initial distance between β and γ.

By definition, wires do not contain any vertices or bends of the drawing or other wires. However, the center link $\overline{\beta\gamma}$ might be crossed by a segment of the drawing or a link of a wire in the other set (see Fig. 2(b) for a crossing with a segment of the drawing). In this case we introduce two virtual bends in the segment or the link on the crossing and symbolically offset one to the right and one to the left. The left bend is thus included in $\mathcal{V}$ while the right bend is not. We can prevent that multiple segments or links cross $\overline{\beta\gamma}$ using so-called *bend-introducing slides* as discussed in [12] (see Fig. 2(c)).

3   Linear Morphs for Disconnected Graphs

Let Γ_I and Γ_O be two equivalent planar orthogonal drawings of a disconnected graph G. For a connected graph there is a unique homotopy class in Γ_I that contains all possible wires that match a given wire w from Γ_O. This statement does not hold for disconnected graphs: there might be more than one homotopy class in Γ_I that matches w (see Fig. 3(a)). If we choose homotopy classes independently for the wires in Γ_I then their union might not be equivalent to the set of wires in Γ_O, for example, wires might cross more than once (see Fig. 3(c)).

Below we show that we can choose homotopy classes for the wires in Γ_I incrementally, first for the lr-wires and then for the tb-wires, while maintaining the correct intersection pattern and hence equivalence with Γ_O. For each of the resulting equivalence classes we add the shortest wire to the set of wires. It remains to argue that the resulting set of wires has spirality $O(n)$ despite the interdependence of the homotopy classes and the fact that the arrangement of drawing and wires can have super-linear complexity (which invalidates the proofs from [12]). Below we consider only $W_\rightarrow$, analogous results hold for $W_\downarrow$.

Lemma 1. *For each right-oriented link $\ell_\rightarrow$ of a wire $w \in W_\rightarrow$ with positive (negative) spirality s there exists a vertical line L and a subsequence of $\Omega(|s|)$ links of w crossing L, such that the absolute spiralities of the links in sequence are $[0, 2, 4, \ldots, |s| - 2, |s|]$, and when ordered top-to-bottom (bottom-to-top) along L form the sequence $[2, 6, 10, \ldots, |s| - 2, |s|, |s| - 4, \ldots, 4, 0]$.*

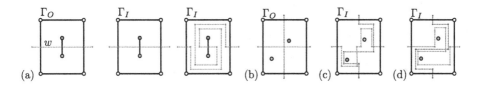

Fig. 3. (a) A (straight-line) wire w in Γ_O (red) and two possible wires in Γ_I from different homotopy classes that both match w. (b) A graph with three connected components. (c) Wires in Γ_I that cross three times. (d) Set of wires equivalent to Γ_O. (Color figure online)

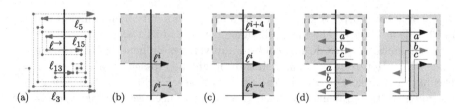

Fig. 4. (a) Lemma 1 for a link $\ell_\rightarrow$ and sequence $S = (\ell_3, \ell_5, \ell_{13}, \ell15, \ell_\rightarrow)$. (b) The i-core of a spiral for a link $\ell^i \in S$ (gray). (c) The i-layer of the spiral (gray). (d) A layer cannot only contain wires as then we can shorten all wires.

Figure 4(a) illustrates Lemma 1. Let $\ell_\rightarrow$ be a right-oriented link on a wire w and w.l.o.g. let $s > 0$ be the spirality of w. Further, let L be a vertical line through $\ell_\rightarrow$ and S a subsequence from w with the properties guaranteed by Lemma 1. Finally, let $\ell^i \in S$ be the unique link with spirality $0 \leq i \leq s$ in S. We define the i-core for S (for $4 \leq i \leq s$ and $i \pmod 4 = 0$) as the region enclosed by the wire w from the intersection between ℓ^{i-4} and L to the intersection between ℓ^i and L and the straight line segment along L connecting them (see Fig. 4(b)). We define the i-layer for S (for $4 \leq i \leq s - 4$ and $i \pmod 4 = 0$) as the difference of the i-core and the $(i + 4)$-core (see Fig. 4(c)).

Lemma 2. *An equivalent set of lr-wires with spirality $O(n)$ exists.*

Proof (Sketch). We prove by induction that we can add a new lr-wire with spirality $O(n)$. If a wire w has $\omega(n)$ layers, then we can argue via shortcuts (see Fig. 4(d)) that w was not shortest with respect to previously inserted wires. □

Lemma 3. *An equivalent set of wires with spirality $O(n)$ exists.*

Proof (Sketch). By Lemma 2 we can insert all lr-wires with spirality $O(n)$. By Lemma 2 from [12] the spirality of intersecting links is the same. Apply Lemma 2 for the tb-wires in the regions between the intersections with lr-wires. □

Theorem 1. *Let Γ_I and Γ_O be two unified planar orthogonal drawings of a (disconnected) graph G. We can morph Γ_I into Γ_O using $\Theta(n)$ linear morphs while maintaining planarity and orthogonality.*

Proof. By Lemma 3 an equivalent set of wires with spirality $s = O(n)$ exists. By Theorem 8 from [12] we can thus morph the drawings into each other using $O(s) = O(n)$ linear morphs. The lower bound of $\Omega(n)$ follows from [1]. □

4   Combining Intermediate Linear Morphs

The proof of Theorem 1 implies a morph between two unified planar orthogonal drawings Γ_I and Γ_O exists using $O(s)$ linear morphs, where s is the spirality of Γ_I. In this section we show how to combine consecutive linear morphs into a total number of only s linear morphs, while maintaining planarity and orthogonality.

The morphs we describe can be encoded by a sequence of drawings, starting with Γ_I and ending with Γ_O, such that every consecutive pair of drawings can be linearly interpolated while maintaining planarity and orthogonality. For notational convenience let $\Gamma_i \rightarrow \Gamma_j$ indicate that Γ_i occurs before Γ_j during the morph and $\Gamma_i \twoheadrightarrow \Gamma_j$ that $\Gamma_i \rightarrow \Gamma_j$ or $\Gamma_i = \Gamma_j$.

Let an *iteration* of the original morph consist of all linear slides that jointly reduce spirality by one. Let the first drawing of iteration s be the first drawing in the original morph with spirality s and the last drawing be the first drawing with spirality $s-1$. Consecutive iterations overlap in exactly one drawing. These drawings in the overlap of iterations are the intermediate steps of the final morph. Within this section let $\Gamma_I \twoheadrightarrow \Gamma_a \rightarrow \Gamma_b \twoheadrightarrow \Gamma_O$, where Γ_a is the first drawing with spirality s and Γ_b is the first drawing with spirality $s - 1$.

4.1 Staircases

Consider two distinct vertices v and w of the drawing. Define an *x-inversion* (*y-inversion*) of v and w between Γ_a and Γ_b when the sign $(+,-,0)$ of $v.x - w.x$ ($v.y - x.y$) differs in Γ_a and Γ_b. We say two vertices are *x-inverted* (*y-inverted*), or simply *inverted*. Two vertices v and w are *separated* in a drawing by a link ℓ when they are both in the vertical (horizontal) strip spanned by ℓ, and v and w are on opposite sides of ℓ.

Lemma 4. *Two vertices v and w can be inverted by a zigzag-removing slide along link ℓ, if and only if v and w are separated by ℓ.*

A *downward staircase* is a sequence of horizontal links where: (1) the left-endpoints are x-monotone increasing and y-monotone decreasing, (2) the projection on the x-axis is overlapping or touching for a pair if and only if they are consecutive in the sequence, and (3) all links have positive spirality. Two vertices v and w are *separated* by a downward staircase if v is in the vertical strip spanned by the first link of the staircase and above it and w is in the vertical strip spanned by the last link and below it. Similar concepts can be defined for upwards staircases and for vertical links.

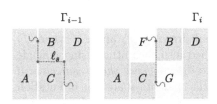

Fig. 5. Regions surrounding ℓ_s in Γ_{i-1} and the matching regions in Γ_i.

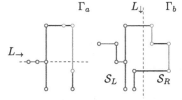

Fig. 6. Sets $\mathcal{S}_L$ and $\mathcal{S}_R$ in Γ_a and Γ_b.

Lemma 5. *Two vertices v and w that are x-inverted (y-inverted) first during a morph from Γ_a to Γ_b, are separated by a horizontal (vertical) staircase of maximum spirality links in Γ_a.*

Proof (Sketch). Assume w.l.o.g. that only one inversion occurs and it occurs from Γ_{b-1} to Γ_b. By Lemma 4, v and w are separated by a link ℓ in Γ_{b-1}. Link ℓ must have maximum absolute spirality as it was selected for the morph. We now prove inductively that a staircase exists in all drawings from Γ_a to Γ_{b-1} by "moving backwards" through the morph. To this end we define four rectangular regions A, B, C, D surrounding ℓ_s in Γ_{i-1} (see Fig. 5). During the linear slide from Γ_{i-1} to Γ_i two new regions F and G are created, which cannot contain vertices. Using these rectangular regions and a case distinction on the type of linear slide, we can argue inductively that a staircase separating v and w must also exist in Γ_{i-1}.   □

4.2   Inversions

We show that every pair of vertices is inverted along at most one axis during the morph from Γ_a to Γ_b. We then prove that Γ_a has spirality one relative to Γ_b.

Lemma 6. *Two vertices v and w can be inverted along only one axis during the morph from Γ_a to Γ_b.*

Lemma 7. *Each vertical (horizontal) line in Γ_b not crossing a vertex, can be matched to a y- (x-)monotone wire in Γ_a.*

Proof (Sketch). Consider a vertical line $L_\downarrow$ in Γ_b not intersecting any vertex. Line $L_\downarrow$ partitions the set of vertices and vertical edges in Γ_b into two subsets $\mathcal{S}_L$ and $\mathcal{S}_R$. Consider a horizontal line $L_\rightarrow$ in Γ_a and consider the maximal intervals formed along it by elements from the same set $\mathcal{S}_L$ or $\mathcal{S}_R$ (see Fig. 6). Set $\mathcal{S}_L$ and $\mathcal{S}_R$ form exactly two maximal intervals along $L_\rightarrow$. Thus a y-monotone line exists correctly splitting $\mathcal{S}_L$ and $\mathcal{S}_R$. We can show that this y-monotone line must intersect horizontal edges in the correct order as well.   □

Lemma 8. *Drawing Γ_a has spirality one relative to Γ_b.*

4.3   Single Linear Morph

We now show that any two planar orthogonal drawings Γ_i and Γ_j, where Γ_i has spirality one relative to Γ_j, can be morphed into each other using a single linear morph while maintaining planarity. Two drawings are *shape-equivalent* if for each edge the sequence of left and right turns is identical and the orientation of the initial segment is identical in both drawings. We say two drawings are *degenerate shape-equivalent* if edges may contain zero-length segments but an assignment of orientations to the segments exists that is consistent with both drawings. Two (degenerate) shape-equivalent drawings are per definition also unified. We can make Γ_a degenerate shape-equivalent to Γ_b by adding zero-length edges whenever maximum absolute spirality links in Γ_a cross an edge.

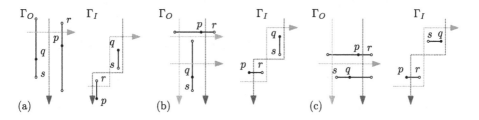

Fig. 7. (a) Two points p and q on vertical segments of the drawing that are inverted along both axes imply wires in Γ_I that are not equivalent to Γ_O. (b) Points p and q on a horizontal and vertical segment. (c) Points p and q on horizontal segments.

We say two *points* p and q on the drawing are *split* by a wire when p and q lie on different sides of the wire.

Lemma 9. *Let Γ_I and Γ_O be two degenerate shape-equivalent drawings, where Γ_I has spirality one. There exists a single linear morph from Γ_I to Γ_O that maintains planarity and orthogonality.*

Proof (Sketch). The partition of the drawing by all wires defines *cells*: regions of the plane not split by any wire. For each cell containing at least one bend or vertex, we can linearly interpolate all vertices and bends in Γ_I to the unique vertex or bend location in Γ_O. This directly defines a linear morph between Γ_I and Γ_O. To argue planarity of this morph, we assume for contradiction that there exist two points p and q on an edge or vertex of the drawing that coincide during the morph (excluding Γ_I and Γ_O). Then p and q must be x- and y-inverted in Γ_O compared to Γ_I and there must be two vertices r and s that are x- and y-inverted and split by at least a tb-wire and a lr-wire. As the lr-wire and the tb-wire are monotone they cross at least three times (see Fig. 7). Contradiction.   □

Theorem 2. *Let Γ_I and Γ_O be two unified planar orthogonal drawings of a (disconnected) graph G, where Γ_I has spirality s. We can morph Γ_I into Γ_O using exactly s linear morphs while maintaining planarity and orthogonality.*

5   Linear Complexity of Intermediate Drawings

Van Goethem and Verbeek [12] describe rerouting and a simplification operations that reduce the complexity of intermediate drawings to $O(n)$. These operations are not compatible with the batched linear morphs we described in Sect. 4. Below we show how to adapt these operations to the batched setting. These adaptations come at the cost of a single additional linear morph.

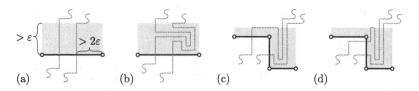

Fig. 8. (a) An ε-band adjacent to the edge. (b) Inserting an s-windmill. (c–d) Reroute wires after linear slide without introducing new crossings.

5.1 Rerouting

To avoid that the linear morphs introduce too many bends in a single iteration of the morph, we show how to route the wires such that only $O(n)$ complexity is added to the drawing in each iteration. The initial rerouting of the wires in Γ_I increases the maximum spirality by one, but it prevents any increase of spirality during the morph. Thus, using Theorem 2, $s+1$ morphs are sufficient to morph two equivalent drawings into each other while maintaining planarity and keeping complexity of the intermediate drawings to $O(n^2)$.

We reroute the wires in $W_{\downarrow}$ and $W_{\rightarrow}$ as follows. Consider an edge e that is crossed by at least two wires in Γ_I. By Lemma 9 from [12] all crossing links have the same spirality. Assume w.l.o.g. that this spirality is positive, otherwise mirror the rotations and replace right by left. Let ε be a small distance such that the ε-band above e is empty except for the links crossing e and that there is more than a 2ε distance between the right-most crossing link and the right-endpoint of e (see Fig. 8(a)).

We insert an s-*windmill* of all crossing wires within the ε-band above e by rerouting the wires as follows. First disconnect all crossing links within the ε-band above e. Then reroute all wires in a parallel bundle to the right, beyond the right-most wire w_r crossing e. Now we spiral the bundle using right turns until the spirality of the links reaches zero. Next we unwind the bundle again within the spiral. Finally we reconnect the wires by routing back parallel to e to maintain the original crossing points (see Fig. 8(b)). This rerouting can be executed without introducing crossings between the wires. It does increase the spirality of the drawing by one.

We now change each iteration as follows. Consider a horizontal edge e crossed by $k > 1$ links of maximum absolute spirality s (assuming $s > 0$) at the start of the iteration. Instead of performing a linear slide on all crossing links, we perform a single linear slide only on the rightmost crossing link. This slide creates a new vertical segment (see Fig. 8(c)). Thanks to the introduction of the s-windmill, we can easily reroute the other crossing wires to intersect the new vertical segment instead of the horizontal segment without introducing other crossings (see Fig. 8(d)). The newly created crossing links must have spirality $s - 1$ as all links crossing the same segment have the same spirality (Lemma 9 from [12]). We can reduce all remaining spirality s links without introducing additional complexity in the drawing.

Lemma 10. *At the start of iteration i of the morph, all wires crossing an edge e with links of spirality i form an i-windmill in an empty ε-band next to e.*

Lemma 11. *Let Γ_s be the first drawing of an iteration and Γ_{s-1}^r the rerouted last drawing. The spirality of Γ_s relative to Γ_{s-1}^r is one.*

Proof (Sketch). We can argue that rerouting wires does not eliminate staircases. A link that is rerouted may have been part of a staircase, but the new links replacing it do not break any staircase properties. As rerouting links maintains staircases, Lemmata 5–8 still apply.   □

Drawing Γ_{s-1}^r compared to Γ_s contains two additional bends in each edge crossed by maximum absolute spirality links in Γ_s. We can make Γ_s and Γ_{s-1}^r degenerate shape-equivalent by inserting an additional zero-length segment at the right-most (left-most for negative spirality) crossing link for each edge crossed by maximum absolute spirality links. By Lemmata 9 and 11 we can morph the resulting Γ_s into Γ_{s-1}^r in a single linear morph while maintaining planarity.

As, independently of how many wires are crossing it, each edge only introduces two new bends, complexity increases by $O(n)$ during each iteration. Thus the overall complexity is $O(s \cdot n)$. We conclude that we can morph two drawings Γ_I and Γ_O, where Γ_I has spirality s, into each other using $s + 1$ linear morphs while maintaining planarity and $O(s \cdot n)$ complexity of the drawing.

5.2   Simplification

By using rerouting we can ensure that the complexity of the drawing increases by at most $O(n)$ in every iteration, but its complexity may still grow to $O(n^2)$ over $O(n)$ iterations. In this section we show how to simplify the intermediate drawings to ensure that the complexity after each iteration is $O(n)$.

We again consider a single iteration starting with Γ_s and ending with Γ_{s-1}. Using rerouting we can find an alternative final drawing Γ_{s-1}^r that also maintains planarity. We now introduce a *redraw* step that further simplifies Γ_{s-1}^r into a straight-line drawing Γ_{s-1}' such that a linear morph from Γ_s to Γ_{s-1}' still maintains planarity. The redraw step works as follows.

For each vertex v in Γ_{s-1}^r, consider a 6ε-sized square box surrounding v that contains only v and a 3ε-part of each outgoing edge from v. If an incident edge e is crossed by a maximum absolute spirality link in Γ_s, then we reroute e inside the 6ε-box around v. Specifically, for an edge e leaving v rightwards, we reroute e within the 6ε-box using the coordinates $(v, v+(0, -\varepsilon), v+(2\varepsilon, -\varepsilon), v+(2\varepsilon, 0), v+(3\varepsilon, 0))$ (see Fig. 9(a)). Analogous rerouting can be done for edges leaving v in other directions. For an edge crossed by a negative spirality link invert the left and right turns.

Lemma 12. *We can redraw all edges in Γ_{s-1} that were crossed by a maximum-spirality link in Γ_s within 6ε-boxes while maintaining planarity of the drawing.*

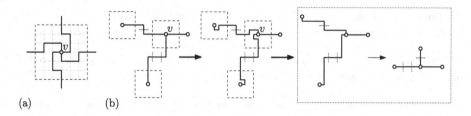

Fig. 9. (a) A 6ε-box surrounding a vertex v (dashed) with four redrawn edges. (b) Original drawing, rerouted drawing, and straightening the drawing.

Proof (Sketch). We can establish a relation between the spiralities of two segments incident at the same vertex. Using this relation we can argue that, after redrawing, no two edges leave a vertex in the same direction. As a result, there are no planarity violations within the 6ε-boxes around vertices. □

Lemma 13. *If Γ_s is a straight-line drawing with spirality $s > 0$ then there exists a straight-line drawing Γ'_{s-1} with spirality $s - 1$.*

Proof (Sketch). Let Γ^r_{s-1} be the drawing obtained by applying rerouting to the last drawing of iteration s. Consider an edge e crossed by maximum absolute spirality links in Γ_s. Edge e has three segments in Γ^r_{s-1} due to the two introduced bends. The first and last segment do not cross any wires. We can apply the redraw step to e, resulting in three more segments at the start and end of e. Finally we eliminate all additional segments of e by performing zigzag-eliminating slides on these segments (see Fig. 9(b)). □

Lemma 14. *The spirality of Γ'_{s-1} relative to Γ_s is one.*

Proof (Sketch). Let the *main wire set* be the set of wires used to compute the morph including rerouting from Γ_s to Γ^r_{s-1}. Consider a *reference wire grid* that is a straight-line wire grid in Γ_s. Using Lemmata 7, 8, and 11 but swapping the roles of Γ_a and Γ_b, we obtain the result that there is an equivalent monotone set of wires in Γ^r_{s-1} matching the reference grid in Γ_s. Thus the spirality of Γ_s relative to Γ^r_{s-1} is one.

When straightening Γ^r_{s-1} to Γ'_{s-1} only zigzag-removing slides are performed on segments not crossed by a wire from the main wire set. As such a segment was not crossed by a wire from the main wire set, the orientation of the segment is unchanged in Γ^r_{s-1}. Specifically, any link of a wire from the reference wire grid that crosses such a segment must have spirality zero. When straightening Γ^r_{s-1} to Γ'_{s-1} the zigzag-removing slides may insert additional bends in these reference wires, but the wires will remain monotone. □

We can make Γ_s degenerate shape-equivalent to Γ'_{s-1} as follows. For each edge e crossed by maximum absolute spirality links, we split e at the crossing with the right-most (or left-most if the links have negative spirality) crossing link and insert a zero-length segment. Furthermore, we add three zero-length segments at the endpoint of each such edge e coincident with the respective endpoint.

Theorem 3. *Let Γ_I and Γ_O be two equivalent drawings of a (disconnected) graph G, where Γ_I has spirality s. We can morph Γ_I into Γ_O using $s + 1$ linear morphs while maintaining planarity, orthogonality, and linear complexity of the drawing during the morph.*

Acknowledgements. Bettina Speckmann and Kevin Verbeek are supported by the Netherlands Organisation for Scientific Research (NWO) under project no. 639.023.208 (B.S.) and no. 639.021.541 (K.V.). We want to thank the anonymous reviewers for their extensive feedback.

References

1. Alamdari, S., Angelini, P., Barrera-Cruz, F., Chan, T., Da Lozzo, G., Di Battista, G., Frati, F., Haxell, P., Lubiw, A., Patrignani, M., Roselli, V., Singla, S., Wilkinson, B.: How to morph planar graph drawings. SIAM J. Comput. **46**(2), 824–852 (2017)
2. Aloupis, G., Barba, L., Carmi, P., Dujmović, V., Frati, F., Morin, P.: Compatible connectivity augmentation of planar disconnected graphs. Discrete Comput. Geom. **54**(2), 459–480 (2015)
3. Angelini, P., Frati, F., Patrignani, M., Roselli, V.: Morphing planar graph drawings efficiently. In: Wismath, S., Wolff, A. (eds.) GD 2013. LNCS, vol. 8242, pp. 49–60. Springer, Cham (2013). https://doi.org/10.1007/978-3-319-03841-4_5
4. Biedl, T., Lubiw, A., Petrick, M., Spriggs, M.J.: Morphing orthogonal planar graph drawings. ACM Trans. Algorithms **9**(4), 29:1–29:24 (2013)
5. Biedl, T., Lubiw, A., Spriggs, M.J.: Morphing planar graphs while preserving edge directions. In: Healy, P., Nikolov, N.S. (eds.) GD 2005. LNCS, vol. 3843, pp. 13–24. Springer, Heidelberg (2006). https://doi.org/10.1007/11618058_2
6. Bläsius, T., Lehmann, S., Rutter, I.: Orthogonal graph drawing with inflexible edges. Comput. Geom. **55**, 26–40 (2016)
7. Cairns, S.: Deformations of plane rectilinear complexes. Am. Math. Monthly **51**(5), 247–252 (1944)
8. Di Battista, G., Liotta, G., Vargiu, F.: Spirality and optimal orthogonal drawings. SIAM J. Comput. **27**(6), 1764–1811 (1998)
9. Didimo, W., Giordano, F., Liotta, G.: Upward spirality and upward planarity testing. SIAM J. Discrete Math. **23**(4), 1842–1899 (2009)
10. Thomassen, C.: Deformations of plane graphs. J. Comb. Theory Ser. B **34**(3), 244–257 (1983)
11. van Goethem, A., Speckmann, B., Verbeek, K.: Optimal Morphs of Planar Orthogonal Drawings II. arXiv e-prints, August 2019. arXiv:1908.08365
12. van Goethem, A., Verbeek, K.: Optimal morphs of planar orthogonal drawings. In: Proceedings of the 34th International Symposium on Computational Geometry (SoCG 2018), pp. 42:1–42:14 (2018)

Computing Stable Demers Cartograms

Soeren Nickel[1], Max Sondag[2]([⊠]), Wouter Meulemans[2],
Markus Chimani[3], Stephen Kobourov[4], Jaakko Peltonen[5],
and Martin Nöllenburg[1]

[1] Algorithms and Complexity Group, TU Wien, Vienna, Austria
soeren.nickel@tuwien.ac.at, noellenburg@ac.tuwien.ac.at
[2] TU Eindhoven, Eindhoven, The Netherlands
{m.f.m.sondag,w.meulemans}@tue.nl
[3] University of Osnabrück, Osnabrück, Germany
markus.chimani@uni-onsabrueck.de
[4] University of Arizona, Tucson, AZ, USA
kobourov@cs.arizona.edu
[5] Tampere University, Tampere, Finland
jaakko.peltonen@tuni.fi

Abstract. Cartograms are popular for visualizing numerical data for
map regions. Maintaining correct adjacencies is a primary quality criterion for cartograms. When there are multiple data values per region (over
time or different datasets) shown as animated or juxtaposed cartograms,
preserving the viewer's mental map in terms of stability between cartograms is another important criterion. We present a method to compute stable Demers cartograms, where each region is shown as a square
and similar data yield similar cartograms. We enforce orthogonal separation constraints with linear programming, and measure quality in terms
of keeping adjacent regions close (cartogram quality) and using similar
positions for a region between the different data values (stability). Our
method guarantees ability to connect most lost adjacencies with minimal
leaders. Experiments show our method yields good quality and stability.

Keywords: Time-varying data · Cartograms · Mental-map
preservation

1 Introduction

Myriad datasets are georeferenced and relate to specific places or regions. A
natural way to visualize such data in their spatial context is by cartographic
maps. A choropleth map is a prominent tool, which colors each region in a map

This research was initiated at NII Shonan Meeting 127 "Reimagining the Mental Map
and Drawing Stability". M. Sondag is supported by The Netherlands Organisation for
Scientific Research (NWO) under project no. 639.023.20. S. Kobourov is supported
by NSF grants CCF-1740858, CCF-1712119 and DMS-1839274. M. Nöllenburg is supported by FWF grant P 31119.

ⓒ Springer Nature Switzerland AG 2019
D. Archambault and C. D. Tóth (Eds.): GD 2019, LNCS 11904, pp. 46–60, 2019.
https://doi.org/10.1007/978-3-030-35802-0_4

by its data value. Such maps have several drawbacks: data may not be correlated to region size and hence the visual salience of large vs small regions is not equal. Moreover, colors are difficult to compare and not the most effective encoding for numeric data [23], requiring a legend to facilitate relative assessment.

Cartograms, also called value-by-area maps, overcome the drawbacks by reducing spatial precision in favor of clearer encoding of data values: the map is deformed such that each region's visual size is proportional to its data value. Attention is then drawn to items with large data values and comparison of relative magnitudes becomes a task of estimating sizes – which relies on more accurate visual variables for numeric data [23]. This also frees up color as a visual variable. Cartogram quality is assessed by criteria [25] including **1. Spatial deformation:** regions should be placed close to their geographic position; **2. Shape deformation:** each region should resemble its geographic shape; **3. Preservation of relative directions:** spatial relations such as north-south and east-west should be maintained. **4. Topological accuracy:** geographically adjacent regions should be adjacent in the cartogram, and vice versa. **5. Cartographic error:** relative region sizes should be close to the data values. Criteria 1–4 describe geographical accuracy of the region arrangement. Maintaining relative directions also helps preserve a viewer's spatial mental model [30] Criterion 5 (also called statistical error) captures how well data values are represented. Often techniques aim at zero cartographic error sacrificing other criteria.

Cartograms can also be effective for showing different datasets of the same regions, arising from time-varying data such as yearly censuses yielding temporally ordered values for each region, or from available measurements of different demographic variables that we want to explore, compare and relate, yielding a vector or set of values for each region. Visualizations for multiple cartograms include animations (especially for time series), small multiples showing a matrix of cartograms, or letting a user interactively switch the mapped value in one cartogram. See for example the interactive Demers cartogram accompanying an article from the New York Times[1]. In such methods, cartograms should be as similar as the data values allow: we thus want cartograms to be *stable* by using similar layouts. This helps retain the viewer's mental map [22], supporting linking and tracking across cartograms. Thus, we obtain an important criterion with multivariate or time-varying data. **Stability:** for high stability, cartograms for the same regions using different data values should have similar layouts. The relative importance of the criteria depends on the tasks to be facilitated. Nusrat and Kobourov's taxonomy of ten tasks [25] can also be considered with multiple cartograms. Many tasks focus on the data values. As such, a representation of a region of low complexity allows for easier estimation and size comparison.

Contribution. We focus on *Demers cartograms* (DC; [3]) which represent each region by a suitably sized square, similar to Dorling cartograms [9] which use circles. Their simplicity allows easy comparison of data values, since aspect ratio is no longer a factor, unlike, e.g., for rectangular cartograms [19]. How-

---

[1] https://archive.nytimes.com/www.nytimes.com/interactive/2008/09/04/business/20080907-metrics-graphic.html, accessed June 2019.

ever, as abstract squares incur shape deformation, in spatial recognition tasks the cartogram embedding as a whole must be informative, so the layout must optimize as much as possible the other geographic criteria: *spatial deformation, preservation of relative directions* and *topological accuracy*. We contribute an efficient linear programming algorithm to compute high-quality stable DCs. Our DCs have no cartographic error, satisfy given constraints on spatial relations, and allow trade-off between topological error and stability. Linear interpolation between different DCs yields no overlap during transformation. Lost adjacencies–satisfying a mild assumption–can be shown as minimal-length planar orthogonal lines. Figure 1 shows examples. Experiments compare settings of our linear program to each other and to a force-directed layout we introduce (also novel for DCs); results show that our linear program efficiently computes stable DCs.

Fig. 1. Cartograms displaying drug poisoning mortality, total GDP and population of the contiguous states of the US in 2016. The layout minimizes distance between adjacent regions. Lost adjacencies are indicated with red leaders. Color is used only to facilitate correspondences between the cartograms. (Color figure online)

Related Work. Cartogram-like representations date to the 1800s. In the 1900s most standard cartogram types were defined, including rectangular value-by-area cartograms [26] and more recent ones [13,19]. The first automatically generated cartograms are continuous deformation ones [29] followed by others [12,15]. Dorling cartograms [9] and DCs [3] exemplify the non-contiguous type representing regions by circles and squares respectively. Layouts representing regions by rectangles and rectilinear polygons have received much attention in algorithmic literature, see e.g. [1,7,11], and typically focus on aspect ratio, topological error and region complexity. Compared to DCs, rectilinear variants have higher visual complexity and added difficulty to assessing areas. No cartogram type can guarantee a both statistically and geographically accurate representation; see a recent survey [25]. Measures exist to evaluate quality of cartogram types and algorithms, see e.g. [2,17].

There is little work on evaluating or computing stable cartograms for time-varying or multivariate data. Yet they are used in such manner, e.g., as a sequence of contiguous cartograms showing the evolution of the Internet [16].

DCs relate to contact representations, encoding adjacencies between neighboring regions as touching squares. The focus in graph theory and graph drawing literature lies on recognizing which graphs can be perfectly represented. Even the unit-disk case is NP-hard [5], though efficient algorithms exist for some restricted graph classes [8]. Klemz et al. [18] consider a vertex-weighted variant using disks, that is, with varying disk sizes. Various other techniques are similar to DCs, using squares or rectangles for geospatial information. Examples include grid maps, see, e.g., [10] for algorithms and [21] for computational experiments. Recently, Meulemans [20] introduced a constraint program to compute optimal solutions under orthogonal order constraints for diamond-shaped symbols. We use similar techniques, but refer to Sect. 2 for a discussion of the differences.

2   Computing a Single DC

First, we consider a DC for a single weight vector. We are given a set of weighted regions with their adjacencies, and a set of directional relations. We compute a layout realizing the weights with disjoint squares that may touch only if adjacent, so that directional relations are "roughly" maintained. We quantify the quality of the layout by considering the distances between any two squares representing adjacent regions. We show that the problem, under appropriate distance measures, can be solved via linear programming in polynomial time.

Formal Setting. We are given an input graph $G = (R, T)$. For each *region* $r \in R$ we are given its centroid in $\mathbb{R}^2$ and its weight $w(r)$, the side length of the square that represents it in output. The graph has an edge in T if and only if the original regions are adjacent, thus their respective squares in the output *should* be adjacent as well. We are also given two sets H, V of ordered region pairs. A pair (r, r') is in H, if r should be horizontally separated from r' such that there exists a vertical line ℓ with the square of r being left of ℓ and r' to its right. Analogously, V encodes vertical separation requirements. If r and r' are adjacent, then (r, r') is either in H or in V (but not in $H \cap V$) and they *should* touch ℓ, otherwise we require a strict separation to avoid false adjacencies; we are given a minimum gap ε to ensure that this non-adjacency can be visually recognized.[2] The sets H and V model the relative directions criterion for DCs and any two regions are paired in at least one of those sets. To ensure a DC exists satisfying the separation constraints, the directed graph $D = (R, H \cup V)$ must be a directed *acyclic* graph (DAG). We consider these relations transitive: if $(r, r') \in H$ and $(r', r'') \in H$, then this enforces that there exists a vertical line separating (r, r'') in any DCs and thus (r, r'') is in H.

The output—a placement of a square for each region—can be stored as a point $P \colon R \to \mathbb{R}^2$ for each region, encoding the center of its square. A placement P is *valid*, if it satisfies the separation constraints of H and V. This implies all squares are pairwise interior disjoint (or fully disjoint for nonadjacent regions). We look for a valid placement where distances between non-touching squares of originally adjacent regions are minimized; this will be made more precise below.

---

[2] In the implementation, ε is the minimum of the side length of the smallest region and 5% of the diagonal of the bounding box of the input regions R.

Deriving Separation Constraints. The regions' weights are given and their adjacencies and centroids easily derived, but separation constraints H and V are not. Various models can determine good directions or separation constraints [6]. We use the following model; it is symmetric and ensures constraints form a DAG.

For two regions (r, r') represented by centroids, we check whether their horizontal or vertical distance is larger. In the former case, we add (r, r') to H if r is left of r' and (r', r) to H otherwise. In the latter case, we add the pair to V in the appropriate order. We call this the *weak setting*. We call constraints added in this setting *primary separation constraints*.

In the *strong setting*, we may add an extra constraint for nonadjacent region pairs whose bounding boxes admit both horizontal and vertical separating lines: if a pair has a primary separation constraint in H or V, we add a *secondary separation constraint* to V or H respectively.

Linear Program. We model optimal solutions to the problem via a polynomially-sized linear program (LP), which lets us solve the problem in polynomial time. For each $r \in R$, we introduce variables x_r and y_r for the center $P(r) = (x_r, y_r)$ of the square. For any originally adjacent regions $\{r, r'\} \in T$ we introduce variables $h_{r,r'}$ and $v_{r,r'}$ for the (non-negative) distance between two squares. For any two regions r, r', we define shorthands: let $w_{r,r'} := \frac{(w(r)+w(r'))}{2}$ and let $gap_{r,r'} = \varepsilon$ if $\{r, r'\} \notin T$, and 0 otherwise.

$$\min \sum_{\{r,r'\}\in T} h_{r,r'} + v_{r,r'} \tag{1}$$

$$x_{r'} - x_r \geq w_{r,r'} + gap_{r,r'} \qquad\qquad \forall(r, r') \in H \quad (2)$$

$$y_{r'} - y_r \geq w_{r,r'} + gap_{r,r'} \qquad\qquad \forall(r, r') \in V \quad (3)$$

$$h_{r,r'} \geq \max\{(x_r - x_{r'}) - w_{r,r'}, (x_{r'} - x_r) - w_{r,r'}\} \qquad \forall\{r, r'\} \in T \quad (4)$$

$$v_{r,r'} \geq \max\{(y_r - y_{r'}) - w_{r,r'}, (y_{r'} - y_r) - w_{r,r'}\} \qquad \forall\{r, r'\} \in T \quad (5)$$

$$h_{r,r'}, v_{r,r'} \geq 0 \qquad\qquad\qquad\qquad \forall\{r, r'\} \in T \quad (6)$$

The objective (1) minimizes a sum of the distances between regions with broken adjacencies in the L_1 metric. Constraints (2) and (3) ensure separation requirements by forcing square centers far enough apart. For nonadjacent regions, the *gap* function assures a recognizable gap of width ε between resulting squares. Constraints (4)–(6) bind distance variables h, v with positional variables x, y. Here (4) and (5) encode two linear constraints per line, one for each term in the 'max' function. As (1) minimizes the distances, it suffices to enforce lower bounds, hence the '$\geq$' in the constraints. In an optimal solution, either one of the two versions, or the non-negativity constraint (6) will be satisfied with equality.

Improving the Gaps. The above model has two minor flaws. *First*, two squares 'touch' even if they only do so at corners; we resolve this by adding ε to the right-hand side of (4) (or (5)) for vertically (or horizontally, respectively) separated region pairs in T. This allows $h_{r,r'} = 0$ ($v_{r,r'} = 0$), when squares share a segment at least ε long. *Second*, in the strong setting the LP asks for a minimum gap ε

along both axes. This is not not needed for visual separation, so we remove the gap requirement from the secondary separation constraint.

Fine-Tuning the Optimization Criteria. The LP minimizes a sum of distances between adjacent regions. Cartogram literature emphasizes counting lost adjacencies between regions, not the distance between them. We prefer our measure since *(1)* there is a big difference if two neighboring countries are set apart by a small or large gap; *(2)* while the LP can be turned to an *integer* linear program to count lost adjacencies, it greatly increases computational complexity—optimizing for adjacencies is typically NP-hard, e.g., for disks [4,5] or segments [14].

Our linear program typically admits several optimal solutions, due to translation invariance and since touching squares may slide freely along each other as long as they touch. We introduce a *secondary* term to the objective to nuance selection of better layouts, multiplied by a *small* constant to not interfere with the original (primary) objective. The secondary term optimizes preservation of relative directions between squares within the freedom of the optimal solution.

Consider regions r and r'. W.l.o.g., assume their original centroids are horizontally farther apart than vertically, and r is left of r', so $(r, r') \in H$. We compute a directional deviation $d_{rr'} = |(y_r + \alpha(x_{r'} - x_r)) - y_{r'}|$, where α is the (finite) slope of the ray from r to r' in the input graph G. Similar to (4), the objective function will minimize $d_{rr'}$; we weigh this term more heavily for adjacent regions. We thus turn the above formula into two linear inequalities.

Alternatives exist for the secondary criterion: displacement from the original location helps find layouts maintaining many adjacencies for grid maps of equal-size squares [10,21]. For each region we measure L_1 displacement from its *origin* (centroid of the original region in the geographic map) to the square center $P(r)$.

Comparison to Overlap Removal. A technique placing disjoint squares exists to remove overlap of diamond (45 degree rotated square) glyphs for spatial point data [20], asking to minimally displace varying-size diamonds to remove all overlap, constrained to keep orthogonal order of their centers. Rotating the scenario to yield squares does not yield axis-parallel order constraints but "diagonal" ones, different from our strong setting. A "weak order constraints" variant is mentioned, related to our LP in the weak setting, if we change our objective to one only optimizing displacement relative to original locations. Figure 2 shows similarities and differences considering the feasibility area between two regions. Extensions in [20] can be applied in our scenario, e.g., reducing actively considered separation constraints by removing transitive relations ("dominance" in [20]). Time-varying data is briefly considered in [20], only conceptualizing a trade-off between origin-displacement and stability for artificial data; we discuss several optimization criteria, also focusing on adjacencies which are not considered in [20], use real-world data experiments, and compare to a baseline DC implementation to move beyond the limits of linear programming.

The lemma below matches an observation from [20] that carries over to our setting. It implies that cartograms for different weight functions but with the

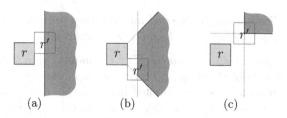

Fig. 2. Feasibility area where r' may be placed w.r.t r, when r' is primarily to the right of r. (a) In terms of feasibility, our weak setting and weak order constraints in [20] coincide. (b) Feasibility using a rotated orthogonal order [20]. (c) Feasibility in the strong setting.

same constraints have a smooth and simple transition between any such DCs helping to retain the user's mental map.

Lemma 1. *Let R be a set of regions with separation constraints H and V. Let A and B be two DCs for R, both satisfying H and V. Then, any linear interpolation between A to B also satisfies H and V and is thus overlap-free.*

3 Computing Stable DCs for Multiple Weights

The method can be extended for regions having multiple weights. We are given a set of weight functions $W = \{w_1, \ldots, w_k\}$. We aim to compute a DC for each $w_i \in W$, i.e., positions $P_i(r)$ for each $r \in R$ and $w_i \in W$. If each weight function represents the same data semantic, say population size, at different times, we consider $W = \{w_1, \ldots, w_k\}$ ordered by the k time steps; we call this setting *time series*. If each weight function represents measurements of different data semantics (possibly at the same time), say population and gross domestic product, we treat W as an unordered set; we call this setting *weight vectors*.

As we focus on cartogram stability over multiple datasets, we combine the weight functions into one LP that computes the set of DCs, with potentially different centers $P_i(r)$ for each region r and weight function w_i. This lets us add constraints and optimization objectives for stability. We change objective (1) and add constraints to minimize displacement between centers of the same region for different weight functions. We re-use notation in Sect. 2 with superscript i denoting respective variables for weight function $w_i \in W$.

$$\min \sum_{i=1}^{k} \sum_{\{r,r'\} \in T} (h_{r,r'}^i + v_{r,r'}^i) + \sum_{\{i,j\} \in I} \sum_{r \in R} (c_r^{i,j} + d_r^{i,j}) \qquad (7)$$

$$c_r^{i,j} \geq \max\{(x_r^i - x_r^j), (x_r^j - x_r^i)\} \qquad \forall r \in R, \{i,j\} \in I \qquad (8)$$

$$d_r^{i,j} \geq \max\{(y_r^i - y_r^j), (y_r^j - y_r^i)\} \qquad \forall r \in R, \{i,j\} \in I \qquad (9)$$

Here set I contains index pairs of weight functions $\{w_i, w_j\}$ for which displacement should be minimized. For each $r \in R$, variables $c_r^{i,j}$ and $d_r^{i,j}$ measure the horizontal and vertical displacement between $P_i(r)$ and $P_j(r)$ due to (8) and (9).

For which weight functions to relate in I, we consider two options: (1) relate *all pairs* of functions so $I = \binom{W}{2}$, which is natural for weight vectors where an analyst may want to compare the DCs for any two weight functions; and (2) relating *consecutive* pairs in a predefined order of the functions so $I = \{(i, i+1) \mid 1 \le i \le k-1\}$, which is natural for time series. An alternative (3) for time series initially computes a DC for w_1 (e.g., minimizing displacement to region centroids in the initial map) and then iteratively solves the LP for one DC and weight function w_i ($i \ge 2$), where we minimize the displacement only with respect to the previously solved DC for weight function w_{i-1}. Due to its restricted solution space (3) is expected to be faster to solve than (2) but with lower stability. In some scenarios another option (4) may be worthwhile: one weight function, say w_1, may be considered central to the dataset and displacements are only considered relative to it, so I contains pairs $\{1, i\}$ for all $2 \le i \le k$.

Not all planar graphs can be represented using touching squares of any size. A real-world example is Luxembourg having three pairwise neighbors; the input graph G is a K_4. Thus any DC may need to break some adjacencies. To show lost adjacencies we use *leaders* – orthogonal polylines connecting the two squares. We want leaders to have minimal length and low complexity which we can guarantee under mild assumptions: (1) leaders can coincide with square boundaries; (2) regions to be connected are *realisable*, i.e., a valid DC (with possibly different weights) exists for each pair of regions such that they are adjacent. Let $L_1^B(r_1, r_2)$ denote the minimal L_1 distance between squares of regions r_1 and r_2 in DC B. The following lemmas are proven in Appendix A in the full version [24] – the proof of the first is constructive and gives a simple $O(n^2)$ algorithm to compute all leaders.

Lemma 2. *Consider DCs with separation constraints H, V and two regions $\{r_1, r_2\} \in T$. Let (r_1, r_2) be a minimal pair in H or V. Then, in any DC B, there is a monotone leader ℓ between r_1 and r_2 with length $L_1^B(r_1, r_2)$.*

Lemma 3. *Let $\{r_1, r_2\} \in T$ and assume a DC A exists with r_1 and r_2 adjacent, from which H and V are derived in the strong setting. Then, for any DC B satisfying H and V, a leader ℓ exists between r_1 and r_2 with at most two bends.*

4   Experimental Setup

We compare 18 variants of our linear programs with each other and to 4 variants of a baseline force-directed DC layout implementation, as described below.

Linear Programs. We categorize our method according to three criteria: (A) optimization term, (B) method of deriving constraints, and (C) how we deal with different time steps. For (A) our linear program admits three primary optimization terms: *TOP* – distance between topologically adjacent regions;

CNT – number of lost adjacencies; ORG – distance to the origin (region's centroid in the geographic map). We use the indicated primary optimization term, complemented by the secondary constraint of maintaining relative directions. For (B), separation constraints are deduced from the input map in one of two ways, S and W, matching the strong and weak case respectively. For (C), we deal with different weight values (time series/weight vectors) in three ways called *stability implementations*: CO – we add an optimization term to minimize distance between layouts of all (complete) weight value pairs; (2) SU – we add an optimization term to minimize distance between layouts of successive weight values; (3) IT – we iteratively solve a linear program including an optimization term to minimize distance to previously calculated layouts. We specify our methods by concatenating the three aspects in order, for example, TOP-S-SU indicates the linear program optimized for distances of topologically adjacent regions with strong separation constraints and with successive weight values linked.

Force-Directed Method. DCs are hard to track down in literature, especially regarding computation. To our knowledge, there is no common baseline for computing a DC; we introduce a simple one. As Dorling cartograms and DCs are similar [3] and Dorling cartograms use a force-directed method, we implement one here, too: FRC. For each pair of regions we define a disjointness force based on Chebyshev distance between their centers, which grows quadratically to push squares apart. We use the same desired distance as in Sect. 2 at which this force becomes zero. We also add a force for cartogram quality, either towards their original locations (FRC-O) or between adjacent regions (FRC-T). We initialize the process with map locations (U; unstable) or the result for previous weights (S; stable). See Appendix B in the full version [24] for more details.

Metrics: Cartogram Quality. Our algorithms inherently yield zero cartographic error, and shape deformation is constant over all possible DCs. To evaluate cartogram quality we use three metrics, each normalized between 0 and 1; smaller values are better. We measure topological accuracy as the number of lost adjacencies (**MADJ**) in each of the k computed layouts, normalized by the number of adjacencies $k|T|$. To measure preservation of relative directions (**MREL**) with respect to the input map, we use the Relative Position Change Metric [28] which captures the preservation of the spatial mental model (orthogonal order) in a fine-grained way. Each rectangle defines eight zones by extending its sides to infinite lines. Between a pair of input map regions (r, r') we consider fractions of the bounding box that fall into each zone; if bounding boxes overlap, we scale values so they sum to 1. We do the same between the corresponding squares in the cartogram layouts. The measure between two regions is half the sum over all absolute differences between fractions per zone; the value is in $[0, 1]$ but is not symmetric. Finally, we take the average over all pairs. For spatial deformation we measure distance to map origins (**MDIS**), average L_1 distance of each region r in the DC to its origin (centroid of r in the geographic map), normalized by dividing with the L_1 distance of the diagonal of the map.

Metrics: Stability. We also want to assess stability, or layout similarity, between the DCs by two quality metrics, based on treemap stability metrics [28], interpreting DCs as special treemaps with added whitespace. The first is based on geometric distances between the layouts: the layout distance (**SDIS**) focuses on the change in position of the squares. The layout distance change function as presented by Shneiderman and Wattenberg [27] is the most common one. It measures Euclidean distance between rectangles r and r'. We take the average over all pairs, and normalize by dividing with the L_1 distance of the largest diagonal of the two DCs. The result is related to our optimization term for quality when dealing with multiple weights (see Sect. 3). The second metric, relative directions between layouts (**SREL**), focuses on changes in relative directions; it is analogous to MREL, but compares two layouts instead.

Datasets. We run experiments on real-world datasets. For time-series data, we expect a gradual change and strong correlation between the different values. For weight-vectors data, we expect more erratic changes and less correlation. We use two maps with rather different geographic structures: the first (**World**) is a map of world countries, having mixed region (country) sizes in a rather unstructured manner; the second (**US**) is a map of the 48 contiguous US states, having relatively high structure in sizes of its states, with large states in the middle and along the west coast and many smaller states along the east coast. We collected five time series for the World and four for the US map of which the details are given in Appendix C in the full version [24]. We transformed these into a weight-vectors dataset by taking the values of 2016 for each of these time series, resulting in five weight vectors for the World map, and four for the US map.

The various datasets have different scales, and need be projected into a reasonable square size to compute a DC. We compute the diagonal Δ of the bounding box of the map. For a time-series dataset, we find the region r with maximal $w_i(r)$ for any i and scale values such that $w_i(r) = \Delta/4$. For a weight-vectors dataset, we do the same, but scale the values for each DC separately.

Running Times. We ran the experiments using IBM ILOG CPLEX 12.8 to solve the (I)LP. We observe the following running times on a normal laptop: *-*-IT and FRC-O-* finished within seconds (USA) or a minute (World); *-*-{SU,CO} took around a minute (USA) or below 5 min (World); FRC-T-* was completed in minutes (USA) or hours (World). CNT-*-* is an *integer* linear program rather than a regular linear program (or force-directed method); its computational complexity is significantly higher, and intractable in many cases. Only CNT-*-IT variants were successfully solved, and only on the US map; for all other cases it ran out of memory (48 GB allocated).

5   Experimental Results

We discuss results and four questions: (1) How much does the strong versus weak setting affect quality? (2) How much does stability implementation matter?

(3) Which optimization criteria perform best? (4) What is the effect of separation constraints in our LP, compared to a force-directed method for DCs? Figure 3 shows the result of two algorithms for the US. Appendix D and the supplementary video in the full version [24] show more DCs for different settings.

Fig. 3. US election turnout data DC in 2016, by TOP-W-CO and CNT-W-IT.

Strong Versus Weak Setting. Figure 4 shows the average metric values for the iterative variants, over all datasets and linear programs. We find that the strong case (additional separation constraints) reduces the error in relative direction for both cartogram quality and stability: the average score for MREL, including CNT variants where possible, reduces from 0.21 to 0.16; similarly, stability (SREL) decreases from 0.059 to 0.045 due to decreased movement freedom of the squares. This is at the expense of topological error (MADJ increases from 0.58 to 0.61) and origin displacement (MDIS increases from 0.16 to 0.17). The effect is present independent of optimization criterion and stability implementation though its strength varies. Effects remain noticeable but of varying strength when we control for type of dataset, except MDIS slightly decreases for US datasets (0.116 to 0.107) in the strong setting. We also see a clear difference between optimization terms (CNT, TOP, ORG), discussed later.

Stability Implementation. In time-series datasets there is little difference in stability over the three settings: time series data change gradually over time so choosing which pairs to optimize does not have much influence. In weight-vectors datasets, even with only few weights per region (five for the World, four for the US), an effect becomes noticeable in the IT setting. CO and SU behave nearly identical, but this might be an artifact of only having a few weights per region. Compared to CO (and SU) setting, the iterative version scores better on MDIS (0.31 versus 0.26) but worse on the stability metric SDIS (0.084 versus 0.10). For weight-vectors datasets it is thus better to use the SU variant as this achieves better stability and is only slightly more expensive to compute compared to IT variants. The added complexity of CO does not seem to pay off.

Optimization Criteria. We use three metrics for cartogram quality: MADJ and MDIS are optimized explicitly with the CNT and ORG objectives respectively,

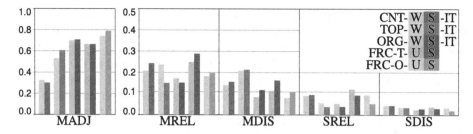

Fig. 4. Bar chart of average metric scores, for IT settings of all linear programs and the FRC directed variants. We see similar effects when switching from the weak version to the strong version of the IT setting for all three optimization settings. We also see a strong effect when choosing different optimization settings for the IT setting. FRC is generally outperformed by the {TOP,ORG}-W-IT variants.

the third metric MREL corresponds to a secondary objective term. To compare the TOP/CNT/ORG objective terms, we consider the IT variant (see Fig. 4), as other stability implementations could not solve the CNT objective; still, we found similar patterns for the SU and CO cases.

For MADJ, CNT finds the optimal value (0.31) under the given constraints. TOP (0.57) does clearly better than ORG (0.70), somewhat in contrast to observations of [10,21]: for grid maps, the MDIS metric that ORG optimizes is a good proxy for maintaining topology; our results suggest this is not so for DCs.

For MDIS and MREL metrics ORG performs best; for MDIS, CNT performs slightly better compared to TOP and vice versa for MREL. Thus, in terms of spatial quality, ORG seems a good objective, except for topological error – which is typically of primary concern for cartograms.

For stability metrics SDIS and SREL, ORG outperforms TOP which outperforms CNT. We explain it by inherent stability of the map which is the same for all DCs. CNT does poorly; it is fairly unconstrained for lost adjacencies whereas TOP aims to keep such pairs close.

ORG scores best on all metrics except MADJ; its MADJ score is high, losing 70% of adjacencies on average. In contrast, CNT optimizes the number of adjacencies, but is clearly worse on other metrics and is computationally expensive. There is thus a trade-off present between topological error and other quality aspects. TOP makes this trade-off, scoring reasonably on most metrics.

Comparison to FRC. Our linear programs enforce separation constraints which help maintain spatial relations and the spatial mental model; they are required for the linear program but not in general. To study their effect, we compare to FRC which does not enforce separation constraints; results are shown in Fig. 4. Comparing FRC-T and FRC-O variants, we see the same behavior as in the TOP versus ORG linear programs: FRC-O performs worse than FRC-T on ADJ, and better on the other metrics. Layout initialization trades off stability versus cartogram quality: FRC-*-S variants have better stability scores and worse quality scores compared to FRC-*-U.

As it has the fewest constraints, we compare ORG-W-IT to FRC methods: FRC-O-* are slightly worse or equal to ORG-W-IT on all metrics; FRC-T-* are worse than ORG-W-IT on all metrics except ADJ where it is a lightly better, but the number of adjacencies lost is still clearly higher compared to TOP-W-IT.

To conclude, in general we outperform FRC for the various metrics by an appropriate setting in our linear program. No single setting outperforms all FRC variants. The large difference with TOP-variants in terms of MADJ suggests TOP variants are a good choice for high-quality stable DCs.

6   Discussion and Future Work

We described a linear program to compute stable Demers cartograms, based on separation constraints and minimizing distance between adjacent regions. It allows overlap-free transitions between weight functions and connecting lost adjacencies with short, low-complexity leaders. Experiments show it offers a good trade-off between topological error and other criteria. It outperforms basic force-directed layouts, though there is not a unique variant that does so, suggesting an interplay between separation constraints, optimization and quality metrics.

In future work we may consider stability in other cartogram styles, and perform human-centered comparisons in addition to computational ones, with methods implemented in interactive systems; such systems can, e.g., emphasize adjacent regions by drawing leaders (at all or more clearly) or link regions back to the geographic map. We focused on Demers cartograms, but there are many different styles of cartograms. Future work may also investigate stable variants of such other cartogram styles and quantitatively or qualitatively compare them.

References

1. Alam, M.J., Biedl, T., Felsner, S., Kaufmann, M., Kobourov, S.G., Ueckerdt, T.: Computing cartograms with optimal complexity. Discrete Comput. Geom. **50**(3), 784–810 (2013). https://doi.org/10.1007/s00454-013-9521-1
2. Alam, M.J., Kobourov, S.G., Veeramoni, S.: Quantitative measures for cartogram generation techniques. Comput. Graph. Forum **34**(3), 351–360 (2015). https://doi.org/10.1111/cgf.12647
3. Bortins, I., Demers, S., Clarke, K.: Cartogram types (2002). http://www.ncgia.ucsb.edu/projects/Cartogram_Central/types.html
4. Bowen, C., Durocher, S., Löffler, M., Rounds, A., Schulz, A., Tóth, C.D.: Realization of simply connected polygonal linkages and recognition of unit disk contact trees. In: Di Giacomo, E., Lubiw, A. (eds.) Graph Drawing and Network Visualization (GD). LNCS, vol. 9411, pp. 447–459. Springer, Heidelberg (2015). https://doi.org/10.1007/978-3-319-27261-0_37
5. Breu, H., Kirkpatrick, D.G.: Unit disk graph recognition is NP-hard. Comput. Geom. **9**(1–2), 3–24 (1998). https://doi.org/10.1016/S0925-7721(97)00014-X
6. Buchin, K., Kusters, V., Speckmann, B., Staals, F., Vasilescu, B.: A splitting line model for directional relations. In: Advances in Geographic Information Systems (SIGSPATIAL), pp. 142–151. ACM (2011). https://doi.org/10.1145/2093973.2093994

7. Buchin, K., Speckmann, B., Verdonschot, S.: Evolution strategies for optimizing rectangular cartograms. In: Xiao, N., Kwan, M.P., Goodchild, M.F., Shekhar, S. (eds.) Geographic Information Science (GIScience). LNCS, vol. 7478, pp. 29–42. Springer, Heidelberg (2012). https://doi.org/10.1007/978-3-642-33024-7_3

8. Di Giacomo, E., Didimo, W., Hong, S.H., Kaufmann, M., Kobourov, S.G., Liotta, G., Misue, K., Symvonis, A., Yen, H.C.: Low ply graph drawing. In: Information, Intelligence, Systems and Applications (IISA). IEEE (2015). https://doi.org/10.1109/IISA.2015.7388020

9. Dorling, D.: Area cartograms: their use and creation. No. 59 in Concepts and Techniques in Modern Geography, University of East Anglia: Environmental Publications (1996). http://www.dannydorling.org/?page_id=1448

10. Eppstein, D., van Kreveld, M., Speckmann, B., Staals, F.: Improved grid map layout by point set matching. Int. J. Comput. Geom. Appl. **25**(02), 101–122 (2015). https://doi.org/10.1142/S0218195915500077

11. Eppstein, D., Mumford, E., Speckmann, B., Verbeek, K.: Area-universal rectangular layouts. In: Symposium on Computational Geometry (SoCG), pp. 267–276. ACM (2009). https://doi.org/10.1145/1542362.1542411

12. Gastner, M., Newman, M.: Diffusion-based method for producing density-equalizing maps. Proc. Natl. Acad. Sci. U.S.A. **101**, 7499–7504 (2004). https://doi.org/10.1073/pnas.0400280101

13. Heilmann, R., Keim, D., Panse, C., Sips, M.: RecMap: rectangular map approximations. In: Information Visualization (InfoVis), pp. 33–40. IEEE (2004). https://doi.org/10.1109/INFVIS.2004.57

14. Hliněný, P.: Contact graphs of line segments are NP-complete. Discrete Math. **235**(1–3), 95–106 (2001). https://doi.org/10.1016/S0012-365X(00)00263-6

15. House, D.H., Kocmoud, C.J.: Continuous cartogram construction. In: IEEE Conference on Visualization, pp. 197–204 (1998). https://doi.org/10.1109/VISUAL.1998.745303

16. Johnson, T., Acedo, C., Kobourov, S., Nusrat, S.: Analyzing the evolution of the Internet. In: Eurographics Conference on Visualization (EuroVis), pp. 43–47. Eurographics Association (2015). https://doi.org/10.2312/eurovisshort.20151123

17. Keim, D.A., North, S.C., Panse, C.: CartoDraw: a fast algorithm for generating contiguous cartograms. IEEE Trans. Vis. Comput. Graph. **10**(1), 95–110 (2004). https://doi.org/10.1109/TVCG.2004.1260761

18. Klemz, B., Nöllenburg, M., Prutkin, R.: Recognizing weighted disk contact graphs. In: Di Giacomo, E., Lubiw, A. (eds.) Graph Drawing and Network Visualization (GD). LNCS, vol. 9411, pp. 433–446. Springer, Heidelberg (2015). https://doi.org/10.1007/978-3-319-27261-0_36

19. van Kreveld, M., Speckmann, B.: On rectangular cartograms. Comput. Geom. **37**(3), 175–187 (2007). https://doi.org/10.1016/j.comgeo.2006.06.002

20. Meulemans, W.: Efficient optimal overlap removal: algorithms and experiments. Comput. Graph. Forum **38**(3), 713–723 (2019). https://doi.org/10.1111/cgf.13722

21. Meulemans, W., Dykes, J., Slingsby, A., Turkay, C., Wood, J.: Small multiples with gaps. IEEE Trans. Vis. Comput. Graph. **23**(1), 381–390 (2017). https://doi.org/10.1109/TVCG.2016.2598542

22. Misue, K., Eades, P., Lai, W., Sugiyama, K.: Layout adjustment and the mental map. J. Vis. Lang. Comput. **6**(2), 183–210 (1995). https://doi.org/10.1006/jvlc.1995.1010

23. Munzner, T.: Visualization Analysis and Design. AK Peters/CRC Press, Natick/Boca Raton (2014)

24. Nickel, S., Sondag, M., Meulemans, W., Chimani, M., Kobourov, S., Peltonen, J., Nöllenburg, M.: Computing stable Demers cartograms. CoRR abs/1908.07291 (2019). http://arxiv.org/abs/1908.07291
25. Nusrat, S., Kobourov, S.: The state of the art in cartograms. Comput. Graph. Forum **35**(3), 619–642 (2016). https://doi.org/10.1111/cgf.12932
26. Raisz, E.: The rectangular statistical cartogram. Geogr. Rev. **24**(2), 292–296 (1934). https://doi.org/10.2307/208794
27. Shneiderman, B., Wattenberg, M.: Ordered treemap layouts. In: Information Visualization (InfoVis), pp. 73–78. IEEE (2001). https://doi.org/10.1109/INFVIS.2001.963283
28. Sondag, M., Speckmann, B., Verbeek, K.: Stable treemaps via local moves. IEEE Trans. Vis. Comput. Graph. **24**(1), 729–738 (2018). https://doi.org/10.1109/TVCG.2017.2745140
29. Tobler, W.R.: A continuous transformation useful for districting. Ann. New York Acad. Sci. **219**, 215–220 (1973). https://doi.org/10.1111/j.1749-6632.1973.tb41401.x
30. Tversky, B.: Cognitive maps, cognitive collages, and spatial mental models. In: Frank, A.U., Campari, I. (eds.) Spatial Information Theory (COSIT). LNCS, vol. 716, pp. 14–24. Springer, Heidelberg (1993). https://doi.org/10.1007/3-540-57207-4_2

Geometric Graph Theory

Bundled Crossings Revisited

Steven Chaplick[1]([⊠]) [iD], Thomas C. van Dijk[1] [iD], Myroslav Kryven[1],
Ji-won Park[2], Alexander Ravsky[3], and Alexander Wolff[1] [iD]

[1] Universität Würzburg, Würzburg, Germany
steven.chaplick@uni-wuerzburg.de
[2] KAIST, Daejeon, Republic of Korea
[3] Pidstryhach Institute for Applied Problems of Mechanics and Mathematics,
National Academy of Sciences of Ukraine, Lviv, Ukraine

Abstract. An effective way to reduce clutter in a graph drawing that
has (many) crossings is to group edges that travel in parallel into *bundles*.
Each edge can participate in many such bundles. Any crossing in this
bundled graph occurs between two bundles, i.e., as a *bundled crossing*.
We consider the problem of bundled crossing minimization: A graph is
given and the goal is to find a bundled drawing with at most k bundled
crossings. We show that the problem is NP-hard when we require a simple
drawing. Our main result is an FPT algorithm (in k) when we require a
simple circular layout. These results make use of the connection between
bundled crossings and graph genus.

1 Introduction

In traditional node–link diagrams, vertices are mapped to points in the plane
and edges are usually drawn as straight-line segments connecting the vertices.
For large and dense graphs, however, such layouts tend to be so cluttered that it
is hard to see any structure in the data. For this reason, Holten [16] introduced
bundled drawings, where edges that are close together and roughly go into the
same direction are drawn using Bézier curves such that the grouping becomes
visible. Due to the practical effectiveness of this approach, it has quickly been
adopted by the InfoVis community [9,15,17,18,24]. However, bundled drawings
have only recently attracted study from a theoretical point of view [1,11,13,14].

Crossing minimization is a fundamental problem in graph drawing [25]. Its
natural generalization in bundled drawings is bundled crossing minimization,
see Definition 1 for the formalization of a bundled crossing. In his survey on
crossing minimization, Schaefer lists the bundled crossing number as a variant
of the crossing number and suggests to study it [25, page 35].

Related Work. Fink et al. [14] considered bundled crossings (which they called
block crossings) in the context of drawing metro maps. A metro network is a

---

The full version of this article is available at ArXiv [5]. M.K. was supported by DAAD;
S.C. was supported by DFG grant WO 758/11-1.

© Springer Nature Switzerland AG 2019
D. Archambault and C. D. Tóth (Eds.): GD 2019, LNCS 11904, pp. 63–77, 2019.
https://doi.org/10.1007/978-3-030-35802-0_5

planar graph where vertices are stations and metro lines are simple paths in this graph. These paths representing metro lines can share edges. They enter an edge at one endpoint in some linear order, follow the edge as x-monotone curves (considering the edge as horizontal), and then leave the edge at the other endpoint in some linear order. In order to improve the readability of metro maps, the authors suggested to bundle crossings. The authors then studied the problem of minimizing bundled crossings in such metro maps. Fink et al. also introduced *monotone* bundled crossing minimization where each pair of lines can intersect at most once. Later, Fink et al. [11] applied the concept of bundled crossings to drawing storyline visualizations. A storyline visualization is a set of x-monotone curves where the x-axis represents time in a story. Given a set of *meetings* (subsets of the curves that must be consecutive at given points in time), the task is to find a drawing that realizes the meetings and minimizes the number of bundled crossings. Fink et al. showed that, in this setting, minimizing bundled crossings is fixed-parameter tractable (FPT) and can be approximated in a restricted case. Our research builds on recent works of Fink et al. [13] and Alam et al. [1], who extended the notion of bundled crossings from sets of x-monotone curves to general drawings of graphs – details below.

Notation and Definitions. In graph drawing, it is common to define a drawing of a graph as a function that maps vertices to points in the plane and edges to Jordan arcs that connect the corresponding points. In this paper, we are less restrictive in that we sometimes allow edges to self-intersect. We will often identify vertices with their points and edges with their curves. Moreover, we assume that each pair of edges shares at most a finite number of points, that edges can touch (that is, be tangent to) each other only at endpoints, and that any point of the plane that is not a vertex is contained in at most two edges. A drawing is *simple* if any two edges intersect at most once and no edge self-intersects. We consider both simple and non-simple drawings; look ahead at Fig. 2 for a simple and a non-simple drawing of $K_{3,3}$.

Definition 1 (Bundled Crossing). *Let D be a drawing, not necessarily simple, and let $I(D)$ be the set of intersection points among the edges (not including the vertices) in D. We say that a* bundling *of D is a partition of $I(D)$ into bundled crossings, where a set $B \subseteq I(D)$ is a* bundled crossing *if the following holds (see Fig. 1).*

- *B is contained in a closed Jordan region $R(B)$ whose boundary consists of four Jordan arcs $\tilde{e}_1$, $\tilde{e}_2$, $\tilde{e}_3$, and $\tilde{e}_4$ that are pieces of edges e_1, e_2, e_3, and e_4 in D (a piece of an edge e is $D[e] \bigcap R(B)$); when the edge pieces are not distinct, we define $R(B)$ not as a Jordan region but as an arc or a point.*
- *The pieces of the edges cut out by the region $R(B)$ can be partitioned into two sets $\tilde{E}_1$ and $\tilde{E}_2$ such that $\tilde{e}_1, \tilde{e}_3 \in \tilde{E}_1$, $\tilde{e}_2, \tilde{e}_4 \in \tilde{E}_2$, and each pair of edge pieces in $\tilde{E}_1 \times \tilde{E}_2$ has exactly one intersection point in $R(B)$, whereas no two edge pieces in $\tilde{E}_1$ (respectively $\tilde{E}_2$) have a common point in $R(B)$.*

Our definition is similar to that of Alam et al. [1] but defines the Jordan region $R(B)$ more precisely. We call the sets $\tilde{E}_1$ and $\tilde{E}_2$ of edge pieces *bundles*

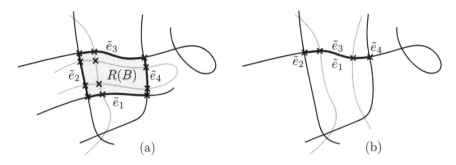

Fig. 1. (a) A non-degenerate bundled crossing B and (b) a degenerate bundled crossing B'; crossings belonging to a bundled crossing are marked with crosses

and the Jordan arcs $\tilde{e}_1, \tilde{e}_3 \in \tilde{E}_1$ and $\tilde{e}_2, \tilde{e}_4 \in \tilde{E}_2$ *frame arcs* of the bundles $\tilde{E}_1$ and $\tilde{E}_2$, respectively. For simple drawings, we accordingly call the edges that bound the two bundles of a bundled crossing *frame edges*. We say that a bundled crossing is *degenerate* if at least one of the bundles consists of only one edge piece; see Fig. 1(b). In this case, the region of the plane associated with the crossing coincides with that edge piece. In particular, any point in $I(D)$ by itself is a degenerate bundled crossing. Hence, every drawing admits a trivial bundling.

We use $\mathrm{bc}(G)$ to denote the *bundled crossing number* of a graph G, i.e., the smallest number of bundled crossings over all bundlings of all simple drawings of G. When we do not insist on simple drawings, we denote the corresponding number by $\mathrm{bc}'(G)$. In the circular setting, where vertices are required to lie on the boundary of a disk and edges inside this disk, we consider the analogous *circular bundled crossing numbers* $\mathrm{bc}^\circ(G)$ and $\mathrm{bc}^{\circ'}(G)$ of a graph G.

Fink et al. [13] showed that it is NP-hard to compute the minimum number of bundled crossings that a given drawing of a graph can be partitioned into. They also showed that this problem generalizes the problem of partitioning a rectilinear polygon with holes into the minimum number of rectangles, and they exploited this connection to construct a 10-approximation for computing the number of bundled crossings in the case of a *fixed circular drawing*. They left open the computational complexity of the general and the circular bundled crossing number for the case that the drawing is not fixed.

Alam et al. [1] showed that $\mathrm{bc}'(G)$ equals the orientable genus of G, which in general is NP-hard to compute [26]. They also showed that there is a graph G with $\mathrm{bc}'(G) \neq \mathrm{bc}(G)$ by proving that $\mathrm{bc}'(K_6) = 1 < \mathrm{bc}(K_6)$. As it turns out, the two problem variants differ in the circular setting, too (see Fig. 2 and Observation 2). For computing $\mathrm{bc}(G)$ and $\mathrm{bc}^\circ(G)$, Alam et al. [1] gave an algorithm whose approximation factor depends on the density of the graph. They posed the existence of an FPT algorithm for $\mathrm{bc}^\circ(G)$ as an open question.

Our Contribution. As some graphs G have $\mathrm{bc}'(G) \neq \mathrm{bc}(G)$ (see Fig. 2), Fink et al. [13] posed the complexity of computing the bundled crossing number $\mathrm{bc}(G)$ of a given graph G as an open problem. We settle this in Sect. 2 as follows:

Theorem 1. *Given a graph G, it is NP-hard to compute* bc(G).

Our main result, which we prove in Sect. 3, resolves an open question of Alam et al. [1] concerning the fixed-parameter tractability of bundled crossing minimization in circular layouts as follows:

Theorem 2. *There is a computable function f such that, for any n-vertex graph G and integer k, we can check, in $O(f(k)n)$ time, whether $bc^\circ(G) \le k$, i.e., whether G admits a circular layout with k bundled crossings. Within the same time bound, such a layout can be computed.*

To prove this, we use an approach similar to that of Bannister and Eppstein [3] for 1-page crossing minimization (that is, edge crossing minimization in a circular layout). Bannister and Eppstein observe that the set of crossing edges of a circular layout with k edge crossings of a graph G forms an arrangement of curves that partition the drawing into $O(k)$ subgraphs, each of which occurs in a distinct face of this arrangement. The subgraphs are obviously outerplanar. This means that G has bounded treewidth (see the full version [5]). So, by enumerating all ways to draw the crossing edges of a circular layout with k edge crossings, and, for each such way, expressing the edge partition problem (into crossing edges and outerplanar components) in extended monadic second order logic (MSO₂), *Courcelle's Theorem* [7] (stated as Theorem 5 in Sect. 3) can be applied (leading to fixed-parameter tractability).

  The difficulty in using this approach for bundled crossing minimization is in showing how to partition the graph into a set of $O(k)$ "crossing edges" (our analogy will be the frame edges) and a collection of $O(k)$ outerplanar graphs. This is where we exploit the connection to genus. Moreover, constructing an MSO₂ formula is somewhat more difficult in our case due to the more complex way our regions interact with our special set of edges.

2   Computing bc(G) Is NP-Hard

For a given graph G, finding a drawing with the fewest bundled crossings resembles computing the *orientable genus*[1] g(G) of G. In fact, Alam et al. [1] showed that $bc'(G) = g(G)$. Thus, deciding $bc'(G) = k$ for some k is NP-hard and that it is FPT in k, since the same holds for deciding g(G) = k [19,23,26].

Theorem 3 [1]. *For every graph G with genus k, it holds that* $bc'(G) = k$.

To show this, Alam et al. [1] first showed that a drawing with k bundled crossings can be lifted onto a surface of genus k, and thus $bc'(G) \ge g(G)$:

Observation 1 [1]. *A drawing D with k bundled crossings can be lifted onto a surface of genus k via a one-to-one correspondence between bundled crossings and handles, i.e., at each bundled crossing, we attach a handle for one of the two edge bundles, thus providing a crossing-free lifted drawing; see Fig. 7.*

---

[1] I.e., computing the fewest *handles* to attach to the sphere so that G can be drawn on the resulting surface without any crossings.

Then, to see that $bc'(G) \leq g(G)$, Alam et al. [1] used the *fundamental poly-gon* representation (or *polygonal schema*) [10] of a drawing on a genus-g sur-face. More precisely, the sides of the polygon are numbered in circular order $a_1, b_1, a_1', b_1', \ldots, a_g, b_g, a_g', b_g'$; for $1 \leq k \leq g$, the pairs (a_k, a_k') and (b_k, b_k') of sides are identified in opposite direction, meaning that an edge leaving side a_k appears on the corresponding position of side a_k'; see Fig. 3 for an example show-ing K_6 drawn in a fundamental square, which models a drawing on the torus. In such a representation, all vertices lie in the interior of the fundamental poly-gon and all edges leave the polygon avoiding vertices of the polygon. Alam et al. [1] showed that such a representation can be transformed into a non-simple bundled drawing with g many bundled crossings. It is not clear, however, when such a representation can be transformed into a simple bundled drawing with g bundled crossings, as this transformation can produce drawings with self-loops and pairs of edges crossing multiple times, e.g., Alam et al. [1, Lemma 1] showed that $bc(K_6) = 2$ while $bc'(K_6) = g(K_6) = 1$.

We show that the problem remains NP-hard for simple drawings.

Proof (of Theorem 1). Let G' be the graph obtained from G by subdividing each edge $O(|E(G)|^2)$ times. We reduce from the NP-hardness of computing the genus $g(G)$ of G by showing that $bc(G') = g(G)$, with Observation 1 in mind.

Consider the embedding of G onto the genus-$g(G)$ surface. By a result of Lazarus et al. [21, Theorem 1], we can construct a fundamental polygon repre-sentation of the embedding so that its boundary intersects with edges of the graph $O(g(G)|E(G)|)$ times. Note that each edge piece outside the polygon intersects each other edge piece at most once; see Fig. 3. We then subdivide the edges by adding a vertex to each intersection of an edge with the bound-ary of the fundamental polygon. This subdividing of edges ensures that no edge intersects itself or intersects another edge more than once in the correspond-ing drawing of the graph on the plane; hence, the drawing is simple. Since $g(G) \leq |E(G)|$, by subdividing edges further whenever necessary, we obtain a drawing of G'. Our subdivisions keep the integrity of all bundled crossings, so $bc(G') \leq g(G)$. On the other hand, since subdividing edges does not affect the genus, $g(G) = g(G') = bc'(G') \leq bc(G')$.   □

3   FPT Algorithms for Computing $bc^{\circ'}(G)$ and $bc^{\circ}(G)$

We now consider circular layouts, where vertices are placed on a circle and edges are routed inside the circle. We note that $bc^{\circ}(G)$ and $bc^{\circ'}(G)$ can be different.

Observation 2. $bc^{\circ'}(K_{3,3}) = 1$ *but* $bc^{\circ}(K_{3,3}) > 1$.

Proof. Let $V(K_{3,3}) = \{a, b, c\} \cup \{a', b', c'\}$. A drawing with $bc^{\circ'}(K_{3,3}) = 1$ is obtained by placing the vertices a, a', b, b', c, c' in clockwise order around a circle; see Fig. 2(b). If a graph G has $bc^{\circ}(G) = 1$ then G is planar because we can embed edges for one bundle outside the circle. Hence, $bc^{\circ}(K_{3,3}) > 1$.

Similarly to computing $bc'(G)$, we use compute $bc^{\circ'}(G)$ via computing genus.

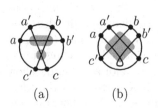

Fig. 2. $\mathrm{bc}^{\circ}(K_{3,3}) \neq \mathrm{bc}^{\circ'}(K_{3,3})$; see Observation 2

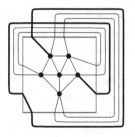

Fig. 3. K_6 drawn in a fundamental square; the self-intersecting edge is bold [1, Fig. 2].

Theorem 4. *Testing whether* $\mathrm{bc}^{\circ'}(G) = k$ *can be done in* $2^{k^{O(1)}} n$ *time.*

Proof (Sketch). This follows from the fact that $\mathrm{bc}^{\circ'}(G) = \mathrm{g}(G^{\star})$ where $G^{\star}$ is a graph with a vertex $v^{\star}$ adjacent to every vertex of G (see the full version [5] and the $2^{g^{O(1)}} n$ time algorithm for genus [19].   □

To prove our main result (Theorem 2) we develop an algorithm that tests whether $\mathrm{bc}^{\circ}(G) = k$ in FPT time with respect to k. Our algorithm is inspired by recent works on circular layouts with at most k crossings [3] and circular layouts where each edge is crossed at most k times [6]. In both of these prior works, it is first observed that the graphs admitting such circular layouts have treewidth $O(k)$, and then algorithms are developed using Courcelle's theorem, which establishes that expressions in MSO_2 logic can be evaluated efficiently. (For basic definitions of both treewidth and MSO_2 logic, see the appendix of the full version.)

Theorem 5 (Courcelle [7,8]). *For any integer* $t \geq 0$ *and any* MSO_2 *formula* ψ *of length* ℓ, *an algorithm can be constructed which takes a graph* G *with* n *vertices,* m *edges, and treewidth at most* t *and decides in* $O(f(t, \ell) \cdot (n + m))$ *time whether* $G \models \psi$ *where the function* f *from this time bound is a computable function of* t *and* ℓ.

We proceed along the lines of Bannister and Eppstein [3], who used a similar approach to show that edge crossing minimization in a circular layout is in FPT (as mentioned in the introduction). We start by very carefully describing a surface (in the spirit of Observation 1) onto which we will lift our drawing. We will then examine the structure of this surface (and our algorithm) for the case of one bundled crossing and finally for k bundled crossings.

3.1   Constructing the Surface Determined by a Bundled Drawing

Consider a bundled circular drawing D. Note that adding parallel edges to the drawing (i.e., making our graph a multi-graph) allows us to assume that every bundled crossing has four distinct frame edges and can be done without modifying the number of bundled crossings; see Fig. 7. Each bundled crossing B defines a Jordan curve made up of the four Jordan arcs $\tilde{e}_1$, $\tilde{e}_2$, $\tilde{e}_3$, $\tilde{e}_4$ in clockwise order taken from its four frame edges $e_1, \ldots, e_4$ respectively (here (e_1, e_3) and (e_2, e_4) frame the two bundles and $e_i = u_i v_i$). Similarly to Observation 1, we can construct a surface S by creating a flat handle (note that this differs from the usual definition of a handle since our flat handles have a boundary) on top of D which connects $\tilde{e}_2$ to $\tilde{e}_4$ and doing so for each bundled crossing. We then lift the drawing D onto S by rerouting the edges of one of the bundles over its corresponding handle for each bundled crossing B obtaining the lifted drawing D_S. To avoid the crossings in D_S of the frame edges that can occur at the foot of the handle of B we can make the handle a bit wider and add *corner-cuts* (as illustrated in Fig. 4) to preserve the topology of the surface. Thus, D_S is crossing-free.

We now cut S into *components* (maximal connected subsets) along the frame edges and corner-cuts of each bundled crossing, resulting in a subdivision Ω of S.

We use D_Ω to denote the sub-drawing of D_S on Ω, i.e., D_Ω is missing the frame edges since these have been cut out. We now consider the components of Ω. Notice that every edge of D_Ω is contained in one component of Ω. In order for a component s of Ω to contain an edge e of D_Ω, s must have both endpoints of e on its boundary. With this in mind we focus on the components of Ω where each one has a vertex of G on its boundary and call such components *regions*. Observe that a crossing in D which does not involve a frame edge corresponds, in D_Ω, to a pair of edges where one goes over a handle and the other goes underneath.

3.2   Recognizing a Graph with One Bundled Crossing

We now discuss how to recognize if an n-vertex graph $G = (V, E)$ can be drawn in a circular layout with one bundled crossing. Consider a bundled circular drawing D of G consisting of one bundled crossing. The bundled crossing consists of two bundles, and so a set F of four frame edges. By $V(F)$ we denote the set of vertices incident to frame edges. Via the construction above, we obtain the subdivided surface Ω; see Fig. 4. Let r_1 and r_2 be the regions that are each bounded by a pair of frame edges corresponding to one of the bundles, and let $r_3, \ldots, r_6$ be the regions each bounded by one edge from one pair and one from the other pair; see Fig. 4. These are all the regions of Ω. Since, as mentioned before, each of the non-frame edges of G (i.e., each $e \in E(G) \setminus F$) along with its two endpoints is contained in exactly one of these regions, each component of $G \setminus V(F)$ including the edges connecting it to vertices of $V(F)$ is drawn in D_Ω in some region of Ω. In this sense, for each region r of Ω, we use G_r to denote the subgraph of G induced by the components of $G \setminus V(F)$ contained in r, including the edges connecting them to vertices in $V(F)$. Additionally, each vertex of G is either incident to an edge in F (in which case it is on the boundary of at least two regions) or it is on the boundary of exactly one region.

Note that there are two types of regions: $\{r_1, r_2\}$ and $\{r_3, r_4, r_5, r_6\}$. Consider a region of the first type, say r_1; see Fig. 4. Observe that r_1 is a topological disk, i.e., G_{r_1} is outerplanar. Moreover, G_{r_1} has a special outerplanar drawing where on the boundary of r_1 (in clockwise order) we see the frame edge e_1,

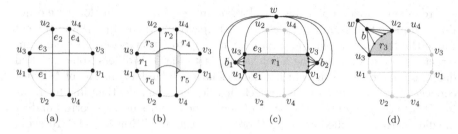

Fig. 4. (a) Bundled crossing; (b) regions, corner-cuts in blue; (c),(d) augmented graphs $G^*_{r_1}$ and $G^*_{r_3}$ consist of the edges of G_{r_1} and G_{r_3} (blue), augmentation vertices and edges (black) (Color figure online)

the vertices mapped to the (u_1, u_3)-arc, the frame edge e_3, then the vertices mapped to the (v_3, v_1)-arc. We now describe how to augment G_{r_1} to a planar graph $G^*_{r_1}$ where in every planar embedding of $G^*_{r_1}$ the sub-embedding of G_{r_1} has this special outerplanar form[2]. The vertex set of $G^*_{r_1}$ is $V(G_{r_1}) \cup \{h, b_1, b_2\}$ where we call h *hub* vertex and b_1 and b_2 *boundary* vertices (one for each arc of the boundary of r_1 to which vertices can be mapped); see Fig. 4. The graph $G^*_{r_1}$ has four types of edges; the edges in $E(G_{r_1})$, edges that make h the hub of a wheel whose cycle is $C = (v_1, b_2, v_3, u_3, b_1, u_1, v_1)$, edges from b_1 to the vertices on the (u_1, u_3)-arc, and edges from b_2 to the vertices on the (v_3, v_1)-arc (both including the end points). Clearly, we can obtain a planar embedding of $G^*_{r_1}$ by drawing the elements of $G^*_{r_1} \setminus G_{r_1}$ "outside" of the outerplanar drawing of G_{r_1} described before. Moreover, every planar embedding of $G^*_{r_1}$ contains an outerplanar embedding of G_{r_1} that can be drawn in the special form needed to "fit" into r_1, in the sense that all of G_{r_1} lies (or can be put) inside the simple cycle C. (For example, if, say, b_1 is a cut vertex, the component hanging off b_1 can be embedded in the face (h, b_1, u_3, h). But then it can easily be moved into C. Similarly, a component that is incident only to u_3 and v_3 can end up in the face (h, u_3, v_3, h), but again, the component can be moved inside C.)

Similarly, for a region of the second type, say r_3, the graph G_{r_3} is outerplanar with a special drawing where all the vertices must be on the (u_3, u_2)-arc of the disk subtended by the two frame edges e_3 and e_2 bounding the region r_3. We augment similarly as for r_1; see Fig. 4. For the augmented graph $G^*_{r_3}$, we add to G_{r_3} a boundary vertex b neighboring all vertices on the (u_3, u_2)-arc and a hub vertex h adjacent to u_2, b, and u_3. Again, $G^*_{r_3}$ is planar since G_{r_3} is outerplanar due to r_3 being a topological disk. Moreover, as b is adjacent to all vertices of G_{r_3}, in every planar embedding of $G^*_{r_3}$, G_{r_3} is embedded outerplanarly and, since b occurs on one side of the triangle $u_3 u_2 h$, the edge $u_3 u_2$ occurs on the boundary of this outerplanar embedding of G_{r_3}. Thus, each planar embedding of $G^*_{r_3}$ provides an outerplanar embedding of G_{r_3} that fits into r_3.

---

[2] This augmentation may sound overly complicated, but is written as to easily generalize to more bundled crossings.

Note that each G_{r_i} fits into r_i because its augmented graph $G^*_{r_i}$ is planar ($\star$). Moreover, as outerplanar graphs have treewidth at most two [22], each graph G_r is outerplanar, and adding the (up to) eight frame vertices raises the treewidth by at most 8, we see that the treewidth of G is at most 10. Namely, in order for G to have $\mathrm{bc}^\circ(G) = 1$, it must have treewidth at most 10 (and this can be checked in linear time using an algorithm of Bodlaender [4]).

To sum up, G has a circular drawing D with at most one bundled crossing because it has treewidth at most 10 and there exist (i) $\beta \leq 4$ frame edges $e_1, e_2, \ldots, e_\beta$ (this set is denoted F) and $v_1, \ldots, v_\xi$ frame vertices (this set is denoted V_F), (ii) a particular circular drawing D_F of frame edges, (iii) the drawing of the one bundled crossing B, and (iv) $\gamma \leq 6$ corresponding regions $r_1, \ldots, r_\gamma$ of the subdivided surface Ω so that the following properties hold (note that the frame vertices partition the boundary of the disk underlying Ω into $\eta \leq 8$ (possibly degenerate) arcs $p_1, \ldots, p_\eta$ where each such p_j is contained in a unique region r_{i_j} of Ω):

1. $E(G)$ is partitioned into $E_0, E_1, \ldots, E_\gamma$, where $E_0 = \{f_1, \ldots, f_\beta\}$.
2. $V(G)$ is partitioned into $V_0, V_1, \ldots, V_\eta$, where $V_0 = \{u_1, \ldots, u_\xi\}$.
3. The mapping $u_i \leftrightarrow v_i$ and $f_i \leftrightarrow e_i$ defines an isomorphism between the subgraph of G formed by (V_0, E_0) and graph (V_F, F).
4. No vertex in $V(G) \setminus V_0$ has incident edges $e \in E_i$, $e' \in E_j$ for $i \neq j$.
5. For each $v \in V_0$, and each edge e incident to v, exactly one of the following is true: (i) $e \in E_0$ or (ii) $e \in E_i$ and v is on the boundary of r_i.
6. For each $v \in V_j$, all edges incident to v belong to E_{i_j}.
7. For each region r_i, let G_i be the graph $(V_0 \cup \bigcup_{j:\, i_j=i} V_j, E_i)$ (i.e., the subgraph that is to be drawn in r_i), and let G^*_i be the corresponding augmented graph (i.e., as in $\star$ above). Each G^*_i is planar.

We now describe the algorithm to test for a simple circular drawing with one bundled crossing. First we check that treewidth of G is at most 10. We then enumerate drawings of up to four edges in the circle. For the drawing D_F that is valid for the set F of frame edges of one bundled crossing, we define our surface and its regions (which makes the augmentation well-defined). We have intentionally phrased these properties so that it is clear that they are expressible in MSO_2 (see the full version [5]). The only property that is not obviously expressible is the planarity of G^*_i. To this end, recall that planarity is characterized by two forbidden minors (i.e., K_5 and $K_{3,3}$) and that, for every fixed graph H, there is an MSO formula MINOR_H so that for all graphs G, it holds that $G \models \mathrm{MINOR}_H$ if and only if G contains H as a minor [8, Corollary 1.14]. Additionally, each G^*_i can be expressed as an *MSO-transduction*[3] of G and our variables (our transduction can be thought of as a kind of 2-copying transduction). Thus, by [8, Theorem 7.10] using the transduction and the MSO formula testing planarity, we can construct an MSO_2 formula ι so that when $G \models \iota$, G^*_i is planar for every i. Therefore, Properties 1–7 can be expressed as an MSO_2 formula ψ and, by

---

[3] For the formalities of transductions, see the book of Courcelle and Engelfriet [8, Section 1.7.1, and Definitions 7.6 and 7.25].

Fig. 5. Configurations for $p = 6$: (a) $D_A(p)$, (b) $D_B(p)$, and (c) induced by a hole

Courcelle's theorem, there is a computable function f such that we can test (in $O(f(\psi, t)n)$ time) whether $G \models \psi$ for an input graph G of treewidth at most t. Thus, since our graph has treewidth at most 10, applying Courcelle's theorem completes our algorithm.

3.3 Recognizing a Graph with k Bundled Crossings

We now generalize the above approach to k bundled crossings. In a drawing D of G together with a solution consisting of k bundled crossings, there are $2k$ bundles making (up to) $4k$ frame edges F. As described above, these bundled crossings provide a surface $\mathcal{S}$, its subdivision Ω, and the corresponding set of regions. The key ingredient above was that every region was a topological disk. However, that is now non-trivial as our regions can go over and under many handles. To show this property, we first consider the following two partial drawings $D_A(p)$ and $D_B(p)$ of a matching with $p + 1$ edges $f_0, f_1 \ldots, f_p$ (see, e.g., Fig. 5) such that

- edge f_i crosses only $f_{i-1 \bmod p+1}$ and $f_{i+1 \bmod p+1}$ for $i = 0, \ldots, p$;
- the endpoints of each edge f_i, $i = 1, \ldots, p - 2$, are inside the cycle C formed by the crossing points and the edge-pieces between these crossing points;
- both endpoints of f_{p-1}, only one endpoint of f_0, and only one endpoint of f_p are contained in C in the drawing $D_A(p)$;
- only one endpoint of f_{p-1}, only one endpoint of f_0, and no endpoints of f_p are contained in C in the drawing $D_B(p)$.

Note that the partial drawings $D_A(p)$ and $D_B(p)$ differ only in how the last edge is drawn with respect to the previous edge. Arroyo et al. [2, Theorem 1.2] showed that such partial drawings are obstructions for pseudolinearity, that is, they cannot be part of any pseudoline arrangement. Therefore, neither of these partial drawings can be *completed* to a simple circular drawing, that is, the endpoints of the edges cannot be extended so that they lie on a circle which contains the drawing. We highlight this fact in the following lemma.

Lemma 1. *For a matching with $p + 1$ edges $f_0, f_1, \ldots, f_p$, neither the partial drawing $D_A(p)$ nor $D_B(p)$ can be completed to a simple circular drawing.*

Using this lemma we can now prove the following statement.

Lemma 2. *Each region r of Ω is a topological disk[4].*

---

[4] We slightly abuse this notion to also mean a simply connected set.

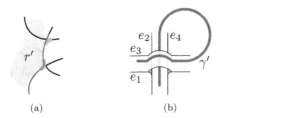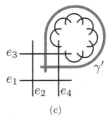

(a)   (b)   (c)

Fig. 6. (a) Projection r' of the region r and its boundary (green, the corner-cuts are in blue) onto the disk of the drawing D (b) projection γ' of a Jordan arc γ that goes over and under the same handle; (c) profile of edges of the projected boundary of r enclosed by the loop made by γ' form a partial drawing $D_A(p)$. (Color figure online)

Proof. First, we show that no region of Ω includes part of both a handle and its *undertunnel*, that is, the part of the surface over which the handle was built. Then we will show that a region also does not include holes.

Let r be a region of the surface subdivision Ω. The boundary of this region is formed by pieces of frame edges that were lifted on the surface S as described above and the additional corner-cuts as illustrated in Fig. 4 in red. Consider the projection r' of r and its boundary on the drawing D in the plane. Note that the projected boundary either follows an edge in D or switches to some another edge via a corner-cut at an intersection point; see Fig. 6(a).

Suppose now, for a contradiction, that r contains both a handle and its undertunnel corresponding to the same bundled crossing $B = ((e_1, e_3), (e_2, e_4))$.

Then there is a Jordan arc $\gamma \subset r$ going over and under this handle making a loop; see Fig. 6(b). Note that the orthogonal projection γ' of γ on the disk of the drawing D self-intersects. The profile of edges along the projected boundary of r that is enclosed by γ' then inevitably contains a partial drawing $D_A(p)$; see Fig. 6(c). And according to Lemma 1, such a partial drawing cannot be completed to a valid simple circular drawing; contradiction.

As for holes, it is easy to see that if r had a hole, the profile of the boundary edges around this hole would give a partial drawing of edges as illustrated in Fig. 5(c). Therefore, the region r is a proper topological disk.   □

The next lemma concerning treewidth is a direct consequence of Lemma 2.

Lemma 3. *If a graph G admits a circular layout with k bundled crossings then its treewidth is at most $8k + 2$.*

Proof. If the graph G can be drawn in a circular layout with k bundled crossings then there exist at most $4k$ frame edges. According to Lemma 2, the removal of their endpoints breaks up the graph into outerplanar components. The treewidth of an outerplanar graph is at most two [22]. Moreover, adding a vertex to a graph raises its treewidth by at most one. Thus, since deleting at most $8k$ frame vertices leaves behind an outerplanar graph, G has treewidth at most $8k + 2$.   □

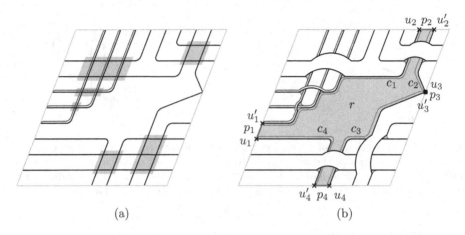

Fig. 7. (a) A bundled drawing D with six bundled crossings (pink); parallel (blue) edges can be inserted to avoid degenerate bundled crossings; (b) the corresponding surface of genus 6; the components of the surface that are not regions are marked in green; the region r (light blue) has a boundary consisting of the arcs of the disk (red) and the arcs c_1, c_2, c_3, and c_4 (traced in orange). (Color figure online)

We now prove Theorem 2, that deciding whether $bc^\circ(G) \leq k$ is FPT in k.

Proof (of Theorem 2). We use Lemma 2 and extend the algorithm of Sect. 3.2.

Suppose G has a circular drawing D with at most k bundled crossings. In D we see the set F of (up to) $4k$ frame edges of these bundled crossings. As before, F together with D defines a subdivided topological surface Ω containing a set of regions R. As in the one bundled crossing case, each edge of G is in exactly one such region, and each vertex of G either is incident to an edge in F (in which case it belongs to at least two regions) or belongs to exactly one region.

Throughout the proof we will refer to Fig. 7 for an example. By Lemma 2, each region r is a topological disk and as such its graph G_r is outerplanar with a quite special drawing D_r described as follows. In particular, if we trace the boundary of r in clockwise order, we see that it is made up of arcs $p_1, \ldots, p_\alpha$ of S, marked in red in Fig. 7(b) (such arcs can degenerate to single points), and Jordan arcs $c_1, \ldots, c_\alpha$, traced in orange in Fig. 7(b), each of which connects two such arcs of the disk. For $i \in \{1, \ldots, \alpha\}$, let u_i and u_i' be the end points of p_i, in clockwise order. So u_i' and u_{i+1} are the endpoints of c_i. No vertex of G_r lies in the interior of c_i.

We now describe G_r^*. First, we add a hub vertex h. Then, for each $i \in \{1, \ldots, \alpha\}$, if u_i' and u_{i+1} (where $u_{\alpha+1}$ is u_1) are not adjacent, we add an edge between them. If p_i is non-degenerate, we add a boundary vertex b_i adjacent to all vertices on p_i (including u_i and u_i') and make h adjacent to u_i, b_i, and u_i'. Otherwise, we make h adjacent to $u_i = u_i'$ and, for technical reasons (see the full version), we identify b_i with u_i and u_i'.

Observe that the resulting graph G_r^* is planar due to the special outerplanar drawing of G_r in r. Moreover, in every planar embedding of G_r^*, there is an outerplanar embedding of G_r where the cyclic order of the arcs c_i and the sets of vertices mapped to the p_i's match their cyclic order in r, implying that G_r fits into r. This is due to the fact that the simple cycle C' around h must be embedded planarly, with all of G_r inside (with the possible and easy-to-fix exceptions described in Sect. 3.2 concerning the cycle C there). Then the order of the vertices in an outerplanar embedding of G_r is the order of the vertices incident to $b_1, \ldots, b_\alpha$ in a planar embedding of G_r^*. So the planarity of G_r^* guarantees that G_r fits into r as needed.

The reason why G has a circular drawing D with at most k bundled crossings is that there is a β-edge k-bundled crossing drawing D_F (of the graph formed by F), whose corresponding surface $\mathcal{S}$ consists of regions $r_1, \ldots, r_\gamma$ (note: $\gamma \leq 2\beta \leq 8k$) so that Properties 1–7 hold.

Our algorithm first checks that the treewidth of G is at most $8k + 2$. Recall that this can be done in linear time (FPT in k) [4]. It then enumerates all possible simple drawings of at most $4k$ edges in the circle[5]. For each drawing, it further enumerates the possible ways to form k bundled crossings so that every edge is a frame edge of at least one bundled crossing. Then, for each such bundled drawing D_F, we build an MSO$_2$ formula φ (see the full version) to express Properties 1–7. Finally, since G has treewidth at most $8k + 2$, we can apply Courcelle's theorem on (G, φ).   □

4 Open Problems

Given our new FPT algorithm for simple circular layouts, it would be interesting to improve its runtime and to investigate whether a similar result can be obtained for general simple layouts. A starting point could be the FPT algorithm of Kawarabayashi et al. [20] for computing the usual crossing number of a graph.

Acknowledgements. We thank Bruno Courcelle for clarifying discussions on the tools available when working with his meta-theorem and in particular MSO$_2$.

References

1. Alam, M.J., Fink, M., Pupyrev, S.: The bundled crossing number. In: Hu, Y., Nöllenburg, M. (eds.) GD 2016. LNCS, vol. 9801, pp. 399–412. Springer, Cham (2016). https://doi.org/10.1007/978-3-319-50106-2_31. http://arxiv.org/abs/1608.08161
2. Arroyo, A., Bensmail, J., Richter, R.B.: Extending drawings of graphs to arrangements of pseudolines. ArXiv report (2018). https://arxiv.org/abs/1804.09317

---

[5] i.e., at most $4k$ curves extending to infinity in both directions where each pair of curves cross at most once. The number of such drawings is proportional to k, and efficient enumeration has been done for the case when every pair of curves cross exactly once [12].

3. Bannister, M.J., Eppstein, D.: Crossing minimization for 1-page and 2-page drawings of graphs with bounded treewidth. J. Graph Algorithms Appl. **22**(4), 577–606 (2018). https://doi.org/10.7155/jgaa.00479

4. Bodlaender, H.L.: A linear-time algorithm for finding tree-decompositions of small treewidth. SIAM J. Comput. **25**(6), 1305–1317 (1996). https://doi.org/10.1137/S0097539793251219

5. Chaplick, S., van Dijk, T.C., Kryven, M., won Park, J., Ravsky, A., Wolff, A.: Bundled crossings revisited. ArXiv report (2019). https://arxiv.org/abs/1812.04263

6. Chaplick, S., Kryven, M., Liotta, G., Löffler, A., Wolff, A.: Beyond outerplanarity. In: Frati, F., Ma, K.-L. (eds.) GD 2017. LNCS, vol. 10692, pp. 546–559. Springer, Cham (2018). https://doi.org/10.1007/978-3-319-73915-1_42

7. Courcelle, B.: The monadic second-order logic of graphs. I. recognizable sets of finite graphs. Inform. Comput. **85**(1), 12–75 (1990). https://doi.org/10.1016/0890-5401(90)90043-H

8. Courcelle, B., Engelfriet, J.: Graph Structure and Monadic Second-Order Logic: A Language-Theoretic Approach. Cambridge Univ Press, Cambridge (2012)

9. Cui, W., Zhou, H., Qu, H., Wong, P.C., Li, X.: Geometry-based edge clustering for graph visualization. IEEE Trans. Vis. Comput. Graph. **14**(6), 1277–1284 (2008). https://doi.org/10.1109/TVCG.2008.135

10. de Verdière, É.C.: Computational topology of graphs on surfaces. In: Tóth, C.D., O'Rourke, J., Goodman, J.E. (eds.) Handbook of Discrete and Computational Geometry, 3rd edn., chap. 23. CRC Press LLC, Boca Raton (2017)

11. van Dijk, T.C., Fink, M., Fischer, N., Lipp, F., Markfelder, P., Ravsky, A., Suri, S., Wolff, A.: Block crossings in storyline visualizations. J. Graph Algorithms Appl. **21**(5), 873–913 (2017). https://doi.org/10.7155/jgaa.00443

12. Felsner, S.: On the number of arrangements of pseudolines. In: SoCG, pp. 30–37. ACM (1996). https://doi.org/10.1145/237218.237232

13. Fink, M., Hershberger, J., Suri, S., Verbeek, K.: Bundled crossings in embedded graphs. In: Kranakis, E., Navarro, G., Chávez, E. (eds.) LATIN 2016. LNCS, vol. 9644, pp. 454–468. Springer, Heidelberg (2016). https://doi.org/10.1007/978-3-662-49529-2_34

14. Fink, M., Pupyrev, S., Wolff, A.: Ordering metro lines by block crossings. J. Graph Algorithms Appl. **19**(1), 111–153 (2015). https://doi.org/10.7155/jgaa.00351

15. Gansner, E.R., Hu, Y., North, S., Scheidegger, C.: Multilevel agglomerative edge bundling for visualizing large graphs. In: Battista, G.D., Fekete, J.D., Qu, H. (eds.) PACIFICVIS, pp. 187–194. IEEE (2011). https://doi.org/10.1109/PACIFICVIS.2011.5742389

16. Holten, D.: Hierarchical edge bundles: visualization of adjacency relations in hierarchical data. IEEE Trans. Vis. Comput. Graph. **12**(5), 741–748 (2006). https://doi.org/10.1109/TVCG.2006.147

17. Hurter, C., Ersoy, O., Fabrikant, S.I., Klein, T.R., Telea, A.C.: Bundled visualization of dynamicgraph and trail data. IEEE Trans. Vis. Comput. Graph. **20**(8), 1141–1157 (2014). https://doi.org/10.1109/TVCG.2013.246

18. Hurter, C., Ersoy, O., Telea, A.: Graph bundling by kernel density estimation. Comput. Graph. Forum **31**, 865–874 (2012). https://doi.org/10.1111/j.1467-8659.2012.03079.x

19. Kawarabayashi, K., Mohar, B., Reed, B.A.: A simpler linear time algorithm for embedding graphs into an arbitrary surface and the genus of graphs of bounded tree-width. In: FOCS, pp. 771–780. IEEE (2008). https://doi.org/10.1109/FOCS.2008.53

20. Kawarabayashi, K., Reed, B.: Computing crossing number in linear time. In: STOC, pp. 382–390. ACM (2007). https://doi.org/10.1145/1250790.1250848

21. Lazarus, F., Pocchiola, M., Vegter, G., Verroust, A.: Computing a canonical polygonal schema of an orientable triangulated surface. In: SoCG, pp. 80–89. ACM (2001). https://doi.org/10.1145/378583.378630

22. Mitchell, S.L.: Linear algorithms to recognize outerplanar and maximal outerplanar graphs. Inform. Process. Lett. **9**(5), 229–232 (1979). https://doi.org/10.1016/0020-0190(79)90075-9

23. Mohar, B.: A linear time algorithm for embedding graphs in an arbitrary surface. SIAM J. Disc. Math. **12**(1), 6–26 (1999)

24. Pupyrev, S., Nachmanson, L., Bereg, S., Holroyd, A.E.: Edge routing with ordered bundles. Comput. Geom. Theory Appl. **52**, 18–33 (2016). https://doi.org/10.1016/j.comgeo.2015.10.005

25. Schaefer, M.: The graph crossing number and its variants: a survey. Electr. J. Combin. Dynamic Survey DS21 (2017). http://www.combinatorics.org/ojs/index.php/eljc/article/view/DS21

26. Thomassen, C.: The graph genus problem is NP-complete. J. Algorithms **10**(4), 568–576 (1989). https://doi.org/10.1016/0196-6774(89)90006-0

Crossing Numbers of Beyond-Planar Graphs

Markus Chimani[1], Philipp Kindermann[2(✉)], Fabrizio Montecchiani[3], and Pavel Valtr[4]

[1] Osnabrück University, Osnabrück, Germany
markus.chimani@uos.de
[2] University of Würzburg, Würzburg, Germany
philipp.kindermann@uni-wuerzburg.de
[3] University of Perugia, Perugia, Italy
fabrizio.montecchiani@unipg.it
[4] Charles University in Prague, Prague, Czech Republic
valtr@kam.mff.cuni.cz

Abstract. We study the 1-planar, quasi-planar, and fan-planar crossing number in comparison to the (unrestricted) crossing number of graphs. We prove that there are n-vertex 1-planar (quasi-planar, fan-planar) graphs such that any 1-planar (quasi-planar, fan-planar) drawing has $\Omega(n)$ crossings, while $\mathcal{O}(1)$ crossings suffice in a crossing-minimal drawing without restrictions on local edge crossing patterns.

1 Introduction

The *crossing number* of a graph G, denoted by $\mathrm{cr}(G)$, is the smallest number of pairwise edge crossings over all possible drawings of G. Many papers are devoted to the study of this parameter, refer to [22,25] for surveys. In particular, minimizing the number of crossings is one of the seminal problems in graph drawing (see, e.g., [2,3,23]), whose importance has been further witnessed by user studies showing how edge crossings may deteriorate the readability of a diagram [20,21,26]. On the other hand, determining the crossing number of a graph is NP-hard [5] and can be solved exactly only on small/medium instances [7]. On the positive side, the crossing number is fixed-parameter tractable in the number of crossings [15] and can be approximated by a constant factor for graphs of bounded degree and genus [10].

A recent research stream studies graph drawings where, rather than minimizing the number of crossings, some edge crossing patters are forbidden; refer to [4,9,11,12] for surveys and reports. A key motivation for the study of so-called *beyond-planar graphs* are recent cognitive experiments showing that already the

Research in this work started at the Bertinoro Workshop on Graph Drawing 2019. MC was supported by DFG under grant CH 897/2-2. FM was supported in part by MIUR under grant 20174LF3T8 AHeAD: efficient Algorithms for HArnessing networked Data. PV was supported by the Czech Science Foundation grant 18-19158S.

D. Archambault and C. D. Tóth (Eds.): GD 2019, LNCS 11904, pp. 78–86, 2019.
https://doi.org/10.1007/978-3-030-35802-0_6

Table 1. Lower and upper bounds the crossing ratio of beyond-planar graphs.

Graph class	Lower bound	Upper bound
1-planar	$n/2 - 1$	$n/2 - 1$
quasi-planar	$\Omega(n)$	$O(n^2)$
k-quasi-planar	$\Omega(n/k^3)$	$f(k) \cdot n^2 \log^2 n$
fan-planar	$\Omega(n)$	$O(n^2)$

absence of specific kinds of edge crossing configurations has a positive impact on the human understanding of a graph drawing [13,18]. Of particular interest for us are three families of beyond-planar graphs that have been extensively studied, namely the k-planar, fan-planar, and k-quasi-planar graphs; refer to [9] for additional families. A *k-planar drawing* is such that each edge is crossed at most $k \geq 1$ times [19] (see also [16] for a survey on 1-planarity). A *k-quasi planar* drawing does not have $k \geq 3$ mutually crossing edges [1]. A *fan-planar drawing* does not contain two independent edges that cross a third one or two adjacent edges that cross another edge from different "sides" [14]. A graph is *k-planar* (*k-quasi-planar, fan-planar*) if it admits a k-planar (k-quasi-planar, fan-planar) drawing; a 3-quasi-planar graph is simply called *quasi-planar*.

In this context, an intriguing question is to what extent edge crossings can be minimized while forbidding such local crossing patterns. In particular, we ask whether avoiding local crossing patterns in a drawing of a graph may enforce an overall large number of crossings, whereas only a few crossings would suffice in a crossing-minimal drawing of the graph. We answer this question in the affirmative for the above-mentioned three families of beyond-planar graphs. Our contribution are summarized in Table 1.

1. In Sect. 2, we prove that there exist n-vertex 1-planar graphs such that the ratio between the minimum number of crossings in a 1-planar drawing of one such graph and its crossing number is $n/2 - 1$. This result can be easily extended to k-planar graphs if we allow parallel edges.
2. In Sect. 3, we prove that there exist n-vertex quasi-planar graphs such that the ratio between the minimum number of crossings in a quasi-planar drawing of one such graph and its crossing number is $\Omega(n)$. Similarly, a $\Omega(n/k^3)$ bound can be proved for k-quasi-planar graphs.
3. In Sect. 4, we prove that there exist n-vertex fan-planar graphs such that the ratio between the minimum number of crossings in a fan-planar drawing of one such graph and its crossing number is $\Omega(n)$.

The lower bound in Result 1 is tight. Since fan-planar and quasi-planar graphs have $\mathcal{O}(n)$ edges, the lower bounds in Results 2 and 3 are a linear factor from the trivial upper bound $\mathcal{O}(n^2)$, and it remains open whether such an upper bound can be achieved (see Sect. 5). All results are based on nontrivial constructions that exhibit interesting structural properties of the investigated graphs.

Notation and Definitions. We assume familiarity with standard definitions about graph drawings and embeddings of planar and nonplanar graphs (see, e.g., [8,9]). In a drawing of a graph, we assume that an edge does not contain a vertex other than its endpoints, no two edges meet tangentially, and no three edges share a crossing. It suffices to only consider *simple* drawings where any two edges intersect in at most one point, which is either a common endpoint or an interior point where the two edges properly cross. Thus, in a simple drawing, any two adjacent edges do not cross and any two non-adjacent edges cross at most once.

We define the *k-planar crossing number* of a k-planar graph G, denoted by $\mathrm{cr}_{k\text{-pl}}(G)$, as the minimum number of crossings over all k-planar drawings of G. The *k-planar crossing ratio* $\varrho_{k\text{-pl}}$ is the supremum of $\mathrm{cr}_{k\text{-pl}}(G)/\mathrm{cr}(G)$ over all k-planar graphs G. Analogously, we define the *quasi-planar* and the *fan-planar crossing number* of a graph G, denoted by $\mathrm{cr}_{\mathrm{quasi}}(G)$ and $\mathrm{cr}_{\mathrm{fan}}(G)$, as well as the *quasi-planar* and the *fan-planar crossing ratio*, denoted by ϱ_{quasi} and ϱ_{fan}.

2   The 1-planar Crossing Ratio

An n-vertex 1-planar graph has at most $4n - 8$ edges and a 1-planar drawing has at most $n - 2$ crossings, that is $\mathrm{cr}_{1\text{-pl}}(G) \leq n - 2$ [16]. Observe that for $\mathrm{cr}(G) < \mathrm{cr}_{1\text{-pl}}(G)$ it has to hold that $\mathrm{cr}(G) \geq 2$. It follows that the 1-planar crossing ratio is $\varrho_{1\text{-pl}} \leq n/2 - 1$. We show that this bound can be achieved.

Theorem 1. *For every $\ell \geq 7$, there exists a 1-planar graph G_ℓ with $n = 11\ell + 2$ vertices such that $\mathrm{cr}_{1\text{-pl}}(G_\ell) = n - 2$ and $\mathrm{cr}(G_\ell) = 2$, which yields the largest possible 1-planar crossing ratio.*

The construction of G_ℓ consists of three parts: a rigid graph P that has to be drawn planar in any 1-planar drawing; its dual P^*; a set of *binding* edges and one *special* edge that force P and P^* to be intertwined in any 1-planar drawing.

To obtain P, we utilize a construction introduced by Korzhik and Mohar [17]. They construct graphs H_ℓ that are the medial extension of the Cartesian product of the path of length 2 and the cycle of length ℓ; see Fig. 1a. They prove that H_ℓ has exactly one 1-planar embedding on the sphere, and that embedding is crossing-free. We choose $P = H_\ell$ as our rigid graph and fix its (1−) planar embedding (when we will refer to P, we will usually mean this embedding).

Let P^* be the dual of P, obtained by placing a dual vertex h^* into each face h of P and connecting two dual vertices if their corresponding faces share an edge; see Fig. 1b. Since P has 5ℓ vertices and 11ℓ edges, by Euler's polyhedra formula it has $6\ell + 2$ faces; thus, P^* has $6\ell + 2$ vertices and 11ℓ edges.

Obviously, $P \cup P^*$ can be drawn planar, as both P and P^* are planar and disjoint. All faces of P have size 3 or 4, except two large (called *polar*) faces f and g of size ℓ. We create a graph G' by adding ℓ binding edges to $P \cup P^*$ between f^* (the vertex of P^* corresponding to face f) and the vertices of P that are incident to f. This forces f^* to be drawn in face f in any 1-planar drawing. In the full version [6] we prove the following lemma, cf. Fig. 1c and d.

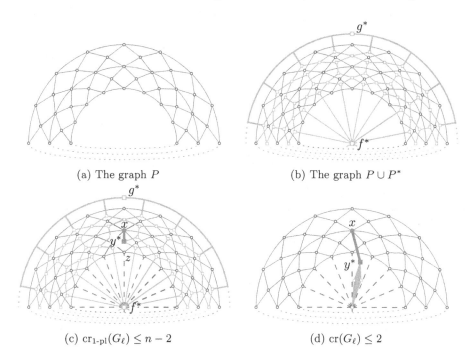

(a) The graph P

(b) The graph $P \cup P^*$

(c) $\mathrm{cr}_{1\text{-pl}}(G_\ell) \leq n - 2$

(d) $\mathrm{cr}(G_\ell) \leq 2$

Fig. 1. Construction of the graph G_ℓ in the proof of Theorem 1. Blue circles and edges are P; red squares and bold edges are P^*; green dashed edges are the binding edges; and the orange very bold edge is the special edge. (Color figure online)

Lemma 2. G' has only two types of 1-planar embeddings (up to the choice of the outer face): a planar one where P^* lies completely inside face f of P; and a 1-planar embedding where f^* lies inside f, g^* lies inside g, and each edge of P crosses an edge of P^* and vice versa.

Let z be a vertex of P on the boundary of f. Let y be the face of size 4 that has z on its boundary. Let x be the degree-6 vertex on the boundary of y. We obtain G_ℓ from G' by adding the *special* edge (x, y^*). In the planar embedding of Lemma 2, P^* and thus y^* lies inside face f of P, so (x, y^*) has to cross at least two edges of P; see Fig. 1d. Choosing the face that corresponds to z as the outer face of P^* gives a non-1-planar drawing of G_ℓ with 2 crossings.

Hence, G' has to be drawn in the second way of Lemma 2; see Fig. 1c. Here, the edge (x, y^*) can be added without further crossings. Graph G_ℓ consists of $n = 11\ell + 2$ vertices in total. Both P and P^* have 11ℓ edges, and each of them is crossed, so there are $n - 2$ crossings in total, which is the maximum possible in a 1-planar drawing. Hence, $\mathrm{cr}_{1\text{-pl}}(G_\ell) = n - 2$ and $\mathrm{cr}(G_\ell) = 2$, so $\varrho_{1\text{-pl}} \leq n/2 - 1$.

The construction used in the proof of Theorem 1 can be generalized to k-planar multigraphs. It suffices to replace each edge of G_ℓ, except the special edge, by a bundle of k parallel edges:

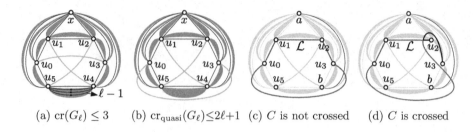

(a) $\mathrm{cr}(G_\ell) \leq 3$ (b) $\mathrm{cr}_{\mathrm{quasi}}(G_\ell) \leq 2\ell{+}1$ (c) C is not crossed   (d) C is crossed

Fig. 2. Illustration for the proof of Theorem 4.

Corollary 3. *For every* $\ell \geq 6$, *there exists a* k-*planar multigraph* $G_{\ell,k}$ *with* $n = 11\ell + 2$ *vertices and maximum edge multiplicity* k *such that* $\mathrm{cr}_k\text{-}pl(G_{\ell,k}) = k^2(n-2)$ *and* $\mathrm{cr}(G_{\ell,k}) = 2k$, *thus* $\varrho_k\text{-}pl \geq k(n-2)/2$.

3   The Quasi-planar Crossing Ratio

An n-vertex quasi-planar graph G has at most $6.5n - 20$ edges, thus $\mathrm{cr}_{\mathrm{quasi}}(G) \in \mathcal{O}(n^2)$ [9]. For $\mathrm{cr}(G) < \mathrm{cr}_{\mathrm{quasi}}$ it has to hold that $\mathrm{cr}(G) \geq 2$, and hence $\varrho_{\mathrm{quasi}} \in \mathcal{O}(n^2)$. We show that the quasi-planar crossing ratio is unbounded, even for $\mathrm{cr}(G) \leq 3$:

Theorem 4. *For every* $\ell \geq 2$, *there exists a quasi-planar graph* G_ℓ *with* $n = 12\ell - 5$ *vertices such that* $\mathrm{cr}_{quasi}(G_\ell) \geq \ell$ *and* $\mathrm{cr}(G_\ell) \leq 3$, *thus* $\varrho_{quasi} \in \Omega(n)$.

In order to prove Theorem 4, we begin with a technical lemma.

Lemma 5. *Let* G *be a graph containing two independent edges* (u,v) *and* (w,z). *Suppose that* u *and* v *(w and z, resp.) are connected by a set* Π_{uv} *(Π_{wz}, resp.) of* $\ell - 1$ *paths of length two. Let* Γ *be a drawing of* G. *If* (u,v) *and* (w,z) *cross in* Γ, *then* Γ *contains at least* ℓ *crossings.*

Proof. Suppose that (u,v) and (w,z) cross. If each of the $\ell - 1$ paths in Π_{wz} crosses (u,v), then the claim follows. Assume otherwise that at least one of these paths does not cross (u,v). This path forms a 3-cycle t with (w,z); the $\ell - 1$ paths of Π_{uv} all cross at least one edge of t, which proves the claim.   □

Proof (of Theorem 4). Let G_ℓ be the graph constructed as follows; cf. Fig. 2a. Start with a 6-cycle $C = \langle u_0, u_1, \ldots, u_5 \rangle$, and a vertex x connected to each of C, yielding graph G'. *Extend* each edge of G' by adding $\ell - 1$ disjoint paths of length two between its endpoints. Finally, add *special* edges (u_i, u_{i+3}), $i = 0, 1, 2$.

The resulting graph G_ℓ has $n = 12(\ell - 1) + 7 = 12\ell - 5$ vertices and admits a drawing with 3 crossings, so $\mathrm{cr}(G_\ell) \leq 3$; see Fig. 2a. Note that G_ℓ admits a quasi-planar drawing with $2\ell + 1$ crossings as shown in Fig. 2b. We prove that $\mathrm{cr}_{\mathrm{quasi}}(G_\ell) \geq \ell$. Let Γ be a quasi-planar drawing of G_ℓ. If there are two edges of G' that cross each other, then the claim follows by Lemma 5.

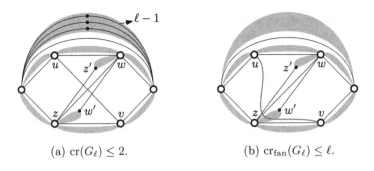

(a) $\mathrm{cr}(G_\ell) \leq 2$.   (b) $\mathrm{cr}_{\mathrm{fan}}(G_\ell) \leq \ell$.

Fig. 3. Illustration for the proof of Theorem 7.

If no special edge would cross G', they would all be drawn within the unique face of size 6 in G'. They would mutually cross, contradicting quasi-planarity.

Thus, at least one special edge, say $s = (u_0, u_3)$, crosses an edge (a, b) of G'. Consider the closed (possibly self-intersecting) curve $\mathcal{L}$ composed of s plus the subpath of C connecting u_0 to u_3 and containing none of the vertices a and b. This curve partitions the plane into two or more regions, and a and b lie in different regions; see Fig. 2c and d for an illustration. Thus (a, b) and the $\ell - 1$ paths connecting a and b cross $\mathcal{L}$, yielding ℓ crossings in Γ, as desired.   □

The above proof can be straight-forwardly extended to k-quasi-planar graphs by using exactly the same construction in which the cycle C has length $2k$. Note that any k-quasi-planar graph has at most $c_k n \log n$ edges, where c_k depends only on k [24], so $\varrho_{\mathrm{quasi}} \leq f(k) \cdot n^2 \log^2 n$.

Corollary 6. *For every $\ell \geq 2$ and $k \geq 3$, there exists a k-quasi-planar graph $G_{\ell,k}$ with $n = 2k(\ell + 1) + 1$ vertices such that $\mathrm{cr}_{quasi}(G_{\ell,k}) \geq \ell$ and $\mathrm{cr}(G_{\ell,k}) \leq k(k-1)/2$, thus $\varrho_{quasi} \in \Omega(n/k^3)$.*

4   The Fan-Planar Crossing Ratio

An n-vertex fan-planar graph G has at most $5n - 10$ edges, thus $\mathrm{cr}_{\mathrm{fan}}(G) \in \mathcal{O}(n^2)$ [9]. For $\mathrm{cr}(G) < \mathrm{cr}_{\mathrm{fan}}(G)$ it has to hold that $\mathrm{cr}(G) \geq 2$, and hence $\varrho_{\mathrm{fan}} \in \mathcal{O}(n^2)$. We show that the fan-planar crossing ratio is unbounded, even for $\mathrm{cr}(G) = 3$.

Theorem 7. *For every $\ell \geq 2$, there exists a fan-planar graph G_ℓ with $n = 9\ell + 1$ vertices such that $\mathrm{cr}_{fan}(G_\ell) = \ell$ and $\mathrm{cr}(G_\ell) = 3$, thus $\varrho_{fan} \in \Omega(n)$.*

Proof. Let G_ℓ be the graph constructed as follows; cf. Fig. 3a. Start with a $K_{3,3}$. Extend each edge of the $K_{3,3}$ by adding $\ell - 1$ disjoint paths of length two between its endpoints, except for two independent edges (u, v) and (w, z). Add vertices

w' and z', edges $\bar{w} = (w, w')$ and $\bar{z} = (z, z')$, ℓ disjoint paths of length two connecting w' and z, and ℓ disjoint paths of length two connecting z' and w.

Graph G_ℓ has $n = 6 + 7(\ell - 1) + 2 + 2\ell = 9\ell + 1$ vertices and admits a drawing with three crossings, see Fig. 3a. Recall that we obtain a subdivision of a graph G by subdividing (even multiple times) any subset of its edges. G_ℓ contains three subdivisions of $K_{3,3}$ sharing only edge (u, v), and thus each subdivision requires at least one distinct crossing in any drawing. It follows that $\mathrm{cr}(G_\ell) = 3$. Note that G_ℓ admits a fan-planar drawing with ℓ crossings, cf. Fig. 3b. We prove that $\mathrm{cr}_{\mathsf{fan}}(G_\ell) = \ell$. Let Γ be a fan-planar drawing of G_ℓ. If any two extended edges cross each other, then the claim follows by Lemma 5. Assume they do not:

G_ℓ contains ℓ subdivions of $K_{3,3}$ that share only (u, v) and $\bar{w}$. Since each $K_{3,3}$ subdivision requires at least one crossing, there are either ℓ crossings in Γ (proving the claim), or (u, v) crosses $\bar{w}$. Similarly, G_ℓ contains ℓ $K_{3,3}$ subdivisions that share only (u, v) and $\bar{z}$, and we can assume that (u, v) crosses $\bar{z}$. But fan-planarity forbids (u, v) to cross both $\bar{w}$ and $\bar{z}$. □

5 Open Problems

The main open question is whether there exist fan-planar and quasi-planar graphs whose crossing ratio is $\Omega(n^2)$. In fact, we conjecture that this bound can be reached, but proving our suspected constructions turns out to be elusive. Another natural research direction is to extend our results to further families of beyond-planar graphs, such as k-gap planar graphs or fan-crossing-free graphs (refer to [9] for definitions). Finally, we may ask whether similar lower bounds can be proved in the geometric setting (i.e., when the edges are drawn as straight-line segments).

References

1. Alon, N., Erdös, P.: Disjoint edges in geometric graphs. Disc. Comput. Geom. **4**, 287–290 (1989). https://doi.org/10.1007/BF02187731
2. Batini, C., Furlani, L., Nardelli, E.: What is a good diagram? A pragmatic approach. In: Proceedings of 4th International Conference on Entity-Relationship Approach (ER 1985), pp. 312–319 (1985). http://dl.acm.org/citation.cfm?id=647510.726382
3. Batini, C., Nardelli, E., Tamassia, R.: A layout algorithm for data flow diagrams. IEEE Trans. Softw. Eng. **12**(4), 538–546 (1986). https://doi.org/10.1109/TSE.1986.6312901
4. Bekos, M.A., Kaufmann, M., Montecchiani, F.: Guest editors' foreword and overview - special issue on graph drawing beyond planarity. J. Graph Algorithms Appl. **22**(1), 1–10 (2018). https://doi.org/10.7155/jgaa.00459
5. Bienstock, D.: Some provably hard crossing number problems. Disc. Comput. Geom. **6**, 443–459 (1991). https://doi.org/10.1007/BF02574701
6. Chimani, M., Kindermann, P., Montecchiani, F., Valtr, P.: Crossing numbers of beyond-planar graphs. Arxiv report (2019). http://arxiv.org/abs/1908.03153

7. Chimani, M., Mutzel, P., Bomze, I.: A new approach to exact crossing minimization. In: Halperin, D., Mehlhorn, K. (eds.) ESA 2008. LNCS, vol. 5193, pp. 284–296. Springer, Heidelberg (2008). https://doi.org/10.1007/978-3-540-87744-8_24

8. Di Battista, G., Eades, P., Tamassia, R., Tollis, I.G.: Graph Drawing. Prentice-Hall, Upper Saddle River (1999)

9. Didimo, W., Liotta, G., Montecchiani, F.: A survey on graph drawing beyond planarity. ACM Comput. Surv. **52**(1), 4:1–4:37 (2019). https://doi.org/10.1145/3301281

10. Hliněný, P., Chimani, M.: Approximating the crossing number of graphs embeddable in any orientable surface. In: Charikar, M. (ed.) Proceedings 21sth Annual ACM-SIAM Symposium Discrete Algorithms (SODA 2010), pp. 918–927. SIAM (2010). https://doi.org/10.1137/1.9781611973075.74

11. Hong, S., Kaufmann, M., Kobourov, S.G., Pach, J.: Beyond-planar graphs: Algorithmics and combinatorics (dagstuhl seminar 16452). In: Dagstuhl Reports, vol. 6, no. 11, pp. 35–62 (2016). https://doi.org/10.4230/DagRep.6.11.35

12. Hong, S., Tokuyama, T.: Algoritihmcs for beyond planar graphs (NII shonan meeting 2016-17). NII Shonan Meeting Report 2016 (2016). http://shonan.nii.ac.jp/shonan/report/no-2016-17/

13. Huang, W., Eades, P., Hong, S.: Larger crossing angles make graphs easier to read. J. Vis. Lang. Comput. **25**(4), 452–465 (2014). https://doi.org/10.1016/j.jvlc.2014.03.001

14. Kaufmann, M., Ueckerdt, T.: The density of fan-planar graphs. Arxiv Report (2014). http://arxiv.org/abs/1403.6184

15. Kawarabayashi, K., Reed, B.A.: Computing crossing number in linear time. In: Johnson, D.S., Feige, U. (eds.) Proceedings 39th Annual ACM Symposium Theory Computing(STOC 2007). pp. 382–390. ACM (2007). https://doi.org/10.1145/1250790.1250848

16. Kobourov, S.G., Liotta, G., Montecchiani, F.: An annotated bibliography on 1-planarity. Comput. Sci. Rev. **25**, 49–67 (2017). https://doi.org/10.1016/j.cosrev.2017.06.002

17. Korzhik, V.P., Mohar, B.: Minimal obstructions for 1-immersions and hardness of 1-planarity testing. J. Graph Theory **72**(1), 30–71 (2013). https://doi.org/10.1002/jgt.21630

18. Mutzel, P.: An alternative method to crossing minimization on hierarchical graphs. SIAM J. Optim. **11**(4), 1065–1080 (2001). https://doi.org/10.1137/S1052623498334013

19. Pach, J., Tóth, G.: Graphs drawn with few crossings per edge. Combinatorica **17**(3), 427–439 (1997). https://doi.org/10.1007/BF01215922

20. Purchase, H.C.: Effective information visualisation: a study of graph drawing aesthetics and algorithms. Interact. Comput. **13**(2), 147–162 (2000). https://doi.org/10.1016/S0953-5438(00)00032-1

21. Purchase, H.C., Carrington, D.A., Allder, J.A.: Empirical evaluation of aesthetics-based graph layout. Empir. Softw. Eng. **7**(3), 233–255 (2002)

22. Schaefer, M.: The graph crossing number and its variants: a survey. Electr. J. Comb., Dynamic Surveys, DS21, 113 p. (2017). https://www.combinatorics.org/ojs/index.php/eljc/article/view/DS21

23. Sugiyama, K., Tagawa, S., Toda, M.: Methods for visual understanding of hierarchical system structures. IEEE Trans. Syst. Man Cybern. **11**(2), 109–125 (1981). https://doi.org/10.1109/TSMC.1981.4308636

24. Suk, A., Walczak, B.: New bounds on the maximum number of edges in k-quasi-planar graphs. Comput. Geom. **50**, 24–33 (2015). https://doi.org/10.1016/j.comgeo.2015.06.001
25. Vrt'o, I.: Crossing numbers of graphs: A bibliography (2014). ftp://ftp.ifi.savba.sk/pub/imrich/crobib.pdf
26. Ware, C., Purchase, H.C., Colpoys, L., McGill, M.: Cognitive measurements of graph aesthetics. Inform. Vis. **1**(2), 103–110 (2002). https://doi.org/10.1057/palgrave.ivs.9500013

On the 2-Colored Crossing Number

Oswin Aichholzer[1] , Ruy Fabila-Monroy[2] , Adrian Fuchs[1],
Carlos Hidalgo-Toscano[3] , Irene Parada[1]([envelope]) , Birgit Vogtenhuber[1] ,
and Francisco Zaragoza[4]

[1] Graz University of Technology, Graz, Austria
{oaich,iparada,bvogt}@ist.tugraz.at, adrian.fuchs@student.tugraz.at
[2] Departamento de Matemáticas, Cinvestav, Mexico City, Mexico
ruyfabila@math.cinvestav.edu.mx
[3] Centro de Investigación e Innovación en Tecnologías de la Información y
Comunicación, Mexico City, Mexico
carlos.hidalgo@infotec.mx
[4] Universidad Autónoma Metropolitana, Mexico City, Mexico
franz@correo.azc.uam.mx

Abstract. Let D be a straight-line drawing of a graph. The rectilinear 2-colored crossing number of D is the minimum number of crossings between edges of the same color, taken over all possible 2-colorings of the edges of D. First, we show lower and upper bounds on the rectilinear 2-colored crossing number for the complete graph K_n. To obtain this result, we prove that asymptotic bounds can be derived from optimal and near-optimal instances with few vertices. We obtain such instances using a combination of heuristics and integer programming. Second, for any fixed drawing of K_n, we improve the bound on the ratio between its rectilinear 2-colored crossing number and its rectilinear crossing number.

Keywords: Complete graph · Rectilinear crossing number · k-colored crossing number

1 Introduction

For a drawing of a non-planar graph G in the plane it is of interest from both a theoretical and practical point of view, to minimize the number of crossings. The minimum such number is known as the *crossing number* $\mathrm{cr}(G)$ of G. There are many variants on crossing numbers, see the comprehensive dynamic survey

This project has received funding from the European Union's Horizon 2020 research and innovation programme under the Marie Skłodowska-Curie grant agreement No. 734922. O.A. and I.P. partially supported by the Austrian Science Fund (FWF) grant W1230. R.F. and C.H. partially supported by CONACYT (Mexico), grant 253261. B.V. partially supported by Austrian Science Fund within the collaborative DACH project *Arrangements and Drawings* as FWF project I 3340-N35. F.Z. partially supported by UAM Azcapotzalco, research grant SI004-12, and SNI Conacyt.

© Springer Nature Switzerland AG 2019
D. Archambault and C. D. Tóth (Eds.): GD 2019, LNCS 11904, pp. 87–100, 2019.
https://doi.org/10.1007/978-3-030-35802-0_7

of Schaefer [25]. In this paper we focus on a version combining two of them: the *k-planar crossing number* and the *rectilinear crossing number*.

The *k-planar crossing number* $cr_k(G)$ of a graph G is the minimum of $cr(G_1) + \cdots + cr(G_k)$ over all sets of k graphs $\{G_1, \ldots, G_k\}$ whose union is G. For $k = 2$, it was introduced by Owens [22] who called it the *biplanar crossing number*; see [13,14] for a survey on biplanar crossing numbers. Shahrokhi et al. [26] introduced the generalization to $k \geq 2$.

A *straight-line drawing* of G is a drawing D of G in the plane in which the vertices are drawn as points in general position, that is, no three points on a line, and the edges are drawn as straight line segments. We identify the vertices and edges of the underlying abstract graph with the corresponding ones in the straight-line drawing. The *rectilinear crossing number* of G, $\overline{cr}(G)$, is the minimum number of pairs of edges that cross in any straight-line drawing of G. Of special relevance is $\overline{cr}(K_n)$, the rectilinear crossing number of the complete graph on n vertices. The current best published bounds on $\overline{cr}(K_n)$ are $0.379972\binom{n}{4} <$ $\overline{cr}(K_n) < 0.380473\binom{n}{4} + \Theta(n^3)$ [3,16]. The upper bound was achieved using a duplication process and has been improved in an upcoming paper [6] to $\overline{cr}(K_n) < 0.38044921\binom{n}{4} + \Theta(n^3)$.

A *k-edge-coloring* of a drawing D of a graph is an assignment of one of k possible colors to every edge of D. The *rectilinear k-colored crossing number* of a graph G, $\overline{cr}_k(G)$, is the minimum number of monochromatic crossings (pairs of edges of the same color that cross) in any k-edge-colored straight-line drawing of G. This parameter was introduced before and called the *geometric k-planar crossing number* [23]. In the same paper, as well as in [26], also the *rectilinear k-planar crossing number* was considered, which asks for the minimum of $\overline{cr}(G_1) + \ldots + \overline{cr}(G_k)$ over all sets of k graphs $\{G_1, \ldots, G_k\}$ whose union is G. We prefer our terminology because the terms geometric and rectilinear are very often used interchangeably and because the term k-planar is extensively used in graph drawing with a different meaning; see for example [15,20]. We remark that in graph drawing, *rectilinear* sometimes also refers to orthogonal grid drawings (which is not the case here).

In this paper we focus on the case where G is the complete graph K_n, and we prove the following lower and upper bounds on $\overline{cr}_2(K_n)$:

$$0.03\binom{n}{4} + \Theta(n^3) < \overline{cr}_2(K_n) < 0.11798016\binom{n}{4} + \Theta(n^3).$$

Our approach is based on theoretical results that guarantee asymptotic bounds from the information of small point sets. Thus, it implies computationally dealing with small sets, both to guarantee a minimum amount of monochromatic crossings (for the lower bound) and to find examples with few monochromatic crossings and some other desired properties (for the upper bound).

From an algorithmic point of view, the decision variant of the crossing number problem was shown to be NP-complete for general graphs already in the 1980s by Garey and Johnson [18]. The version for straight-line drawings is also known to be NP-hard, and actually, computing the rectilinear crossing number is

∃ℝ-complete [19]. So whenever considering crossing numbers, it is rather likely that one faces computationally difficult problems.

In our case the challenge is twofold. On the one hand, we need to optimize the point configuration (order type) to obtain a small number of crossings, which is the original question about the rectilinear crossing number of K_n. On the other hand, we need to determine a coloring of the edges of K_n that minimizes the colored crossing number for a fixed point set.

For the first problem there is not even a conjecture of point configurations that minimize the rectilinear crossing number of K_n for any n. The latter problem corresponds to finding a maximum cut in a segment intersection graph, which in general is NP-complete [9]. Moreover, these two problems are not independent. There exist examples where a point set with a non-minimal number of uncolored crossings allows for a coloring of the edges so that the resulting colored crossing number is smaller than the best colored crossing number obtained from a set minimizing the uncolored crossing number. Thus, the two optimization processes need to interleave if we want to guarantee optimality. But, as we will see in Sect. 2, even this combined optimization does not guarantee to yield the best asymptotic result. There are sets of fixed cardinality and with larger 2-colored crossing number which—due to an involved duplication process—give a better asymptotic constant than the best minimizing sets. This is in contrast to the uncolored setting [2,3], where for any fixed cardinality, sets with a smaller crossing number always give better asymptotic constants. Also, it clearly indicates that our extended duplication process for 2-colored crossings differs essentially from the original version.

As mentioned, drawings with few crossings do not necessarily admit a coloring with few monochromatic crossings. This observation motivates the following question: given a fixed straight-line drawing D of K_n, what is the ratio between the number of monochromatic crossings for the best 2-edge-coloring of D and the number of (uncolored) crossings in D? A simple probabilistic argument shows that this ratio is less than $1/2$. In Sect. 4, we improve that bound, showing that for sufficiently large n, it is less than $1/2 - c$ for some positive constant c.

In a slight abuse of notation, we denote with $\overline{cr}(D)$ the number of pairs of edges in D that cross and call it the rectilinear crossing number of D. The (rectilinear) 2-colored crossing number of a straight-line drawing D, $\overline{cr}_2(D)$, is then the minimum of $\overline{cr}(D_1) + \overline{cr}(D_2)$, over all pairs of straight-line drawings $\{D_1, D_2\}$ whose union is D. For a given 2-edge-coloring χ of D, we denote with $\overline{cr}_2(D, \chi)$ the number of monochromatic crossings in D. Thus, $\overline{cr}_2(D)$ is the minimum of $\overline{cr}_2(D, \chi)$ over all 2-edge-colorings χ of D.

Outline. In Sect. 2 we prove that, given a 2-colored straight-line drawing D of K_n, there is a duplication process that allows us to obtain a 2-colored straight-line drawing D_k of $K_{2^k n}$ for any $k \geq 1$ whose 2-colored crossing number $\overline{cr}_2(D_k)$ can be easily calculated. Moreover, we can obtain the asymptotic value when $k \to \infty$. By finding good sets of constant size as a seed for the duplication process, we obtain an asymptotic upper bound for $\overline{cr}_2(K_n)$. In Sect. 3 we obtain a lower bound for $\overline{cr}_2(K_n)$ using the crossing lemma, and we improve it with an approach again using small drawings. For sufficiently large n, we show in Sect. 4 that for any straight-line drawing D of K_n, $\overline{cr}_2(D)/\overline{cr}(D) < 1/2 - c$ for a positive constant c, that is, using two colors saves more than half of the crossings. Finally, in Sect. 5 we present some open problems.

2   Upper Bounds on $\overline{cr}_2(K_n)$

For the rectilinear crossing number $\overline{cr}(K_n)$, the best upper bound [6] comes from finding examples of straight-line drawings of K_n (for a small value of n) with few crossings which are then used as a seed for the duplication process in [2,3]. To be able to apply this duplication process, the starting set P with m points has to contain a halving matching. If m is even (odd), a *halving line* of P is a line that passes exactly through two (one) points of P and leaves the same number of points of P to each side. If it is possible to match each point p of P with a halving line of P through this point in such a way that no two points are matched with the same line, P is said to have a *halving matching*. It is then shown in [2] that every point of P can be substituted by a pair of points in its close neighborhood such that the resulting set Q with $2m$ points contains again a halving matching. Iterating this process leads to the mentioned upper bound for $\overline{cr}(K_n)$, where this bound depends only on m and the number of crossings of the starting set P.

In this section, we prove that a significantly more involved but similar approach can be adopted for the 2-colored case. Unlike the original approach, we cannot always get a matching which simultaneously halves both color classes. Moreover, even for sets where such a halving matching exists, it cannot be guaranteed that this property is maintained after the duplication step. We will see below that we need a more involved approach, where the matchings are related to the distribution of the colored edges around a vertex. Consequently, the number of crossings which are obtained in the duplication, and thus, the asymptotic bound we get, not only depends on the 2-colored crossing number of the starting set, but also on the specific distribution of the colors of the edges. In that sense, both the heuristics for small drawings and the duplication process for the 2-colored crossing number differ significantly from the uncolored case.

Throughout this section, P is a set of m points in general position in the plane, where m is even. Let p be a point in P. By slight abuse of notation, in the following we do not distinguish between a point set and the straight-line drawing of K_n it induces. Given a 2-coloring χ of the edges induced by P, we denote by $L(p)$ and $S(p)$ the edges incident to $p \in P$ of the larger and smaller color class at p, respectively. An edge $e = (p, q)$ incident to p is called a χ-*halving edge of* p if the number of edges of $L(p)$ to the right of the line ℓ_e spanned by e (and directed from q to p) and the number of edges of $L(p)$ to the left of ℓ_e differ by at most one. A matching between the points of P and their χ-halving edges is called a χ-*halving matching for* P.

Theorem 1. *Let P be a set of m points in general position and let χ be a 2-coloring of the edges induced by P. If P has a χ-halving matching, then the 2-colored rectilinear crossing number of K_n can be bounded by*

$$\overline{cr}_2(K_n) \leq \frac{24A}{m^4} \binom{n}{4} + \Theta(n^3)$$

where A is a rational number that depends on P, χ, and the χ-halving matching for P.

Proof. First we describe a process to obtain from P a set Q of $2m$ points, a 2-edge-coloring χ' of the edges that Q induces, and a χ'-halving matching for Q. The set Q is constructed as follows. Let p be a point in P and $e = (p, q)$ its χ-halving edge in the matching. We add to Q two points p_1, p_2 placed along the line spanned by e and in a small neighborhood of p such that:

(i) if f is an edge different from e that is incident to p, then p_1 and p_2 lie on different sides of the line spanned by f;

(ii) if f is an edge different from e that is not incident to p, then p_1 and p_2 lie on the same side of the line spanned by f as p; and

(iii) the point p_1 is further away from q than p_2.

The set Q has $2m$ points and the above conditions ensure that they are in general position.

Next, we define a coloring χ' and a χ'-halving matching for Q. For every edge (p, q) of P, we color the four edges $(p_i, q_j), i, j \in \{1, 2\}$ with the same color as (p, q). Hence, the only edges remaining to be colored are the edges (p_1, p_2) between the duplicates of a point $p \in P$. Let ℓ_e be the line spanned by e and directed from q to p. Further, let q_1 and q_2 be the points that originated from duplicating q, such that q_1 lies to the left of ℓ_e and q_2 lies to the right of ℓ_e. Denote by $L_l(p)$ and $L_r(p)$ the number of edges in $L(p)$ to the left and right of e, respectively. Analogously, denote by $S_l(p)$ and $S_r(p)$ the number edges in $S(p)$ to the left and right of e. For the following case distinction, we assume that the colors are red and blue and that the larger color class at p is blue.

There are six cases in which p can fall, depending on the color of the edge e and on the relation between the numbers $L_l(p)$ and $L_r(p)$ of blue edges incident to p on the left and the right side of ℓ_e; see Fig. 1. The edge e of P has color red in the first three cases and color blue in the last three cases. The edge (p_1, p_2) receives color blue in Cases 1 and 3, and color red in the remaining cases. The thick edges in Fig. 1 represent the matching edges for p_1 and p_2 in Q, where the arrow points to the point it is matched with. For each of p_1 and p_2, the resulting numbers of incident red and blue edges that are to the left and to the right of the line spanned by the matching edge are written next to those lines in the figure. They also show that the matching edges are indeed χ'-halving edges in each case. A detailed case distinction can be found in the full version [7].

Having completed the coloring χ' for the edges induced by Q, we next consider the number of monochromatic crossings in the resulting drawing on Q. We claim the following for $\overline{cr}_2(Q, \chi')$:

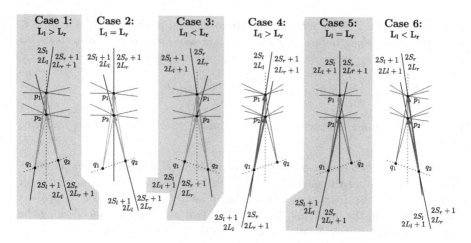

Fig. 1. The cases in the duplication process of Theorem 1 when the larger color class at p is blue.The dotted lines represent the lines spanned by the χ-halving matching edges for P. The numbers of blue (red) edges at p to the left and right of l_e, is denoted with L_l and L_r (S_l and S_r), respectively. (Color figure online)

Claim 1. *The pair* (Q, χ') *satisfies*

$$\overline{\mathrm{cr}}_2(Q, \chi') = 16 \, \overline{\mathrm{cr}}_2(P, \chi) + \binom{m}{2} - m$$

$$+ 4 \sum_p \left(\binom{L_l(p)}{2} + \binom{L_r(p)}{2} + \binom{S_l(p)}{2} + \binom{S_r(p)}{2} \right)$$

$$+ 2 \sum_p (H_l(p) + H_r(p)).$$

The proof of this claim follows the same counting technique used in [2]. The proof can be found in the full version [7].

We now apply the duplication process multiple times. To this end, consider again the six different cases for a point $p \in P$ when obtaining a coloring and a matching for Q. Note that if one of the Cases 1, 2, 3, 4 and 6 applies for p, then the same case applies for its duplicates $p_1, p_2 \in Q$ (and will apply in all further duplication iterations). If p falls in Case 5, then for p_1 and p_2 we have Case 2 and 4, respectively. As no point in Q falls in Case 5, from now on, we assume that P is such that no point of P falls in Case 5 either.

Let $k \geq 1$ be an integer and let (Q_k, χ_k) be the pair obtained by iterating the duplication process k times, with $(Q_0, \chi_0) = (P, \chi)$. We claim the following on $\overline{\mathrm{cr}}_2(Q_k, \chi_k)$, the number of monochromatic crossings in the 2-edge-colored drawing of K_n induced by Q_k and χ_k:

Claim 2. *After k iterations of the duplication process, the following holds*

$$\overline{\mathrm{cr}}_2(Q_k, \chi_k) = A \cdot 2^{4k} + B \cdot 2^{3k} + C \cdot 2^{2k} + D \cdot 2^k$$

where A, B, C and D are rational numbers that depend on P and its χ-halving matching.

The proof of this claim uses a careful analysis of the structure of (Q_k, χ_k) in dependence of (P, χ) and the χ-halving matching for P. This analysis, followed by involved calculations to obtain the statement of Claim 2, can be found in the full version [7]. Applying Claim 2 to an initial drawing on m vertices and letting $n = 2^k m$, we get:

$$\overline{\text{cr}}_2(K_n) \leq \overline{\text{cr}}_2(Q_k, \chi_k) = \frac{24A}{m^4} \binom{n}{4} + \Theta(n^3)$$

which completes the proof of Theorem 1 when n is of the form $2^k m$. The proof for $2^k m < n < 2^{k+1} m$ then follows from the fact that $\overline{\text{cr}}_2(K_n)$ is an increasing function.   □

We remark that the duplication process described in the proof of Theorem 1 can also be applied if the initial set P has odd cardinality. However, then it might happen that the resulting matching is not χ'-halving for the resulting set Q. Moreover, a similar process can even be applied with any matching between the points of P and the edges induced by P, where in that situation one needs to specify how the colors for the edges between duplicates of points (and possibly a matching for the resulting set) is chosen.

In the uncolored duplication process for obtaining bounds on $\overline{\text{cr}}(K_n)$, halving matchings always yield the best asymptotic behavior, which only depends on $|P|$ and $\overline{\text{cr}}(P)$. This is not the case for the 2-colored setting, where we ideally would like to achieve simultaneously for every point $p \in P$ that (i) both color classes are of similar size, (ii) both color classes are evenly split by the matching edge, and (iii) $\overline{\text{cr}}_2(P)$ is small. Yet, this is in general not possible. Starting with a χ-halving matching for P we obtain (ii) at least for the larger color class at every point of P. Moreover, this is hereditary by the design of our duplication process.

The results of this section imply that for large cardinality we can obtain straight-line drawings of the complete graph with a reasonably small 2-colored crossing number by starting from *good* sets of constant size. Similar as in [6] we apply a heuristic combining different methods to obtain straight-line drawings of the complete graph with low 2-colored crossing number. Our heuristic iterates three steps of (1) locally improving a set, (2) generating larger good sets, and (3) extracting good subsets, where also after steps (2) and (3) a local optimization is done. The currently best (with respect to the crossing constant, see below) straight-line drawing D with 2-edge coloring χ we found in this way[1] has $n = 135$ vertices, a 2-colored crossing number of $\overline{\text{cr}}_2(D, \chi) = 1470756$, and contains a χ-halving matching.

Let $\overline{\text{cr}}_2$ be the *rectilinear 2-colored crossing constant*, that is, the constant such that the best straight-line drawing of K_n for large values of n has at most

---

[1] The interested reader can get a file with the coordinates of the points, the colors of the edges, and a χ-halving matching from http://www.crossingnumbers.org/projects/monochromatic/sets/n135.php.

$\overline{cr}_2\binom{n}{4}$ monochromatic crossings. Its existence follows from the fact that the limit $\lim_{n\to\infty} \overline{cr}_2(K_n)/\binom{n}{4}$ exits and is a positive number (the proof goes along the same lines as for the uncolored case [24]). Using the above-mentioned currently best straight-line 2-edge colored drawing and plugging it into the machinery developed in the proof of Theorem 1 we get

Theorem 2. *The rectilinear 2-colored crossing constant satisfies*

$$\overline{cr}_2 \leq \frac{182873519}{1550036250} < 0.11798016.$$

In [3] a lower bound of $\overline{cr} \geq \frac{277}{729} > 0.37997267$ has been shown for the rectilinear crossing constant. We can thus give an upper bound on the asymptotic ratio between the best rectilinear 2-colored drawing of K_n and the best rectilinear drawing of K_n of $\overline{cr}_2/\overline{cr} \leq 0.31049652$.

3   Lower Bounds on $\overline{cr}_2(K_n)$

In this section we consider lower bounds for the 2-colored crossing number and the biplanar crossing number of K_n.

In related work [23], the authors present lower and upper bounds on the $\sup \overline{cr}_k(G)/\overline{cr}(G)$ where the supremum is taken over all non-planar graphs. We remark that this lower bound does not yield a lower bound for $\overline{cr}_2(K_n)$ as their bound is obtained for "midrange" graphs (graphs with a subquadratic but superlinear number of edges). Czabarka et al. mention a lower bound on the biplanar crossing number of general graphs depending on the number of edges [14, Equation 3]. For the complete graph, this yields a lower bound of $\overline{cr}_2(K_n) \geq 1/1944 n^4 - O(n^3)$. A better bound of $\overline{cr}_2 \geq \frac{24}{29 \cdot 32} = 3/116 > 1/39$ can be obtained from (an improved version of) the crossing lemma [4,21], which states that for an undirected simple graph with n vertices and e edges with $e > 7n$, the crossing number of the graph is at least $\frac{e^3}{29n^2}$.

Alternatively, the following result shows that from the 2-colored rectilinear crossing number of small sets we can obtain lower bounds for larger sets.

Lemma 1. *Let $\overline{cr}_2(m) = \hat{c}$ for some $m \geq 4$. Then for $n > m$ we have $\overline{cr}_2(K_n) \geq \frac{24\hat{c}}{m(m-1)(m-2)(m-3)}\binom{n}{4}$ which implies $\overline{cr}_2 \geq \frac{24\hat{c}}{m(m-1)(m-2)(m-3)}$.*

Proof. Every subset of m points of K_n induces a drawing with at least $\hat{c}$ crossings, and thus we have $\hat{c}\binom{n}{m}$ crossings in total. In this way every crossing is counted $\binom{n-4}{m-4}$ times. This results in a total of $\frac{24\hat{c}}{m(m-1)(m-2)(m-3)}\binom{n}{4}$ crossings.   □

As K_8 can be drawn such that $\overline{cr}_2(K_8) = 0$ (see Fig. 2 left) we next determine $\overline{cr}_2(K_9)$. We use the optimization heuristic mentioned from Sect. 2 to obtain good colorings for all 158 817 order types of K_9 (which are provided by the order type data base [5]). In this way, it is guaranteed that all (crossing-wise) different straight-line drawings of K_9 (uncolored) are considered.

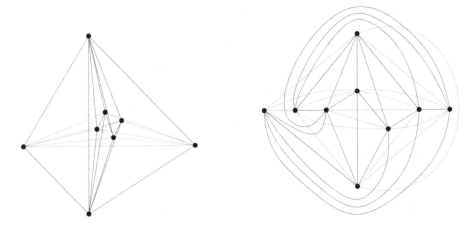

Fig. 2. Left: a 2-colored rectilinear drawing of K_8 without monochromatic crossings.Right: a 2-colored drawing of K_9 with only one monochromatic (red) crossing. (Color figure online)

To prove that the heuristics indeed found the best colorings we consider the intersection graph for each drawing D. In the intersection graph every edge in D is a vertex, and two vertices are connected if their edges in D cross. Note that each odd cycle in the intersection graph of D gives rise to a monochromatic crossing in D. On the other hand, several odd cycles might share a crossing and only one monochromatic crossing is forced by them. We thus set up an integer linear program, where for every crossing of D we have a non-negative variable and for each odd cycle the sum of the variables corresponding to the crossings of the cycle has to be at least 1. The objective function aims to minimize the sum of all variables, which by construction is a lower bound for the number of monochromatic crossings in D.

With that program and some additional methods for speedup (see [17] for details), we have been able to obtain matching lower bounds and hence determine the 2-colored crossing numbers for all order types of K_9 within a few hours. The best drawings we found have 2 monochromatic crossings, and thus $\overline{cr}_2(K_9) = 2$. Using Lemma 1 for $m = 9$ and $\hat{c} = 2$ we get a bound of $\overline{cr}_2 \geq 1/63$, which is worse than what we obtained from the crossing lemma. Repeating the process of computing lower bounds for sets of small cardinality, we checked all order types of size up to 11 [8] and obtained $\overline{cr}_2(K_{10}) = 5$ and $\overline{cr}_2(K_{11}) = 10$. By Lemma 1, the latter gives the improved lower bound of $\overline{cr}_2 \geq 1/33$.

3.1 Straight-Line Versus General Drawings

The best straight-line drawings of K_n with $n \leq 8$ have no monochromatic crossing, see again Fig. 2 left. In [23, Section 3], the authors state that no graph is known were the k-planar crossing number is strictly smaller than the rectilinear k-planar crossing number for any $k \geq 2$. Moreover, according to personal

communication [27], the similar question whether a graph exists where the
k-planar crossing number is strictly smaller than the rectilinear k-colored crossing number was open. We next argue that K_9 is such an example. From the
previous section we know that $\overline{cr}_2(K_9) = 2$. Inspecting rotation systems for
$n = 9$ [1] which have the minimum number of 36 crossings, we have been able to
construct a drawing of K_9 which has only one monochromatic crossing, see Fig. 2
right. As the graph thickness of K_9 is 3 [12,28], we cannot draw K_9 with just
two colors without monochromatic crossings. Thus, we get the following result.

Observation 1. *The biplanar crossing number for K_9 is 1 and is thus strictly
smaller than the rectilinear 2-colored crossing number $\overline{cr}_2(K_9) = 2$.*

4   Upper Bounds on the Ratio $\overline{cr}_2(D)/\overline{cr}(D)$

In this section we study the extreme values that $\overline{cr}_2(D)/\overline{cr}(D)$ can attain for
straight-line drawings D of K_n. Using a simple probabilistic argument as in [23],
2-coloring the edges uniformly at random, it can be shown that $\overline{cr}_2(D)/\overline{cr}(D) <
1/2$ for every straight-line drawing D, even if the underlying graph is not K_n.

In the following, we show that for K_n this upper bound on $\overline{cr}_2(D)/\overline{cr}(D)$
can be improved. To obtain our improved bound, we find subdrawings of D and
colorings such that many of the crossings in these drawings are between edges of
different colors. To this end, we need to find large subsets of vertices of D with
identical geometric properties. We use the following definition and theorem. Let
$(Y_1, ..., Y_k)$ be a tuple of finite subsets of points in the plane. A *transversal* of
$(Y_1, ..., Y_k)$ is a tuple of points $(y_1, \ldots, y_k)$ such that $y_i \in Y_i$ for all i.

Theorem 3 (Positive fraction Erdős-Szekeres theorem). *For every integer $k \geq 4$ there is a constant $c_k > 0$ such that every sufficiently large finite point
set $X \subset \mathbb{R}^2$ in general position contains k disjoint subsets $Y_1, \ldots, Y_k$, of size at
least $c_k|X|$ each, such that each transversal of $(Y_1, \ldots, Y_k)$ is in convex position.*

The Positive Fraction Erdős-Szekeres theorem was proved by Bárány and
Valtr [11], see also Matoušek's book [21]. Although it is not stated in the theorem,
every transversal of the Y_i has the same (labelled) order type. Making use of that
result we obtained the following theorem.

Theorem 4. *There exists an integer $n_0 > 0$ and a constant $c > 0$ such that for
any straight-line drawing D of K_n on $n \geq n_0$ vertices, $\overline{cr}_2(D)/\overline{cr}(D) < \frac{1}{2} - c$.*

Proof. Let c_4 be as in Theorem 3 and let n_0 be such that Theorem 3 holds for
$k = 4$ and for point sets with at least n_0 points. Let D be a straight-line drawing
of K_n, where $n \geq n_0$.

Our general strategy is as follows. We first find subsets of edges of D that can
be 2-colored such that many of the crossings between these edges are between
pairs of edges of different colors. We remove these edges and search for a subset
of edges with the same property. We repeat this process as long as possible. We

2-color the remaining edges so that at most half of the crossings are monochromatic. Afterwards, we put back the edges we removed while 2-coloring them in a convenient way.

We define a sequence of subsets $V = X_0 \supset X_1 \supset \cdots \supset X_m$ of vertices of D, where $V = X_0$ is the set of vertices of D, and tuples $(F_1, F_1'), \ldots, (F_m, F_m')$ of sets of edges of D as follows. Suppose that X_i has been defined. If $|X_i| < n_0$, we stop the process. Otherwise we apply Theorem 3 to X_i, to obtain a tuple (Y_1, Y_2, Y_3, Y_4) of disjoint subsets of points X_i, each with exactly $\lfloor c_4 |X_i| \rfloor$ vertices, such that every transversal (y_1, y_2, y_3, y_4) of (Y_1, Y_2, Y_3, Y_4) is a convex quadrilateral. Without loss of generality we assume that (y_1, y_2, y_3, y_4) appear in clockwise order around this quadrilateral. This implies that the edge (y_1, y_3) crosses the edge (y_2, y_4). Let F_i be the set of edges with an endpoint in Y_1 and an endpoint in Y_3; let F_i' be the set of edges with an endpoint in Y_2 and an endpoint in Y_4; and finally, let $X_{i+1} = X_i \setminus (Y_1 \cup Y_2)$. Note that every edge in F_i crosses every edge in F_i'.

We now consider the remaining edges. Let $\overline{F}$ be the set of edges of D that are not contained in any F_i nor in any F_i' for $1 \le i \le m$. Let H be the straight-line drawing with the same vertices as D and with edge set equal to $\overline{F}$. By a probabilistic argument 2-coloring the edges uniformly at random, there is a coloring χ' of the edges of H so that $\overline{cr}(H)/\overline{cr}_2(H, \chi') \ge 2$.

We now 2-color the edges in F_i and F_i'. We define a sequence of straight-line drawings $H = D_{m+1}, \subset D_m \subset \cdots \subset D_0 = D$ and a corresponding sequence of 2-edge-colorings $\chi' = \chi_{m+1}, \chi_m, \ldots, \chi_0 = \chi$ that satisfies the following. Each χ_i is a 2-edge-coloring of D_i. Also χ_{i-1} when restricted to D_i equals χ_i. Suppose that D_i and χ_i have been defined and that $0 < i \le m + 1$. Let D_{i-1} be the straight-line drawing with the same vertices as D and with edge set E_{i-1} equal to $E_i \cup F_{i-1} \cup F_{i'-1}$ (where E_i is the edge set of D_i). Since χ_{i-1} coincides with χ_i in the edges of E_i, we only need to specify the colors of F_{i-1} and F_{i-1}'. We color the edges of F_i with the same color and the edges of F_{i-1}' with the other color. There are two options for doing this, and one of them guarantees that at most half of the crossings between an edge of $F_{i-1} \cup F_{i-1}'$ and an edge of D_i are monochromatic. We choose this option to define χ_{i-1}.

In what follows we assume that D has been colored by χ. Let C be the set of pairs of edges of D that cross. Of these, let C_1 be the subset of pairs of edges such that both of them are contained in $F_i \cup F_i'$ for some $1 \le i \le m$. Let $C_2 := C \setminus C_1$. Note that, by construction of χ, at most half of the pairs of edges in C_2 are of edges of the same color. For a given i, let E_i' be the subset of pairs of edges in C_1 such that both edges are in $F_i \cup F_i'$. Let (Y_1, Y_2, Y_3, Y_4) be the tuple of disjoint subsets of points X_i used to define F_i and F_i'. Recall that each Y_i consists of $\lfloor c_4 |X_i| \rfloor$ points. Every pair of crossing edges defines a convex quadrilateral and, conversely, every convex quadrilateral defines a unique pair of crossing edges. Therefore, by construction there at most $c_4^4 \lfloor |X_i| \rfloor^4 / 2$ pairs of edges in E_i' such that both edges are of the same color; and there are exactly $\lfloor c_4 |X_i| \rfloor^4$ pairs of edges in E_i' such that the edges are of different color. Thus, at most $\frac{1}{3}$ of the pairs of edges in E_i' are edges of the same color.

Therefore, $\frac{\overline{cr}_2(D,\chi)}{\overline{cr}(D)} \leq \frac{\frac{1}{2}|C_1|+\frac{1}{3}|C_2|}{|C_1|+|C_2|}$. This is maximized when C_1 is as large as possible. Since there are in total at most $\binom{n}{4}$ pairs of edges that cross, we have $|C_1| \leq \binom{n}{4} - |C_2|$. Thus,

$$\frac{\overline{cr}_2(D,\chi)}{\overline{cr}(D)} \leq \frac{\frac{1}{2}\binom{n}{4} - \frac{1}{6}|C_2|}{\binom{n}{4}}.$$

We now obtain a lower bound for the size of C_2. Note that $|X_0| = n$ and $|X_i| \geq (1 - 4c_4)|X_{i-1}|$. This implies that $|X_i| \geq (1 - 4c_4)^i n$ and that $|E_i| \geq c_4{}^4(1 - 4c_4)^{4i}n^4$. Therefore,

$$|C_2| = \sum_{i=1}^{m}|E_i| \geq \sum_{i=1}^{m} c_4{}^4(1-4c_4)^{4i}n^4 = 24c_4{}^4 \left(\frac{1}{1 - (1 - 4c_4)^4} - 1 - o(1) \right) \binom{n}{4},$$

which completes the proof.   □

In the full version [7] we explore the ratio $\overline{cr}_2(D)/\overline{cr}(D)$ for certain classes of straight-line drawings of K_n.

5   Conclusion and Open Problems

In this paper we have shown lower and upper bounds on the rectilinear 2-colored crossing number for K_n as well as its relation to the rectilinear crossing number for fixed drawings of K_n. Besides improving the given bounds, some open problems arise from our work.

(1) How fast can the best edge-coloring of a given straight-line drawing of K_n be computed? This problem is related to the max-cut problem of segment intersection graphs, which has been shown to be NP-complete for general graphs [9]. But for the intersection graph of K_n the algorithmic complexity is still unknown.

(2) What can we say about the structure of 2-colored crossing minimal sets? For the rectilinear crossing number it is known that optimal sets have a triangular convex hull [10]. For $n = 8, 9$ we have optimal sets with 3 and 4 extreme points, but so far all minimal sets for $n \geq 10$ have a triangular convex hull.

(3) We have seen that for convex sets asymptotically, the ratio $\overline{cr}_2(D)/\overline{cr}(D)$ approaches $3/8$ from below when $n \to \infty$. It can be observed that among all point sets (order types) of size 10, the convex drawing D of K_{10} is the only one that provides the largest ratio of $\overline{cr}_2(D)/\overline{cr}(D) = 2/7$, while the best factor $5/76$ is reached by sets minimizing $\overline{cr}_2(D)$. Is it true that the convex set has the worst (i.e., largest) factor? And is the best (smallest) factor always achieved by optimizing sets, that is, sets with $\overline{cr}_2(D) = \overline{cr}_2(K_n)$?

References

1. Ábrego, B.M., Aichholzer, O., Fernández-Merchant, S., Hackl, T., Pammer, J., Pilz, A., Ramos, P., Salazar, G., Vogtenhuber, B.: All good drawings of small complete graphs. In: Proceedings of the 31st European Workshop on Computational Geometry (EuroCG 2015), pp. 57–60 (2015)
2. Ábrego, B.M., Fernández-Merchant, S.: Geometric drawings of K_n with few crossings. J. Comb. Theory Ser. A **114**(2), 373–379 (2007). https://doi.org/10.1016/j.jcta.2006.05.003
3. Ábrego, B.M., Fernández-Merchant, S., Leaños, J., Salazar, G.: A central approach to bound the number of crossings in a generalized configuration. Electron. Notes Discret. Math. **30**, 273–278 (2008). https://doi.org/10.1016/j.endm.2008.01.047
4. Ackerman, E.: On topological graphs with at most four crossings per edge. ArXiv e-Prints (2013). https://arxiv.org/abs/1509.01932
5. Aichholzer, O.: The order type data base. http://www.ist.tugraz.at/aichholzer/research/rp/triangulations/ordertypes/. Accessed 1 Oct 2018
6. Aichholzer, O., Duque, F., García-Quintero, O.E., Fabila-Monroy, R., Hidalgo-Toscano, C.: An ongoing project to improve the rectilinear and pseudolinear crossing constants. ArXiv e-Prints (2018). https://arxiv.org/abs/1907.07796
7. Aichholzer, O., Fabila-Monroy, R., Fuchs, A., Hidalgo-Toscano, C., Parada, I., Vogtenhuber, B., Zaragoza, F.: On the 2-colored crossing number. ArXiv e-Prints (2019). http://arxiv.org/abs/1908.06461
8. Aichholzer, O., Krasser, H.: Abstract order type extension and new results on the rectilinear crossing number. Comput. Geom.: Theory Appl. **36**(1), 2–15 (2006). https://doi.org/10.1016/j.comgeo.2005.07.005
9. Aichholzer, O., Mulzer, W., Schnider, P., Vogtenhuber, B.: NP-completeness of max-cut for segment intersection graphs. In: Proceedings of the 34th European Workshop on Computational Geometry (EuroCG 2018), pp. 1–6 (2018)
10. Aichholzer, O., Orden, D., Ramos, P.: On the structure of sets attaining the rectilinear crossing number. In: Proceedings of the 22nd European Workshop on Computational Geometry (EuroCG 2006), pp. 43–46 (2006)
11. Bárány, I., Valtr, P.: A positive fraction Erdős-Szekeres theorem. Discret. Comput. Geom. **19**(3), 335–342 (1998). https://doi.org/10.1007/PL00009350
12. Battle, J., Harary, F., Kodama, Y.: Every planar graph with nine points has a nonplanar complement. Bull. Am. Math. Soc. **68**, 569–571 (1962). https://doi.org/10.1090/S0002-9904-1962-10850-7
13. Czabarka, É., Sýkora, O., Székely, L.A., Vrt'o, I.: Biplanar crossing numbers I: a survey of results and problems. In: Győri, E., Katona, G.O.H., Lovász, L., Fleiner, T. (eds.) More Sets, Graphs and Numbers: A Salute to Vera Sós and András Hajnal. BSMS, vol. 15, pp. 57–77. Springer, Heidelberg (2006). https://doi.org/10.1007/978-3-540-32439-3_4
14. Czabarka, E., Sýkora, O., Székely, L.A., Vrt'o, I.: Biplanar crossing numbers II. Comparing crossing numbers and biplanar crossing numbers using the probabilistic method. Random Struct. Algorithms **33**(4), 480–496 (2008). https://doi.org/10.1002/rsa.20221
15. Didimo, W., Liotta, G., Montecchiani, F.: A survey on graph drawing beyond planarity. ACM Comput. Surv. **52**(1), 4:1–4:37 (2019). https://doi.org/10.1145/3301281
16. Fabila-Monroy, R., López, J.: Computational search of small point sets with small rectilinear crossing number. J. Graph Algorithms Appl. **18**(3), 393–399 (2014). https://doi.org/10.7155/jgaa.00328

17. Fuchs, A.: On the number of monochromatic crossings in rectilinear embeddings of complete graphs. Master's thesis, University of Technology Graz, Austria (2019)
18. Garey, M., Johnson, D.S.: Crossing number is NP-complete. SIAM J. Algebr. Discret. Methods **4**(3), 312–316 (1983)
19. Hernández-Vélez, C., Leaños, J., Salazar, G.: On the pseudolinear crossing number. J. Graph Theory **84**(3), 297–310 (2016). https://doi.org/10.1002/jgt.22027
20. Hong, S.H., Kaufmann, M., Kobourov, S.G., Pach, J.: Beyond-planar graphs: algorithmics and combinatorics (Dagstuhl seminar 16452). Dagstuhl Rep. **6**(11), 35–62 (2017). https://doi.org/10.4230/DagRep.6.11.35
21. Matoušek, J.: Lectures on Discrete Geometry. Graduate Texts in Mathematics, vol. 212. Springer, New York (2002). https://doi.org/10.1007/978-1-4613-0039-7
22. Owens, A.: On the biplanar crossing number. IEEE Trans. Circuit Theory **18**(2), 277–280 (1971). https://doi.org/10.1109/TCT.1971.1083266
23. Pach, J., Székely, L.A., Tóth, C.D., Tóth, G.: Note on k-planar crossing numbers. Comput. Geom. **68**, 2–6 (2018). https://doi.org/10.1016/j.comgeo.2017.06.015
24. Richter, R.B., Thomassen, C.: Relations between crossing numbers of complete and complete bipartite graphs. Am. Math. Mon. **104**(2), 131–137 (1997). https://doi.org/10.1080/00029890.1997.11990611
25. Schaefer, M.: The graph crossing number and its variants: a survey. Electron. J. Comb. Dyn. Surv. 21 (2013/2017)
26. Shahrokhi, F., Sýkora, O., Székely, L.A., Vrt'o, I.: On k-planar crossing numbers. Discret. Appl. Math. **155**(9), 1106–1115 (2007). https://doi.org/10.1016/j.dam.2005.12.011. Advances in Graph Drawing: The 11th International Symposium on Graph Drawing
27. Tóth, C.D.: Personal communication (2018)
28. Tutte, W.T.: The non-biplanar character of the complete 9-graph. Can. Math. Bull. **6**(3), 319–330 (1963). https://doi.org/10.4153/CMB-1963-026-x

Minimal Representations of Order Types by Geometric Graphs

Oswin Aichholzer[1], Martin Balko[2], Michael Hoffmann[3], Jan Kynčl[2],
Wolfgang Mulzer[4], Irene Parada[1]([✉]), Alexander Pilz[1],
Manfred Scheucher[5], Pavel Valtr[2], Birgit Vogtenhuber[1], and Emo Welzl[3]

[1] Institute of Software Technology, Graz University of Technology, Graz, Austria
{oaich,iparada,apilz,bvogt}@ist.tugraz.at
[2] Department of Applied Mathematics, Charles University,
Prague, Czech Republic
{balko,kyncl,valtr}@kam.mff.cuni.cz
[3] Department of Computer Science, ETH Zürich, Zürich, Switzerland
{hoffmann,emo}@inf.ethz.ch
[4] Institut für Informatik, Freie Universität Berlin, Berlin, Germany
mulzer@inf.fu-berlin.de
[5] Institute of Mathematics, Technische Universität Berlin, Berlin, Germany
scheucher@math.tu-berlin.de

Abstract. In order to have a compact visualization of the order type of a given point set S, we are interested in geometric graphs on S with few edges that unequivocally display the order type of S. We introduce the concept of *exit edges*, which prevent the order type from changing under continuous motion of vertices. Exit edges have a natural dual characterization, which allows us to efficiently compute them and to bound their number.

Keywords: Geometric graph · Straight-line drawing · Order type · Pseudoline arrangement · Triangular cell

1 Introduction

Let $S, T \subset \mathbb{R}^2$ be two sets of n labeled points in general position (no three collinear). We say that S and T have *the same order type* if there is a bijection

Research supported by the German Science Foundation (DFG), the Austrian Science Fund (FWF), and the Swiss National Science Foundation (SNSF) within the collaborative DACH project *Arrangements and Drawings*. O.A., I.P., and B.V. were partially supported by Austrian Science Fund (FWF) grant W1230. M.B., J.K., and P.V. are supported by grant no. 18-19158S of the Czech Science Foundation (GAČR). M.B. and J.K. are supported by Charles University project UNCE/SCI/004. M.B. has received funding from European Research Council (ERC) under the European Union's Horizon 2020 research. M.H. and E.W. are supported by SNSF Project 200021E-171681. A.P. was supported by a Schrödinger fellowship of the Austrian Science Fund (FWF): J-3847-N35. M.S. was partially supported by DFG Grant FE 340/12-1. W.M. was partially supported by ERC StG 757609 and DFG Grant 3501/3-1.

© Springer Nature Switzerland AG 2019
D. Archambault and C. D. Tóth (Eds.): GD 2019, LNCS 11904, pp. 101–113, 2019.
https://doi.org/10.1007/978-3-030-35802-0_8

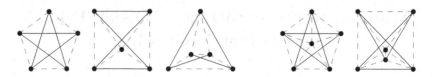

Fig. 1. Left: representatives of the three order types of five points in general position. Right: representatives of two order types of six points. Exit edges are drawn in black.

$\varphi : S \to T$ such that any triple $(p, q, r) \in S^3$ of three distinct points has the same orientation (clockwise or counterclockwise) as the image $(\varphi(p), \varphi(q), \varphi(r)) \in T^3$. The resulting equivalence relation on planar n-point sets has a finite number of equivalence classes, the *order types* [9]. Representatives of several distinct order types of five or six points are illustrated in Fig. 1. Among other things, the order type determines which geometric graphs can be drawn on a point set without crossings. Thus, order types appear ubiquitously in the study of extremal problems on geometric graphs.

Now, suppose we have discovered an interesting order type, and we would like to illustrate it in a publication. One solution is to give explicit coordinates of a representative point set S; see Fig. 2 left. This is unlikely to satisfy most readers. We could also present S as a set of dots in a figure. For some point sets (particularly those with extremal properties), the reader may find it difficult to discern the orientation of an almost collinear point triple. To mend this, we could draw all lines spanned by two points in S. In fact, it suffices to present only the *segments* between the point pairs (the complete geometric graph on S). The orientation of a triple can then be obtained by inspecting the corresponding triangle; see Fig. 2 middle. However, such a drawing is rather dense, and we may have trouble following an edge from one endpoint to the other. Therefore, we want to reduce the number of edges in the drawing as much as possible, but so that the order type remains uniquely identifiable; see Fig. 2 right.

Results. We introduce the concept of *exit edges* to capture which edges are sufficient to uniquely describe a given order type in a robust way under continuous motion of vertices. More precisely, in a geometric drawing of a representative point set with all exit edges, at least one vertex needs to move across an (exit) edge in order to change the order type. We give an alternative characterization of exit edges in terms of the dual line arrangement, where an exit edge corresponds to one or two empty triangular cells. This allows us to efficiently compute the set of exit edges for a given set of n points in $O(n^2)$ time and space.

Using the more general framework of abstract order types and their dual pseudoline arrangements, we prove that every set of $n \geq 4$ points has at least $(3n-7)/5$ exit edges. We also describe a family of n points with $n-3$ exit edges, showing that this bound is asymptotically tight. An upper bound of $n(n-1)/3$ follows from known results on the number of triangular cells in line arrangements [10]. Thus, compared to the complete geometric graph with $n(n-1)/2$ edges, using only exit edges we save at least one third of the edges.

(-1,1)
(1,1)
(-1,-1)
(1,-1)
(-0.6,0.4)
(-0.6,-0.4)

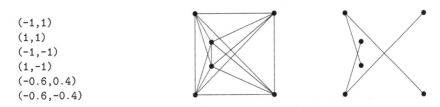

Fig. 2. Three different representations of an order type of six points.

Identification of Order Types. Let S be a set of n labeled points in the plane. A *geometric graph* on S is a graph with vertex set S whose edges are represented as line segments between their endpoints. A geometric graph is thus a drawing of an abstract graph. Two geometric graphs G and H are *isomorphic* if there is an orientation-preserving homeomorphism of the plane transforming G into H. Each class of this equivalence relation may be described combinatorially by the cyclic orders of the edge segments around vertices and crossings, and by the incidences of vertices, crossings, edge segments, and faces. In the following, we will consider topology-preserving deformations. An *ambient isotopy* of the Euclidean plane is a continuous map $f : \mathbb{R}^2 \times [0,1] \to \mathbb{R}^2$ such that $f(\cdot, t)$ is a homeomorphism for every $t \in [0,1]$ and $f(\cdot, 0) = \mathrm{Id}$. Note that if there is an ambient isotopy transforming a geometric graph G into another geometric graph H, then G and H are isomorphic.

Definition 1. *Let G be a geometric graph on a point set S. We say that G is* supporting *for S if every ambient isotopy f of $\mathbb{R}^2$ that keeps the images of the edges of G straight (thus, transforming G into another geometric graph) and that allows at most three points of $f(S,t)$ to be collinear for every $t \in [0,1]$, also preserves the order type of the vertex set.*

Related Work. The connection between order types and straight-line drawings has been studied intensively, both for planar drawings and for drawings minimizing the number of crossings. For example, it is NP-complete to decide whether a planar graph can be embedded on a given point set [5]. Continuous movements of the vertices of plane geometric graphs have also been considered [2]. The continuous movement of points maintaining the order type was considered by Mnëv [7,14]. He showed that there are point sets with the same order type such that there is no ambient isotopy between them preserving the order type, settling a conjecture by Ringel [15]. The orientations of triples that have to be fixed to determine the order type are strongly related to the concept of *minimal reduced systems* [4].

Outline. We introduce the concept of *exit edges* for a given point set. The resulting *exit graphs* are always supporting, though they are not necessarily minimal. In Sect. 2 we show that some exit edges are rendered unnecessary by non-stretchability of certain pseudoline arrangements. Despite being non-minimal in general, we argue that exit graphs are good candidates for supporting graphs by

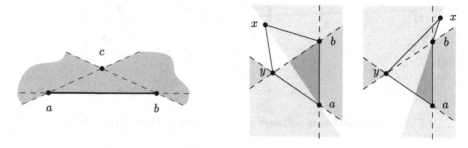

Fig. 3. Characterizing exit edges. Left: If the gray region is empty of points, then the edge ab is an exit edge. Right: An illustration of the proof of Proposition 3.

discussing their dual representation in pseudoline arrangements (Sect. 3). This connection allows us to both compute exit edges efficiently and give bounds on their number (Sect. 4). Supporting graphs in general need not be connected, and two minimal geometric graphs that are supporting for point sets with different order types can be drawings of the same abstract graph; see Fig. 1 right. Thus, the structure of the drawing is crucial. In Sect. 5 we provide some further properties of the exit graphs. We conjecture that graphs based on exit edges are not only supporting but also they encode the order type, as discussed in Sect. 6.

2   Exit Edges

Clearly, every complete geometric graph is supporting. To obtain a supporting graph with fewer edges, we select edges so that no vertex of the resulting geometric graph can be moved to change the order type while preserving isomorphism.

Definition 2. *Let $S \subset \mathbb{R}^2$ be finite and in general position. Let $a, b, c \in S$ be distinct. Then, ab is an* exit edge *with* witness c *if there is no $p \in S$ such that the line $\overline{ap}$ separates b from c or the line $\overline{bp}$ separates a from c. The geometric graph on S whose edges are the exit edges is called the* exit graph *of S.*

Equivalently, ab is an *exit edge* with *witness* c if and only if the double-wedge through a between b and c and the double-wedge through b between a and c contain no point of S in their interior; see Fig. 3 left.

An exit edge has at most two witnesses. If $|S| \geq 4$ and ab is an exit edge in S with witness c, neither ac nor bc can be an exit edge with witness b or a, respectively. We illustrate the set of exit edges for sets of 5 points in Fig. 1 left.

Exit edges can be characterized via 4-holes. For an integer $k \geq 3$, a *(general) k-hole* in S is a simple polygon $\mathcal{P}$ spanned by k points of S whose interior contains no point of S. If $\mathcal{P}$ is convex, we call $\mathcal{P}$ a *convex k-hole*. A point $a \in S$ or an edge ab with $a, b \in S$ is *extremal* for S if it lies on the boundary of the convex hull of S. A point or an edge in S that is not extremal in S is *internal* to S.

Proposition 3. *Let $S \subset \mathbb{R}^2$ be a set of points in general position and let a, $b \in S$. Then, ab is not an exit edge of S if and only if the following conditions hold:*

1. *If ab is extremal in S, then ab is an edge of at least one convex 4-hole in S.*
2. *If ab is internal in S, then there are two 4-holes $abxy$ and $bauv$, in counter-clockwise order, such that their reflex angles (if any) are incident to ab.*

We remark that an internal exit edge either has a witness on both sides or is incident to at least one general 4-hole on one side.

Proof. Let ab be an exit edge with a witness c that lies, without loss of generality, to the left of $\overrightarrow{ab}$. Suppose there is a general 4-hole $abxy$, traced counterclockwise, such that the reflex angle of $abxy$ (if it exists) is incident to ab. We can assume that y lies to the left of $\overrightarrow{ab}$, as in Fig. 3 right. First, suppose that $abxy$ is convex (this must hold if ab is extremal). Since ab is an exit edge with witness c, the line $\overline{ax}$ does not separate c from b and the line $\overline{by}$ does not separate c from a. Thus, c must be inside the 4-hole $abxy$, which is impossible. Second, suppose that $abxy$ is not convex (then, ab is internal), and x is to the right of $\overline{ab}$. Since ab is an exit edge with witness c, the line $\overline{bx}$ does not separate a from c and the line $\overline{ay}$ does not separate b from c, so c lies inside the 4-hole $abxy$, again a contradiction.

Conversely, assume that ab is not an exit edge. First, let ab be extremal, and let p be the closest point in $S \setminus \{a, b\}$ to the line $\overline{ab}$. The triangle abp is a 3-hole in S. Since p is not a witness for ab, there is a point $q \in S \setminus \{a, b, p\}$ such that, without loss of generality, the line $\overline{bq}$ separates a from p. Since ab is extremal, q lies on the same side of $\overline{ab}$ as p and, in particular, the polygon $abpq$ is convex. If we choose q so that it is the closest such point to the line $\overline{ap}$, the triangles bpq and abq are 3-holes in S. Altogether, we obtain a convex 4-hole $abpq$ in S.

Second, let ab be internal. Let p be closest in $S \setminus \{a, b\}$ to the line $\overline{ab}$ such that p lies to the left of $\overline{ab}$. The triangle abp is a 3-hole in S. Since p is not a witness for ab, there is a point $q \in S \setminus \{a, b, p\}$ such that either the line $\overline{bq}$ separates a from p or the line $\overline{aq}$ separates b from p. If q lies to the left of $\overline{ab}$, we obtain a convex 4-hole as in the previous case. Thus, we can assume that all such points q lie to the right of $\overline{ab}$. We choose the point q so that it is (one of the) closest to the line $\overline{ab}$ among all points that prevent ab from being an exit edge with witness p. Without loss of generality, we assume that the line $\overline{bq}$ separates a from p. The choice of q guarantees that bpq is a 3-hole in S. Thus, $abqp$ is a 4-hole in S incident to ab from the left. An analogous argument with a point p' from $S \setminus \{a, b\}$ that is closest to $\overline{ab}$ such that p' lies to the right of $\overline{ab}$ shows that there is an appropriate 4-hole in S incident to ab from the right. □

Proposition 4. *Let $S \subset \mathbb{R}^2$ be finite and in general position and, for every $t \in [0, 1]$, let $S(t)$ be a continuous deformation of S at time t. More formally, let $f : \mathbb{R}^2 \times [0, 1] \to \mathbb{R}^2$ be an ambient isotopy and $S(t) = \{f(s, t) \mid s \in S\}$, for $t \in [0, 1]$. Let $S^c \subseteq S$ be the first subset of at least three points to become collinear. Let (a, b, c) be the first triple to become collinear, at time $t_0 > 0$. If c lies on the segment ab in $S(t_0)$, then ab is an exit edge of $S(0)$ with witness c.*

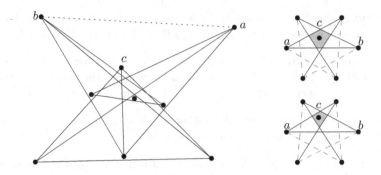

Fig. 4. Left: moving c over ab to orient (a, b, c) clockwise, without changing the orientation of other triples, would contradict Pappus's theorem [15]. Right: it is not always possible to move a witness c continuously to the corresponding exit edge ab.

Proof. For $t \in [0, t_0)$, the triple orientations in $S(t)$ remain unchanged, and in $S(t_0)$, the point c lies on ab. Thus, for $t \in [0, t_0)$, there is no line through two points of $S(t)$ that strictly separates the relative interior of ab from c. In particular, there is no such separating line through a or b in $S(0)$. Hence, ab is an exit edge with witness c.   □

Corollary 5. *The exit graph of every point set is supporting.*

The proof of Proposition 4 also shows that if a line separates c from the relative interior of ab, then there is such a line through a or b. This may suggest that the exit edges are necessary for a supporting graph. However, this is not true in general. For example, in Fig. 4 left, we see a construction by Ringel [15]: ab is an exit edge with witness c, but c cannot move over ab without violating Pappus' theorem. We note that in this situation, we might consider the *abstract order type* for the triple orientations we would obtain after moving c over ab. Since there is no planar point set with this set of triple orientations, this abstract order type is not *realizable*. Deciding realizability is (polynomial-time-)equivalent to the existential theory of the reals [14]. We will revisit these concepts in Sect. 4.

We note that there are point sets where two or more other exit edges prevent a witness c from crossing its corresponding exit edge ab; see, for example, Fig. 4 bottom right. Since the two geometric graphs in Fig. 4 right are not isomorphic, they cannot be transformed into each other by a continuous deformation as the one used in Definition 1. However, in this example, while c cannot move to ab without changing the order type in Fig. 4 bottom right, if ab were not present, we could first change the point set to the one in Fig. 4 top right and then move c over ab. Thus, ab indeed has to be in a supporting graph.

3   Exit Edges and Empty Triangular Cells

The (real) *projective plane* $\mathbb{P}^2$ is a non-orientable surface obtained by augmenting the Euclidean plane $\mathbb{R}^2$ by a *line at infinity*. This line has one *point at infinity* for each direction, where all parallel lines with this direction intersect. Thus, in $\mathbb{P}^2$, each pair of parallel lines intersects in a unique point.

For a point set S in the Euclidean plane, add a line ℓ_∞ to obtain the projective plane. We use a duality transformation that maps a point s of $\mathbb{P}^2$ to a line s^* in $\mathbb{P}^2$. In this way, we get a set of lines S^* dual to S, giving a projective line arrangement $\mathcal{A}$. The removal of a line from $\mathcal{A}$ does not disconnect $\mathbb{P}^2$. Since $\mathbb{P}^2$ has non-orientable genus 1, removing any two lines ℓ_1 and ℓ_2 from $\mathbb{P}^2$ disconnects it into two components. We call the closure of each of the two components a *halfplane* determined by ℓ_1 and ℓ_2. The *marked cell* c_∞ is the cell of $\mathcal{A}$ that contains the point ℓ_∞^* dual to the line ℓ_∞. By appropriately choosing the duality transformation, we can assume that ℓ_∞^* lies at vertical infinity.

The combinatorial structure of $\mathcal{A}$, together with the marked cell, determines the order type of S. We show how to identify exit edges and their witnesses in dual line arrangements.

We use the marked cell c_∞ to orient the lines from S^*: first, we orient the lines on the boundary of c_∞ in one direction. Then, we iteratively remove lines that have already been oriented, and we define the orientation for the remaining lines from S^* by considering the new lines on the boundary of c_∞. Then, c_∞ is the only cell whose boundary is *oriented consistently*, that is, it can be traversed completely along the resulting orientation. In particular, for an unmarked triangular cell $\triangle$ in $\mathcal{A}$, the directed edges of $\triangle$ form a transitive order on its vertices, with a unique vertex of $\triangle$ in the middle. We call this vertex the *exit vertex* of $\triangle$ and the line through the other two vertices of $\triangle$ the *witness line* of $\triangle$.

Note that if we consider the duality mapping a point $p = (p_x, p_y)$ from the real plane to the (non-vertical) line $p^* : y = p_x x - p_y$, then the described orientation procedure corresponds to orienting these dual lines from left to right.

Theorem 6. *Let $S \subset \mathbb{R}^2$ be in general position, and let $a, b, c \in S$. Then, ab is an exit edge with witness c if and only if the lines a^*, b^*, and c^* bound an unmarked triangular cell $\triangle$ in the arrangement $\mathcal{A}$ of lines from S^* so that c^* is the witness line of $\triangle$ and the point $\overline{ab}^* = a^* \cap b^*$ is the exit vertex of $\triangle$.*

Proof. For two points $p, q \in S$ and their dual lines $p^*, q^* \in S^*$, we denote by $w(p^*, q^*)$ the halfplane determined by p^* and q^* that does not contain the marked cell. Thus, the boundary of $w(p^*, q^*)$ is not oriented consistently. Since projective duality preserves incidences, the condition that no line spanned by two points of S intersects the edge pq is equivalent in S^* to $w(p^*, q^*)$ not containing any vertex of $\mathcal{A}$.

Let $\triangle$ be the triangular region determined by the intersection of the two halfplanes $w(a^*, c^*)$ and $w(b^*, c^*)$. By the projective duality, ab is an exit edge with witness c in S if and only if no line of S^* intersects a^* inside $w(b^*, c^*)$ or b^* inside $w(a^*, c^*)$. In other words, if and only if two sides of $\triangle$, lying on a^* and b^*, contain no intersection with lines from S^*. This is equivalent to $\triangle$ being a cell of the arrangement $\mathcal{A}$. Moreover, a^* and b^* share the exit vertex of $\triangle$; see Fig. 5. Consequently, the exit vertex $a^* \cap b^*$ is the dual of the line containing the exit edge ab. $\qquad\square$

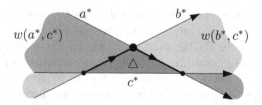

Fig. 5. An illustration of the proof of Theorem 6. If ab is an exit edge with witness c in S, then the two bold drawn segments of the corresponding triangular cell are unintersected, and thus, bound an unmarked triangular cell in S^*. The exit vertex is represented with a black disk.

Corollary 7. *Let S be a set of n points in general position. Then the exit edges of S can be enumerated in $O(n^2)$ time by constructing the dual line arrangement of S and checking which cells are unmarked triangular cells.*

4   On the Number of Exit Edges

Line arrangements can be generalized to so-called pseudoline arrangements. A *pseudoline* is a closed curve in the projective plane $\mathbb{P}^2$ whose removal does not disconnect $\mathbb{P}^2$. A set of pseudolines in $\mathbb{P}^2$, where any two pseudolines cross exactly once, determines a (projective) *pseudoline arrangement*. If no three pseudolines intersect in a common point, the pseudoline arrangement is *simple*. All notions that we have introduced for line arrangements, such as consistent orientations, exit vertices, or witness lines, naturally extend to pseudolines.

A pseudoline arrangement is *stretchable* if it is isomorphic to a line arrangement, that is, the corresponding cell complexes into which the two arrangements partition $\mathbb{P}^2$ are isomorphic. The combinatorial dual analogues of line arrangements and pseudoline arrangements are order types and abstract order types, respectively. Thus, deciding if a pseudoline arrangement is stretchable is (polynomial-time-)equivalent to the existential theory of the reals [7,14].

As discussed in Sect. 3, the maximum number of *triangular cells* in a simple projective pseudoline arrangement gives an upper bound on the number of exit edges of a point set. However, one triangular cell could be c_∞, and there could be pairs of triangular cells with the same exit vertex. We call a configuration of the latter type an *hourglass*; see Fig. 6. We say that the two pseudolines p and q that define the exit vertex of the two triangular cells of an hourglass H *slice* H and that H is *sliced* by p and by q.

Observation 8. *A triangular cell can be a part of at most one hourglass.*

Observation 9. *An exit edge ab with two witness points is dual to an hourglass with exit vertex $\overline{ab}^*$.*

Any projective arrangement of $n \geq 4$ lines has at least n triangular cells, as each line is incident to at least three triangular cells [12]. This is known to be

Fig. 6. Left: the two triangular cells $\triangle_1$ and $\triangle_2$ do not form an hourglass, because they share a vertex that is not an exit vertex. Right: the two triangular cells $\triangle_1$ and $\triangle_2$ form an hourglass because they share an exit vertex.

tight. Therefore, taking into account the marked cell c_∞ and possible hourglasses, any set of $n \geq 4$ points has at least $\lceil \frac{n-1}{2} \rceil$ exit edges. We improve this lower bound by bounding from below the difference between the number of triangular cells and the number of hourglasses.

Proposition 10. *Any set of $n \geq 4$ points in the plane has at least $(3n-7)/5$ exit edges.*

For the proof of Proposition 10 we use the following two lemmas. The first is a theorem by Grünbaum [10, Theorem 3.7 on p. 50], and the second can be derived from the proof of that theorem.

Lemma 11 (Grünbaum [10]). *In a simple pseudoline arrangement L every pseudoline from L is incident to at least three triangular cells.*

Lemma 12 (Grünbaum [10]). *Let L be a simple arrangement of pseudolines, and let H be a closed halfplane determined by two pseudolines $\ell_1, \ell_2 \in L$. If two other pseudolines of L cross in the interior of H, then there is a triangular cell in H that is incident to ℓ_1 but not to ℓ_2.*

Proof (of Proposition 10). Let L be a simple projective line arrangement of $n \geq 4$ pseudolines $\ell_1, \ell_2, \ldots, \ell_n$. For each pseudoline $\ell_i \in L$, let t_i be the number of triangular cells incident to ℓ_i and h_i the number of hourglasses sliced by ℓ_i. Set $x_i = t_i - h_i/2$. For each pseudoline $\ell_i \in L$, there are three possible cases.

Case (i): there is no hourglass sliced by ℓ_i. By Lemma 11, every pseudoline is incident to at least three triangular cells. Thus, we have $x_i = t_i \geq 3$.

Case (ii): the pseudoline ℓ_i slices an hourglass together with some pseudoline ℓ_j and the interior of each of the two halfplanes determined by ℓ_i and ℓ_j contains at least one crossing of some other pair of pseudolines. By Lemma 12, ℓ_i is incident to the two triangular cells of the hourglass plus at least two other triangular cells, one in each closed halfplane. (We ignore here that a cell might be the marked one.) Thus, $t_i \geq 4$. Observation 8 implies $h_i \leq t_i/2$. Overall we get $x_i = t_i - h_i/2 \geq t_i - t_i/4 \geq (3/4) \cdot 4 = 3$.

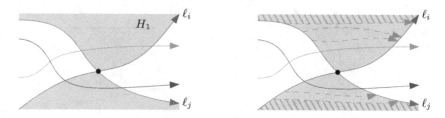

Fig. 7. In case (iii), both ℓ_1 and ℓ_2 must bound the marked cell, shown striped on the right picture. Moreover, that cell is bounded by four pseudolines.

Case (iii): the pseudoline ℓ_i slices an hourglass together with some pseudoline ℓ_j, and one of the two closed halfplanes H_1 and H_2 determined by ℓ_i and ℓ_j contains no crossing of any other pair of pseudolines in its interior. Suppose the closed halfplane that contains no further crossing is H_1. Then, the hourglass sliced by ℓ_i and ℓ_j is in H_1, as the other two lines defining the hourglass do not cross in that halfplane; see Fig. 7 (left). Since H_1 contains no crossing in its interior, it is divided by the other pseudolines into 4-gons and the two triangular cells of the hourglass. In particular, the marked cell is bounded by only four pseudolines, two of them being ℓ_i and ℓ_j; see Fig. 7, right. Thus, there can be at most four pseudolines for which case (iii) applies. Notice that in this case $h_i = 1$, since any other hourglass sliced by ℓ_i would have one triangular cell in each of the two halfplanes H_1 and H_2 and the two triangular cells in H_1 form the already-counted hourglass (and by Observation 8 they cannot be part of another hourglass). Thus, we can only guarantee that $x_i \geq 3 - 1/2 = 5/2$. However, as we showed, this case can happen at most for two pairs of pseudolines.

Let T be the total number of triangular cells in L and let H be the total number of hourglasses. Summing the contributions of cases (i)–(iii), we have

$$3T - H = \sum_{i=1}^{n} t_i - \frac{1}{2} \sum_{i=1}^{n} h_i = \sum_{i=1}^{n} x_i \geq 3 \cdot (n-4) + 4 \cdot \left(\frac{5}{2}\right) = 3n - 2.$$

By Observation 8, we have $T \geq 2H$. Combining these inequalities, we get

$$T - H = \frac{3T - H + 2(T - 2H)}{5} \geq \frac{3T - H}{5} \geq \frac{3n - 2}{5}.$$

By Theorem 6, the number of exit edges in a point set is equal to the number of exit vertices in its dual line arrangement. In general, the number of exit vertices in a pseudoline arrangement is bounded from below by $T - H - 1$. Therefore, there are at least $\frac{3}{5}n - \frac{7}{5}$ exit edges. □

We do not know if the lower bound in Proposition 10 is tight. The smallest number of exit edges we could achieve is $n - 3$ for $n \geq 9$; see Fig. 8.

The number of triangular cells in a simple arrangement of n lines in the projective plane $\mathbb{P}^2$ is at most $n(n-1)/3$ [10], so there are at most $n^2/3 + O(n)$ exit edges. This means that representing an order type with the exit graph instead

Fig. 8. Construction with $n - 3$ exit edges.

of the complete geometric graph saves at least one third of the edges. Palásti and Füredi [17] showed that for every value of n there are simple arrangement of n lines in $\mathbb{P}^2$ with $n(n - 3)/3$ triangular cells. Moreover, Roudneff [16] and Harborth [11] proved that the upper bound $n(n-1)/3$ is tight for infinitely many values of n (see also [3]). The point sets that are dual to the currently-known arrangements that maximize the number of triangular cells have $n^2/6+O(n)$ exit edges, since most of their exit edges have two witnesses. This gives a quadratic lower bound in the worst case, but the leading coefficient remains unknown. It is worth noting that there are line arrangements with no pair of adjacent triangular cells [13], which implies the existence of point sets where every exit edge has precisely one witness.

5   Properties of Exit Graphs

We present some further results on supporting graphs and exit graphs.

Theorem 13. *Any geometric graph supporting a point set S, with $|S| \geq 9$, contains a crossing.*

Proof. Let G be a geometric graph with vertex set S without crossings. There is a point set S' with a different order type that also admits G: Dujmović [6] showed that every plane graph admits a plane straight-line embedding with at least $\sqrt{n/2}$ points on a line; as we have a point set with a collinear triple that admits G, there are at least two point sets in general position with a different order type that admit G. Moreover, one can continuously morph S to S' while keeping the corresponding geometric graph planar and isomorphic to G (see, for example, [2]). Therefore, G does not support S. $\square$

Proposition 14. *Let S be a point set in general position in $\mathbb{R}^2$ and let G be its exit graph. Every vertex in the unbounded face of G is extremal, that is, it lies on the boundary of the convex hull of S.*

Note that, as shown in Fig. 4 left, an analogous statement does not hold for general supporting graphs. The proof can be found in the full version [1].

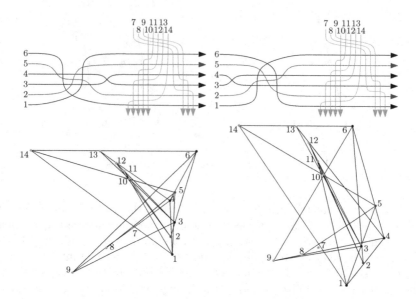

Fig. 9. Top: two arrangements of 14 pseudolines with the same set of triangular cells (extending [8, Figure 3]). No triangular cell crossed by the line at infinity. Bottom: corresponding dual point sets. The order types are not the same (see for example the number of extremal points).

6 Concluding Remarks

We conjecture that the geometric graph G of exit edges not only is support-ing for S, but also that any point set S' that is the vertex set of a geometric graph isomorphic to G has the same order type as S. One might conjecture that already knowing all exit edges and their witnesses (in the dual line arrangement, all triangular cells and their orientations) is sufficient to determine the order type. Surprisingly, this turns out to be wrong. A counterexample is sketched in Fig. 9 as a dual (stretchable) pseudoline arrangement of 14 lines in the projective plane, based on an example by Felsner and Weil [8]. It consists of two arrange-ments of six lines in the Euclidean plane that are combinatorially different, but share the set of triangular cells and their orientations. While the exit edges are the same for the two different order types, the corresponding exit graphs are not isomorphic. In the dual of that example the order of the triangular cells along each pseudoline differs, but that extra information is not enough to dis-tinguish the two order types: We can modify the pseudoline arrangements in Fig. 9 by, essentially, duplicating pseudolines 1–6 and making a pseudoline and its duplication cross between the crossings with two green pseudolines (7–14). An illustration is presented in the full version [1].

Acknowledgments. This work was initiated during the *Workshop on Sidedness Queries*, October 2015, in Ratsch, Austria. We thank Thomas Hackl, Vincent Kusters, and Pedro Ramos for valuable discussions.

References

1. Aichholzer, O., Balko, M., Hoffmann, M., Kynčl, J., Mulzer, W., Parada, I., Pilz, A., Scheucher, M., Valtr, P., Vogtenhuber, B., Welzl, E.: Minimal representations of order types by geometric graphs (2019). http://arxiv.org/abs/1908.05124
2. Alamdari, S., Angelini, P., Barrera-Cruz, F., Chan, T.M., Da Lozzo, G., Di Battista, G., Frati, F., Haxell, P., Lubiw, A., Patrignani, M., Roselli, V., Singla, S., Wilkinson, B.T.: How to morph planar graph drawings. SIAM J. Comput. **46**(2), 824–852 (2017). https://doi.org/10.1137/16M1069171
3. Blanc, J.: The best polynomial bounds for the number of triangles in a simple arrangement of n pseudo-lines. Geombinatorics **21**, 5–17 (2011). https://edoc.unibas.ch/47402
4. Bokowski, J., Sturmfels, B.: On the coordinatization of oriented matroids. Discret. Comput. Geom. **1**, 293–306 (1986). https://doi.org/10.1007/BF02187702
5. Cabello, S.: Planar embeddability of the vertices of a graph using a fixed point set is NP-hard. J. Graph Algorithms Appl. **10**(2), 353–363 (2006). https://doi.org/10.7155/jgaa.00132
6. Dujmović, V.: The utility of untangling. J. Graph Algorithms Appl. **21**(1), 121–134 (2017). https://doi.org/10.7155/jgaa.00407
7. Felsner, S., Goodman, J.E.: Pseudoline arrangements. In: Tóth, C.D., O'Rourke, J., Goodman, J.E. (eds.) Handbook of Discrete and Computational Geometry, pp. 125–157, 3rd edn. CRC Press (2017). https://doi.org/10.1201/9781315119601
8. Felsner, S., Weil, H.: A theorem on higher Bruhat orders. Discret. Comput. Geom. **23**(1), 121–127 (2000). https://doi.org/10.1007/PL00009485
9. Goodman, J.E., Pollack, R.: Multidimensional sorting. SIAM J. Comput. **12**(3), 484–507 (1983). https://doi.org/10.1137/0212032
10. Grünbaum, B.: Arrangements and spreads. AMS (1972). https://bookstore.ams.org/cbms-10/
11. Harborth, H.: Some simple arrangements of pseudolines with a maximum number of triangles. Ann. N. Y. Acad. Sci. **440**(1), 31–33 (1985). https://doi.org/10.1111/j.1749-6632.1985.tb14536.x
12. Levi, F.: Die Teilung der projektiven Ebene durch Gerade oder Pseudogerade. Ber. Math.-Phys. Kl. Sächs. Akad. Wiss. Leipzig **78**, 256–267 (1926). (in German)
13. Ljubić, D., Roudneff, J.P., Sturmfels, B.: Arrangements of lines and pseudolines without adjacent triangles. J. Comb. Theory. Ser. A **50**(1), 24–32 (1989). https://doi.org/10.1016/0097-3165(89)90003-4
14. Mnev, N.E.: The universality theorems on the classification problem of configuration varieties and convex polytopes varieties. In: Viro, O.Y., Vershik, A.M. (eds.) Topology and Geometry — Rohlin Seminar. LNM, vol. 1346, pp. 527–543. Springer, Heidelberg (1988). https://doi.org/10.1007/BFb0082792
15. Ringel, G.: Teilungen der Ebene durch Geraden oder topologische Geraden. Math. Z. **64**, 79–102 (1956)
16. Roudneff, J.P.: On the number of triangles in simple arrangements of pseudolines in the real projective plane. Discret. Math. **60**, 243–251 (1986). https://doi.org/10.1016/0012-365X(86)90016-6
17. Füredi, Z., Palásti, I.: Arrangements of lines with a large number of triangles. Proc. Am. Math. Soc. **92**(4), 561–566 (1984). https://doi.org/10.2307/2045427

Balanced Schnyder Woods for Planar Triangulations: An Experimental Study with Applications to Graph Drawing and Graph Separators

Luca Castelli Aleardi[✉][iD]

LIX, Ecole Polytechnique, Institut Polytechnique de Paris, Palaiseau, France
amturing@lix.polytechnique.fr

Abstract. In this work we consider balanced Schnyder woods for planar graphs, which are Schnyder woods where the number of incoming edges of each color at each vertex is balanced as much as possible. We provide a simple linear-time heuristic leading to obtain well balanced Schnyder woods in practice. As test applications we consider two important algorithmic problems: the computation of Schnyder drawings and of small cycle separators. While not being able to provide theoretical guarantees, our experimental results (on a wide collection of planar graphs) suggest that the use of balanced Schnyder woods leads to an improvement of the quality of the layout of Schnyder drawings, and provides an efficient tool for computing short and balanced cycle separators.

1 Introduction

Schnyder woods [27] and its generalizations are a deep tool for dealing with the combinatorics of planar [13] and surface maps [10,11,18]. They lead to efficient algorithmic and combinatorial solutions for a broad collection of problems, arising in several domains, from enumerative combinatorics to graph drawing and computational geometry. For instance, the use of Schnyder woods has led to linear-time algorithms for grid drawing [3,18,27], to the optimal encoding and uniform sampling of planar maps [26], to the design of compact data structures [9] and to deal with geometric spanners [5]. Schnyder woods lead to fast implementations (also integrated in open source libraries [25]) and provide strong tools for establishing rigorous theoretical guarantees that hold in the worst case, even for irregular, random or pathological cases. The main idea motivating this work is that, in practice, most real-word graphs exhibit strong regularities which make them far from the random and pathological cases. Based on this remark, many geometry processing algorithms try to exploit this regularity in order to obtain better results in practice. For instance, when applied to regular graphs, many mesh compression schemes [19] achieve good compression rates, well below

This work is supported by the French ANR GATO (ANR-16-CE40-0009-01).

D. Archambault and C. D. Tóth (Eds.): GD 2019, LNCS 11904, pp. 114–121, 2019.
https://doi.org/10.1007/978-3-030-35802-0_9

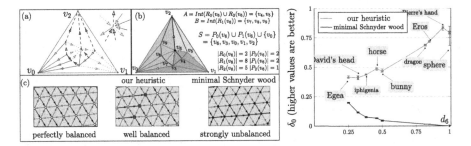

Fig. 1. (a) A planar triangulation endowed with a Schnyder wood. (b) a separator (A, B, S) obtained from the Schnyder wood. (c) three Schnyder woods of the same portion of a spherical grid: our heuristic leads to a majority of balanced vertices (white circles), while the minimal Schnyder wood is strongly unbalanced. (Right chart) Evaluation of the balance of Schnyder woods (tests are repeated with 500 random seeds) (Color figure online).

the worst-case optimal bound guaranteed by [26]. As far as we know, the problem of providing an adaptive analysis of Schnyder woods taking into account the graph regularity has not been investigated so far. This work provides empirical evidence about the fact that balanced Schnyder woods can lead to fast solutions achieving good results in practice, especially for real-world graphs. As test applications, we evaluate the layout quality of a Schnyder drawing depending on the balance of the underlying Schnyder wood, and we consider the problem of computing small separators for planar graphs, which has been extensively investigated [22–24,28], due to its relevance for many graph algorithms.

Preliminaries and Related Works. In this work we deal with *planar triangulations*, which are genus 0 simple maps where every face is triangulated (we will denote by n the number of vertices and by m the number of edges). Given a planar triangulation with a distinguished root (outer) face (v_0, v_1, v_2), a Schnyder wood [27] is defined as a coloring (with colors 0, 1 or 2) and orientation of the inner edges such that each inner vertex has exactly one outgoing incident edge for each color, and the remaining incident edges must satisfy the local Schnyder rule (see Fig. 1(a)). A given rooted triangulation may admit many Schnyder woods [1,2,14]: among them, the *minimal* one (without ccw oriented triangles) plays a fundamental role [9,26]. Here we focus on balanced Schnyder woods, for which the ingoing edges are evenly distributed around inner vertices. A related problem concerns the computation of egalitarian orientations: unfortunately the results in [6] only apply to unconstrained orientations. Schnyder woods have led to a linear-time algorithm providing an elegant solution to the grid drawing problem (solved independently also in [17]): in its pioneristic work [27] Schnyder showed that a planar graph with n vertices admits a straight-line drawing on a grid of size $O(n) \times O(n)$. Schnyder drawings have a number of nice properties that make them useful for addressing problems [4,5,12] involving planar graphs in several distinct domains. While recent works [21] provide a probabilistic study of the converge for uniformly sampled triangulations endowed with a Schnyder wood, as far as we know there are no theoretical or empirical evaluations of the quality of Schnyder drawings for regular graphs. Given a graph G we consider *small separators* which are defined by a partition (A, B, S) of all vertices such that S is a separating vertex set of small size (usually $|S| = O(\sqrt{m})$), and the remaining vertices in $G \setminus S$ belong to a balanced partition (A, B) satisfying $|A| \leq \alpha n$, $|B| \leq \alpha n$ (usually, for planar graphs, the *balance ratio* is $\alpha = \frac{2}{3}$). Here we focus on *simple cycle separators* [24], for which fast implementations [16,20] have been recently proposed (some of them [16] are provided with a worst-case bound of $\sqrt{8m}$ on the cycle size).

2   Contribution

2.1   Balanced Schnyder Woods

Our first step is to measure the balance of a Schnyder wood: given an inner vertex v of degree $deg(v)$ having $indeg_i(v)$ incoming edges of color i (for $i \in \{0, 1, 2\}$), we define its *defect* as $\delta(v) = \max_i indeg_i(v) - \min_i indeg_i(v)$ if $deg(v)$ is a multiple of 3, and $\delta(v) = \max_i indeg_i(v) - \min_i indeg_i(v) - 1$ otherwise. We say that a vertex is *balanced* if $\delta(v) = 0$ and a Schnyder wood is *well balanced* if a majority of vertices have a small defect. For regular graphs is possible, in principle, to get a Schnyder wood that is perfectly balanced ($\delta(v) = 0$ everywhere) as shown in Fig. 1(c). In practice many Schnyder woods are unbalanced and we are not aware of existing theoretical or empirical results on the balance of Schnyder woods.

An Heuristic for Well Balanced Schnyder Woods. We make use of the well known incremental vertex shelling procedure [7] that computes a Schnyder wood with a sequence of vertex removals. This procedure has many degrees of freedom: at each step the choice of the vertex to be removed can possibly lead to a different Schnyder wood. In order to get as much as possible balanced vertices, we retard the removal of some vertices according to a balance priority, defined as the total number of ingoing edges incident to a vertex during the shelling procedure. The balance can be further improved by performing the reversal of oriented triangles in a post-processing[1] step. We refer to [8] for more details.

2.2   From Schnyder Drawings to Small Simple Cycle Separators

Schnyder woods provides a very fast procedure for partitioning, given an arbitrary inner vertex v, the set of inner faces of a triangulation into three sets $R_0(v)$, $R_1(v)$ and $R_2(v)$ (respectively blue, red and gray triangles in Fig. 1(b)), whose boundaries consist of the three disjoint paths $P_0(v)$, $P_1(v)$ and $P_2(v)$ emanating from v. The computation of simple cycle separators can be done as follows: for each vertex v check whether the two sets $A = Int(R_i(v) \cup R_{i+1}(v))$ and $B = Int(R_{i+2}(v))$ satisfy the prescribed balance ratio for at least one index

---

[1]   The results presented in Sect. 2.3 are obtained without post-processing step.

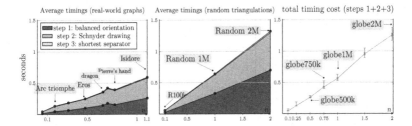

Fig. 2. Evaluation of timing costs over 100 executions (allocating 1 GB of RAM for the JVM): timings are expressed as a function of the size (millions of vertices).

$i \in \{0, 1, 2\}$ (indices are modulo 3, and $Int(R)$ denotes the set of inner vertices of a region R): then select the vertex for which the corresponding cycle length $|P_i(v)| + |P_{i+1}(v)| + 1$ is minimal. All this steps can be performed almost instantaneously, since all the quantities above are encoded in the Schnyder drawing itself (see [27] for more details). As far as we know there are no theoretical guarantees on both the partition balance and boundary size: as observed in practice, most vertices lead to unbalanced partitions whose boundary size can be very large.

2.3 Experimental Results

Datasets and Experimental Setting. We run our experimental evaluations[2] on a broad variety of graphs[3], including real-world meshes used in geometry processing (made available by the **aim@shape** and **Thingi10k** repositories), synthetic regular graphs with different shapes (**sphere, cylinder, ...**), random planar triangulations (generated with uniform random sampling [26]), and Delaunay triangulations of random points. As done in geometric modeling, we use the proportion of degree 6 vertices, denoted by d_6, to measure the regularity of a graph: d_6 is close to 1 for regular meshes, while is usually below 0.3 for irregular and random graphs. To evaluate the balance of a Schnyder woods we use the proportion of balanced vertices, denoted by δ_0, and the average defect computed on all vertices, denoted by δ_{avg}. As for previous works [16,20], the results (e.g. the size of the separator) can depend on the choice of the initial seed (the root face in our case). We perform tests with hundreds of random seeds: for each choice of the seed, we adopt whisker plots to show the entire range of computed values, while each box represents the middle 50% of values (as in Figs. 1 and 4).

Balance of Schnyder Woods. To evaluate the balance quality of the Schnyder woods we plot the value δ_0 as a function of d_6: our balanced Schnyder woods are compared to minimal ones in Fig. 1. Experimental results strongly suggests that our heuristic leads to well balanced Schnyder woods. Our heuristic performs

---

[2] Our datasets and code can be found at http://www.lix.polytechnique.fr/~amturing/ software.html.

[3] Previous works [16,20] triangulate the input graph in a preprocessing phase.

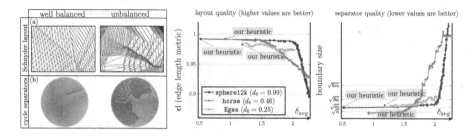

Fig. 3. For a fixed initial seed, we generate a sequence of Schnyder woods by starting from a well balanced Schnyder wood (computed with our heuristic) and by randomly reversing ccw oriented triangles. In the charts we plot the layout and separator quality as functions of the average defect δ_{avg} of the corresponding Schnyder wood.

particularly well for regular graphs, for which a large majority of vertices are balanced (79% in average for the `sphere` graph). The results are good also for irregular graphs (`egea`), where about 45% of vertices are balanced. Also observe that the choice of the initial seed has a limited effect on the balance of the resulting Schnyder wood. Minimal Schnyder woods represent a bad case, especially for regular graphs: most vertices have a large defect and the resulting paths $P_0(v)$, $P_1(v)$ and $P_2(v)$ resemble very long spirals.

Runtime Performances. The algorithmic solutions relying on Schnyder woods are simple to implement and extremely fast. As observed in practice (see Fig. 2), our `Java` implementation allows processing between $1.43M$ and $1.92M$ vertices per second: we run our tests on an EliteBook with a core i7-5600U 2.60 GHz (with Ubuntu 16.04 and 1 GB of RAM allocated for the JVM). This has to be compared to the `C` implementations of previous results on cycle separators [16,20], running on an Intel Xeon X5650 2.67 GHz (with 48.4 GB of RAM): the fastest variant of the procedures tested in [16] allows processing between $0.54M$ and $0.62M$ vertices per second for the case of square `grids`. Our timing costs are little affected by the choice of the initial seed and the structural properties of the graph. Observe that once the Schnyder drawing is given, the extraction of the cycle separator is instantaneous ($0.01\,$s for a $1M$ vertices graph).

Layout Quality. A qualitative evaluation of the graph layouts based on the balance Schnyder woods is provided by the pictures in Fig. 3(a) showing two portions of the Schnyder drawings of a regular sphere graph. When starting from a our well balanced Schnyder woods the shape of triangles is much more balanced, and the resulting drawing partially captures the regularity of the grid. When starting from an unbalanced Schnyder wood the drawing exhibits many long edges and flat triangles, a typical drawback of Schnyder drawings. In order to provide a quantitative measure of the layout quality we consider the *edge lengths aesthetic metric* defined by $\mathfrak{el} = 1 - d_{el}$, where d_{el} is the average percent deviation of edge lengths: values close to 1 mean that most edges have the same length (see [15] for more details). For a fixed initial seed, we start from a

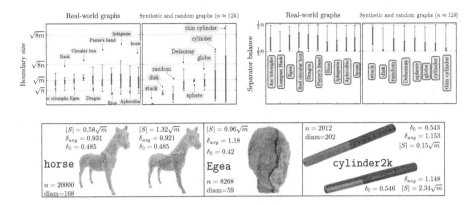

Fig. 4. We evaluate the quality of our simple cycle separators obtained from our balanced Schnyder woods (tests are repeated using 200 random seeds). The left charts report the boundary sizes, while the right charts show the plots of the separator balance (the normalized size of the smallest of the two sets A and B). The graphs are listed from left to right according to the increasing values of their relative diameter.

balanced Schnyder wood obtained with our heuristic and we randomly reverse ccw oriented triangles, obtaining a sequence of Schnyder woods which are more and more unbalanced. The middle chart in Fig. 3 reports the values of $\mathfrak{e}l$ as a function of the average defect: the layout quality tends to deteriorate as soon as Schnyder woods get more unbalanced (high values of δ_{avg}).

Length and Balance of Separators. We look for separators with a balance ratio $\alpha = \frac{2}{3}$ that are *short*: the boundary size is at most $|S| \leq \sqrt{8m}$, as required in [16]. We plot in the charts of Fig. 4 the boundary sizes and partition balances of the separators obtained from a Schnyder drawing as described in Sect. 2.1. Our tests, repeated over several tens of graphs, confirm our intuition that balanced Schnyder woods lead to good separators for a large majority of classes of graphs. As for the layout quality, the separator size and balance strongly depend on the balance of the underlying Schnyder wood (right chart in Fig. 3). The boundary size of the separator is affected by the choice of the seed for graphs with large diameter: a good choice of the seed would prevent from getting too long cycles. For graphs with small diameter (e.g. random triangulations) Schnyder woods lead to very short separators, while the size is closed to $\sqrt{m}$ for most real-world graphs. Our separators are often longer when compared with the results obtained in [16], but well below the prescribed bound of $\sqrt{8m}$.

References

1. Bernardi, O., Bonichon, N.: Catalan's intervals and realizers of triangulations. J. Comb. Theory Ser. A **116**(1), 55–75 (2009). https://hal.archives-ouvertes.fr/hal-00143870(22 pages)

2. Bonichon, N.: A bijection between realizers of maximal plane graphs and pairs of non-crossing Dyck paths. Discret. Math. **298**, 104–114 (2005). https://hal.archives-ouvertes.fr/hal-00307593

3. Bonichon, N., Felsner, S., Mosbah, M.: Convex drawings of 3-connected plane graphs. Algorithmica **47**(4), 399–420 (2007). https://doi.org/10.1007/s00453-006-0177-6

4. Bonichon, N., Gavoille, C., Hanusse, N., Ilcinkas, D.: Connections between theta-graphs, delaunay triangulations, and orthogonal surfaces. In: Thilikos, D.M. (ed.) WG 2010. LNCS, vol. 6410, pp. 266–278. Springer, Heidelberg (2010). https://doi.org/10.1007/978-3-642-16926-7_25

5. Bonichon, N., Gavoille, C., Hanusse, N., Perković, L.: Plane spanners of maximum degree six. In: Abramsky, S., Gavoille, C., Kirchner, C., Meyer auf der Heide, F., Spirakis, P.G. (eds.) ICALP 2010. LNCS, vol. 6198, pp. 19–30. Springer, Heidelberg (2010). https://doi.org/10.1007/978-3-642-14165-2_3

6. Borradaile, G., Iglesias, J., Migler, T., Ochoa, A., Wilfong, G.T., Zhang, L.: Egalitarian graph orientations. J. Graph Algorithms Appl. **21**(4), 687–708 (2017). https://doi.org/10.7155/jgaa.00435

7. Brehm, E.: 3-orientations and Schnyder 3-tree-decompositions. Master's thesis, FB Mathematik und Informatik, Freie Universität Berlin (2000)

8. Castelli Aleardi, L.: Balanced Schnyder woods for planar triangulations: an experimental study with applications to graph drawing and graph separators (2019). https://arxiv.org/abs/1908.06688

9. Castelli Aleardi, L., Devillers, O.: Array-based compact data structures for triangulations: practical solutions with theoretical guarantees. JoCG **9**(1), 247–289 (2018). https://doi.org/10.20382/jocg.v9i1a8

10. Castelli Aleardi, L., Fusy, É., Lewiner, T.: Schnyder woods for higher genus triangulated surfaces, with applications to encoding. Discret. Comput. Geom. **42**(3), 489–516 (2009). https://hal.inria.fr/hal-00712046v1

11. Despré, V., Gonçalves, D., Lévêque, B.: Encoding toroidal triangulations. Discret. Comput. Geom. **57**(3), 507–544 (2017). https://doi.org/10.1007/s00454-016-9832-0

12. Dhandapani, R.: Greedy drawings of triangulations. Discret. Comput. Geom. **43**(2), 375–392 (2010). https://doi.org/10.1007/s00454-009-9235-6

13. Felsner, S.: Lattice structures from planar graphs. Electr. J. Comb. **11**(1) (2004)

14. Felsner, S., Zickfeld, F.: On the number of planar orientations with prescribed degrees. Electr. J. Comb. **15**(1) (2008)

15. Fowler, J.J., Kobourov, S.G.: Planar preprocessing for spring embedders. In: 20th International Symposium Graph Drawing, pp. 388–399 (2012)

16. Fox-Epstein, E., Mozes, S., Phothilimthana, P.M., Sommer, C.: Short and simple cycle separators in planar graphs. ACM J. Exp. Algorithm **21**(1), 2:2:1–2:2:24 (2016)

17. de Fraysseix, H., Pach, J., Pollack, R.: How to draw a planar graph on a grid. Combinatorica **10**(1), 41–51 (1990)

18. Gonçalves, D., Lévêque, B.: Toroidal maps: Schnyder woods, orthogonal surfaces and straight-line representations. Discret. Comput. Geom. **51**(1), 67–131 (2014). https://doi.org/10.1007/s00454-013-9552-7

19. Gotsman, C.: On the optimality of valence-based connectivity coding. Comput. Graph. Forum **22**(1), 99–102 (2003). https://doi.org/10.1111/1467-8659.t01-1-00649

20. Holzer, M., Schulz, F., Wagner, D., Prasinos, G., Zaroliagis, C.D.: Engineering planar separator algorithms. ACM J. Exp. Algorithm **14** (2009). https://doi.org/10.1145/1498698.1571635
21. Li, Y., Sun, X., Watson, S.S.: Schnyder woods, sle(16), and liouville quantum gravity. Technical report arXiv:1705.03573v1 [math.PR], ArXiV, May 2016
22. Lipton, R.J., Tarjan, R.E.: A separator theorem for planar graphs. SIAM J. Appl. Math. **36**(2), 177–189 (1979)
23. Lipton, R.J., Tarjan, R.E.: Applications of a planar separator theorem. SIAM J. Comput. **9**(3), 615–627 (1980). https://doi.org/10.1137/0209046
24. Miller, G.L.: Finding small simple cycle separators for 2-connected planar graphs. J. Comput. Syst. Sci. **32**(3), 265–279 (1986). https://doi.org/10.1016/0022-0000(86)90030-9
25. PIGALE, Public Implementation of a Graph Algorithm Library and Editor. http://pigale.sourceforge.net/
26. Poulalhon, D., Schaeffer, G.: Optimal coding and sampling of triangulations. Algorithmica **46**(3–4), 505–527 (2006)
27. Schnyder, W.: Embedding planar graphs on the grid. In: Proceedings of the Annual ACM-SIAM Symposium on Discrete Algorithms, vol. 90, pp. 138–148 (1990). http://departamento.us.es/dma1euita/PAIX/Referencias/schnyder.pdf
28. Spielman, D.A., Teng, S.: Disk packings and planar separators. In: Proceedings of the Twelfth Annual Symposium on Computational Geometry, pp. 349–358 (1996). https://doi.org/10.1145/237218.237404

Clustering

A Quality Metric for Visualization of Clusters in Graphs

Amyra Meidiana[1(✉)], Seok-Hee Hong[1], Peter Eades[1], and Daniel Keim[2]

[1] University of Sydney, Sydney, Australia
amei2916@uni.sydney.edu.au, {seokhee.hong,peter.eades}@sydney.edu.au
[2] University of Konstanz, Konstanz, Germany
keim@uni-konstanz.de

Abstract. Traditionally, graph quality metrics focus on readability, but recent studies show the need for metrics which are more specific to the discovery of patterns in graphs. Cluster analysis is a popular task within graph analysis, yet there is no metric yet explicitly quantifying how well a drawing of a graph represents its cluster structure.

We define a clustering quality metric measuring how well a node-link drawing of a graph represents the clusters contained in the graph. Experiments with deforming graph drawings verify that our metric effectively captures variations in the visual cluster quality of graph drawings. We then use our metric to examine how well different graph drawing algorithms visualize cluster structures in various graphs; the results confirm that some algorithms which have been specifically designed to show cluster structures perform better than other algorithms.

1 Introduction

Clustering is an important task in graph analysis. Visualization can be a useful tool in this task, where a good drawing of a network should be able to highlight important group structures within the network and allow a user to accurately answer group-level analytical tasks. To this end, a number of graph layout algorithms specifically focused on faithfully depicting clusters within a graph have been introduced.

The quality of a drawing of a graph is often measured using *aesthetic criteria* which rate the readability of the visualization, such as the number of edge crossings or symmetry. However, these measures become less significant when working with large graphs (e.g. [19]). More recent work considers quality metrics more extensible to large graphs, such as *shape-based metrics* which compare the original topology of a graph to one derived from the positioning of vertices in its drawing [9]. Newly introduced is also the concept of more specific quality metrics concerned with the discovery of specific patterns with visualizations [5]. Although general quality metrics are still necessary, these more specific metrics are useful when developing visualizations geared for a more specific

This work is supported by ARC DP grant.

D. Archambault and C. D. Tóth (Eds.): GD 2019, LNCS 11904, pp. 125–138, 2019.
https://doi.org/10.1007/978-3-030-35802-0_10

purpose - for example, clustered graph visualizations which can be used to support various classes of group-level tasks [34].

Despite a longstanding recognition of cluster discovery as one important goal in graph visualization and the definition of quality metrics that regard the depiction or discovery of specific structures, there is yet to be defined a metric that explicitly quantifies how well a visualization represents the underlying clustering structure of the graph. We therefore introduce a *clustering quality metric* which scores a drawing of a graph based on how well the clustering structure of the graph is displayed within it. We present the following contributions:

1. We define the *clustering quality metric*, a new metric to measure the visual cluster quality of node-link graph drawings. In our framework, we compare the ground truth clustering provided for the vertices a graph to the geometric clustering derived from the graph's drawing, and the similarity of both clusterings denotes the quality of the visualization of clusters within the drawing.
2. We validate the metric through deformation experiments of graph drawings. Results of the experiment confirm that as the graphs are distorted resulting in the clusters to become visually less distinct from each other in the drawings, the scores computed using our metric decrease.
3. We compare various graph drawing algorithms using our metric to discover which methods perform better in visualizing cluster structures. We compare drawing algorithms of different types, including layouts that have been designed specifically to emphasize clusters. Our experiments confirm that these layouts perform better than others not explicitly geared towards cluster visualization, especially for real world graphs.

2   Related Work

2.1   Graph Drawing Quality Metrics

Aesthetics have been described as one criterion to be achieved by graph drawing algorithms [3]. The concept of aesthetics is concerned with the readability of graphs and include standards such as the minimization of edge crossing and bends, and minimization of drawing area used. A number of studies have verified the correlation of such aesthetic metrics with the ability of users to execute tasks on the graph (e.g. [17,30,31]). However, these studies tend to focus on smaller graphs, and newer studies (e.g. [19]) have discovered that the effects of these aesthetic criteria are not as apparent in larger graphs.

Shape-based metrics [9] attempt to address this limitation by computing a shape graph based on the drawing of a graph, where two vertices are connected with an edge if they are "close" to each other, and comparing it to the topology of the original graph - a good drawing is expected to have a shape graph similar to its actual topology. For recent work on visualization quality metrics, Behrisch et al. [5] provides a survey covering various visualization techniques, including but not limited to node-link drawings, and notes that measuring the effectiveness of node-link drawings in supporting analytical tasks is an open research question.

2.2 Clustering Comparison Metrics

Clustering refers to the division of a set of items into *clusters*, where items in the same cluster are more similar to each other than to items in a different cluster [1]. Despite the seemingly simple definition, the notions of "similarity" and what constitutes a "cluster" differ between contexts, leading to the birth of various clustering algorithms and thus multiple ways to cluster the same set [11]. To compare two clusterings C and C' of the same set, a number of metrics exist:

- *Rand Index (RI)* measures the similarity of C and C' based on the number of pairs of elements classified into the same group in both C and C' and the number of pairs of elements classified into different groups in both C and C' [32]. *Adjusted Rand Index (ARI)* [18] is a version corrected for chance.
- *Mutual Information (MI)*, when applied to two random variables, measures how much information of one can be gathered from the other, and is also applicable to comparisons between two clusterings C and C' [7]. *Normalized Mutual Information (NMI)* [36] is a normalized version, while *Adjusted Mutual Information (AMI)* [38] is a version adjusted for chance.
- *Fowlkes-Mallows Index (FMI)* compares a clustering C' to a target clustering C using the number of true positives, false positives, and false negatives [12].
- *Homogeneity (HOM)* and *completeness (CMP)* have been described as desirable outcomes of a cluster assignment C' compared to a target clustering C, where homogeneity measures to what extent each cluster in C' only contains members of the same cluster in C, and completeness refers to the extent that all members of a cluster in C are assigned to the same cluster in C' [33].

2.3 Graph Drawing Algorithms

In this section, we briefly describe a number of types of algorithms used to compute graph layouts:

- *Force-directed* layouts model a graph as a system where repulsive forces exist between all pairs of vertices and neighboring vertices attract each other [13].
- *Multi-level* layouts improve the time efficiency of force-directed layouts through steps of *coarsening* the graph into a smaller graph such as through clustering, applying the layout on the smaller graph, and using it as an initial layout to draw the less coarse graph until a layout for the original is computed [15].
- *Multi-dimensional scaling (MDS)* methods are based on dimension reduction techniques that aim to display high-dimensional data in fewer dimensions while preserving the distances between the data points [37].
- *Stress-based* layouts utilize the stress function found in the MDS literature. These methods compute a layout by minimizing an adapted stress function that considers the geometric and theoretical distances between vertices [14].
- *Spectral* methods computes the layout of a graph using the eigenvectors of matrices related to the graph, such as adjacency or Laplacian matrices [20].

3   Clustering Metric for Graph Visualization

We propose a new task-specific metric for graph visualization, the *clustering quality metric*, for measuring how well a drawing of a graph represents its underlying clustering structure. We compute the similarity between a ground truth clustering of a graph's vertices to a geometric clustering derived from its drawing and compute the clustering quality using the similarity of the two clusterings. Figure 1 summarizes the framework used for our proposed metric.

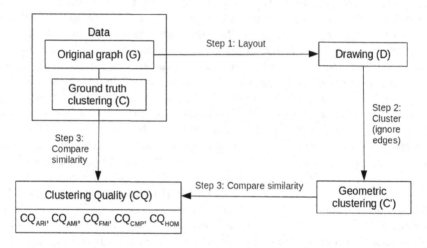

Fig. 1. The framework for the clustering quality metric. The framework takes as input a graph G with a predefined ground truth clustering C. A drawing D is produced by applying a layout algorithm to G, from which a geometric clustering C' of the vertices is computed. Computing the similarity of C and C' produces the clustering quality CQ score, which can be done using a variety of clustering comparison metrics.

Let $G = (V, E)$ be a graph and $C = \{C_i, i = 1...k\}$ be the ground truth clustering of V, the vertex set of G. Although in some applications a vertex may belong to multiple clusters, in this study, we focus on non-overlapping clusters as a starting point in developing the metric.

Step 1: We apply a layout algorithm to G to obtain a graph drawing D, which provides geometric positions for each node in G. A node-link drawing of a graph with no additional visual variables implicitly denotes groupings of vertices through the proximity of vertices to each other and a user is more likely to perceive two vertices drawn close together as belonging to the same group rather than two vertices drawn further apart.

Step 2: We compute a geometric clustering $C' = \{C'_i, i = 1...k\}$ purely based on the geometric positions of vertices in D. Any geometric clustering algorithm can be used, but in this work, we use *k-means clustering*, which partitions a set into k subsets that minimize the within-class variance [25]. We use k-means clustering as it is a widely used method applicable to geometric clustering with existing fast and efficient heuristic approximations and because for our experiments, we know the number of ground truth clusters.

Step 3: Using C', we compute the clustering quality of D by computing the similarity of C with C' to produce a clustering quality score CQ. Any clustering comparison metrics can be used with our framework, however we use the following metrics discussed in Sect. 2.2: Adjusted Rand Index (CQ_{ARI}), Adjusted Mutual Information (CQ_{AMI}), Fowlkes-Mallows Index (CQ_{FMI}), Homogeneity (CQ_{HOM}), and Completeness (CQ_{CMP}). These metrics have been established for measuring a clustering's quality when a target ground truth is available. In the cases of CQ_{ARI} and CQ_{AMI}, they were taken over other variants of RI and MI as they are adjusted for chance. All these metrics produce a score of 1 for perfect clustering, while independent clusterings attain values close to 0.

4   Validation Experiments

4.1   Experiment Design

To validate our metric, we designed deformation experiments for graph drawings. We start with a drawing of a graph that displays its clusters such that the number of visible clusters and their respective sizes accurately represent the ground truth clusters and the clusters are well-separated from each other with no overlap.

We then progressively deform the drawing. In each experiment, we performed 10 steps of deformation, where in each step, the coordinates of each vertex from the previous step are perturbed by a small value in the range $[0, \delta]$, with δ being in the range of 0.05-0.1 multiplied by the drawing area. We compute the clustering quality score and compare the scores across all steps of the deformation.

Based on the clustering comparison metrics, we expect our approach to produce scores in the range of $[0, 1]$ where a higher value denotes a closer similarity between the geometrical clustering C' derived from the drawing D and the ground truth clustering C. Therefore, we formulate the following hypothesis in order to validate our metric:

Hypothesis 1: The clustering quality metric scores will decrease as the graph drawings are deformed.

To create the initial layout, we used the Backbone layout from Visone [4] as this layout produced drawings scoring 1 or nearly 1 on our metric for our datasets. The exception is that we used sfdp from Graphviz [10] for $cv - many - verydense - mid$ and $gnm - many - mid - verysparse$, where sfdp produces drawings with higher clustering quality metric scores than backbone. We used cluster comparison metrics implementations from scikit-learn [29].

Each dataset for our validation experiment is created by first creating a small graph. Each vertex is replaced with a larger graph of a specified internal density - each will become a cluster of the dataset. Then, each edge is replaced with inter-cluster edges with a specified external density. Table 1 shows the dataset details. $|c|$ stands for the number of clusters and $avg(cd)$ denotes the average internal density of the clusters, as opposed to the global density denoted in the previous column.

Each graph is named in the format $[name] - [no. of clusters] - [internal density] - [external density]$, where we vary the parameters to increase generality. The prefixes denote the structure used to generate the clustered graph - c stands for a complete graph, b denotes a bipartite graph, s denotes a star graph, t denotes a tree, p denotes a path, rn denotes an r-regular graph, cv is a complete graph with variable cluster sizes, and gnm denotes a $G_{n,m}$ random graph.

Table 1. Validation datasets

Name	$|V|$	$|E|$	$|c|$	Density	$avg(cd)$
$c - few - verydense - mid$	439	9552	9	0.0497	0.399
$s - few - verydense - dense$	2051	256108	10	0.122	0.400
$t - few - dense - mid$	2082	164180	15	0.0379	0.400
$c - few - dense - mid$	898	31516	9	0.0391	0.349
$p - few - verydense - verysparse$	3002	230055	15	0.0511	0.759
$c - mid - verydense - mid$	815	18674	15	0.0563	0.797
$r3 - mid - dense - verysparse$	1773	53103	20	0.0338	0.670
$cv - many - verydense - mid$	2000	54749	30	0.0274	0.788
$r3 - many - verydense - sparse$	3045	124727	30	0.0269	0.801
$gnm - many - mid - verysparse$	2685	26098	30	0.00724	0.214

4.2   Results

Figure 2 displays one deformation experiment example, where vertices are colored based on their combinatorial cluster membership. In step 0 (Fig. 2 (a)), vertices of the same cluster are positioned close to each other, there is minimal overlap between each cluster and the layout produces CQ scores of 1. As the

(a) Step 0   (b) Step 2   (c) Step 5   (d) Step 9

Fig. 2. Deformation experiment for $r3 - mid - dense - verysparse$, drawn using Backbone layout, showing how each subsequent step further deforms the clusters in the drawing.

positions are perturbed, vertices of the same cluster grow further apart. The clusters also continue mixing with each other, until vertices are no less likely to be placed closer to members of other clusters than vertices in its own cluster.

Figure 3 shows the clustering metric scores for each deformation step, with the scores averaged for all datasets in Table 1. We expect to see the CQ scores decreasing after each deformation step, which is indeed what the figure shows, confirming Hypothesis 1 for a wide variety of clustered graphs.

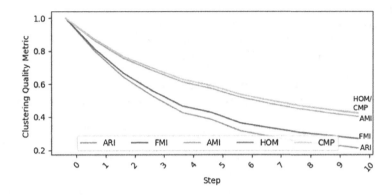

Fig. 3. Average of clustering quality scores for all validation experiments. The decreasing trend for all clustering comparison metrics show that our metric successfully captures the deteriorating visual clustering quality and validates Hypothesis 1. We also see that CQ_{AMI} and CQ_{FMI} are more sensitive to changes in the visual cluster quality, from the steeper curves. Also note that CQ_{HOM} and CQ_{CMP} produce highly similar results such that their curves overlap.

4.3   Discussion and Summary

Figure 3 shows that the plots of the clustering quality metric scores produce a downward slope. This validates our metric and the usage of all selected clustering comparison metrics with our framework. It can also be seen that the scores of our metric deteriorate at different rates when different clustering comparison metrics are used: CQ_{ARI} deteriorates at the fastest rate, followed closely by CQ_{FMI}. CQ_{HOM} and CQ_{CMP} obtains very similar scores with their curves overlapping, while CQ_{AMI} degrades at a slightly faster rate. Therefore, we conclude that CQ_{ARI} and CQ_{FMI} are more sensitive to changes in clustering visualisation quality than the other metrics.

In summary, the validation experiments have shown that our metric reflects the visual clustering quality of drawings of clustered graphs. Furthermore, from the different rates of change of the clustering quality scores when different clustering comparison metrics are used, we conclude that CQ_{ARI} and CQ_{FMI} are better at capturing changes in visual cluster quality and are recommended for use with our framework.

5   Layout Comparison Experiments

5.1   Experiment Design

After the validation experiments have shown that our metric effectively measures visual cluster quality, we compare the performance of a number of graph drawing algorithms against our metric. We selected layouts of different types:

- Force-directed: *Fruchterman-Reingold (FR)* [13] and *Organic* from yfiles [39].
- Multi-level: *FM3* [15] and *sfdp* [10,16].
- MDS: *Metric MDS* based on classical scaling [37] and *Pivot MDS* [6].
- Stress-based: *Stress Majorization* [14] and *Sparse Stress Minimization* [28].
- Spectral: spectral layout with graph laplacian.

We also selected a few layouts which purport to focus on the discovery of clusters or important community structures in a graph to test their claims:

- *LinLog* [26] modifies the force-directed model to emphasize clusters.
- *Backbone* [27] utilizes triadic or quadratic Simmelian backbones to extract important community structures from "hairball" graphs.
- *tsNET* [22] is based on t-distributed Stochastic Neighbor Embedding (t-SNE), a dimensionality reduction technique [24], and aims to preserve point neighborhoods.

Based on the selection of algorithms, we formulate the following hypothesis:

Hypothesis 2: LinLog, backbone, and tsNET will score higher on our metric than other selected layouts in visualizing clusters in graphs.

We used implementations provided from Tulip [8] (FR, FM3, Pivot MDS, Stress Majorization, LinLog), visone [4] (Backbone, Metric MDS, Sparse Stress Minimization, Spectral), yEd [39] (Organic), Graphviz [10] (sfdp), and Kruiger's implementation of tsNET [21]. We re-used some datasets from the validation experiments and created some new ones, listed in Table 2. We also selected real

Table 2. Additional layout comparison datasets

Name	$|V|$	$|E|$	$|c|$	Density	avg(cd)
$b - many - dense - sparse$	1797	49210	30	0.0305	0.560
$cv - mid - verydense - mid$	939	21798	20	0.0495	0.162
$s - mid - mid - sparse$	2116	24175	20	0.0108	0.216
$w - many - mid - verysparse$	2485	68844	25	0.0223	0.554
$r4 - many - verydense - verysparse$	3045	124727	30	0.0269	0.801
$revije - 90$	124	1334	14	0.127	0.377
$SS - Butterfly - 0 - 85$	832	13009	10	0.0376	0.258
$email - Eu - core - lcc$	986	16687	34	0.0344	0.490

world graph datasets with existing vertex categorization, which are listed under the double line in Table 2. The datasets were taken from Pajek [2] and Stanford Network Analysis Project's (SNAP) repository [23,40].

5.2 Results

Tables 3 and 4 show layout comparison examples, with colours representing ground truth clusters, with CQ scores displayed in Figs. 4 and 5 respectively. LinLog, tsNET, and Backbone score higher than other layouts for both datasets, supporting Hypothesis 2. In Table 3 and Fig. 4, where the number of clusters are small, other layouts such as sfdp, FR, FM3, and spectral also score close to 1. Meanwhile, in the example in Table 4 and Fig. 5 displaying a real world graph with a larger number of clusters, LinLog, tsNET, and backbone's performances more clearly surpass the other layouts.

Table 3. Layout comparison for $c - few - verydense - mid$

FR	Organic	Stress Maj.	Metric MDS	Backbone	FM3
Spectral	S. Stress Min.	tsNET	Pivot MDS	sfdp	LinLog

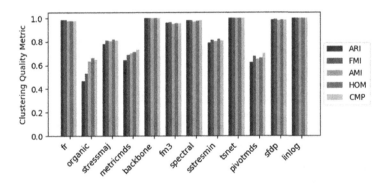

Fig. 4. Clustering quality metrics for $c - few - verydense - mid$. LinLog, tsNET, and Backbone produces scores of 1 on our metrics, in line with Hypothesis 2. For this dataset, sfdp, FR, FM3, and spectral also score highly, close to 1.

Table 4. Layout comparison for *email − Eu − core − lcc*

FR	Organic	Stress Maj.	Metric MDS
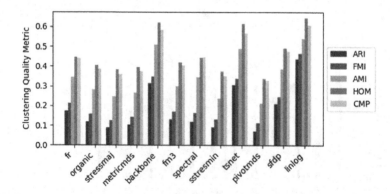			
Backbone	FM3	Spectral	S. Stress Min.
tsNET	Pivot MDS	sfdp	LinLog

Fig. 5. Clustering quality metrics for *email − Eu − core − lcc*. LinLog, backbone, and tsNET clearly outperform other layouts, as expected from Hypothesis 2. Among non-cluster-focused layouts, sfdp produces the highest scores.

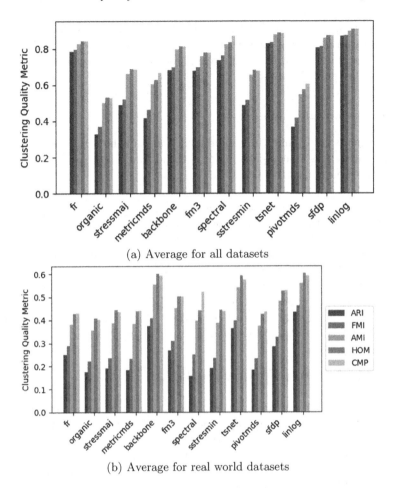

(a) Average for all datasets

(b) Average for real world datasets

Fig. 6. Clustering quality metrics averaged per layout for all layout comparison datasets (a) and for real world datasets only (b). In (a), we see that tsNET and LinLog produce the highest scores, validating Hypothesis 2 for the two layouts. Meanwhile in (b), we see that on real world datasets, LinLog, tsNET, and Backbone outperforms other layout algorithms in accordance to Hypothesis 2.

Figure 6(a) shows the scores averaged across all layout comparison datasets and Fig. 6(b) show the scores averaged across real world datasets. Averaged across all datasets, LinLog scores the highest, with tsNET close behind, confirming Hypothesis 2 for these two layouts. Backbone scores well on many graphs, but sometimes deteriorates in quality when the number of clusters becomes larger compared to the total size of the graph, causing it to score lower than tsNET and LinLog on average (see Fig. 6(a)). Even so, it still outperforms the other algorithms on real world datasets as seen in Fig. 6(b), which supports Hypothesis 2 for Backbone on real world graphs.

In the case of synthetic datasets, sfdp also tends to perform well, as seen in the overall averaged clustering quality metric scores in Fig. 6(a). LinLog, backbone, and tsNET still outperforms it with real world datasets as seen from Fig. 6(b), however, in line with Hypothesis 2.

5.3 Discussion and Summary

Our experiments verify that LinLog and tsNET attains the highest average scores on our metrics across all comparison datasets and Backbone attains equally high average scores on real world datasets.

A point of note is that LinLog often has issues with excessive node overlaps, especially when the internal cluster density is high - this can be seen in Table 3, where the nodes of each cluster are positioned very close together such that they almost appear as only one node, and to a lesser extent in Table 4 where the red cluster is packed quite closely together. Backbone does not have this problem on any tested graphs. Thus, we can conclude that Backbone also has its advantages for practical applications of clustered graph visualization.

In summary, our experiments have confirmed Hypothesis 2 for LinLog and tsNET, which consistently obtained the highest scores across all datasets, while for Backbone it is more supported on real world structures.

6 Conclusion and Future Work

We have introduced a new graph drawing quality metric for the visualization of clusters in graph. Deformation experiments has shown the effectiveness of the metric in measuring how well a drawing of a graph depicts the clusters in the graph. We have also compared graph drawings produced by layouts emphasizing cluster structures to non-cluster-focused layouts and validated the claims of these cluster-focused layouts especially on real world structures.

A direction for future work is to refine the metric by combining it with readability metrics, such as to address node overlaps, and further validating it with human evaluation. Other geometric clustering algorithms besides k-means can also be tested, including fuzzy clustering algorithms that accomodate overlaps between clusters, and concepts of visual cluster separations for scatterplots [35] can also be considered.

References

1. Aldenderfer, M.S., Blashfield, R.: Cluster Analysis. Beverly Hills: Sage Publications, Thousand Oaks (1984)
2. Batagelj, V., Mrvar, A.: Pajek data sets (2003). http://pajek.imfm.si/doku.php?id=data:index
3. Battista, G.D., Eades, P., Tamassia, R., Tollis, I.G.: Graph Drawing: Algorithms for the Visualization of Graphs. Prentice Hall PTR, Upper Saddle River (1998)

4. Baur, M., Benkert, M., Brandes, U., Cornelsen, S., Gaertler, M., Köpf, B., Lerner, J., Wagner, D.: Visone Software for Visual Social Network Analysis. In: Mutzel, P., Jünger, M., Leipert, S. (eds.) GD 2001. LNCS, vol. 2265, pp. 463–464. Springer, Heidelberg (2002). https://doi.org/10.1007/3-540-45848-4_47

5. Behrisch, M., Blumenschein, M., Kim, N.W., Shao, L., El-Assady, M., Fuchs, J., Seebacher, D., Diehl, A., Brandes, U., Pfister, H., Schreck, T., Weiskopf, D., Keim, D.A.: Quality metrics for information visualization. In: Computer Graphics Forum, vol. 37, pp. 625–662. Wiley Online Library (2018)

6. Brandes, U., Pich, C.: Eigensolver methods for progressive multidimensional scaling of large data. In: Kaufmann, M., Wagner, D. (eds.) GD 2006. LNCS, vol. 4372, pp. 42–53. Springer, Heidelberg (2007). https://doi.org/10.1007/978-3-540-70904-6_6

7. Cover, T.M., Thomas, J.A.: Elements of Information Theory. Wiley-Interscience, New York (1991)

8. David, A.: Tulip. In: Mutzel, P., Jünger, M., Leipert, S. (eds.) GD 2001. LNCS, vol. 2265, pp. 435–437. Springer, Heidelberg (2002). https://doi.org/10.1007/3-540-45848-4_34

9. Eades, P., Hong, S.H., Nguyen, A., Klein, K.: Shape-based quality metrics for large graph visualization. J. Graph Algorithms Appl. **21**(1), 29–53 (2017)

10. Ellson, J., Gansner, E., Koutsofios, L., North, S.C., Woodhull, G.: Graphviz—open source graph drawing tools. In: Mutzel, P., Jünger, M., Leipert, S. (eds.) GD 2001. LNCS, vol. 2265, pp. 483–484. Springer, Heidelberg (2002). https://doi.org/10.1007/3-540-45848-4_57

11. Estivill-Castro, V.: Why so many clustering algorithms: a position paper. SIGKDD Explor. Newsl. **4**(1), 65–75 (2002). https://doi.org/10.1145/568574.568575

12. Fowlkes, E.B., Mallows, C.L.: A method for comparing two hierarchical clusterings. J. Am. Stat. Assoc. **78**(383), 553–569 (1983). https://doi.org/10.1080/01621459.1983.10478008

13. Fruchterman, T.M.J., Reingold, E.M.: Graph drawing by force-directed placement. Softw.: Practice Exp. **21**(11), 1129–1164 (1991). https://doi.org/10.1002/spe.4380211102

14. Gansner, E.R., Koren, Y., North, S.: Graph drawing by stress majorization. In: Pach, J. (ed.) GD 2004. LNCS, vol. 3383, pp. 239–250. Springer, Heidelberg (2005). https://doi.org/10.1007/978-3-540-31843-9_25

15. Hachul, S., Jünger, M.: Drawing large graphs with a potential-field-based multilevel algorithm. In: Pach, J. (ed.) GD 2004. LNCS, vol. 3383, pp. 285–295. Springer, Heidelberg (2005). https://doi.org/10.1007/978-3-540-31843-9_29

16. Hu, Y.: Efficient, high-quality force-directed graph drawing. Math. J. **10**(1), 37–71 (2005)

17. Huang, W., Hong, S.H., Eades, P.: Effects of crossing angles. In: 2008 IEEE Pacific Visualization Symposium, pp. 41–46. IEEE (2008)

18. Hubert, L., Arabie, P.: Comparing partitions. J. Classif. **2**(1), 193–218 (1985). https://doi.org/10.1007/BF01908075

19. Kobourov, S.G., Pupyrev, S., Saket, B.: Are crossings important for drawing large graphs? In: Duncan, C., Symvonis, A. (eds.) GD 2014. LNCS, vol. 8871, pp. 234–245. Springer, Heidelberg (2014). https://doi.org/10.1007/978-3-662-45803-7_20

20. Koren, Y.: Drawing graphs by eigenvectors: theory and practice. Comput. Math. Appl. **49**(11–12), 1867–1888 (2005). https://doi.org/10.1016/j.camwa.2004.08.015

21. Kruiger, J.F.: tsnet (2017). https://github.com/HanKruiger/tsNET/

22. Kruiger, J.F., Rauber, P.E., Martins, R.M., Kerren, A., Kobourov, S., Telea, A.C.: Graph layouts by t-SNE. Comput. Graph. Forum **36**(3), 283–294 (2017). https://doi.org/10.1111/cgf.13187

138   A. Meidiana et al.

23. Leskovec, J., Krevl, A.: SNAP Datasets: Stanford large network dataset collection, June 2014. http://snap.stanford.edu/data
24. Maaten, L.V.D., Hinton, G.: Visualizing data using t-SNE. J. Mach. Learn. Res. **9**(Nov), 2579–2605 (2008)
25. MacQueen, J., et al.: Some methods for classification and analysis of multivariate observations. In: Proceedings of the Fifth Berkeley Symposium on Mathematical Statistics and Probability, vol. 1, pp. 281–297. University of California Press (1967)
26. Noack, A.: An energy model for visual graph clustering. In: Liotta, G. (ed.) GD 2003. LNCS, vol. 2912, pp. 425–436. Springer, Heidelberg (2004). https://doi.org/10.1007/978-3-540-24595-7_40
27. Nocaj, A., Ortmann, M., Brandes, U.: Untangling the hairballs of multi-centered, small-world online social media networks. J. Graph Algorithms Appl. **19**(2), 595–618 (2015). https://doi.org/10.7155/jgaa.00370
28. Ortmann, M., Klimenta, M., Brandes, U.: A sparse stress model. In: Hu, Y., Nöllenburg, M. (eds.) GD 2016. LNCS, vol. 9801, pp. 18–32. Springer, Cham (2016). https://doi.org/10.1007/978-3-319-50106-2_2
29. Pedregosa, F., Varoquaux, G., Gramfort, A., Michel, V., Thirion, B., Grisel, O., Blondel, M., Prettenhofer, P., Weiss, R., Dubourg, V., et al.: Scikit-learn: machine learning in python. J. Mach. Learn. Res. **12**(Oct), 2825–2830 (2011)
30. Purchase, H.: Which aesthetic has the greatest effect on human understanding? In: DiBattista, G. (ed.) GD 1997. LNCS, vol. 1353, pp. 248–261. Springer, Heidelberg (1997). https://doi.org/10.1007/3-540-63938-1_67
31. Purchase, H.C., Cohen, R.F., James, M.: Validating graph drawing aesthetics. In: Brandenburg, F.J. (ed.) GD 1995. LNCS, vol. 1027, pp. 435–446. Springer, Heidelberg (1996). https://doi.org/10.1007/BFb0021827
32. Rand, W.M.: Objective criteria for the evaluation of clustering methods. J. Am. Stat. Assoc. **66**(336), 846–850 (1971). https://doi.org/10.1080/01621459.1971.10482356
33. Rosenberg, A., Hirschberg, J.: V-measure: a conditional entropy-based external cluster evaluation measure. In: Proceedings of the 2007 Joint Conference on Empirical Methods in Natural Language Processing and Computational Natural Language Learning (EMNLP-CoNLL), pp. 410–420 (2007)
34. Saket, B., Simonetto, P., Kobourov, S.: Group-level graph visualization taxonomy. CoRR abs/1403.7421 (2014)
35. Sedlmair, M., Tatu, A., Munzner, T., Tory, M.: A taxonomy of visual cluster separation factors. Comput. Graph. Forum **31**(3pt4), 1335–1344 (2012). https://doi.org/10.1111/j.1467-8659.2012.03125.x
36. Strehl, A., Ghosh, J.: Cluster ensembles–a knowledge reuse framework for combining multiple partitions. J. Mach. Learn. Res. **3**(Dec), 583–617 (2002)
37. Torgerson, W.S.: Multidimensional scaling: I. Theory and method. Psychometrika **17**(4), 401–419 (1952). https://doi.org/10.1007/BF02288916
38. Vinh, N.X., Epps, J., Bailey, J.: Information theoretic measures for clusterings comparison: variants, properties, normalization and correction for chance. J. Mach. Learn. Res. **11**(Oct), 2837–2854 (2010)
39. Wiese, R., Eiglsperger, M., Kaufmann, M.: yfiles - visualization and automatic layout of graphs. In: Jünger, M., Mutzel, P. (eds.) Graph Drawing Software. Mathematics and Visualization, pp. 173–191. Springer, Heidelberg (2004). https://doi.org/10.1007/978-3-642-18638-7_8
40. Zitnik, M., Sosič, R., Maheshwari, S., Leskovec, J.: BioSNAP Datasets: Stanford biomedical network dataset collection, August 2018. http://snap.stanford.edu/biodata

Multi-level Graph Drawing Using Infomap Clustering

Seok-Hee Hong[1]([⊠]), Peter Eades[1], Marnijati Torkel[1], Ziyang Wang[1],
David Chae[1], Sungpack Hong[2], Daniel Langerenken[2], and Hassan Chafi[2]

[1] University of Sydney, Sydney, Australia
{seokhee.hong,peter.eades,mtor0581,zwan0130,min.chae}@sydney.edu.au
[2] Oracle Research Lab, Belmont, USA
{sungpack.hong,daniel.langerenken,hassan.chafi}@oracle.com

Abstract. *Infomap clustering* finds the community structures that minimize the expected description length of a random walk trajectory; algorithms for infomap clustering run fast in practice for large graphs. In this paper we leverage the effectiveness of Infomap clustering combined with the multi-level graph drawing paradigm. Experiments show that our new Infomap based multi-level algorithm produces good visualization of large and complex networks, with significant improvement in quality metrics.

1 Introduction

The *multi-level graph drawing* is a popular approach to visualize large and complex graphs to improve the quality of drawings. It recursively coarsens the graph and then uncoarsens the drawing using layout refinement. There are a number of multi-level graph drawing algorithms available [7,9,11,14,17,18,20]. They mainly differ in the *coarsening* method.

Clustering is a widely used analysis method for identifying groups with strong similarity, or communities in the data. Graph clustering is to partition a graph such that vertices in the same cluster are more interconnected. *Infomap clustering* computes clusters by translating a graph into a map, which decomposes the myriad nodes and links into modules that represent the graph [19]. It maximizes an objective function called the minimum description length of a random walk trajectory, where the approximation to the optimal solution can be computed quickly. Infomap performed the best in community finding experiments [15].

In this paper, we present a new multi-level graph drawing algorithm based on Infomap clustering. More specifically, we leverage the effectiveness of Infomap clustering, combined with the multi-level graph drawing paradigm. Experiments with real-world large and complex networks such as protein-protein interaction networks, Facebook graph, Autonomous Systems (AS) graphs as well as benchmark graphs show that our new multi-level algorithms produce good visualization with significant improvement in quality metrics, including shape-based

Research supported by ARC Linkage Grant with Oracle labs.

D. Archambault and C. D. Tóth (Eds.): GD 2019, LNCS 11904, pp. 139–146, 2019.
https://doi.org/10.1007/978-3-030-35802-0_11

metrics [4], edge crossing and stress. It also requires a small number of coarsening steps for medium to large graphs, which makes it fast to run.

2   Related Work

Hadany and Harel presented the multi-scale method using an edge contraction based coarsening method and a force-directed layout preserving topological properties such as cluster size and vertex degree [10]. Koren and Harel presented FMS, which used a k-center approximation based coarsening method and a force-directed layout with a beautification [14].

Walshaw presented a multi-level algorithm using a matching, by repeatedly collapsing maximal independent subsets of graph edges, and a grid variant of Fruchterman-Reingold [6] layout [20]. Gajer et al. presented GRIP using a maximum independent set filtration based coarsening method, and an intelligent initial placement of vertices based on both graph and Euclidean distances [7].

Quigley and Eades presented FADE using the quad tree, and Barnes-Hut n-body method [1] for approximation of the repulsive force computation in a force-directed layout [18]. Hachul and Junger presented FM^3 using similar method to compute the repulsive forces between vertices, where subgraphs with small diameter, called solar system, are partitioned and collapsed to obtain a multi-level representation [9]. Hu presented the sfdp layout, also using the Barnes-Hut approximation method [11]. Frishman and Tal [5] presented a multi-level force directed graph layout on the GPU, based on spectral partitioning and Kamada-Kawai layout [12]. Bartel et al. presented an experimental study for extensive comparison of various multi-level algorithms, using a combination of coarsening methods, initial placement and graph layout methods [2].

More recently, Meyerhenke et al. presented a multi-level algorithm using a label propagation method for the coarsening step, and Maxent stress optimisation layout [8] on shared memory parallelization [16]. Nguyen and Hong used fast k-core coarsening method, which can be computed in linear time [17].

3   Infomap Based Multi-level Algorithm

The multi-level graph drawing algorithm is an iterative process consisting of the following three steps: coarsening, initialization (or placement), and graph layout (or refinement). Roughly speaking, the coarsening step is to cluster vertices to define a smaller graph, recursively until the size of the graph falls below the threshold, resulting in a coarse graph hierarchy, $G_0, G_1, \ldots, G_L$. The layout of graph G_L is then extended to the layout of graph G_{L-1} by placement (i.e., add vertices back to the layout) and refinement step. Recursively, these steps extend the layout of graph G_L to G_0 by repetitively interpolating from G_i to G_{i-1}. In each iteration, the layout of G_i is used to compute an initial placement of G_{i-1}, and then the layout algorithm is applied to refine the layout.

3.1   Coarsening: Infomap Clustering

Let $G = (V, E)$ be a graph with vertex set V and edge set E. The coarsening step computes a graph level hierarchy by iteratively computing a sequence of smaller graphs $G_0, G_1, G_2, \ldots, G_L$, where the original graph $G = G_0$. At each level, a coarser graph (or clustered graph) is computed by combining a sets of vertices belong to the same cluster in G_i and replacing into a single vertex in G_{i+1}, recursively until the predefined stop criterion is satisfied.

Infomap clustering finds community structure that minimizes the expected description length of a random walk trajectory [19]. It computes clusters by translating a graph into a map, which decomposes the myriad nodes and links into modules that represent the graph. The algorithm maximizes an objective function called the Minimum Description Length.

We first compute the Infomap clustering of G, and partition the vertex set V into V_i based on the clusters. More specifically, we define a clustered graph with a weighted vertex set (i.e., the number of vertices belong to each cluster) and a weighted edge set (i.e., the number of edges between the partitioned vertex set). The vertices $u_1, u_2, \ldots, u_k \in V_i$ are merged to form a new cluster vertex $v \in V_{i+1}$, where the weight of v is computed as $|v| = |u_1| + |u_2| + \ldots + |u_k|$. Similarly, the weight of the collapsed edges are computed as the sum of the weights of the edges that it replaces. This coarsening phase stops when the resulting clustered graph has a small size (say 50) or there is no reduction in terms of size.

3.2   Initialization: Placement

This step aims to compute a good initial layout of G_{i-1} using the layout of G_i. Let $v_i \in V_i$ of G_i corresponds to a cluster of vertices $u_1, u_2, \ldots, u_k \in V_{i-1}$ of G_{i-1}. We add back vertices $u_j, j = 1, \ldots, k$ to the layout of G_i by initializing the positions of u_j using the position of v_i. Here we use the following three variations.

- Circle placement: It places all $u_j, j = 1, \ldots, k$ at the circle with a small radius, where the center of the circle is the location of v_i.
- Barycenter placement: It places each vertex at the barycenter of its neighbors [7].
- Zero placement: It places all $u_j, j = 1, \ldots, k$ at the same position as v_i with small perturbation [20].

3.3   Refinement: Force-Directed Layout

The initial layout of G_{i-1} is recursively refined at each level using a force-directed algorithm. We use layout algorithms, previously used in other multi-level graph drawing algorithm experiments [2]:

- FR: Fruchterman and Reingold layout [6].
- FRG: grid variant of Fruchterman and Reingold layout, used in [20].
- FME (Fast-Multipole Embedder): an improvement of NME (New Multipole Method) layout of FM^3 [9], designed for a multi-level method in [2].

4   Experiments

We implemented Infomap clustering based multi-level algorithm using OGDF [3], which was used in the comparison experiments of multi-level algorithms [2]. We used a standard Dell laptop with Intel Core i7, 16 GB RAM.

We first experimented with three different placement methods, and found that there is no significant difference in terms of layout quality. We choose the barycenter placement, which shows slightly better performance, with three layouts FR, FRG and FME for comparison.

More specifically, we have the following variations for comparison:

– InfomapFR: Infomap multi-level algorithm with FR layout
– InfomapFRG: Infomap multi-level algorithm with FR grid variant layout
– InfomapFME: Infomap multi-level algorithm with FME layout.

The experiment was conducted with real-world benchmark data sets including social networks such as facebook, biological networks such as protein-protein interaction networks, and benchmark graphs used in previous work [2,17,20].

Table 1 shows the details of the data sets, the number of coarsening levels and runtime (seconds), where D represents the density of a graph G and L represents the number of levels. We can clearly see that the Infomap coarsening method

Table 1. Data sets, size, number of levels (L) and runtime.

Graph G	$|V_0|$	$|E_0|$	D	L	Time	$|V_1|$	$|E_1|$	$|V_2|$	$|E_2|$	$|V_3|$	$|E_3|$
G_15_0	1785	20459	11.5	2	0.02	59	100	9	8		
nasa1824	1824	18692	10.3	2	0.02	53	217	5	7		
G_4_0	2075	4769	2.3	2	0.02	89	326	8	11		
yeastppi	2361	7182	3.0	2	0.04	302	1923	101	0		
soc_h	2426	11630	4.8	2	0.02	301	1088	149	1		
oflights	2939	15677	5.3	2	0.03	170	477	19	24		
ecolippi	3796	78120	20.6	2	0.03	245	2453	53	1		
facebook	4039	88234	21.9	2	0.02	93	272	7	11		
3elt	4720	13722	2.9	2	0.05	189	489	17	35		
USpowerGrid	4941	6594	1.3	2	0.18	489	963	44	104		
as19990606	5188	10974	2.1	2	0.17	368	2034	12	38		
commanche_dual	7920	19800	2.5	2	0.24	503	1365	34	71		
p2p-Gnutella05	8846	31839	3.6	2	0.20	830	18154	3	0		
astroph2001	16046	121251	7.6	3	0.61	1219	9333	395	68	369	0
condmat2001	16264	47594	2.9	3	1.33	1720	4574	798	774	726	0
crack-dual	20141	30043	1.5	3	1.16	1357	3633	84	216	10	18
bcsstk31	35588	608502	17.1	2	0.36	453	2295	25	44		
shock-9	36476	71290	2.0	3	1.17	1351	3852	74	191	8	14
del16	65536	196575	3.0	3	1.95	1981	5921	101	290	8	16

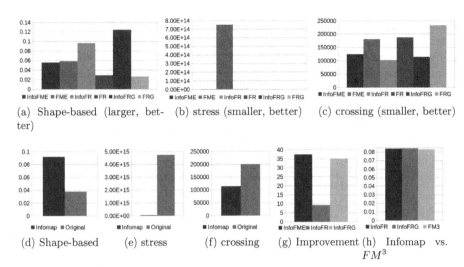

(a) Shape-based (larger, better)   (b) stress (smaller, better)   (c) crossing (smaller, better)

(d) Shape-based   (e) stress   (f) crossing   (g) Improvement   (h) Infomap vs. FM^3

Fig. 1. Significant improvement in quality metrics: (a) (b) (c) Average of metrics per layout: InfomapFME (blue), FME (red), InfomapFR (yellow), FR (green), InfomapFRG (brown), FRG (cyan); (d) (e) (f) Average of metrics: Infomap (blue) vs. Original (red); (g) Average of improvement by Infomap over Original in Shape-based metrics: InfomapFME (blue), InfomapFR (red), InfomapFRG (yellow); (h) Average of shape-based metrics: InfomapFR (blue), InfomapFRG (red), FM^3 (yellow). (Color figure online)

produced small number of levels such as 2 or 3 for most of data sets. Overall, Infomap clustering runs quite fast for medium size graphs.

Comparison of Quality Metrics: For large and complex graphs, edge crossing may not a suitable metric to measure the quality of drawings [4,13]. We used the *shape-based metrics* [4]; this is a new graph drawing quality measure specially designed for large graphs. Roughly speaking, the shape-based metrics measure the *faithfulness* of graph drawing, i.e., how well the *shape* of the drawing represents the structure (or shape) of the graph.

Figures 1(a), (b) and (c) show the comparison of *average metrics* between six layouts (i.e., Infomap multi-level vs. FME, FRG, FR original layouts) using *shape-based quality metrics* (Q), *stress* and *edge crossings*. Clearly, we can see that Infomap multi-level layouts perform significantly better than the original layouts. In general, InfomapFR and InfomapFRG perform better than InfomapFME. Figures 1(d), (e) and (f) show the *average metrics* between Infomap multi-level and original layouts. Overall, we can see that Infomap multilevel layouts outperform original layouts. Figure 1(g) shows the *average improvement* by Infomap multi-level layouts over original layouts in shape-based metrics (i.e., $(Q_{Infomap}/Q_{Original} - 1)$). Clearly, significant improvement was achieved by InfomapFME and InfomapFRG.

Visual Comparison: Overall, Infomap multi-level layouts perform significantly better than original layouts. In general, InfomapFR and InfomapFRG perform significantly better than other layouts, and InfomapFME achieved the most significant improvement over FME. For example, Fig. 2 shows visual comparison between layouts of $3elt$.

Comparison with FM^3: Figure 1(h) shows *average shape-based metrics* between InfomapFR, InfomapFRG and FM^3, excluding the outlier. Clearly, we can see that InfomapFR and Infomap FRG perform similar to FM^3 in shape-based metrics. For *layout comparison*, see Fig. 3. We can see that InfomapFR perform similar to FM^3, and for some instances perform better than FM^3.

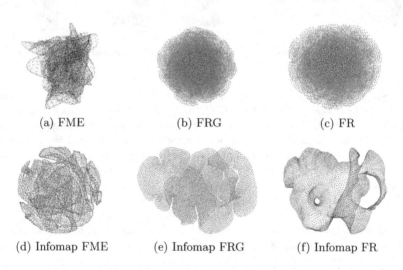

(a) FME	(b) FRG	(c) FR

(d) Infomap FME	(e) Infomap FRG	(f) Infomap FR

Fig. 2. Visual comparison of $3elt$.

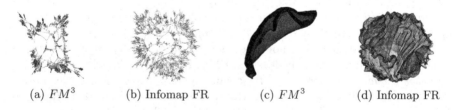

(a) FM^3	(b) Infomap FR	(c) FM^3	(d) Infomap FR

Fig. 3. Comparison with FM^3: (a) (b) *USpowerGrid*; (c) (d) *shock-9*

Summary: Our experimental results provide strong evidence that our Infomap based multi-level algorithm performs considerably well for real-world social networks, biological networks and benchmark graphs.

- Overall, Infomap multi-level layouts perform significantly better than original layouts in terms of quality metrics and visualisation.

- Metric wise, InfomapFR and InfomapFRG perform better than InfomapFME.
- InfomapFME achieved the most significant improvement.
- InfomapFR and InfomapFRG perform similar to FM^3 in terms of shape-based metrics and visual comparison.

References

1. Barnes, J., Hut, P.: A hierarchical O (N log N) force-calculation algorithm. Nature **324**, 446 (1986)
2. Bartel, G., Gutwenger, C., Klein, K., Mutzel, P.: An experimental evaluation of multilevel layout methods. In: Brandes, U., Cornelsen, S. (eds.) GD 2010. LNCS, vol. 6502, pp. 80–91. Springer, Heidelberg (2011). https://doi.org/10.1007/978-3-642-18469-7_8
3. Chimani, M., Gutwenger, C., Jünger, M., Klau, G.W., Klein, K., Mutzel, P.: The open graph drawing framework (OGDF). In: Handbook on Graph Drawing and Visualization, pp. 543–569 (2013)
4. Eades, P., Hong, S.-H., Klein, K., Nguyen, A.: Shape-based quality metrics for large graph visualization. In: Di Giacomo, E., Lubiw, A. (eds.) GD 2015. LNCS, vol. 9411, pp. 502–514. Springer, Cham (2015). https://doi.org/10.1007/978-3-319-27261-0_41
5. Frishman, Y., Tal, A.: Multi-level graph layout on the GPU. IEEE Trans. Vis. Comput. Graph. **13**(6), 1310–1319 (2007)
6. Fruchterman, T.M., Reingold, E.M.: Graph drawing by force-directed placement. Softw.: Pract. Exper. **21**(11), 1129–1164 (1991)
7. Gajer, P., Kobourov, S.G.: GRIP: graph drawing with intelligent placement. J. Graph Algorithms Appl. **6**(3), 203–224 (2002)
8. Gansner, E.R., Hu, Y., North, S.C.: A maxent-stress model for graph layout. IEEE Trans. Vis. Comput. Graph. **19**(6), 927–940 (2013)
9. Hachul, S., Jünger, M.: Drawing large graphs with a potential-field-based multilevel algorithm. In: Pach, J. (ed.) GD 2004. LNCS, vol. 3383, pp. 285–295. Springer, Heidelberg (2005). https://doi.org/10.1007/978-3-540-31843-9_29
10. Hadany, R., Harel, D.: A multi-scale algorithm for drawing graphs nicely. Discrete Appl. Math. **113**(1), 3–21 (2001)
11. Hu, Y.: Efficient, high-quality force-directed graph drawing. Mathematica J. **10**(1), 37–71 (2005)
12. Kamada, T., Kawai, S.: An algorithm for drawing general undirected graphs. Inf. Process. Lett. **31**(1), 7–15 (1989)
13. Kobourov, S.G., Pupyrev, S., Saket, B.: Are crossings important for drawing large graphs? In: Duncan, C., Symvonis, A. (eds.) GD 2014. LNCS, vol. 8871, pp. 234–245. Springer, Heidelberg (2014). https://doi.org/10.1007/978-3-662-45803-7_20
14. Koren, D., Harel, Y.: A fast multi-scale method for drawing large graphs. J. Graph Algorithms Appl. **6**(3), 179–202 (2002)
15. Lancichinetti, A., Fortunato, S.: Community detection algorithms: a comparative analysis. Phys. Rev. E **80**, 056117 (2009)
16. Meyerhenke, H., Nöllenburg, M., Schulz, C.: Drawing large graphs by multilevel maxent-stress optimization. In: Di Giacomo, E., Lubiw, A. (eds.) GD 2015. LNCS, vol. 9411, pp. 30–43. Springer, Cham (2015). https://doi.org/10.1007/978-3-319-27261-0_3

146   S.-H. Hong et al.

17. Nguyen, A., Hong, S.: k-core based multi-level graph visualization for scale-free networks. In: 2017 IEEE Pacific Visualization Symposium, PacificVis 2017, Seoul, South Korea, 18–21 April 2017, pp. 21–25 (2017)
18. Quigley, A., Eades, P.: FADE: graph drawing, clustering, and visual abstraction. In: Marks, J. (ed.) GD 2000. LNCS, vol. 1984, pp. 197–210. Springer, Heidelberg (2001). https://doi.org/10.1007/3-540-44541-2_19
19. Rosvall, M., Bergstrom, C.T.: Maps of random walks on complex networks reveal community structure. Proc. Nat. Acad. Sci. **105**(4), 1118–1123 (2008)
20. Walshaw, C., et al.: A multilevel algorithm for force-directed graph-drawing. J. Graph Algorithms Appl. **7**(3), 253–285 (2003)

On Strict (Outer-)Confluent Graphs

Henry Förster[1] , Robert Ganian[2] , Fabian Klute[2(✉)] ,
and Martin Nöllenburg[2]

[1] University of Tübingen, Tübingen, Germany
foersth@informatik.uni-tuebingen.de
[2] Algorithms and Complexity Group, TU Wien, Vienna, Austria
{rganian,fklute,noellenburg}@ac.tuwien.ac.at

Abstract. A strict confluent (SC) graph drawing is a drawing of a graph
with vertices as points in the plane, where vertex adjacencies are represented not by individual curves but rather by unique smooth paths
through a planar system of junctions and arcs. If all vertices of the graph
lie in the outer face of the drawing, the drawing is called a strict outerconfluent (SOC) drawing. SC and SOC graphs were first considered by
Eppstein et al. in Graph Drawing 2013. Here, we establish several new
relationships between the class of SC graphs and other graph classes,
in particular string graphs and unit-interval graphs. Further, we extend
earlier results about special bipartite graph classes to the notion of strict
outerconfluency, show that SOC graphs have cop number two, and establish that tree-like (Δ-)SOC graphs have bounded cliquewidth.

1 Introduction

Confluent drawings of graphs are geometric graph representations in the
Euclidean plane, in which vertices are mapped to points, but edges are not drawn
as individually distinguishable geometric objects. Instead, an edge between two
vertices u and v is represented by a smooth path between the points of u and v
through a crossing-free system of arcs and junctions. Since multiple edge representations may share some arcs and junctions of the drawing, this allows dense
and non-planar graphs to be drawn in a plane way (e.g., see Fig. 2 for a confluent
drawing of K_5). Hence confluent drawings can be seen as theoretical counterpart of heuristic edge bundling techniques, which are frequently used in network
visualizations to reduce visual clutter in layouts of dense graphs [2, 25].

More formally, a *confluent drawing* D of a graph $G = (V, E)$ consists of a
set of points representing the vertices of G, a set of junction points, and a set
of smooth arcs, such that each arc starts and ends at either a vertex point or
a junction, no two arcs intersect (except at common endpoints), and all arcs
meeting in a junction share the same tangent line in the junction point. There

A poster containing some of the results of this paper was presented at GD 2017. Robert
Ganian acknowledges support by the Austrian Science Fund (FWF, project P31336)
and is also affiliated with FI MUNI, Brno, Czech Republic.

D. Archambault and C. D. Tóth (Eds.): GD 2019, LNCS 11904, pp. 147–161, 2019.
https://doi.org/10.1007/978-3-030-35802-0_12

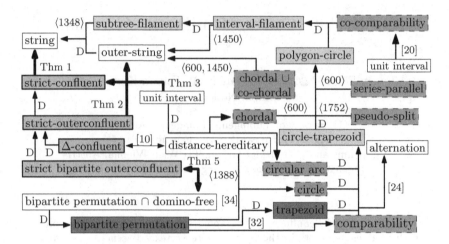

Fig. 1. Inclusions among graph classes related to SOC graphs. Arrows point from sub- to superclass, where edge label 'D' marks an inclusion by definition. Fat arrows are inclusions shown in this paper and are labelled with the corresponding theorem. Green boxes are confluent graph classes. Red, dashed boxes are classes that are incomparable to SOC graphs. Orange boxes are classes that are potential superclasses of SOC graphs. Blue boxes are potential subclasses of the SOC graphs. The numbers in ⟨·⟩ indicate references of graphclasses.org. (Color figure online)

is an edge $(u, v) \in E$ if and only if there is a smooth path from u to v in D not passing through any other vertex.

Confluent drawings were introduced by Dickerson et al. [8], who identified classes of graphs that admit or do not admit confluent drawings. Subsequently, the notions of strong and tree confluency have been introduced [27], as well as Δ-confluency [10]. Confluent drawings have further been used for drawings of layered graphs [11] and Hasse diagrams [13].

Eppstein et al. [12] defined the class of strict confluent (SC) drawings, which require that every edge of the graph must be represented by a unique smooth path and that there are no self-loops. They showed that for general graphs it is NP-complete to decide whether an SC drawing exists. An SC drawing is called *strict outerconfluent* (SOC) if all vertices lie on the boundary of a (topological) disk that contains the SC drawing. For graphs with a given cyclic vertex order, Eppstein et al. [12] presented a constructive efficient algorithm for testing the existence of an SOC drawing. Without a given vertex order, neither the recognition complexity nor a characterization of such graphs is known.

We approach the characterization problem by comparing the SOC graph class with a hierarchy of classes of intersection graphs. In general a *geometric intersection graph* $G = (V, E)$ is a graph with a bijection between the vertices V and a set of geometric objects such that two objects intersect if and only if the corresponding vertices are adjacent. Common examples include interval graphs, string graphs [9] and circle graphs [15]. Since confluent drawings make heavy use of intersecting curves to represent edges in a planar way, it seems natural to ask what kind of geometric intersection models can represent a confluent graph.

Contributions. After introducing basic definitions and properties in Sect. 2, we show in Sect. 3 that SC and SOC graphs are, respectively, string and outerstring graphs [28]. Section 4 shows that every unit interval graph [30,33] can be drawn strict confluent. In Sect. 5, we consider the so-called strict bipartite-outerconfluent drawings: by following up on an earlier result of Hui et al. [27], we show that graphs which admit such a drawing are precisely the domino-free bipartite permutation graphs. Inspired by earlier work of Gavenčiak et al. [16], we examine in Sect. 6 the cop number of SOC graphs and show that it is at most two. In [14], we show additionally that many natural subclasses of outer-string graphs are incomparable to SOC graphs (see red, dashed boxes in Fig. 1). More specifically, we show that circle [15], circular-arc [26], series-parallel [31], chordal [17], co-chordal [4], and co-comparability [22] graphs are all incomparable to SOC graphs. This list may help future research by excluding a series of natural candidates for sub- and super-classes of SOC graphs. Finally, in Sect. 7, we show that the cliquewidth of so-called tree-like Δ-SOC graphs is bounded by a constant, generalizing a previous result of Eppstein et al. [10].

Due to space constraints some proofs are omitted; we refer to [14] for full details.

2   Preliminaries

A *confluent diagram* $D = (N, J, \Gamma)$ in the plane $\mathbb{R}^2$ consists of a set N of points called *nodes*, a set J of points called *junctions* and a set Γ of simple smooth curves called *arcs* whose endpoints are in $J \cup N$. Further, two arcs may only intersect at common endpoints. If they intersect in a junction they must share the same tangent line, see Fig. 2.

Let $D = (N, J, \Gamma)$ be a confluent diagram and let $u, v \in N$ be two nodes. A uv-path $p = (\gamma_0, \ldots, \gamma_k)$ in D is a sequence of arcs $\gamma_0 = (u, j_1), \gamma_1 = (j_1, j_2), \ldots, \gamma_k = (j_k, v) \in \Gamma$ such that $j_1, \ldots j_k$ are junctions and p is a smooth curve. In Fig. 2 the unique uy-path passes through junctions i, j, k. If there is at most one uv-path for each pair of nodes u, v in N and if there are no self-loops, i.e., no uu-path for any $u \in N$, we say that D is a *strict* confluent diagram. The uniqueness of uv-paths and the absence of self-loops imply that every uv-path is actually a path in the graph-theoretic sense, where no vertex is visited twice. We further define $P(D)$ as the set of all smooth paths between all pairs of nodes in N. Let $p \in P(D)$ be a path and $j \in J$ a junction in D, then we write $j \in p$, if p passes through j.

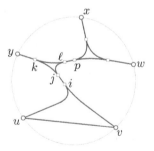

Fig. 2. A strict outerconfluent diagram representing K_5. Nodes are disks, junctions are squares.

As observed by Eppstein et al. [12], we may assume that every junction is a *binary* junction, where exactly three arcs meet such that the three enclosed angles are $180°, 180°, 0°$. In other words two arcs from the same direction merge into the third arc, or, conversely, one arc splits into two arcs. A (strict) confluent diagram with higher-degree junctions can easily be transformed into an equivalent (strict) one with only binary junctions.

Let $j \in J$ be a binary junction with the three incident arcs $\gamma_1, \gamma_2, \gamma_3$. Let the angle enclosed by γ_1 and γ_2 be $0°$ and the angle enclosed by γ_3 and γ_1 (or γ_2) be $180°$. Then we say that j is a *merge-junction* for γ_1 and γ_2 and a *split-junction* for γ_3. We also say that γ_1 and γ_2 *merge* at j and that γ_3 *splits* at j. Given two nodes $u, v \in N$ and a junction $j \in J$ we say that j is a merge-junction for u and v if there is a third node $w \in N$, a uw-path p and a vw-path q such that $j \in p$ and $j \in q$, the respective incoming arcs $\gamma_p = (j_p, j)$ and $\gamma_q = (j_q, j)$ are distinct and the suffix paths of p and q from j to w are equal. Conversely, we say that a junction $j \in J$ is a split-junction for a node $u \in N$ if there are two nodes $v, w \in N$, a uv-path p, and a uw-path q such that $j \in p$ and $j \in q$, the prefix paths of p and q from u to j are equal and the respective subsequent arcs $\gamma_p = (j, j_p)$ and $\gamma_q = (j, j_q)$ are distinct. In Fig. 2, junction i is a merge-junction for u and v, while it is a split junction for each of w, x, y. Two junctions $i, j \in J$ are called a *merge-split pair* if i and j are connected by an arc γ and both i and j are split-junctions for γ; in Fig. 2, junctions i and j form a merge-split pair, as well as junctions ℓ and p.

We call an arc $\gamma \in \Gamma$ *essential* if we cannot delete γ without changing adjacencies in the represented graph. We call a confluent diagram D *reduced*, if every arc is essential. Notice that this is a different notion than strictness, since it is possible that in a confluent diagram we find two essential arcs between a pair of nodes. Without loss of generality we can assume that the nodes of an outerconfluent diagram are placed on a circle with all arcs and junctions inside the circle. We can infer a *cyclic order* π from an outerconfluent diagram D by walking clockwise around the boundary of the unbounded face and adding the nodes to π in the order they are visited.

From a confluent diagram $D = (N, J, \Gamma)$ we derive a simple, undirected graph $G_D = (V_D, E_D)$ with $V_D = N$ and $E_D = \{(u, v) \mid \exists uv\text{-path } p \in P(D)\}$. We say D is a confluent drawing of a graph G if G is isomorphic to G_D and that G is a (strict) (outer-)confluent graph if it admits a (strict) (outer-)confluent drawing.

3   Strict (Outer-)Confluent $\subset$ (Outer-)String

The class of *string graphs* [28] contains all graphs $G = (V, E)$ which can be represented as the intersection graphs of open curves in the plane. We show that they form a superclass of SC graphs and that every SOC graph is an outer-string graph [28]. *Outer-string* graphs are string graphs that can be represented so that strings lie inside a disk and intersect the boundary of the disk in one endpoint. Note that strings are allowed to self-intersect and cross each more than once.

Let $D = (N, J, \Gamma)$ be a strict confluent diagram. For every node $u \in N$ we construct the *junction tree* T_u of u, with root u and a leaf for each neighbor v of u in G_D. The interior vertices of T_u are the junctions which lie on the (unique) uv-paths. The strictness of D implies that T_u is a tree. Observe that every internal node of T_u has at most two children. Further, every merge-junction for u is a vertex with one child in T_u, and every split-junction for u has two children. For every junction j in T_u we can define the sub-tree $T_{u,j}$ of T_u with root j.

Lemma 1. *Let $D = (N, J, \Gamma)$ be a strict confluent diagram, let $u, v \in N$ be two nodes and let i, j be two distinct merge-junctions for u, v. Then i is neither an ancestor nor a descendant of j in T_u (and, by symmetry, in T_v).*

To create a string representation of an SC graph we trace the paths of a strict confluent diagram $D = (N, J, \Gamma)$, starting from each node $u \in N$ and combine them into a string representation. Figure 3 shows an example. We traverse the junction tree for each $u \in N$ on the left-hand side of each arc (seen from its root u) and create a string $t(u)$, the *trace* of u, with respect to T_u as follows.

Start from u and traverse T_u in left-first DFS order. Upon reaching a leaf ℓ make a clockwise U-turn and backtrack to the previous split-junction of T_u. When returning to a split-junction we have two cases. (a) coming from the left subtree: cross the arc from the left subtree at the junction and descend into the right subtree. (b) coming from the right subtree: cross the arc to the left subtree again and backtrack upward in the tree along the existing trace to the previous split-junction of T_u.

Finally, at a merge-junction i with at least one trace from the other arc merging into i already drawn: Let $v \in N$ such that u and v merge at i and $t(v)$ is already tracing the subtree $T_{u,i} = T_{v,i}$. In this case we temporarily cut open the part of trace $t(v)$ closest to $t(u)$, route $t(u)$ through the gap and let it follow $t(v)$ along $T_{u,i}$ until it returns to junction i, where $t(u)$ passes through the gap again. Since $T_{u,i} = T_{v,i}$ this is possible without $t(u)$ intersecting $t(v)$.

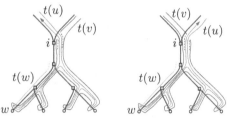

Fig. 3. Two possible configurations for inserting a new trace $t(u)$ that meets an existing trace $t(v)$ at a merge junction i; $t(v)$ is cut and re-routed.

Now it remains to reconnect the two open ends of $t(v)$, but this can again be done without any new intersections by winding $t(v)$ along the "outside" of $t(u)$. See Fig. 3 for an illustration. If there are multiple traces with this property, they can all be treated as a single "bundled" trace within $T_{u,i}$.

Theorem 1. *Every SC graph is a string graph.*

Proof. Given an SC graph $G = (V, E)$ with a strict confluent drawing $D = (N, J, \Gamma)$ we construct the traces as described above for every node $u \in N$. In the following let u, v be two nodes of D. We distinguish three cases.

Case 1 (*uv-path in $P(D)$*): We draw $t(u)$ and $t(v)$ as described above. Since there is a uv-path in $P(D)$ we have to guarantee that $t(u)$ and $t(v)$ intersect at least once. We introduce crossings at the leaves corresponding to u and v in T_u

and T_v when $t(u)$ and $t(v)$ make a U-turn; see how the trace $t(u)$ intersects $t(w)$ near the leaf w in Fig. 3.

Case 2 (No uv-path in $P(D)$ and u, v share no merge-junction): In this case T_u and T_v are disjoint trees. Traces can meet only at shared junctions and around leaves, but since $t(u)$ and $t(v)$ trace disjoint trees intersections are impossible.

Case 3 (No uv-path in $P(D)$ and u, v share a merge-junction): First assume u and v share a single merge-junction $i \in J$ and assume $t(v)$ is already drawn when creating trace $t(u)$. We have to be careful that $t(v)$ and $t(u)$ do not intersect. If we route the traces at the merge-junction i as depicted in Fig. 3, they visit the shared subtree $T_{u,i} = T_{v,i}$ without intersecting each other.

Now assume u and v share $k > 1$ merge-junctions $j_1, \ldots, j_k \in J$ and u and v merge at each j_i. Consequently we find k shared subtrees $T^1, \ldots, T^k$. By Lemma 1, however, we know that the intersection of these subtrees is empty. Hence we can treat every merge-junction and its subtree independently as in the case of a single merge-junction.

These are all the cases how two junction trees can interact. Hence the traces $t(u)$ and $t(v)$ for nodes $u, v \in N$ intersect if and only if there is a uv-path in $P(D)$ and, equivalently, the edge $(u, v) \in E_D$. Further, every trace is a continuous curve, so this set of traces yields a string representation of G.   □

A construction following the same principle can in fact be used to show:

Theorem 2. *Every SOC graph is an outer-string graph.*

4   Unit Interval Graphs and SC

In this section we consider so-called unit interval graphs. Let $G = (V, E)$ be a graph, then G is a unit interval graph if there exists a *unit-interval* layout Γ_{UI} of G, i.e. a representation of G where each vertex $v \in V$ is represented as an interval of unit length and edges are given by the intersections of the intervals.

Theorem 3. *Every unit-interval graph is an SC graph.*

Proof (Sketch). Our proof technique is constructive and describes how to compute a strict confluent diagram D for a given graph G based on its unit-interval layout Γ_{UI}. Based on the ordering of intervals in Γ_{UI}, we first greedily compute a set of cliques which are subgraphs of G. In particular, we ensure that the left-to-right-ordered set of cliques has the property that vertices in a clique are only incident to vertices in the same clique and to the two neighboring cliques; see Fig. 4(a). We then create an SOC diagram for each clique; see the red, blue and green layouts of the three cliques in Fig. 4(b).

In order to realize the remaining edges we first make the following useful observation. Let $(v_1, \ldots, v_k)$ denote the vertices of some clique C ordered from left to right according to Γ_{UI}. Then since all vertices are represented by unit intervals, if v_i is incident to a vertex w in the subsequent clique, also v_j must be incident to w for $i < j \le k$. We use this observation to insert a split junction b_i

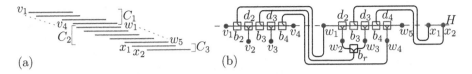

Fig. 4. (a) A unit interval graph G with a decomposition of its vertices into a set of cliques as described in the proof of Theorem 3; and (b) a strict confluent layout of G computed by the algorithm described in the proof of Theorem 3. (Color figure online)

in the SOC diagram of C such that all vertices with index at least i can access a smooth arc that connects them with w; see the black arcs in Fig. 4(b). We route arcs between cliques C_i and C_{i+1} first above clique C_i, then let it intersect with a line H that passes through all the cliques (which intuitively inverts the ordering of such arcs) and then finish the drawing below clique C_{i+1}; refer to Fig. 4(b) for an illustration. By adopting this scheme for each pair of consecutive cliques, intersections can be prevented.   □

5   Strict Bipartite-Outerconfluent Drawings

Let G be a bipartite graph with bipartition (X, Y). An outerconfluent drawing of G is *bipartite-outerconfluent* if the vertices in X (and hence also Y) occur consecutively on the boundary. Graphs admitting such a drawing are called *bipartite-outerconfluent*. The *bipartite permutation* graphs are just the graphs that are bipartite and *permutation* graphs, where a permutation graph is a graph that has an intersection model of straight lines between two parallel lines [29].

Theorem 4 (Hui et al. [27]). *The class of bipartite-permutation graphs is equal to the class of bipartite-outerconfluent graphs, i.e., the class of bipartite graphs admitting an intersection representation of straight-line segments between two parallel lines.*

It is natural to consider the idea of bipartite drawings also in the strict outerconfluent setting. We call a strict outerconfluent drawing D of G *bipartite* if it is bipartite-outerconfluent and strict. The graphs admitting such a drawing are called *strict bipartite-outerconfluent graphs*. In this section we extend Theorem 4 to the notion of strictness. The next lemma and observation are required in the proof of our theorem. The *domino* graph is the graph resulting from gluing two 4-cycles together at an edge.

Lemma 2. *Suppose that a reduced confluent diagram $D = (N, J, \Gamma)$ contains two distinct uv-paths. Then we can find in $G_D = (V_D, E_D)$ a set $V' \subseteq V_D$ such that $G[V']$ is isomorphic to C_6 with at least one chord.*

Observation 1. *Let $G = (V, E)$ be a graph and $V' \subseteq V$ a subset of six vertices such that $G[V']$ is isomorphic to a domino graph and let $X \cup Y = V'$ be the* *corresponding bipartition. Now let* π *be a cyclic order of* V' *in which the vertices in* X *and in* Y *are contiguous, respectively. Then there is no strict outerconfluent diagram* $D = (N, J, \Gamma)$ *with order* π *and* $G_D = G[V']$ *or, consequently,* $G_D = G$.

Theorem 5. *The (bipartite-permutation* $\cap$ *domino-free)-graphs are exactly the strict bipartite-outerconfluent graphs.*

Proof (Sketch). Let $G = (V, E)$ be a (bipartite-permutation $\cap$ domino-free) graph. By Theorem 4 we can find a bipartite-outerconfluent diagram $D = (N, J, \Gamma)$ which has $G_D = G$. Now assume that D is reduced but not strict. In this case we find six nodes $N' \subseteq N$ corresponding to a vertex set $V' \subseteq V_D$ in G_D such that $G_D[V'] = (V', E')$ is a C_6 with at least one chord by Lemma 2. In addition, since D (and hence also G_D) is bipartite and domino-free, we know there are two or three chords. Then $G_D[V']$ is a $K_{3,3}$ minus one edge $e \in E'$ or $K_{3,3}$. In a bipartite diagram these can always be drawn in a strict way.

For the other direction, consider a strict bipartite-outerconfluent diagram $D = (N, J, \Gamma)$. By Theorem 4, G_D is a bipartite permutation graph, and by Observation 1, it must be domino-free. Thus, G_D must be as described.   □

6   Strict Outerconfluent Graphs Have Cop Number Two

The *cops-and-robbers* game [1] on a graph $G = (V, E)$ is a two-player game with perfect information. The *cop-player* controls k *cop tokens*, while the *robber-player* has one *robber token*. In the first move the cop-player places the cop tokens on vertices of the graph, and then the robber places his token on another vertex. In the following the players alternate, in each turn moving their tokens to a neighboring vertex or keeping them at the current location. The cop-player is allowed to move all cops at once and multiple cops may be at the same vertex. The goal of the cop-player is to catch the robber, i.e., place one of its tokens on the same vertex as the robber.

The *cop number* $cop(G)$ of a graph G is the smallest integer k such that the cop-player has a winning strategy using k cop tokens. Gavenčiak et al. [16] showed that the cop number of outer-string graphs is between three and four, while the cop-number of many other interesting classes of intersection graphs, such as circle graphs and interval filament graphs, is two. We show that the cop number of SOC graphs is two as well.

Consider a SOC drawing $D = (N, J, \Gamma)$ of a graph $G = (V, E)$, which we can assume to be connected. For nodes $u, v \in N$, let the node interval $N[u, v] \subset N$ be the set of nodes in clockwise order between u and v on the outer face, excluding u and v. Let the cops be located on nodes $C \subseteq N$ and the robber be located on $r \in N$. We say that the robber is *locked* to a set of nodes $N' \subset N$ if $r \in N'$ and every path from r to $N \setminus N'$ contains at least one node that is either in C or adjacent to a node in C; in other words, a robber is locked to N' if it can be prevented from leaving N' by a cop player who simply remains stationary unless the robber can be caught in a single move. The following lemma establishes that a single cop can lock the robber to one of two "sides" of a SOC drawing.

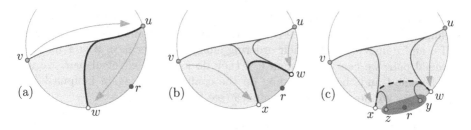

Fig. 5. Moves of the cops to confine the robber to a strictly smaller range.

Lemma 3. *Let $D = (N, J, \Gamma)$ be a SOC diagram of a graph G. Let a cop be placed on node u, the robber on node $r \neq u$ and not adjacent to u, and let $v \neq r$ be an arbitrary neighbor of u. Then the robber is either locked to $N[u, v]$ or locked to $N[v, u]$.*

Let $u, v \in N$ be two nodes of a SOC diagram $D = (N, J, \Gamma)$. We call a neighbor w of u in $N[u, v]$ *cw-extremal* (resp. *ccw-extremal*) for u, v (assuming such a neighbor exists), if it is the last neighbor of u in the clockwise (resp. counterclockwise) traversal of $N[u, v]$. Now let u, v be two neighboring nodes in N, $w \in N[u, v]$ be the cw-extremal node for u and $x \in N[u, v]$ be the ccw-extremal node for v. If w appears before x in the clockwise traversal of $N[u, v]$ we call w, x the *extremal pair* of the uv-path, see Fig. 5(b) and (c). In the case where only one node of u, v has an extremal neighbor w, say u, we define the extremal pair as v, w. In the following we assume that for a given uv-path the extremal pair exists.

Lemma 4. *Let $D = (N, J, \Gamma)$ be a SOC diagram of a graph G, $u, v \in N$ be two nodes connected by a uv-path in $P(D)$ and $w, x \in N[u, v]$ the extremal pair of the uv-path. If the cops are placed at u and v and the robber is at $r \in N[u, v]$, $r \neq w$, $r \neq x$, there is a move that locks the robber to $N[u, w]$, $N[w, x]$ or $N[x, v]$.*

Lemma 5. *Let $D = (N, J, \Gamma)$ be a SOC diagram of a graph G, $u, v \in N$ be two nodes connected by a uv-path in $P(D)$ and $w, x \in N[u, v]$ be the extremal pair of the uv-path such that there is no wx-path in $P(D)$. If the robber is at $r \in N[w, x]$ and the cops are placed on w, x we can find $y, z \in N[w, x] \cup \{w, x\}$ such that the yz-path exists in $P(D)$ and the robber is locked to $N[y, z]$.*

Combining Lemmas 3, 4 and 5 yields the result.

Theorem 6. *SOC graphs have cop number two.*

Proof (Sketch). Let $D = (N, J, \Gamma)$ be a strict-outerconfluent diagram of a (connected) graph G. Choose any uv-path in $P(D)$ and place the cops on u and v as the initial move. The robber must be placed on a node r that is either in $N[u, v]$ or in $N[v, u]$; by symmetry, let us assume the former. By Lemma 3, the robber is now locked to $N[u, v] \neq \emptyset$.

In every move we will shrink the locked interval until eventually the robber is caught. We distinguish three cases, based on the extremal neighbors w and x of u and v in $N[u,v]$ and their ordering along the outer face. If w, x form no extremal pair, we can use Lemma 3, if they do form an extremal pair, we use first Lemma 4 and then, depending on the configuration, again Lemma 3 (see Fig. 5(b)) or go into the case of Lemma 5 (see Fig. 5(c)).   □

Theorem 6 suggests a closer link between SOC graphs and interval-filament graphs [18], another subclass of outer-string graphs with cop number two.

7   Clique-Width of Tree-Like Strict Outerconfluent Graphs

In 2005, Eppstein et al. [10] showed that every strict confluent graph whose arcs in a strict confluent drawing topologically form a tree is distance hereditary and hence exhibits certain well-understood structural properties—in particular, every such graph has bounded *clique-width* [6]. These graphs are called Δ-*confluent* graphs. In their tree like confluent drawings an additional type of 3-way junction is allowed, the Δ-*junction*, which smoothly links together all three incident arcs. See Fig. 6, where the junctions j' and k' now form a single Δ-junction instead of three separate merge or split junctions.

In this section, we lift the result of Eppstein et al. [10] to the class of strict outerconfluent graphs: in particular, we show that as long as the arcs incident to junctions (including Δ-junctions) topologically form a tree, strict outerconfluent graphs also have bounded clique-width. Equivalently, we show that "extending" any drawing covered by Eppstein et al. [10] through the addition of outerplanar drawings of subgraphs in order to produce a strict outerconfluent drawing does not substantially increase the clique-width of the graph. Since the notion of clique-width will be central to this section, we formally introduce it below (see also the work of Courcelle et al. [6]). A k-graph is a graph whose

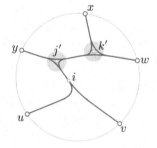

Fig. 6. A Δ-confluent diagram representing $K_5 - (u,v)$. Nodes are disks, junctions are squares. Δ-junctions are marked with a grey circle.

vertices are labeled by $[k] = \{1, 2, \ldots, k\}$; formally, the graph is equipped with a labeling function $\gamma \colon V(G) \to [k]$, and we also use $\gamma^{-1}(i)$ to denote the set of vertices labeled i for $i \in [k]$. We consider an arbitrary graph as a k-graph with all vertices labeled by 1. We call the k-graph consisting of exactly one vertex v (say, labeled by i) an *initial* k-graph and denote it by $i(v)$. The clique-width of a graph G is the smallest integer k such that G can be constructed from *initial* k-graphs by means of repeated application of the following three operations:

1. Disjoint union (denoted by $\oplus$);
2. Relabeling: changing all labels i to j (denoted by $p_{i \to j}$);
3. Edge insertion: adding an edge between every vertex labeled by i and every vertex labeled by j, where $i \neq j$ (denoted by $\eta_{i,j}$ or $\eta_{j,i}$).

The construction sequence of a k-graph G using the above operations can be represented by an algebraic term composed of $i(v)$, $\oplus$, $p_{i \to j}$ and $\eta_{i,j}$ (where $v \in V(G)$, $i \neq j$ and $i, j \in [k]$). Such a term is called a k-*expression* defining G, and the *clique-width* of G is the smallest integer k such that G can be defined by a k-expression. Distance-hereditary graphs are known to have clique-width at most 3 [23] and outerplanar graphs have clique-width at most 5 due to having treewidth at most 2 [3,7].

Let *(tree-like)* Δ *-SOC graphs* be the class of all graphs which admit strict outerconfluent drawings (including Δ-junctions) such that the union of all arcs incident to at least one junction topologically forms a tree. Clearly, the edge set E of every tree-like Δ-SOC graph $G = (V, E)$ with confluent diagram D_G can be partitioned into sets E_s and E_c, where E_s (the set of *simple edges*) contains all edges represented by single-arc paths in D not passing through any junction and E_c (the set of *confluent edges*) contains all remaining edges in G. Let $G_c = G[E_c] = (V_c, E_c)$ be the subgraph of G induced by E_c, i.e., V_c is obtained from V by removing all vertices without incident edges in E_c.

We note that even though G_c is known to be distance-hereditary [10] and $G - E_c$ is easily seen to be outerplanar, this does not imply that tree-like Δ-SOC graphs have bounded clique-width—indeed, the union of two graphs of bounded clique-width may have arbitrarily high clique-width (consider, e.g., the union of two sets of disjoint paths that create a square grid). Furthermore, one cannot easily adapt the proof of Eppstein et al. [10] to tree-like Δ-SOC graphs, as that explicitly uses the structure of distance-hereditary graphs; note that there exist outerplanar graphs which are not distance-hereditary, and hence tree-like Δ-SOC graphs are a strict superclass of distance hereditary graphs. Before proving the desired theorem, we introduce an observation which will later allow us to construct parts of G in a modular manner.

Observation 2. *Let $H = (V, E)$ be a graph of clique-width $k \geq 2$, let V_1, V_2 be two disjoint subsets of V, and let $s \in V \setminus (V_1 \cup V_2)$. Then there exists a $(3k + 1)$-expression defining H so that in the final labeling all vertices in V_1 receive label 1, all vertices in V_2 receive label 2, s receives label 3 and all remaining vertices receive label 4.*

Theorem 7. *Every tree-like Δ-SOC graph has clique-width at most 16.*

Proof (Sketch). We begin by partitioning the edge set of the considered Δ-SOC graph into E_c and E_s, as explained above, and by setting an arbitrary arc incident to a junction as the root r. Given a tree-like Δ-SOC drawing of the graph, our aim will be to pass through the confluent arcs of the drawing in a leaves-to-root manner so that at each step we construct a 16-expression for a certain circular segment of the outer face. This way, we will gradually build up the 16-expression for G from modular parts, and once we reach the root we will have a complete 16-expression for G.

At its core, the proof partitions nodes in the drawing into *regions*, delimited by arcs connecting nodes and junctions (such nodes are *not* part of any region). Each region is an outerplanar graph (which has clique-width at most 5), and furthermore the nodes in a region can only be adjacent to the nodes on the boundary of that region. Hence, by Observation 2 using $k = 5$, each region can be constructed by a 16-expression which also uses separate labels to capture the neighborhood of that region to its border. See Fig. 7 for an illustration.

The second ingredient used in the proof is tied to the tree-like structure of the drawing. In particular, one cannot construct a 16-expression (and even any k-expression for constant k) by simply joining the regions together in the order they appear along the outer face. Instead, to handle the adjacencies imposed by the paths in the drawing, one needs to process regions (and their bordering vertices) in an order which respects the structure of the tree. To do so, we introduce a notion of *depth*: nodes have a depth of 0, while junctions have depth equal to the largest depth of its "children" plus 1. Regions are then processed in an order which matches the depth of the corresponding junctions: for instance, if in Fig. 7 one of the junctions a_1 and a_2 has depth d then junction j' has depth $d + 1$,

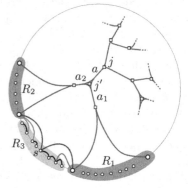

Fig. 7. Sketch of a tree-like Δ-SOC graph G with its regions. (Color figure online)

and so the blue regions will be constructed by modular 16-expressions before the yellow one. Afterwards, all three regions R_1, R_2, R_3 will be merged together into a blue region with a single 16-expression. By iterating this process, upon reaching the root r we obtain a 16-expression that constructs the whole Δ-SOC graph.   □

8   Conclusion

While this work provides the first in-depth study of SC and SOC graphs, a number of interesting open questions remain. One such question is motivated by our results on the cop-number of SOC graphs: we showed that SOC graphs are incomparable to most classes identified to have cop number two by Gavenčiak et al. [16], but we could not show such a result for the class of interval-filament graphs [18]. It seems likely that SOC graphs are contained in this class. Similarly, it is open whether SC graphs are contained in subtree-filament graphs. Furthermore, it is conceivable that a similar construction for the inclusion in string graphs, Sect. 3, could be used to show similar results for non-strict confluent graphs. Finally, investigating the curve complexity of our construction might provide insight into the curve complexity of SC and SOC diagrams.

On the algorithmic side, Sect. 7 raises the question of whether clique-width might be used to recognize SOC graphs, and perhaps even for finding SOC drawings. Another decomposition-based approach would be to use so-called split-decompositions [19], which we did not consider here. It is also open whether all bipartite permutation and trapezoid graphs [5, 21] are SOC graphs. Since bipartite permutation graphs are equivalent to bipartite trapezoid graphs [5, 21], the former represents a promising first step in this direction. It also remains open if it is possible to drop the unit length condition on the intervals in Sect. 4. We did not see an obvious way of adapting the construction for confluent drawings of interval graphs [8].

References

1. Aigner, M., Fromme, M.: A game of cops and robbers. Discrete Appl. Math. **8**(1), 1–12 (1984). https://doi.org/10.1016/0166-218X(84)90073-8
2. Bach, B., Riche, N.H., Hurter, C., Marriott, K., Dwyer, T.: Towards unambiguous edge bundling: investigating confluent drawings for network visualization. IEEE Trans. Vis. Comput. Graph. **23**(1), 541–550 (2017). https://doi.org/10.1109/TVCG.2016.2598958
3. Baker, B.S.: Approximation algorithms for NP-complete problems on planar graphs. J. ACM **41**(1), 153–180 (1994). https://doi.org/10.1145/174644.174650
4. Benzaken, C., Crama, Y., Duchet, P., Hammer, P.L., Maffray, F.: More characterizations of triangulated graphs. J. Graph Theory **14**(4), 413–422 (1990). https://doi.org/10.1002/jgt.3190140404
5. Brandstädt, A., Spinrad, J., Stewart, L.: Bipartite permutation graphs are bipartite tolerance graphs. Congressus Numerantium **58**, 165–174 (1987)
6. Courcelle, B., Makowsky, J.A., Rotics, U.: Linear time solvable optimization problems on graphs of bounded clique-width. Theory Comput. Syst. **33**(2), 125–150 (2000). https://doi.org/10.1007/s002249910009
7. Courcelle, B., Olariu, S.: Upper bounds to the clique width of graphs. Discrete Appl. Math. **101**(1–3), 77–114 (2000). https://doi.org/10.1016/S0166-218X(99)00184-5
8. Dickerson, M., Eppstein, D., Goodrich, M.T., Meng, J.Y.: Confluent drawings: visualizing non-planar diagrams in a planar way. J. Graph Algorithms Appl. **9**(1), 31–52 (2005). https://doi.org/10.7155/jgaa.00099
9. Ehrlich, G., Even, S., Tarjan, R.E.: Intersection graphs of curves in the plane. J. Comb. Theory Ser. B **21**(1), 8–20 (1976). https://doi.org/10.1016/0095-8956(76)90022-8
10. Eppstein, D., Goodrich, M.T., Meng, J.Y.: Delta-confluent drawings. In: Healy, P., Nikolov, N.S. (eds.) GD 2005. LNCS, vol. 3843, pp. 165–176. Springer, Heidelberg (2006). https://doi.org/10.1007/11618058_16
11. Eppstein, D., Goodrich, M.T., Meng, J.Y.: Confluent layered drawings. Algorithmica **47**, 439–452 (2007). https://doi.org/10.1007/s00453-006-0159-8
12. Eppstein, D., Holten, D., Löffler, M., Nöllenburg, M., Speckmann, B., Verbeek, K.: Strict confluent drawing. J. Comput. Geom. **7**(1), 22–46 (2016). https://doi.org/10.20382/jocg.v7i1a2
13. Eppstein, D., Simons, J.A.: Confluent Hasse diagrams. J. Graph Algorithms Appl. **17**(7), 689–710 (2013). https://doi.org/10.1007/978-3-642-25878-7_2

14. Förster, H., Ganian, R., Klute, F., Nöllenburg, M.: On strict (outer-)confluent graphs. CoRR abs/1908.05345 (2019). http://arxiv.org/abs/1908.05345
15. Gabor, C.P., Supowit, K.J., Hsu, W.L.: Recognizing circle graphs in polynomial time. J. ACM **36**(3), 435–473 (1989). https://doi.org/10.1145/65950.65951
16. Gavenčiak, T., Jelínek, V., Klavík, P., Kratochvíl, J.: Cops and robbers on intersection graphs. In: Cai, L., Cheng, S.-W., Lam, T.-W. (eds.) ISAAC 2013. LNCS, vol. 8283, pp. 174–184. Springer, Heidelberg (2013). https://doi.org/10.1007/978-3-642-45030-3_17, https://doi.org/10.1016/j.ejc.2018.04.009
17. Gavril, F.: Algorithms for minimum coloring, maximum clique, minimum covering by cliques, and maximum independent set of a chordal graph. SIAM J. Comput. **1**(2), 180–187 (1972). https://doi.org/10.1137/0201013
18. Gavril, F.: Maximum weight independent sets and cliques in intersection graphs of filaments. Inf. Process. Lett. **73**(5–6), 181–188 (2000). https://doi.org/10.1016/S0020-0190(00)00025-9
19. Gioan, E., Paul, C.: Split decomposition and graph-labelled trees: characterizations and fully dynamic algorithms for totally decomposable graphs. Discrete Appl. Math. **160**(6), 708–733 (2012). https://doi.org/10.1016/j.dam.2011.05.007
20. Golumbic, M.C.: Algorithmic Graph Theory and Perfect Graphs, vol. 57. Elsevier, Amsterdam (2004). https://doi.org/10.1002/net.3230130214
21. Golumbic, M.C., Monma, C.L., Trotter Jr., W.T.: Tolerance graphs. Discrete Appl. Math. **9**(2), 157–170 (1984). https://doi.org/10.1016/0166-218X(84)90016-7
22. Golumbic, M.C., Rotem, D., Urrutia, J.: Comparability graphs and intersection graphs. Discrete Math. **43**(1), 37–46 (1983). https://doi.org/10.1016/0012-365X(83)90019-5
23. Golumbic, M.C., Rotics, U.: On the clique-width of some perfect graph classes. Int. J. Found. Comput. Sci. **11**(3), 423–443 (2000). https://doi.org/10.1142/S0129054100000260
24. Halldórsson, M.M., Kitaev, S., Pyatkin, A.: Alternation graphs. In: Kolman, P., Kratochvíl, J. (eds.) WG 2011. LNCS, vol. 6986, pp. 191–202. Springer, Heidelberg (2011). https://doi.org/10.1007/978-3-642-25870-1_18
25. Holten, D.: Hierarchical edge bundles: visualization of adjacency relations in hierarchical data. IEEE Trans. Vis. Comput. Graph. **12**(5), 741–748 (2006). https://doi.org/10.1109/TVCG.2006.147
26. Hsu, W.L.: Maximum weight clique algorithms for circular-arc graphs and circle graphs. SIAM J. Comput. **14**(1), 224–231 (1985). https://doi.org/10.1137/0214018
27. Hui, P., Pelsmajer, M.J., Schaefer, M., Stefankovic, D.: Train tracks and confluent drawings. Algorithmica **47**(4), 465–479 (2007). https://doi.org/10.1007/s00453-006-0165-x
28. Kratochvíl, J.: String graphs. I. The number of critical nonstring graphs is infinite. J. Comb. Theory Ser. B **52**(1), 53–66 (1991). https://doi.org/10.1016/0095-8956(91)90090-7
29. Pnueli, A., Lempel, A., Even, S.: Transitive orientation of graphs and identification of permutation graphs. Can. J. Math. **23**(1), 160–175 (1971). https://doi.org/10.4153/CJM-1971-016-5
30. Roberts, F.S.: Indifference graphs. In: Proof Techniques in Graph Theory, pp. 139–146 (1969)
31. Takamizawa, K., Nishizeki, T., Saito, N.: Linear-time computability of combinatorial problems on series-parallel graphs. J. ACM **29**(3), 623–641 (1982). https://doi.org/10.1145/322326.322328
32. Trotter, W.T.: Combinatorics and Partially Ordered Sets: Dimension Theory, vol. 6. JHU Press, Baltimore (2001). https://doi.org/10.1137/1035116

33. Wegner, G.: Eigenschaften der Nerven homologisch-einfacher Familien im Rn. Ph.D. thesis, Universität Göttingen (1967)
34. Yu, C.W., Chen, G.H.: Efficient parallel algorithms for doubly convex-bipartite graphs. Theoret. Comput. Sci. **147**(1–2), 249–265 (1995). https://doi.org/10.1016/0304-3975(94)00220-D

Quality Metrics

On the Edge-Length Ratio
of Planar Graphs

Manuel Borrazzo and Fabrizio Frati[(✉)]

University Roma Tre, Rome, Italy
{borrazzo,frati}@dia.uniroma3.it

Abstract. The edge-length ratio of a straight-line drawing of a graph is the ratio between the lengths of the longest and of the shortest edge in the drawing. The planar edge-length ratio of a planar graph is the minimum edge-length ratio of any planar straight-line drawing of the graph.

In this paper, we study the planar edge-length ratio of planar graphs. We prove that there exist n-vertex planar graphs whose planar edge-length ratio is in $\Omega(n)$; this bound is tight. We also prove upper bounds on the planar edge-length ratio of several families of planar graphs, including series-parallel graphs and bipartite planar graphs.

1 Introduction

The reference book for the graph drawing research field "Graph Drawing: Algorithms for the Visualization of Graphs", by Di Battista, Eades, Tamassia, and Tollis [6], mentions that the minimization of the maximum edge length, provided that the minimum edge length is a fixed value, is among the most important aesthetic criteria that one should aim to satisfy in order to guarantee the readability of a graph drawing. A measure that naturally captures this concept is the *edge-length ratio* of a drawing; this is defined as the ratio between the lengths of the longest and shortest edge in the drawing.

In this paper we are interested in the construction of planar straight-line drawings with small edge-length ratio. From an algorithmic point of view, it has long been known that deciding whether a graph admits a planar straight-line drawing with edge-length ratio equal to 1 is an NP-hard problem. This was first proved by Eades and Wormald [7] for biconnected planar graphs and then by Cabello et al. [3] for triconnected planar graphs. From a combinatorial point of view, the study of planar straight-line drawings with small edge-length ratio started only recently, when Lazard, Lenhart, and Liotta [11] proved that every outerplanar graph admits a planar straight-line drawing with edge-length ratio smaller than 2 and that, for every fixed $\epsilon > 0$, there exist outerplanar graphs whose every planar straight-line drawing has edge-length ratio larger than $2 - \epsilon$.

Adopting the notation and the definitions from [10,11], we denote by $\rho(\Gamma)$ the edge-length ratio of a straight-line drawing Γ of a graph G, i.e., $\rho(\Gamma) = \max\limits_{e_1,e_2 \in E(G)} \frac{\ell_\Gamma(e_1)}{\ell_\Gamma(e_2)}$, where $\ell_\Gamma(e)$ denotes the length of the segment representing an

© Springer Nature Switzerland AG 2019
D. Archambault and C. D. Tóth (Eds.): GD 2019, LNCS 11904, pp. 165–178, 2019.
https://doi.org/10.1007/978-3-030-35802-0_13

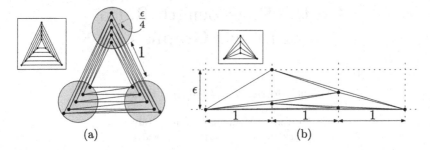

Fig. 1. (a) A drawing with edge-length ratio less than $1 + \varepsilon$ of the nested-triangle graph. (b) A drawing with edge-length ratio less than 3 of the plane 3-tree obtained as the join of a path with an edge.

edge e in Γ. The *planar edge-length ratio* $\rho(G)$ of G is the minimum edge-length ratio of any planar straight-line drawing of G. We prove the following results.

First, we prove that there exist n-vertex planar graphs whose planar edge-length ratio is in $\Omega(n)$. This bound is asymptotically tight, as every planar graph admits a planar straight-line drawing on an $O(n) \times O(n)$ grid [5,13]; such a drawing has edge-length ratio in $O(n)$. While our lower bound is not surprising, it was unexpectedly challenging to prove it. Some classes of graphs which are often used in order to prove lower bounds for graph drawing problems turn out to have constant planar edge-length ratio; see Fig. 1.

Second, we provide upper bounds for the planar edge-length ratio of several families of planar graphs. Namely, we prove that plane 3-trees have planar edge-length ratio bounded by their "depth" and that, for every fixed $\epsilon > 0$, bipartite planar graphs have planar edge-length ratio smaller than $1+\epsilon$. Most interestingly, we prove that every n-vertex graph with treewidth at most two, including 2-trees and series-parallel graphs, has sub-linear planar edge-length ratio; our upper bound is $O(n^{\log_2 \phi}) \subseteq O(n^{0.695})$, where $\phi = \frac{1+\sqrt{5}}{2}$ is the golden ratio. Lazard et al. [11] asked whether the planar edge-length ratio of 2-trees is bounded by a constant; recently, at the 14th Bertinoro Workshop on Graph Drawing, Fiala announced a negative answer to the above question. Thus, our upper bound provides a significant counterpart to Fiala's result; further, our result sharply contrasts with the fact that there exist n-vertex 2-trees whose every planar straight-line *grid* drawing requires an edge to have length in $\Omega(n)$ [9].

The paper is organized as follows. In Sect. 2, we introduce some definitions; in Sect. 3, we prove a lower bound for the planar edge-length ratio of planar graphs; in Sect. 4, we prove upper bounds for the planar edge-length ratio of families of planar graphs; finally, in Sect. 5, we conclude and present some open problems. The proofs marked (*) are deferred to the full version of the paper [1].

2 Definitions and Preliminaries

A *drawing* of a graph represents each vertex as a point in the plane and each edge as an open curve between its end-vertices. A drawing is *straight-line* if each

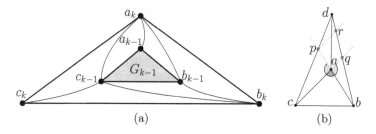

Fig. 2. (a) Construction of the graph G_k from the graph G_{k-1}. (b) Illustration for the proofs of Lemmata 2 and 3.

edge is represented by a straight-line segment. A drawing is *planar* if no two edges intersect, except at common end-vertices. A planar drawing of a graph defines connected regions of the plane, called *faces*. The only unbounded face is the *outer face*, while the other faces are *internal*. Two planar drawings of a (connected) graph are *equivalent* if: (i) the clockwise order of the edges incident to each vertex is the same in both drawings; and (ii) the clockwise order of the edges along the boundary of the outer face is the same in both drawings. A *plane embedding* is an equivalence class of planar drawings and a *plane graph* is a graph with a prescribed plane embedding. Throughout the paper, whenever we talk about a planar drawing of a plane graph G, we always assume, even when not explicitly stated, that it respects the plane embedding associated to G.

For any two distinct points a and b in the plane, we denote by $\overline{ab}$ the straight-line segment between a and b and by $||\overline{ab}||$ the Euclidean length of such a segment. For any three distinct and non-collinear points a, b, and c in the plane, we denote by abc the triangle whose vertices are a, b, and c. Further, for a triangle Δ, we denote by $p(\Delta)$ its perimeter and by $\angle_a(\Delta)$ the angle at a vertex a of Δ.

We will use the following lemma more than once.

Lemma 1 (*). *For a planar graph G and a subgraph G' of G, we have $\rho(G') \leq \rho(G)$.*

3   A Lower Bound for Planar Graphs

In this section we prove the following result.

Theorem 1. *For every $n = 3k$ with $k \in \mathbb{N}_{>0}$, there exists an n-vertex planar graph whose planar edge-length ratio is in $\Omega(n)$.*

We start by defining the class of planar graphs that we use in order to prove the theorem. For a 3-cycle $\mathcal{C}$ in a plane graph G, we denote by abc the clockwise order in which the vertices a, b, and c of $\mathcal{C}$ occur along $\mathcal{C}$. For any integer $k \geq 1$, we define a $3k$-vertex plane graph G_k as follows; refer to Fig. 2(a). Let G_1 coincide with a 3-cycle $\mathcal{C}_1 = a_1 b_1 c_1$. Now suppose that, for some integer $k \geq 2$, a plane graph G_{k-1} has been defined so that its outer face is delimited by a

3-cycle $\mathcal{C}_{k-1} = a_{k-1}b_{k-1}c_{k-1}$. Let G_k consist of (i) a 3-cycle $\mathcal{C}_k = a_k b_k c_k$, (ii) the plane graph G_{k-1}, embedded inside $\mathcal{C}_k$, and (iii) the edges $a_k a_{k-1}$, $a_k b_{k-1}$, $a_k c_{k-1}$, $b_k b_{k-1}$, $b_k c_{k-1}$, $c_k c_{k-1}$. Note that G_k has $3k$ vertices.

We first prove a lower bound for the edge-length ratio $\rho(\Gamma)$ of any planar straight-line drawing Γ of G_k in which the outer face is delimited by $\mathcal{C}_k$. Assume, without loss of generality up to a scaling of Γ, that the length of the shortest edge is 1. We prove that, for $i = 1, 2, \dots, k$, the perimeter $p(\Delta_i)$ of the triangle Δ_i representing $\mathcal{C}_i$ in Γ is at least $\gamma \cdot i$, for a constant γ to be determined later. This implies that $p(\Delta_k) \in \Omega(k)$, hence the longest of the three segments composing Δ_k has length in $\Omega(k)$, and the edge-length ratio $\rho(\Gamma)$ of Γ is in $\Omega(k)$.

The perimeter $p(\Delta_1)$ of Δ_1 is at least 3, given that each of the three segments composing Δ_1 has length greater than or equal to 1. Now assume that $p(\Delta_{i-1}) \geq \gamma \cdot (i-1)$, for some integer $i \geq 2$ and some constant $\gamma \leq 3$. We prove that $p(\Delta_i) \geq p(\Delta_{i-1}) + \gamma$, which implies that $p(\Delta_i) \geq \gamma \cdot i$.

We introduce two geometric lemmata; refer to Fig. 2(b). Let $\Delta = abc$ be a triangle and let d be a point outside Δ such that a either lies inside the triangle $\Delta' = bcd$, or it lies in the interior of $\overline{bd}$, or it lies in the interior of $\overline{cd}$.

Lemma 2 (*). $p(\Delta') > p(\Delta)$.

Lemma 3. If $\|\overline{ad}\| \geq 1$ and $\angle_a(\Delta) \leq 90°$, then $p(\Delta') > p(\Delta) + 1$.

Proof: Suppose first that a lies in the interior of $\overline{cd}$. Since $\angle_a(\Delta) \leq 90°$, we have that $\angle_a(bad) \geq 90°$, hence $\|\overline{bd}\| > \|\overline{ba}\|$. It follows that $p(\Delta') - p(\Delta) = \|\overline{bd}\| + \|\overline{ad}\| - \|\overline{ba}\| > 1$. The case in which a lies in the interior of $\overline{bd}$ is analogous.

Suppose next that a lies inside Δ'. Let p (q) be the intersection point of the straight line through a and b (through a and c) with $\overline{cd}$ (with $\overline{bd}$). Since $\angle_a(\Delta) \leq 90°$, we have that $\angle_a(cap) = \angle_a(baq) \geq 90°$, hence $\|\overline{cp}\| > \|\overline{ca}\|$ and $\|\overline{bq}\| > \|\overline{ba}\|$. It follows that $p(\Delta') - p(\Delta) > \|\overline{dp}\| + \|\overline{dq}\|$.

We claim that $\|\overline{dp}\| > \|\overline{aq}\|$. Let r be the intersection point between $\overline{bd}$ and the line passing through p that is parallel to the line through a and c. The triangles baq and bpr are similar, hence $\angle_p(bpr) = \angle_a(baq) \geq 90°$. Thus, $\angle_r(bpr) < 90°$ and $\angle_r(dpr) > 90°$. It follows that $\|\overline{dp}\| > \|\overline{pr}\|$; further, $\|\overline{pr}\| > \|\overline{aq}\|$, again by the similarity of the triangles baq and bpr. This proves the claim. It can be analogously proved that $\|\overline{dq}\| > \|\overline{ap}\|$.

By the triangular inequality, we have $\|\overline{ap}\| + \|\overline{dp}\| > \|\overline{ad}\|$ and $\|\overline{aq}\| + \|\overline{dq}\| > \|\overline{ad}\|$, hence $\|\overline{ap}\| + \|\overline{dp}\| + \|\overline{aq}\| + \|\overline{dq}\| > 2\|\overline{ad}\| \geq 2$. Since $\|\overline{dp}\| > \|\overline{aq}\|$ and $\|\overline{dq}\| > \|\overline{ap}\|$, it follows that $\|\overline{dp}\| + \|\overline{dq}\| > 1$. □

We now return the proof that $p(\Delta_i) \geq p(\Delta_{i-1}) + \gamma$; refer to Fig. 3. Assume, w.l.o.g. that $\overline{b_{i-1}c_{i-1}}$ is horizontal, with b_{i-1} to the right of c_{i-1} and with a_{i-1} above them. Let Δ'_{i-1} and Δ''_{i-1} be the triangles $a_i b_{i-1} c_{i-1}$ and $a_i b_i c_{i-1}$ in Γ, respectively. By Lemma 2, we have $p(\Delta_i) > p(\Delta''_{i-1}) > p(\Delta'_{i-1}) > p(\Delta_{i-1})$. If $\angle_{a_{i-1}}(\Delta_{i-1}) \leq 90°$, then by Lemma 3 we have $p(\Delta'_{i-1}) > p(\Delta_{i-1}) + 1$, thus $p(\Delta_i) > p(\Delta_{i-1}) + 1$ and we are done, as long as $\gamma \leq 1$. Assume hence that $\angle_{a_{i-1}}(\Delta_{i-1}) > 90°$; this implies that a_{i-1} is to the left of the vertical line ℓ_b through b_{i-1}. Further, if $\angle_{b_{i-1}}(\Delta'_{i-1}) \leq 90°$, then by Lemma 3 we have

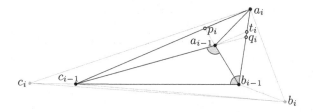

Fig. 3. Illustration for the proof that $p(\Delta_i) \geq p(\Delta_{i-1}) + \gamma$.

$p(\Delta''_{i-1}) > p(\Delta'_{i-1}) + 1$, thus $p(\Delta_i) > p(\Delta_{i-1}) + 1$ and we are done, as long as $\gamma \leq 1$. Assume hence that $\angle_{b_{i-1}}(\Delta'_{i-1}) > 90°$; this implies that a_i is to the right of ℓ_b.

Let p_i (q_i) be the intersection point of the straight line through a_{i-1} and b_{i-1} (through a_{i-1} and c_{i-1}) with $\overline{c_{i-1}a_i}$ (with $\overline{b_{i-1}a_i}$).

Assume first that $||\overline{a_iq_i}|| \geq 0.4$. By Lemma 2, we have that $p(b_{i-1}c_{i-1}q_i) > p(\Delta_{i-1})$; further, since $\angle_{q_i}(c_{i-1}q_ia_i) > \angle_{b_{i-1}}(c_{i-1}b_{i-1}a_i) > 90°$, we have that $||\overline{c_{i-1}a_i}|| > ||\overline{c_{i-1}q_i}||$, hence $p(\Delta'_{i-1}) = p(b_{i-1}c_{i-1}q_i) + ||\overline{c_{i-1}a_i}|| + ||\overline{a_iq_i}|| - ||\overline{c_{i-1}q_i}|| > p(\Delta_{i-1}) + 0.4$ and we are done, as long as $\gamma \leq 0.4$.

Assume next that $||\overline{a_iq_i}|| < 0.4$. We show that this implies that $||\overline{a_ip_i}|| \geq 0.4$. Suppose, for a contradiction, that $||\overline{a_ip_i}|| < 0.4$. Consider the intersection point t_i of $\overline{b_{i-1}a_i}$ with the line through a_{i-1} parallel to the line through c_{i-1} and a_i. Since $\angle_{q_i}(a_{i-1}q_it_i) = \angle_{q_i}(c_{i-1}q_ia_i) > 90°$, we have $||\overline{a_{i-1}q_i}|| < ||\overline{a_{i-1}t_i}||$. Further, by the similarity of the triangles $b_{i-1}a_ip_i$ and $b_{i-1}t_ia_{i-1}$, we have $||\overline{a_{i-1}t_i}|| < ||\overline{a_ip_i}||$. Hence, $||\overline{a_{i-1}q_i}|| < 0.4$. Then the triangular inequality implies that $||\overline{a_{i-1}a_i}|| < ||\overline{a_{i-1}q_i}|| + ||\overline{a_iq_i}|| < 0.8$, while $||\overline{a_{i-1}a_i}|| \geq 1$, given that $a_{i-1}a_i$ is an edge of G_i, a contradiction. We can hence assume that $||\overline{a_ip_i}|| \geq 0.4$. In order to conclude our argument, we are going to use the following.

Lemma 4 (*). *Let T be a triangle with vertices u, v, and w, where $\angle_{u}vw < 90°$. Then $||\overline{vw}|| < \sqrt{||\overline{uv}||^2 + ||\overline{uw}||^2}$.*

Let $x = ||\overline{b_{i-1}a_i}||$, $y = ||\overline{a_ip_i}||$, and $z = ||\overline{b_{i-1}p_i}||$. By Lemma 2, we have $p(b_{i-1}c_{i-1}p_i) > p(\Delta_{i-1})$, hence $p(\Delta'_{i-1}) - p(\Delta_{i-1}) > x + y - z$. Note that $x \geq 1$, since $b_{i-1}a_i$ is an edge of G_i, and $y \geq 0.4$, by assumption. By Lemma 4, we have that $z < \sqrt{x^2 + y^2}$, hence $p(\Delta'_{i-1}) - p(\Delta_{i-1}) > x + y - \sqrt{x^2 + y^2}$. The derivative $\frac{\partial(x+y-\sqrt{x^2+y^2})}{\partial x} = \frac{\sqrt{x^2+y^2}-x}{\sqrt{x^2+y^2}}$ is positive for every value of x and y; the same is true for the derivative $\frac{\partial(x+y-\sqrt{x^2+y^2})}{\partial y}$. Hence, the minimum value of $x + y - \sqrt{x^2 + y^2}$ is achieved when x and y are minimum, that is, when $x = 1$ and $y = 0.4$. With such values we get $x + y - \sqrt{x^2 + y^2} > 0.32$. Hence, $p(\Delta'_{i-1}) > p(\Delta_{i-1}) + 0.32$ and we are done, as long as $\gamma \leq 0.32$.

By picking $\gamma = 0.3$, we conclude the proof that $p(\Delta_i) \geq p(\Delta_{i-1}) + \gamma$, which implies that $p(\Delta_k) \in \Omega(k)$ and hence that $\rho(\Gamma) \in \Omega(k)$.

Finally, we remove the assumption that the outer face of Γ is delimited by $\mathcal{C}_k$. This is done as follows. Consider the complete graph K_4 on four vertices, say a, b, c, and d; further, consider two copies G'_k and G''_k of G_k, where $\mathcal{C}'_k$ and $\mathcal{C}''_k$ denote the copies of the cycle $\mathcal{C}_k$ in G'_k and G''_k, respectively. Glue G'_k and G''_k with K_4 by identifying the 3-cycle abc with $\mathcal{C}'_k$ and the 3-cycle abd with $\mathcal{C}''_k$. Denote by G the resulting n-vertex planar graph. In any planar drawing Γ of G, the planar drawing of G'_k has its outer face delimited by $\mathcal{C}'_k$ or the planar drawing of G''_k has its outer face delimited by $\mathcal{C}''_k$, hence $\rho(\Gamma) \in \Omega(k)$. The proof of Theorem 1 is concluded by observing that $k \in \Omega(n)$.

4   Upper Bounds for Planar Graph Classes

In this section we prove upper bounds for the planar edge-length ratio of various families of planar graphs.

4.1   Plane 3-Trees

A *plane 3-tree* is a maximal plane graph that can be constructed as follows. The only plane 3-tree with 3 vertices is a plane 3-cycle. For $n \geq 4$, an n-vertex plane 3-tree G is obtained from an $(n-1)$-vertex plane 3-tree G' by inserting a vertex v inside an internal face f of G' and by connecting v to the three vertices of G' incident to f. An n-vertex plane 3-tree G is naturally associated with a rooted ternary tree T_G whose internal nodes represent the internal vertices of G and whose leaves represent the internal faces of G (T_G is called *representative tree* of G in [12]). Formally, T_G is defined as follows. If $n = 3$, then T_G is a single node, representing the unique internal face of G. If $n > 3$, then G can be obtained by inserting a vertex v inside an internal face f of a plane 3-tree G' and by connecting v to the three vertices of G' incident to f. Let t_f be the leaf representing f in $T_{G'}$. Then T_G is obtained from $T_{G'}$ by inserting three new leaves as children of t_f. In T_G, t_f represents v and its children represent the faces of G incident to v. The *depth* of T_G is the maximum number of nodes in any root-to-leaf path in T_G. The *depth* of G is the depth of T_G. We have the following.

Theorem 2. *Every plane 3-tree with depth k has planar edge-length ratio in $O(k)$.*

Proof: Let G be any plane 3-tree with depth k. Fix any constant $\epsilon > 0$ and represent the 3-cycle $\mathcal{C}$ delimiting the outer face of G as any triangle Δ whose y-extension is ϵ and whose three sides have x-extension equal to 1, k, and $k+1$.

Now assume that we have constructed a drawing Γ' of a plane 3-tree G' which is a subgraph of G that includes $\mathcal{C}$. Assume that Γ' satisfies the following invariant: every internal face f of G' is delimited by a triangle whose three sides have x-extension equal to 1, greater than or equal to k_f, and greater than or equal to $k_f + 1$, where k_f is the depth of the subtree of T_G rooted at the node

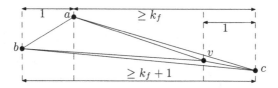

Fig. 4. Inserting a vertex v of G in a face f of G'.

corresponding to f. Initially, this is the case with $G' = \mathcal{C}$ and $\Gamma' = \Delta$; note that the only internal face of G' corresponds to the root of T_G, which has depth k.

Let t_f be any leaf of $T_{G'}$ which is not a leaf of T_G. Let f be the internal face of G' represented by t_f in $T_{G'}$, let Δ_f be the triangle representing f in Γ', let abc be the 3-cycle delimiting f in G, and let v be the internal vertex of G represented by t_f in T_G; see Fig. 4. By the invariant, we can assume that the x-extensions of $\overline{ab}$, $\overline{ac}$, and $\overline{bc}$ are equal to 1, greater than or equal to k_f, and greater than or equal to $k_f + 1$, respectively. Place v inside f in Γ' so that the x-extension of $\overline{vc}$ is equal to 1 and draw the edges va, vb, and vc as straight-line segments. This results in a planar straight-line drawing Γ'' of a plane 3-tree G'' which is a subgraph of G and which has one more vertex than G'. The invariant is satisfied by Γ''; in particular, $\overline{av}$ and $\overline{bv}$ have x-extension greater than or equal to $k_f - 1$ and greater than or equal to k_f, respectively, hence each face f' of G'' incident to v is delimited by a triangle whose sides have the desired x-extension, since the subtree of T_G rooted at the node corresponding to f' has depth $k_f - 1$.

Eventually, we get a planar straight-line drawing of G such that every edge has length at least 1, given that it has x-extension greater than or equal to 1, and at most $k + 1 + \epsilon \in O(k)$, given that it has x-extension smaller than or equal to $k + 1$ and y-extension smaller than or equal to ϵ. □

The bound in Theorem 2 is tight, as Theorem 1 shows that a plane 3-tree with depth k might have planar edge-length ratio in $\Omega(k)$. Further, Theorem 2 implies that any *balanced* n-vertex plane 3-tree, i.e., a plane 3-tree G such that T_G is a balanced tree, has planar edge-length ratio in $O(\log n)$.

4.2 2-Trees

For any integer $n \geq 2$, an n-vertex 2-*tree* G is a graph whose vertex set has an ordering $v_1, v_2, \ldots, v_n$ such that $v_1 v_2$ is an edge of G, called *root* of G, and, for $i = 3, \ldots, n$, the vertex v_i has exactly two neighbors $p(v_i)$ and $q(v_i)$ in $\{v_1, v_2, \ldots, v_{i-1}\}$, where $p(v_i)$ and $q(v_i)$ are adjacent in G. The vertices $v_3, v_4, \ldots, v_n$, i.e., the vertices of G not in its root are called *internal*. For an edge $v_i v_j$ of G, an *apex* of $v_i v_j$ is a vertex v_k, with $k > i$ and $k > j$, such that $p(v_k) = v_i$ and $q(v_k) = v_j$; further, the *side edges* of $v_i v_j$ are all the edges $v_i v_k$ and $v_j v_k$ such that v_k is an apex of $v_i v_j$; finally, an edge $v_i v_j$ is *trivial* if it has no apex, otherwise it is *non-trivial*. In this section we prove the following theorem.

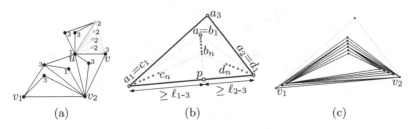

Fig. 5. (a) A linear 2-tree H. The apexes and the side edges of the edge uv are gray; the only non-trivial side edge of uv is thicker. The numbers show the classes of the vertices. (b) The points $b_1, b_2, \ldots, b_n, c_1, c_2, \ldots, c_n, d_1, d_2, \ldots, d_n$ inside $a_1 a_2 a_3$ (for the sake of readability, there are fewer points than there should be). (c) The drawing of H constructed by the algorithm L2T-drawer.

Theorem 3. *Every n-vertex 2-tree has planar edge-length ratio in $O(n^{\log_2 \phi}) \subseteq O(n^{0.695})$, where $\phi = \frac{1+\sqrt{5}}{2}$ is the golden ratio.*

In the following, we first define a family of 2-trees, which we call *linear 2-trees*, and show that they admit drawings with constant edge-length ratio. We will later show how to find, in any 2-tree G, a subgraph which is a linear 2-tree and whose removal splits G into "small" components. This decomposition, together with the drawing algorithm for linear 2-trees, will be used in order to construct a planar straight-line drawing of G with sub-linear edge-length ratio.

A *linear* 2-*tree* is a 2-tree such that every edge has at most one non-trivial side edge; see Fig. 5(a). We now classify the vertices of a linear 2-tree H into vertices of class 1, class 2, and class 3, so that every edge of H has its end-vertices in different classes. First, v_1 and v_2 are vertices of class 1 and class 2, respectively, where $v_1 v_2$ is the root of H. Now we repeatedly consider an edge uv of H such that u and v have been already classified and the apexes of uv have not been classified yet. We let every apex be in the unique class different from the classes of u and v. Based on the classification of the vertices of H, we also classify the edges of H into edges of class 1-2, class 1-3, and class 2-3, where an edge is of class a-b if its end-vertices are of classes a and b.

We now show an algorithm, called **L2T-drawer**, that constructs a planar straight-line drawing Γ_H of a linear 2-tree H. The algorithm L2T-drawer receives in input a triangle $a_1 a_2 a_3$ and three real values $\ell_{1\text{-}2}, \ell_{1\text{-}3}, \ell_{2\text{-}3} \geq 1$ such that $\ell_{1\text{-}3} + \ell_{2\text{-}3} \leq ||\overline{a_1 a_2}||$ and $\ell_{1\text{-}2} < ||\overline{a_1 a_2}||$. The algorithm constructs a planar straight-line drawing Γ_H of H with the following properties: (L1) for $i = 1, 2$, the vertex v_i lies at a_i; (L2) every internal vertex of H lies inside $a_1 a_2 a_3$; and (L3) the length of every edge of class x-y is at least $\ell_{x\text{-}y}$, for each x-$y \in \{1\text{-}2, 1\text{-}3, 2\text{-}3\}$.

Refer to Fig. 5(b). Let a be a point inside $a_1 a_2 a_3$ such that the line through a orthogonal to $\overline{a_1 a_2}$ intersects $\overline{a_1 a_2}$ in a point p with $||\overline{a_1 p}|| \geq \ell_{1\text{-}3}$ and $||\overline{a_2 p}|| \geq \ell_{2\text{-}3}$; this exists since $\ell_{1\text{-}3} + \ell_{2\text{-}3} \leq ||\overline{a_1 a_2}||$. Let $\epsilon = \min\{||\overline{a a_1}|| - \ell_{1\text{-}3}, ||\overline{a a_2}|| - \ell_{2\text{-}3}, ||\overline{a_1 a_2}|| - \ell_{1\text{-}2}\}$ and note that $\epsilon > 0$. Let $b_1, b_2, \ldots, b_n$ be n points on $\overline{ap}$, in this order from a to p, with $||\overline{a b_n}|| \leq \frac{\epsilon}{3}$. Further, let $c_1 = a_1, c_2, \ldots, c_n$ be n points on $\overline{a_1b_n}$, in this order from a_1 to b_n, with $\|\overline{a_1c_n}\| \leq \frac{\epsilon}{3}$, and let $d_1 = a_2, d_2, \ldots, d_n$ be n points on $\overline{a_2b_n}$, in this order from a_2 to b_n, with $\|\overline{a_2d_n}\| \leq \frac{\epsilon}{3}$.

The algorithm L2T-drawer is as follows. Refer to Fig. 5(c). We initialize Γ_H by drawing the root v_1v_2 of H as the straight-line segment $\overline{a_1a_2}$, where a_i represents v_i, for $i = 1, 2$. Now L2T-drawer proceeds in steps. During one step, all the apexes and side edges of a single non-trivial edge of H are drawn. The algorithm maintains the invariant that, before each step, Γ_H is a planar straight-line drawing of an m-vertex subgraph H_m of H such that the following properties are satisfied for some integers j, k, l with $m = j + k + l$: (i) the vertices of H_m of classes 1, 2, and 3 are drawn at the points $c_1, \ldots, c_k$, at the points $d_1, \ldots, d_l$, and at the points $b_1, \ldots, b_j$, respectively; further, if H_m does not coincide with H, then (ii) there is exactly one edge e_m that is a non-trivial edge of H, that is in H_m, and whose apexes are not in H_m, and (iii) the end-vertices of e_m lie at b_j and c_k, or at c_k and d_l, or at b_j and d_l. The invariant is indeed satisfied after the initialization of Γ_H to a drawing of v_1v_2, with $m = 2$, $k = l = 1$, and $j = 0$.

We now perform one step. Assume that e_m is a 1-2 edge, hence its end-vertices lie at c_k and d_l; the other cases are analogous. Draw the $x \geq 1$ apexes of e_m, which are vertices of class 3, at the points $b_{j+1}, \ldots, b_{j+x}$, so that the only non-trivial side edge e_{m+x} of e_m, if any, is incident to the apex drawn at b_{j+x}. Draw the side edges of e_m as straight-line segments. After this step, Γ_H is a planar straight-line drawing of an $(m+x)$-vertex subgraph H_{m+x} of H satisfying the invariant; in particular, at most one side edge e_{m+x} of e_m is non-trivial in H, given that H is a linear 2-tree; this implies property (ii).

Eventually, the algorithm constructs a planar straight-line drawing Γ_H of H. By construction, the vertices v_1 and v_2 are placed at a_1 and a_2, respectively, hence Γ_H satisfies property (L1). Further, the internal vertices of H are placed at the points $b_1, b_2, \ldots, b_n, c_2, c_3, \ldots, c_n, d_2, d_3, \ldots, d_n$, which are inside $a_1a_2a_3$, by construction, hence Γ_H satisfies property (L2). Finally, consider any edge of class 1-3, which is represented by a straight-line segment $\overline{c_kb_j}$. By the triangular inequality we have $\|\overline{c_kb_j}\| \geq \|\overline{aa_1}\| - \|\overline{a_1c_k}\| - \|\overline{ab_j}\| \geq \ell_{1\text{-}3} + \epsilon - \frac{2\epsilon}{3} > \ell_{1\text{-}3}$. Analogously, any edge of class 2-3 has length larger than $\ell_{2\text{-}3}$ and any edge of class 1-2 has length larger than $\ell_{1\text{-}2}$ in Γ_H; hence Γ_H satisfies property (L3).

We now deal with general 2-trees. Let G be a 2-tree with root v_1v_2. Let H be any subgraph of G that is a linear 2-tree and that has v_1v_2 as its root. For any edge uv of H we define an H-component G_{uv} of G as follows. Remove from G the vertices of H and their incident edges; this splits G into connected components and we let G_{uv} be the 2-tree which is the subgraph of G induced by u, by v, and by the vertex sets of the connected components containing a vertex adjacent to both u and v; see Fig. 6(a). The edge uv is the root of G_{uv}. An H-component of G is of class 1-2, 1-3, or 2-3 if its root is of class 1-2, 1-3, or 2-3, respectively.

For technical reasons, we let n be the number of vertices of G minus one. The plan is: (1) to find a subgraph H of G that is a linear 2-tree with root v_1v_2 such that every H-component of G has "few" internal vertices; (2) to construct a planar straight-line drawing Γ_H of H by means of the algorithm L2T-drawer; and (3) to recursively draw each H-component, plugging such drawings into Γ_H,

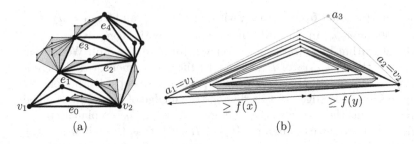

Fig. 6. (a) A 2-tree G and a subgraph H of G which is a linear 2-tree; the vertices and edges of H are represented by larger disks and thicker lines, respectively. The H-components of G are shown within shaded regions. (b) The planar straight-line drawing Γ of G constructed by the algorithm in the proof of Theorem 3.

thus obtaining a drawing of G. We start with the following lemma, which draws inspiration from a technique for decomposing binary trees proposed by Chan [4].

Lemma 5 (*). *There exists a subgraph H of G that is a linear 2-tree, that has v_1v_2 as its root, and that satisfies the following property. Let x, y, and z be the maximum number of vertices of an H-component of G of class 1-3, 2-3, and 1-2, respectively, minus one. Then $z \leq \frac{n}{2}$; further (i) $x \leq \frac{n}{2}$ and $y \leq \frac{n-x}{2}$, or (ii) $y \leq \frac{n}{2}$ and $x \leq \frac{n-y}{2}$, or (iii) $x+y \leq \frac{2n}{3}$.*

We now show an algorithm to construct a planar straight-line drawing Γ of G. Let $f(n) = n^{\log_2 \phi}$, where $\phi = \frac{1+\sqrt{5}}{2}$. The algorithm receives in input a triangle $a_1a_2a_3$, whose hypotenuse $\overline{a_1a_2}$ is such that $\|\overline{a_1a_2}\| \geq f(n)$, and constructs a planar straight-line drawing Γ of G satisfying the following properties: (T0) the length of every edge is at least 1 and at most $\|\overline{a_1a_2}\|$; (T1) for $i = 1, 2$, the vertex v_i lies at a_i; and (T2) every internal vertex of G lies inside $a_1a_2a_3$.

If $n = 1$, that is, G coincides with the edge v_1v_2, then Γ is the straight-line segment $\overline{a_1a_2}$. Then property (T1) is trivially satisfied, property (T2) is vacuous, and property (T0) is satisfied since $\|\overline{a_1a_2}\| \geq n^{\log_2 \phi} = 1$.

Assume next that $n > 1$; refer to Fig. 6(b). Let H be a subgraph of G satisfying the properties of Lemma 5; in particular (i) $x \leq \frac{n}{2}$ and $y \leq \frac{n-x}{2}$, or (ii) $y \leq \frac{n}{2}$ and $x \leq \frac{n-y}{2}$, or (iii) $x+y \leq \frac{2n}{3}$, where x and y are the maximum number of vertices of an H-component of G of class 1-3 and 2-3, respectively, minus one. We construct a planar straight-line drawing Γ_H of H by applying the algorithm L2T drawer with input the triangle $a_1a_2a_3$ and the real values $\ell_{1\text{-}3} = f(x)$, $\ell_{2\text{-}3} = f(y)$, and $\ell_{1\text{-}2} = f(z)$; note that $\ell_{1\text{-}2} = f(z) < f(n) \leq \|\overline{a_1a_2}\|$, as $z < n$; we will prove later that $f(n) \geq f(x) + f(y)$, which implies that $\|\overline{a_1a_2}\| \geq \ell_{1\text{-}3} + \ell_{2\text{-}3}$.

Let $G_1, \ldots, G_k$ be the H-components of G; for $i = 1, \ldots, k$, let u_iv_i be the root of G_i. Note that u_iv_i is an edge of H, hence it is represented by a straight-line segment $\overline{u_iv_i}$ in Γ_H. For $i = 1, \ldots, k$, let w_i be a point such that the triangle $\Delta_i = u_iv_iw_i$ lies inside $a_1a_2a_3$, does not intersect Γ_H other than at $\overline{u_iv_i}$, and does not intersect any distinct triangle Δ_j, except at common vertices. Since

Γ_H is planar, choosing w_i sufficiently close to $\overline{u_i v_i}$ suffices to accomplish these objectives. For $i = 1, \ldots, k$, we recursively draw G_i so that u_i and v_i lie at the same points as in Γ_H and so that every internal vertex of G_i lies inside Δ_i. This concludes the construction of a planar straight-line drawing Γ of G.

We prove that Γ satisfies properties (T0)–(T2). Property (T1) is satisfied since Γ_H satisfies property (L1); further, property (T2) is satisfied since Γ_H satisfies property (L2), since the internal vertices of G_i lie inside the triangle Δ_i, and since Δ_i lies inside $a_1 a_2 a_3$, by construction. We now deal with property (T0). The length of every edge of H in Γ is at least $\min\{f(x), f(y), f(z)\}$ by property (L3) of Γ_H; further, $f(x) = x^{\log_2 \phi} \geq 1$, $f(y) = y^{\log_2 \phi} \geq 1$, and $f(z) = z^{\log_2 \phi} \geq 1$, given that $x, y, z \geq 1$. The length of every edge of H in Γ is at most $\|\overline{a_1 a_2}\|$, given that every vertex of H lies inside or on the boundary of $a_1 a_2 a_3$, by properties (L1) and (L2) of Γ_H, and given that $\overline{a_1 a_2}$ is the hypotenuse of $a_1 a_2 a_3$. The length of every edge of G not in H is at least 1 and at most $\|\overline{a_1 a_2}\|$ by induction and since every triangle Δ_i lies inside $a_1 a_2 a_3$.

We now prove that $f(n) \geq f(x) + f(y)$. In the case in which (i) $x \leq \frac{n}{2}$ and $y \leq \frac{n-x}{2}$, or (ii) $y \leq \frac{n}{2}$ and $x \leq \frac{n-y}{2}$, the inequality $f(n) \geq f(x) + f(y)$ has been already proved by Chan [4]. Assume hence that (iii) $x + y \leq \frac{2n}{3}$.

We use Hölder's inequality, that is, $\sum_{i=1}^{k} r_i s_i \leq (\sum_{i=1}^{k} r_i^p)^{\frac{1}{p}} (\sum_{i=1}^{k} s_i^q)^{\frac{1}{q}}$, for every real $p, q > 1$ with $\frac{1}{p} + \frac{1}{q} = 1$ and every vectors $(r_1, \ldots, r_k), (s_1, \ldots, s_k) \in \mathbb{R}^k$.

By employing the values $\frac{1}{p} = \log_2 \phi$, $\frac{1}{q} = 1 - \log_2 \phi$, $r_1 = x^{\log_2 \phi}$, $r_2 = y^{\log_2 \phi}$, and $s_1 = s_2 = 1$, we get $f(x) + f(y) = x^{\log_2 \phi} + y^{\log_2 \phi} \leq (x+y)^{\log_2 \phi} \cdot 2^{(1-\log_2 \phi)} \leq (\frac{2n}{3})^{\log_2 \phi} \cdot \frac{2}{2^{\log_2 \phi}} = \frac{2}{3^{\log_2 \phi}} n^{\log_2 \phi} < 0.933 \cdot n^{\log_2 \phi} < n^{\log_2 \phi} = f(n)$.

Applying the described algorithm with a triangle $a_1 a_2 a_3$ whose hypotenuse has length $\|\overline{a_1 a_2}\| = f(n)$ results in a planar straight-line drawing of G with edge-length ratio at most $f(n) = n^{\log_2 \phi}$. This concludes the proof of Theorem 3.

We remark that $f(n) = n^{\log_2 \phi}$ is the smallest possible function when using the decomposition of Lemma 5, as an example in which $x = \frac{n}{2}$ and $y = \frac{n}{4}$ shows.

We also remark that a graph has treewidth at most 2 if and only if it is a subgraph of a 2-tree; hence, Lemma 1 and Theorem 3 imply the following.

Corollary 1. *Every graph with treewidth at most 2 has planar edge-length ratio in $O(n^{\log_2 \phi}) \subseteq O(n^{0.695})$, where $\phi = \frac{1+\sqrt{5}}{2}$ is the golden ratio.*

The bound on the treewidth in the above result is the best possible, as the proof of Theorem 1 shows that an n-vertex planar graph with treewidth 3 might have planar edge-length ratio in $\Omega(n)$. Observe that graphs with treewidth 1, i.e., trees, have planar edge-length ratio equal to 1.

4.3 Bipartite Planar Graphs

In this section we deal with bipartite planar graphs.

Theorem 4. *For every $\epsilon > 0$, every n-vertex bipartite planar graph has planar edge-length ratio smaller than $1 + \epsilon$.*

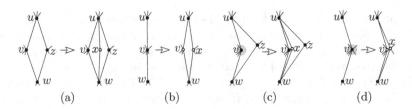

Fig. 7. (a) The operation P_0. (b) The operation P_1. (c) and (d) show how to transform Γ' into Γ by applying P_0 or P_1, respectively. The gray disk is D.

Proof: First, by Lemma 1 and since any bipartite planar graph can be augmented to maximal by adding edges to it, it suffices to prove the statement for maximal bipartite planar graphs. Second, Brinkmann et al. [2] proved that every n-vertex maximal bipartite plane graph G is either a plane 4-cycle, or can be obtained from an $(n-1)$-vertex maximal bipartite plane graph G' by applying either the operation P_0 shown in Fig. 7(a), in which a path uxw is inserted in a face f of G' delimited by a 4-cycle $uvwz$, or the operation P_1 shown in Fig. 7(b), in which a path uvw of G' is transformed into a 4-cycle $uvwx$.

We now prove that, for every $\epsilon > 0$, any n-vertex maximal bipartite plane graph G admits a planar straight-line drawing Γ in which every edge has length larger than 1 and smaller than $1 + \epsilon$. The proof is by induction on n. If $n = 4$, then G is a 4-cycle embedded in the plane, and the desired drawing Γ of G is any square with side length equal to $1 + \delta$, with $0 < \delta < \epsilon$.

If $n > 4$, then let G' be an $(n-1)$-vertex maximal bipartite plane graph such that G can be obtained from G' by applying either the operation P_0 or the operation P_1. Fix any δ such that $0 < \delta < \epsilon$; inductively construct a planar straight-line drawing Γ' of G' in which every edge has length larger than 1 and smaller than $1 + \delta$. Let $\ell_1 = \min_e\{\ell_{\Gamma'}(e) - 1\}$, $\ell_2 = \min_e\{1 + \epsilon - \ell_{\Gamma'}(e)\}$, and $\ell = \min\{\ell_1, \ell_2\}$. Let D be a disk with radius ℓ centered at v in Γ'.

Both the operations P_0 and P_1 correspond to the *expansion* of a vertex v into an edge vx, followed by the removal of vx. Hence, by standard continuity arguments (see, e.g., the proof of Fáry's theorem [8]), a planar straight-line drawing Γ of G can be obtained from Γ' by suitably replacing the vertex v with the edge vx, so that the position of v in Γ is the same as in Γ', and so that x is arbitrarily close to v. Thus, both if P_0 or if P_1 transforms G' into G, we can obtain a planar straight-line drawing Γ of G in which every vertex other than x is at the same position as in Γ', and in which x is inside the disk D; see Figs. 7(c) and (d). Note that, for every edge e of G that is not incident to x, we have $1 < \ell_\Gamma(e) < 1 + \epsilon$, given that $1 < \ell_{\Gamma'}(e) < 1 + \delta$. Further, consider any edge $e = tx$ of G and note that $e' = tv$ is an edge of G'. By the triangular inequality we have $\|\overline{tx}\| < \|\overline{tv}\| + \|\overline{vx}\| < \ell_{\Gamma'}(e') + \ell \leq \ell_{\Gamma'}(e') + (1 + \epsilon - \ell_{\Gamma'}(e')) = 1 + \epsilon$, and $\|\overline{tx}\| > \|\overline{tv}\| - \|\overline{vx}\| > \ell_{\Gamma'}(e') - \ell \geq \ell_{\Gamma'}(e') - (\ell_{\Gamma'}(e') - 1) = 1$. This concludes the induction and hence the proof of the theorem. $\qquad\square$

Note that the bound in Theorem 4 is the best possible, as there exist bipartite planar graphs (for example any complete bipartite graph $K_{2,m}$ with $m \geq 3$) that admit no planar straight-line drawing with edge-length ratio equal to 1.

5 Conclusions and Open Problems

In this paper we have proved that there exist n-vertex planar graphs whose planar edge-length ratio is in $\Omega(n)$; that is, in any planar straight-line drawing of one of such graphs, the ratio between the length of the longest edge and the length of the shortest edge is in $\Omega(n)$. Further, we have proved upper bounds for the planar edge-length ratio of several graph classes, most notably an $O(n^{0.695})$ upper bound for the planar edge-length ratio of 2-trees.

Several problems remain open; we mention some of them. First, what is the asymptotic behavior of the planar edge-length ratio of 2-trees? In particular, we wonder whether our geometric construction can lead to a better upper bound if coupled with a decomposition technique better than the one in Lemma 5. Second, is the planar edge-length ratio of cubic planar graphs sub-linear? The proof of Theorem 1 shows that this question has a negative answer when extended to all bounded-degree planar graphs. Finally, is the planar edge-length ratio of k-outerplanar graphs bounded by some function of k? The results from [11] show that this is indeed the case for $k = 1$.

References

1. Borrazzo, M., Frati, F.: On the edge-length ratio of planar graphs. CoRR abs/1908.03586 (2019)
2. Brinkmann, G., Greenberg, S., Greenhill, C.S., McKay, B.D., Thomas, R., Wollan, P.: Generation of simple quadrangulations of the sphere. Discrete Math. **305**(1–3), 33–54 (2005)
3. Cabello, S., Demaine, E.D., Rote, G.: Planar embeddings of graphs with specified edge lengths. J. Graph Algorithms Appl. **11**(1), 259–276 (2007)
4. Chan, T.M.: A near-linear area bound for drawing binary trees. Algorithmica **34**(1), 1–13 (2002)
5. de Fraysseix, H., Pach, J., Pollack, R.: How to draw a planar graph on a grid. Combinatorica **10**(1), 41–51 (1990)
6. Di Battista, G., Eades, P., Tamassia, R., Tollis, I.G.: Graph Drawing: Algorithms for the Visualization of Graphs. Prentice-Hall, Upper Saddle River (1999)
7. Eades, P., Wormald, N.C.: Fixed edge-length graph drawing is NP-hard. Discrete Appl. Math. **28**(2), 111–134 (1990)
8. Fáry, I.: On straight line representations of planar graphs. Acta Sci. Math. (Szeged) **11**, 229–233 (1948)
9. Frati, F.: Lower bounds on the area requirements of series-parallel graphs. Discrete Math. Theoret. Comput. Sci. **12**(5), 139–174 (2010)
10. Lazard, S., Lenhart, W., Liotta, G.: On the edge-length ratio of outerplanar graphs. In: Frati, F., Ma, K.-L. (eds.) GD 2017. LNCS, vol. 10692, pp. 17–23. Springer, Cham (2018). https://doi.org/10.1007/978-3-319-73915-1_2

11. Lazard, S., Lenhart, W.J., Liotta, G.: On the edge-length ratio of outerplanar graphs. Theoret. Comput. Sci. **770**, 88–94 (2019)
12. Mondal, D., Nishat, R.I., Rahman, M.S., Alam, M.J.: Minimum-area drawings of plane 3-trees. J. Graph Algorithms Appl. **15**(2), 177–204 (2011)
13. Schnyder, W.: Embedding planar graphs on the grid. In: Johnson, D.S. (ed.) ACM-SIAM Symposium on Discrete Algorithms (SODA 1990), pp. 138–148 (1990)

Node Overlap Removal Algorithms:
A Comparative Study

Fati Chen[1,2]([⊠]) [iD], Laurent Piccinini[2], Pascal Poncelet[1],
and Arnaud Sallaberry[1,2] [iD]

[1] LIRMM - CNRS - Université de Montpellier, Montpellier, France
{fati.chen,pascal.poncelet,arnaud.sallaberry}@lirmm.fr
[2] Université Paul-Valéry Montpellier 3, Montpellier, France
{fati.chen,laurent.piccinini,arnaud.sallaberry}@univ-montp3.fr

Abstract. Many algorithms have been designed to remove node overlapping, and many quality criteria and associated metrics have been proposed to evaluate those algorithms. Unfortunately, a complete comparison of the algorithms based on some metrics that evaluate the quality has
never been provided and it is thus difficult for a visualization designer to
select the algorithm that best suits his needs. In this paper, we review 21
metrics available in the literature, classify them according to the quality criteria they try to capture, and select a representative one for each
class. Based on the selected metrics, we compare 8 node overlap removal
algorithms. Our experiment involves 854 synthetic and real-world graphs.

Keywords: Graph drawing · Layout adjustment · Node overlap
removal

1 Introduction

Graph drawing algorithms are good at creating rich expressive graph layouts but
often consider nodes as points with no dimensions. After changing the size of
nodes in the case of annotation or evolving graphs, it causes node overlap which
hides information. Post-process algorithms, named *layout adjustment* [15], have
been proposed to remove node overlap.

The objective of these algorithms is, given an initial positioning of the nodes
and a size for each one, to provide a new embedding so that there are no overlapping nodes any more. A classical zoom-in function maintaining the sizes of
the nodes (i.e. uniform scaling) provides such an embedding, but it expands
the visualisation, resulting in large areas without any objects. Therefore, a node
overlap removal algorithm must take into account the area of the drawing, and
try to minimise it. Positioning the nodes evenly on a grid meets this objective
but will result in the loss of the user's mental picture of the original embedding.
Thus, it is also important to minimise the change on the layout.

Since a preliminary work in 1995 [15], many algorithms have been designed to
reach these purposes, and many quality criteria have been proposed to evaluate

© Springer Nature Switzerland AG 2019
D. Archambault and C. D. Tóth (Eds.): GD 2019, LNCS 11904, pp. 179–192, 2019.
https://doi.org/10.1007/978-3-030-35802-0_14

them. Unfortunately, a complete comparison of the algorithms based on the different criteria has never been provided and it is thus difficult for a visualisation designer to select the one that best suits his needs.

In this paper, our contribution comes in two forms: (1) We propose a classification of 21 quality metrics, grouping them according to the quality criterion they try to capture. We also discuss their relevance and we select a representative one for each class. (2) We compare state-of-the-art node overlapping approaches in regards to the previously selected metrics. This experiment involves 854 graphs, including synthetic ones (random, tree, scale-free, small-world) and real world ones.

The paper is organised as follows: after a brief reminder in Sect. 2 of the definitions and the notations used in this paper, we present and discuss the quality criteria and the metrics in Sect. 3. Then we compare the algorithms in Sect. 4. Finally we conclude in Sect. 5.

2   Preliminaries

In this paper, we use the following definitions and notations.

$G = (V, E)$ denotes a graph where V is the set of nodes and E the set of edges. The number of nodes $|V|$ is denoted by n and the number of edges $|E|$ by m. We consider each node as a rectangle. Thus, for a node $v \in V$, its width and its height are denoted by the couple (w_v, h_v) which is not impacted by the layout adjustment.

The initial embedding is defined as an injection $\mathcal{E}_G : V \rightarrow \mathbb{R}^2$ such that $\forall v \in V$, $\mathcal{E}_G(v) = (x_v, y_v)$ where (x_v, y_v) are the coordinates of the center of the node v. The overlapping-free embedding is denoted by $\mathcal{E}'_G$. To simplify notations, we denote $v = (x_v, y_v)$ instead of $\mathcal{E}_G(v)$, and $v' = (x'_v, y'_v)$ instead of $\mathcal{E}'_G(v)$. Remark that two nodes $(u, v) \in V^2$ are overlapping when :

$$|x_v - x_u| < \frac{w_v + w_u}{2} \quad \text{and} \quad |y_v - y_u| < \frac{h_v + h_u}{2}$$

The bounding box Bb of an embedding $\mathcal{E}_G$ is defined as the smallest rectangle containing all the nodes of G; w_{bb} (resp. h_{bb}) denotes the width (resp. the height) of the initial embedding, w'_{bb} (resp. h'_{bb}) denotes the width (resp. the height) of the overlapping-free one. They are determined as follows:

$$w_{bb} = \left| \max_{v \in V} \left(x_v + \frac{w_v}{2} \right) - \min_{u \in V} \left(x_u - \frac{w_u}{2} \right) \right| \tag{1}$$

$$h_{bb} = \left| \max_{v \in V} \left(y_v + \frac{h_v}{2} \right) - \min_{u \in V} \left(y_u - \frac{h_u}{2} \right) \right| \tag{2}$$

The position of the center of the bounding box is denoted by $c_{bb} = (x_{bb}, y_{bb})$ in the initial embedding, and $c'_{bb} = (x'_{bb}, y'_{bb})$ in the overlapping-free embedding.

The convex hull of an embedding $\mathcal{E}_G$ is defined as the smallest convex region containing all the nodes of G. Note that it is computed by using the 4 corners of the nodes, and not only their center, in a way that the rectangles representing the nodes are fully included into it. In the following, Ch denotes the convex hull of the original embedding, Ch' the convex hull of the free-overlapping one, c_{ch} the center of mass of Ch, c'_{ch} the center of mass of Ch'.

3   Quality Criteria

Many criteria have been proposed in the literature to evaluate the quality of the embeddings resulting from adjustment algorithms. Unfortunately, the experiments provided by the authors of the different approaches are not always based on the same metrics. With a view to provide a uniform protocol of experiment and a complete comparison of the algorithms, we need to review the quality criteria and the metrics used to evaluate them. We also need to select a representative metric for each criterion.

We identified 5 classes of metrics (*Orthogonal Ordering preservation, Spread minimisation, Global Shape preservation, Node Movement minimisation* and *Edge Length preservation*), each of them depicting a quality criterion. Table 1 shows the metrics assigned into the classes. The formulas are given in the discussion below.

The following subsections contain the metrics of a specific class. In each of them, we select one representative metric, based on the corresponding quality criterion and the properties that the metrics aim at capturing. Our discussion also sometimes involves the coefficient of correlation of two metrics run following the protocol described in the comparison section, Sect. 4.

3.1   Orthogonal Ordering Preservation

The orthogonal ordering class groups the metrics which try to quantify how much an adjustment algorithm preserves the initial orthogonal ordering. We recall that the orthogonal ordering is respected when all nodes satisfy the following conditions:

$$\begin{cases} x_u < x_v \Leftrightarrow x'_u < x'_v \\ y_u < y_v \Leftrightarrow y'_u < y'_v \\ x_u = x_v \Leftrightarrow x'_u = x'_v \\ y_u = y_v \Leftrightarrow y'_u = y'_v \end{cases}$$

The first metric of this class, oo_o [15], is equal to 1 if the overlapping-free graph embedding preserves the initial orthogonal ordering, 0 otherwise. Also, if only one couple of nodes does not satisfy those conditions, the value of oo_o is the same as when many ones do not satisfy it.

To overcome this issue, Huang *et al.* [11] proposed a metric based on the *Kendall's Tau distance*. For each couple of nodes, they first compute an inversion number $inv(u, v)$ corresponding to 0 if the orthogonal ordering is preserved

Table 1. List of metrics classified by the quality criteria they try to capture: metrics selected for the comparison appear in bold italics. The *Abbreviations* are based on some initials of the names. For example, *sp_bb_a* means that the metric is in the class *Spread minimisation*, it uses the embedding *Bounding Box* to quantify the *Area* spreading. The *Range* column contains the set of values that the metric can take. The *Target* column refers to the target value to meet the corresponding criterion.

Abbreviation	Name	Range	Target
	Orthogonal Ordering preservation		
oo_o	Original [15]	$\{0,1\}$	1
oo_kt	Kendall's Tau Distance [11]	$[0,1]$	0
oo_ni	Number of Inversions [17]	$[0, n(n-1)]$	0
oo_nni	***Normalised Number of Inversions***	$[0,1]$	0
	Spread minimisation		
sp_bb_l1ml	Bounding Box L1 Metric Length [12]	$[1,+\infty[$	1
sp_bb_a	Bounding Box Area [15]	$[1,+\infty[$	1
sp_bb_na	Bounding Box Normalised Area [11]	$[0,1[$	0
sp_ch_a	***Convex Hull Area*** [17]	$[1,+\infty[$	1
	Global Shape preservation		
gs_bb_ar	Bounding Box Aspect Ratio [12]	$]0,+\infty[$	1
gs_bb_iar	***Bounding Box Improved Aspect Ratio***	$[1,+\infty[$	1
gs_ch_sd	Convex Hull Standard Deviation [17]	$[0,+\infty[$	0
	Node Movement minimization		
nm_mn	Moved Nodes [11]	$[0,1]$	0
nm_dm_me	Distance Moved Mean Euclidean [17]	$[0,+\infty[$	0
nm_dm_ne	Distance Moved Normalized Euclidean [13]	$[0,1]$	0
nm_dm_h	Distance Moved Hamiltonian [10,11]	$[0,+\infty[$	0
nm_dm_se	Distance Moved Squared Euclidean [14]	$[0,+\infty[$	0
nm_dm_imse	***Distance Moved Improved Mean Squared Euclidean***	$[0,+\infty]$	0
nm_d	Displacement [5]	$]0,+\infty[$	0
nm_knn	K-Nearest Neighbours [16]	$[0,+\infty[$	0
	Edge Length preservation		
el_r	Ratio [12]	$[1,+\infty[$	1
el_rsdd	***Relative Standard Deviation Delaunay*** [5]	$[0,+\infty]$	0

between them, 1 otherwise. The metric is then defined as the normalised sum of the inversion numbers:

$$\text{oo\_kt} = \frac{\sum_{u \neq v} inv(u,v)}{n(n-1)}$$

Strobelt *et al.* [17] introduced the number of inversions:

$$\text{oo\_ni} = \sum_{\substack{(u,v)\in V^2 \\ x_u > x_v}} \begin{cases} 1 & \text{if } x_u' < x_v' \\ 0 & \text{otherwise} \end{cases}$$

$$+ \sum_{\substack{(u,v)\in V^2 \\ x_u > x_v}} \begin{cases} 1 & \text{if } y_u' < y_v' \\ 0 & \text{otherwise} \end{cases}$$

This metric has the drawback of providing non-normalized values. However, it holds the benefit of penalizing inversions occurring on each axis independently ($x-$ and $y-axis$), instead of penalizing in the same manner an inversion occurring in only one axis and an inversion occurring in the two axes. Thus, in our study, we combine the two metrics by using a normalised version of the latter:

$$\text{oo\_nni} = \frac{oo\_ni}{n(n-1)}$$

3.2   Spread Minimisation

A classical zoom-in function maintaining the sizes of the nodes (i.e. uniform scaling) provides an overlapping-free embedding, but it expands the visualisation, resulting in large areas without any objects. To avoid this issue, quality metrics have been introduced to quantify embedding spreading. Their purpose is to favour algorithms inducing low spreading.

The L1 metric length [12] is the ratio:

$$\text{sp\_bb\_l1ml} = \frac{\max(w_{bb}', h_{bb}')}{\max(w_{bb}, h_{bb})}$$

The drawback of this technique is to consider only one dimension of the embedding, width or height. For instance, considering an example where $w_{bb} = 4$, $h_{bb} = 2$, $w_{bb}' = 4$, $h_{bb}' = 4$, the value of the L1 metric length is 1 (which is the target value), whereas the area of the overlapping-free embedding is twice as large as in the initial embedding. The ratio between the bounding box areas of the two embeddings [15] overcomes this issue:

$$\text{sp\_bb\_a} = \frac{w_{bb}' \times h_{bb}'}{w_{bb} \times h_{bb}}$$

While the result gives an unbounded value greater than 1, Huang *et al.* [11] proposes a normalised version producing values in the interval $[0, 1[$:

$$\text{sp\_bb\_na} = 1 - \frac{w_{bb} \times h_{bb}}{w_{bb}' \times h_{bb}'}$$

Unfortunately, this criterion is poorly intuitive and it is hard to figure out what the values represent.

In our comparison, we selected another version of the ratio of areas involving convex hulls [17], as it better captures the concrete area of the drawing:

$$sp\_ch\_a = \frac{area(Ch')}{area(Ch)}$$

3.3 Global Shape Preservation

This class contains metrics that try to capture the ability of the algorithms to preserve the global shape of the initial embedding. The first one was proposed by Li et al. [12]:

$$gs\_bb\_ar = \begin{cases} \text{if } w'_{bb} > h'_{bb} & \dfrac{w'_{bb} \times h_{bb}}{h'_{bb} \times w_{bb}} \\ \text{otherwise} & \dfrac{h'_{bb} \times w_{bb}}{w'_{bb} \times h_{bb}} \end{cases}$$

The underlying idea is to capture the variation of the aspect ratio (w_{bb}/h_{bb}) between the initial and the overlapping-free embedding. For instance, let us consider an example where $w_{bb} = 3$, $h_{bb} = 2$, $w'_{bb} = 6$, $h'_{bb} = 4$. In this case, the overlapping-free embedding is twice as large as the initial one but the aspect ratio remains the same 3/2. The gs_bb_ar is 1, which is the target value. Now let us consider another example where $w_{bb} = 3$, $h_{bb} = 2$, $w'_{bb} = 4$, $h'_{bb} = 6$. In this case, the initial aspect ratio is 3/2 whereas the overlapping-free one is 2/3. The gs_bb_ar is now 2.25, which is not the target value; it reveals a distortion of the initial embedding during the overlap removal process. The main drawback of this metric is that it can reach values in the interval $]0, +\infty[$ while the target value is 1. Thus, it is hard to decide, for instance, which algorithm is the best between two of them if the first one obtains a score of 0.67 and the second one a score of 4.56. To overcome this issue, we propose to refine it as follows:

$$gs\_bb\_iar = \max\left(\frac{w'_{bb} \times h_{bb}}{h'_{bb} \times w_{bb}}, \frac{h'_{bb} \times w_{bb}}{w'_{bb} \times h_{bb}}\right)$$

In this case, the target value is 1 and the metric cannot reach values below it. This criterion is the one we selected for our study.

An alternative to this approach based on the convex hull has been proposed by Strobelt et al. [17]. The idea is to evaluate the distortion of the convex hull by comparing, between both embeddings, the distances of convex hull points to their center. Let ℓ_θ (resp. ℓ'_θ) be the euclidean distance between the center of mass c_{ch} (resp. c'_{ch}) of the convex hull Ch (resp. Ch') and the intersection of the convex hull with the line going through c_{ch} (resp. c'_{ch}) and with an angle θ (θ varying from 0° to 350° in 10° steps). Then, the difference is defined as the ratio $d_\theta = \ell'_\theta/\ell_\theta$. The metric is the standard deviation of the 36 measures of d_θ:

$$gs\_ch\_sd = \sqrt{\frac{1}{36} \sum_{\substack{\theta=10k \\ k=0,\cdots,35}} (d_\theta - \bar{d})^2}$$

$$\text{where } \overline{d} = \frac{1}{36} \sum_{\substack{\theta=10k \\ k=0,\cdots,35}} d_\theta \text{ is the mean value}$$

Based on the experiments presented below in Sect. 4, we observed that gs_bb_iar and gs_ch_sd have a correlation coefficient of 0.77, showing that they both tend to capture similar aspects of the adjustment process. We selected the former for its simplicity and its ease of interpretation.

3.4 Node Movement Minimisation

This class contains the metrics quantifying the changes in node positions after running an adjustment algorithm. The underlying intuition is that an algorithm involving high node movements will provide an overlapping-free configuration different from the original one, and thus may result in a substantial loss of the mental model.

The simplest metric of this class was presented by Huang et al. [11]:

$$nm\_mn = \frac{nb}{n}$$

Here, nb represents the number of nodes which have moved between the initial and the overlapping-free embedding. The main drawback of this approach is that a node overlap removal algorithm may induce very small changes in most nodes, which does not affect the mental model preservation, while inducing a very bad result. To tackle this problem and add more granularity over the evaluation of node movements, a series of metrics have been proposed, based on the same underlying quality function:

$$nm\_dm = f(n) \times \sum_{v \in V} \text{dist}(v, v')$$

where f is a normalising function of $n = |V|$ and $dist$ is a distance between v and v'. Table 2 sums up the ones used in the literature.

Table 2. Functions used to tune the distances moved metric.

$\text{dist}(v, v')$ \ $f(n)$	1	$1/n$	$\frac{1}{k\sqrt{2}\times n}$
$\|v' - v\|$		nm_dm_me [17]	nm_dm_ne [13]
$\|v' - v\|^2$	nm_dm_se [14]	nm_dm_imse	
$\|x'_v - x_v\| + \|y'_v - y_v\|$	nm_dm_h [10]		

The function f comes in three different forms. Marriott et al. [14] and Huang et al. [10] do not include any f, which is similar to having $f(n) = 1$. The drawback is that the resulting value highly depends on the number of nodes in the graph. That is why Strobelt et al. [17] proposed to use the mean of the distances, which corresponds to $f(n) = 1/n$. Finally, Lyons *et al.* [13] proposed $f(n) = 1/(k\sqrt{2} \times n)$, where k is the maximum between w'_{bb} and h'_{bb}. In this case, $k\sqrt{2}$ is the diagonal of a square containing the embedding, thus a maximum distance available for a node. Thus, this f function normalises the values of the metric. Unfortunately, this normalisation induces very small values that are hard to interpret. That is why we preferred using $f(n) = 1/n$ for our study.

Three *dist* functions have been proposed in the literature. The most intuitive one is the Euclidean distance $\|v' - v\|$ [13,17]. The squared Euclidean distance $\|v' - v\|^2$ [14] avoids the square root computation and discriminates high changes better. It is the one we selected for our study. The Manhattan distance $|x'_v - x_v| + |y'_v - y_v|$ has also been used [10], but it is less intuitive and has close results (*nm_dm_se* and *nm_dm_h* have a correlation coefficient of 0.9).

Let us consider an adjustment algorithm that pushes nodes on the x-axis. The preservation of the global shape is not optimal but the preservation of the configuration should reach a good score, as a node on right-top in the initial embedding would remain on right-top in the overlapping-free embedding. In order to better capture the relative movement of a node between the two embeddings, a *shift* function can be applied to align the center of the initial bounding box with the center of the final one, and a *scale* function to align the size of the initial bounding box to the size of the final one:

$$shift(v) = (x_v + x'_{bb} - x_{bb}, y_v + y'_{bb} - y_{bb})$$

$$scale(v) = (x_v \times \frac{w'_{bb}}{w_{bb}}, y_v \times \frac{h'_{bb}}{h_{bb}})$$

Considering this, we selected the following node movement metric:

$$nm\_dm\_imse = \frac{1}{n} \times \sum_{v \in V} \|v' - scale(shift(v))\|^2$$

nm_d [5] (the complete formula is available in the paper) is also based on the idea that the metric should be based on modified initial positions to better capture the relative movement of the nodes between the two embeddings. Besides including the *shift* and the *scale* functions, it also rotates the initial embedding with an angle θ that minimizes the distances between the nodes of the initial embedding and the ones of the overlapping-free embedding:

$$rotation(v) = (x_v \cos\theta - y_v \sin\theta, x_v \sin\theta + y_v \cos\theta)$$

We have not included the rotation in our experiment as we consider that it can induce a loss of the mental model (think about the recognition of a map turned inside down).

An alternative to quantify how much an overlapping-free configuration may result in a substantial loss of the mental model is to look at the neighbourhoods at the nodes and compare them before and after the adjustment. Based on a *KNN* approach, Nachmanson *et al.* [16] proposed the following metric:

$$nm\_knn(k) = \sum_{v \in V} (k - |N_k(v) \cap N_k(v')|)^2$$

where $N_k(v)$ (resp. $N_k(v')$) denotes the k nearest neighbours of v (resp. v'), in terms of Euclidean distance, in the initial (resp. overlapping-free) embedding. We did not select this metric because, unlike the other metrics of the class, it requires to fix a parameter (k).

3.5 Edge Length Preservation

This class contains the two metrics based on edge lengths. The set of edges can be E or can be another set derived from the graph.

Standard force-based layout algorithms tend to produce uniform lengths of edges. Indeed, the first metric of this class captures whether the edge lengths of a graph remain uniform or not after applying an adjustment algorithm [12]:

$$el\_r = \frac{\max_{(u,v)\in E^2} \|u' - v'\|}{\min_{(u,v)\in E^2} \|u' - v'\|}$$

While many layout algorithms are not designed to produce uniform edge lengths, we did not select this approach, which is not related to the mental model preservation for these algorithms. We preferred the next one, based on a Delaunay triangulation.

Let E_{dt} be the set of edges of a Delaunay triangulation performed on the nodes of the initial embedding. The second metric of this class, el_rsdd, is based on computing the coefficient of variation, also known as the relative standard deviation, of the edge lengths ratio as follows [5]:

$$r_{uv} = \frac{\|u' - v'\|}{\|u - v\|}, \quad (u,v) \in E_{dt}^2$$

$$\overline{r} = \frac{1}{|E_{dt}|} \sum_{(u,v)\in E_{dt}^2} r_{uv}$$

$$el\_rsdd = \frac{\sqrt{\frac{1}{|E_{dt}|}\sum_{(u,v)\in E_{dt}^2} (r_{uv} - \overline{r})^2}}{\overline{r}}$$

4 Algorithms Comparison

In this section, we compare 8 algorithms of the literature in terms of quality and running time: uniform *Scaling*, *PFS* [15], *PFS'* [8], *FTA* [11], *VPSC* [3], *PRISM* [5], *RWordle-L* [17], and *GTREE* [16]. The quality of an overlapping-free embedding is evaluated with the metrics identified in the last section, by following a 3 steps procedure: **Step 1: Datasets** We generate 840 synthetic graphs containing 10 to 1,000 nodes. These graphs are provided by 4 generation models available on the OGDF library [2]: random graphs [4], random trees, small world graphs [18], and scale-free graphs [1]. We also use 14 real-world graphs selected from the Graphviz test suite[1] [6], previously used by the authors of *PRISM* [5] and *GTREE* [16]. **Step 2: Overlapping-free embedding computation** Synthetic graphs resulting from the first step are initially positioned by the FM^3 layout algorithm [7]. Then, we apply the 8 node overlap removal algorithms, thus providing a set of 6,720 overlapping-free graph embeddings. Graphviz test suite graphs are initially positioned by the *SFDP* layout algorithm [9] to follow the same baseline embedding as Gansner *et al.* [5]. We then apply the 8 node overlap removal algorithms thus providing 112 overlapping-free graph embeddings. **Step 3: Metrics computation** We finally compute the values of the 5 selected metrics on the 6.832 overlapping-free synthetic and real-world graph embeddings. We also measure the computation time of the algorithms.

4.1   Quality

Figure 1 and 2 show the aggregated metrics values on the synthetic and real-world datasets. Unsurprisingly, *Scaling*, *PFS* and *PFS'* obtain the best scores at *oo_nni* as it is proved that they maintain the original orthogonal ordering. Though, all the algorithms tested got good results for this criterion.

		Scaling	PFS	PFS'	FTA	VPSC	PRISM	RWordle-L	GTREE
	Q1	0.00	0.00	0.00	0.00	0.00	0.00	0.00	0.00
oo_nni	Median	0.00	0.00	0.00	0.01	0.00	0.01	0.01	0.02
	Q3	0.00	0.00	0.00	0.04	0.03	0.02	0.10	0.03
	Q1	1.06	1.04	1.00	1.00	1.00	1.01	1.00	1.12
sp_ch_a	Median	8.70	1.69	1.22	1.02	1.00	1.12	1.01	1.49
	Q3	54.29	17.67	5.04	4.21	2.30	4.22	1.99	5.94
	Q1	1.00	1.01	1.00	1.00	1.00	1.00	1.00	1.01
gs_bb_iar	Median	1.01	1.19	1.04	1.01	1.00	1.04	1.00	1.04
	Q3	1.06	1.65	1.14	1.66	1.94	1.23	1.07	1.08
	Q1	0.00	6.59	1.65	0.94	0.42	9.82	2.18	28.22
nm_dm_imse	Median	0.00	694.83	116.06	37.87	9.71	131.57	27.60	292.56
	Q3	0.00	7899.2	1782.8	2966.0	283.6	688.1	597.7	2221.7
	Q1	0.00	0.05	0.05	0.04	0.03	0.13	0.07	0.13
eb_rsdd	Median	0.00	0.25	0.21	0.22	0.17	0.31	0.25	0.28
	Q3	0.00	0.33	0.26	0.60	0.47	0.38	0.68	0.36

Fig. 1. Aggregated values of the selected metrics among the synthetic graphs: first quartile, median and third quartile.

Scaling highly increases the size of the embedding, which induces a bad score for *sp_ch_a*. *PFS* also obtains a bad score for this criterion. *VPSC* and *RWordle-L* produce the most compact embeddings, while the other algorithms give intermediary results.

---

[1] https://gitlab.com/graphviz/graphviz/blob/master/rtest/graphs/ (accessed: 2019-07).

	Scaling	PFS	PFS'	FTA	VPSC	PRISM	RWordle-L	GTREE
oo_nni	0.00	0.00	0.00	0.07	0.01	0.02	0.04	0.02
sp_ch_a	217.26	210.28	6.92	6.03	2.36	2.18	1.53	4.02
gs_bb_iar	1.04	1.97	1.22	1.95	2.20	1.33	1.04	1.17
nm_dm_imse	0.00	9768157.94	63594.74	4297355.84	34611.40	42919.66	35928.91	37331.83
eb_rsdd	0.00	0.39	0.28	0.27	0.34	0.37	0.50	0.35

Fig. 2. Mean values of the selected metrics among the real-world graphs.

Scaling preserves the initial global shape[2] (*gs_bb_iar* score). *PFS* is the worst algorithm on this criterion. The other algorithms obtained good median scores on synthetic graphs, but the third quartile scores show that *FTA* and *VPSC* can produce a certain amount of distorted embeddings. This is confirmed by the tests on real-world graphs, where they obtain worse results.

Scaling obtains the best results for the node movement minimisation criterion, followed by *VPSC* and *RWordle-L*. *FTA* also obtained a good median score on synthetic graphs, but its third quartile value shows that it can generate a certain amount of embeddings with high changes, as also illustrated by the bad score obtained on the real-world graphs. *PFS'* and *PRISM* obtained intermediary results. Finally, *GTREE* had bad results on the synthetic graphs, while it obtained pretty good ones on the real-world graphs.

Scaling preserves relative edge lengths. All the other criteria obtained comparable median score between 0.17 and 0.31. However, the third quartile on the synthetic graphs shows that *FTA*, *VPSC* and *RWordle-L* generate a certain amount of embeddings with high variations. This observation is confirmed by the results on the real-world graphs for *FTA* and *RWordle-L*.

4.2 Computation Time

Figure 3 and 4 show the aggregated running time values on the synthetic and real-world datasets. *Scaling*, *PFS*, *PFS'* and *VPSC* require lower running time. *FTA* and *GTREE* induce intermediate running time, but the third quartile shows that *FTA* can induce a certain amount of time consuming embedding computations. Finally, *PRISM* is time consuming for small graphs, but have intermediate results for larger graphs, while *RWordle-L* has good results for small graphs but is very time-consuming for larger ones.

---

[2] The global shape preservation score for *Scaling* is not 1 because of the size of the nodes that remains the same between the initial and the overlapping-free embeddings.

		Scaling	PFS	PFS'	FTA	VPSC	PRISM	RWordle-L	GTREE
	Q1	0.00	0.00	0.00	0.00	0.00	3.00	0.00	0.00
10	Median	0.00	0.00	0.00	0.00	0.00	4.00	0.00	1.00
	Q3	0.00	0.00	0.00	0.00	0.00	12.00	0.00	3.00
	Q1	0.00	0.00	0.00	0.00	0.00	4.00	0.00	1.00
20	Median	0.00	0.00	0.00	0.00	0.00	6.00	0.00	1.00
	Q3	0.00	0.00	0.00	0.00	1.00	5.00	1.00	4.00
	Q1	0.00	0.00	0.00	0.00	1.00	7.00	0.00	2.00
50	Median	0.00	0.00	0.00	1.00	1.00	16.00	3.00	5.00
	Q3	0.00	0.00	1.00	2.00	2.00	41.25	9.25	8.25
	Q1	0.00	0.00	1.00	1.00	2.00	25.75	0.00	8.00
100	Median	0.00	1.00	1.00	4.00	3.00	44.00	36.50	13.00
	Q3	1.00	1.00	1.00	12.00	4.00	69.00	84.25	26.25
	Q1	1.00	2.00	4.00	4.00	4.00	40.75	2.00	19.00
200	Median	1.00	2.00	4.00	25.50	10.50	81.50	274.00	30.00
	Q3	2.00	3.00	5.00	89.00	15.00	134.25	589.50	47.00
	Q1	2.00	15.00	25.00	26.00	25.00	196.75	21.00	59.50
500	Median	10.00	17.00	29.50	207.50	93.00	308.00	3410.00	109.00
	Q3	13.00	20.00	35.00	1392.00	140.75	487.25	7246.33	174.75
	Q1	4.75	57.00	113.50	87.75	125.50	623.75	112.00	165.00
1000	Median	34.50	67.50	133.00	1022.00	496.00	1053.00	26877.50	297.00
	Q3	58.25	77.00	176.25	8694.00	1065.75	1440.00	55173.25	418.75

Fig. 3. Aggregated running times among the synthetic graphs, function of number of nodes (10 to 1,000): first quartile, median and third quartile.

	Scaling	PFS	PFS'	FTA	VPSC	PRISM	RWordle-L	GTREE
b100	41	82	126	19823	321	3750	191109	285
b102	5	16	24	82	15	229	96	33
b124	0	0	0	3	1	28	1	6
b143	0	0	1	9	2	42	6	11
badvoro	20	43	63	32593	383	1512	56155	213
dpd	0	0	0	1	1	1	0	1
mode	0	1	2	93	4	172	281	34
NaN	0	1	1	2	0	17	0	7
ngk10_4	0	0	0	0	0	9	0	2
root	7	24	49	10866	85	2301	43169	192
rowe	1	0	0	1	0	3	0	1
size	0	0	0	1	0	18	1	2
unix	0	0	0	1	0	5	0	1
xx	1	3	5	27	7	158	60	35

Fig. 4. Mean values of running times among the real-world graphs.

5   Conclusion

As a conclusion, even if *Scaling* optimises 4 out of 5 criteria and is very fast to compute on the graphs of our datasets, it does not represent a satisfying solution as it increases the size of the embedding too much. *PFS* is also not satisfying as it got poor results on 3 criteria. *FTA* obtained intermediate results over all the criteria, which is less good than all its remaining competitors. *PFS'* and *PRISM* obtained comparable results but the latter is more time-consuming. Both have intermediate results for shape preservation and node movement minimisation, which might be considered as two essential criteria. *GTREE* suffers from inducing high node movements on our datasets. Overall, *VPSC* and *RWordle-L* obtained the best quality results. While *RWordle-L* outperforms *VPSC* on global shape preservation and is comparable on the other criteria, *VPSC* outperforms *RWordle-L* in terms of running time. Finally, considering the different types of graphs (random graphs, random trees, small world graphs, and scale-free graphs), we did not observe any significant differences in terms of results.

Acknowledgement. This research has been partly funded by a national French grant (ANR Daphne 17-CE28-0013-01).

References

1. Barabaśi, A.L., Albert, R.: Emergence of scaling in random networks. Science **286**(5439), 509–512 (1999)
2. Chimani, M., Gutwenger, C., Jünger, M., Klau, G.W., Klein, K., Mutzel, P.: The open graph drawing framework (OGDF). In: Tamassia, R. (ed.) Handbook on Graph Drawing and Visualization, pp. 543–569. Chapman and Hall/CRC, London (2013)
3. Dwyer, T., Marriott, K., Stuckey, P.J.: Fast node overlap removal. In: Healy, P., Nikolov, N.S. (eds.) GD 2005. LNCS, vol. 3843, pp. 153–164. Springer, Heidelberg (2006). https://doi.org/10.1007/11618058_15
4. Erdös, P., Rényi, A.: On random graphs. Publicationes Mathematicae Debrecen **6**, 290–291 (1959)
5. Gansner, E., Hu, Y.: Efficient, proximity-preserving node overlap removal. J. Graph Algorithms Appl. **14**(1), 53–74 (2010)
6. Gansner, E.R., North, S.C.: An open graph visualization system and its applications to software engineering. Softw. Pract. Exp. **30**(11), 1203–1233 (2000)
7. Hachul, S., Jünger, M.: Drawing large graphs with a potential-field-based multilevel algorithm. In: Pach, J. (ed.) GD 2004. LNCS, vol. 3383, pp. 285–295. Springer, Heidelberg (2005). https://doi.org/10.1007/978-3-540-31843-9_29
8. Hayashi, K., Inoue, M., Masuzawa, T., Fujiwara, H.: A layout adjustment problem for disjoint rectangles preserving orthogonal order. In: Whitesides, S.H. (ed.) GD 1998. LNCS, vol. 1547, pp. 183–197. Springer, Heidelberg (1998). https://doi.org/10.1007/3-540-37623-2_14
9. Hu, Y.: Efficient, high-quality force-directed graph drawing. Math. J. **10**(1), 37–71 (2005)
10. Huang, X., Lai, W.: Force-transfer: a new approach to removing overlapping nodes in graph layout. In: Proceedings of the 26th Australasian computer science conference, vol. 16, pp. 349–358. Australian Computer Society, Inc. (2003)
11. Huang, X., Lai, W., Sajeev, A., Gao, J.: A new algorithm for removing node overlapping in graph visualization. Inf. Sci. **177**(14), 2821–2844 (2007)
12. Li, W., Eades, P., Nikolov, N.: Using spring algorithms to remove node overlapping. In: Proceedings of the 2005 Asia-Pacific Symposium on Information Visualisation, vol. 45, pp. 131–140. APVis 2005. Australian Computer Society Inc, Darlinghurst (2005)
13. Lyons, K.A., Meijer, H., Rappaport, D.: Algorithms for cluster busting in anchored graph drawing. J. Graph Algorithms Appl. **2**(1), 1–24 (1998)

14. Marriott, K., Stuckey, P., Tam, V., He, W.: Removing node overlapping in graph layout using constrained optimization. Constraints **8**(2), 143–171 (2003)
15. Misue, K., Eades, P., Lai, W., Sugiyama, K.: Layout adjustment and the mental map. J. Vis. Lang. Comput. **6**(2), 183–210 (1995)
16. Nachmanson, L., Nocaj, A., Bereg, S., Zhang, L., Holroyd, A.: Node overlap removal by growing a tree. In: Hu, Y., Nöllenburg, M. (eds.) GD 2016. LNCS, vol. 9801, pp. 33–43. Springer, Cham (2016). https://doi.org/10.1007/978-3-319-50106-2_3
17. Strobelt, H., Spicker, M., Stoffel, A., Keim, D., Deussen, O.: Rolled-out wordles: a heuristic method for overlap removal of 2D data representatives. Comput. Graph. Forum **31**(3), 1135–1144 (2012)
18. Watts, D.J., Strogatz, S.H.: Collective dynamics of 'small-world' networks. Nature **393**, 440–442 (1998)

Graphs with Large Total Angular Resolution

Oswin Aichholzer[1], Matias Korman[2], Yoshio Okamoto[3], Irene Parada[1],
Daniel Perz[1(✉)], André van Renssen[4], and Birgit Vogtenhuber[1]

[1] Graz University of Technology, Graz, Austria
{oaich,iparada,daperz,bvogt}@ist.tugraz.at
[2] Tufts University, Medford, MA, USA
matias.korman@tufts.edu
[3] The University of Electro-Communications and RIKEN Center for Advanced
Intelligence Project, Tokyo, Japan
okamotoy@uec.ac.jp
[4] The University of Sydney, Sydney, Australia
andre.vanrenssen@sydney.edu.au

Abstract. The total angular resolution of a straight-line drawing is the minimum angle between two edges of the drawing. It combines two properties contributing to the readability of a drawing: the angular resolution, which is the minimum angle between incident edges, and the crossing resolution, which is the minimum angle between crossing edges. We consider the total angular resolution of a graph, which is the maximum total angular resolution of a straight-line drawing of this graph. We prove that, up to a finite number of well specified exceptions of constant size, the number of edges of a graph with n vertices and a total angular resolution greater than $60°$ is bounded by $2n - 6$. This bound is tight. In addition, we show that deciding whether a graph has total angular resolution at least $60°$ is NP-hard.

Keywords: Graph drawing · Total angular resolution · Angular resolution · Crossing resolution · NP-hardness

1 Introduction

The *total angular* resolution of a drawing D, or short $\text{TAR}(D)$, is the smallest angle occurring in D, either between two edges incident to the same vertex or between two crossing edges. In other words, $\text{TAR}(D)$ is the minimum of the

This work started during the Japan-Austria Joint Seminar *Computational Geometry Seminar with Applications to Sensor Networks* supported by the Japan Society for the Promotion of Science (JSPS) and the Austrian Science Fund (FWF) under grant AJS 399. O.A., I.P., D.P., and B.V. are partially supported by the FWF grants W1230 (Doctoral Program Discrete Mathematics) and I 3340-N35 (Collaborative DACH project *Arrangements and Drawings*).

D. Archambault and C. D. Tóth (Eds.): GD 2019, LNCS 11904, pp. 193–199, 2019.
https://doi.org/10.1007/978-3-030-35802-0_15

angular resolution AR(D) and the crossing resolution CR(D) of the same drawing. Furthermore, the total angular resolution of a graph G is defined as the maximum of TAR(D) over all drawings D of G. Similarly, the angular resolution and the crossing resolution of G are the maximum of AR(D) and CR(D), respectively, over all drawings D of G. The total angular resolution of a graph is in general smaller than the minimum of its crossing resolution and its angular resolution. Note that all drawings considered in this work are straight-line.

Formann et al. [8] were the first to introduce the angular resolution of graphs and showed that finding a drawing of a graph with angular resolution at least $90°$ is NP-hard. Fifteen years later experiments by Huang et al. [9,11] showed that the crossing resolution plays a major role in the readability of drawings. Consequently research in that direction was intensified. In particular right angle crossing drawings (or short RAC drawings) were studied [6,12], and NP-hardness of the decision version for right angles was proven [3].

The upper bound for the number of edges of αAC drawings (drawings with crossing resolution α) is $\frac{180°}{\alpha}(3n-6)$ [7]. For the two special classes of RAC drawings and $60°$AC drawings better upper bounds are known. More precisely, RAC drawings have at most $4n-10$ edges [6] and αAC drawings with $\alpha > 60°$ have at most $6.5n - 20$ edges [1].

Argyriou et al. [4] were the first to study the total angular resolution, calling it just *total resolution*. They presented drawings of complete and complete bipartite graphs with asymptotically optimal total angular resolution. Recently Bekos et al. [5] presented a new algorithm for finding a drawing of a given graph with high total angular resolution which was performing superior to earlier algorithms like [4,10] on the considered test cases.

2   Upper Bound on the Number of Edges

We say a drawing D is *planarized* if we replace every crossing by a vertex so that this new vertex splits both crossing edges into two edges. We denote this planarized drawing by $P(D)$. Furthermore, every edge in $P(D)$ has two sides and every side is incident to exactly one cell of D. Note that both sides of an edge can be incident to the same cell. We define the size of a cell of a connected drawing D as the number of sides in $P(D)$ incident to this cell.

In this section we show that for almost all graphs with TAR(G) > $60°$ the number of edges is bounded by $2n-6$. We start by showing a bound for the number of edges in a connected drawing D depending on the size of the unbounded cell of D.

Lemma 1. *Let D be a connected drawing with $n \geq 1$ vertices and m edges. If the unbounded cell of D has size k and TAR(D) > $60°$, then $m \leq 2n-2-\lceil k/2 \rceil$.*

Proof. If at least three edges cross each other in a single point, then there exists an angle with at most $60°$ at this crossing point. Therefore every crossing is incident to two edges. We planarize the drawing D and get $n' = n + \mathrm{cr}(D)$ and $m' = m + 2\,\mathrm{cr}(D)$ where $\mathrm{cr}(D)$ is the number of crossings in D, n' is the number of vertices of $P(D)$, and m' is the number of edges of $P(D)$. Since we have a planar graph, we can use Euler's formula to compute the number f of faces in $P(D)$ as

$$f = -n + m + \mathrm{cr}(D) + 2. \tag{1}$$

Moreover, every bounded cell of D has at least size 4, as otherwise $P(D)$ contains a triangle which implies an angle of at most $60°$. By definition, the unbounded cell of D has size k and we obtain the following inequality

$$2m' \geq 4(f-1) + k. \tag{2}$$

Combining Equation (1) and Inequality (2) gives $m \leq 2n - 2 - \lceil k/2 \rceil$.   □

From Lemma 1 it follows directly that a connected drawing D on $n \geq 3$ vertices and with $\mathrm{TAR}(D) > 60°$ fulfills $m \leq 2n - 4$.

Observation 1, which will be useful to prove Lemma 2, follows from the fact that the sum of interior angles in a simple polygon is $180°(p-2)$.

Observation 1. *Let D be a plane drawing where the boundary of the unbounded cell is a simple polygon P with $p > 3$ vertices. Let the inner degree of a vertex v_i of P be the number d'_i of edges incident to v_i that lie in the interior of P. If $\mathrm{TAR}(D) > 60°$, then $\sum_{v_i \in V(P)} d'_i \leq 2p - 7$ holds.*

Lemma 2. *Let D be a connected plane drawing on $n \geq 3$ vertices, where D is not a path on 3 vertices and not a 4-gon. If $\mathrm{TAR}(D) > 60°$, then $m \leq 2n - 5$.*

Proof. The unbounded cell of D cannot have size 3, as in this case the convex hull of the drawing is a triangle and we have $\mathrm{TAR}(D) \leq 60°$. If the drawing D has an unbounded cell of size at least 5 and $\mathrm{TAR}(D) > 60°$, then $m \leq 2n - 5$ follows directly from Lemma 1. Otherwise, the unbounded cell of D has size 4, which, as D is not a path on 3 vertices, implies that the boundary of D is a 4-gon F. By Observation 1 and the fact that D is not a 4-cycle, there is precisely one edge e in the interior of and incident to F. Let D' be the drawing we get by deleting all vertices and edges of F and also the edge e. The drawing D' is connected and has $n' \geq 1$ vertices and m' edges, where $n = n' + 4$ and $m = m' + 5$. By Lemma 1 we know that $m' \leq 2n' - 2$ and we derive $m = m' + 5 \leq 2n' - 2 + 5 \leq 2n - 5$. □

Two drawings are *combinatorially equivalent* if all cells are bounded by the same edges, all crossing edge pairs are the same, and the order of crossings along an edge are the same. We can extend Lemma 2 in the following way.

Lemma 3. *Let D be a connected plane drawing on $n \geq 3$ vertices with $\mathrm{TAR}(D) > 60°$. If D is not combinatorially equivalent to one of the exceptions E1–E9 as listed below and depicted in Appendix B of [2], then $m \leq 2n - 6$.*

E1 *A tree on at most 4 vertices.*
E2 *An empty 4-gon.*
E3 *A 4-gon with one additional vertex connected to one vertex of the 4-gon.*

Fig. 1. (a) The drawings of exception E9. (b) A drawing D of a graph with $m=2n-6$ and $\mathrm{TAR}(D) > 60°$.

E4 *An empty 5-gon.*

E5 *A 5-gon with one inner vertex connected to two non-neighboring vertices of the 5-gon.*

E6 *A 5-gon with an edge inside, connected with 3 edges to the 5-gon such that the 5-gon is partitioned into two empty 4-gons and one empty 5-gon.*

E7 *A 6-gon with an additional diagonal between opposite vertices.*

E8 *A 6-gon with an additional vertex or edge inside, connected with 3 or 4, respectively, edges to the 6-gon such that the 6-gon is partitioned into 3 or 4, respectively, empty 4-gons.*

E9 *A 6-gon with either a path on 3 vertices or a 4-cycle inside, connected as depicted also in Fig. 1(a).*

The proof of Lemma 3 is similar to the one of Lemma 2 and can be found in Appendix A of [2]. Note that Lemma 3 considers plane drawings. If D has a crossing, then $P(D)$ has a vertex of degree 4. The only drawings in the exceptions with a vertex with degree 4 are shown in Fig. 1(a). It can be shown that, when replacing the vertices of degree 4 in any of them by a crossing, the resulting drawings have $\mathrm{TAR}(D) \leq 60°$. A detailed proof of this fact can be found in Appendix C of [2] and will be useful for the proof of the next theorem.

Theorem 1. *Let G be a graph with $n \geq 3$ vertices, m edges and $\mathrm{TAR}(G) > 60°$. Then $m \leq 2n - 6$ except if G is either a graph of an exception for Lemma 3 or only consists of three vertices and one edge.*

Proof. Assume there exists a graph which is not in the list of exceptions for Lemma 3 with $\mathrm{TAR}(G) > 60°$. Consider a drawing D of G with $\mathrm{TAR}(D) > 60°$ and its planarization $P(D)$.

Applying Lemma 1 to every component gives $m \leq 2m - 6$, with the only exception consisting of three vertices and one edge (Exception E0). For details see Appendix D of [2]. So for the rest of the proof only consider connected graphs.

If three edges cross in a single point, then in $P(D)$ this point has degree 6 and therefore an angle with at most 60°. Hence $P(D)$ has $m_P = m + 2\,\mathrm{cr}(D)$ edges and $n_P = n + \mathrm{cr}(D)$ vertices. Let $m = 2n - c$. This is equivalent to $m_P = 2n_P - c$. Since $\mathrm{TAR}(P(D)) \geq \mathrm{TAR}(D) > 60°$, by applying Lemma 3 we get that $m_P \leq 2n_P - 6$ or $P(D)$ is in the exceptions. If $m_P \leq 2n_P - 6$, then also $m \leq 2n - 6$. If $P(D)$ is in the exceptions, then, as observed before, D is in the exceptions. □

The bound of Theorem 1 is the best possible in the sense that there are infinitely many graphs with $m = 2n - 6$ and $\text{TAR}(G) > 60°$. Consider for example the layered 8-gon with two edges in the middle depicted in Fig. 1(b), which can be generalized to any $n = 8k$ with $k \in \mathbb{N}$. In the full version of this work we present examples for every $n \geq 9$ and also discuss plane drawings of planar graphs.

3   NP-hardness

Formann et al. [8] showed that the problem of determining whether there exists a drawing of a graph with angular resolution of $90°$ is NP-hard. Their proof, which is by reduction from 3SAT with exactly three different literals per clause, also implies NP-hardness of deciding whether a graph has a drawing with total angular resolution of $90°$. We adapt their reduction to show NP-hardness of the decision problem for $\text{TAR}(G) \geq 60°$. A full version of the proof of Theorem 2 can be found in Appendix E of [2].

Theorem 2. *It is* NP-*hard to decide whether a graph G has* $\text{TAR}(G) \geq 60°$.

Proof (sketch). Given a 3SAT formula with variables $x_1, x_2, \ldots, x_n$ and clauses $c_1, c_2, \ldots, c_m$, where every clause contains exactly three different literals, we first construct a graph G for it. The basic building blocks of G consist of triangles, which must be equilateral in any drawing with total angular resolution $60°$.

We use three types of gadgets; see Fig. 2(a). The clause gadget has a designated *clause vertex* C_j and the variable gadget has two *literal vertices* $X_{i,j}, \overline{X}_{i,j}$ per clause c_j. For each gadget, the embedding with total angular resolution $60°$ is unique up to rotation, scaling and reflection.

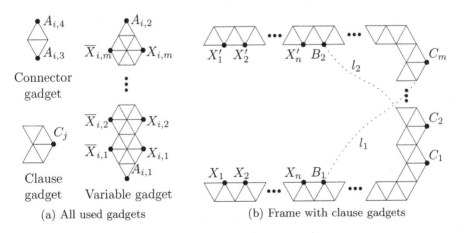

(a) All used gadgets   (b) Frame with clause gadgets

Fig. 2. Gadgets and frame of the NP-hardness proof.

For connecting the gadgets, we build a 3-sided frame; see Fig. 2(b). It consists of a straight *bottom path* of $2n + 2m - 1$ triangles alternatingly facing up and

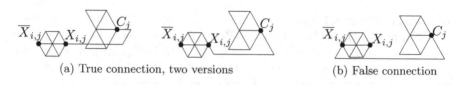

(a) True connection, two versions   (b) False connection

Fig. 3. Connections between clause and literal vertices in the NP-hardness proof.

down, a sequence of m clause gadgets stacked on top of each other to the right
(one for each clause, with the clause vertices $C_1, \ldots, C_m$ facing to the right),
and a *top path* of $2n + 2m - 1$ triangles alternatingly facing down and up.
The leftmost n vertices of degree three on the upper side of the bottom path
and the lower side of the top path ($X_1, \ldots, X_n$ and $X_1', \ldots, X_n'$) are used for
the variables: For each variable x_i, we add a variable gadget and a connector
gadget by identifying $A_{i,1}$ with X_i, $A_{i,2}$ with $A_{i,3}$, and $A_{i,4}$ with X_i', respectively.
Finally, a clause-literal path consisting of three consecutive edges between $X_{i,j}$
($\overline{X}_{i,j}$) and C_j is added whenever x_i ($\overline{x}_i$) is a literal of clause c_j.

The following holds for any drawing D of the graph G with $TAR(D) \geq 60°$.
(1) The embedding of the frame is unique up to rotation, scaling, and reflection.
Hence we can assume that it is embedded as in Fig. 2(b). *(2)* Each variable
gadget together with its connector gadget must be drawn vertically between its
X_i and X_i', either with all $X_{i,j}$ to the right of the $\overline{X}_{i,j}$ or the other way around.
(3) All clause-literal paths leave from their clause vertices to the right, and one
path per clause leaves horizontally to the right.

We claim that $TAR(G) \geq 60°$ if and only if the initial 3SAT formula is
satisfiable. For the one direction, consider a satisfying truth assignment of the
formula. We draw the variable gadgets with all true literal sides to the right and
scaled (via the connector gadgets) such that different gadgets have their vertices
at different heights, and we draw the clause-literal paths as indicated in Fig. 3.
For the other direction, consider a drawing of C with $TAR(D) = 60°$. Using the
straight lines ℓ_1 and ℓ_2 sketched in Fig. 2(b), one can show that every clause-
literal path that leaves the clause vertex horizontally must end at a literal vertex
facing to the right. Setting the according literals to true gives a non-contradicting
variable assignment that in turn fulfills all clauses.   □

4   Conclusion

In this work we have shown that, up to a finite number of well specified exceptions of constant size, any graph G with $TAR(G) > 60°$ has at most $2n - 6$
edges. In addition we have been able to obtain similar bounds for graphs with
$TAR(G) \geq 90°$ and $TAR(G) > 120°$: For graphs with $TAR(G) \geq 90°$ we have
$m \leq 2n - 2\sqrt{n}$ and for $TAR(G) > 120°$ we have $m \leq n$ for $n \geq 7$, which is best
possible. We conjecture that almost all graphs with $TAR(G) > \frac{k-2}{k}90°$ have at
most $2n - 2 - \lfloor \frac{k}{2} \rfloor$ edges.

From a computational point of view, we have proven that finding a drawing of a given graph with total angular resolution at least 60° is NP-hard. The same was known before for at least 90° [8]. On the other hand, for large angles, the recognition problem eventually becomes easy (for example, G can be drawn with TAR(G) > 120° if and only if it is the union of cycles of at least 7 vertices and arbitrary paths). This yields the following open problem: At which angle(s) does the decision problem change from NP-hard to polynomially solvable?

References

1. Ackerman, E., Tardos, G.: On the maximum number of edges in quasi-planar graphs. J. Comb. Theory Ser. A **114**, 563–571 (2007). https://doi.org/10.1016/j.jcta.2006.08.002

2. Aichholzer, O., Korman, M., Okamoto, Y., Parada, I., Perz, D., van Renssen, A., Vogtenhuber, B.: Graphs with large total angular resolution (2019). https://arxiv.org/abs/1908.06504v1

3. Argyriou, E.N., Bekos, M.A., Symvonis, A.: The straight-line RAC drawing problem Is NP-hard. In: Černá, I., et al. (eds.) SOFSEM 2011. LNCS, vol. 6543, pp. 74–85. Springer, Heidelberg (2011). https://doi.org/10.1007/978-3-642-18381-2_6

4. Argyriou, E.N., Bekos, M.A., Symvonis, A.: Maximizing the total resolution of graphs. Comput. J. **56**(7), 887–900 (2013). https://doi.org/10.1093/comjnl/bxs088

5. Bekos, M.A., Förster, H., Geckeler, C., Holländer, L., Kaufmann, M., Spallek, A.M., Splett, J.: A heuristic approach towards drawings of graphs with high crossing resolution. In: Biedl, T., Kerren, A. (eds.) GD 2018. LNCS, vol. 11282, pp. 271–285. Springer, Cham (2018). https://doi.org/10.1007/978-3-030-04414-5_19

6. Didimo, W., Eades, P., Liotta, G.: Drawing graphs with right angle crossings. Theoret. Comput. Sci. **412**(39), 5156–5166 (2011). https://doi.org/10.1016/j.tcs.2011.05.025

7. Dujmovic, V., Gudmundsson, J., Morin, P., Wolle, T.: Notes on large angle crossing graphs. Chic. J. Theor. Comput. Sci. **4**, 1–14 (2011). https://doi.org/10.4086/cjtcs.2011.004

8. Formann, M., Hagerup, T., Haralambides, J., Kaufmann, M., Leighton, F.T., Symvonis, A., Welzl, E., Woeginger, G.J.: Drawing graphs in the plane with high resolution. SIAM J. Comput. **22**, 1035–1052 (1993). https://doi.org/10.1137/0222063

9. Huang, W.: Using eye tracking to investigate graph layout effects. In: 2007 6th International Asia-Pacific Symposium on Visualization, pp. 97–100. IEEE (2007). https://doi.org/10.1109/APVIS.2007.329282

10. Huang, W., Eades, P., Hong, S.H., Lin, C.: Improving multiple aesthetics produces better graph drawings. J. Vis. Lang. Comput. **24**(4), 262–272 (2013). https://doi.org/10.1016/j.jvlc.2011.12.002

11. Huang, W., Hong, S.H., Eades, P.: Effects of crossing angles. In: 2008 IEEE Pacific Visualization Symposium, pp. 41–46 (2008). https://doi.org/10.1109/PACIFICVIS.2008.4475457

12. Kreveld, M.: The quality ratio of RAC drawings and planar drawings of planar graphs. In: Brandes, U., Cornelsen, S. (eds.) GD 2010. LNCS, vol. 6502, pp. 371–376. Springer, Heidelberg (2011). https://doi.org/10.1007/978-3-642-18469-7_34

Arrangements

Computing Height-Optimal Tangles Faster

Oksana Firman[1]($\boxtimes$), Philipp Kindermann[1], Alexander Ravsky[2],
Alexander Wolff[1], and Johannes Zink[1]

[1] Institut für Informatik, Universität Würzburg, Würzburg, Germany
oksana.firman@uni-wuerzburg.de
[2] National Academy of Sciences of Ukraine, Lviv, Ukraine
alexander.ravsky@uni-wuerzburg.de

Abstract. We study the following combinatorial problem. Given a set
of n y-monotone *wires*, a *tangle* determines the order of the wires on
a number of horizontal *layers* such that the orders of the wires on any
two consecutive layers differ only in swaps of neighboring wires. Given
a multiset L of *swaps* (that is, unordered pairs of numbers between 1
and n) and an initial order of the wires, a tangle *realizes* L if each pair of
wires changes its order exactly as many times as specified by L. The aim
is to find a tangle that realizes L using the smallest number of layers.
We show that this problem is NP-hard, and we give an algorithm that
computes an optimal tangle for n wires and a given list L of swaps in
$O((2|L|/n^2 + 1)^{n^2/2} \cdot \varphi^n \cdot n)$ time, where $\varphi \approx 1.618$ is the golden ratio.
We can treat lists where every swap occurs at most once in $O(n!\varphi^n)$
time. We implemented the algorithm for the general case and compared
it to an existing algorithm. Finally, we discuss feasibility for lists with a
simple structure.

1 Introduction

The subject of this paper is the visualization of so-called *chaotic attractors*,
which occur in chaotic dynamic systems. Such systems are considered in physics,
celestial mechanics, electronics, fractals theory, chemistry, biology, genetics, and
population dynamics. Birman and Williams [3] were the first to mention tangles
as a way to describe the topological structure of chaotic attractors. They investigated how the orbits of attractors are knotted. Later Mindlin et al. [6] showed
how to characterize attractors using integer matrices that contain numbers of
swaps between the orbits. Our research is based on a recent paper of Olszewski
et al. [7]. In the framework of their paper, one is given a set of wires that hang
off a horizontal line in a fixed order, and a multiset of swaps between the wires;
a tangle then is a visualization of these swaps, i.e., an order in which the swaps
are performed, where only adjacent wires can be swapped and disjoint swaps
can be done simultaneously. For an example of a list of swaps (described by an

© Springer Nature Switzerland AG 2019
D. Archambault and C. D. Tóth (Eds.): GD 2019, LNCS 11904, pp. 203–215, 2019.
https://doi.org/10.1007/978-3-030-35802-0_16

$(n \times n)$-matrix) and a tangle that realizes this list, see Fig. 1. Olszewski et al. gave an algorithm for minimizing the height of a tangle. They didn't analyze the asymptotic running time of their algorithm (which we estimate below), but tested it on a benchmark set.

Wang [8] used the same optimization criterion for tangles, given only the final permutation. She showed that there is always a height-optimal tangle where no swap occurs more than once. She used odd-even sort, a parallel variant of bubble sort, to compute tangles with at most one layer more than the minimum. Bereg et al. [1,2] considered a similar problem. Given a final permutation, they showed how to minimize the number of bends or *moves* (which are maximal "diagonal" segments of the wires).

Framework, Terminology, and Notation. We modify the terminology of Olszewski et al. [7] in order to introduce a formal algebraic framework for the problem. Given n wires, a *(swap) list* $L = (l_{ij})$ of *order* n is a symmetric $n \times n$ matrix with non-negative entries and zero diagonal. The *length* of L is $|L| = \sum_{i<j} l_{ij}$. A list $L' = (l'_{ij})$ is a *sublist* of L if $l'_{ij} \leq l_{ij}$ for each $i, j \in [n]$. A list is *simple* if all its entries are zeros or ones.

A *permutation* is a bijection of the set $[n] = \{1, \ldots, n\}$ onto itself. The set S_n of all permutations of the set $[n]$ is a group whose multiplication is a composition of maps (i.e., $(\pi\sigma)(i) = \pi(\sigma(i))$ for each pair of permutations $\pi, \sigma \in S_n$ and each $i \in [n]$). The identity of the group S_n is the identity permutation id_n. We write a permutation $\pi \in S_n$ as the sequence of numbers $\pi^{-1}(1)\pi^{-1}(2)\ldots\pi^{-1}(n)$. For instance, the permutation π of [4] with $\pi(1) = 3$, $\pi(2) = 4$, $\pi(3) = 2$, and $\pi(4) = 1$ is written as 4312. We denote the set of all permutations of order 2 in S_n by $S_{n,2}$, that is, $\pi \in S_{n,2}$ if and only if $\pi\pi = \mathrm{id}_n$ and $\pi \neq \mathrm{id}_n$. For example, $2143 \in S_{4,2}$.

For $i, j \in [n]$ with $i \neq j$, the *swap* ij is the permutation that exchanges i and j, whereas the other elements of $[n]$ remain fixed. A set S of swaps is *disjoint* if each element of $[n]$ participates in at most one swap of S. Therefore, the product $\prod S$ of all elements of a disjoint set S of swaps does not depend on the order of factors and belongs to $S_{n,2}$. Conversely, for each permutation $\varepsilon \in S_{n,2}$ there exists a unique disjoint set $S(\varepsilon)$ of swaps such that $\varepsilon = \prod S(\varepsilon)$.

A permutation $\pi \in S_n$ *supports* a permutation $\varepsilon \in S_{n,2}$ if, for each swap $ij \in S(\varepsilon)$, i and j are neighbors in the sequence π. By induction with respect to n, we can easily show that any permutation $\pi \in S_n$ supports exactly $F_{n+1} - 1$ permutations of order 2, where F_n is the n-th number in the Fibonacci sequence.

Permutations π and σ are *adjacent* if there exists a permutation $\varepsilon \in S_{n,2}$ such that π supports ε and $\sigma = \pi\varepsilon$. In this case, $\sigma\varepsilon = \pi\varepsilon\varepsilon = \pi$ and σ supports ε, too. A *tangle* T of *height* h is a sequence $\langle \pi_1, \pi_2, \ldots, \pi_h \rangle$ of permutations in which every two consecutive permutations are adjacent. A tangle can also be viewed as a sequence of $h - 1$ *layers*, each of which is a set of disjoint swaps. A *subtangle* of T is a sequence $\langle \pi_k, \pi_{k+1}, \ldots, \pi_\ell \rangle$ of consecutive permutations of T. For a tangle T, we define $L(T) = (l_{ij})$ as the symmetric $n \times n$ matrix with zero diagonal, where l_{ij} is the number of occurrences of swap ij in T. We say

$$L_n = \begin{pmatrix} 0 & 1 & 1 & \dots & 1 & 0 & 2 \\ 1 & 0 & 1 & \dots & 1 & 2 & 0 \\ 1 & 1 & 0 & \dots & 1 & 0 & 2 \\ \vdots & \vdots & \vdots & \ddots & \vdots & \vdots & \vdots \\ 1 & 1 & 1 & \dots & 0 & \mathbf{0} & 2 \\ 0 & 2 & 0 & \dots & \mathbf{0} & 0 & n-1 \\ 2 & 0 & 2 & \dots & 2 & n-1 & 0 \end{pmatrix}$$

(The bold zeros and twos must be
swapped if n is even.)

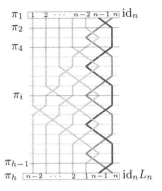

Fig. 1. A list L_n for n wires and a tangle of minimum height $h = 3n - 4$ realizing L_n for id_n. Here, $n = 7$. The tangle is not simple because $\pi_2 = \pi_4$.

that T *realizes* the list $L(T)$. To make the reader familiar with our formalism, we observe the following.

Observation 1. *The tangle in Fig. 1 realizes the list L_n specified there; all tangles that realize L_n have the same order of swaps along each wire.*

Proof. For $i, j \in [n-2]$ with $i \neq j$, the wires i and j swap exactly once, so their order reverses. Additionally, each wire $i \in [n-2]$ swaps twice with the wire $k \in \{n-1, n\}$ that has the same parity as i. Observe that wire $i \in [n-2]$ must first swap with each $j \in [n-2]$ with $j > i$, then twice with the correct $k \in \{n-1, n\}$, say $k = n$, and finally with each $j' \in [n-2]$ with $j' < i$. Otherwise, if i swaps with $i - 1$ before swapping with n, then i cannot reach n because $i - 1$ swaps only with $n - 1$ among the two wires $\{n-1, n\}$ and thus separates i from n. This establishes the unique order of swaps along each wire. □

A list is π-*feasible* if it can be realized by a tangle starting from a permutation π. An id_n-feasible list is *feasible*. For example, the list defined by the two swaps 13 and 24 is *not* feasible.

By E, we denote the (simple) list $E = (e_{ij})$ with $e_{ij} = 1$ if $i \neq j$, and $e_{ij} = 0$ otherwise. This list is feasible for any permutation; a tangle realizing E is commonly known as *pseudo-line arrangement*. So tangles can be thought of as generalizations of pseudo-line arrangements where the numbers of swaps are prescribed and even feasibility becomes a difficult question.

A list $L = (l_{ij})$ can also be considered a multiset of swaps, where l_{ij} is the multiplicity of swap ij. By $ij \in L$ we mean $l_{ij} > 0$. A tangle is *simple* if all its permutations are distinct. Note that the height of a simple tangle is at most $n!$.

The *height* $h(L)$ of a feasible list L is the minimum height of a tangle that realizes L. A tangle T is *optimal* if $h(T) = h(L(T))$. In the TANGLE-HEIGHT MINIMIZATION problem, we are given a swap list L and the goal is to compute an optimal tangle T realizing L. As initial wire order, we always assume id_n.

Our Contribution. We show that TANGLE-HEIGHT MINIMIZATION is NP-hard (see Sect. 2). We give an exact algorithm for simple lists running in $O(n!\varphi^n)$ time, where $\varphi = \frac{\sqrt{5}+1}{2} \approx 1.618$ is the golden ratio, and an exact algorithm for general lists running in $O((2|L|/n^2+1)^{n^2/2}\varphi^n n)$ time, which is polynomial in $|L|$ for fixed $n \geq 2$ (see Sect. 3). We implemented the algorithm for general lists and compared it to the algorithm of Olszewski et al. [7] using their benchmark set (see Sect. 4). We show that the asymptotic runtimes of the algorithms of Olszewski et al. [7] for simple and for general lists are $O(\varphi^{2|L|}5^{-|L|/n}n)$ and $2^{O(n^2)}$, respectively. In Sect. 5, we discuss feasibility for lists with simple structure.

2   Complexity

We show the NP-hardness of TANGLE-HEIGHT MINIMIZATION by reduction from 3-PARTITION. An instance of 3-PARTITION is a multiset A of $3m$ positive integers $n_1, \ldots, n_{3m}$, and the task is to decide whether A can be partitioned into m groups of three elements each that all sum up to the same value $B = \sum_{i=1}^{3m} n_i/m$. 3-PARTITION remains NP-hard if restricted to instances where B is polynomial in m, and $B/4 < n_i < B/2$ for each $i \in [3m]$ [5]. We reduce from this version.

Theorem 1. *The decision version of* TANGLE-HEIGHT MINIMIZATION *is NP-hard.*

Proof. Given an instance A of 3-PARTITION, we construct in polynomial time a list L of swaps such that there is a tangle T realizing L with height at most $H = 2m^4B + 7m^2$ if and only if A is a yes-instance of 3-PARTITION.

In L, we use two inner wires ω and ω' with $\omega' = \omega + 1$ that swap $2m$ times. Thus, in a tangle realizing L, ω and ω' provide a twisted structure with $m + 1$ "loops" of ω and ω' (ω on the left side and ω' on the right side) and m "loops" of ω' and ω (ω' on the left side and ω on the right side). We call them ω–ω' *loops* and ω'–ω *loops*, respectively. The first ω–ω' loop is *open*, that is, it is bounded by the start permutation and the first ω–ω' swap. Symmetrically, the last ω–ω' loop is open. All other ω–ω' loops and all ω'–ω loops are *closed*, that is, they are bounded by two consecutive ω–ω' swaps. Apart from ω and ω', the list L uses three different types of wires. Refer to Fig. 2 for an illustration.

We use the first type of wires of L to represent the numbers in A. To this end, we introduce wires $\alpha_1, \alpha_2, \ldots, \alpha_{3m}$, which we call α-wires, and wires $\alpha'_1, \alpha'_2, \ldots, \alpha'_{3m}$, which we call α'-wires. Initially, these wires are ordered $\alpha_{3m} < \cdots < \alpha_1 < \omega < \omega' < \alpha'_1 < \cdots < \alpha'_{3m}$. For each $i \in [3m]$, we have $2m^3n_i$ swaps α_i–α'_i. We use the factor 2 in the number of α_i–α'_i swaps to make the initial permutation and the final permutation of this part the same. The factor m^3 helps us to prove the correctness because it dominates the number of *intermediate swaps*, which are swaps that cannot occur on the same layer as any α_i–α'_i swap. The intermediate swaps together will require a total height of only $O(m^2)$. Clearly, all ω–ω' swaps are intermediate swaps, but we will identify more below.

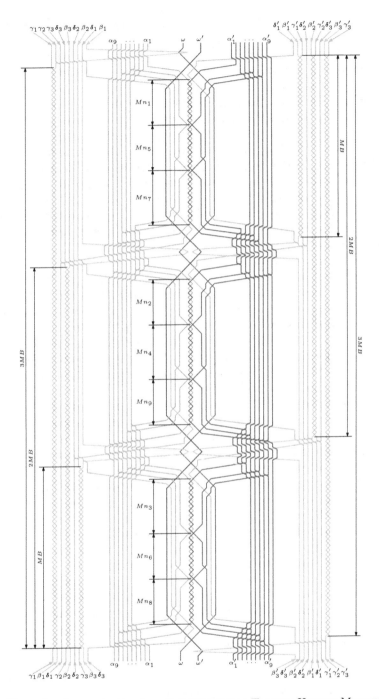

Fig. 2. Example of our reduction from 3-PARTITION to TANGLE-HEIGHT MINIMIZATION with $A_1 = \{n_1, n_5, n_7\}$, $A_2 = \{n_2, n_4, n_9\}$, $A_3 = \{n_3, n_6, n_8\}$, $m = 3$, $B = \sum_{i=1}^{3m} n_i/m$, and $M = 2m^3$.

We now argue why no two α_i–α_i' swaps can appear on the same layer. Clearly, the same swap cannot appear multiple times on the same layer. Also, there cannot be two swaps α_i–α_i' and α_j–α_j' with $i \neq j$ on the same layer because L does not contain any swap α_i–α_j' or α_j–α_i'. For the α-wires and the α'-wires to swap with each other, for each $i \in [3m]$, L has two α_i–ω' swaps and two α_i'–ω swaps, but no α_i–ω swaps and no α_i'–ω' swaps. Therefore, α_i–α_i' swaps can only occur within ω'–ω loops. Every pair of α-wires swaps twice, and so does every pair of α'-wires. This allows each α-wire to once pass all α-wires to its right in order to reach an ω'–ω loop, and then to go back. Observe that the order in which the α-wires do this is not fixed. Note that some of the α–ω' and α'–ω swaps are intermediate swaps that are needed for the α- and α'-wires to enter and to leave the ω'–ω loops.

Using the second type of wires, we now build a rigid structure around the ω–ω' loops. We use the construction of Fig. 1 on both sides of the wires ω and ω', as follows. For each $i \in [m]$, we introduce wires β_i, δ_i and β_i', δ_i' such that $\delta_m < \beta_m < \cdots < \delta_1 < \beta_1 < \alpha_{3m}$ and $\alpha_{3m}' < \delta_1' < \beta_1' < \cdots < \delta_m' < \beta_m'$. On each side, every pair of wires of the second type swaps exactly once – as the green wires in Fig. 1. Hence, in the final permutation, their order is reversed on both sides. For every $i \in [m]$, each of the wires β_i and δ_i' has two swaps with ω and each of the wires δ_i and β_i' has two swaps with ω'. To allow them to pass the α-wires, each β- and each δ-wire swaps twice with each α-wire. The same holds on the right-hand side for the α'-, β'- and δ'-wires. Note that this does not restrict the choice of the ω'–ω loops where the α_i–α_i' swaps take place. This is important for the correctness of our reduction.

Further note that some of the swaps of the β- and δ-wires with the wires ω, ω', and the α-wires are intermediate swaps. For example, β_1 has to swap with all α-wires and twice with the wire ω before any swap of an α- and an α'-wire can occur. Accordingly, some of the swaps of the β'- and δ'-wires with ω, ω', and the α'-wires are intermediate swaps as well. Still, it is obvious that the number of layers needed to accommodate all intermediate swaps is $O(m^2)$.

We denote the third type of wires by γ_i, γ_i' for $i \in [m]$. On the left side, the γ-wires are initially on the far left, that is, we set $\gamma_1 < \cdots < \gamma_m < \delta_m$. In the final permutation π, these γ-wires end up in between the β- and δ-wires in the order $\pi(\gamma_1) < \pi(\beta_1) < \pi(\delta_1) < \cdots < \pi(\gamma_m) < \pi(\beta_m) < \pi(\delta_m)$. On the right side, the γ'-wires start in a similarly interwoven configuration: $\delta_1' < \beta_1' < \gamma_1' < \cdots < \delta_m' < \beta_m' < \gamma_m'$. The γ'-wires end up in order on the far right; see Fig. 2.

To ensure that each ω'–ω loop has a fixed minimum height, we introduce many swaps between the γ- and β-wires, and between the γ'- and β'-wires: For $i \in [m]$, every γ_i has $(m-i+1) \cdot 2m^3 B$ swaps with β_i, and every γ_i' has $i \cdot 2m^3 B$ swaps with β_i'. Additionally, every γ_i has one swap with every β_j and δ_j with $j < i$, and every γ_i' has one swap with every β_j' and δ_j' with $j > i$. Recall that the subinstance of L induced by $\delta_m, \beta_m, \ldots, \delta_1, \beta_1, \omega, \omega'$ is the same as the instance L_n with wires $1, 2, \ldots, n$ in Fig. 1. Observe that, for any realization of the list L_n, the order of the swaps along each wire is the same as in the tangle on the right side. Therefore, by Observation 1, no γ_i–β_i swap is above the i-th ω–ω'

loop; see Fig. 2. (Recall that we start counting from the first open ω–ω' loop.) Accordingly, no γ_i'–β_i' swap is below the $(i + 1)$-th ω–ω' loop. Since there are $(m - i + 1) \cdot 2m^3 B$ swaps of γ_i–β_i, occurring on different layers, the subtangle below and including the i-th ω–ω' loop has height at least $(m - i + 1) \cdot 2m^3 B$. Accordingly, since there are $i \cdot 2m^3 B$ swaps of γ_i'–β_i', occurring on different layers, the subtangle above and including the $(i + 1)$-th ω–ω' loop has height at least $i \cdot 2m^3 B$. Thus, the whole tangle has height at least $2m^4 B$.

It remains to prove that there is a tangle T realizing L with height at most $H = 2m^4 B + 7m^2$ if and only if A is a yes-instance of 3-PARTITION.

First, assume that A is a yes-instance. Let L be a tangle constructed in the same way as the example given in Fig. 2. Then it is clear that T realizes L. We now estimate the height of T. For each partition of three elements n_i, n_j, n_k of a solution of A, we assign exactly one ω'–ω loop, in which we let the swaps of the pairs (α_i, α_i'), (α_j, α_j'), (α_k, α_k') occur. Therefore, every ω'–ω loop has height $2m^3 B + c$, where c is a small constant for the involved wires to enter and leave the loop. Observe that the additional height for the intermediate swaps we need at the beginning, at the end, and between each two consecutive ω'–ω loops is always at most $6m + k$ for some small constant k. So in total, the height of the constructed tangle is $m \cdot (2m^3 B + c) + (m + 1) \cdot (6m + k) = 2m^4 B + 6m^2 + (c + k + 6)m + k$. This is at most H for $m > c + 2k + 6$.

Now, assume that A is a no-instance. This means that any tangle realizing L has an ω'–ω loop of height at least $2m^3(B + 1)$ because there is no 3-Partition of A and, for each unit of an item in A, there are $2m^3$ swaps. Assume that the i-th ω'–ω loop has height at least $2m^3(B + 1)$. We know that the subtangle from the very beginning to the end of the i-th ω–ω' loop has height at least $(i - 1) \cdot 2m^3 B$ and the subtangle from the beginning of the $(i + 1)$-th ω–ω' loop to the very end has height at least $(m - i) \cdot 2m^3 B$. In between, there is the i-th ω'–ω loop with height $2m^3(B + 1)$. Summing these three values up, we have a total height of at least $2m^4 B + 2m^3$. Since this is greater than H for $m > 3.5$, we conclude that L cannot be realized by a tangle of height at most H, and thus our reduction is complete. $\qquad\square$

3   Exact Algorithms

The two algorithms that we describe in this section test whether a given list is feasible and, if yes, construct an optimal tangle realizing the list.

For a permutation $\pi \in S_n$ and a list $L = (l_{ij})$, we define a map $\pi L \colon [n] \to [n]$, $i \mapsto \pi(i) + |\{j \colon \pi(i) < \pi(j) \le n \text{ and } l_{ij} \text{ odd}\}| - |\{j \colon 1 \le \pi(j) < \pi(i) \text{ and } l_{ij} \text{ odd}\}|$. For each wire $i \in [n]$, $\pi L(i)$ is the position of the wire after all swaps in L have been applied to π. A list L is called π-consistent if $\pi L \in S_n$, or, more rigorously, if πL induces a permutation of $[n]$. An id_n-consistent list is consistent. For example, the list $\{12, 23, 13\}$ is consistent, whereas the list $\{13\}$ is not. If L is not consistent, then it is clearly not feasible. However, not all consistent lists are feasible e.g., the list $\{13, 13\}$ is consistent but not feasible. For a list $L = (l_{ij})$, we define $1(L) = (l_{ij} \bmod 2)$. Since $\mathrm{id}_n L = \mathrm{id}_n 1(L)$, the list

L is consistent if and only if $1(L)$ is consistent. We can compute $1(L)$ and check its consistency in $O(n + |1(L)|) = O(n^2)$ time. Hence, in the sequel we assume that all lists are consistent. For any permutation $\pi \in S_n$, we define the simple list $L(\pi) = (l_{ij})$ such that for $0 \le i < j \le n$, $l_{ij} = 0$ if $\pi(i) < \pi(j)$, and $l_{ij} = 1$ otherwise.

We use the following two lemmas which are proved in the full version [4].

Lemma 1. *For every permutation $\pi \in S_n$, $L(\pi)$ is the unique simple list with* $\mathrm{id}_n L(\pi) = \pi$.

Lemma 2. *For every tangle $T = \langle \pi_1, \pi_2, \ldots, \pi_h \rangle$, we have $\pi_1 L(T) = \pi_h$.*

Simple lists. Let L be a consistent simple list. Wang's algorithm [8] creates a simple tangle from $\mathrm{id}_n L$, so L is feasible. Let $T = (\mathrm{id}_n = \pi_1, \pi_2, \ldots, \pi_h = \mathrm{id}_n L)$ be any tangle such that $L(T)$ is simple. Then, by Lemma 2, $\mathrm{id}_n L(T) = \pi_h$. By Lemma 1, $L(\pi_h)$ is the unique simple list with $\mathrm{id}_n L(\pi_h) = \pi_h = \mathrm{id}_n L$, so $L(T) = L(\pi_h) = L$ and thus T is a realization of L.

We compute an optimal tangle realizing $L = (l_{ij})$ as follows. Consider the graph G_L whose vertex set $V(G_L)$ consists of all permutations $\pi \in S_n$ with $L(\pi) \le L$ (componentwise). A directed edge (π, σ) between vertices $\pi, \sigma \in V(G_L)$ exists if and only if π and σ are adjacent as permutations and $L(\pi) \cap L(\pi^{-1}\sigma) = \varnothing$; the latter means that the set of (disjoint) swaps whose product transforms π to σ cannot contain swaps from the set whose product transforms id_n to π. The graph G_L has at most $n!$ vertices and maximum degree $F_{n+1} - 1$, see introduction (page 2). Notice, that $F_n = (\varphi^n - (-\varphi)^{-n})/\sqrt{5} \in \Theta(\varphi^n)$. Furthermore, for each $h \ge 0$, there is a natural bijection between tangles of height $h+1$ realizing L and paths of length h in the graph G_L from the initial permutation id_n to the permutation $\mathrm{id}_n L$. A shortest such path can be found by BFS in $O(E(G_L)) = O(n!\varphi^n)$ time.

Theorem 2. *For a simple list of order n,* Tangle-Height Minimization *can be solved in $O(n!\varphi^n)$ time.*

General lists. W.l.o.g., assume that $|L| \ge n/2$; otherwise, there is a wire $k \in [n]$ that doesn't belong to any swap. This wire splits L into smaller lists with independent realizations. (If there is a swap ij with $i < k < j$, then L is infeasible.)

Let $L = (l_{ij})$ be the given list. We compute an optimal tangle realizing L (if it exists) as follows. Let λ be the number of distinct sublists of L. We consider them ordered non-decreasingly by their length. Let L' be the next list to consider. We first check its consistency by computing the map $\mathrm{id}_n L'$. If L' is consistent, we compute an optimal realization $T(L')$ of L' (if it exists), adding a permutation $\mathrm{id}_n L'$ to the end of a shortest tangle $T(L'') = \langle \pi_1, \ldots, \pi_h \rangle$ with π_h adjacent to $\mathrm{id}_n L'$ and $L'' + L(\langle \pi_h, \mathrm{id}_n L' \rangle) = L'$. This search also checks the feasibility of L' because such a tangle $T(L')$ exists if and only if the list L' is feasible. Since there are $F_{n+1} - 1$ permutations adjacent to $\mathrm{id}_n L'$, we have to check at most $F_{n+1} - 1$ lists L''. Hence, in total we spend $O(\lambda(F_{n+1} - 1)n)$ time for L. Assuming that $n \ge 2$, we bound λ as follows, where we obtain the first inequality from the inequality between arithmetic and geometric means, the second one from Bernoulli's inequality, and the third one from $1 + x \le e^x$.

$$\lambda = \prod_{i<j}(l_{ij}+1) \le \left(\frac{\sum_{i<j}(l_{ij}+1)}{\binom{n}{2}}\right)^{\binom{n}{2}} = \left(\frac{|L|}{\binom{n}{2}}+1\right)^{\binom{n}{2}} \le \left(\frac{2|L|}{n^2}+1\right)^{\frac{n^2}{2}} \le e^{|L|}.$$

Theorem 3. *For a list L of order n,* TANGLE-HEIGHT MINIMIZATION *can be solved in $O((2|L|/n^2 + 1)^{n^2/2} \cdot \varphi^n \cdot n)$ time.*

4 Theoretical and Experimental Comparison

In order to be able to compare the algorithm of Olszewski et al. [7] to ours, we first analyze the asymptotic runtime behavior of the algorithm of Olszewski et al. Their algorithm constructs a search tree whose height is bounded by the height $h(L)$ of an optimal tangle for the given list L. The tree has $1 + d + d^2 + \cdots + d^{h(L)-1} = (d^{h(L)} - 1)/(d - 1)$ vertices, where $d = F_{n+1} - 1$ is a bound on the number of edges leaving a vertex. Neglecting the time it takes to deal with each vertex, the total running time is $\Omega(\varphi^{(n+1)(h(L)-1)} \cdot 5^{-(h(L)-1)/2})$. Since $2|L|/n \le h(L)-1 \le |L|$, this is at least $\Omega(\varphi^{2|L|} \cdot 5^{-|L|/n} \cdot n)$, which is exponential in $|L|$ for fixed $n \ge 2$ and, hence, slower than our algorithm for the general case if we assume that $|L| \ge n/2$ (see Theorem 3).

It is known (see, e.g., Wang [8]) that, for any simple list L, $h(L) \le n + 1$. This implies that, on simple lists, the algorithm of Olszewski et al. runs in $O(\varphi^{(n+1)n} \cdot 5^{-n} \cdot n) = 2^{O(n^2)}$ time, whereas our algorithm for simple lists runs in $O(n!\varphi^n) = 2^{O(n \log n)}$ time.

We implemented the algorithm for general lists (see Theorem 3) and compared the running time of our implementation with the one of Olszewski et al. [7]. Their code and a database of all possible elementary linking matrices (most of them non-simple) of 5 wires (14 instances), 6 wires (38 instances), and 7 wires (115 instances) are available at https://gitlab.uni.lu/PCOG. We used their code and their benchmarks to compare our implementations. Both their and our code is implemented in Python3.

The matrices in the benchmark are quite small: the largest instance for 5 wires has 8 swaps, the largest instance for 6 wires has 15 swaps, and the largest instance for 7 wires has 27 swaps. Further, the algorithm of Olszewski et al. could not solve any of the six instances with 7 wires and ≥ 22 swaps within two hours (while our algorithm solved four of them within 10 s and the other two within 50 s), so we removed them from the data set. For better comparisons, we additionally created 10 random matrices each for $n = 5$ and $|L| = 9, \ldots, 49$, for $n = 6$ and $|L| = 16, \ldots, 49$, and for $n = 7$ and $|L| = 22, \ldots, 49$. To this end, we randomly and uniformly generated vectors of length $n(n + 1)/2$ and sum $|L|$ by drawing samples from a multinomial distribution and rejecting them if the corresponding swap list is not feasible. This gave us 414 instances for 5 wires, 358 instances for 6 wires, and 379 instances for 7 wires in total. Our source code,

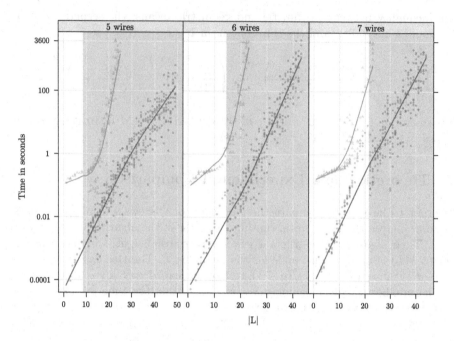

Fig. 3. Comparison of our algorithm (blue circles) with the algorithm of Olszewski et al. [7] (red triangles). The means are plotted as a trend curve. The elapsed time is plotted on a log-scale. The shaded regions correspond to randomly generated instances. (Color figure online)

the benchmarks, and the experimental data are available at https://github.com/PhKindermann/chaotic-attractors.

We ran our experiments on a single compute node of the High Performance Computing Cluster of the University of Würzburg[1]. This node consists of two Intel Xeon Gold 6134 processors, both with eight cores of 3.20 GHz. The node runs under Debian 4.9.144-3 and has 384 GB of memory. Both algorithms used only a single core. We gave both algorithms 12 h of computation time for each $n = 5, 6, 7$ to solve as many instances as possible (ordered by the number of swaps), stopping an instance after 1 h if no solution has been found yet. To avoid noise, we repeated these experiments five times and took the arithmetic mean. The results are summarized in Fig. 3.

Among the benchmark instances of Olszewski et al., our algorithm could solve almost all in less than 1 s, and the maximum running time was 8 s for one instance. The benchmark instances that could not be solved within 2 h by the algorithm of Olszewski et al. could also all be solved in less than 1 min by our algorithm. We solved all 414 instances for 5 wires within 2 h. Within the 12-hour time, we solved 303 instances with 6 wires and 333 instances with 7 wires. The

---

[1] https://www.rz.uni-wuerzburg.de/dienste/rzserver/high-performance-computing/.

algorithm by Olszewski et al. solved 163 instances with 5 wires, 97 instances with 6 wires, and 120 instances with 8 wires, within 12 h each. Our algorithm used at most 2 GB memory, whereas for the algorithm of Olszewski et al. the 384 GB RAM did not suffice for many instances.

5   Deciding Feasibility

Since computing a tangle of minimum height realizing a given list turned out to be NP-hard, the question arises whether it is already NP-hard to decide if a given list is feasible. As we could not answer this question in its full generality, we are investigating the feasibility for special classes of lists in this section.

Recall that for a list $L = (l_{ij})$, we defined $1(L) = (l_{ij} \bmod 2)$, see Sect. 3. Now we also define $2(L) = (l'_{ij})$, where $l'_{ij} = 0$ if $l_{ij} = 0$, $l'_{ij} = 1$ if l_{ij} is odd, and $l'_{ij} = 2$ otherwise. Clearly, $\pi 1(L) = \pi 2(L) = \pi L$ for each $\pi \in S_n$. A list (l_{ij}) is *even* if all l_{ij} are even, and *odd* if all non-zero l_{ij} are odd. A list L is even if and only if the list $1(L)$ is the zero list. A list L is odd if and only if $1(L) = 2(L)$.

Simple Lists. If we restrict our study to simple lists, we can easily decide feasibility. We use the following lemma, which is well-known (see, e.g., Wang [8]).

Lemma 3 (Wang [8]). *For any $n \geq 2$ and permutations $\pi, \sigma \in S_n$, there is a tangle T of height at most $n + 1$ that starts from π, ends at σ, and the list $L(T)$ is simple.*

Proposition 1. *A simple list L is feasible if and only if L is consistent. Thus, we can check the feasibility of L in $O(n + |L|)$ time.*

Proof. Clearly, if L is feasible, then L is also consistent. If L is consistent, then $\mathrm{id}_n L$ is a permutation. By Lemma 3, there exists a tangle T which starts from id_n, ends at $\mathrm{id}_n L$, and the list $L(T)$ is simple. By Lemma 2, $\pi L(T) = \pi L$. By Lemma 1, $L(T) = L$. So L is also feasible. As discussed in the beginning of Sect. 3, we can check the consistency of L in $O(n + |L|)$ time, which is equivalent to checking the feasibility of L.   □

Odd Lists. For odd lists, feasibility reduces to that of simple lists. For $A \subseteq [n]$, let L_A be the list that consists of all swaps ij of L such that $i, j \in A$. We prove the following Proposition 2 in the full version [4].

Proposition 2. *For $n \geq 3$ and an odd list L, the following statements are equivalent:*

1. *The list L is feasible.*
2. *The list $1(L)$ is feasible.*
3. *For each triple $A \subseteq [n]$, the list L_A is feasible.*
4. *For each triple $A \subseteq [n]$, the list $1(L_A)$ is feasible.*
5. *The list L is consistent.*
6. *The list $1(L)$ is consistent.*

7. *For each triple $A \subseteq [n]$, the list L_A is consistent.*
8. *For each triple $A \subseteq [n]$, the list $1(L_A)$ is consistent.*

Note that, for any feasible list L, it does not necessarily hold that $2(L)$ is feasible; see, e.g. list L_n from Observation 1.

Even Lists. For even lists, it is not as clear as for odd lists whether we can decide feasibility efficiently. An even list is always consistent, since it does not contain odd swaps and the final permutation is the same as the initial one. We conjecture that the following characterization is true, and we give some alternative formulations (see Proposition 3).

We say that a list (l_{ij}) is *non-separable* if, for every $1 \le i < k < j \le n$, $l_{ik} = l_{kj} = 0$ implies $l_{ij} = 0$. Clearly, non-separability is a necessary condition for a list to be feasible. For even lists, we conjecture that this is also sufficient. Note that any triple $A \subseteq [n]$ of an even list is feasible if and only if it is non-separable (which is not true for general lists, e.g., $L = \{12, 23\}$ is not feasible).

Conjecture 1. *Every non-separable even list L is feasible.*

We have verified the correctness of Conjecture 1 for $n \le 8$ by testing all lists using a computer. Moreover, Conjecture 1 is true for sufficiently "rich" lists according to the following lemma, which we prove in the full version [4].

Lemma 4. *Every even non-separable list $L = (l_{ij})$ with $l_{ij} \ge n$ or $l_{ij} = 0$ for every $1 \le i, j \le n$ is feasible.*

We now give some alternative formulations of Conjecture 1. To this end, we define a *minimal feasible* (even) list to be a(n even) list where we cannot remove swaps to obtain another feasible (even) list without creating new zero-entries. We say that a list is 0–2 if all its entries are either 0 or 2.

Proposition 3. *The following claims are equivalent:*

1. *Every non-separable even list L is feasible. (Conjecture 1)*
2. *Every non-separable 0–2 list L is feasible.*
3. *For each feasible even list L, the list $2(L)$ is feasible.*
4. *Every minimal feasible even list L is a 0–2 list.*

Proof. 1 $\Rightarrow$ 2. By definition.

2 $\Rightarrow$ 3. Since the list L is feasible, it is non-separable and, thus, also the list $2(L)$ is non-separable. Since $2(L)$ is non-separable and 0–2 (because L is even), $2(L)$ is feasible.

3 $\Rightarrow$ 4. Clearly, a list L never has fewer swaps than $2(L)$. Therefore, all minimal feasible lists are 0–2.

4 $\Rightarrow$ 1. Let $L = (l_{ij})$ be an even non-separable list. By Lemma 4, the list $nL := (n \cdot l_{ij})$ is feasible. Let L' be a minimal feasible even list that we obtain from nL by removing swaps without creating new zero-entries. Since every minimal feasible even list is 0–2 by assumption, we have $L' = 2(L)$. Hence, any tangle realizing L' can be extended to a tangle realizing L using the same procedure as in the proof of Proposition 2 (2 $\Rightarrow$ 1), so L is feasible. $\qquad\square$

6   Conclusions and Open Problems

Inspired by the practical research of Olszewski et al. [7], we have considered tangle-height minimization. We have shown that the problem is NP-hard, but we note that membership in NP is not obvious because the minimum height can be exponential in the size of the input. We leave open the complexity of the feasibility problem for general lists. Even if feasibility turns out to be NP-hard, can we decide it faster than finding optimal tangles?

For the special case of simple lists, we have a faster algorithm, but its running time of $O(n!\varphi^n)$ is still depressing given that odd-even sort [8] can compute a solution of height at most one more than the optimum in $O(n^2)$ time. This leads to the question whether height-minimization is NP-hard for simple lists.

Our most tantalizing open problem, however, is whether Conjecture 1 holds.

Acknowledgments. We thank Thomas C. van Dijk for stimulating discussions and the anonymous reviewers for helpful comments.

References

1. Bereg, S., Holroyd, A., Nachmanson, L., Pupyrev, S.: Representing permutations with few moves. SIAM J. Disc. Math. **30**(4), 1950–1977 (2016). https://doi.org/10. 1137/15M1036105. http://arxiv.org/abs/1508.03674
2. Bereg, S., Holroyd, A.E., Nachmanson, L., Pupyrev, S.: Drawing permutations with few corners. In: Wismath, S., Wolff, A. (eds.) GD 2013. LNCS, vol. 8242, pp. 484–495. Springer, Cham (2013). https://doi.org/10.1007/978-3-319-03841-4_42. http://arxiv.org/abs/1306.4048
3. Birman, J.S., Williams, R.F.: Knotted periodic orbits in dynamical systems–I: Lorenz's equation. Topology **22**(1), 47–82 (1983). https://doi.org/10.1016/0040-9383(83)90045-9
4. Firman, O., Kindermann, P., Ravsky, A., Wolff, A., Zink, J.: Computing height-optimal tangles faster. Arxiv report (2019). http://arxiv.org/abs/1901.06548
5. Garey, M.R., Johnson, D.S.: Computers and Intractability: A Guide to the Theory of NP-Completeness. Freeman, New York (1979)
6. Mindlin, G., Hou, X.J., Gilmore, R., Solari, H., Tufillaro, N.B.: Classification of strange attractors by integers. Phys. Rev. Lett. **64**, 2350–2353 (1990). https://doi. org/10.1103/PhysRevLett.64.2350
7. Olszewski, M., Meder, J., Kieffer, E., Bleuse, R., Rosalie, M., Danoy, G., Bouvry, P.: Visualizing the template of a chaotic attractor. In: Biedl, T., Kerren, A. (eds.) GD 2018. LNCS, vol. 11282, pp. 106–119. Springer, Cham (2018). https://doi.org/ 10.1007/978-3-030-04414-5_8. http://arxiv.org/abs/1807.11853
8. Wang, D.C.: Novel routing schemes for IC layout part I: two-layer channel routing. In: Proceedings of 28th ACM/IEEE Design Automation Conference (DAC 1991), pp. 49–53 (1991). https://doi.org/10.1145/127601.127626

On Arrangements of Orthogonal Circles

Steven Chaplick[1], Henry Förster[2], Myroslav Kryven[1][✉],
and Alexander Wolff[1]

[1] Universität Würzburg, Würzburg, Germany
myroslav.kryven@uni-wuerzburg.de
[2] Universität Tübingen, Tübingen, Germany
foersth@informatik.uni-tuebingen.de

Dedicated to Honza Kratochvíl on his 60th birthday.

Abstract. In this paper, we study arrangements of *orthogonal circles*, that is, arrangements of circles where every pair of circles must either be disjoint or intersect at a right angle. Using geometric arguments, we show that such arrangements have only a linear number of faces. This implies that *orthogonal circle intersection graphs* have only a linear number of edges. When we restrict ourselves to orthogonal *unit* circles, the resulting class of intersection graphs is a subclass of penny graphs (that is, contact graphs of unit circles). We show that, similarly to penny graphs, it is NP-hard to recognize orthogonal unit circle intersection graphs.

1 Introduction

For the purpose of this paper, an *arrangement* is a (finite) collection of curves such as lines or circles in the plane. The study of arrangements has a long history; for example, Grünbaum [15] studied arrangements of lines in the projective plane. Arrangements of circles and other closed curves have also been studied extensively [1,2,13,19,22]. An arrangement is *simple* if no point of the plane belongs to more than two curves and every two curves intersect. A *face* of an arrangement $\mathcal{A}$ in the projective or Euclidean plane P is a connected component of the subdivision induced by the curves in $\mathcal{A}$, that is, a face is a component of $P \setminus \bigcup \mathcal{A}$.

For a given type of curves, people have investigated the maximum number of faces that an arrangement of such curves can form. In 1826, Steiner [23] showed that a simple arrangement of straight lines can have at most $\binom{n}{2} + \binom{n}{1} + \binom{n}{0}$ faces while an arrangement of circles can have at most $2\left(\binom{n}{2} + \binom{n}{0}\right)$ faces.

Alon et al. [2] and Pinchasi [22] studied the number of *digonal* faces, that is, faces that are bounded by two edges, for various kinds of arrangements of

The full version of this article is available at ArXiv [5]. M.K. was supported by DAAD; S.C. was supported by DFG grant WO 758/11-1, H.F. was supported by DFG grant Ka812/17-1.

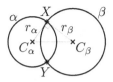

Fig. 1. Circles α and β are orthogonal if and only if $\triangle C_\alpha X C_\beta$ is orthogonal.

circles. For example, any arrangement of n unit circles has $O(n^{4/3} \log n)$ digonal faces [2] and at most $n + 3$ digonal faces if every pair of circles intersects [22], whereas arrangements of circles with arbitrary radii have at most $20n - 2$ digonal faces if every pair of circles intersects [2].

The same arrangements can, however, have quadratically many *triangular* faces, that is, faces that are bounded by three edges. A lower bound example with quadratically many triangular faces can be constructed from a simple arrangement $\mathcal{A}$ of lines by projecting it on a sphere (disjoint from the plane containing $\mathcal{A}$) and having each line become a great circle. This is always possible since the line arrangement is simple; for more details see [12, Section 5.1]. In this process we obtain $2p_3$ triangular faces, where p_3 is the number of triangular faces in the line arrangement. The great circles on the sphere can then be transformed into a circle arrangement in a different plane using the stereographic projection. This gives rise to an arrangement of circles with $2p_3$ triangular faces in this plane. Füredi and Palásti [14] provided simple line arrangements with $n^2/3 + O(n)$ triangular faces. With the argument above, this immediately yields a lower bound of $2n^2/3 + O(n)$ on the number of triangular faces of arrangements of circles. Felsner and Scheucher [13] showed that this lower bound is tight by proving that an arrangement of *pseudocircles* (that is, closed curves that can intersect at most twice and no point belongs to more than two curves) can have at most $2n^2/3 + O(n)$ triangular faces.

One can also specialize circle arrangements by fixing an angle (measured as the angle between the two tangents at either intersection point) at which each pair of intersecting circles intersect; this was recently discussed by Eppstein [10]. In this paper, we consider arrangements of circles with the restriction that each pair of circles must intersect at a right angle. An arrangement of circles in which each intersecting pair intersect at a right angle is called *orthogonal*. We make the following simple observation regarding orthogonal circles; see Fig. 1.

Observation 1. *Let α and β be two circles with centers C_α, C_β and radii r_α, r_β, respectively. Then α and β are orthogonal if and only if $r_\alpha^2 + r_\beta^2 = |C_\alpha C_\beta|^2$.*

We discuss further basic properties of orthogonal circles in Sect. 2. In particular, in an arrangement of orthogonal circles no two circles can touch and no three circles can intersect at the same point.

The main result of our paper is that arrangements of n orthogonal circles have at most $14n$ intersection points and at most $15n + 2$ faces; see Theorem 1 (in Sect. 3). This is different from arrangements of orthogonal circular arcs, which can have quadratically many quadrangular faces; see the arcs inside the blue square in Fig. 5. In Sect. 3.2 we also consider small (that is, digonal and triangular) faces and provide bounds on the number of such faces in arrangements of orthogonal circles.

Given a set of geometric objects, their *intersection graph* is a graph whose vertices correspond to the objects and whose edges correspond to the pairs of intersecting objects. Restricting the geometric objects to a certain shape restricts the class of graphs that admit a representation with respect to this shape. For example, graphs represented by disks in the Euclidean plane are called *disk intersection graphs*. The special case of *unit disk graphs*—intersection graphs of unit disks—has been studied extensively. Recognition of such graphs as well as many combinatorial problems restricted to these graphs such as coloring, independent set, and domination are all NP-hard [6]; see also the survey of Hliněný and Kratochvíl [17]. Instead of restricting the radii of the disks, people have also studied restrictions of the type of intersection. If the disks are only allowed to touch, the corresponding graphs are called *coin graphs*. Koebe's classical result says that the coin graphs are exactly the planar graphs. If all coins have the same size, the represented graphs are called *penny graphs*. These graphs have been studied extensively, too [4,8,11]. For example, they are NP-hard to recognize [3,7].

As with the arrangements above, we again consider a restriction on the intersection angle. We define the *orthogonal circle intersection graphs* as the intersection graphs of arrangements of orthogonal circles. In Sect. 4, we investigate properties of these graphs. For example, similar to the proof of our linear bound on the number of intersection points for arrangements of orthogonal circles (Theorem 1), we observe that such graphs have only a linear number of edges.

We also consider *orthogonal unit circle intersection graphs*, that is, orthogonal circle intersection graphs with a representation that consists only of unit circles. We show that these graphs are a proper subclass of penny graphs. It is NP-hard to recognize penny graphs [9]. We modify the NP-hardness proof of Di Battista et al. [7, Section 11.2.3], which uses the *logic engine*, to obtain the NP-hardness of recognizing orthogonal unit circle intersection graphs (Theorem 4).

2   Preliminaries

We will use the following type of Möbius transformation [20]. Let α be a circle having center at C_α and radius r_α. The *inversion* with respect to α is a mapping that maps any point $P \neq C_\alpha$ to a point P' on the ray $C_\alpha P$ so that $|C_\alpha P'| \cdot |C_\alpha P| = r_\alpha^2$. Inversion maps each circle not passing through C_α to another circle and a circle passing through C_α to a line; see Fig. 2. Inversion and orthogonal circles are closely related. For example, in order to construct the image P' of some point P that lies inside the inversion circle α, consider the intersection points X and Y of α and the line that is orthogonal to the line through C_α and P in P; see Fig. 2c. The point P' then is simply the center of the circle β that is orthogonal to α and goes through X and Y. This follows from the similarity of the orthogonal triangles $\triangle C_\alpha X P'$ and $\triangle C_\alpha X P$. A useful property of inversion,

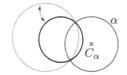
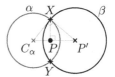

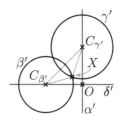

(a) a circle passing through C_α is mapped to a line (and vice versa)

(b) a circle not passing through C_α is mapped to another circle

(c) constructing the inversion P' of a point P w.r.t. α via a circle β orthogonal to α

Fig. 2. Examples of inversion

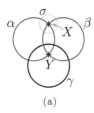

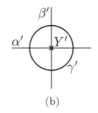

(a)

(b)

Fig. 3. (a) Three pairwise intersecting circles, the red inversion circle is centered at X; (b) image of the inversion. (Color figure online)

Fig. 4. Illustration for the proof of Lemma 2

as of any other Möbius transformation, is that it preserves angles. Using inversion we can easily show several properties of orthogonal circles.

Lemma 1. *No orthogonal circle intersection graph contains a K_4. In other words, in an arrangement of orthogonal circles there cannot be four pairwise orthogonal circles.*

Proof. Assume that there are four pairwise orthogonal circles α, β, γ, and δ. Let X and Y be the intersection points of α and β. Consider the inversion with respect to a circle σ centered at X. The images of α and β are orthogonal lines α' and β' that intersect at Y', which is the image of Y; see Fig. 3. The image of γ is a circle γ' centered at Y' but so is the image δ' of δ. Thus γ' and δ' are either disjoint or equal, but not orthogonal to each other, a contradiction.   □

Lemma 2. *No orthogonal circle intersection graph contains an induced C_4. In other words, in an arrangement of orthogonal circles there cannot be two pairs of circles such that each circle of one pair is orthogonal to each circle of the other pair and the circles within the pairs are not orthogonal.*

Proof. Assume there are two pairs (α, β) and (γ, δ) of circles such that the circles within each pair do not intersect each other and each circle of one pair intersects both circles of the other pair. Consider an inversion via a circle σ centered at one of the intersection points of the circles α and δ. In the image they will

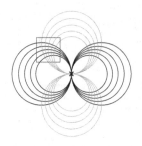

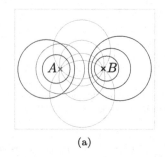

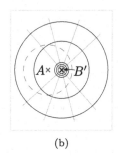

(a)   (b)

Fig. 5. Apollonian circles consisting of two parabolic pencils of circles (one in black, the
other in gray). (Color figure
online)

Fig. 6. (a) Apollonian circles consisting of an elliptic
(in gray) and hyperbolic (in black) pencil of circles;
(b) its inversion via a circle centered at A (in red).
(Color figure online)

become lines α' and δ'. The image β' of the circle β must intersect δ' but not
α', therefore, its center must lie on the line δ' and it should be to one side of the
line α'; see Fig. 4. Similarly the center of the image γ' of the circle γ must lie
on the line α' and γ' should be to one side of the line δ'. Shift the drawing so
that the intersection of α' and δ' is at the origin O and observe that the triangle
$\triangle C_{\beta'}OC_{\gamma'}$ is orthogonal, where $C_{\beta'}$ and $C_{\gamma'}$ are the centers of the circles β' and
γ'. Let X be the intersection point of these circles that is closer to the origin.
This point X is contained in the triangle $\triangle C_{\beta'}OC_{\gamma'}$. Therefore the triangle
$\triangle C_{\beta'}XC_{\gamma'}$ cannot be orthogonal—a contradiction.   □

A *pencil* is a family of circles who share a certain characteristic. In a *parabolic*
pencil all circles have one point in common, and thus are all tangent to each
other; see Fig. 5. In an *elliptic* pencil all circles go through two given points; see
the gray circles in Fig. 6a. In a *hyperbolic* pencil all circles are orthogonal to a
set of circles that go through two given points, that is, to some elliptic pencil;
see the black circles in Fig. 6a.

For an elliptic pencil whose circles share two points A and B and the corresponding hyperbolic pencil, the circles in the hyperbolic pencil possess several
properties useful for our purposes [20]. Their centers are collinear and they consist of non-intersecting circles that form two nested structures of circles, one
containing A, the other one containing B in its interior; see Fig. 6a.

Two pencils of circles such that each circle in one pencil is orthogonal to
each circle in the other are called *Apollonian circles*. There can be two such
combinations of pencils, that is, one with two parabolic pencils and one with an
elliptic and a hyperbolic pencil. We focus on the latter since such Apollonian
circles contain arbitrarily large arrangements of orthogonal circles, that is, two
orthogonal circles from the elliptic pencil and arbitrary many circles from the
hyperbolic pencil. Equivalently, such Apollonian circles are an inversion image
of a family of concentric circles centered at some point X and concurrent lines
passing through X; see Fig. 6b. We use this equivalence in the next proof.

Lemma 3. *Three circles such that one is orthogonal to the two others belong to the same family of Apollonian circles. Two sets of circles such that each circle in one set is orthogonal to each circle in the other set and each set has at least two circles belong to the same family of Apollonian circles. In particular the set belonging to the elliptic pencil can contain at most two circles.*

Proof. Consider three circles such that one is orthogonal to two others. If all three are pairwise orthogonal, then their inversion via a circle centered at one of their intersection points (see Fig. 3a) is two perpendicular lines and a circle centered at their intersection point (see Fig. 3b), therefore, they belong to the same family of Apollonian circles. If two circles do not intersect, then by [20, Theorem 13], it is always possible to invert them into two concentric circles. Since inversion preserves angles, the image of the third circle must be orthogonal to both concentric circles and therefore it must be a straight line passing through the center of both circles. Therefore, the three circles belong to the same family of Apollonian circles.

Consider now two sets S_1 and S_2 of circles such that each circle in one set is orthogonal to each circle in the other set and each set has at least two circles. By Lemma 2 there must be two circles α and β in one of the sets, say S_1, that are orthogonal. Consider an inversion via a circle σ centered at one of the intersection points X of the circles α and β. In the image they will become orthogonal lines α' and β' intersecting at a point Y. Because inversion preserves angles, the image of each circle in S_2 is a circle centered at Y. Since S_2 contains at least two circles, the image of each circle in S_1 must be orthogonal to two circles centered at Y, therefore, it must be a straight line passing through Y. Thus, the circles in S_1 and S_2 belong to the same family of Apollonian circles and S_1 contains at most two circles. □

Because each triangular or quadrangular face consists of either three circles such that one is orthogonal to two others or two pairs of circles such that each circle in one pair is orthogonal to each circle in the other pair, we obtain the following observation from Lemma 3.

Observation 2. *In any arrangement of orthogonal circles, each triangular and each quadrangular face is formed by Apollonian circles.*

3   Arrangements of Orthogonal Circles

In this section we study the number of faces of an arrangement of orthogonal circles. In Sect. 3.1, we give a bound on the total number of faces. In Sect. 3.2, we separately bound the number of faces formed by two and three edges.

Let $\mathcal{A}$ be an arrangement of orthogonal circles in the plane. By a slight abuse of notation, we will say that a circle α *contains* a geometric object o and mean that the disk bounded by α contains o. We say that a circle $\alpha \in \mathcal{A}$ is *nested* in a circle $\beta \in \mathcal{A}$ if α is contained in β. We say that a circle $\alpha \in \mathcal{A}$ is nested *consecutively* in a circle $\beta \in \mathcal{A}$ if α is nested in β and there is no other

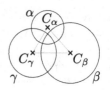

Fig. 7. Deepest circles in bold

Fig. 8. $\angle C_\beta C_\alpha C_\gamma \geq \pi/3$

circle $\gamma \in \mathcal{A}$ such that α is nested in γ and γ is nested in β. Consider a subset $S \subseteq \mathcal{A}$ of maximum cardinality such that for each pair of circles one is nested in the other. The innermost circle α in S is called a *deepest* circle in $\mathcal{A}$; see Fig. 7.

Lemma 4. *Let α be a circle of radius r_α, and let S be a set of circles orthogonal to α. If S does not contain nested circles and each circle in S has radius at least r_α, then $|S| \leq 6$. Moreover, if $|S| = 6$, then all circles in S have radius r_α and α is contained in the union of the circles in S.*

Proof. Let C_α be the center of α. Consider any two circles β and γ in S with centers C_β and C_γ and with radii r_β and r_γ, respectively. Since $r_\beta \geq r_\alpha$ and $r_\gamma \geq r_\alpha$, the edge $C_\beta C_\gamma$ is the longest edge of the triangle $\triangle C_\beta C_\alpha C_\gamma$; see Fig. 8. So the angle $\angle C_\beta C_\alpha C_\gamma$ is at least $\pi/3$. Thus, $|S| \leq 6$.

Moreover, if $|S| = 6$ then, for each pair of circles β and γ in S that are consecutive in the circular ordering of the circle centers around C_α, it holds that $\angle C_\beta C_\alpha C_\gamma = \pi/3$. This is only possible if $r_\beta = r_\gamma = r_\alpha$. Thus, all the circles in S have radius r_α and α is contained in the union of the circles in S; see Fig. 9b. □

3.1   Bounding the Number of Faces

Theorem 1. *Every arrangement of n orthogonal circles has at most $14n$ intersection points and $15n + 2$ faces.*

The above theorem (whose formal proof is at the end of the section) follows from the fact that any arrangement of orthogonal circles contains a circle α with at most seven *neighbors* (that is, circles that are orthogonal to α).

Lemma 5. *Every arrangement of orthogonal circles has a circle that is orthogonal to at most seven other circles.*

Proof. If no circle is nested within any other, Lemma 4 implies that the smallest circle has at most six neighbors, and we are done.

So, among the deepest circles in $\mathcal{A}$, consider a circle α with the smallest radius. Let r_α be the radius of α. Note that α is nested in at least one circle. Let β be a circle such that α and β are consecutively nested. Denote the set of all circles in $\mathcal{A}$ that are orthogonal to α but not to β by S_α. All circles in S_α are nested in β. Since α is a deepest circle, S_α contains no nested circles; see Fig. 9a. Since the radius of every circle in S_α is at least r_α, Lemma 4 ensures that S_α

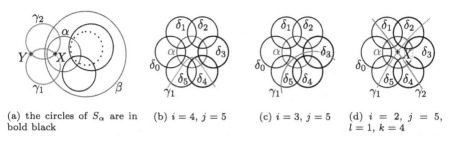

(a) the circles of S_α are in bold black   (b) $i = 4, j = 5$ (c) $i = 3, j = 5$ (d) $i = 2, j = 5, l = 1, k = 4$

Fig. 9. Illustrations to the proof of Lemma 5

contains at most six circles. Given the structure of Apollonian circles (Lemma 3), there can be at most two circles that intersect both α and β. This together with Lemma 4 immediately implies that α cannot be orthogonal to more than eight circles. In the following we show that there can be at most seven such circles.

If there is only one circle intersecting both α and β, then α is orthogonal to at most seven circles in total, and we are done.

Otherwise, there are two circles orthogonal to both α and β. Let these circles be γ_1 and γ_2. We assume that S_α contains exactly six circles. Hence, by Lemma 4, all circles in S_α have radius r_α. Let $S_\alpha = (\delta_0, \ldots, \delta_5)$ be ordered clockwise around α so that every two circles δ_i and δ_j with $i \equiv j + 1 \bmod 6$ are orthogonal.

Let X and Y be the intersection points of γ_1 and γ_2; see Fig. 9a. Note that, by the structure of Apollonian circles, one of the intersection points, say X, must be contained inside α, whereas the other intersection point Y must lie in the exterior of β. Since the circles in S_α are contained in β, none of them contains Y. Further, no circle δ_i in S_α contains X, as otherwise the circles δ_i, α, γ_1, and γ_2 would be pairwise orthogonal, contradicting Lemma 1. Recall that, by Lemma 4, α is contained in the union of the circles in S_α. Since X is not contained in this union, γ_1 intersects two different circles δ_i and δ_j, and γ_2 intersects two different circles δ_k and δ_l. Note that γ_1 and γ_2 cannot intersect the same circle ε in S_α, because ε, α, γ_1, and γ_2 would be pairwise orthogonal, contradicting Lemma 1. Therefore, the indices i, j, k, and l are pairwise different.

We now consider possible values of the indices i, j, k, and l, and show that in each case we get a contradiction to Lemma 1 or Lemma 2. If $j \equiv i + 1 \bmod 6$, then γ_1, α, δ_i, and δ_j would be pairwise orthogonal, contradicting Lemma 1; see Fig. 9b. If $j \equiv i + 2 \bmod 6$, then γ_1, δ_i, δ_{i+1}, and δ_j would form an induced C_4 in the intersection graph; see Fig. 9c. This would contradict Lemma 2. If $j \equiv i + 3 \bmod 6$ and $k \equiv l + 3 \bmod 6$, then either $k \equiv i + 1 \bmod 6$ or $i \equiv l + 1 \bmod 6$; see Fig. 9d. W.l.o.g., assume the latter and observe that then γ_2, δ_i, γ_1, δ_l would form an induced C_4, again contradicting Lemma 2.

We conclude that S_α contains at most five circles. Together with γ_1 and γ_2, at most seven circles are orthogonal to α.   □

Using the lemma above and Euler's formula, we now can prove Theorem 1.

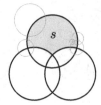

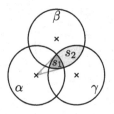

Fig. 10. Region s is a face in the arrangement of the bold circles

Fig. 11. Angles subtended by the regions s_1 and s_2 in the circle α; $\angle(s_1, \alpha) = -\angle(s_2, \alpha)$

Proof (of Theorem 1). Let $\mathcal{A}$ be an arrangement of orthogonal circles. By Lemma 5, $\mathcal{A}$ contains a circle α orthogonal to at most seven circles. The circle α yields at most 14 intersection points. By induction, the whole arrangement has at most $14n$ intersection points.

Consider the planarization G' of $\mathcal{A}$, and let n', m', f', and c' denote the numbers of vertices, edges, faces, and connected components of G', respectively. Since every vertex in the planarization corresponds to an intersection, the resulting graph is 4-regular and therefore $m' = 2n'$. By Euler's formula, we obtain $f' = n' + 1 + c'$. This yields $f' \leq 15n + 1$ since $n' \leq 14n$ and $c' \leq n$. □

3.2 Bounding the Number of Small Faces

In the following we study the number of faces of each type, that is, the number of digonal, triangular, and quadrangular faces. We begin with some notation. Let $\mathcal{A}$ be an arrangement of orthogonal circles in the plane. Let S be some subset of the circles of $\mathcal{A}$. A face in S is called a region in $\mathcal{A}$ formed by S; see for instance Fig. 10. Note that each face of $\mathcal{A}$ is also a region.

Let s be the region formed by some circular arcs $a_1, a_2, \ldots, a_k$ enumerated in counterclockwise order around s. For an arc a_i with $i \in \{1, \ldots, k\}$, let α be the circle that supports a_i. If $C_\alpha = (x_\alpha, y_\alpha)$ is the center of α and r_α its radius, we can write α as $\{C_\alpha + r_\alpha(\cos t, \sin t) : t \in [0, 2\pi]\}$. Let u and v be the endpoints of a_i so that we meet u first when we traverse s counterclockwise when starting outside of a_i. Let $u = C_\alpha + r_\alpha(\cos t_1, \sin t_1)$ and $v = C_\alpha + r_\alpha(\cos t_2, \sin t_2)$. We say that the region s *subtends* an angle in the circle α of size $\angle(s, a_i) = t_2 - t_1$ with respect to the arc a_i. Note that $\angle(s, a_i)$ is negative if a_i forms a concave side of s; see Fig. 11. If the circle α forms only one side of the region s, then we just say that the region s *subtends* an angle in the circle α of size $\angle(s, \alpha) = t_2 - t_1$. Moreover, if s is a digonal region, that is, it is formed by only two circles α and β, then we simply say that β subtends an angle of $\angle(\beta, \alpha) = t_2 - t_1$ in α to mean $\angle(s, \alpha)$.

By *total angle* we denote the sum of subtended angles by s with respect to all the arcs that form its sides, that is, $\sum_{i=1}^{k} \angle(s, a_i)$.

We now give an upper bound on the number of digonal and triangular faces in an arrangement $\mathcal{A}$ of n orthogonal circles. The tool that we utilize in this section is the Gauss–Bonnet formula [24] which, in the restricted case of orthogonal circles in the plane, states that, for every region s formed by some circular arcs $a_1, a_2, \ldots, a_k$, it holds that

$$\sum_{i=1}^{k} \angle(s, a_i) + \frac{k\pi}{2} = 2\pi.$$

This formula implies that each digonal or triangular face subtends a total angle of size π and of size $\pi/2$, respectively. Thus, we obtain the following bounds.

Theorem 2. *Every arrangement of n orthogonal circles has at most $2n$ digonal faces and at most $4n$ triangular faces.*

Proof. Because faces do not overlap, each digonal or triangular face uses a unique convex arc of a circle bounding this face. Therefore, the sum of angles subtended by digonal or triangular faces formed by the same circle must be at most 2π. Analogously, the sum of total angles over all digonal or triangular faces cannot exceed $2n\pi$. By the Gauss–Bonnet formula each digonal or triangular face subtends a total angle of size π or $\pi/2$, respectively. This gives an upper bound of $2n$ on the number of digonal faces and an upper bound of $4n$ on the number of triangular faces. □

Theorem 2 can be generalized to all convex orthogonal closed curves since the Gauss–Bonnet formula does not require curves to be circular. In contrast to this, for example, a grid made of axis parallel rectangles has quadratically many quadrangular faces. This makes circles a special subclass of convex orthogonal closed curves. We refer to the full version for more details [5].

The Gauss–Bonnet formula does not help us to get an upper bound on the number of quadrangular faces. Using Observation 2, however, it is possible to restrict the types of quadrangular faces to several shapes and obtain bounds on the number of faces of each type. Apart from being interesting in its own right, such a bound also provides a bound on the total number of faces in an arrangement of orthogonal circles. Namely, since the average degree of a face in an arrangement of orthogonal circles is 4, a bound on the number of faces of degree at most 4 gives a bound on the number of all faces in the arrangement (via Euler's formula). Unfortunately, the bound on the number of quadrangular faces that we achieved was $17n$ and thus higher than the bound $15n + 2$ that we now have for the number of *all* faces in an arrangement of n orthogonal circles.

4 Intersection Graphs of Orthogonal Circles

Given an arrangement $\mathcal{A}$ of orthogonal circles, consider its *intersection* graph, which is the graph with vertex set $\mathcal{A}$ that has an edge between any pair of intersecting circles in $\mathcal{A}$. Lemmas 1 and 2 imply that such a graph does not contain any K_4 and any induced C_4. We show that such graphs can be non-planar (Lemma 6), then we bound their edge density (Theorem 3), and finally we consider the intersection graphs arising from orthogonal *unit* circles (Theorem 4).

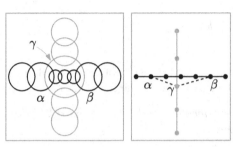

(a) a chain crossing and its intersection graph

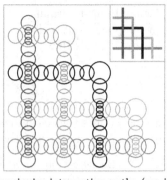

(b) pairwise intersecting paths (see inset) and the corresponding chains in an orthogonal circle representation

Fig. 12. Construction of an orthogonal circle intersection graph that contains K_n as a minor (here $n = 5$).

Lemma 6. *For every n, there is an intersection graph of orthogonal circles that contains K_n as a minor. The representation uses circles of three different radii.*

Proof. Let a *chain* be an arrangement of orthogonal circles whose intersection graph is a path. We say that two chains C_1 and C_2 *cross* if two disjoint circles α and β of one chain, say C_1, are orthogonal to the same circle γ of the other chain C_2; see Fig. 12a (left). If two chains cross, their paths in the intersection graph are connected by two edges; see the dashed edges in Fig. 12a (right).

Consider an arrangement of n rectilinear paths embedded on a grid where each pair of curves intersect exactly once; see the inset in Fig. 12b. We convert the arrangement of paths into an arrangement of chains such that each pair of chains crosses; see Fig. 12b. Now consider the intersection graph of the orthogonal circles in the arrangement of chains. If we contract each path in the intersection graph that corresponds to a chain, we obtain K_n. □

Next, we discuss the density of orthogonal circle intersection graphs. Gyárfás et al. [16] have shown that any C_4-free graph on n vertices with average degree at least a has clique number at least $a^2/(10n)$. Due to Lemma 1, we know that orthogonal circle intersection graphs have clique number at most 3. Thus, their average degree is bounded from above by $\sqrt{30n}$, leading to at most $\sqrt{7.5}n^{\frac{3}{2}}$ edges in total. However, Lemma 5 implies the following stronger bound.

Theorem 3. *The intersection graph of a set of n orthogonal circles has at most $7n$ edges.*

Proof. The geometric representation of an orthogonal circle intersection graph is an arrangement of orthogonal circles. By Lemma 5, an arrangement of n orthogonal circles always has a circle orthogonal to at most seven circles. Therefore, the corresponding intersection graph always has a vertex of degree at most seven. Thus, it has at most $7n$ edges. □

(a) all orthogonal unit circle intersection graphs are penny graphs   (b) penny graphs that aren't orthogonal unit circle intersection graphs

Fig. 13. Penny graphs vs. orthogonal unit circle intersection graphs

(a) $1.5n$ digonal faces   (b) $2n$ triangular faces   (c) $4(n-3)$ quadrangular faces

Fig. 14. Arrangements of n orthogonal circles with many digonal, triangular, and quadrangular faces.

The remainder of this section concerns a natural subclass of orthogonal circle intersection graphs, the orthogonal *unit* circle intersection graphs. Recall that these are orthogonal circle intersection graphs with a representation that consists of unit circles only. As Fig. 13a shows, every representation of an orthogonal unit circle intersection graph can be transformed (by scaling each circle by a factor of $\sqrt{2}/2$) into a representation of a *penny graph*, that is, a contact graph of equal-size disks. Hence, every orthogonal unit circle intersection graph is a penny graph – whereas the converse is not true. For example, C_4 or the 5-star are penny graphs but not orthogonal unit circle intersection graphs (see Fig. 13b).

Orthogonal unit circle intersection graphs being penny graphs implies that they inherit the properties of penny graphs, e.g., their maximum degree is at most six and their edge density is at most $\lfloor 3n - \sqrt{12n-6} \rfloor$, where n is the number of vertices [21, Theorem 13.12, p. 211]. Because triangular grids are orthogonal unit circle intersection graphs, this upper bound is tight.

As it turns out, orthogonal unit circle intersection graphs share another feature with penny graphs: their recognition is NP-hard. The hardness of penny-graph recognition can be shown using the *logic engine* [7, Section 11.2], which simulates an instance of the Not-All-Equal-3-Sat (NAE3SAT) problem. We establish a similar reduction for the recognition of orthogonal unit circle intersection graphs; the details are in the full version [5].

Theorem 4. *It is NP-hard to recognize orthogonal unit circle intersection graphs.*

5   Discussions and Open Problems

In Sect. 3 we have provided upper bounds for the number of faces of an orthogonal circle arrangement. As for lower bounds on the number of faces, we found only very simple arrangements containing $1.5n$ digonal, $2n$ triangular, and $4(n-3)$

quadrangular faces; see Figs. 14a, b, and c, respectively. Can we construct better lower bound examples or improve the upper bounds?

Recognizing (unit) disk intersection graphs is $\exists\mathbb{R}$-complete [18]. But what is the complexity of recognizing (general) orthogonal circle intersection graphs?

Acknowledgments. We thank Alon Efrat for useful discussions and an anonymous reviewer for pointing us to the Gauss-Bonnet formula.

References

1. Agarwal, P.K., Aronov, B., Sharir, M.: On the complexity of many faces in arrangements of pseudo-segments and circles. In: Aronov, B., Basu, S., Pach, J., Sharir, M. (eds.) Discrete and Computational Geometry: The Goodman-Pollack Festschrift, pp. 1–24. Springer, Heidelberg (2003). https://doi.org/10.1007/978-3-642-55566-4_1

2. Alon, N., Last, H., Pinchasi, R., Sharir, M.: On the complexity of arrangements of circles in the plane. Discret. Comput. Geom. **26**(4), 465–492 (2001). https://doi.org/10.1007/s00454-001-0043-x

3. Breu, H., Kirkpatrick, D.G.: Unit disk graph recognition is NP-hard. Comput. Geom. Theory Appl. **9**(1–2), 3–24 (1998). https://doi.org/10.1016/S0925-7721(97)00014-X

4. Cerioli, M.R., Faria, L., Ferreira, T.O., Protti, F.: A note on maximum independent sets and minimum clique partitions in unit disk graphs and penny graphs: complexity and approximation. RAIRO Theor. Inf. Appl. **45**(3), 331–346 (2011). https://doi.org/10.1051/ita/2011106

5. Chaplick, S., Förster, H., Kryven, M., Wolff, A.: On arrangements of orthogonal circles. ArXiv report (2019). https://arxiv.org/abs/1907.08121

6. Clark, B.N., Colbourn, C.J., Johnson, D.S.: Unit disk graphs. Discret. Math. **86**(1–3), 165–177 (1990). https://doi.org/10.1016/0012-365X(90)90358-O

7. Di Battista, G., Eades, P., Tamassia, R., Tollis, I.G.: Graph Drawing: Algorithms for the Visualization of Graphs. Prentice Hall, Upper Saddle River (1999)

8. Dumitrescu, A., Pach, J.: Minimum clique partition in unit disk graphs. Graphs Combin. **27**(3), 399–411 (2011). https://doi.org/10.1007/s00373-011-1026-1

9. Eades, P., Whitesides, S.: The logic engine and the realization problem for nearest neighbor graphs. Theoret. Comput. Sci. **169**(1), 23–37 (1996). https://doi.org/10.1016/S0304-3975(97)84223-5

10. Eppstein, D.: Circles crossing at equal angles (2018). https://11011110.github.io/blog/2018/12/22/circles-crossing-equal.html. Accessed 11 May 2019

11. Eppstein, D.: Triangle-free penny graphs: degeneracy, choosability, and edge count. In: Frati, F., Ma, K.L. (eds.) GD 2017. LNCS, vol. 10692. Springer, Heidelberg (2018). https://doi.org/10.1007/978-3-319-73915-1_39. https://arxiv.org/abs/1708.05152

12. Felsner, S.: Geometric Graphs and Arrangements: Some Chapters from Combinatorial Geometry. Springer, Heidelberg (2004). https://doi.org/10.1007/978-3-322-80303-0

13. Felsner, S., Scheucher, M.: Arrangements of pseudocircles: triangles and drawings. In: Frati, F., Ma, K.-L. (eds.) GD 2017. LNCS, vol. 10692, pp. 127–139. Springer, Cham (2018). https://doi.org/10.1007/978-3-319-73915-1_11

14. Füredi, Z., Palásti, I.: Arrangements of lines with a large number of triangles. Proc. Amer. Math. Soc. **92**(4), 561–566 (1984). https://doi.org/10.1090/S0002-9939-1984-0760946-2
15. Grünbaum, B.: Arrangements and spreads. In: CBMS Regional Conference Series in Mathmatics, vol. 10. AMS, Providence (1972)
16. Gyárfás, A., Hubenko, A., Solymosi, J.: Large cliques in C_4-free graphs. Combinatorica **22**(2), 269–274 (2002). https://doi.org/10.1007/s004930200012
17. Hliněný, P., Kratochvíl, J.: Representing graphs by disks and balls (a survey of recognition complexity results). Discret. Math. **229**(1–3), 101–124 (2001). https://doi.org/10.1016/S0012-365X(00)00204-1
18. Kang, R.J., Müller, T.: Sphere and dot product representations of graphs. Discret. Comput. Geom. **47**(3), 548–568 (2012). https://doi.org/10.1007/s00454-012-9394-8
19. Kang, R.J., Müller, T.: Arrangements of pseudocircles and circles. Discret. Comput. Geom. **51**(4), 896–925 (2014). https://doi.org/10.1007/s00454-014-9583-8
20. Ogilvy, C.S.: Excursions in Geometry. Oxford University Press, New York (1969)
21. Pach, J., Agarwal, K.P.: Combinatorial Geometry. Wiley-Interscience Series in Discrete Mathematics and Optimization. Wiley, Hoboken (1995)
22. Pinchasi, R.: Gallai-Sylvester theorem for pairwise intersecting unit circles. Discret. Comput. Geom. **28**(4), 607–624 (2002). https://doi.org/10.1007/s00454-002-2892-3
23. Steiner, J.: Einige Gesetze über die Theilung der Ebene und des Raumes. Journal für die reine und angewandte Mathematik **1**, 349–364 (1826). https://doi.org/10.1515/crll.1826.1.349
24. Weisstein, E.W.: Gauss-Bonnet formula (2019). http://mathworld.wolfram.com/Gauss-BonnetFormula.html. Accessed 27 July 2019

Extending Simple Drawings

Alan Arroyo[1] , Martin Derka[2], and Irene Parada[3]([✉])

[1] IST Austria, Klosterneuburg, Austria
alanmarcelo.arroyoguevara@ist.ac.at
[2] University of Waterloo, Waterloo, ON, Canada
mderka@uwaterloo.ca
[3] Graz University of Technology, Graz, Austria
iparada@ist.tugraz.at

Abstract. Simple drawings of graphs are those in which each pair of edges share at most one point, either a common endpoint or a proper crossing. In this paper we study the problem of extending a simple drawing $D(G)$ of a graph G by inserting a set of edges from the complement of G into $D(G)$ such that the result is a simple drawing. In the context of rectilinear drawings, the problem is trivial. For pseudolinear drawings, the existence of such an extension follows from Levi's enlargement lemma. In contrast, we prove that deciding if a given set of edges can be inserted into a simple drawing is NP-complete. Moreover, we show that the maximization version of the problem is APX-hard. We also present a polynomial-time algorithm for deciding whether one edge uv can be inserted into $D(G)$ when $\{u, v\}$ is a dominating set for the graph G.

Keywords: Simple drawings · Edge insertion · NP-hardness · APX-hardness

1 Introduction

A *simple drawing* of a graph G (also known as *good drawing* or as *simple topological graph* in the literature) is a drawing $D(G)$ of G in the plane such that every pair of edges share at most one point that is either a proper crossing (no tangent edges allowed) or an endpoint. Moreover, no three edges intersect in the same point and edges must neither self-intersect nor contain other vertices than their endpoints. Simple drawings, despite often considered in the study of crossing numbers, have basic aspects that are yet unknown.

The long-standing conjectures on the crossing numbers of K_n and $K_{n,m}$, known as the Harary-Hill and Zarankiewicz's conjectures, respectively, have

This work was started at the Crossing Numbers Workshop 2016 in Strobl (Austria). M.D. was partially supported by NSERC. I.P. is supported by the Austrian Science Fund (FWF): W1230. This project has received funding from the European Union's Horizon 2020 research and innovation programme under the Marie Skłodowska-Curie grant agreement No. 754411.

© Springer Nature Switzerland AG 2019
D. Archambault and C. D. Tóth (Eds.): GD 2019, LNCS 11904, pp. 230–243, 2019.
https://doi.org/10.1007/978-3-030-35802-0_18

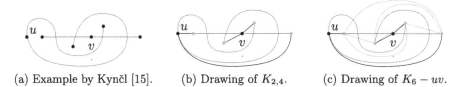

(a) Example by Kynčl [15].   (b) Drawing of $K_{2,4}$.   (c) Drawing of $K_6 - uv$.

Fig. 1. Drawings in which the edge uv cannot be inserted.

drawn particular interest in the study of simple drawings of complete and complete bipartite graphs. The intensive study of these conjectures has produced deep results about simple drawings of K_n [14,18] and $K_{n,m}$ [8].

In contrast to our knowledge about K_n, little is known about simple drawings of general graphs. In [16] it was observed that, when studying simple drawings of general graphs, it is natural to try to extend them, by inserting the missing edges between non-adjacent vertices. One of the main results in this paper suggests that there is no hope for efficiently deciding when such operation can be performed.

The complement $\overline{G}$ of a graph G is the graph with the same vertex set as G and where two distinct vertices are adjacent if and only they are not adjacent in G. Given a simple drawing $D(G)$ of a graph $G = (V, E)$ and a subset M of *candidate edges* from $\overline{G}$, an *extension* of $D(G)$ with M is a simple drawing $D'(H)$ of the graph $H = (V, E \cup M)$ that contains $D(G)$ as a subdrawing. If such an extension exists, then we say that M can be *inserted* into $D(G)$.

Given a simple drawing, an extension with one given edge is not always possible, as shown by Kynčl [15] (in Fig. 1a the edge uv cannot be inserted, because uv would cross an edge incident either to u or to v). We can extend this example to a simple drawing of $K_{2,4}$ (Fig. 1b) and we can then use it to construct drawings of $K_{n,m}$ with larger values of m and n in which an edge uv cannot be inserted. Moreover, Kynčl's drawing can be extended to a simple drawing of K_6 minus one edge where the missing edge cannot be inserted (Fig. 1c). From this drawing one can construct drawings of K_n with $n \geq 6$ minus one edge where the only missing edge cannot be inserted.

Extensions, by inserting both vertices and edges, have received a great deal of attention in the last decade, specially for (different classes of) plane drawings [2,4,6,9,13,17,19]. It has also been of interest to study crossing number questions on planar graphs with one additional edge [7,11,20]. Note that the term *augmentation* has also been used in the literature for the similar problem of inserting edges and/or vertices to a graph [10]. Extensions of simple drawings have been previously considered in the context of *saturated* drawings, that is, drawings where no edge can be inserted [12,16].

Our Contribution. We study the computational complexity of extending a simple drawing $D(G)$ of a graph G. In Sect. 2, we show that deciding if $D(G)$ can be extended with a set M of candidate edges is NP-complete. Moreover, in Sect. 3, we prove that finding the largest subset of edges from M that extend $D(G)$ is

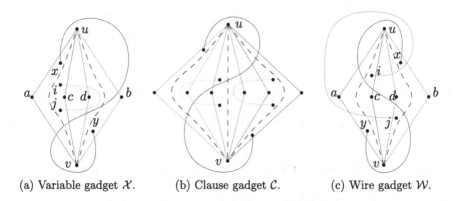

(a) Variable gadget $\mathcal{X}$.   (b) Clause gadget $\mathcal{C}$.   (c) Wire gadget $\mathcal{W}$.

Fig. 2. Basic gadgets for the proof of Theorem 1.

APX-hard. Finally, in Sect. 4, we present a polynomial-time algorithm to decide whether an edge uv can be inserted into $D(G)$ when $\{u,v\}$ is a dominating set for G.

2   Inserting a Given Set of Edges Is NP-complete

In this section we prove the following result:

Theorem 1. *Given a simple drawing $D(G)$ of a graph $G = (V, E)$ and a set M of edges of the complement of G, it is* NP-*complete to decide if $D(G)$ can be extended with the set M.*

Notice first that the problem is in NP, since it can be described combinatorially. Our proof of Theorem 1 is based on a reduction from *monotone* 3SAT [5]. An instance of that problem consists of a Boolean formula ϕ in 3-CNF with a set of variables $X = \{x_1, \ldots, x_n\}$ and a set of clauses $K = \{C_1, \ldots, C_m\}$. Moreover, in each clause either all the literals are positive (*positive clause*) or they are all negative (*negative clause*). The *bipartite graph $G(\phi)$ associated to ϕ* is the graph with vertex set $X \cup K$ and where a variable x_i is adjacent to a clause C_j if and only if $x_i \in C_j$ or $\overline{x_i} \in C_j$.

We now show how to construct a simple drawing from a given formula. We start by introducing our three basic gadgets, the *variable gadget*, the *clause gadget*, and the *wire gadget*, shown in Fig. 2.

The variable gadget contains two nested cycles, $avbu$ on the outside and $cvdu$ on the inside, drawn in the plane without any crossings. Two additional vertices x and y are drawn in the interior of $avcu$ and $dvbu$, respectively. They are connected with an edge that, starting in x, crosses the edges au, ub, dv, cv, av, and vb, in this order, and ends in y. Another two vertices i and j are drawn inside the region in the interior of $avcu$ that is incident to x. They are connected with an edge that, starting in i, crosses the edges uc, ud, vd, and vc, in this order,

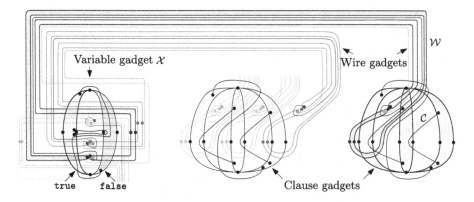

Fig. 3. Reduction from monotone 3SAT. (Color figure online)

and ends in j; see Fig. 2a. Notice that the edge uv can be inserted only in two possible regions: either inside the cycle $avcu$ or inside the cycle $dvbu$. Drawing the edge uv in any other region would force it to cross uj or xy more than once. The clause gadget and the wire gadget are similarly defined; see Fig. 2b–c.

In each of these three gadgets shown in Fig. 2, the edge uv can only be inserted in the regions where the dashed arcs are drawn. In the rest of the paper, when we refer to the *regions* in a gadget we mean these regions where the edge uv can be inserted.

In a variable gadget, these regions encode the truth assignment of the corresponding variable x_i: inserting the edge uv in the left region corresponds to the assignment $x_i = \mathtt{true}$, while inserting it in the right region corresponds to $x_i = \mathtt{false}$. We call these left and right regions in a variable gadget the **true** and **false** regions, respectively. In a clause gadget, each of the three regions is associated to a literal in the corresponding clause. Wire gadgets propagate the truth assignment of the variables to the clauses. They are drawn between the gadgets corresponding to clauses and variables that are incident in $G(\phi)$. The idea is that if an assignment makes a literal not satisfy a clause, then the edge uv in the wire gadget blocks the region in the clause gadget corresponding to that literal by forcing uv to cross that region twice.

Let $w^{(\mathcal{G})}$ denote vertex w in gadget $\mathcal{G}$. The following lemma shows that we can get the desired behavior with a wire gadget connecting a variable gadget and a clause gadget. The precise placement of a wire gadget with respect to the variable gadget and the clause gadget that it connects is illustrated in Fig. 3.

Lemma 2. *We can combine a variable gadget $\mathcal{X}$, a clause gadget $\mathcal{C}$, and a wire gadget $\mathcal{W}$ to produce a simple drawing with the following properties.*

- *If $u^{(\mathcal{X})}v^{(\mathcal{X})}$ is inserted in the **false** region in $\mathcal{X}$, then inserting $u^{(\mathcal{W})}v^{(\mathcal{W})}$ prevents $u^{(\mathcal{C})}v^{(\mathcal{C})}$ from being inserted in one specified target region in $\mathcal{C}$.*
- *If $u^{(\mathcal{X})}v^{(\mathcal{X})}$ is inserted in the **true** region in $\mathcal{X}$, then we can insert $u^{(\mathcal{W})}v^{(\mathcal{W})}$ in a way such that $u^{(\mathcal{C})}v^{(\mathcal{C})}$ can then be inserted in any region in $\mathcal{C}$.*

Proof. We start with a drawing of the variable gadget $\mathcal{X}$ and the clause gadget $\mathcal{C}$ such that the two gadgets are drawn on a line and they are disjoint. A representation of how the wire gadget is then inserted is shown in Fig. 3. In this proof we focus on the wire gadget drawn with blue edges and vertices.

In Fig. 3, gadget $\mathcal{X}$ lies to the left of gadget $\mathcal{C}$. The true and false regions in $\mathcal{X}$ are shaded in green and red, respectively. We assume that the target region in $\mathcal{C}$ is the leftmost one, shaded in yellow. The left and right regions in the wire gadget are shaded in red and yellow, respectively.

If the edge $u^{(\mathcal{X})}v^{(\mathcal{X})}$ is inserted in the false region in $\mathcal{X}$ then the edge $u^{(\mathcal{W})}v^{(\mathcal{W})}$ cannot be inserted in the yellow region in $\mathcal{W}$, since it would cross $u^{(\mathcal{X})}v^{(\mathcal{X})}$ twice. Thus, $u^{(\mathcal{W})}v^{(\mathcal{W})}$ can only be inserted in the red region in $\mathcal{W}$. If inserted in that region, $u^{(\mathcal{C})}v^{(\mathcal{C})}$ cannot be inserted in the yellow region in $\mathcal{C}$, since it would cross $u^{(\mathcal{W})}v^{(\mathcal{W})}$ twice. In contrast, if the edge $u^{(\mathcal{X})}v^{(\mathcal{X})}$ is inserted in the true (green) region in $\mathcal{X}$, then $u^{(\mathcal{W})}v^{(\mathcal{W})}$ can be inserted in either of the two regions in $\mathcal{W}$. In particular, it can be inserted in the yellow region in a way such that $u^{(\mathcal{C})}v^{(\mathcal{C})}$ can then be inserted in any region in $\mathcal{C}$.

Finally, notice that if the target region in $\mathcal{C}$ is not the leftmost one, we can adapt the construction by leaving the region(s) to the left in $\mathcal{C}$ uncrossed by the wire gadget $\mathcal{W}$; see the clause gadget in the middle of Fig. 3. $\square$

Let ϕ be an instance of monotone 3SAT and let $G(\phi)$ be the bipartite graph associated to ϕ. Let $D(\phi)$ be a 2-page book drawing of $G(\phi)$ in which (i) all vertices lie on an horizontal line, and from left to right, first the ones corresponding to negative clauses, then to variables, and finally to positive clauses; and (ii) the edges incident to vertices corresponding to positive clauses are drawn as circular arcs above that horizontal line, while the ones incident to vertices corresponding to negative clauses are drawn as circular arcs below it. In an slight abuse of notation, we refer to the vertices in $D(\phi)$ corresponding to variables and clauses simply as variables and clauses, respectively.

We construct a simple drawing D' from $D(\phi)$ by first replacing the variables and clauses by variable gadgets and clause gadgets, respectively, and drawn in disjoint regions. Moreover, the clause gadgets corresponding to negative clauses are rotated 180°. We then insert the wire gadgets. The edges in $D(\phi)$ connecting variables to positive clauses are replaced by wire gadgets drawn as in the proof of Lemma 2; see Fig. 3. Similarly, the edges in $D(\phi)$ connecting variables to negative clauses are replaced by wire gadgets drawn as the ones before, but rotated 180°.

We now describe how to draw the wire gadgets with respect to each other, so that the result is a simple drawing; see Fig. 3 for a detailed illustration. First, we focus on the drawing locally around the variable gadgets. Consider a set of edges in $D(\phi)$ connecting a variable with some positive clauses. The drawing $D(\phi)$ defines a clockwise order of these edges around the common vertex starting from the horizontal line. We insert the corresponding wire gadgets locally around the variable gadget following this order. Each new gadget is inserted shifted up and to the right with respect to the previous one (as the blue and green gadgets depicted in Fig. 3). Edges in $D(\phi)$ connecting a variable with some negative clauses are replaced by wire gadgets in an analogous manner with a 180° rotation. We assign the three different regions in a clause gadget to the target regions in the wire gadgets following the rotation of the edges around the clause in $D(\phi)$. (Note that we can assume without loss of generality, by possibly duplicating variables, that each clause in ϕ contains three literals.) Thus, locally around a clause gadget, it is then possible to draw the different wire gadgets connecting to it without crossing. Since $D(\phi)$ is a 2-page book drawing, the constructed drawing D' is a simple drawing.

Let M be the set of uv edges of all the gadgets. The fact that ϕ is satisfiable if and only if M can be inserted into D' follows now from Lemma 2, finishing the proof of Theorem 1.

3   Maximizing the Number of Edges Inserted Is **APX**-hard

In this section we show that the maximization version of the problem of inserting missing edges from a prescribed set into a simple drawing is APX-hard. This implies that, if P $\neq$ NP, then no PTAS exists for this problem. We start by showing that this maximization problem is NP-hard.

Theorem 3. *Given a simple drawing $D(G)$ of a graph $G = (V, E)$ and a set M of edges in the complement $\overline{G}$, it is* NP-*hard to find a maximum subset of edges $M' \subseteq M$ that extends $D(G)$.*

Our proof of Theorem 3 is based on a reduction from the maximum independent set problem (MIS). By showing that the reduction when the input graph has vertex degree at most three is actually a PTAS-reduction we will then conclude that the problem is APX-hard.

An independent set of a graph $G = (V, E)$ is a set of vertices $S \subseteq V$ such that no two vertices in S are incident with the same edge. The problem of determining the maximum independent set (MIS) of a given graph is APX-hard even when the graph has vertex degree at most three [1]. We first describe the construction of a simple drawing $D'(G')$ from the graph G of a given MIS instance. Then we argue that for a well-selected set of edges M that are not present in $D'(G')$, finding a maximum subset $M' \subseteq M$ that can be inserted into $D'(G')$ is equivalent to finding a maximum independent set of G.

3.1   Constructing a Drawing from a Given Graph

We begin by introducing our two basic gadgets, the vertex gadget $\mathcal{V}$ and the edge gadget $\mathcal{E}$, shown in Fig. 4. They are reminiscent of the gadgets in the previous section, but adapted to this different reduction. Similarly as in the previous gadgets, there is only one region in which the edge uv can be inserted into $\mathcal{V}$ and only two regions in which the edge uv can be inserted into $\mathcal{E}$. These regions are the ones in which the dashed arcs in Fig. 4b are drawn.

In Fig. 4c we combined an edge gadget and two vertex gadgets. This figure shows a copy $\mathcal{E}^{(e)}$ of the gadget $\mathcal{E}$ (that corresponds to an edge $e = wz$) drawn

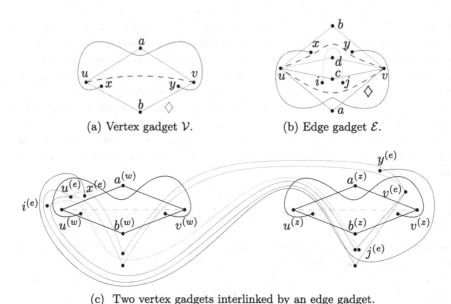

(a) Vertex gadget $\mathcal{V}$.

(b) Edge gadget $\mathcal{E}$.

(c) Two vertex gadgets interlinked by an edge gadget.

Fig. 4. Basic gadgets and drawings for the proof of Theorem 3.

over two different copies, $\mathcal{V}^{(w)}$ and $\mathcal{V}^{(z)}$, of the gadget $\mathcal{V}$ (that correspond to vertices w and z, respectively). We relabel the vertices in the copies of these gadgets by using the vertex or edge to which they correspond as their superscripts. Since there is only one region in which $v^{(w)}u^{(w)}$ and $v^{(z)}u^{(z)}$ can be drawn, inserting both of these edges prevents $v^{(e)}u^{(e)}$ from being inserted. Inserting either only $v^{(w)}u^{(w)}$ or only $v^{(z)}u^{(z)}$ leaves exactly one possible region where $v^{(e)}u^{(e)}$ can be inserted.

We have all the ingredients needed for our construction. Suppose that we are given a simple graph $G = (V, E)$. This graph admits a 1-page book drawing $D(G)$ in which the vertices are placed on a horizontal line and the edges are drawn as circular arcs in the upper halfplane. Since the edge gadget does not interlink the vertex gadgets symmetrically, we consider the edges in $D(G)$ with an orientation from their left endpoint to their right one.

The following lemma shows that is possible to replace each vertex $w \in V$ in the drawing by a vertex gadget $\mathcal{V}^{(w)}$ and each edge $e \in E$ by an edge gadget $\mathcal{E}^{(e)}$, and obtain simple drawing $D'(G')$ (where G' is the disjoint union of the underlying graphs of the vertex- and edge-gadgets).

Lemma 4. *Given a 1-page book drawing $D(G)$ of a graph $G = (V, E)$, then we can replace every vertex by a vertex gadget and every edge by an edge gadget to obtain a simple drawing.*

Proof. We show that the copies $\{\mathcal{E}^{(e)} : e \in E\}$ can be inserted into $\bigcup_{w \in V} \mathcal{V}^{(w)}$ such that such that vertex gadgets corresponding to different vertices are drawn

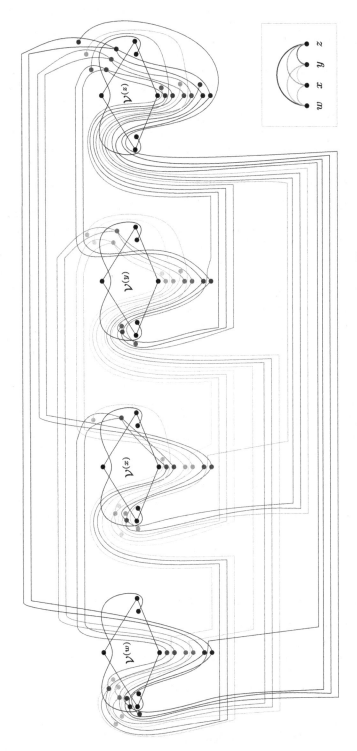

Fig. 5. Drawing obtained by a reduction from K_4.

in disjoint regions and for every edge $e = wz \in E$, $\mathcal{V}^{(w)} \cup \mathcal{V}^{(z)} \cup \mathcal{E}^{(e)}$ is as in Fig. 4c (up to interchanging the indices w and z), and such that the resulting drawing is simple.

First, for each vertex $w \in V$ we place the gadget $\mathcal{V}^{(w)}$ in its position, so all the copies of $\mathcal{V}$ lie (equidistant) on a horizontal line and do not cross each other. For the edges of G, since the drawing in Fig. 4c is not symmetric, we choose an orientation. We orient all the edges in the 1-page book drawing $D(G)$ of G from left to right. We start by inserting the corresponding $\mathcal{E}$ gadgets from left to right and from the shortest edges in $D(G)$ to the longest. For an edge wz, the intersections of the gadget $\mathcal{E}^{(wz)}$: (i) with the edges $u^{(w)}a^{(w)}$ and $u^{(w)}b^{(w)}$ are placed to the left of all the previous intersections of other edge gadgets with that edge; (ii) with the edge $v^{(w)}b^{(w)}$ are placed to the right of all the previous intersections with that edge; (iii) with the edge $v^{(w)}a^{(w)}$ are placed to the right of previous intersections with gadgets $\mathcal{E}^{(wt)}$ and to the left of previous intersections with gadgets $\mathcal{E}^{(tw)}$; (iv) with the edges $u^{(z)}a^{(z)}$ and $u^{(z)}b^{(z)}$ are placed to the left of the previous intersections with gadgets $\mathcal{E}^{(tz)}$; (v) with the edge $v^{(z)}b^{(z)}$ are placed to the left of all previous intersections; and (vi) with the edge $v^{(z)}a^{(z)}$ are placed to the left of all previous intersections with gadgets $\mathcal{E}^{(tz)}$; see Fig. 5.

Moreover, the arcs of an edge gadget connecting two vertex gadgets are drawn either completely in the upper half-plane or completely in the lower one with respect to the horizontal line and two arcs cross at most twice. If they are part of edges in edge gadgets connected to the same vertex gadget, they might cross locally around this vertex gadget. However, after this crossing, they follow the circular-arc routing induced by $D(G)$ (or its mirror image) and do not cross again. Otherwise, with respect to each other, they follow the circular-arc routing induced by $D(G)$ (or its mirror image) and thus cross at most once; see Fig. 5.

Since in neither of the gadgets two incident edges cross, and edges of different gadgets are vertex-disjoint, we only have to worry about edges from different gadgets crossing more than once. By construction, no edge in an edge gadget intersects more than once with an edge in a vertex gadget. Thus, it remains to show that any two edges from two distinct edge gadgets cross at most once. Such two edges are included in a subgraph H of G with exactly four vertices. The drawing induced by the four vertex gadgets and the at most six edge gadgets is homeomorphic to a subdrawing of the drawing in Fig. 5. It is routine to check that it is a simple drawing, and thus any two edges cross at most once.   □

3.2   Reduction from Maximum Independent Set

Proof (of Theorem 3). Given a graph $G = (V, E)$, we reduce the problem of deciding whether G has an independent set of size k to the problem of deciding whether the simple drawing $D'(G')$ constructed as in Lemma 4 with a candidate set of edges M (where $M = \{u^{(w)}v^{(w)} : w \in V\} \cup \{u^{(e)}v^{(e)} : e \in E\}$) can be extended with a set of edges $M' \subseteq M$ of cardinality $|M'| = |E| + k$.

To show the correctness of the (polynomial) reduction, we first show that if G has an independent set I of size k, then we can extend $D'(G')$ with a set M' of $|E| + k$ edges of M. Clearly, the k edges $\{u^{(w)}v^{(w)} : w \in I\}$ can be inserted into

$D'(G')$ by the construction of the drawing. Since I is an independent set, each edge has at most one endpoint in I. Thus, in every edge gadget $\mathcal{E}^{(e)}$ at most one of the two possibilities for inserting the edge $u^{(e)}v^{(e)}$ is blocked by the previous k inserted edges. We therefore can also insert the $|E|$ edges $\{u^{(e)}v^{(e)} : e \in E\}$.

Conversely, let $M' \subset M$ be a set of $|E| + k$ edges can be inserted into $D'(G')$ and that contains the minimum number of uv edges from vertex gadgets. If the set of vertices $\{w \in V : u^{(w)}v^{(w)} \in M'\}$ is an independent set of G, then we are done, since at most $|E|$ edges of M' can be from edge gadgets, so at least k are from vertex gadgets. Otherwise, there are two edges $u^{(w)}v^{(w)}$ and $u^{(z)}v^{(z)}$ in M' such that the corresponding vertices $w, z \in V$ are connected by the edge $wz \in E$. By the construction of $D'(G')$ this implies that the edge $u^{(wz)}v^{(wz)}$ belongs to M, but it cannot be in M'. By removing the edge $u^{(w)}v^{(w)}$ and inserting the edge $u^{(wz)}v^{(wz)}$ into $D'(G')$, we obtain another valid extension with the same cardinality but one less uv edge from a vertex gadget. This contradicts our assumption. $\square$

The presented reduction can be further analyzed to show that the problem is actually APX-hard. Note that the problem we are reducing from, maximum independent set in simple graphs, is APX-hard [1] even in graphs with vertex degree at most three. Our reduction can be shown to be an L-reduction in that case, implying a PTAS-reduction. This shows the following result (details are provided in the full version [3]):

Corollary 5. *Given a simple drawing $D(G)$ of a graph G and a set of edges M of the complement of G, finding the size of the largest subset of edges from M extending $D(G)$ is APX-hard.*

4   Inserting One Edge in a Simple Drawing

In this section, we consider the problem of extending a simple drawing of a graph by inserting exactly one edge uv for a given pair of non-adjacent vertices u and v. We start by rephrasing our problem as a problem of finding a certain path in the dual of the planarization of the drawing.

Given a simple drawing $D(G)$ of a graph $G = (V, E)$, the *dual graph* $G^*(D)$ has a vertex corresponding to each cell of $D(G)$ (where a cell is a component of $\mathbb{R}^2 \setminus D(G)$). There is an edge between two vertices if and only if the corresponding cells are separated by the same segment of an edge in $D(G)$. Notice that $G^*(D)$ can also be defined as the plane dual of the planarization of $D(G)$, where crossings are replaced by vertices so that the resulting drawing is plane.

We define a coloring χ of the edges of $G^*(D)$ by labeling the edges of the original graph G using numbers from 1 to $|E|$, and assigning to each edge of $G^*(D)$ the label of the edge that separates the cells corresponding to its incident vertices. Given two vertices $u, v \in V$, let $G^*(D, \{u, v\})$ be the subgraph of $G^*(D)$ obtained by removing the edges corresponding to connections between cells separated by an (arc of an) edge incident to u or to v, and let χ' be the coloring of the edges coinciding with χ in every edge. The problem of extending $D(G)$

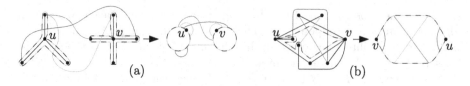

Fig. 6. Reduction to the path problem with holes.

with one edge uv is equivalent to the existence of a heterochromatic path in $G^*(D, \{u, v\})$ (i.e., no color is repeated) with respect to χ, between two vertices that corresponds to a cell incident to u and a cell incident to v, respectively.

We remark that, from this dual perspective, it is clear that the problem of deciding if a simple drawing can be extended with a given set of edges is in NP.

The general problem of finding an heterochromatic path in an edge-colored graph is NP-complete, even when each color is assigned to at most two edges. The proof can be found in the full version [3].

Theorem 6. *Given a (multi)graph G with an edge-coloring χ and two vertices x and y, it is NP-complete to decide whether there is a heterochromatic path in G from x to y, even when each color is assigned to at most two edges.*

However, in our setting the multigraph and the coloring come from a simple drawing. The following theorem shows a particular case in which we can decide in polynomial time if an edge can be inserted.

Theorem 7. *Let $D(G)$ be a simple drawing of a graph $G = (V, E)$ and let u, $v \in V(G)$ be non-adjacent vertices. If $\{u, v\}$ is a dominating set for G, that is, every vertex in $V \setminus \{u, v\}$ is a neighbor of u or v, then the problem of extending $D(G)$ with the edge uv can be decided in polynomial time.*

An algorithm proving this result can be found in the full version [3]. We sketch here the idea. The first step is to reduce our problem to the *path problem with holes* (PPH): Given two open disks $h_1, h_2 \subseteq \mathbb{R}^2$ whose closures (called holes) are either disjoint or they coincide $h_1 = h_2$, a set $\mathcal{J}$ of colored Jordan curves in $\Gamma = \mathbb{R}^2 \setminus (h_1 \cup h_2)$, and two distinct points $p, q \in \Gamma \setminus \bigcup \mathcal{J}$, we want to decide if there is a pq-arc intersecting at most one arc in $\mathcal{J}$ from each color. If $h_1 = h_2$, we say that the instance of the PPH has one hole.

Consider the subdrawing $D_{u,v}$ of $D(G)$ consisting of u, v, all vertices adjacent to them and all the edges incident to u or to v. Figure 6 illustrates the reduction from the problem of inserting uv in $D_{u,v}$ to the PPH. Based on our reduction, one can make further assumptions on any instance $(\Gamma, \mathcal{J}, p, q)$ that we consider of the PPH problem: (i) for every two different arcs $\alpha_1, \alpha_2 \in \mathcal{J}$, $|\alpha_1 \cap \alpha_2| \leq 1$; (ii) pairs of arcs in $\mathcal{J}$ with the same color do not cross; and (iii) each arc in $\mathcal{J}$ has both ends on the union of the boundaries of the holes $\partial h_1 \cup \partial h_2$.

Given an instance $(\Gamma, \mathcal{J}, p, q)$ of the PPH, an arc $\alpha \in J$ is *separating* if p and q are on different connected components of $\Gamma \setminus \alpha$. We divide the arcs in $\mathcal{J}$

Fig. 7. Transforming an instance of PPH: (a) enlarging a hole along an arc and (b) cutting through an arc.

into three different types: (T1) arcs with ends on different holes; (T2) separating arcs with ends on the same hole; and (T3) non-separating arcs with ends on the same hole.

Arcs of type T3 can be preprocessed with the operation that we denote *enlarging one hole using* α, as showed in Fig. 7(a). Once all the arcs in $\mathcal{J}$ are of type either T1 or T2, the algorithm determines the existence of a feasible pq-arc based on the colors of the arcs in $\mathcal{J}$. If all the arcs have different colors we have a solution. Otherwise we consider two arcs of the same color. If both arcs are of type T2, then there is no valid pq-arc and our algorithm stops. For handling the cases in which at least one of these arcs is of type T1, the idea is to try to find a solution that does not cross it. To do so, we use the operation denoted *cutting through an arc* illustrated in Fig. 7(b). If of the two arcs of the same color is of type T1 and the other is of type T2, there is a valid pq-arc if and only if there is a valid pq-arc after cutting through the T1 arc. Otherwise, if both are of type T1, there is a solution if and only if either there is a solution after cutting through the first arc or there is a solution after cutting through the second one. Note that the operation of cutting through an arc produces an instance with only one from an instance with two holes. This guarantees that the algorithm runs in polynomial time.

5   Conclusions

In this paper we showed that given a simple drawing $D(G)$ of a graph $G = (V, E)$ and a prescribed set M of edges of the complement of G, it is NP-complete to decide whether M can be inserted into $D(G)$. Moreover, it is APX-hard to find the maximum subset of edges in M that can be inserted into $D(G)$. We remark that the reduction showing APX-hardness cannot replace the one showing NP-hardness of inserting the whole set M of edges, since, by construction, in the APX-hardness reduction some of the edges in M cannot be inserted.

Focusing on the case $|M| = 1$, we showed that a generalization of this problem is NP-complete and we found sufficient conditions guaranteeing a polynomial time decision. We hope that this paves the way to solve the following question.

Problem 1. Given a simple drawing $D(G)$ of a graph G and a pair u, v of non-adjacent edges, what is the computational complexity of deciding whether we can insert uv into $D(G)$ such that the result is a simple drawing?

Acknowledgments. We want to thank the anonymous reviewers for their insightful comments.

References

1. Alimonti, P., Kann, V.: Some APX-completeness results for cubic graphs. Theoret. Comput. Sci. **237**(1), 123–134 (2000). https://doi.org/10.1016/S0304-3975(98)00158-3

2. Angelini, P., Di Battista, G., Frati, F., Jelínek, V., Kratochvíl, J., Patrignani, M., Rutter, I.: Testing planarity of partially embedded graphs. ACM Trans. Algorithms **11**(4), 32 (2015). https://doi.org/10.1145/2629341

3. Arroyo, A., Derka, M., Parada, I.: Extending simple drawings. ArXiv e-Prints (2019). http://arxiv.org/abs/1908.08129

4. Bagheri, A., Razzazi, M.: Planar straight-line point-set embedding of trees with partial embeddings. Inf. Process. Lett. **110**(12–13), 521–523 (2010). https://doi.org/10.1016/j.ipl.2010.04.019

5. de Berg, M., Khosravi, A.: Optimal binary space partitions in the plane. Int. J. Comput. Geom. Appl. **22**(03), 187–205 (2010). https://doi.org/10.1142/S0218195912500045

6. Brückner, G., Rutter, I.: Partial and constrained level planarity. In: Klein, P.N. (ed.) Proceedings of the 28th Annual ACM-SIAM Symposium on Discrete Algorithms (SODA 2017), pp. 2000–2011 (2017). https://doi.org/10.1137/1.9781611974782.130

7. Cabello, S., Mohar, B.: Adding one edge to planar graphs makes crossing number and 1-planarity hard. SIAM J. Comput. **42**(5), 1803–1829 (2013). https://doi.org/10.1137/120872310

8. Cardinal, J., Felsner, S.: Topological drawings of complete bipartite graphs. J. Comput. Geom. **9**(1), 213–246 (2018). https://doi.org/10.20382/jocg.v9i1a7

9. Da Lozzo, G., Di Battista, G., Frati, F.: Extending upward planar graph drawings. In: Friggstad, Z., Sack, J.-R., Salavatipour, M.R. (eds.) WADS 2019. LNCS, vol. 11646, pp. 339–352. Springer, Cham (2019). https://doi.org/10.1007/978-3-030-24766-9_25

10. Eswaran, K.P., Tarjan, R.E.: Augmentation problems. SIAM J. Comput. **5**(4), 653–665 (1976)

11. Gutwenger, C., Mutzel, P., Weiskircher, R.: Inserting an edge into a planar graph. Algorithmica **41**(4), 289–308 (2005). https://doi.org/10.1007/s00453-004-1128-8

12. Hajnal, P., Igamberdiev, A., Rote, G., Schulz, A.: Saturated simple and 2-simple topological graphs with few edges. J. Graph Algorithms Appl. **22**(1), 117–138 (2018). https://doi.org/10.7155/jgaa.00460

13. Jelínek, V., Kratochvíl, J., Rutter, I.: A Kuratowski-type theorem for planarity of partially embedded graphs. Comput. Geom.: Theory Appl. **46**(4), 466–492 (2013). https://doi.org/10.1016/j.comgeo.2012.07.005

14. Kynčl, J.: Simple realizability of complete abstract topological graphs simplified. In: Di Giacomo, E., Lubiw, A. (eds.) GD 2015. LNCS, vol. 9411, pp. 309–320. Springer, Cham (2015). https://doi.org/10.1007/978-3-319-27261-0_26

15. Kynčl, J.: Improved enumeration of simple topological graphs. Discret. Comput. Geom. **50**(3), 727–770 (2013). https://doi.org/10.1007/s00454-013-9535-8

16. Kynčl, J., Pach, J., Radoičić, R., Tóth, G.: Saturated simple and k-simple topological graphs. Comput. Geom. **48**(4), 295–310 (2015). https://doi.org/10.1016/j.comgeo.2014.10.008

17. Mchedlidze, T., Nöllenburg, M., Rutter, I.: Extending convex partial drawings of graphs. Algorithmica **76**(1), 47–67 (2015). https://doi.org/10.1007/s00453-015-0018-6

18. Pach, J., Solymosi, J., Tóth, G.: Unavoidable configurations in complete topological graphs. Discret. Comput. Geom. **30**(2), 311–320 (2003). https://doi.org/10.1007/s00454-003-0012-9
19. Patrignani, M.: On extending a partial straight-line drawing. Int. J. Found. Comput. Sci. **17**(5), 1061–1070 (2006). https://doi.org/10.1142/S0129054106004261
20. Riskin, A.: The crossing number of a cubic plane polyhedral map plus an edge. Studia Scientiarum Mathematicarum Hungarica **31**(4), 405–414 (1996)

Coloring Hasse Diagrams
and Disjointness Graphs of Curves

János Pach[1,2] and István Tomon[1(✉)]

[1] École Polytechnique Fédérale de Lausanne, Lausanne, Switzerland
{janos.pach,istvan.tomon}@epfl.ch
[2] Rényi Institute, Budapest, Hungary

Abstract. Given a family of curves $\mathcal{C}$ in the plane, its disjointness graph is the graph whose vertices correspond to the elements of $\mathcal{C}$, and two vertices are joined by an edge if and only if the corresponding sets are disjoint. We prove that for every positive integer r and n, there exists a family of n curves whose disjointness graph has girth r and chromatic number $\Omega(\frac{1}{r}\log n)$. In the process we slightly improve Bollobás's old result on Hasse diagrams and show that our improved bound is best possible for uniquely generated partial orders.

Keywords: String graph · Hasse diagram · Chromatic number

1 Introduction

There are two important, seemingly unrelated, concepts that play important roles in Geometric Graph Theory and in Graph Drawing: *Hasse diagrams* and *string graphs*.

Hasse diagrams were introduced by Vogt [26] at the end of the 19th century for concise representation of partial orders. Today they are widely used in graph drawing algorithms. Let P be a partially ordered set with partial ordering $\prec$. For any $x, y \in P$, we say that y *covers* x if $x \prec y$ and there is no $z \in P$ such that $x \prec z \prec y$. The *Hasse diagram* of P is the directed graph on the elements of P, where there is an edge from x to y if and only if y covers x. If we disregard the direction of the edges, we obtain the *cover graph* of P. The graph on P whose two elements are connected by an edge if and only if they are related by $\prec$ is the *comparability graph* of P. The cover graph is a subgraph of the comparability graph.

The *intersection graph* of a family of sets $\mathcal{C}$ is the graph whose vertices correspond to the elements of $\mathcal{C}$ and two vertices are joined by an edge if and only if the corresponding sets have a nonempty intersection. The *disjointness graph* of $\mathcal{C}$ is the complement of the intersection graph of $\mathcal{C}$. A *string*, or *curve*, γ is the image of a continuous function $f : [0,1] \to \mathbb{R}^2$. A curve γ is *grounded* if one of its

Research partially supported by Swiss National Science Foundation grants no. 200020-162884 and 200021-175977.

endpoints is on the y-axis, and γ lies in the nonnegative half-plane. See Fig. 1 for an illustration of a grounded family of curves and its disjointness graph. A *string graph* is the intersection graph of curves. The notion was introduced by Benzer [2] and Sinden [24] to describe the incidence structures of intervals in chromosomes and metallic layers in printed networks, respectively. The systematic study of string graphs was initiated in [5] and [10].

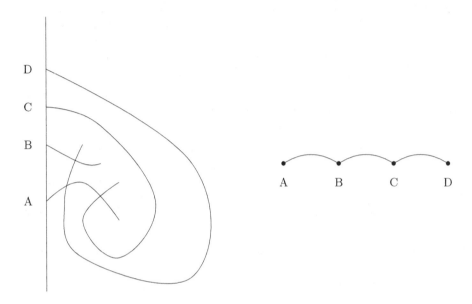

Fig. 1. A family of grounded curves and its disjointness graph.

The first sign that the above concepts are intimately related was the following simple fact discovered by Golumbic, Rotem, Urrutia [9], and Lovász [15]: Every comparability graph is the disjointness graph of a collection of curves in the plane. A partial converse of this statement was established in [8].

A useful characterization of cover graphs in terms of strings follows directly from Corollary 2.7 of Middendorf and Pfeiffer [16] and Theorem 1 in [24]. See also [13] and [23] (page 2).

Theorem 1. *[16], [24] A triangle-free graph is a cover graph of a partially ordered set if and only if it is isomorphic to the disjointness graph of a family of grounded curves.*

The *girth* of a graph G is the length of the shortest cycle in G. Obviously, every triangle-free graph has girth at least *four*. According to a classical result of Erdős [6], for every $r \geq 3$, there exist graphs with n vertices and girth at least r which have arbitrarily large chromatic numbers. Erdős's construction is probabilistic and does not posses any geometric structure.

For geometrically defined graphs, the situation is more complicated. The chromatic number of intersection graphs of axis-parallel rectangles [1] or chords of a cycle [4,11,12] and disjointness graphs of segments in the plane [21,25] can be bounded from above by a function of their clique numbers. In sharp contrast to this, Pawlik, Kozik, Krawczyk, Lasoń, Micek, Trotter, and Walczak [22] proved that there exist triangle-free intersection graphs of n segments with chromatic number $\Omega(\log \log n)$, which disproved a longstanding conjecture of Erdős. In [19], triangle-free disjointness graphs of n curves were constructed, with chromatic number $\Omega(\log n)$, cf. [17]. This construction is based on shift graphs, defined by Erdős and Hajnal [7]. It appears to be difficult to extend this method to obtain disjointness graphs of curves with high girth and high chromatic number.

The aim of the present note is to construct such graphs.

Theorem 2. *For every positive integer r and for every sufficiently large n, there exists a family of n curves whose disjointness graph has girth at least r and chromatic number at least $\Omega(\frac{1}{r} \log n)$.*

This result does not remain true if we are allowed to use only *x-monotone* curves, that is, if every vertical line meets each curve in at most one point. In this case, the chromatic number of the cover graph is bounded from above by a constant [20,21].

In view of Theorem 1, in order to prove Theorem 2, it is sufficient to establish the following.

Theorem 3. *For every positive integer r and for every sufficiently large n, there exists a poset on n vertices whose cover graph has girth at least r and chromatic number $\Omega(\frac{1}{r} \log n)$.*

The study of combinatorial properties of cover graphs (Hasse diagrams) is an extensive area of research in the theory of partial orders. Bollobás [3] was the first to show the existence of partial orders (actually, lattices) whose cover graphs have arbitrarily large girth and chromatic number. Alternative constructions were found by Nešetřil *et al.* [14,18]. Bollobás's proof, which gives the best known asymptotic bound, builds on Erdős's probabilistic construction [6] mentioned above. It shows that for a fixed girth r and $n \to \infty$, the chromatic number of a cover graph with n vertices can be as large as $\Omega(\frac{\log n}{\log \log n})$. Our Theorem 3 improves on this bound.

It is possible that Theorem 3 can be further improved. However, we can show that our bound is tight for an interesting family of cover graphs. A partially ordered set P is called *uniquely generated* if for every comparable pair of vertices $x \prec y$, there exists a unique sequence of vertices $x = v_1 \prec \cdots \prec v_k = y$ such that v_{i+1} covers v_i for $i = 1, \ldots, k - 1$. Obviously, if there is no chain with 3 elements in P, then P is uniquely generated and its cover graph is bipartite.

Theorem 4. *(i) If P is a uniquely generated poset on n vertices, then the chromatic number of its cover graph is at most $\lfloor \log_2 n \rfloor + 1$.*
(ii) For every integer $r > 3$ and for every sufficiently large n, there exists a uniquely generated poset on n vertices whose cover graph has girth at least r and chromatic number at least $\Omega(\frac{1}{r} \log n)$.

2   Cover Graphs with Large Chromatic Number

In this section, we prove Theorem 4. Note that then Theorem 3 is an immediate consequence of part (ii) of Theorem 4. We omit floors an ceilings for easier readability.

Proof of Theorem 4, part (i). Let G be the cover graph of P, let $<_P$ be the partial ordering on P, and let $\prec$ be a linear extension of $<_P$. For any $x \in P$, let $C(x)$ denote the set of vertices of P covered by x.

We prove that the greedy coloring of G with respect to $\prec$ uses at most $1 + \lfloor \log_2 n \rfloor$ colors. Let $v_1 \prec \cdots \prec v_n$ be the vertices of G. Color them with the elements of $\mathbb{Z}^+$, as follows. For $i = 1, \ldots, n$, if $v_1, \ldots, v_{i-1}$ have already been colored, then color v_i with the smallest positive integer k that does not appear among the colors of $C(v_i)$.

For each vertex $v \in V(G)$, let $T(v)$ denote the set of vertices $u \in V(G)$ such that $u \leq_P v$. Note that, as P is uniquely generated, the subgraph of G induced by $T(v)$ is a tree. We claim that if v received color k, then $|T(v)| \geq 2^{k-1}$. This clearly implies (i), because if the total number of colors used by our coloring is K, then we have $n \geq 2^{K-1}$.

We prove the claim by induction on k. For $k = 1$, the statement is trivial. Suppose that $k \geq 2$ and that the claim is true for all positive integers smaller than k. As v received color k, we can find $k - 1$ vertices $u_1, \ldots, u_{k-1} \in C(v)$ such that the color of u_i is i, for $i = 1, \ldots, k - 1$. By the induction hypothesis, we have $|T(u_i)| \geq 2^{i-1}$. Since the trees $T(u_1), \ldots, T(u_{k-1}) \subset T(v)$ are pairwise disjoint, we obtain $|T(v)| \geq 1 + \sum_{i=1}^{k-1} 2^{i-1} = 2^{k-1}$, as required.   □

For the proof of part (ii) of Theorem 4, we need the following technical lemma.

Lemma 1. *Let A and B be two m-element sets and let G be the random graph on $A \cup B$ in which every $a \in A$ and $b \in B$ are joined by an edge independently with probability $p = \frac{d}{m}$.*

Then the probability that there exist $X \subset A$ and $Y \subset B$ such that $|X||Y| \geq 3m^2/d$ and there is no edge between X and Y is at most 2^{-m}.

Proof. Let $N = \frac{3m^2}{d}$. For any $X \subset A$ and $Y \subset B$, let $I(X,Y)$ denote the event that there exists no edge between X and Y. Obviously, we have $\mathbb{P}(I(X,Y)) = (1-p)^{|X||Y|} \leq e^{-p|X||Y|}$. This yields

$$\mathbb{P}\left(\bigcup_{\substack{X \subset A, Y \subset B \\ |X||Y| \geq N}} I(X,Y)\right) \leq \sum_{\substack{X \subset A, Y \subset B \\ |X||Y| \geq N}} e^{-p|X||Y|} \leq 2^{2m} e^{-pN} < 2^{-m}.$$

□

Proof of Theorem 4, part (ii). Assume that $n \geq 2^{10r}$, and let $N = 3n$, $k = \frac{\log_2 N}{10r}$, and $m = \frac{N}{k}$. If G is a graph whose vertex set is a subset of the integers, a *monotone path* in G is a path with vertices $c_0 < c_1 < \cdots < c_t$ and edges $c_i c_{i+1}$ for $i = 1, \ldots, t-1$. A pair of vertices $\{a, b\}$ of G is called *bad*, if there exist two edge-disjoint monotone paths whose endpoints are a and b.

Our goal is to construct a graph G on the vertex set $\{1, \ldots, N\}$ satisfying the following three conditions:

1. G has no independent set of size larger than $7m$,
2. G has at most $\frac{N}{3}$ bad pairs of vertices,
3. the number of cycles in G of length smaller than r is at most $\frac{N}{3}$.

Suppose that such a graph G exists. Let G' denote the graph obtained from G by deleting $\frac{2N}{3}$ vertices: at least one vertex from every bad pair and at least one vertex from every cycle of length smaller than r. Then G' has n vertices and girth at least r. Condition 1 implies that the chromatic number of G' is at least $\frac{n}{7m} > \frac{1}{10^3 r} \log_2 n$. Define a partially ordered set P with partial ordering $<_P$ on $V(G')$ in such a way that $a <_P b$ if and only if $a < b$ and there exists a monotone path in G' with endpoints a and b. Then P meets all the requirements of part (ii) of the theorem. Indeed, as G' has no bad pair of vertices, the cover graph of P is equal to G', and P is uniquely generated.

We construct a graph G with the above three properties, as follows. Divide $\{1, \ldots, N\}$ into k intervals of size m, denoted by $A_1, \ldots, A_k$. For every $1 \leq i < j \leq k$ and for any $x \in A_i$, $y \in A_j$, join x and y by an edge independently with probability $p_{ij} = \frac{2^{j-i}}{m}$. Denote the resulting graph by G.

First, we show that, with probability larger than $\frac{2}{3}$, condition 1 is satisfied: G does not contain an independent set of size larger than $7m$. Let $\mathcal{A}$ denote the event that for every pair (i, j) with $1 \leq i < j \leq k$, and for every pair of subsets $X \subset A_i$ and $Y \subset A_j$ with no edge running between X and Y, we have $|X||Y| < 3m^2 2^{i-j}$. By Lemma 1, for a fixed pair (i, j) with $1 \leq i < j \leq k$, with probability at least $1 - 2^{-m}$ there exists no $X \subset A_i$ and $Y \subset A_j$ such that $|X||Y| \geq 3m^2 2^{i-j}$ and there is no edge between X and Y. As there are fewer than k^2 different pairs (i, j) with $1 \leq i < j \leq k$, we have $\mathbb{P}(\mathcal{A}) \geq 1 - k^2 2^{-m} > \frac{2}{3}$.

We show that if $\mathcal{A}$ happens, then G has no independent set of size larger than $7m$. Suppose for contradiction that $I \subset V(G)$ is an independent set with $|I| > 7m$. For $i = 1, \ldots, k$, let $I_i = I \cap A_i$. Clearly, there exists an index $1 \leq h \leq k$ such that $\sum_{i=1}^{h} |I_i| \geq 3m$ and $\sum_{i=h+1}^{k} |I_i| \geq 3m$. Then we have

$$9m^2 \leq \left(\sum_{i=1}^{h} |I_i| \right) \left(\sum_{i=h+1}^{k} |I_i| \right) = \sum_{i=1}^{h} \sum_{j=h+1}^{k} |I_i||I_j| \leq \sum_{i=1}^{h} \sum_{j=h+1}^{k} 3m^2 2^{i-j}, \quad (1)$$

where the last inequality holds if $\mathcal{A}$ occurs. However,

$$\sum_{i=1}^{h} \sum_{j=h+1}^{k} 2^{i-j} \leq \sum_{l=1}^{k} l 2^{-l} < 2,$$

which contradicts the left-hand side of (1).

Next, we prove that the probability that G satisfies condition 2 is larger than $\frac{2}{3}$. Let X stand for the number of bad pairs of vertices in G, and let $\mathcal{B}$ denote the event that $X \leq \frac{N}{3}$. Let $x \in A_i$ and $y \in A_j$, where $1 \leq i < j \leq k$. Let $x = v_0, v_2, \ldots, v_l = y$ such that $v_t \in A_{i_t}$ for $t = 0, \ldots, l$, where $i = i_0 < \cdots < i_l = j$. The probability that $v_0, \ldots, v_l$ is a monotone path in G is

$$\prod_{t=0}^{l-1} \frac{2^{i_{l+1}-i_l}}{m} = \frac{2^{j-i}}{m^l} < \frac{2^k}{m^l}.$$

There are $\binom{j-i-1}{l-1} m^{l-1} < 2^k m^{l-1}$ ways to choose the vertices of a monotone path of length l with endpoints x and y. Hence, the probability that there exist two edge-disjoint monotone paths with endpoints x and y, where one of these paths has length l and the other has length l', is at most

$$(2^k m^{l-1})(2^k m^{l'-1}) \frac{2^k}{m^l} \frac{2^k}{m^{l'}} = \frac{2^{4k}}{m^2}.$$

There are fewer than k^2 ways to choose (l, l'), so the probability that $\{x, y\}$ is a bad pair of vertices is less than $\frac{k^2 2^{4k}}{m^2} < \frac{1}{9n}$. Therefore, we have $\mathbb{E}(X) < N^2 \frac{1}{9N} = \frac{N}{9}$. Applying Markov's inequality, we obtain that $1 - \mathbb{P}(\mathcal{B}) = \mathbb{P}(X > \frac{N}{3}) < \frac{1}{3}$.

Finally, we show that G satisfies condition 3, with probability larger than $\frac{2}{3}$. Let Y be the number of cycles of length at most $r - 1$ in G, and let $\mathcal{C}$ denote the event that $Y \leq \frac{N}{3}$. Let $p = n^{-(r-1)/r}$. Note that each pair of vertices in G is joined by an edge with probability at most $\frac{2^k}{m} < p$. Then we have

$$\mathbb{E}(Y) < \sum_{l=3}^{r-1} N^l p^l < r N^{\frac{r-1}{r}} < \frac{N}{9}.$$

Indeed, there are $\frac{(l-1)!}{2}\binom{N}{l} < N^l$ possible copies of the cycle of length l, and the probability that a fixed copy of such a cycle appears in G is at most p^l. Applying Markov's inequality, we get $1 - \mathbb{P}(\mathcal{C}) = \mathbb{P}(Y > \frac{N}{3}) < \frac{1}{3}$.

In conclusion, we proved that $\mathbb{P}(\mathcal{A}), \mathbb{P}(\mathcal{B}), \mathbb{P}(\mathcal{C}) > \frac{2}{3}$. Thus, the probability that the event $\mathcal{A} \wedge \mathcal{B} \wedge \mathcal{C}$ occurs is nonzero. This means that there exists a graph G satisfying conditions 1, 2, and 3, which completes the proof of the theorem.□

Acknowledgements. We are grateful to Bartosz Walczak for valuable discussions. He gave a direct construction proving Theorem 1 and pointed out where the components of the statement appeared in the literature.

References

1. Asplund, E., Grünbaum, B.: On a coloring problem. Math. Scand. **8**, 181–188 (1960)
2. Benzer, S.: On the topology of the genetic fine structure. Proc. Natl. Acad. Sci. U.S.A. **45**(11), 1607–1620 (1959)

3. Bollobás, B.: Colouring lattices. Algebra Universalis **7**, 313–314 (1977)
4. Davies, J., McCarty, R.: Circle graphs are quadratically χ-bounded. arXiv:1905.11578
5. Ehrlich, G., Even, S., Tarjan, R.E.: Intersection graphs of curves in the plane. J. Comb. Theory Ser. B **21**(1), 8–20 (1976)
6. Erdős, P.: Graph theory and probability. Can. J. Math. **11**, 34–38 (1959)
7. Erdős, P., Hajnal, A.: Some remarks on set theory. IX. Combinatorial problems in measure theory and set theory. Mich. Math. J. **11**(2), 107–127 (1964)
8. Fox, J., Pach, J.: String graphs and incomparability graphs. Adv. Math. **230**(3), 1381–1401 (2012)
9. Golumbic, M., Rotem, D., Urrutia, J.: Comparability graphs and intersection graphs. Discrete Math. **43**, 37–46 (1983)
10. Graham, R.L.: Problem 1, Open Problems at 5th Hungarian Colloquium on Combinatorics (1976). In: Hajnal, A., Sós, V.T. (eds.) Combinatorics, vol. II. North-Holland, Amsterdam, p. 1195 (1978)
11. Gyárfás, A.: On the chromatic number of multiple interval graphs and overlap graphs. Discrete Math. **55**, 161–166 (1985)
12. Kostochka, A., Kratochvíl, J.: Covering and coloring polygon-circle graphs. Discrete Math. **163**(1–3), 299–305 (1997)
13. Kratochvíl, J.: String graphs. I. The number of critical nonstring graphs is infinite. J. Comb. Theory Ser. B **52**(1), 53–66 (1991)
14. Kříž, I., Nešetřil, J.: Chromatic number of Hasse diagrams, eyebrows and dimension. Order **8**(1), 41–48 (1991)
15. Lovász, L.: Perfect graphs. In: Selected Topics in Graph Theory, vol. 2, pp. 55–87. Academic Press, London (1983)
16. Middendorf, M., Pfeiffer, F.: Weakly transitive orientations, Hasse diagrams and string graphs. Discrete Math. **111**(1–3), 393–400 (1993). In: Graph theory and combinatorics (Marseille-Luminy 1990)
17. Mütze, T., Walczak, B., Wiechert, V.: Realization of shift graphs as disjointness graphs of 1-intersecting curves in the plane. arXiv:1802.09969
18. Nešetřil, J., Rödl, V.: A short proof of the existence of highly chromatic hypergraphs without short cycles. J. Comb. Theory Ser. B **27**(2), 225–227 (1979)
19. Pach, J., Tardos, G., Tóth, G.: Disjointness graphs of segments. In: 33rd International Symposium on Computational Geometry, SoCG 2017, 77, Leibniz Zentrum, Dagstuhl, pp. 59:1–59:15 (2017)
20. Pach, J., Tomon, I.: On the chromatic number of disjointness graphs of curves. In: 35th International Symposium on Computational Geometry, SoCG 2019, 129 Leibniz Zentrum, Dagstuhl, pp. 54:1–54:17 (2019)
21. Pach, J., Törőcsik, J.: Some geometric applications of Dilworth's theorem. Discrete Comput. Geom. **12**(1), 1–7 (1994)
22. Pawlik, A., Kozik, J., Krawczyk, T., Lasoń, M., Micek, P., Trotter, W.T., Walczak, B.: Triangle-free intersection graphs of line segments with large chromatic number. J. Comb. Theory Ser. B **105**, 6–10 (2014)
23. Rok, A., Walczak, B.: Outerstring graphs are χ-bounded. https://arxiv.org/abs/1312.1559
24. Sinden, F.W.: Topology of thin film RC-circuits. Bell System Tech. J. **45**, 1639–1662 (1966)
25. Tóth, G.: Note on geometric graphs. J. Comb. Theory Ser. A **89**(1), 126–132 (2000)
26. Vogt, H. G.: Leçons sur la résolution algébrique des équations, Nony, p. 91 (1895)

A Low Number of Crossings

Efficient Generation of Different Topological Representations of Graphs Beyond-Planarity

Patrizio Angelini⬤, Michael A. Bekos$^{(\boxtimes)}$⬤, Michael Kaufmann⬤, and Thomas Schneck⬤

Institut für Informatik, Universität Tübingen, Tübingen, Germany
{angelini,bekos,mk,schneck}@informatik.uni-tuebingen.de

Abstract. Beyond-planarity focuses on combinatorial properties of classes of non-planar graphs that allow for representations satisfying certain local geometric or topological constraints on their edge crossings. Beside the study of a specific graph class for its maximum edge density, another parameter that is often considered in the literature is the size of the largest complete or complete bipartite graph belonging to it.

Overcoming the limitations of standard combinatorial arguments, we present a technique to systematically generate all non-isomorphic topological representations of complete and complete bipartite graphs, taking into account the constraints of the specific class. As a proof of concept, we apply our technique to various beyond-planarity classes and achieve new tight bounds for the aforementioned parameter.

Keywords: Beyond planarity · Complete (bipartite) graphs · Generation of topological representations ·

1 Introduction

Beyond-planarity is an active research area concerned with combinatorial properties of non-planar graphs that lie in the "neighborhood" of planar graphs. More concretely, these graphs allow for non-planar drawings in which certain geometric or topological crossing configurations are forbidden. The most studied beyond-planarity classes, with early results dating back to 60's [10,43], are the *k-planar* graphs [40], which forbid an edge to be crossed more than k times, and the *k-quasiplanar* graphs [4], which forbid k mutually crossing edges; see Fig. 1a–b.

More recently, several other classes have been suggested (e.g., [6,12]), also motivated by cognitive experiments [33,38] indicating that the absence of certain types of crossings helps in improving the readability of a drawing; for a survey, refer to [25]. Some of the most studied are: (i) *fan-planar* graphs, in which no edge can be crossed by two independent edges or by two adjacent edges from

This project was supported by DFG grant KA812/18-1.

D. Archambault and C. D. Tóth (Eds.): GD 2019, LNCS 11904, pp. 253–267, 2019.
https://doi.org/10.1007/978-3-030-35802-0_20

different directions [13,14,34], (ii) *fan-crossing free* graphs, in which no edge can be crossed by two adjacent edges [17,20], (iii) *gap-planar* graphs, in which each crossing is assigned to one of its two involved edges, such that each edge can be assigned at most one crossing [12], and (iv) *RAC graphs*, in which edge crossings occur only at right angles [23,24,26]; see Fig. 1c–e. Note that all the aforementioned graph classes are *topological*, i.e., each edge is represented as a simple curve, with the only exception of the class of RAC graphs, which is a purely *geometric* graph class, i.e., each edge must be represented as a straight-line segment. In this work, we refer to the aforementioned topological graph classes as *beyond-planarity classes of topological graphs*.

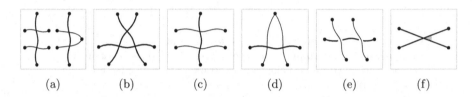

| (a) | (b) | (c) | (d) | (e) | (f) |

Fig. 1. Different forbidden crossing configurations in: (a) 1-planar, (b) 3-quasiplanar, (c) fan-planar, (d) fan-crossing free, (e) gap-planar, and (f) RAC graphs.

A common characteristic of these graph classes is that their edge density is at most linear in the number of vertices, e.g., 1-planar graphs with n vertices have at most $4n-8$ edges [40]; see Table 1. Another common measure to determine the extent of a specific class is the size of the largest complete or complete bipartite graph belonging to it [12,15,21,22], which also provides a lower bound on their chromatic number [31] and has been studied in related fields (e.g., [9,18,27,32]).

For 1-planar graphs, Czap and Hudák [21] proved that the complete graph K_n is 1-planar if and only if $n \leq 6$, and that the complete bipartite graph $K_{a,b}$, with $a \leq b$, is 1-planar if and only if $a \leq 2$, or $a = 3$ and $b \leq 6$, or $a = b = 4$. An analogous characterization is known for the class of RAC graphs by Didimo et al. [22,23], who proved that K_n is a RAC graph if and only if $n \leq 5$, while $K_{a,b}$, with $a \leq b$, is a RAC graph if and only if $a \leq 2$, or $a = 3$ and $b \leq 4$. For the classes of 3-quasiplanar (also known as *quasiplanar*), gap-planar, and fan-crossing free graphs, characterizations exist only for complete graphs, i.e., K_n is quasiplanar if and only if $n \leq 10$ [3,15], gap-planar if and only if $n \leq 8$ [12], and fan-crossing free if and only if $n \leq 6$ [20,21]; Table 1 gives more details.

To prove the "if part" of these characterizations, one has to provide a certificate drawing of the respective graph. The proof for the "only if part" is generally more complex, as it requires arguments to show that no such drawing exists.

One of the main techniques is provided by the linear edge density of the graph classes; e.g., K_7 is neither 1-planar nor fan-crossing free, as it has more than $4n - 8$ edges [20,40]. However, this technique has a limited applicability; e.g., for 2-planar and fan-planar graphs, which have at most $5n-10$ edges, it only ensures that K_9 is not a member of these classes. Proving that K_8 is also not a member requires a different approach. The limitations are even more evident for complete bipartite graphs, as they are sparser than the complete ones (see Sect. 4).

Table 1. Known results and our findings. For each class, we present the largest complete and complete bipartite graphs that belong to this class (col. "∈"), and the smallest ones that do not (col. "∉"). Color gray indicates weaker results that follow from other entries.

Class	Density	Complete				Complete bipartite			
		∈	Ref	∉	Ref	∈	Ref	∉	Ref
1-planar	$4n-8$	K_6	[21, Fig. 1]	K_7	[40, Theorem 1]	$K_{3,6}$	[21, Fig. 2]	$K_{3,7}$	[21, Lemma 4.2]
						$K_{4,4}$	[21, Fig. 3]	$K_{4,5}$	[21, Lemma 4.3]
2-planar	$5n-10$	K_7	[14, Fig. 7]	K_8	Char. 2	$K_{3,10}$	[6, Lemma 1]	$K_{3,11}$	[6, Lemma 1]
						$K_{4,6}$	Char. 3	$K_{4,7}$	Char. 3
						$K_{4,5}$		$K_{5,5}$	Char. 3 [35]
3-planar	$\frac{11}{2}n-11$	K_8	Char. 2	K_9	Char. 2	$K_{3,14}$	[6, Lemma 1]	$K_{3,15}$	[6, Lemma 1]
						$K_{4,9}$	Char. 4	$K_{4,10}$	Char. 4
						$K_{5,6}$	Char. 4	$K_{5,7}$	Char. 4
						$K_{5,6}$		$K_{6,6}$	Char. 4
4-planar	$6n-12$	K_9	Char. 2	K_{10}	Char. 2	$K_{3,18}$	[6, Lemma 1]	$K_{3,19}$	[6, Lemma 1]
						$K_{4,11}$	Obs. 5	$K_{4,19}$	
						$K_{5,8}$	Obs. 5	$K_{5,19}$	
						$K_{6,6}$	Obs. 5	$K_{6,19}$	
fan-planar	$5n-10$	K_7	[14, Fig. 7]	K_8	Char. 6	$K_{4,n}$	[34, Fig. 3]	$K_{5,5}$	Char. 7
fan-crossing free	$4n-8$	K_6	[21, Fig. 1]	K_7	[20, Theorem 1]	$K_{3,6}$		$K_{3,7}$	Char. 9
						$K_{4,6}$	Char. 9	$K_{4,7}$	
						$K_{4,5}$		$K_{5,5}$	Char. 9
gap-planar	$5n-10$	K_8	[12, Fig. 7]	K_9	[12, Theorem 23]	$K_{3,12}$	[12, Fig. 7]	$K_{3,14}$	[11, Theorem 1]
						$K_{4,8}$	[12, Fig. 9]	$K_{4,9}$	Obs. 11
						$K_{5,6}$	[12, Fig. 9]	$K_{5,7}$	[12]
						$K_{5,6}$		$K_{6,6}$	[11, Theorem 1]
RAC	$4n-10$	K_5	[26, Fig. 5]	K_6	[23, Theorem 1]	$K_{3,4}$	[22, Fig. 4]	$K_{3,5}$	[22, Theorem 2]
						$K_{3,4}$		$K_{4,4}$	[22, Theorem 2]
quasiplanar	$\frac{13}{2}n-20$	K_{10}	[15, Fig. 1]	K_{11}	[3, Theorem 5]	$K_{4,n}$	[34, Fig. 3]	$-$	
						$K_{5,18}$	Obs. 13	?	
						$K_{6,10}$	Obs. 13	?	
						$K_{7,7}$	Obs. 13	$K_{7,52}$	[3, Theorem 5]

Another technique consists of showing that the minimum number of crossings required by *any* drawing of a certain graph (as derived by, e.g., the Crossing Lemma [2,5,39] or closed formulas [44]) exceeds the maximum number of crossings allowed in the considered graph class. However, this technique only applies to classes that impose such restrictions, e.g., gap- and 1-planar graphs [11,21].

This difficulty in finding combinatorial arguments to prove that certain complete (bipartite) graphs do not belong to specific classes often results in the need of a large case analysis on the different topological representations of the graph. Beside the proofs in [22,35], we give in [8] another example of a combinatorial proof that, based on a tedious case analysis, yields a characterization of the complete bipartite fan-crossing free graphs. The range of the cases in these proofs justifies the need of a tailored approach to systematically explore them.

Our Contribution. We suggest a technique to engineer the analysis of all topological representations of a graph that satisfy certain beyond-planarity constraints. Our technique does not extend to classes of geometric graphs, and is tailored for complete and complete bipartite graphs, as we exploit their symmetry to reduce the search space, by discarding equivalent topological representations.

In Sect. 3, we present an algorithm to generate all possible representations of such graphs under different topological constraints on the crossing configurations. Our algorithm builds on two key ingredients, which allow to drastically reduce the search space. First, the representations are constructed by adding a vertex at a time, directly taking into account the topological constraints, thus avoiding constructing unnecessary representations. Second, at each intermediate step, the produced drawings are efficiently tested for equivalence (up to a relabeling of the vertices), which usually allows to discard a large set of them. Using this algorithm, we derived characterizations for several classes, as described in Sect. 4; Table 1 positions our results with respect to the state of the art. We give preliminary definitions in Sect. 2 and discuss future directions in Sect. 5.

2   Preliminaries

We assume familiarity with standard definitions on planar graphs and drawings (see, e.g., [8]). We assume *simple* drawings, in which there are no self-crossing edges, two edges cross at most once, and adjacent edges do not cross; note this assumption is not without loss of generality [3]. Given a planarization Γ of a graph G, a *half-pathway for a vertex* u in Γ is a path in the dual of Γ from a face incident to u to some face in Γ, called its *destination*; see Fig. 2a. The *length* of a half-pathway is the number of edges in this path. A half-pathway for u is *valid* with respect to a beyond-planarity class $\mathcal{C}$ of topological graphs, if Γ can be augmented such that (i) a vertex v is placed in its destination, (ii) edge (u, v) is drawn as a curve from u to v that crosses only the edges that are dual to the edges in this half-pathway, in the same order, and (iii) the drawing of (u, v) violates neither the simplicity of the resulting drawing nor the crossing restrictions of class $\mathcal{C}$. Accordingly, a *pathway for an edge* (u, v) is a half-pathway for vertex u in Γ, whose destination is a face incident to vertex v. A *valid pathway* is defined analogously, with the only exception that v is already part of Γ.

Another ingredient of our algorithm is an equivalence-relationship between different drawings of a graph G, i.e, drawings D_1 and D_2 of G are *isomorphic* [36] if there exists a homeomorphism of the sphere transforming D_1 into D_2. Namely, D_1 and D_2 are isomorphic if D_1 can be transformed into D_2 by relabeling vertices, edges, and faces of D_1, and by moving vertices and edges of D_1, so that at no time of this process new crossings are introduced, existing crossings are eliminated, or the order of the crossings along an edge is modified. We define a *valid* bijective mapping between vertices, crossings, edges, and faces of the planarizations Γ_1 and Γ_2 of D_1 and D_2 such that: **(P.1)** if an edge (v_1, w_1) is mapped to an edge (v_2, w_2) in Γ_1 and Γ_2, respectively, and v_1 is mapped to v_2, then w_1 is mapped to w_2; **(P.2)** if a face f_1 is mapped to a face f_2 in Γ_1 and Γ_2, respectively, and an edge e_1 incident to f_1 is mapped to an edge e_2 incident to f_2, then the predecessor (successor) of e_1 is mapped to the predecessor (successor) of e_2 when walking along the boundaries of f_1 and f_2 in clockwise direction. Also, the face incident to the other side of e_1 is mapped to the face incident to the other side of e_2. Clearly, Properties P.1 and P.2 are sufficient for D_1 and D_2 to be isomorphic. We believe they are also necessary, but this is beyond the scope of this work. Note that Property P.2 guarantees that two vertices are mapped to each other only if they have the same degree.

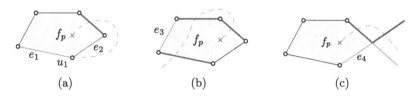

(a)   (b)   (c)

Fig. 2. The prohibited edges (blue solid) for a half-pathway (red dashed) that ends in a face f_p. The thick blue edges are prohibited, because they are crossed by the half-pathway. In (a) edges e_1 and e_2 are prohibited, since they are incident to u_1. In (b) edge e_3 is prohibited, since, in order to cross this edge, the half-pathway would make a self-crossing. In (c) edge e_4 is prohibited since it is part of a crossed edge. (Color figure online)

Several works [1, 30, 42] that generate simple drawings of complete graphs adopt a weaker definition of isomorphism; two drawings D_1 and D_2 are *weakly isomorphic* [36], if there exists an incidence preserving bijection between their vertices and edges, such that two edges cross in D_1 if and only if they do in D_2. Weakly isomorphic drawings that are non-isomorphic differ in the order in which their edges cross [29]. Two simple drawings of a complete graph with the same cyclic order of the edges around each vertex (called *rotation system*) are weakly isomorphic, and vice versa [29,41]; hence, generating all simple drawings of a complete graph reduces to finding all rotation systems that determine simple drawings [37]. However, this property holds only for complete graphs [1], while for the complete bipartite graphs, which are more difficult to handle, only partial results exist in this direction [19]. Thus, we decided not to follow this approach.

3   Generation Procedure

Let $\mathcal{C}$ be a beyond-planarity class of topological graphs and let G be a graph with $n \geq 3$ vertices. Assuming that G is either complete or complete bipartite, we describe in this section an algorithm to generate all non-isomorphic simple drawings of G that are certificates that G belongs to $\mathcal{C}$ (if any). We stress that, if G is neither complete nor complete bipartite, then it is a more involved task to recognize isomorphic drawings [28], and thus to eliminate them, which is a key point in the efficiency of our approach (we provide more details in Sect. 4).

Our algorithm aims at computing a set S containing all non-isomorphic simple drawings of G. In the base of the recursion, graph G is a cycle of length 3 or 4, depending on whether G is the complete graph K_3 or the complete bipartite graph $K_{2,2}$. In the former case, set S only contains a planar drawing of K_3, while in the latter case, S contains a planar drawing and one with a crossing between two non-adjacent edges. This is because, in both cases, any other drawing is either isomorphic to one of these, or non-simple.

In the recursive step, we consider a vertex v of G and assume that we have recursively computed the set S for $G \setminus \{v\}$. We may assume w.l.o.g. that $S \neq \emptyset$, as otherwise G would not belong to C. Then, we consider each drawing of S and our goal is to report all non-isomorphic simple drawings of G that have it as a subdrawing. In other words, we aim at reporting all non-isomorphic simple drawings that can be derived by all different placements of vertex v and the routing of its incident edges in the drawings of S. To this end, let Γ be the planarization of one of the drawings in S, and let $u_1, \ldots, u_k$ be the neighbors of v in G, where $k = \deg(v)$. If G is a complete graph, then $k = n - 1$; otherwise, G is a complete bipartite graph $K_{a,b}$ with $a + b = n$, and $k = a$ or $k = b$ holds.

We start by computing all possible valid half-pathways for u_1 in Γ with respect to C, which corresponds to constructing all possible drawings of edge (v, u_1) that respect simplicity and the restrictions of class C. To compute these half-pathways, we again use recursion. For each half-pathway, we maintain a list of so-called *prohibited* edges, which are not allowed to be crossed when inserting edge (u_1, v), as otherwise either the simplicity or the crossing restrictions of class C would be violated; see Fig. 2. This list is initialized with all edges incident to u_1 and is updated at every recursive step.

In the base of this inner recursion, we determine all valid half-pathways for u_1 of length zero; this means that, for each face f incident to u_1, we create a half-pathway that starts at f and has its destination also at f, which corresponds to placing v in f and drawing edge (v, u_1) crossing-free. Assume now that we have computed all valid half-pathways of some length $i \geq 0$ in Γ. We show how to compute all valid half-pathways for u_1 of length $i + 1$ (if any). Consider a half-pathway p of length i. Let f_p be its destination. Every non-prohibited edge e of f_p implies a new half-pathway of length $i + 1$, composed of p followed by the edge that is dual to e in Γ. Note that this process will eventually terminate, since the length of a half-pathway is bounded by the number of edges of Γ.

For each valid half-pathway p computed by the procedure above, we obtain a new drawing by inserting (u_1, v) into Γ following p and by inserting v into the destination of p. It remains to insert the remaining edges incident to v, i.e., $(v, u_2), \ldots, (v, u_k)$, into each of these drawings – again in all possible ways. For this, we proceed mostly as above with one difference. Instead of half-pathways, we search for valid pathways for each edge (v, u_i), $2 \leq i \leq k$, i.e., we only consider pathways that start in a face incident to v and end in a face incident to u_i.

If we find an edge (v, u_i) for which no valid pathway exists, we declare that Γ cannot be extended to a simple drawing of G that respects the crossing restrictions of $\mathcal{C}$. Otherwise, the computed drawings of G are added to $\mathcal{S}$, once all the drawings of $G \setminus \{v\}$ have been removed from it. To maintain our initial invariant, however, once a new drawing is to be added to $\mathcal{S}$, it will be first checked for isomorphism against all previously added drawings. If there is an isomorphic one, then the current drawing is discarded; otherwise, it is added to $\mathcal{S}$.

We stress that we test isomorphism using Properties P.1 and P.2 of a valid bijection. Since these properties are sufficient but we do not know whether they are also necessary, set $\mathcal{S}$ might contain some isomorphic drawings. However, our experiments indicate that the vast majority of them will be discarded.

Testing for Isomorphism. We describe a procedure to test whether the planarizations Γ_1 and Γ_2 of two drawings of G comply with Properties P.1 and P.2 of a valid bijection. We start by selecting two edges $e_1 = (v_1, w_1)$ and $e_2 = (v_2, w_2)$ in Γ_1 and Γ_2, respectively, whose end-vertices have compatible types (i.e., v_1 and v_2 are both real vertices or both crossings, and the same holds for w_1 and w_2). We bijectively map e_1 to e_2, v_1 to v_2, and w_1 to w_2, which complies with Property P.1. We call this a *base mapping* and try to extend it to a valid bijection.

We map to each other the face f_1 of Γ_1 that is "left" of e_1 (when walking along e_1 from v_1 to w_1) and the face f_2 of Γ_2 that is "left" of e_2 (when walking along e_2 from v_2 to w_2). If the degrees of f_1 and f_2 are different, then the base mapping cannot be extended. Otherwise, both f_1 and f_2 have degree δ, and we walk simultaneously along their boundaries, starting at e_1 and e_2 respectively; in view of Property P.2, for each $i = 1, \ldots, \delta$, we bijectively map the i-th vertex (either real or crossing) of f_1 to the i-th vertex of f_2, and the i-th edge of f_1 to the i-th edge of f_2. If a crossing is mapped to a real vertex, or if the degrees of two mapped vertices are different, then the base mapping cannot be extended.

If the vertices and edges of f_1 and f_2 have been mapped successfully, we proceed by considering the two maximal connected subdrawings Γ_1' and Γ_2' of Γ_1 and Γ_2, respectively, such that each edge of Γ_1' and Γ_2' has at least one face incident to it that is already mapped. Consider an edge e_1' of Γ_1' that is incident to only one mapped face f_1' (such an edge exists, as long as the base mapping has not been completely extended). Let e_2' be the edge of Γ_2' mapped to e_1'; note that e_2' must be incident to a face f_2' that is mapped to f_1' and to a face that is not mapped yet. We map to each other the faces incident to e_1' end e_2' that are not mapped yet, and we proceed by applying the procedure described above (i.e., we walk along the boundaries of f_1' and f_2' simultaneously, while ensuring that the mapping remains valid). If this procedure can be performed successfully, then we have computed two subdrawings Γ_1'' and Γ_2'', such that $\Gamma_1' \subseteq \Gamma_1''$, $\Gamma_2' \subseteq \Gamma_2''$, and each edge of them has at least one face incident to it that is already mapped. Hence, we can recursively apply the aforementioned procedure to Γ_1'' and Γ_2''.

Drawings Γ_1 and Γ_2 are isomorphic, if the base mapping can be eventually extended. If not, then we have to consider another base mapping and check whether this can be extended. Note that the case where e_1 is bijectively mapped to e_2, v_1 to w_2, and w_1 to v_2 defines a different base mapping than the one we were currently considering. If none of the base mappings can be extended, then we consider Γ_1 and Γ_2 as non-isomorphic. To reduce the number of base mappings that we have to consider, we first count the number of edges of Γ_1 and Γ_2 whose endpoints are both real vertices, both crossings, and those consisting of one real vertex and one crossing. These numbers have to be the same in Γ_1 and Γ_2. Since it is enough to consider base mappings only restricted to one of the three types of edges, we choose the type with the smallest positive number of occurrences. We summarize the above discussion in the following theorem.

Theorem 1. *Let G be a complete (or a complete bipartite) graph and let C be a beyond-planarity class of topological graphs. Then, G belongs to C if and only if, under the restrictions of class C, our algorithm returns a valid drawing of G.*

4   Proof of Concept - Applications

In this section we use the algorithm described in Sect. 3 to test whether certain complete or complete bipartite graphs belong to specific beyond-planarity graph classes. We give corresponding characterizations and discuss how our findings are positioned within the literature. Our lower bound examples are drawings that certify membership to particular beyond-planarity graph classes, computed by an implementation (https://github.com/beyond-planarity/complete-graphs) of our algorithm; for typesetting reasons we redrew them. Our upper bounds are the smallest corresponding instances reported as negative by our algorithm.

The Class of k-planar Graphs. We start our discussion with the case of complete graphs. As already mentioned in the introduction, the complete graph K_n is 1-planar if and only if $n \leq 6$ [21].

For the case of complete 2-planar graphs, the fact that a 2-planar graph with n vertices has at most $5n - 10$ edges [40] implies that K_9 is not a member of this class. Figure 7 in [14], on the other hand, shows that K_7 is 2-planar. We close this gap by showing, with our implementation, that even K_8 is not 2-planar.

For the cases of complete 3-, 4-, and 5-planar graphs, the application of a similar density argument as above proves that K_{10}, K_{11}, and K_{19} are not 3-, 4-, and 5-planar, respectively [2,39]. With our implementation, we could show that even K_9 is not 3-planar, while K_{10} is neither 4- nor 5-planar. On the other hand, our algorithm was able to construct 3- and 4-planar drawings of K_8 and K_9, respectively; see Fig. 3a and b. Note that a 6-planar drawing of K_{10} can be easily derived from the 4-planar drawing of K_9 in Fig. 3b by adding one extra vertex inside the red colored triangle. We have the following characterization.

Characterization 2. *For $k \in \{1, 2, 3, 4\}$, the complete graph K_n is k-planar if and only if $n \leq 5 + k$. Also, K_n is 5-planar if and only if $n \leq 9$.*

Note that the 3-planarity of K_8 implies that the chromatic number of 3-planar graphs is lower bounded by 8. Analogous implications can be derived for the classes of 4-, 5-, and 6-planar graphs. Another observation that came out from our experiments is that, up to isomorphism, K_6 has a unique 1-planar drawing,

K_7 has only two 2-planar drawings, and K_8 has only three 3-planar drawings, while the number of non-isomorphic 4-planar drawings of K_9 is significantly larger, namely 35. For more details, refer to Table 2, and to [8].

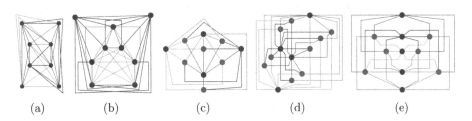

(a)   (b)   (c)   (d)   (e)

Fig. 3. Illustration of (a) a 3-planar drawing of K_8, (b) a 4-planar drawing of K_9, (c) a drawing of $K_{4,6}$ that is both 2-planar and fan-crossing free, (d) a 3-planar drawing of $K_{4,9}$, and (e) a 3-planar drawing of $K_{5,6}$.

Consider now a complete bipartite graph $K_{a,b}$ with $a \leq b$. Note that $a \leq 2$ implies that $K_{a,b}$ is planar; thus, it trivially belongs to all beyond-planarity graph classes. Also, recall that $K_{a,b}$ is 1-planar if and only if $a \leq 2$, or $a = 3$ and $b \leq 6$, or $a = b = 4$ [21]. Further, a recent combinatorial result states that $K_{3,b}$ is k-planar if and only if $b \leq 4k + 2$ [6]. So, in the following we assume $a \geq 4$.

For complete bipartite 2-planar graphs, the fact that a bipartite 2-planar graph with n vertices has at most $3.5n - 7$ edges [7] implies that neither $K_{4,15}$ nor $K_{5,8}$ is 2-planar. With our implementation, we could show that $K_{4,7}$ and $K_{5,5}$ are not 2-planar, while $K_{4,6}$ is (see Fig. 3c), yielding the following characterization.

Characterization 3. *The complete bipartite graph $K_{a,b}$ (with $a \leq b$) is 2-planar if and only if (i) $a \leq 2$, or (ii) $a = 3$ and $b \leq 10$, or (iii) $a = 4$ and $b \leq 6$.*

As opposed to the corresponding 2-planar case, there exists no upper bound on the edge density of 3-planar graphs tailored for the bipartite setting. The upper bound of $5.5n - 11$ edges [39] for general 3-planar graphs with n vertices does not provide any negative instance for $a \leq 5$, and only proves that $K_{6,b}$, with $b \geq 45$, is not 3-planar. With our implementation, we could provide significant improvements, by showing that $K_{4,10}$, $K_{5,7}$, and $K_{6,6}$ are not 3-planar, while $K_{4,9}$ and $K_{5,6}$ are (see Fig. 3d and e), which yields the following characterization.

Characterization 4. *The complete bipartite graph $K_{a,b}$ (with $a \leq b$) is 3-planar if and only if (i) $a \leq 2$, or (ii) $a = 3$ and $b \leq 14$, or (iii) $a = 4$ and $b \leq 9$, or (iv) $a = 5$ and $b \leq 6$.*

For complete bipartite 4-planar graphs, we were unable to derive a characterization, but only some partial results, because the search space becomes drastically larger and, as a consequence, our generation technique could not terminate. To give an intuition, note that $K_{4,4}$ has 81817 non-isomorphic 4-planar drawings, which makes the computation of the corresponding non-isomorphic drawings of $K_{4,5}$ infeasible in reasonable time; for more details refer to [8].

However, we were at least able to report some positive certificate drawings by slightly refining our generation technique. Instead of computing *all* possible non-isomorphic simple drawings of graph $K_{a-1,b}$ or $K_{a,b-1}$, in order to compute the corresponding ones for $K_{a,b}$, we only computed few *samples*, hoping that we will eventually find a positive certificate drawing. With this so-called *DFS-like* approach, we managed to derive 4-planar drawings for $K_{4,11}$, $K_{5,8}$, and $K_{6,6}$; see Fig. 4. We summarize these findings in the following observation.

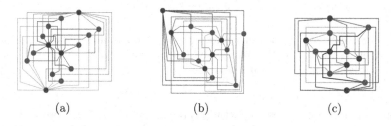

| (a) | (b) | (c) |

Fig. 4. Illustration of 4-planar drawings of (a) $K_{4,11}$, (b) $K_{5,8}$ and (c) $K_{6,6}$.

Observation 5. *The complete bipartite graph $K_{a,b}$ (with $a \leq b$) is 4-planar if (i) $a \leq 2$, or (ii) $a = 3$ and $b \leq 18$, or (iii) $a = 4$ and $b \leq 11$, or (iv) $a = 5$ and $b \leq 8$, or (v) $a = 6$ and $b = 6$. Further, $K_{a,b}$ is not 4-planar if $a \geq 3$ and $b \geq 19$.*

The Class of Fan-Planar Graphs. We start our discussion with complete graphs. The fact that a fan-planar graph with n vertices has at most $5n - 10$ edges [34] implies that K_9 is not fan-planar, while Fig. 7 in [14] shows that K_7 is. With our implementation, we showed that K_8 is not fan-planar, even relaxing the requirement that an edge crossed by two or more adjacent edges must be crossed from the same direction; see, e.g., [16]. This yields the following characterization.

Characterization 6. *The complete graph K_n is fan-planar if and only if $n \leq 7$.*

Consider now a complete bipartite graph $K_{a,b}$ with $a \leq b$. For $a \leq 4$, $K_{a,b}$ is fan-planar for any value of b [34]. On the other hand, the fact that a bipartite fan-planar graph has at most $4n-12$ edges [7] implies that $K_{5,9}$ is not fan-planar. Using our implementation, we could show that even $K_{5,5}$ is not fan-planar (again by relaxing the requirement of having the crossings from the same direction). These two results together imply the following characterization.

Characterization 7. *The complete bipartite graph $K_{a,b}$ (with $a \leq b$) is fan-planar if and only if $a \leq 4$.*

The Class of Fan-Crossing Free Graphs. A characterization for the case of complete graphs can be derived by combining two known results. First, K_6 is fan-crossing free, as it is 1-planar. We additionally show in [8] that, up to isomorphism, K_6 has a unique fan-crossing free drawing. Second, the fact that a fan-crossing free graph with n vertices has at most $4n - 8$ edges [20] implies that K_7 is not fan-crossing free. Hence, we have the following characterization.

Characterization 8 (Cheong et al. [20], Czap et al. [21]). *The complete graph K_n is fan-crossing free if and only if $n \leq 6$.*

As already stated, for the complete bipartite fan-crossing free graphs, we provide in [8] a combinatorial proof of their characterization. The same result was also obtained by our implementation.

Characterization 9. *The complete bipartite graph $K_{a,b}$ (with $a \leq b$) is fan-crossing free if and only if (i) $a \leq 2$, or (ii) $a \leq 4$ and $b \leq 6$.*

The Class of Gap-Planar Graphs. A characterization of the complete gap-planar graphs has already been provided [12] as follows.

Characterization 10 (Bae et al. [12]). *The complete graph K_n is gap-planar if and only if $n \leq 8$.*

For the case of complete bipartite graphs, Bae et al. [12] proved that $K_{3,12}$, $K_{4,8}$, and $K_{5,6}$ are gap-planar, while $K_{3,15}$, $K_{4,11}$, and $K_{5,7}$ are not. These negative results were derived using the technique discussed in Sect. 1 that compares the crossing number of these graphs with their number of edges, which is an upper bound to the number of crossings allowed in a gap-planar drawing. By refining this technique, Bachmaier et al. [11] proved that even $K_{3,14}$, $K_{4,10}$, and $K_{6,6}$ are not gap-planar. Hence, towards a characterization the cases that are left open are $K_{3,13}$ and $K_{4,9}$. Here, we address one of these two open cases by showing that $K_{4,9}$ is not gap-planar, thus yielding the following observation.

Observation 11. *The complete bipartite graph $K_{a,b}$ (with $a \leq b$) is gap-planar if (i) $a \leq 2$, or (ii) $a = 3$ and $b \leq 12$, or (iii) $a = 4$ and $b \leq 8$, or (iv) $a = 5$ and $b \leq 6$. Further, $K_{a,b}$ is not gap-planar if (i) $a = 3$ and $b \geq 14$, or (ii) $a = 4$ and $b \geq 9$, or (iii) $a = 5$ and $b \geq 7$, or (iv) $a \geq 6$ and $b \geq 6$.*

The Class of Quasiplanar Graphs. A characterization for the complete quasiplanar graphs can be also derived by combining two known results. Namely, the fact that a quasiplanar graph with n vertices has at most $6.5n - 20$ edges [3] implies that K_{11} is not quasiplanar, while K_{10} is in fact quasiplanar [15].

Characterization 12 (Ackerman et al. [3], Brandenburg [15]). *The complete graph K_n is quasiplanar if and only if $n \leq 10$.*

Consider now a complete bipartite graph $K_{a,b}$ with $a \leq b$. First, we observe that for $a \leq 4$, graph $K_{a,b}$ is quasiplanar for any value of b, since it is even fan-planar [34]. On the other hand, the fact that a quasiplanar graph with n vertices has at most $6.5n - 20$ edges [3] does not provide any negative answer for $a \leq 6$, while for $a = 7$ it only implies that $K_{7,52}$ is not quasiplanar. We stress that we were not able to find any improvement on the latter result. The reason is the same as the one that we described for the class of complete bipartite 4-planar graphs. To give an intuition, we note that $K_{4,4}$ has in total 46711 non-isomorphic quasiplanar drawings, which makes the computation of the corresponding non-isomorphic drawings of $K_{4,5}$ infeasible in reasonable time; refer to [8] for details. Notably, using the DFS-like variant of our algorithm, we were able to derive at least positive certificate drawings for $K_{5,18}$, $K_{6,10}$, and $K_{7,7}$, which are given in [8]. We summarize these findings in the following observation.

Observation 13. *The complete bipartite graph $K_{a,b}$ (with $a \leq b$) is quasiplanar if (i) $a \leq 4$, or (ii) $a = 5$ and $b \leq 18$, or (iii) $a = 6$ and $b \leq 10$, or (iv) $a = 7$ and $b \leq 7$. Further, $K_{a,b}$ is not quasiplanar if $a \geq 7$ and $b \geq 52$.*

Table 2. A comparison of the number of drawings reported by our algorithm with the elimination of isomorphic drawings (col. "Non-Iso") and without it (col. "All") for the classes of 1- and 2-planar graphs; the corresponding execution times (in sec.) to compute these drawings are reported next to them.

Class	Complete					Complete bipartite				
	Graph	Non-Iso	Time	All	Time	Graph	Non-Iso	Time	All	Time
1-planar	K_4	2	0.043	8	0.043	$K_{2,3}$	3	0.061	34	0.061
	K_5	1	0.043	30	0.206	$K_{3,3}$	2	0.049	84	0.539
	K_6	1	0.020	120	0.737	$K_{3,4}$	3	0.065	960	5.642
	K_7	0	0.006	0	0.448	$K_{4,4}$	2	0.044	1584	10.871
						$K_{4,5}$	0	0.010	0	7.198
	Total:	4	0.112	158	1.434	total:	10	0.229	2662	24.311
2-planar	K_4	2	0.028	8	0.028	$K_{2,3}$	6	0.090	76	0.090
	K_5	4	0.105	294	2.661	$K_{3,3}$	19	0.254	2352	10.571
	K_6	6	0.233	2664	3.292	$K_{3,4}$	71	1.458	52248	244.964
	K_7	2	0.119	8400	55.323	$K_{4,4}$	38	1.152	168624	1128.457
	K_8	0	0.029	0	51.321	$K_{4,5}$	37	1.826	1200384	8135.843
						$K_{5,5}$	0	0.357	0	12639.293
	Total:	14	0.514	11366	112.625	total:	171	5.137	1423684	22159.218

5   Conclusions and Open Problems

We conclude this work by noting that our results also have some theoretical implications. In particular, $K_{5,5}$ was conjectured in [7] not to be fan-planar; Characterization 7 settles in the positive this conjecture. By Characterization 7 and Observation 11, we deduce that $K_{5,5}$ is a certificate that there exist graphs which are gap-planar but not fan-planar. Since $K_{4,9}$ is fan-planar but not gap-planar, the two classes are incomparable, which answers a related question posed in [12] about the relationship between 1-gap-planar graphs and fan-planar graphs.

We stress that the elimination of isomorphic drawings is a key step in our algorithm, as shown in Table 2. For example, to test whether $K_{5,5}$ is 2-planar without the elimination of intermediate isomorphic drawings, one would need to investigate 1423684 drawings, while in the presence of this step only 171. This significantly reduced the required time to roughly 5 seconds, including the time to perform all isomorphism tests and eliminations. We provide further insights in [8], where we broaden our description to the other classes.

Our work leaves two open problems. Is it possible to extend our approach to graphs that are neither complete nor complete bipartite, e.g., to k-trees or to k-degenerate graphs (for small values of k)? A major difficulty is that, in the absence of symmetry, discarding isomorphic drawings becomes more complex. A general observation from our proof of concept is that our approach was of limited applicability on the classes of complete bipartite k-planar graphs, for $k > 3$, and complete bipartite quasiplanar graphs, for which we could report partial results. So, is it possible to broaden these results by deriving improved upper bounds on the edge densities of these classes tailored for the bipartite setting (see, e.g., [7]).

References

1. Ábrego, B.M., Aichholzer, O., Fernández-Merchant, S., Hackl, T., Pammer, J., Pilz, A., Ramos, P., Salazar, G., Vogtenhuber, B.: All good drawings of small complete graphs. In: EuroCG, pp. 57–60 (2015)
2. Ackerman, E.: On topological graphs with at most four crossings per edge. CoRR abs/1509.01932 (2015)
3. Ackerman, E., Tardos, G.: On the maximum number of edges in quasi-planar graphs. J. Comb. Theory Ser. A **114**(3), 563–571 (2007). https://doi.org/10.1016/j.jcta.2006.08.002
4. Agarwal, P.K., Aronov, B., Pach, J., Pollack, R., Sharir, M.: Quasi-planar graphs have a linear number of edges. Combinatorica **17**(1), 1–9 (1997). https://doi.org/10.1007/BF01196127
5. Aigner, M., Ziegler, G.M.: Proofs from THE BOOK, 3rd edn. Springer, Heidelberg (2004). https://doi.org/10.1007/978-3-662-05412-3
6. Angelini, P., Bekos, M.A., Kaufmann, M., Kindermann, P., Schneck, T.: 1-fan-bundle-planar drawings of graphs. Theor. Comput. Sci. **723**, 23–50 (2018). https://doi.org/10.1016/j.tcs.2018.03.005
7. Angelini, P., Bekos, M.A., Kaufmann, M., Pfister, M., Ueckerdt, T.: Beyond-planarity: Turán-type results for non-planar bipartite graphs. In: ISAAC. LIPIcs, vol. 123, pp. 28:1–28:13. Schloss Dagstuhl (2018). https://doi.org/10.4230/LIPIcs.ISAAC.2018.28
8. Angelini, P., Bekos, M.A., Kaufmann, M., Schneck, T.: Efficient generation of different topological representations of graphs beyond-planarity. CoRR 1908.03042v2 (2019)

9. Arleo, A., Binucci, C., Di Giacomo, E., Evans, W.S., Grilli, L., Liotta, G., Meijer, H., Montecchiani, F., Whitesides, S., Wismath, S.K.: Visibility representations of boxes in 2.5 dimensions. Comput. Geom. **72**, 19–33 (2018). https://doi.org/10.1016/j.comgeo.2018.02.007

10. Avital, S., Hanani, H.: Graphs. Gilyonot Lematematika **3**, 2–8 (1966)

11. Bachmaier, C., Rutter, I., Stumpf, P.: 1-gap planarity of complete bipartite graphs. In: Biedl, T.C., Kerren, A. (eds.) Graph Drawing and Network Visualization. LNCS, vol. 11282, pp. 646–648. Springer, Heidelberg (2018)

12. Bae, S.W., Baffier, J., Chun, J., Eades, P., Eickmeyer, K., Grilli, L., Hong, S., Korman, M., Montecchiani, F., Rutter, I., Tóth, C.D.: Gap-planar graphs. Theor. Comput. Sci. **745**, 36–52 (2018). https://doi.org/10.1016/j.tcs.2018.05.029

13. Bekos, M.A., Cornelsen, S., Grilli, L., Hong, S., Kaufmann, M.: On the recognition of fan-planar and maximal outer-fan-planar graphs. Algorithmica **79**(2), 401–427 (2017)

14. Binucci, C., Di Giacomo, E., Didimo, W., Montecchiani, F., Patrignani, M., Symvonis, A., Tollis, I.G.: Fan-planarity: properties and complexity. Theor. Comp. Sci. **589**, 76–86 (2015)

15. Brandenburg, F.J.: A simple quasi-planar drawing of K_{10}. In: Graph Drawing. LNCS, vol. 9801, pp. 603–604. Springer (2016)

16. Brandenburg, F.J.: A first order logic definition of beyond-planar graphs. J. Graph Algorithms Appl. **22**(1), 51–66 (2018)

17. Brandenburg, F.J.: On fan-crossing and fan-crossing free graphs. Inf. Process. Lett. **138**, 67–71 (2018). https://doi.org/10.1016/j.ipl.2018.06.006

18. Bruckdorfer, T., Cornelsen, S., Gutwenger, C., Kaufmann, M., Montecchiani, F., Nöllenburg, M., Wolff, A.: Progress on partial edge drawings. J. Graph Algorithms Appl. **21**(4), 757–786 (2017). https://doi.org/10.7155/jgaa.00438

19. Cardinal, J., Felsner, S.: Topological drawings of complete bipartite graphs. JoCG **9**(1), 213–246 (2018)

20. Cheong, O., Har-Peled, S., Kim, H., Kim, H.: On the number of edges of fan-crossing free graphs. Algorithmica **73**(4), 673–695 (2015). https://doi.org/10.1007/s00453-014-9935-z

21. Czap, J., Hudák, D.: 1-planarity of complete multipartite graphs. Discrete Appl. Math. **160**(4–5), 505–512 (2012). https://doi.org/10.1016/j.dam.2011.11.014

22. Didimo, W., Eades, P., Liotta, G.: A characterization of complete bipartite RAC graphs. Inf. Process. Lett. **110**(16), 687–691 (2010)

23. Didimo, W., Eades, P., Liotta, G.: Drawing graphs with right angle crossings. Theor. Comput. Sci. **412**(39), 5156–5166 (2011)

24. Didimo, W., Liotta, G.: The crossing-angle resolution in graph drawing. In: Pach, J. (ed.) Thirty Essays on Geometric Graph Theory, pp. 167–184. Springer, New York (2013). https://doi.org/10.1007/978-1-4614-0110-0_10

25. Didimo, W., Liotta, G., Montecchiani, F.: A survey on graph drawing beyond planarity. ACM Comput. Surv. **52**(1), 4:1–4:37 (2019)

26. Eades, P., Liotta, G.: Right angle crossing graphs and 1-planarity. Discrete Appl. Math. **161**(7–8), 961–969 (2013)

27. Eppstein, D., Kindermann, P., Kobourov, S.G., Liotta, G., Lubiw, A., Maignan, A., Mondal, D., Vosoughpour, H., Whitesides, S., Wismath, S.K.: On the planar split thickness of graphs. Algorithmica **80**(3), 977–994 (2018). https://doi.org/10.1007/s00453-017-0328-y

28. Garey, M., Johnson, D.S.: Computers and Intractability: A Guide to the Theory of NP-Completeness. W. H. Freeman & Co., New York (1979)

29. Gioan, E.: Complete graph drawings up to triangle mutations. In: Kratsch, D. (ed.) WG 2005. LNCS, vol. 3787, pp. 139–150. Springer, Heidelberg (2005). https://doi.org/10.1007/11604686_13

30. Gronau, H.D.O., Harborth, H.: Numbers of nonisomorphic drawings for small graphs. Congressus Numerantium **71**, 105–114 (1990)

31. Hadwiger, H.: Über eine Klassifikation der Streckenkomplexe. Vierteljschr. Naturforsch. Ges. Zürich **88**, 133–143 (1943)

32. Hartsfield, N., Jackson, B., Ringel, G.: The splitting number of the complete graph. Graphs Comb. **1**(1), 311–329 (1985). https://doi.org/10.1007/BF02582960

33. Huang, W., Hong, S., Eades, P.: Effects of crossing angles. In: PacificVis 2008, pp. 41–46. IEEE (2008)

34. Kaufmann, M., Ueckerdt, T.: The density of fan-planar graphs. CoRR 1403.6184 (2014)

35. Kehribar, Z.: $K_{5,5}$ kann nicht 2-planar gezeichnet werden: Analyse und Beweis, Bachelor Thesis, Universität Tübingen (2018)

36. Kynčl, J.: Simple realizability of complete abstract topological graphs in P. Discrete Comput. Geom. **45**(3), 383–399 (2011). https://doi.org/10.1007/s00454-010-9320-x

37. Kynčl, J.: Improved enumeration of simple topological graphs. Discrete Comput. Geom. **50**(3), 727–770 (2013). https://doi.org/10.1007/s00454-013-9535-8

38. Mutzel, P.: An alternative method to crossing minimization on hierarchical graphs. SIAM J. Optimiz. **11**(4), 1065–1080 (2001). https://doi.org/10.1137/S1052623498334013

39. Pach, J., Radoičić, R., Tardos, G., Tóth, G.: Improving the crossing lemma by finding more crossings in sparse graphs. Discrete Comput. Geom. **36**(4), 527–552 (2006)

40. Pach, J., Tóth, G.: Graphs drawn with few crossings per edge. Combinatorica **17**(3), 427–439 (1997)

41. Pach, J., Tóth, G.: How many ways can one draw a graph? Combinatorica **26**(5), 559–576 (2006). https://doi.org/10.1007/s00493-006-0032-z

42. Rafla, N.H.: The good drawings D_n of the complete graph K_n. Ph.D. thesis, McGill. University, Montreal, Quebec (1988)

43. Ringel, G.: Ein Sechsfarbenproblem auf der Kugel. Abh. Math. Sem. Univ. Hamb. **29**, 107–117 (1965)

44. Zarankiewicz, K.: On a problem of P. Turán concerning graphs. Fundamenta Mathematicae **41**, 137–145 (1954)

The QuaSEFE Problem

Patrizio Angelini[1]([✉])[iD], Henry Förster[1][iD], Michael Hoffmann[2][iD],
Michael Kaufmann[1][iD], Stephen Kobourov[3][iD], Giuseppe Liotta[4],
and Maurizio Patrignani[5]

[1] University of Tübingen, Tübingen, Germany
angelini@informatik.uni-tuebingen.de
[2] ETH Zürich, Zürich, Switzerland
[3] University of Arizona, Tucson, USA
[4] University of Perugia, Perugia, Italy
[5] University Roma Tre, Rome, Italy

Abstract. We initiate the study of Simultaneous Graph Embedding
with Fixed Edges in the beyond planarity framework. In the `QuaSEFE`
problem, we allow edge crossings, as long as each graph individually is
drawn quasiplanar, that is, no three edges pairwise cross. We show that
a triple consisting of two planar graphs and a tree admit a `QuaSEFE`.
This result also implies that a pair consisting of a 1-planar graph and a
planar graph admits a `QuaSEFE`. We show several other positive results
for triples of planar graphs, in which certain structural properties for
their common subgraphs are fulfilled. For the case in which simplicity is
also required, we give a triple consisting of two quasiplanar graphs and a
star that does not admit a `QuaSEFE`. Moreover, in contrast to the planar
`SEFE` problem, we show that it is not always possible to obtain a `QuaSEFE`
for two matchings if the quasiplanar drawing of one matching is fixed.

Keywords: Quasiplanar · SEFE · Simultaneous graph drawing

1 Introduction

Simultaneous Graph Embedding is a family of problems where one is given a set
of graphs $G_1, \ldots, G_k$ with shared vertex set V and is required to produce drawings $\Gamma_1, \ldots, \Gamma_k$ of them, each satisfying certain readability properties, so that
each vertex has the same position in every Γ_i. The readability property that is
usually pursued is the planarity of the drawing, and a large body of research has
been devoted to establish the complexity of the corresponding decision problem,
or to determine whether such embeddings always exist, given the number and
the types of the graphs; for a survey refer to [9].

Work started at Dagstuhl Seminar 19092, "Beyond-Planar Graphs: Combinatorics,
Models and Algorithms". Research supported by MIUR Project "MODE" under PRIN
20157EFM5C, by MIUR Project "AHeAD" under PRIN 20174LF3T8, by Roma Tre
University Azione 4 Project "GeoView", by DFG grant Ka812/17-1, by NSF under
grants CCF-1740858 and CCF-1712119, and by SNSF Project 200021E-171681.

D. Archambault and C. D. Tóth (Eds.): GD 2019, LNCS 11904, pp. 268–275, 2019.
https://doi.org/10.1007/978-3-030-35802-0_21

These problems have been studied both from a geometric (*Geometric Simultaneous Embedding* - GSE) [6,16] and from a topological point of view (*Simultaneous Embedding with Fixed Edges* - SEFE) [10,12,19]. In particular, in GSE the edges are straight-line segments, while in SEFE they are topological curves, but the edges shared between two graphs G_i and G_j have to be drawn in the same way in Γ_i and Γ_j. Unless otherwise specified, we focus on the topological setting.

We study a relaxation of the SEFE problem, where the graphs can be drawn with edge crossings. However, we prohibit certain crossing configurations in the drawings $\Gamma_1, \ldots, \Gamma_k$, to guarantee their readability, i.e., we require that they satisfy the conditions of a graph class in the area of *beyond-planarity*; see [15] for a survey on this topic. We initiate this study with the class of *quasiplanar* graphs [2,3,18], by requiring that no Γ_i contains three mutually crossing edges.

Definition 1 (QuaSEFE). *Given a set of graphs $G_1 = (V, E_1), \ldots, G_k = (V, E_k)$ with shared vertex set V, we say that $\langle G_1, \ldots, G_k \rangle$ admits a QuaSEFE if there exist quasiplanar drawings $\Gamma_1, \ldots, \Gamma_k$ of $G_1, \ldots, G_k$, respectively, so that each vertex of V has the same position in every Γ_i and each edge shared between two graphs G_i and G_j is drawn in the same way in Γ_i and Γ_j. Further, the QuaSEFE problem asks whether an instance $\langle G_1, \ldots, G_k \rangle$ admits a QuaSEFE.*

It may be worth mentioning that the problem of computing quasiplanar simultaneous embeddings of graph pairs has been studied in the geometric setting [13,14]. Also, simultaneous embeddings have been considered in relation to another beyond-planarity geometric graph class, namely *RAC graphs* [7,8,17,20].

We prove in Sect. 2 that any triple of two planar graphs and a tree admits a QuaSEFE, which also implies that any pair consisting of a 1-planar graph[1] and a planar graph admits a QuaSEFE. Recall that, for the original SEFE problem, there exist even negative instances composed of two outerplanar graphs [19]. Further, we investigate triples of planar graphs in which the common subgraphs have specific structural properties. Finally, we show negative results in more specialized settings in Sect. 3 and conclude with open problems in Sect. 4.

2 Sufficient Conditions for QuaSEFEs

In this section, we provide several sufficient conditions for the existence of a QuaSEFE, mainly focusing on instances composed of three planar graphs G_1, G_2, and G_3. We start with a theorem relating the existence of a SEFE of two of the input graphs to the existence of a QuaSEFE of the three input graphs.

Theorem 1. *Let $G_1 = (V, E_1)$, $G_2 = (V, E_2)$, and $G_3 = (V, E_3)$ be planar graphs with shared vertex set V. If $\langle G_1 \setminus G_3, G_2 \setminus G_3 \rangle$ admits a SEFE, then $\langle G_1, G_2, G_3 \rangle$ admits a QuaSEFE, in which the drawing of G_3 is planar.*

---

[1] A graph is k-planar if it admits a drawing where each edge has at most k crossings.

Proof. First construct a SEFE of $\langle G_1 \setminus G_3, G_2 \setminus G_3 \rangle$, and then construct a planar drawing of G_3, whose vertices have already been placed, but whose edges have not been drawn yet, using the algorithm by Pach and Wenger [23].

The drawing of G_3 is planar, by construction. The drawing of G_1 is quasiplanar, as it is partitioned into two subgraphs, $G_1 \setminus G_3$ and $G_1 \cap G_3$, each of which is drawn planar. Analogously, the drawing of G_2 is quasiplanar.   □

Since every pair composed of a planar graph and a tree admits a SEFE [19], we derive from Theorem 1 the following positive result for the QuaSEFE problem.

Corollary 1. *Let $G_1 = (V, E_1)$ and $G_3 = (V, E_3)$ be planar graphs and $T_2 = (V, E_2)$ be a tree with shared vertex set V. Then $\langle G_1, T_2, G_3 \rangle$ admits a QuaSEFE, in which the drawing of G_3 is planar.*

Corollary 1 already shows that allowing quasiplanarity significantly enlarges the set of positive instances. We further strengthen this result, by additionally guaranteeing that even the tree is drawn planar. For this, we use a result on the *partially embedded planarity* [5] problem (PEP): Given a planar graph G, a subgraph H of G, and a planar embedding $\mathcal{H}$ of H, is there a planar embedding of G whose restriction to H coincides with $\mathcal{H}$? In particular, we will exploit the following characterization, which is the core of a linear-time algorithm for PEP.

Lemma 1 ([5]). *Let $(G, H, \mathcal{H})$ be an instance of PEP. A planar embedding $\mathcal{G}$ of G is a solution for $(G, H, \mathcal{H})$ if and only if the following conditions hold:* **(C.1)** *for every vertex $v \in V$, the edges incident to v in H appear in the same cyclic order in the rotation schemes of v in $\mathcal{H}$ and in $\mathcal{G}$; and* **(C.2)** *for every cycle C of H, and for every vertex v of $H \setminus C$, we have that v lies in the interior of C in $\mathcal{G}$ if and only if it lies in the interior of C in $\mathcal{H}$.*

Theorem 2. *Let $G_1 = (V, E_1)$ and $G_3 = (V, E_3)$ be planar graphs and $T_2 = (V, E_2)$ be a tree with shared vertex set V. Then $\langle G_1, T_2, G_3 \rangle$ admits a QuaSEFE, in which the drawings of G_1 and T_2 are planar.*

Proof. Consider planar embeddings $\mathcal{G}_1$ and $\mathcal{G}_3^*$ of G_1 and $G_3 \setminus G_1$, respectively. We draw G_1 according to $\mathcal{G}_1$. This fixes the embedding of the subgraph $T_2 \cap G_1$ of T_2, thus resulting in an instance of the PEP problem. Since T_2 is acyclic, Condition C.1 of Lemma 1 is trivially fulfilled. Also, since every rotation scheme of T_2 is planar, we choose for the edges of $(T_2 \cap G_3) \setminus G_1$ an order compatible with $\mathcal{G}_3^*$, still satisfying Condition C.1. Finally, we draw the remaining edges of G_3 by considering the instance of PEP defined by its embedded subgraph $(T_2 \cap G_3) \setminus G_1$. Condition C.1 is trivially satisfied, and Condition C.1 is satisfied by construction, if we add the edges of G_3 according to $\mathcal{G}_3^*$. Since crossings edges of the same graph belong to $G_3 \setminus G_1$ and $G_3 \cap G_1$, the drawing of G_3 is quasiplanar.   □

The additional property guaranteed by Theorem 2 is crucial to infer the first result in the simultaneous embedding setting for a class of beyond-planar graphs.

Theorem 3. *Let $G_1 = (V, E_1)$ be a 1-planar graph and $G_2 = (V, E_2)$ be a planar graph. Then $\langle G_1, G_2 \rangle$ admits a QuaSEFE.*

Proof. As G_1 is 1-planar, it is the union of a planar graph G_1' and a forest F_1 [1]. We augment F_1 to a tree T_1. By Theorem 2, there is a QuaSEFE of $\langle G_1', T_1, G_2 \rangle$ where G_1' and T_1 are drawn planar. Thus, G_1 is drawn quasiplanar.   □

We now study properties of the subgraphs induced by the edges that belong to one, to two, or to all the input graphs. We denote by H_i the subgraph induced by the edges only in G_i; by $H_{i,j}$ the subgraph induced by the edges only in G_i and G_j; and by H the subgraph induced by the edges in all graphs; see Fig. 1a.

The following two corollaries of Theorem 1 list sufficient conditions for $G_1 \setminus G_3$ and $G_2 \setminus G_3$ to have a SEFE. In the first case, $H_{1,2}$ has a unique embedding, which fulfills the conditions of Lemma 1 with respect to any planar embedding of G_1 and of G_2. In the second case, this is because $G_1 \setminus G_3$ is a subgraph of $G_2 \setminus G_3$.

Corollary 2. *Let* $G_1 = (V, E_1)$, $G_2 = (V, E_2)$, $G_3 = (V, E_3)$ *be planar graphs with shared vertex set* V. *If* $H_{1,2}$ *is acyclic and has maximum degree 2, then* $\langle G_1, G_2, G_3 \rangle$ *admits a QuaSEFE.*

Corollary 3. *Let* $G_1 = (V, E_1)$, $G_2 = (V, E_2)$, $G_3 = (V, E_3)$ *be planar graphs with shared vertex set* V. *If* $H_1 = \emptyset$, *then* $\langle G_1, G_2, G_3 \rangle$ *admits a QuaSEFE.*

Contrary to the previous corollaries, Theorem 1 has no implication for the graph H, as there are instances with $H = \emptyset$ where no pair of graphs has a SEFE. However, we show that a simple structure of H is still sufficient for a QuaSEFE.

Theorem 4. *Let* $G_1 = (V, E_1)$, $G_2 = (V, E_2)$, $G_3 = (V, E_3)$ *be planar graphs with shared vertex set* V. *If* H *has a planar embedding that can be extended to a planar embedding* $\mathcal{G}_i$ *of each graph* G_i, *then* $\langle G_1, G_2, G_3 \rangle$ *admits a QuaSEFE.*

Proof. We draw the graph $G_1 \setminus H_{1,3} = H_1 \cup H_{1,2} \cup H$ with embedding $\mathcal{G}_1$, the graph $G_2 \setminus H_{1,2} = H_2 \cup H_{2,3} \cup H$ with embedding $\mathcal{G}_2$, and the graph $G_3 \setminus H_{2,3} = H_3 \cup H_{1,3} \cup H$ with embedding $\mathcal{G}_3$. Then, the edges of G_1 are partitioned into two sets, one belonging to $G_1 \setminus H_{1,3}$ and one to $G_3 \setminus H_{2,3}$, each of which is drawn planar. As the same holds for the edges of G_2 and G_3, the statement follows. □

Corollary 4. *Let* $G_1 = (V, E_1)$, $G_2 = (V, E_2)$, $G_3 = (V, E_3)$ *be planar graphs with shared vertex set* V. *If* H *is acyclic and has maximum degree 2, then* $\langle G_1, G_2, G_3 \rangle$ *admits a QuaSEFE.*

The above discussion shows that, if one of the seven subgraphs in Fig. 1a is empty, or has a sufficiently simple structure, $\langle G_1, G_2, G_3 \rangle$ admits a QuaSEFE. Most notably, this is always the case in the *sunflower* setting [4, 21, 24], in which every edge belongs either to a single graph or to all graphs, i.e., $H_{1,2} = H_{1,3} = H_{2,3} = \emptyset$. We extend this result to any set of planar graphs. We remark that SEFE is NP-complete in the sunflower setting for three planar graphs [4, 24].

Theorem 5. *Let* $G_1 = (V, E_1), \ldots, G_k = (V, E_k)$ *be planar graphs with shared vertex set* V *in the sunflower setting. Then* $\langle G_1, \ldots, G_k \rangle$ *admits a QuaSEFE.*

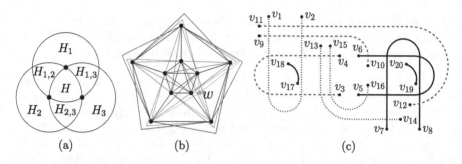

Fig. 1. (a) Subgraphs induced by the edges in one, two, or three graphs. (b) A simple quasiplanar drawing of Q_1 in Theorem 6, obtained by adding w to the drawing of K_{10} by Brandenburg [11]. (c) Theorem 7: Edge (v_{18}, v_{20}) crosses either all dotted blue or all dashed red edges, making (v_5, v_6) and (v_7, v_8) uncrossable (Color figure online).

Proof. Let H be the graph induced by the edges belonging to all graphs. We independently draw planar the graph H and every subgraph $G_i \setminus H$, for $i = 1, \ldots, k$. This guarantees that each G_i is drawn quasiplanar.   □

We remark that all our proofs are constructive. Moreover, the corresponding algorithms run in linear time, as they exploit linear-time algorithms for constructing planar embeddings of graphs [22], for extending their partial embeddings [5], and for partitioning 1-planar graphs into planar graphs and forests [1].

3   Counterexamples for QuaSEFE

In this section we complement our positive results, by providing negative instances of the QuaSEFE problem in two specific settings. We start with a negative result about the existence of a *simple* QuaSEFE for two quasiplanar graphs and one star. Here *simple* means that a pair of independent edges in the same graph is allowed to cross at most once and a pair of adjacent edges in the same graph is not allowed to cross. Note that our algorithms in Sect. 2 may produce non-simple drawings. Also, the maximum number of edges in a quasiplanar graph with n vertices depends on whether simplicity is required or not [2].

Theorem 6. *There exist two quasiplanar graphs* $Q_1 = (V, E_1)$, $Q_2 = (V, E_2)$ *and a star* $S_3 = (V, E_3)$ *with shared vertex set* V *such that* $\langle Q_1, Q_2, S_3 \rangle$ *does not admit a simple QuaSEFE.*

Proof. Let $V = \{v_1, \ldots, v_{10}, w\}$ and let E_{10} be the edges of the complete graph on $V \setminus \{w\}$. Further, let $E_1 = E_{10} \cup \{(w, v_1), \ldots, (w, v_6)\}$, let $E_2 = E_{10} \cup \{(w, v_7)\}$, and let $E_3 = \{(w, v_1), \ldots, (w, v_{10})\}$. By construction, S_3 is the star on all eleven vertices with center w, while Fig. 1b shows that there is a simple quasiplanar drawing of Q_1 (and of Q_2, which is a subgraph of Q_1, up to vertex relabeling).

Suppose that $\langle Q_1, Q_2, S_3 \rangle$ has a simple QuaSEFE, and let $\Gamma_{1,2}$ be the drawing of the union of Q_1 and Q_2 that is part of it. Since the union of Q_1 and Q_2 has 52 edges, which exceeds the upper bound of $6.5n - 20$ edges in a simple quasiplanar graph [2], $\Gamma_{1,2}$ is not simple or not quasiplanar. Since (w, v_7) is the only edge in $\Gamma_{1,2}$ that is not in Q_1, edge (w, v_7) is involved in every crossing violating simplicity or quasiplanarity. Analogously, one of $(w, v_1), \ldots, (w, v_6)$, say (w, v_1), is involved in every crossing violating simplicity or quasiplanarity; in particular, (w, v_1) crosses (w, v_7). Since both (w, v_1) and (w, v_7) belong to S_3, the drawing of S_3 that is part of the simple QuaSEFE is not simple, a contradiction.   □

The second special setting is the one in which one of the input graphs is already drawn in a quasiplanar way, and the goal is to draw the other input graphs so that the resulting simultaneous drawing is a QuaSEFE. This setting is motivated by the natural approach, for an instance $\langle G_1, \ldots, G_k \rangle$, of first constructing a solution for $\langle G_1, \ldots, G_{k-1} \rangle$ and then adding the remaining edges of G_k. Note that, since the drawing of the first graph partially fixes a drawing of the second graph, this can be seen as a version of the PEP problem for quasiplanarity.

For the original SEFE problem, this setting always has a solution when the graph that is already drawn (in a planar way) is a general planar graph, and the other graph is a tree [19]. In a surprising contrast, we construct negative instances for the QuaSEFE problem that are composed of two matchings only.

Theorem 7. *Let $M_1 = (V, E_1)$ and $M_2 = (V, E_2)$ be two matchings on the same vertex set V and let Γ_1 be a quasiplanar drawing of M_1. Instance $\langle M_1, M_2 \rangle$ does not always admit a QuaSEFE in which the drawing of M_1 is Γ_1.*

Proof. First recall that the edges in $E_1 \cap E_2$ have to be drawn in the quasiplanar drawing Γ_2 of G_2 as they are in Γ_1. Consider the quasiplanar drawing Γ_1 of the matching (v_{2i-1}, v_{2i}), with $i = 1, \ldots, 10$, in Fig. 1c, and let E_2 contain the edges (v_{17}, v_{19}) and (v_{18}, v_{20}). Since v_{17} is enclosed in a region bounded by the crossing edges (v_1, v_2) and (v_3, v_4), in any quasiplanar drawing of M_2 edge (v_{17}, v_{19}) crosses exactly one of (v_1, v_2) and (v_3, v_4). In the first case, (v_{17}, v_{19}) crosses also (v_{13}, v_{14}) and (v_{15}, v_{16}) (dotted blue). In the second case, (v_{17}, v_{19}) crosses also (v_9, v_{10}) and (v_{11}, v_{12}) (dashed red). In both cases, (v_5, v_6) and (v_7, v_8) cannot be crossed, and thus (v_{17}, v_{19}) cannot be drawn so that Γ_2 is quasiplanar.   □

4   Conclusions and Open Problems

We initiated the study of simultaneous embeddability in the beyond planar setting, which is a fertile and almost unexplored research direction that promises to significantly enlarge the families of representable graphs when compared with the planar setting. We conclude the paper by listing a few open problems.

- A natural question is whether two 1-planar graphs, a quasiplanar graph and a matching, three outerplanar graphs, or four paths admit a QuaSEFE. All our algorithms construct drawings with a stronger property than quasiplanarity, namely that they are composed of two sets of planar edges. Exploiting quasiplanarity in full generality may lead to further positive results.

- Motivated by Theorem 6, we ask whether some of the constructions presented in Sect. 2 can be modified to guarantee the simplicity of the drawings.
- Another intriguing direction is to determine the computational complexity of the QuaSEFE problem, both in its general version and in the two restrictions studied in Sect. 3. In particular, the setting in which one of the graphs is already drawn can be considered as a quasiplanar version of the PEP problem, which is known to be linear-time solvable in the planar case [5].
- Extend the study to other beyond-planarity classes. For example, do any two planar graphs admit a k-planar SEFE for some constant k?

References

1. Ackerman, E.: A note on 1-planar graphs. Discret. Appl. Math. **175**, 104–108 (2014). https://doi.org/10.1016/j.dam.2014.05.025
2. Ackerman, E., Tardos, G.: On the maximum number of edges in quasi-planar graphs. J. Comb. Theor. Ser. A **114**(3), 563–571 (2007). https://doi.org/10.1016/j.jcta.2006.08.002
3. Agarwal, P.K., Aronov, B., Pach, J., Pollack, R., Sharir, M.: Quasi-planar graphs have a linear number of edges. Combinatorica **17**(1), 1–9 (1997). https://doi.org/10.1007/BF01196127
4. Angelini, P., Da Lozzo, G., Neuwirth, D.: Advancements on SEFE and partitioned book embedding problems. Theor. Comput. Sci. **575**, 71–89 (2015). https://doi.org/10.1016/j.tcs.2014.11.016
5. Angelini, P., Di Battista, G., Frati, F., Jelínek, V., Kratochvíl, J., Patrignani, M., Rutter, I.: Testing planarity of partially embedded graphs. ACM Trans. Algorithms **11**(4), 32:1–32:42 (2015). https://doi.org/10.1145/2629341
6. Angelini, P., Geyer, M., Kaufmann, M., Neuwirth, D.: On a tree and a path with no geometric simultaneous embedding. J. Graph Algorithms Appl. **16**(1), 37–83 (2012)
7. Argyriou, E.N., Bekos, M.A., Kaufmann, M., Symvonis, A.: Geometric RAC simultaneous drawings of graphs. J. Graph Algorithms Appl. **17**(1), 11–34 (2013). https://doi.org/10.7155/jgaa.00282
8. Bekos, M.A., van Dijk, T.C., Kindermann, P., Wolff, A.: Simultaneous drawing of planar graphs with right-angle crossings and few bends. J. Graph Algorithms Appl. **20**(1), 133–158 (2016). https://doi.org/10.7155/jgaa.00388
9. Bläsius, T., Kobourov, S.G., Rutter, I.: Simultaneous embedding of planar graphs. In: Tamassia, R. (ed.) Handbook on Graph Drawing and Visualization, pp. 349–381. Chapman and Hall/CRC, London (2013)
10. Bläsius, T., Rutter, I.: Simultaneous PQ-ordering with applications to constrained embedding problems. ACM Trans. Algorithms **12**(2), 16:1–16:46 (2016). https://doi.org/10.1145/2738054
11. Brandenburg, F.J.: A simple quasi-planar drawing of K_{10}. In: Hu, Y., Nöllenburg, M. (eds.) Graph Drawing. LNCS, vol. 9801, pp. 603–604 (2016)
12. Braß, P., Cenek, E., Duncan, C.A., Efrat, A., Erten, C., Ismailescu, D., Kobourov, S.G., Lubiw, A., Mitchell, J.S.B.: On simultaneous planar graph embeddings. Comput. Geom. **36**(2), 117–130 (2007). https://doi.org/10.1016/j.comgeo.2006.05.006

13. Di Giacomo, E., Didimo, W., Liotta, G., Meijer, H., Wismath, S.K.: Planar and quasi-planar simultaneous geometric embedding. Comput. J. **58**(11), 3126–3140 (2015). https://doi.org/10.1093/comjnl/bxv048

14. Didimo, W., Kaufmann, M., Liotta, G., Okamoto, Y., Spillner, A.: Vertex angle and crossing angle resolution of leveled tree drawings. Inf. Process. Lett. **112**(16), 630–635 (2012). https://doi.org/10.1016/j.ipl.2012.05.006

15. Didimo, W., Liotta, G., Montecchiani, F.: A survey on graph drawing beyond planarity. ACM Comput. Surv. **52**(1), 4:1–4:37 (2019). https://doi.org/10.1145/3301281

16. Estrella-Balderrama, A., Gassner, E., Jünger, M., Percan, M., Schaefer, M., Schulz, M.: Simultaneous geometric graph embeddings. In: Hong, S.-H., Nishizeki, T., Quan, W. (eds.) GD 2007. LNCS, vol. 4875, pp. 280–290. Springer, Heidelberg (2008). https://doi.org/10.1007/978-3-540-77537-9_28

17. Evans, W.S., Liotta, G., Montecchiani, F.: Simultaneous visibility representations of plane st-graphs using L-shapes. Theor. Comput. Sci. **645**, 100–111 (2016). https://doi.org/10.1016/j.tcs.2016.06.045

18. Fox, J., Pach, J., Suk, A.: The number of edges in k-quasi-planar graphs. SIDMA **27**(1), 550–561 (2013). https://doi.org/10.1137/110858586

19. Frati, F.: Embedding graphs simultaneously with fixed edges. In: Kaufmann, M., Wagner, D. (eds.) GD 2006. LNCS, vol. 4372, pp. 108–113. Springer, Heidelberg (2007). https://doi.org/10.1007/978-3-540-70904-6_12

20. Grilli, L.: On the NP-hardness of GRacSim drawing and k-SEFE problems. J. Graph Algorithms Appl. **22**(1), 101–116 (2018). https://doi.org/10.7155/jgaa.00456

21. Haeupler, B., Jampani, K.R., Lubiw, A.: Testing simultaneous planarity when the common graph is 2-connected. J. Graph Algorithms Appl. **17**(3), 147–171 (2013). https://doi.org/10.7155/jgaa.00289

22. Hopcroft, J.E., Tarjan, R.E.: Efficient planarity testing. J. ACM **21**(4), 549–568 (1974). https://doi.org/10.1145/321850.321852

23. Pach, J., Wenger, R.: Embedding planar graphs at fixed vertex locations. Graphs Comb. **17**(4), 717–728 (2001). https://doi.org/10.1007/PL00007258

24. Schaefer, M.: Toward a theory of planarity: Hanani-tutte and planarity variants. J. Graph Algorithms Appl. **17**(4), 367–440 (2013). https://doi.org/10.7155/jgaa.00298

CHORDLINK: A New Hybrid Visualization Model

Lorenzo Angori[1], Walter Didimo[1], Fabrizio Montecchiani[1],
Daniele Pagliuca[1,2], and Alessandra Tappini[1(✉)]

[1] Dipartimento di Ingegneria, Università degli Studi di Perugia, Perugia, Italy
{lorenzo.angori,alessandra.tappini}@studenti.unipg.it
{walter.didimo, fabrizio.montecchiani}@unipg.it
[2] Agenzia delle Entrate, Arezzo, Italy
daniele.pagliuca@agenziaentrate.it

Abstract. Many real-world networks are globally sparse but locally dense. Typical examples are social networks, biological networks, and information networks. This double structural nature makes it difficult to adopt a homogeneous visualization model that clearly conveys an overview of the network and the internal structure of its communities at the same time. As a consequence, the use of hybrid visualizations has been proposed. For instance, NODETRIX combines node-link and matrix-based representations (Henry et al., 2007). In this paper we describe CHORDLINK, a hybrid visualization model that embeds chord diagrams, used to represent dense subgraphs, into a node-link diagram, which shows the global network structure. The visualization is intuitive and makes it possible to interactively highlight the structure of a community while keeping the rest of the layout stable. We discuss the intriguing algorithmic challenges behind the CHORDLINK model, present a prototype system, and illustrate case studies on real-world networks.

1 Introduction

The challenges in the design of effective visualizations for the analysis of real-world networks are not only related to the size of these networks, but also to the complexity of their structure. In particular, many networks in a variety of application domains are globally sparse but locally dense, i.e., they contain *communities* (or *clusters*) of highly connected nodes, and such communities are loosely connected to each other (see, e.g., [17,20,34]). Typical examples are social networks such as collaboration and financial networks [6,12,33,42]. Other examples include biological networks (e.g., metabolic and protein-protein interaction networks) and information networks; see, e.g., [16,24,31]. A visual exploration

Work partially supported by: (*i*) MIUR, under grant 20174LF3T8 "AHeAD: efficient Algorithms for HArnessing networked Data", (*ii*) Dipartimento di Ingegneria - Università degli Studi di Perugia, under grants RICBASE2017WD and RICBA18WD: "Algoritmi e sistemi di analisi visuale di reti complesse e di grandi dimensioni".

D. Archambault and C. D. Tóth (Eds.): GD 2019, LNCS 11904, pp. 276–290, 2019.
https://doi.org/10.1007/978-3-030-35802-0_22

of these networks should allow users to perform two main tasks [37]: (T1) getting an overview of the high-level structure of the network; (T2) identifying and analyzing in detail the communities of the network. However, the heterogeneity of the network connectivity level makes it difficult to adopt a homogeneous visualization that supports both the aforementioned tasks simultaneously.

This scenario naturally motivates the use of *hybrid visualizations* that combine different drawing styles, depending on the connectivity degree of the various portions of the network. A notable example is NODETRIX [22], which adopts a node-link diagram to represent the (sparse) global structure of the network and the more compact matrix representation to visualize denser subgraphs; the user can select the portions of the diagram to be represented as adjacency matrices.

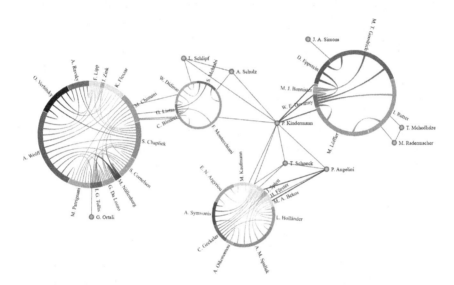

Fig. 1. A CHORDLINK visualization of a co-authorship network. The drawing has four clusters, represented as chord diagrams. In each chord diagram, circular arcs of the same color are copies of the same author. For example, in the smallest cluster, F. Montecchiani has two (green) copies, each connected to some nodes external to the cluster. (Color figure online)

Contribution. Inspired by NODETRIX, we aim to design a hybrid visualization model that supports tasks (T1) and (T2), and that can be integrated into an interactive visual analytics system. In particular, our design is driven by two main requirements: (R1) the model must support the drawing stability throughout the user interaction, so to maintain the user's mental map during an interactive analysis of the network; (R2) the drawing styles to convey the different portions of the network should be intuitive for non-expert users, as for a node-link representation. Our contribution is as follows:

(i) We propose CHORDLINK, a new model that embeds *chord diagrams*, used for the visualization of dense subgraphs (communities), into a node-link diagram, which shows the global network structure (Sect. 3). Chord diagrams are an extension of circular drawings, where nodes are represented as circular arcs instead of points (see, e.g., [29]). Figure 1 shows a CHORDLINK visualization.

(ii) As a proof-of-concept of our model, we describe a prototype system that implements it and we discuss some case studies on different kinds of real-world networks, namely fiscal networks and co-authorship networks (Sect. 4). A short video of the system can be found at https://youtu.be/ezphnPEdA8Y.

(iii) Finally, our model introduces new optimization problems (Sect. 3.2) that are of independent interest, and that may inspire future research (Sect. 5).

For space reasons some details have been omitted and can be found in [2].

Methodology. The CHORDLINK model represents a community C selected in a node-link diagram Γ as a specific type of chord diagram, which we denote as $\Gamma(C)$. Regarding (R1), a suitable replication of the nodes of C allows us to preserve the geometry of the nodes and edges outside $\Gamma(C)$; this avoids new edge crossings out of the cluster and supports the user's mental map during an interactive analysis of the network. Such a node-replication also gives additional freedom to reduce the number of edge crossings in $\Gamma(C)$. Regarding (R2), the representation $\Gamma(C)$ remains intuitive for users who are familiar with the node-link style, because an edge in C is still represented as a geometric curve. This makes it easy, for example, to recognize paths in C, a basic task that is sometimes difficult to perform in a matrix-based representation [19, 22].

2   Related Work

Early works in graph visualization propose hybrid models that combine Euler/Venn Diagrams, used to represent inclusion relationships between sets of objects, with Jordan arcs, which convey other types of relationships between these sets [21, 38]. Similar drawing styles are extensively used to represent *compound graphs*, where the nodes are hierarchically grouped into clusters and where there can be binary relationships between clusters other than between nodes (see, e.g., [14, 28, 40] for surveys on the subject). Hybrid visualizations that mix node-link and treemaps are also studied [15, 43], sometimes in terms of algorithmic techniques for quick computation of clustered layouts [13, 32].

The NODETRIX model is the first attempt to visually convey both the global structure of a sparse network and its locally dense subgraphs by combining node-link and matrix-based representations [22]. This work has inspired a subsequent array of papers, either devoted to the development of visual analytics systems for complex graphs or focused on the theoretical properties of visualizations in the NODETRIX model. In the first direction, an interesting variant of the NODE-TRIX model is proposed in [5]; while in NODETRIX the clusters represented as an adjacency matrix are selected by the user, in [5] the set of clusters is computed by the drawing algorithm so that the resulting graph of clusters (drawn as an orthogonal layout) is planar; the user can choose the drawing style inside each cluster region, including the possibility of using a matrix-based representation. In the second direction, several papers study the so-called *hybrid planarity* testing problem, both in the NODETRIX model [8,10] and in a different model where clusters are intersection graphs of geometric objects [1]. This problem asks whether a given graph admits a hybrid visualization such that the edges represented as geometric links do not cross any cluster region and do not cross each other. Also, complexity results on a relaxation of the hybrid planarity testing problem are given in [9]; similar to CHORDLINK, this relaxation allows for a limited replication of the nodes of a cluster, but in [9] the clusters are defined by the algorithm and intra-cluster edges are not considered.

Our CHORDLINK model uses a specific type of chord diagram to represent clusters. Chord diagrams are effectively adopted in several visualization systems to analyze dense networks in various contexts, including comparative genomics [29], urban mobility trajectories [18], and software profiling on distributed graph processing systems [4]. Other applications of chord diagrams can be found at http://www.circos.ca/. They have also been extended to support hierarchical data sets (see, e.g., [3,25]). We finally remark that the use of circular layouts for visualizing clustered graphs is proposed in [39]. In that approach, the node set of the input network is partitioned into user-defined clusters, and each cluster is represented as a circular layout with nodes drawn as points and edges drawn as straight segments; hence, each node of the network belongs to a circular layout and the whole drawing of the network is computed by knowing in advance the set of clusters. In the CHORDLINK model we assume that the user can define the clusters interactively, and that the drawing of the network must be updated accordingly, while controlling the drawing stability.

3   The CHORDLINK Model

Let $G = (V, E)$ be a network and let Γ be a node-link diagram of G. The CHORDLINK model is conceived to work in an interactive system, in which the user can iteratively select a cluster C of nodes in Γ and the system automatically redraws the subgraph $G[C]$ induced by C as a chord diagram $\Gamma(C)$. The nodes of C are required to lie within a topologically connected region of the plane (e.g., within a circular or a rectangular region); the drawing of nodes and edges of Γ out of $G[C]$ should change as little as possible to enforce stability.

If a node $w \in C$ is connected to a node outside C, we say that w is *extrovert*, else w is *introvert*. To maintain the drawing outside $\Gamma(C)$ stable, the CHORDLINK model allows for a suitable replication of the nodes. Namely, every extrovert node $w \in C$ can have multiple occurrences in $\Gamma(C)$, while an introvert node of C will occur exactly once in $\Gamma(C)$. The occurrences of w are called *copies* of w. A copy of w is represented in $\Gamma(C)$ by a circular arc c_w, coinciding with a portion of the circumference of $\Gamma(C)$. The set of arcs c_w, over all copies of the nodes w of C, partitions the circumference of $\Gamma(C)$. An edge $(u, w) \notin G[C]$, with $u \notin C$ and $w \in C$, is drawn as a straight-line segment incident to one of the circular arcs c_w. An edge $(w, z) \in G[C]$ is drawn as a simple curve, called *chord*, connecting one of the circular arcs c_w to one of the circular arcs c_z.

3.1   General Strategy

Assume that all nodes of a selected cluster C in Γ lie in a circular region $R(C)$ and that all the other nodes of Γ are outside $R(C)$; also, assume that no node of C is located exactly at the center of $R(C)$ (otherwise slightly perturb the region). According to the CHORDLINK model, we locally redraw Γ so that the boundary of the chord diagram $\Gamma(C)$ coincides with the boundary of $R(C)$. This is done through a general strategy that consists of the following phases (see Fig. 2):

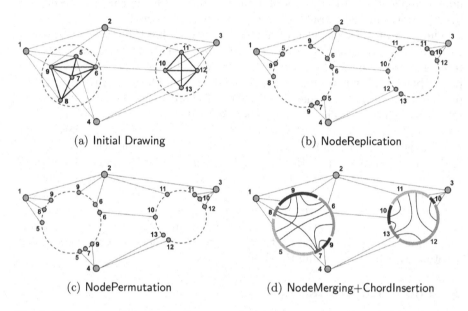

(a) Initial Drawing   (b) NodeReplication

(c) NodePermutation   (d) NodeMerging+ChordInsertion

Fig. 2. Illustration of the general strategy for the CHORDLINK model. (a) An initial node-link diagram with two selected clusters (dashed regions). (b) Drawing after the NodeReplication phase. (c) Output of the NodePermutation phase; for example, in the left cluster the copies of the nodes adjacent to 1 and to 4 are permuted so to reduce the number of non-consecutive copies of 5 and 9. (d) Final drawing after the NodeMerging and ChordInsertion phases; chords are inserted so to minimize their number of crossings.

NodeReplication. For each extrovert node $w \in C$ connected to a node $u \notin C$, create a copy v of w at the intersection point between (u, w) and the boundary of $R(C)$, and replace the segment $\overline{uw}$ with its subsegment $\overline{uv}$. For each introvert node $w \in C$, create a unique copy of w at the intersection point between the boundary of $R(C)$ and the radius of $R(C)$ passing through w. Then, remove all the elements of Γ that are properly inside $R(C)$. At the end we have a circular sequence of copies of the nodes of C along the boundary of $R(C)$; two copies of the same node may not be consecutive in this sequence.

NodePermutation. Permute the copies of the nodes of C along the boundary of $R(C)$ in such a way to minimize the total number of non-consecutive copies of the same node. To preserve the geometry of the drawing outside $R(C)$, two copies can be permuted only if they are adjacent to the same node $u \notin C$.

NodeMerging. For each maximal subsequence of consecutive copies of a node w (possibly a single copy) along the boundary of $R(C)$, replace all these copies by a circular arc c_w that spans at least the whole subsequence.

ChordInsertion. For each edge $(w, z) \in G[C]$, select one of the copies c_w and one of the copies c_z, and insert a chord inside $R(C)$ connecting c_w and c_z. This selection can be done in order to optimize some desired function; for example, one can try to minimize the total number of crossings between chords and/or to maximize the angles formed by two crossing chords.

3.2   Algorithms

In the following we describe specific algorithms to solve the optimization problems posed by the NodePermutation and ChordInsertion phases. In the full version [2], we explain how to handle the NodeMerging phase and the case in which for a selected cluster there is not a circular region that includes exactly its nodes.

Algorithm for the NodePermutation Phase. Let C be a selected cluster in the current drawing Γ. The optimization problem in the NodePermutation phase asks to find a permutation of the copies of the nodes of C along the boundary of $R(C)$ such that the total number of non-consecutive copies of the same node is minimized. However, to preserve the geometry of the links outside $R(C)$ (thus avoiding the introduction of edge crossings), two copies can be permuted only if they have a common neighbor $u \notin C$. Formally, we model the problem as follows.

Let $u_1, u_2, \ldots, u_k$ be the set of nodes not in C that are adjacent to some node of C. For each u_i ($i = 1, \ldots, k$), denote by $\langle v_{i,1}, v_{i,2}, \ldots, v_{i,h_i} \rangle$ the clockwise sequence of copies of extrovert nodes of C along $R(C)$ attached to u_i. For example, assume that C is the left-side cluster in Fig. 2(b); if we set $u_1 = 1$, $u_2 = 2$, $u_3 = 10$, and $u_4 = 4$ then we have: $\langle v_{1,1} = 8, v_{1,2} = 9, v_{1,3} = 5 \rangle$; $\langle v_{2,1} = 9, v_{2,2} = 6 \rangle$; $\langle v_{3,1} = 6 \rangle$; $\langle v_{4,1} = 5, v_{4,2} = 9 \rangle$. The sequence $\langle v_{i,1}, v_{i,2}, \ldots, v_{i,h_i} \rangle$ is called the *group of* u_i. Clearly, two elements of the same group never represent copies of the same node of C. Denote by $\mathcal{E}$ the set of copies of the extrovert nodes of C on the boundary of $R(C)$. Suppose that $v \in \mathcal{E}$ is a copy of a node $w \in C$ and that $n(v)$ is the next copy of w encountered by walking clockwise on the boundary of $R(C)$. We denote by $\chi(v, n(v))$ the *cost of* $\{v, n(v)\}$ and we define it as follows: $\chi(v, n(v)) = 0$ if no copies of nodes of C are encountered between v and $n(v)$ while walking clockwise on the boundary of $R(C)$; $\chi(v, n(v)) = 1$ otherwise. Our optimization problem asks to find a permutation of the copies in the group of u_i (for each $i = 1, \ldots, k$) that minimizes the objective function $\sum_{v \in \mathcal{E}} \chi(v, n(v))$.

We describe a dynamic programming algorithm that we designed with the aim of computing an exact solution for this optimization problem when all the copies in each group are consecutive along the boundary of $R(C)$ (like in Fig. 2); if this is not the case, our algorithm is used as a heuristic for the problem. If all the copies of each group are consecutive, two node permutations π and π'

yield the same cost if for each group the first element is the same in both π and π' and the same holds for the last element. Hence, it suffices to minimize the pairs of consecutive groups such that their two neighboring elements are copies of different nodes. More formally, let $B_0, B_1, \ldots, B_{k-1}$ be the clockwise sequence of groups along $R(C)$, starting from an arbitrary group B_0. For each group B_i, let f_i and l_i be its first and its last element, respectively, i.e., l_i and f_{i+1} (indexes taken modulo k) are consecutive along $R(C)$. Our dynamic programming formulation considers the cost of choosing the first and the last element of B_i assuming that this choice has been already done for the groups $B_{i+1}, \ldots, B_{k-1}$. Namely, denote by $O_i(v_{i,j}, v_{i,z})$ the cost of choosing $f_i = v_{i,j}$ and $l_i = v_{i,z}$. For each possible pairs of elements $v_{i,j}, v_{i,z}$ in B_i and $v_{i+1,j'}, v_{i+1,z'}$ in B_{i+1}, the following holds:

$$O_i(v_{i,j}, v_{i,z}) = O_{i+1}(v_{i+1,j'}, v_{i+1,z'}) + \begin{cases} 0, & \text{if } v_{i+1,j'} = v_{i,z} \\ 1, & \text{if } v_{i+1,j'} \neq v_{i,z} \end{cases} \tag{1}$$

The optimal solution is then $\chi_{opt} = \min_{v_{0,j}, v_{0,z} \in B_0} O_0(v_{0,j}, v_{0,z})$. To solve the above recurrence we fix f_0 and compute a table of size $\sum_{i=0}^{k-1} \binom{h_i}{2} \leq m^2$, where m is the number of edges of G. We repeat this procedure for each of the $h_0 \leq m$ possible values of f_0 and we select the optimal solution among them; this algorithm takes $O(m^3)$ time. Note that, to speed up the algorithm, the elements $v_{i,j}$ such that there is no element $v_{i+1,j'} = v_{i,j}$ in B_{i+1} (resp. $v_{i-1,j'} = v_{i,j}$ in B_{i-1}) can be ignored, since selecting them as first or last element of B_i always increases the cost of the solution. In particular, we first remove them in a preprocessing step, and then reinsert them in any position between f_i and l_i.

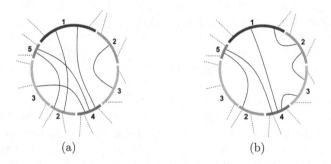

(a)   (b)

Fig. 3. Example of different choices in the ChordInsertion phase. The set of chords in each drawing represents the edges $(1, 2)$, $(1, 4)$, $(2, 3)$, $(2, 5)$, $(3, 4)$, $(4, 5)$. In (a) the chords form 3 crossings, while in (b) they do not cross, due to a more convenient choice of the representative pair of arcs for the edges $(1, 2)$ and $(3, 4)$. The dashed lines represent stubs of possible outside edges incident to the cluster.

Algorithm for the ChordInsertion Phase. In this phase, for each edge $(w, z) \in G[C]$ we have to select one of the circular arcs c_w associated with w and one of the circular arcs c_z associated with z, and we add a chord connecting c_w

to c_z. The specific selection of a pair $\{c_w, c_z\}$ for each edge (w, z) determines the total number of crossings between chords. For example, Fig. 3 shows a schematic illustration of two different chord diagrams for a cluster C. The cluster has seven circular arcs, associated with nodes 1, 2, 3, 4, 5; the edges of $G[C]$ are $(1, 2)$, $(1, 4)$, $(2, 3)$, $(2, 5)$, $(3, 4)$, and $(4, 5)$. The chords representing these edges cause in total 3 crossings in Fig. 3(a), while they do not cross in the drawing of Fig. 3(b), where we have chosen a different pair of arcs for the edges $(1, 2)$ and $(3, 4)$.

Our algorithm for selecting the set of chords aims to minimize the number of crossings and to maximize the minimum angle at a crossing point of two crossing chords. This optimization goal is motivated by several works that show the negative impact of the number of crossings (e.g., [35, 36, 41]) and in particular of small crossing angles (e.g., [26, 27]) in graph layouts.

We model the above optimization problem as follows. We assume that each circular arc c_w is collapsed into a single point p_w, coinciding with the center of c_w. Once the set of chords incident to p_w is decided by the algorithm, we expand back p_w to c_w and equally distribute the chords incident to p_w along c_w. Note that, the number of crossings between non-adjacent chords only depends on the circular order of their end-points along $R(C)$ and not on their exact position. Hence, two non-adjacent chords (p_w, p_z), (p_x, p_y) cross if and only the corresponding chords (c_w, c_z), (c_x, c_y) cross, independent of the position of the end-points of the chords along c_w, c_z, c_x, and c_y. Also, two adjacent chords (p_w, p_z) and (p_w, p_x) never cross, and therefore the corresponding chords (c_w, c_z) and (c_w, c_x) will not cross if we use the same circular order. Moreover, if (p_w, p_z) and (p_x, p_y) are two crossing chords, we denote by $a(\overline{wz}, \overline{xy})$ the minimum angle formed by the segments $\overline{wz}$ and $\overline{xy}$ at their crossing point; this gives an estimation of the crossing angular resolution of the two chords if each chord is drawn as a monotone curve approximating the straight segment between its end-points. For any two chords $e_{wz} = (p_w, p_z)$ and $e_{xy} = (p_x, p_y)$, we define the *cost of* the unordered pair $\{e_{wz}, e_{xy}\}$ as a function $\alpha(e_{wz}, e_{xy})$ such that: $\alpha(e_{wz}, e_{xy}) = 0$ if e_{wz} and e_{xy} do not cross; $\alpha(e_{wz}, e_{xy}) = 1 - a(\overline{wz}, \overline{xy})/\pi$ otherwise. Since $a(\overline{wz}, \overline{xy}) \in (0, \pi/2]$, we have $\alpha(e_{wz}, e_{xy}) \in [0.5, 1)$. We aim to select a set S of chords for the edges of $G[C]$ that minimizes the cost function $\alpha(S) = \sum_{\{e_{wz}, e_{xy}\} \in S \times S} \alpha(e_{wz}, e_{xy})$.

To solve this problem we use a heuristic algorithm based on a greedy strategy. Let $E(C)$ be the set of edges of $G[C]$ and let $E_1(C) \subseteq E(C)$ be the subset of edges having one representative chord (p_w, p_z), i.e., $(w, z) \in E_1(C)$ if and only if w and z have a unique copy on the boundary of $R(C)$. Also, let $E_2(C) = E(C) \setminus E_1(C)$ be the remaining subset of edges of $G[C]$. For example, in the cluster C of Fig. 3 we have $E_1(C) = \{(4, 5)\}$ and $E_2(C) = \{(1, 2), (1, 4), (2, 3), (2, 5), (3, 4)\}$. Our algorithm first adds to the drawing $\Gamma(C)$ the chords representing the edges of $E_1(C)$ (in any order), because for these edges there are no alternative choices. After that, the algorithm executes $|E_2(C)|$ iterations. Each iteration i ($1 \leq i \leq |E_2(C)|$) removes an edge (w, z) from $E_2(C)$ and adds to the drawing one of its representative chords (p_w, p_z). More precisely, let S_0 be the set of chords added for the edges in $E_1(C)$ and let S_i denote the set of chords added at the end of iteration i. At the beginning of iteration i, for each edge $(w, z) \in E_2(C)$ and for each chord (p_w, p_z) that is representative of (w, z), the algorithm computes the cost of inserting (p_w, p_z) in the current drawing, i.e., the cost $\alpha(S_{i-1} \cup \{(p_w, p_z)\})$; then it selects the chord that yields the minimum cost and removes from $E_2(C)$ the corresponding edge. Denote by S' the whole set of representative chords for the edges of $E(C)$. Since the cost $\alpha(S_{i-1} \cup \{(p_w, p_z)\})$ can be easily computed in $O(|S_{i-1}|)$ time from the cost $\alpha(S_{i-1})$ and from the set of chords in S_{i-1}, and since $|S_{i-1}| = O(|E(C)|)$, the whole greedy algorithm takes $O(|S'||E(C)|^2)$ time.

4   A Prototype System

As a proof-of-concept of the CHORDLINK model, we realized a prototype system that implements it. The system is developed in Javascript (so to run in a Web browser) and the implementation uses the D3.JS library [7], https://d3js.org. We first describe the main features of the system interface and its interaction functionalities. Then, we discuss two case studies that show how the system can be used to perform the analysis tasks (T1) and (T2) on different kinds of real networks, namely a fiscal network and a co-authorship network.

Interface and Interaction. Through the interface of our system, the user can import a network in the GML file format [23]. The system initially computes a node-link diagram of the network using a force-directed algorithm; we exploit an implementation available in the D3.JS library. The interface supports the visualization of weighted edges by using different levels of edge thickness to convey this information. The user can execute some common operations, like node movement, zooming, and panning. Node labels can be displayed according to different policies. One can show/hide all labels at the same time or enable/disable each label individually. Alternatively, the system can automatically manage the visualization of labels based on node-degrees and on the current zoom level of the layout (labels of low-degree nodes are hidden after a zoom-out operation). Regardless of the labeling policy, a mouse-hover operation on a node or on an edge causes the display of a tooltip that reports the label of that element.

In order to represent a desired cluster C as a chord diagram $\Gamma(C)$, the user can select the nodes of C in the layout (e.g., through a rectangular region selection). The visualization of $\Gamma(C)$ is such that: (i) All the circular arcs c_w associated with the same node $w \in C$ are assigned the same color; the label of w is displayed near to one of its corresponding arcs, namely the longest one. (ii) Each chord between two arcs c_w and c_z has a color that gradually goes from the color of c_w to that of c_z; this helps to visually detect the end-nodes of the chord. (iii) The size of each chord reflects the weight of the corresponding edge (the maximum thickness for the chords in $\Gamma(C)$ depends on the minimum length of the circular arcs and on their inner degree). A mouse-hover operation on a circular arc c_w of $\Gamma(C)$ highlights all the arcs associated with w, as well as all the edges incident to c_w (see Fig. 6(a) in [2]). The user can move a chord diagram $\Gamma(C)$ or drag a node $u \notin C$ to drop it in $\Gamma(C)$; this operation adds u to C and causes an immediate update of the drawing. The user can click on $\Gamma(C)$ to collapse it into a single *cluster-node* (whose size is proportional to the number of nodes in C); a click operation on a cluster-node expands back it into the original chord diagram. Collapsing/expanding each cluster individually helps focusing on specific portions of the network without losing the general context where they are embedded (see Fig. 6(b) in [2]).

Case Studies: Fiscal Networks. The first case study falls into the domain of fiscal risk analysis. We considered a real network of taxpayers and their economic transactions. The network is provided by the IRV (Italian Revenue Agency) and refers to a portion of data for the fiscal year 2014, consisting of 174 subjects with high fiscal risk and 200 economic transactions between them [11]. Figure 4 depicts a CHORDLINK visualization of this network computed by our system after the selection of six clusters (Fig. 7 in [2] reports the initial node-link diagram). The thickness of an edge (u, v) reflects the amount of transactions between u and v in the considered year (we discretized the range of amounts into 5 values of thickness). For privacy reasons data are anonymized; a node's label reports the ID number and the geographic area of the corresponding taxpayer.

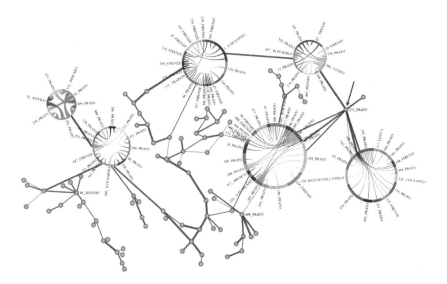

Fig. 4. A visualization obtained by selecting some communities in a node-link diagram.

Regarding task (T1), we observe that the network consists of several communities and of few nodes with high degree. A visual analysis of the network reveals that the node with ID 272 (marked with an arrow in the figure) acts as a broker between three communities, since it has strong connections with them. Regarding task (T2), the chord diagram of each community makes it possible to analyze the connections between its nodes, by overcoming the node overlaps in the node-link diagram. The position of nodes and the geometry of edges outside the chord diagrams do not change with respect to the initial node-link diagram, since all nodes of every selected community lie in a circular region not containing other nodes of the network. Focusing on the rightmost chord diagram $\Gamma(C)$ in Fig. 4, we can see that the node with ID 272 is connected to two nodes of high degree inside $\Gamma(C)$ (those with IDs 195 and 198), which belong to the same geographic area. An analyst of the IRV identified this subgraph as a suspicious scheme characterized by several economic transactions, where the seller is a so-called "missing trader" with serious tax irregularities (omitted VAT payments or tax declarations); nodes with IDs 195 and 198 are missing traders. From a deepest inspection of the connections in $\Gamma(C)$ and from additional attributes of its taxpayers, the analyst confirmed the presence of a tax evasion pattern. Similar conclusions were derived from the analysis of other communities in the network.

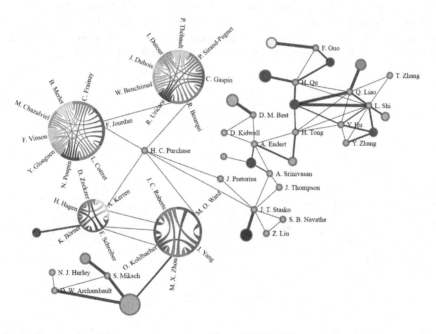

Fig. 5. A co-authorship network extracted from DBLP. Bigger nodes are cluster-nodes. (Color figure online)

Case Studies: Co-authorship Networks. The second case study considers co-authorship networks extracted from the DBLP dataset [30], which contains publication data in computer science. Through a query consisting of keywords and Boolean operators, one can retrieve a set of publications on a desired topic. We use the results returned by DBLP to construct networks where nodes are authors and edges indicate co-authorships, weighted by the number of papers shared by their end-nodes. Nodes are labeled with authors' names and edges with the titles of the corresponding publications.

We performed the query "network AND visualization" and limited to 500 the number of search results (i.e., publications) to be returned. The resulting network consists of 1766 nodes, 3780 edges, and 382 connected components. The largest of these components contains 118 nodes and 322 edges. A CHORDLINK visualization of this component is shown in Fig. 5, where several dense portions of the original node-link layout have been identified as communities. To make the diagram easier to read, some communities (on the left side) have been expanded and some others (on the right side) have been collapsed. We now discuss some findings that involve tasks (T1) and (T2) in an interleaved manner.

From the general structure of the clustered network one can clearly distinguish several central actors. For example, on the left side of the drawing we can observe that H. C. Purchase is connected to four distinct communities. Following the links incident to this author and the connections between the related authors inside the clusters, we can see that H. C. Purchase forms a 3-cycle with A. Kerren and M. O. Ward (this author has two copies in his cluster), who fall into two distinct communities. By exploring the edge labels, we see that this cycle originates from a work titled "Introduction to Multivariate Network Visualization", while the communities to which A. Kerren and M. O. Ward belong mainly derive from the works "Heterogeneous Networks on Multiple Levels" and "Novel Visual Metaphors for Multivariate Networks", respectively. By analyzing the literature more in detail, one can observe that these three works appear in the same book, referring to the Dagstuhl Seminar Multivariate Network Visualization. The orange cluster-node in the bottom of the drawing, call it C, seems to be strongly related to nodes S. Miksch, D. W. Archambault, and M. X. Zhou. Indeed, the links of these three authors with C refer to a common work, "Temporal Multivariate Networks". Since D. W. Archambault has only two connections with nodes outside C, it seems reasonable to move it inside C by a drag operation.

If we analyze this community in detail (Fig. 8 in [2] shows its chord diagram), the connections reveal that the aforementioned work has other 5 authors in addition to the 3 already cited. Two of them, K. Ma and C. Muelder, have a connection thicker than the other pairs of nodes, which indicates a stronger cooperation. Also, there are two nodes of C, namely S. Diehl and F. Tzeng, that are loosely connected in this cluster. We deduce that it would be convenient to keep them out of the community, even if the original node-link diagram locates them very close to the other nodes of C.

5   Final Remarks and Future Work

The CHORDLINK model proposed in this paper is a new kind of hybrid visualization. It can complement previous models conceived for the visual analysis of networks that are globally sparse but locally dense. Among its advantages, CHORDLINK makes it possible to keep the visualization stable during the interaction. This is especially true when the nodes of a community, that is going to be represented as a chord diagram, are close to each other in the node-link layout (which is most often the case if it is computed by a force-directed algorithm). Nonetheless, CHORDLINK has also some clear limits. In particular, the readability of a chord diagram may degrade when the size of a cluster increases; our current visualization can be effectively used for clusters up to 20–25 nodes, while it becomes less effective for bigger clusters.

Besides these considerations, we believe that the CHORDLINK model opens the way for intriguing research directions: (i) We conjecture that the optimization problems at the core of a CHORDLINK visualization are computationally hard. It would be interesting to prove NP-hardness and to design new algorithms to be compared with our heuristics. (ii) It may be worth developing a system that combines the CHORDLINK and the NODETRIX models, allowing users to switch from a visualization to the other for each cluster. This would merge the advantages of both models. (iii) One can exploit an automatic clustering algorithm for the CHORDLINK model, e.g., one that guarantees the planarity of the inter-cluster graph [5].

References

1. Angelini, P., Da Lozzo, G., Di Battista, G., Frati, F., Patrignani, M., Rutter, I.: Intersection-link representations of graphs. J. Graph Algorithms Appl. **21**(4), 731–755 (2017). https://doi.org/10.7155/jgaa.00437
2. Angori, L., Didimo, W., Montecchiani, F., Pagliuca, D., Tappini, A.: ChordLink: a new hybrid visualization model. CoRR abs/1908.08412 (2019). http://arxiv.org/abs/1908.08412
3. Argyriou, E.N., Symvonis, A., Vassiliou, V.: A fraud detection visualization system utilizing radial drawings and heat-maps. In: Laramee, R.S., Kerren, A., Braz, J. (eds.) IVAPP 2014, pp. 153–160. SciTePress (2014). https://doi.org/10.5220/0004735501530160
4. Arleo, A., Didimo, W., Liotta, G., Montecchiani, F.: Profiling distributed graph processing systems through visual analytics. Future Gener. Comput. Syst. **87**, 43–57 (2018). https://doi.org/10.1016/j.future.2018.04.067
5. Batagelj, V., Brandenburg, F., Didimo, W., Liotta, G., Palladino, P., Patrignani, M.: Visual analysis of large graphs using (X, Y)-clustering and hybrid visualizations. IEEE Trans. Vis. Comput. Graph. **17**(11), 1587–1598 (2011). https://doi.org/10.1109/TVCG.2010.265
6. Bedi, P., Sharma, C.: Community detection in social networks. Wiley Interdiscip. Rev. Data Min. Knowl. Discov. **6**(3), 115–135 (2016). https://doi.org/10.1002/widm.1178
7. Bostock, M., Ogievetsky, V., Heer, J.: D^3 data-driven documents. IEEE Trans. Vis. Comput. Graph. **17**(12), 2301–2309 (2011). https://doi.org/10.1109/TVCG.2011.185
8. Da Lozzo, G., Di Battista, G., Frati, F., Patrignani, M.: Computing NodeTrix representations of clustered graphs. J. Graph Algorithms Appl. **22**(2), 139–176 (2018). https://doi.org/10.7155/jgaa.00461
9. Di Giacomo, E., Lenhart, W.J., Liotta, G., Randolph, T.W., Tappini, A.: (k, p)-planarity: a relaxation of hybrid planarity. In: Das, G.K., Mandal, P.S., Mukhopadhyaya, K., Nakano, S. (eds.) WALCOM 2019. LNCS, vol. 11355, pp. 148–159. Springer, Cham (2019). https://doi.org/10.1007/978-3-030-10564-8_12
10. Di Giacomo, E., Liotta, G., Patrignani, M., Rutter, I., Tappini, A.: NodeTrix planarity testing with small clusters. Algorithmica (2019). https://doi.org/10.1007/s00453-019-00585-6

11. Didimo, W., Giamminonni, L., Liotta, G., Montecchiani, F., Pagliuca, D.: A visual analytics system to support tax evasion discovery. Decis. Support Syst. **110**, 71–83 (2018). https://doi.org/10.1016/j.dss.2018.03.008

12. Didimo, W., Liotta, G., Montecchiani, F.: Network visualization for financial crime detection. J. Vis. Lang. Comput. **25**(4), 433–451 (2014). https://doi.org/10.1016/j.jvlc.2014.01.002

13. Didimo, W., Montecchiani, F.: Fast layout computation of clustered networks: algorithmic advances and experimental analysis. Inf. Sci. **260**, 185–199 (2014). https://doi.org/10.1016/j.ins.2013.09.048

14. Dogrusöz, U., Giral, E., Cetintas, A., Civril, A., Demir, E.: A layout algorithm for undirected compound graphs. Inf. Sci. **179**(7), 980–994 (2009). https://doi.org/10.1016/j.ins.2008.11.017

15. Fekete, J.D., Wang, D., Dang, N., Aris, A., Plaisant, C. (eds.): Overlaying graph links on treemaps. In: IEEE Symposium on Information Visualization Conference Compendium (demonstration) (2003)

16. Flake, G.W., Lawrence, S., Giles, C.L., Coetzee, F.: Self-organization and identification of web communities. IEEE Comput. **35**(3), 66–71 (2002). https://doi.org/10.1109/2.989932

17. Fortunato, S.: Community detection in graphs. Phys. Rep. **486**(3–5), 75–174 (2010). https://doi.org/10.1016/j.physrep.2009.11.002

18. Gabrielli, L., Rinzivillo, S., Ronzano, F., Villatoro, D.: From tweets to semantic trajectories: mining anomalous urban mobility patterns. In: Nin, J., Villatoro, D. (eds.) CitiSens 2013. LNCS (LNAI), vol. 8313, pp. 26–35. Springer, Cham (2014). https://doi.org/10.1007/978-3-319-04178-0_3

19. Ghoniem, M., Fekete, J., Castagliola, P.: On the readability of graphs using node-link and matrix-based representations: a controlled experiment and statistical analysis. Inf. Visual. **4**(2), 114–135 (2005)

20. Girvan, M., Newman, M.E.J.: Community structure in social and biological networks. Proc. Natl. Acad. Sci. USA **99**(12), 7821–7826 (2002). https://doi.org/10.1073/pnas.122653799

21. Harel, D.: On visual formalisms. Commun. ACM **31**(5), 514–530 (1988). https://doi.org/10.1145/42411.42414

22. Henry, N., Fekete, J., McGuffin, M.J.: NodeTrix: a hybrid visualization of social networks. IEEE Trans. Vis. Comput. Graph. **13**(6), 1302–1309 (2007). https://doi.org/10.1109/TVCG.2007.70582

23. Himsolt, M.: GML: a portable graph file format (technical report Universität Passau) (2010)

24. Holme, P., Huss, M., Jeong, H.: Subnetwork hierarchies of biochemical pathways. Bioinformatics **19**(4), 532–538 (2003). https://doi.org/10.1093/bioinformatics/btg033

25. Holten, D.: Hierarchical edge bundles: visualization of adjacency relations in hierarchical data. IEEE Trans. Vis. Comput. Graph. **12**(5), 741–748 (2006). https://doi.org/10.1109/TVCG.2006.147

26. Huang, W., Eades, P., Hong, S.: Larger crossing angles make graphs easier to read. J. Vis. Lang. Comput. **25**(4), 452–465 (2014). https://doi.org/10.1016/j.jvlc.2014.03.001

27. Huang, W., Hong, S., Eades, P.: Effects of sociogram drawing conventions and edge crossings in social network visualization. J. Graph Algorithms Appl. **11**(2), 397–429 (2007). https://doi.org/10.7155/jgaa.00152

28. Kaufmann, M., Wagner, D. (eds.): Drawing Graphs, Methods and Models (The Bookgrow out of a Dagstuhl Seminar, April 1999). LNCS, vol. 2025. Springer, Heidelberg (2001). https://doi.org/10.1007/3-540-44969-8

29. Krzywinski, M., Schein, J., Birol, I., Connors, J., Gascoyne, R., Horsman, D., Jones, S.J., Marra, M.A.: Circos: an information aesthetic for comparative genomics. Genome Res. **19**(9), 1639–1645 (2009). https://doi.org/10.1101/gr.092759.109

30. Ley, M.: The DBLP computer science bibliography. https://dblp.uni-trier.de

31. Mahmoud, H., Masulli, F., Rovetta, S., Russo, G.: Community detection in protein-protein interaction networks using spectral and graph approaches. In: Formenti, E., Tagliaferri, R., Wit, E. (eds.) CIBB 2013 2013. LNCS, vol. 8452, pp. 62–75. Springer, Cham (2014). https://doi.org/10.1007/978-3-319-09042-9_5

32. Muelder, C., Ma, K.: A treemap based method for rapid layout of large graphs. In: PacificVis, pp. 231–238. IEEE Computer Society (2008). https://doi.org/10.1109/PACIFICVIS.2008.4475481

33. Onnela, J., Kaski, K., Kertész, J.: Clustering and information in correlationbased financial networks. Eur. Phys. J. B-Condens. Matter Complex Syst. **38**(2), 353–362 (2004). https://doi.org/10.1140/epjb/e2004-00128-7

34. Porter, M.A., Onnela, J.P., Mucha, P.J.: Communities in networks. Not. Am. Math. Soc. **56**(1082–1097), 1164–1166 (2009)

35. Purchase, H.C.: Effective information visualisation: a study of graph drawing aesthetics and algorithms. Interact. Comput. **13**(2), 147–162 (2000). https://doi.org/10.1016/S0953-5438(00)00032-1

36. Purchase, H.C., Carrington, D.A., Allder, J.: Empirical evaluation of aesthetics-based graph layout. Empir. Softw. Eng. **7**(3), 233–255 (2002)

37. Shneiderman, B.: The eyes have it: a task by data type taxonomy for information visualizations. In: Proceedings of the 1996 IEEE Symposium on Visual Languages, Boulder, Colorado, USA, 3–6 September 1996, pp. 336–343 (1996). https://doi.org/10.1109/VL.1996.545307

38. Sindre, G., Gulla, B., Jokstad, H.G.: Onion graphs: asthetics and layout. In: VL, pp. 287–291. IEEE Computer Society (1993). https://doi.org/10.1109/VL.1993.269613

39. Six, J.M., Tollis, I.Y.G.: A framework for user-grouped circular drawings. In: Liotta, G. (ed.) GD 2003. LNCS, vol. 2912, pp. 135–146. Springer, Heidelberg (2004). https://doi.org/10.1007/978-3-540-24595-7_13

40. Sugiyama, K.: Graph Drawing and Applications for Software and Knowledge Engineers, Series on Software Engineering and Knowledge Engineering, vol. 11. World Scientific (2002). https://doi.org/10.1142/4902

41. Ware, C., Purchase, H.C., Colpoys, L., McGill, M.: Cognitive measurements of graph aesthetics. Inf. Visual. **1**(2), 103–110 (2002). https://doi.org/10.1057/palgrave.ivs.9500013

42. Wu, H., He, J., Pei, Y., Long, X.: Finding research community in collaboration network with expertise profiling. In: Huang, D.-S., Zhao, Z., Bevilacqua, V., Figueroa, J.C. (eds.) ICIC 2010. LNCS, vol. 6215, pp. 337–344. Springer, Heidelberg (2010). https://doi.org/10.1007/978-3-642-14922-1_42

43. Zhao, S., McGuffin, M.J., Chignell, M.H.: Elastic hierarchies: Combining treemaps and node-link diagrams. In: INFOVIS, pp. 57–64. IEEE Computer Society (2005). https://doi.org/10.1109/INFVIS.2005.1532129

Stress-Plus-X (SPX) Graph Layout

Sabin Devkota$^{(\boxtimes)}$ (ID), Reyan Ahmed(ID), Felice De Luca(ID),
Katherine E. Isaacs(ID), and Stephen Kobourov(ID)

Department of Computer Science, University of Arizona, Tucson, USA
{devkotasabin,abureyanahmed,felicedeluca}@email.arizona.edu,
{kisaacs,kobourov}@cs.arizona.edu

Abstract. Stress, edge crossings, and crossing angles play an important role in the quality and readability of graph drawings. Most standard graph drawing algorithms optimize one of these criteria which may lead to layouts that are deficient in other criteria. We introduce an optimization framework, Stress-Plus-X (SPX), that simultaneously optimizes stress together with several other criteria: edge crossings, minimum crossing angle, and upwardness (for directed acyclic graphs). SPX achieves results that are close to the state-of-the-art algorithms that optimize these metrics individually. SPX is flexible and extensible and can optimize a subset or all of these criteria simultaneously. Our experimental analysis shows that our joint optimization approach is successful in drawing graphs with good performance across readability criteria.

1 Introduction

Several criteria have been proposed for evaluating the quality of graph layouts [37], including minimizing stress, minimizing the number of edge crossings, minimizing drawing area, as well as maximizing the angle between edge crossings, maintaining separation between marks ("resolution"), and preserving highly connected neighborhoods. In the case of directed acyclic graphs (DAGs), maintaining consistent edge direction, i.e., upwardness, is preferable. While these criteria have been shown to improve human performance for graph tasks, automatic layout approaches actively target at most one from the list.

We propose a framework, *Stress-Plus-X (SPX)*, for automatic layout of node-link diagrams that targets multiple graph layout criteria simultaneously. SPX formulates the layout as an optimization problem that combines stress minimization with penalty terms representing other criteria. Composing and weighting the terms in the objective function provides the flexibility and extensibility.

With the adage *"Don't let perfect be the enemy of good"* in mind, the goal of SPX is not to optimize any one particular criterion at the cost of all others, but to find a balance across the criteria as optimizing only one criterion can lead to poor quality drawings [21]. As an extreme example, for minimum drawing area we can

---

This work is supported in part by NSF grants CCF-1740858, CCF-1712119 and DMS-1839274, DMS-1839307.

D. Archambault and C. D. Tóth (Eds.): GD 2019, LNCS 11904, pp. 291–304, 2019.
https://doi.org/10.1007/978-3-030-35802-0_23

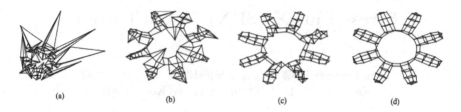

Fig. 1. Different layouts of the same graph from the crossing angle maximization Graph Drawing Contest: (a) from the Tübingen algorithm that won in 2018 [3]; (b) from the KIT algorithm that won in 2017 [8]; (c-d) from SPX with different balance in the optimization of stress, crossing angle, and edge crossings.

place all vertices on top of each other, yet perform poorly in the other quality criteria. A similar example is shown in Fig. 1 where (a–b) show the outputs on a Graph Drawing Contest graph produced by two state-of-the-art algorithms for crossing angle maximization [3,8] while (c–d) show the outputs of SPX with different balance in the optimization of stress, crossing angle, and edge crossings. Note that the SPX approach better preserves topology and produces visually appealing results and although (d) has the lowest crossing angle, it arguably provides the most recognizable drawing. Delving further into this observation, we examined the contest graphs across several metrics, as shown in Fig. 2, noting that optimizing for one criterion could yield extreme drawings. Graph 2018-8 in the middle row is a case where optimal crossing angle (center) requires a very large drawing area. Graph 2017-2 in the last row is a case where the best crossing angle (left) exhibits poor vertex resolution. These observations motivated us to seek a balance of criteria to improve drawings.

To demonstrate our framework, we formulate optimization terms for three criteria: minimizing edge crossings, maximizing the crossing angle, and upwardness (all of which have been used in Graph Drawing Contests). We compare our edge crossing formulation to state-of-the-art approaches on a corpus of community graphs. SPX achieves better edge crossing results than just optimizing stress, and frequently outperforms several of the crossing-centric algorithms. Similarly, we show that aiming only at the optimization of crossing angle tends to significantly impact the quality of the layout for other criteria. Although the angle-centric algorithms outperform SPX, our algorithm generates comparable crossing angle values and sometimes outperforms the angle-centric ones, while still achieving better performance on other drawing aspects. Finally, we compare our upwardness preserving approach to existing directed graph layout approaches [6,14,18,27].

In summary, our contributions are: (1) Stress-Plus-X (SPX), a framework for optimizing multiple graph drawing criteria simultaneously (Sect. 3); (2) Optimization terms for maximizing edge crossing angles (Sect. 3.2) and upwardness preservation (Sect. 3.3); and (3) An evaluation of our optimization terms in comparison to state-of-the-art single criterion approaches (Sect. 4). An extended version of this paper is available at arxiv [11].

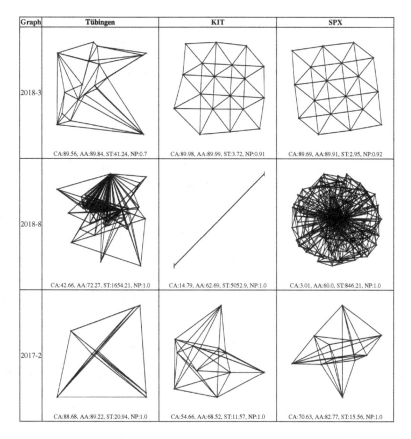

Graph	Tübingen	KIT	SPX
2018-3	CA:89.56, AA:89.84, ST:41.24, NP:0.7	CA:89.98, AA:89.99, ST:3.72, NP:0.91	CA:89.69, AA:89.91, ST:2.95, NP:0.92
2018-8	CA:42.66, AA:72.27, ST:1654.21, NP:1.0	CA:14.79, AA:62.69, ST:5052.9, NP:1.0	CA:3.01, AA:60.0, ST:846.21, NP:1.0
2017-2	CA:88.68, AA:89.22, ST:20.94, NP:1.0	CA:54.66, AA:68.52, ST:11.57, NP:1.0	CA:70.63, AA:82.77, ST:15.56, NP:1.0

Fig. 2. Graphs from the 2017-18 Graph Drawing Contests. In graph 2018-3, the crossing angles are all within 1% of the optimal, yet SPX best shows the underlying graph structure. The best crossing angle layout for Graph 2018-8 (center) yields a large drawing area. The best crossing angle layout for Graph 2017-2 (left) yields poor vertex resolution. We report crossing angle (CA), average angle (AA), stress (ST), and neighborhood preservation (NP).

2   Background and Related Work

Existing graph layout algorithms usually optimize a single drawing criterion, e.g., minimizing stress or maximizing the minimum edge crossing angle. We define these criteria formally and discuss layout approaches that focus on them.

Stress: Stress measures the difference between node-pair distances in a layout and their graph-theoretic distances, based on an all-pairs shortest path computation. It is a natural measure of how well the layout captures the structure in the underlying graph. Let $\mathbf{C}_i$ be the position of the ith node in a layout $\mathbf{C}$ and d_{ij} be the graph distance between node pair i, j. Then stress$(\mathbf{C}) = \sum_{i<j}(w_{ij}\|\mathbf{C}_i - \mathbf{C}_j\| - d_{ij})^2$. A typical normalization value is $w_{ij} = d_{ij}^{-2}$.

Kamada and Kawai [22] formulate the graph layout problem as that of minimizing stress and use energy-based optimization. Gansner *et al.* [17] use stress majorization instead. Stress-based graph visualization can also be seen as a special case of a multi-dimensional scaling (MDS) [24,34], which is a powerful dimensionality reduction technique. Variants of MDS are used in many graph layout systems, including [5,17,28]. None of these methods aim to optimize other criteria such as minimizing edge crossings or maximizing crossing angles.

Wang *et al.* [36] reformulate stress to incorporate target edge directions and lengths and propose constraints to reduce crossings or improve crossing angle in given subgraphs, but not in the entire graph. Constrained layout algorithms [13,14] combine stress minimization or force-directed layout with separation constraints between node pairs. Constrained layouts, however, do not optimize for edge crossings or crossing angles. When used with force-directed layout algorithms (such as Fruchterman-Reingold [16]) instead of with stress minimization, stress is also not optimized.

Edge Crossings and Crossing Angles: Minimizing the number of crossings between edges in a graph layout has been shown to be an important heuristic in readability of graphs [29], prompting interest in several graph drawing contests [1,4]. Other than recent works by Radermacher *et al.* [30] and Shabbeer *et al.* [33] (discussed in Sect. 2), there is little work on directly minimizing edge crossings in general graphs.

The crossing angle of a straight-line drawing of a graph is the smallest angle between two crossing edges in the layout. Large crossing angles have been shown [2,20,21] to improve graph readability and several heuristics have been proposed to maximize crossing angles. Demel *et al.* (KIT) [8] propose a greedy heuristic to select the best position for a single vertex from a random set of points. Bekos *et al.* (Tübingen) [3] propose selecting a vertex arbitrarily from a set of vertices, called the *vertex-pool*, which contains a subset of the vertices which are adjacent to the pairs of edges that have the minimum crossing angle. Both approaches above performed very well in crossing angle maximization, but neither is concerned with stress minimization or other criteria.

Upward Drawing: A drawing of a directed acyclic graph is upward if the target vertex of each directed edge has a strictly higher y-coordinate than the source vertex. Upward drawing is used to show ordering or precedence between entities in a variety of settings [14,18]. Sugiyama layout [35] is the most common approach for creating upward drawings. The layout algorithm assigns ranks to the vertices to determine their y-coordinates followed by computing their x-coordinates to minimize crossings between consecutive layers. Examples include *dot* [18], *dagre* [6], and *OGDF* [27]. Mixed graphs, where only subgraphs are drawn upward, have also been drawn using this approach [32].

Neighborhood Preservation and Drawing Area: While stress captures how well *global* graph distances are realized in the layout, neighborhood preservation captures how well *local* neighborhoods are preserved in the layout. This is the optimization goal of more recent dimensionality reduction techniques such as t-SNE [25] and UMap [26]. Specifically, in the context of graph drawing, neighborhood preservation is defined as the Jaccard similarity between the adjacent nodes in the graph and the nearest nodes in the layout, averaged over all nodes in the graph [23].

Drawing area refers to the size of the canvas used to layout the graph and is implicit when nodes are placed on an integer grid. Large drawing area is undesirable due to difficulties navigating the visualization or resolving the marks. Minimizing drawing area has also been used in Graph Drawing Contest challenges [12, 19].

Joint Optimization: Our work aims to jointly optimize several graph drawing heuristics simultaneously. Huang *et al.* [21] previously optimized for two criteria simultaneously, namely crossing angle and angular resolution of the graph in a force-directed setting. Shabbeer *et al.* [33] minimized stress and edge crossings simultaneously using an optimization-based approach.

The objective function of Shabbeer *et al.* contains penalties for edge crossings. Edge crossings can be expressed as a system of non-linear constraints. Consider two edges $\mathbf{A} = \begin{pmatrix} a_1^x & a_1^y \\ a_2^x & a_2^y \end{pmatrix}$ and $\mathbf{B} = \begin{pmatrix} b_1^x & b_1^y \\ b_2^x & b_2^y \end{pmatrix}$ where the two nodes of $\mathbf{A}$ are (a_1^x, a_1^y), and (a_2^x, a_2^y) and similarly for $\mathbf{B}$. Farkas' Theorem can be used to state that the edges $\mathbf{A}$ and $\mathbf{B}$ do not cross if and only if there exists $\mathbf{u}$, and γ, such that

$$\mathbf{Au} + \gamma\mathbf{e} \geq \mathbf{0}, \mathbf{Bu} + (1 + \gamma)\mathbf{e} \leq \mathbf{0} \tag{1}$$

where $\mathbf{e}$ is a 2-dimensional vector of ones. Intuitively, Eq. 1 states that for a pair of edges $\mathbf{A}$ and $\mathbf{B}$ to not cross, there must exist a line that strictly separates the edges A and B, i.e., there is a non-zero margin between them. Here, $\mathbf{u}$ refers to a vector that is perpendicular to the direction of the separating line and γ is a scalar value that ensures the non-zero margin of separation between the edges.

This set of inequalities can be transformed into a penalty term, $penalty(\mathbf{A}, \mathbf{B})$, for edge pair $\mathbf{A}, \mathbf{B}$ such that it is zero for non-crossing edge pairs and strictly positive for crossing edge pairs.

$$penalty(\mathbf{A}, \mathbf{B}) = \min_{u,v} ||(-\mathbf{Au} - \gamma\mathbf{e})_+||_1 + ||(\mathbf{Bu} + (1 + \gamma)\mathbf{e})_+||_1 \tag{2}$$

where $(z)_+ = max(0, z)$. The penalty term is combined with stress as a cost function and then iterative optimization is used to compute a layout. They demonstrate their approach on small biological networks.

Our approach differs in that our goal is a framework for balancing multiple criteria to achieve good results across them. We introduce penalties and constraints for crossing angle maximization and upward drawings. We further introduce a weighting to the edge crossings. Finally, we introduce a hyperparameter to directly balance across criteria.

3   SPX Algorithm

Stress-Plus-X (SPX) is a unified framework that can simultaneously optimize stress along with other graph drawing criteria. The "X" in SPX refers to the constraints that encode the additional criteria. We describe cost functions for encoding the number of edge crossings and crossing angle respectively, as well as constraints for preserving upwardness. The general SPX model is as follows:

$$cost(\mathbf{C}, \mathbf{u}, \gamma, \boldsymbol{\rho}) = stress(\mathbf{C}) + K \times \sum Penalties(\mathbf{C}, \mathbf{u}, \gamma, \mathbf{P}) \qquad (3)$$

with node coordinates $\mathbf{C}$, balancing hyperparameter K, optional penalty parameters P (e.g., ρ_i in Sect. 3.1), and γ and $\mathbf{u}$ as described in Sect. 2.

Intuitively, decreasing stress, decreasing the penalty term for X, or decreasing both results in a decrease in the objective function. Hence, minimizing the objective function simultaneously optimizes for both stress and "X."

Modifying the value of K allows us to control the balance between the stress and the "X" terms. Figure 1(c) and (d) show two layouts of the same graph created with different K parameterizations. Adjusting K to better balance criteria can result in a more intuitive drawing.

Optimization Procedure: We optimize the cost function iteratively in two phases. We first compute the optimal $\mathbf{u}$ and γ for each pair of edges $(\mathbf{A}, \mathbf{B})$ via linear programming to minimize the penalties, $penalty(\mathbf{A}, \mathbf{B})$. Then, keeping the $\mathbf{u}'s$ and γ's constant, for all edge pairs, we optimize the cost function by modifying $\mathbf{C}$ using gradient descent; see Algorithm 1.

Algorithm 1. Stress-plux-X(G)

Compute initial layout $\mathbf{C}_0$ (using stress majorization, force-directed layout, or random initialization)
for Number-of-iterations **do**
    Keeping the node coordinates $\mathbf{C}$ constant, find optimal $\mathbf{u}$ and γ for each edge pair $(\mathbf{A}, \mathbf{B})$ using linear programming to minimize $penalty(\mathbf{A}, \mathbf{B})$
    Keeping $\mathbf{u}$'s and γ's constant, minimize $cost(\mathbf{C}, \mathbf{U}, \gamma, \boldsymbol{\rho})$ by updating $\mathbf{C}$ using gradient descent
end for

3.1   Stress Plus Crossing Minimization

The edge crossings penalty is: $\sum_{i=1}^{l}(\rho_i/2) * \{||(-\mathbf{A}^i(\mathbf{C})u^i - \gamma^i\mathbf{e})_+||_1 + ||(\mathbf{B}^i(\mathbf{C})u^i + (1 + \gamma^i)\mathbf{e})_+||_1\}$ where l is the number of edge pairs, $\mathbf{A}^i(\mathbf{C})$ and $\mathbf{B}^i(\mathbf{C})$ are the first and second edges of edge pair i as matrices $\mathbf{A}$ and $\mathbf{B}$, u^i, γ^i are the $\mathbf{u}, \gamma$ terms for edge pair i, and ρ_i is a weight on the penalty for edge pair i.

Shabbeer *et al.* [33] use a compounding weight where each edge crossing gets penalized more the longer it persists through the optimization iterations. We found that such a penalty can result in the introduction of new edge crossings for graphs that are larger and denser. With this in mind, we use a binary weight for ρ_i: the value is 1 when edges intersect and 0 otherwise.

The cost function for stress plus crossing minimization further differs from Shabbeer *et al.* in the criteria weighting parameter K. Figure 4 (Sect. 4) shows that the use of binary weights and hyperparameter K helps SPX achieve better results compared to Shabbeer *et al.*

3.2   Stress Plus Crossing Angle Maximization

Our crossing angle maximization penalty is the edge crossing penalty with an additional factor of $cos^2(\theta_i)$ in each factor of the summation, where θ_i is the angle between a pair of crossing edges. We use cos^2 to constrain to positive values and give a heavier weight to smaller crossing angles. Note this modified penalty function explicitly maximizes the minimum crossing angle and implicitly minimizes the number of crossings, as when a crossings is removed altogether it cannot contribute to the minimum crossing angle.

3.3   Stress Plus Upward Crossing Minimization

We add the upwardness criteria to SPX by adding constraints to the model. Let (u, v) be a directed edge. Then, in the drawing of the graph the y coordinate of v should be strictly larger than the y coordinate of u. We enforce this directly with a linear constraint $(y_v > y_u)$. If the input graph is a DAG then we add this constraint for all edges. If the graph is mixed then we add the constraints only for the directed edges.

3.4   Implementation

We implemented SPX in Python. It uses the stress majorization formulation of Gansner *et al.* [17] to minimize stress and the edge crossing detection code from Demel *et al.* [8]. SPX source code and experimental material are available at https://github.com/devkotasabin/SPX-graph-layout.

Initial Layouts: We ran our experiments using 3 different layout algorithms as input to the SPX algorithm: stress majorization (`neato`), force-directed layout (`sfdp`), and random initialization. Both `neato` and `sfdp` are available in the GraphViz package [15]. To ameliorate the effects of sensitivity to initial layout, we employ random starts of SPX, using each method multiple times and choosing the layout that maximizes the objective.

Gradient Descent Algorithms: We experimented with the following algorithms for gradient descent (GD) [31]: bfgs, l-bfgs, vanilla GD, momentum-based GD, Nesterov momentum-based GD, Adagrad, RMSprop, and Adam. We found that for different types of graphs, different GD variants yielded better results and we kept all but bfgs and l-bfgs in our parameter sweep based on their performance in our pilot experiments. Section 3.5 contains further analysis of different GD variants and their convergence plots.

Parallelization: Each combination of random initial layout, gradient descent algorithm, and value of K is independent and thus can be run in parallel. Operations on each edge pair, such as computing $\mathbf{u}$ and γ, as well as summing the penalties, can also be parallelized. However, running edge pairs fully in parallel would incur significant overhead. We leave the implementation of this approach as future work.

3.5 Convergence Analysis

Figure 3 illustrates the convergence behavior of SPX using the six variants of gradient descent from Sect. 3.4 on two graphs, graph 5 from the community graphs of Sect. 4 (top row) and graph 9 from 2018 Graph Drawing contest (bottom row). Convergence behavior of the variants differ depending on graph. Figure 3 shows the values for number of crossings, stress, and crossing angle over 100 iterations for a fixed value of K(= 2) for both graphs.

For both graphs, at least one gradient descent variant converges within 100 iterations.

In the first graph, Momentum and Nesterov converge rapidly and then get stuck in local minima. In the first graph, they overcome the local minima to continue convergence, while on the second graph they diverge after the minima. We hypothesize convergence per variant is dependent on graph properties and thus use all six. Further analysis and optimization is left for future work.

4 Results

As SPX is designed to be a flexible framework, we evaluate it in three different contexts. First, we compare SPX to Shabbeer *et al.* [33] on stress and number of crossings showing SPX performs better.

Second, we compare SPX to two state-of-the-art algorithms for crossing angle optimization: Demel *et al.* from KIT [8] and Bekos *et al.* from Tübingen [3]. We compare across five readability metrics discussed earlier: stress (ST), number of crossings (NC), crossing angle (CA), drawing area (DA), and neighborhood preservation (NP). We show SPX balanced multiple criteria simultaneously rather than optimizing one at the expense of others.

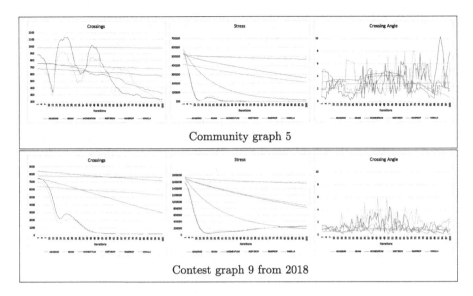

Fig. 3. Number of crossings, stress, and crossing angle over 100 iterations for 6 variants of GD algorithms on 2 graphs run with fixed K(= 2). The 2 graphs are graph 5 from random subset of 25 community graphs (top row) and graph 9 from 2018 graph drawing contest (bottom row).

Third, we compare SPX to existing approaches that directly optimize crossings [6,14,18,27] for upward drawings of DAGs. Our results show that SPX can preserve upwardness while performing better across other readability criteria.

4.1 Datasets and Experimental Settings

For the first two evaluations we used the 2017 and 2018 graph drawing contest graphs [9,10], as well as a collection of 400 graphs used in a crossings minimization study by Radermacher *et al.* [30]. For the third evaluation (upward drawing of DAGs) we generated 4 trees and 30 DAGs of different sizes.

We ran our experiments using all six gradient descent variants discussed in Sect. 3.4. We swept the values of K in the range of 2^{-5} to 2^5 in exponential increments. We used three different initial layout algorithms as input: `neato`, `sfdp`, and random initialization with five different starts each. Metrics were calculated using the `graphmetrics` library of De Luca [7].

4.2 Comparison to Shabbeer *et al.*

We compare SPX with two algorithms - Shabbeer *et al.* [33] and stress majorization [17] on the corpus of 100 community graphs. The crossings value for stress is taken from Radermacher *et al.* [30] and the stress value calculated as lowest from five random `neato` [15] layouts. We run the SPX variant that performs stress-plus-crossing minimization only and compare using two metrics, number of crossings and stress, because Shabbeer *et al.* minimizes only for these two metrics. We do not perform the same two-metric comparison with the crossing minimization algorithms of Radermacher *et al.* [30] because they are not concerned with stress. We provide details about crossing minimization only for SPX and the algorithms of Radermacher *et al.*

Figure 4a shows that on average SPX produces fewer crossings than both other approaches. Figure 4b shows that on average SPX produces layouts with lower stress than both other approaches. We hypothesize that SPX performs better than stress majorization on stress because of SPX's multiple random starts and the use of **neato** as one of the initializations.

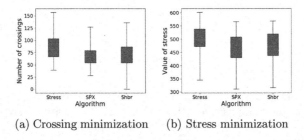

(a) Crossing minimization   (b) Stress minimization

Fig. 4. Comparing SPX, Shabbeer *et al.* and stress majorization in terms of the number of crossing and stress minimization using 100 community graphs.

4.3   Comparison Across Several Criteria

We examine several readability criteria across the layouts obtained by the three algorithms designed to minimize crossing angle: KIT [8], Tübingen [3], and SPX. In particular, we consider stress, neighborhood preservation, edge crossings, drawing area, and crossing angle.

Though our impetus was the graph drawing contest graphs, they are diverse in structure, making it difficult to compare across them. To perform a bulk comparison, we randomly select a subset of 25 graphs from community graphs described above.

Figure 5 shows the results for the 25 graphs, presented in a pairwise fashion of metrics. We plot the metrics so that points in the lower left corner indicate good performance in the two metrics. From the plots we can see that most of the SPX drawings are in the well-performing corner.

Figure 6 shows an example of a community graph, drawn by all three algorithms. SPX achieves best stress and crossing angle while performing very close to the winner, KIT, in terms of number of crossings.

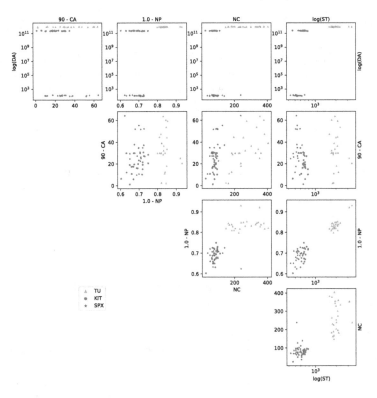

Fig. 5. Pairwise metric evaluation of the KIT, Tübingen, and SPX algorithms using stress (ST), number of crossings (NC), crossing angle (CA), neighborhood preservation (NP), and drawing area (DA).

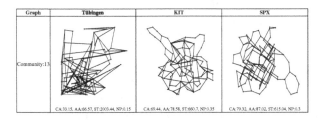

Fig. 6. Outputs of the Tübingen, KIT, SPX algorithms on a community graph.

4.4 Comparison of Upward Drawings

To evaluate SPX for upward drawing, we compare it to several state-of-the-art directed graph algorithms across several metrics on a corpus of 4 trees and 30 DAGs, described in [11].

We compared SPX to *dot* [18]; *dagre* [6] and both variants of Sugiyama in *OGDF* [27]: the barycenter heuristic ("*ogdfb*") and the median heuristic ("*ogdfm*"). We verified all algorithms, including SPX, produced completely upward drawings. We measured drawing area (A), stress (ST), and number of crossings (CR). We also measured height and width separately, but found their behavior to be the same as those for drawing area. The results of the experiment are reported in Table 1. Each cell indicates the number of times each algorithm had the best value for the metric, with ties being attributed to both algorithms.

Table 1. The number of times each algorithm had the best metric value for upward drawings of 4 complete balanced binary trees (left) and 30 DAGs (right).

	dagre	dot	ogdfb	ogdfm	SPX
ST	0	0	0	0	4
A	0	0	0	0	4
CR	4	4	4	4	4

4 binary trees

	dagre	dot	ogdfb	ogdfm	SPX
ST	0	0	0	0	30
A	0	0	0	0	30
CR	2	5	8	11	14

30 directed acyclic graphs

Table 1 shows that SPX consistently produces the best drawings across the metrics, although all other algorithms also produce planar layouts for the complete binary trees. However, there is a caveat in the measure of area. We do not impose any resolution to the upwardness of the drawings. The SPX drawings are very small in area compared to those generated by the other algorithms. Imposing a resolution constraint could increase crossings and stress, indicating a post-processing to enforce resolution may be a better option. We experimented with a naïve scaling parameter which results in very large area. We leave a more appropriate post-processing algorithm as future work.

5   Conclusions and Future Work

As some of the drawings in this paper show, optimizing just one layout criterion can result in unreadable drawings. It seems like a natural idea to consider approaches that balance multiple layout criteria. SPX is an example of such a graph layout framework that balances the optimization of multiple criteria and achieves quality that is close to one criterion state-of-the-art algorithms. Currently SPX considers stress minimization, crossing minimization, crossing angle maximization, and upwardness. A natural direction for future work is to incorporate additional layout criteria. Our current implementation of SPX relies on a combination of stress minimization and a linear program solver. As a result the algorithm is prohibitively slow for large graphs. Possible ways to speed up the algorithm, such as multi-level computation, are worth exploring.

References

1. Ábrego, B.M., Fernández-Merchant, S., Salazar, G.: The rectilinear crossing number of k_n: closing in (or are we?). Thirty Essays Geom. Gr. Theory (2012). https://doi.org/10.1007/978-1-4614-0110-0_2

2. Argyriou, E.N., Bekos, M.A., Symvonis, A.: Maximizing the total resolution of graphs. In: Brandes, U., Cornelsen, S. (eds.) GD 2010. LNCS, vol. 6502, pp. 62–67. Springer, Heidelberg (2011). https://doi.org/10.1007/978-3-642-18469-7_6

3. Bekos, M.A., Förster, H., Geckeler, C., Holländer, L., Kaufmann, M., Spallek, A.M., Splett, J.: A heuristic approach towards drawings of graphs with high crossing resolution. CoRR abs/1808.10519 (2018). http://arxiv.org/abs/1808.10519

4. Buchheim, C., Chimani, M., Gutwenger, C., Jünger, M., Mutzel, P.: Crossings and planarization. In: Handbook of Graph Drawing and Visualization, pp. 43–85 (2013)

5. Chen, L., Buja, A.: Local multidimensional scaling for nonlinear dimension reduction, graph drawing, and proximity analysis. J. Am. Stat. Assoc. **104**(485), 209–219 (2009)

6. Dagrejs: dagrejs/dagre. https://github.com/dagrejs/dagre/wiki

7. De Luca, F.: graphmetrics library (2019). https://github.com/felicedeluca/graphmetrics

8. Demel, A., Dürrschnabel, D., Mchedlidze, T., Radermacher, M., Wulf, L.: A greedy heuristic for crossing-angle maximization. In: Biedl, T., Kerren, A. (eds.) GD 2018. LNCS, vol. 11282, pp. 286–299. Springer, Cham (2018). https://doi.org/10.1007/978-3-030-04414-5_20

9. Devanny, W., Kindermann, P., Löffler, M., Rutter, I.: Graph drawing contest report. In: Frati, F., Ma, K.-L. (eds.) GD 2017. LNCS, vol. 10692, pp. 575–582. Springer, Cham (2018). https://doi.org/10.1007/978-3-319-73915-1_44

10. Devanny, W., Kindermann, P., Löffler, M., Rutter, I.: Graph drawing contest report. In: Biedl, T., Kerren, A. (eds.) GD 2018. LNCS, vol. 11282, pp. 609–617. Springer, Cham (2018). https://doi.org/10.1007/978-3-030-04414-5_43

11. Devkota, S., Ahmed, R., De Luca, F., Isaacs, K., Kobourov, S.: Stress-Plus-X (SPX) graph layout (2019)

12. Duncan, C.A., Gutwenger, C., Nachmanson, L., Sander, G.: Graph drawing contest report. In: Didimo, W., Patrignani, M. (eds.) GD 2012. LNCS, vol. 7704, pp. 575–579. Springer, Heidelberg (2013). https://doi.org/10.1007/978-3-642-36763-2_58

13. Dwyer, T.: Scalable, versatile and simple constrained graph layout. Comput. Graph. Forum **28**, 991–998 (2009)

14. Dwyer, T., Koren, Y., Marriott, K.: Ipsep-cola: an incremental procedure for separation constraint layout of graphs. IEEE Trans. Vis. Comput. Gr. **12**, 821–828 (2006). https://doi.org/10.1109/TVCG.2006.156

15. Ellson, J., Gansner, E., Koutsofios, L., North, S.C., Woodhull, G.: Graphviz—open source graph drawing tools. In: Mutzel, P., Jünger, M., Leipert, S. (eds.) GD 2001. LNCS, vol. 2265, pp. 483–484. Springer, Heidelberg (2002). https://doi.org/10.1007/3-540-45848-4_57

16. Fruchterman, T.M.J., Reingold, E.M.: Graph drawing by force-directed placement. Softw. Pract. Exp. **21**(11), 1129–1164 (1991). https://doi.org/10.1002/spe.4380211102

17. Gansner, E.R., Koren, Y., North, S.: Graph drawing by stress majorization. In: Pach, J. (ed.) GD 2004. LNCS, vol. 3383, pp. 239–250. Springer, Heidelberg (2005). https://doi.org/10.1007/978-3-540-31843-9_25

18. Gansner, E.R., North, S.C., Vo, K.P.: Technique for drawing directed graphs (1990). uS Patent 4,953,106

19. Gutwenger, C., Löffler, M., Nachmanson, L., Rutter, I.: Graph drawing contest report. In: Duncan, C., Symvonis, A. (eds.) GD 2014. LNCS, vol. 8871, pp. 501–506. Springer, Heidelberg (2014). https://doi.org/10.1007/978-3-662-45803-7_42

20. Huang, W., Eades, P., Hong, S.H.: Larger crossing angles make graphs easier to read. J. Vis. Lang. Comput. **25**(4), 452–465 (2014). https://doi.org/10.1016/j.jvlc. 2014.03.001

21. Huang, W., Eades, P., Hong, S.H., Lin, C.C.: Improving multiple aesthetics produces better graph drawings. J. Vis. Lang. Comput. **24**(4), 262–272 (2013). https://doi.org/10.1016/j.jvlc.2011.12.002. http://www.sciencedirect.com/science/article/pii/S1045926X11000814

22. Kamada, T., Kawai, S.: An algorithm for drawing general undirected graphs. Inf. Process. Lett. **31**(1), 7–15 (1989). https://doi.org/10.1016/0020-0190(89)90102-6. http://www.sciencedirect.com/science/article/pii/0020019089901026

23. Kruiger, J.F., Rauber, P.E., Martins, R.M., Kerren, A., Kobourov, S., Telea, A.C.: Graph layouts by t-sne. Comput. Graph. Forum **36**(3), 283–294 (2017). https://doi.org/10.1111/cgf.13187

24. Kruskal, J.B.: Multidimensional scaling by optimizing goodness of fit to a nonmetric hypothesis. Psychometrika **29**(1), 1–27 (1964)

25. Maaten, L.V.D., Hinton, G.: Visualizing data using t-sne. J. Mach. Learn. Res. **9**(Nov), 2579–2605 (2008)

26. McInnes, L., Healy, J., Melville, J.: Umap: uniform manifold approximation and projection for dimension reduction. arXiv preprint arXiv:1802.03426 (2018)

27. Mutzel, P., Chimani, M., Gutwenger, C., Klein, K.: OGDF an open graph drawing framework. In: 15th International Symposium on Graph Drawing (2007). https://doi.org/10.17877/DE290R-7670

28. Pich, C.: Applications of multidimensional scaling to graph drawing. Ph.D. thesis (2009)

29. Purchase, H.: Which aesthetic has the greatest effect on human understanding? In: DiBattista, G. (ed.) GD 1997. LNCS, vol. 1353, pp. 248–261. Springer, Heidelberg (1997). https://doi.org/10.1007/3-540-63938-1_67

30. Radermacher, M., Reichard, K., Rutter, I., Wagner, D.: A geometric heuristic for rectilinear crossing minimization. In: Pagh, R., Venkatasubramanian, S. (eds.) The 20th Workshop on Algorithm Engineering and Experiments (ALENEX 2018), pp. 129–138 (2018). https://doi.org/10.1137/1.9781611975055.12

31. Ruder, S.: An overview of gradient descent optimization algorithms. CoRR abs/1609.04747 (2016). http://arxiv.org/abs/1609.04747

32. Seemann, J.: Extending the sugiyama algorithm for drawing uml class diagrams: towards automatic layout of object-oriented software diagrams. In: DiBattista, G. (ed.) GD 1997. LNCS, vol. 1353, pp. 415–424. Springer, Heidelberg (1997). https://doi.org/10.1007/3-540-63938-1_86

33. Shabbeer, A., Ozcaglar, C., Gonzalez, M., Bennett, K.P.: Optimal embedding of heterogeneous graph data with edge crossing constraints. In: NIPS Workshop on Challenges of Data Visualization (2010)

34. Shepard, R.N.: The analysis of proximities: multidimensional scaling with an unknown distance function. Psychometrika **27**(2), 125–140 (1962)

35. Sugiyama, K., Tagawa, S., Toda, M.: Methods for visual understanding of hierarchical system structures. IEEE Trans. Syst. Man Cybern. **11**(2), 109–125 (1981)

36. Wang, Y., Wang, Y., Sun, Y., Zhu, L., Lu, K., Fu, C., Sedlmair, M., Deussen, O., Chen, B.: Revisiting stress majorization as a unified framework for interactive constrained graph visualization. IEEE Trans. Vis. Comput. Gr. **24**(1), 489–499 (2018). https://doi.org/10.1109/TVCG.2017.2745919

37. Ware, C., Purchase, H., Colpoys, L., McGill, M.: Cognitive measurements of graph aesthetics. Inf. Vis. **1**(2), 103–110 (2002)

Best Paper in Track 1

Exact Crossing Number Parameterized by Vertex Cover

Petr Hliněný[1]([⊠])[iD] and Abhisekh Sankaran[2]

[1] Faculty of Informatics of Masaryk University, Brno, Czech Republic
hlineny@fi.muni.cz
[2] Department of Computer Science and Technology, University of Cambridge,
Cambridge, UK
abhisekh.sankaran@cl.cam.ac.uk

Abstract. We prove that the exact crossing number of a graph can be efficiently computed for simple graphs having bounded vertex cover. In more precise words, CROSSING NUMBER is in FPT when parameterized by the vertex cover size. This is a notable advance since we know only *very few* nontrivial examples of graph classes with unbounded and yet efficiently computable crossing number. Our result can be viewed as a strengthening of a previous result of Lokshtanov [arXiv, 2015] that OPTIMAL LINEAR ARRANGEMENT is in FPT when parameterized by the vertex cover size, and we use a similar approach of reducing the problem to a tractable instance of INTEGER QUADRATIC PROGRAMMING as in Lokshtanov's paper.

Keywords: Graph drawing · Crossing number · Parameterized complexity · Vertex cover

1 Introduction

The crossing number $cr(G)$ of a graph G is the minimum number of pairwise edge crossings in a drawing of G in the plane. We refer to Sect. 2 for the definitions of a drawing and edge crossings. Finding the crossing number of a graph is one of the most prominent combinatorial optimization problems in graph theory and is NP-hard already in very special cases, e.g., even when considering a planar graph with one added edge [5]. Moreover, we know that computing the crossing number is APX-hard [3], i.e., there does not exist a PTAS (unless P=NP).

On the other hand, there is an algorithm [1] that approximates not directly the crossing number, but the quantity $n + cr(G)$ where $n = |V(G)|$; its currently best incarnation does so within a factor of $\mathcal{O}(\log^2 n)$ [11]. The first sublinear approximation factor of $\tilde{\mathcal{O}}(n^{0.9})$ has been achieved by [9]. Concerning rectilinear

---

P. Hliněný—Supported by the Czech Science Foundation, project no. 17-00837S.
A. Sankaran—Supported by the Leverhulme Trust through a Research Project Grant on 'Logical Fractals'.

© Springer Nature Switzerland AG 2019
D. Archambault and C. D. Tóth (Eds.): GD 2019, LNCS 11904, pp. 307–319, 2019.
https://doi.org/10.1007/978-3-030-35802-0_24

drawings of dense graphs there is another recent approximation result [12]. Much better crossing number approximation results are known for some restricted graph classes, such as for graphs embeddable in a fixed surface [15,17] and for graphs from which few edges or vertices can be removed to make them planar [4,6,7,18].

Despite this recent progress in crossing number approximations in special cases, there are nearly no nontrivial formulas or efficient algorithms for computing the exact crossing number in sufficiently "rich" graph classes. Even for very nicely structured classes such as the complete graphs, the complete bipartite graphs and the toroidal grids (Cartesian products of cycles), their exact crossing numbers are only conjectured, but not proved (e.g., we do not know $\mathrm{cr}(K_{13})$).

One notable exception (to near-impossibility of computing efficiently the exact crossing number) is that the exact crossing number can be efficiently computed (even in linear time) when it is bounded [16,19]; more precisely, that CROSSING NUMBER is in linear-time FPT when parameterized by the solution value. However, considering *nontrivially rich* graph classes with unbounded crossing number, there currently seems to be only one such further efficient result by Biedl et al. [2]; computing the exact crossing number for maximal graphs of pathwidth exactly 3.

Our paper brings one more small piece to this crossing-number puzzle. A *vertex cover* in a graph G is a set X of vertices of G such that every edge of G has at least one end in X. We prove that it is possible to compute in FPT the exact crossing number of a simple graph G when the minimum vertex cover size of G is bounded as a parameter.

Theorem 1. *Given a simple graph G, the problem to compute the crossing number of G and the corresponding optimal drawing of G is fixed-parameter tractable with respect to the parameter $k = |X|$ where $X \subseteq V(G)$ is a vertex cover of G.*

We remark that computing the minimum vertex cover X is itself in FPT when parameterized by $|X|$ [10], and so we do not need X on the input.

Although bounding the vertex cover also bounds the pathwidth of a graph, our Theorem 1 is incomparable with [2] since their result gives exact values only for pathwidth 3 (for higher values of pathwidth [2] gives an approximation). Another notable point is that the classes of graphs of bounded vertex cover are monotone (closed under taking subgraphs), while the exact algorithm in [2] requires maximal graphs of pathwidth 3 (again, for non-maximal such graphs there is only a 2-approximation).

In the algorithm of Theorem 1 we follow the approach of Lokshtanov [20], who showed that INTEGER QUADRATIC PROGRAMMING is in FPT when parameterized by the number of variables and the maximum of the absolute values of the coefficients. Lokshtanov then used his IQP algorithm to show that the problem OPTIMAL LINEAR ARRANGEMENT of a graph G is in FPT when parameterized by the minimum vertex cover size of G. In the OPTIMAL LINEAR ARRANGEMENT problem of a graph G, the task is to find a linear ordering of the vertex set of G which minimizes the sum of "lengths" of the edges of G. With respect to this, it is worth to note that the first NP-hardness proof for CROSSING NUMBER [14] used a simple reduction from OPTIMAL LINEAR ARRANGEMENT, one which asymptotically preserves almost any reasonable graph parameter including vertex cover. Although we cannot directly apply our algorithm to the result of that reduction (due to a presence of parallel edges, as explained below), a simple modification along the lines of that reduction allows to deduce Lokshtanov's result for OPTIMAL LINEAR ARRANGEMENT from our algorithm.

2 Basic Definitions

We use the standard terminology of graph theory. A special attention has to be paid to simplicity of graphs – while (non-)simplicity is usually not an issue for the crossing number (just subdivide parallel edges), it becomes important with respect to the minimum vertex cover. Therefore, we will consider *simple graphs* throughout the paper by default, and we will use the term *multigraph* otherwise.

A *drawing* of a graph $G = (V, E)$ is a mapping of the vertices V to distinct points in the plane, and of the edges E to simple curves connecting their respective end points but not containing any other vertex point. When convenient, we will refer to the elements (vertices and edges) of the drawing as to the corresponding elements of G. A *crossing* is a common point of two distinct edge curves, other than their common end point. It is well established that the search for an optimal solution to the crossing number problem can be restricted to so called *good drawings*: any pair of edges crosses at most once, adjacent edges do not cross, and there is no crossing point in common of three or more edges.

Definition 2. *The problem* CROSSING NUMBER *asks for a* good drawing D *of a given graph G with the least possible number of crossings.*

The number of crossings in a particular drawing D is denoted by cr(D) *and the minimum over all good drawings D of a graph G by* cr(G). *We call* cr(D) *and* cr(G) the crossing number *of the drawing D and the graph G, respectively.*

We will also need to deal with *weighted crossing number*. Consider a graph H with a weight assignment $w : E(H) \rightarrow \mathbb{N}$. Then a crossing between edges $e, f \in E(H)$ naturally counts as $w(e) \cdot w(f)$ crossings (as if they were bunches of $w(e)$ and $w(f)$ parallel edges). The weighted crossing number cr(H) of weighted H is defined as in Definition 2 while counting crossings this weighted way.

Following [20], we introduce the problem INTEGER QUADRATIC PROGRAMMING (IQP) in a generalized form. [1] Its input consists of a $k \times k$ integer matrix $\boldsymbol{Q}$, an $m \times k$ and $m' \times k$ integer matrices $\boldsymbol{A}$ and $\boldsymbol{C}$, a k-dimensional integer vector $\boldsymbol{p}$, and m- and m'-dimensional integer vectors $\boldsymbol{b}$ and $\boldsymbol{d}$. The task is to find an optimal solution $\boldsymbol{z}^\circ$ to the following optimization problem:

---

[1] The stated generalized form comes from page 4 of [20], formula (2) and below.

$$\begin{aligned}
\text{Minimize} \quad & z^T Q z + p^T z \\
\text{subject to} \quad & A z \;\le\; b \\
& C z \;=\; d \\
& z \;\in\; \mathbb{Z}^k
\end{aligned} \tag{1}$$

Note that "finding a solution" of an IQP instance means exactly one of the following three outcomes: the instance is infeasible and we correctly detect that, or the instance is feasible and unbounded and we again detect that, or the instance is feasible and bounded and we output an optimal solution z°.

Theorem 3. (Lokshtanov [20]). *Consider the* INTEGER QUADRATIC PROGRAMMING *problem as above* (1), *where the input consists of the integer matrices* A, C, Q *and the integer vectors* b, d, p. *Let L denote the length of the combined bit-representation of this input, and let λ be the largest absolute value of the entries in the matrices* A, C *and* Q, *and the entries in the vector* p. *There exists an algorithm which finds a solution of this instance of IQP in time* $f(k, \lambda) \cdot L^{\mathcal{O}(1)}$ *for some computable function f (that is, fixed-parameter tractable with input size L and parameters k and λ).*

3   Clustered Optimal Drawings

We start with a high-level idea of our solution. Consider a simple graph G and a vertex cover $X \subseteq V(G)$ of fixed size $k = |X|$. Then $V(G) \setminus X$ is an independent set and every vertex of $V(G) \setminus X$ can be classified by its neighbourhood in X (and this classification is unique up to automorphisms). At the first sight it thus appears natural to form "uniform" clusters of the vertices with the same neighbourhood (to be treated the same way, whatever this means), and so seemingly "reduce" the input size to $\mathcal{O}(2^k)$ and then solve it in FPT by brute force. This is, unfortunately, not at all sufficient.[2]

As we will see, while solving the crossing number problem, it would be enough to additionally classify the vertices of $V(G) \setminus X$ by the cyclic ordering of their edges in the (yet to be found) optimal drawing of G. Furthermore, it will also be useful to restrict the arguments to the aforementioned good drawings (pairs of edges crosses at most once and adjacent pairs do not cross). In particular, in a good drawing the edges incident to one common vertex always form an uncrossed star. We give the following core definition (see also Fig. 1):

---

[2]   In the exemplary case of OPTIMAL LINEAR ARRANGEMENT [20], one may easily see the problem on the graph $K_{k,n}$, whose smaller part X of size k is the minimum vertex cover and all vertices of the larger part form one cluster with the same neighbourhood X. Yet, an optimal linear arrangement solution for $K_{k,n}$ has to alternate the vertices of X and those of the large part in the middle of the arrangement. Hence in this particular case of OLA, one has to consider at least the relative position of vertices with respect to X in addition to their neighbourhoods.

Definition 4. *Let G be a graph with a vertex cover X, and D be a good drawing of G. Then two vertices $x, y \in V(G) \setminus X$ belong to the same topological cluster in D (with implicit respect to X) if their neighbourhood in X is the same, and the clockwise cyclic order of the neighbours of x within D is the same as the clockwise cyclic order of the neighbours of y.*

(Note that a vertex of X does not belong to any topological cluster in D.)

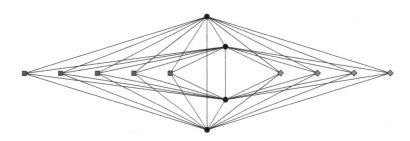

Fig. 1. An illustration of topological clusters with respect to the vertex cover formed by the middle four black vertices: any two blue (square) vertices on the left belong to the same topological cluster, and likewise for any two red (diamond) vertices. Any blue and a red vertex belong to different topological clusters. (Color figure online)

We aim to show that an optimal drawing of our graph G can be obtained in such a form that the topological clusters of vertices and the incident edges are "drawn closely together". The same idea, only for the special case of the complete bipartite graph $G = K_{k,n}$, has already been used by Christian, Richter and Salazar [8] (their research goal, though, was different). Our paper can be considered a generalization of (a part of) [8]. To achieve our goal, we separate two kinds of crossings and rigorously describe a topological clustering of a drawing of G.

Assume a good drawing D of G. Having two edges $e, f \in E(G)$ such that one end of e belongs to the same topological cluster in D as one end of f, we say that a (possible) crossing of e and f is a *cluster crossing*. All other edge crossings occurring in D are called *non-cluster crossings* (and they include all crossings on edges with both ends in X). Here we denote by $\mathrm{cr}^n(D)$ the number of non-cluster crossings (possibly weighted) in the drawing D.

In the following definition we, informally, select one weighted "representative" of each topological cluster of a drawing D.

Definition 5. *A drawing D_X is called a topological clustering of the drawing D (of a graph G) with respect to its vertex cover X if the following hold:*

- *D_X is an induced subdrawing of D and $V(D_X) \supseteq X$,*
- *every vertex of $V(D) \setminus X$ belongs to the same topological cluster in D as some vertex of $V(D_X) \setminus X$,*
- *no two vertices of $V(D_X) \setminus X$ are in the same topological cluster in D_X, and*

– D_X is equipped with a weight function $c : V(D_X) \setminus X \to \mathbb{N}$ such that, for every $t \in V(D_X) \setminus X$, the size of the topological cluster in D containing t equals $c(t)$.

Note that there can be many (topologically) different topological clusterings D_X of the same drawing D, depending on how the "representative" vertices from $V(D_X) \setminus X$ are chosen. See also Fig. 2

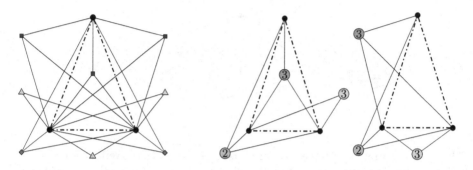

Fig. 2. An illustration: the graph on the left has a vertex cover X formed by the three black vertices. There are three topological clusters wrt. X, depicted by the blue (square), red (diamond) and yellow (triangle) vertices. On the right, we can see two different topological clusterings of this graph, with the weights in the node circles. (Color figure online)

In view of Definition 5, it will be useful to consider the crossing number of the following *independently weighted* graphs. For a graph H, an independent set $Y \subseteq V(H)$, and a weight function $c : Y \to \mathbb{N}$, let the edge weights in H be as follows: for $e \in E(H)$ having one end $x \in Y$ we set $c'(e) := c(x)$, and for edges e with both ends in $V(H) \setminus Y$ we set $c'(e) = 1$. This defines the weighted crossing number of H and, in particular for $Y = V(D_X) \setminus X$, the weighted crossing number $\mathrm{cr}(D_X)$ of D_X with respect to its weight function c.

We can now formulate and prove the core claim:

Lemma 6. *For every good drawing D of a graph G with a vertex cover X, there exists its topological clustering D_X such that the number of non-cluster crossings in D is at least $\mathrm{cr}(D_X)$.*

Proof. We start by setting $D' := D$, and then we will inductively choose suitable representatives of the topological clusters of D' until we arrive at desired D_X.

Let the *cost* of a vertex x of D' be the sum of all non-cluster(!) crossings carried by the edges of D' incident to x. We pick any (nonempty) topological cluster $S \subseteq V(D') \setminus X$ and choose a vertex $s_0 \in S$ having the least cost among those of S. If $|S| > 1$, we iterate the following over all $s \in S \setminus \{s_0\}$;

- remove s and its incident edges from the drawing D',
- choose a new vertex s' in a tiny neighbourhood of s_0 in D', and draw the edges of s' in a tiny strip along the edges of s_0 while making only such non-cluster crossings as those that exist on the edges of s_0.

Since s_0 and s have had the same neighbourhood, the new drawing D'' as a graph (with s') is isomorphic to D' and the vertices s_0 and s' still belong to the same topological cluster. Moreover, since s_0 has been chosen with the least cost, we have $\mathrm{cr}^n(D'') \leq \mathrm{cr}^n(D')$. (New cluster crossings can be simply ignored.)

At the end of the previous procedure, we get a drawing D° (isomorphic to D') which has no more non-cluster crossings than D', and all edges of the cluster S are drawn "the same way closely together" in D°. From this it follows that the number of non-cluster crossings carried by the edges incident to the cluster S in D° equals $|S|$-times this number on the edges incident to s_0 in D'. We therefore define D_1 as the induced subdrawing of D' obtained by deleting the vertices of $S \setminus \{s_0\}$ and assigning the weight $c(s_0) = |S|$. In the setting of independently weighted graph underlying D_1, we get $\mathrm{cr}^n(D_1) = \mathrm{cr}^n(D^\circ) \leq \mathrm{cr}^n(D')$.

We finish the proof inductively. Let r be the number of topological clusters of given D. For $i = 1, 2, \ldots, r-1$, we now repeat the previous steps for $D' := D_i$, obtaining a subdrawing D_{i+1} such that $\mathrm{cr}^n(D_{i+1}) \leq \mathrm{cr}^n(D_i)$. Finally, D_r is a topological clustering of D with respect to X by Definition 4 and we conclude $\mathrm{cr}(D_r) = \mathrm{cr}^n(D_r) \leq \ldots \leq \mathrm{cr}^n(D_1) \leq \mathrm{cr}^n(D)$.   □

4   Counting the Crossings in Clusters and Between

In order to complement Lemma 6, we need to estimate also the number of cluster crossings in a drawing D. This is actually quite easy using the fact that two vertices in the same topological cluster have the same cyclic ordering of their neighbours. We use the following simple claim (cf. Fig. 3):

Lemma 7 ([8, Lemma 2.1]). *Let x, y be the two vertices of degree m in $K_{2,m}$ for $m \geq 3$. Consider any good drawing D of $K_{2,m}$ such that the clockwise cyclic order of the neighbours of x within D is the same as the clockwise cyclic order of the neighbours of y. Then $\mathrm{cr}(D) \geq \lfloor \frac{m}{2} \rfloor \cdot \lfloor \frac{m-1}{2} \rfloor := Z(m)$.*

Corollary 8. *Consider a good drawing D of a graph G with a vertex cover X, and a topological cluster $S \subseteq V(D) \setminus X$ of size $c = |S|$. Let the degree of vertices in S be m. Then the number of cluster crossings in D between the edges incident with S is at least $\binom{c}{2} \cdot \lfloor \frac{m}{2} \rfloor \cdot \lfloor \frac{m-1}{2} \rfloor = \binom{c}{2} \cdot Z(m)$.*

Readers may notice that the formula $\binom{c}{2} \cdot \lfloor \frac{m}{2} \rfloor \cdot \lfloor \frac{m-1}{2} \rfloor$ in the lemma is not symmetric in c and m – it grows on one hand with $c^2/2$ and on the other hand with $m^2/4$. This is correct since the setting is also not symmetric. The vertices in S are required to have the same cyclic order of neighbours in D, but the neighbours of S do not have this property.

Proof. Let $Z(m) = \lfloor \frac{m}{2} \rfloor \cdot \lfloor \frac{m-1}{2} \rfloor$. For $s \in S$, denote by $R_s \subseteq G$ the subgraph (a star) induced by s and the incident edges of s in G.

There is nothing to prove (the bound equals 0) for $m \leq 2$ or $c = 1$. Otherwise, for every pair $s_1, s_2 \in S$, $s_1 \neq s_2$, we apply Lemma 7 to the subdrawing D_{s_1,s_2} of D induced by $R_{s_1} \cup R_{s_2}$, getting at least $Z(m)$ crossings within D_{s_1,s_2}. If $s_3 \in S$ is different from s_1 and s_2, then the crossings in D_{s_1,s_3} are all distinct from the crossings in D_{s_1,s_2}; this is since $E(D_{s_1,s_2}) \cap E(D_{s_1,s_3}) = E(R_{s_1})$, but the edges on R_{s_1} are all incident to s_1 and so they cannot mutually cross in a good drawing. Consequently, each of the $\binom{c}{2}$ invocations of Lemma 7 contributes a collection of at least $Z(m)$ new crossings, providing the overall lower bound of $\binom{c}{2} \cdot Z(m)$ cluster crossings between the edges incident with S. $\square$

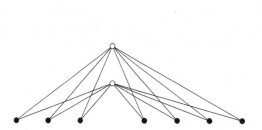 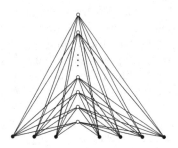

Fig. 3. Left: an optimal drawing of $K_{2,7}$ achieving the minimum number of $Z(7) = 9$ crossings among all drawings in which the two vertices of degree 7 have the same cyclic order of their neighbours (as in Lemma 7). Right: "stacking" the left subdrawings such that the total number of cluster crossings here matches the lower bound given by Corollary 8.

The next step is to introduce an "abstract level" of a topological clustering. Simply put, a drawing D is an *abstract topological clustering* of a graph G with respect to its vertex cover X if D is a topological clustering of *some* drawing of G, but without the weight function. More precisely:

Definition 9. *A drawing C_X is an* abstract topological clustering *of a graph G with respect to its vertex cover X if the following hold:*

- *C_X is a good drawing of an induced subgraph of G containing X,*
- *for every vertex $w \in V(G) \setminus X$ there is a vertex in $V(C_X) \setminus X$ having the same neighbourhood as w in G, and*
- *no two vertices of $V(C_X) \setminus X$ are in the same topological cluster in C_X.*

We will further use the term of *planarization* of a drawing D, which is the plane graph obtained from D by turning every crossing into a new degree-4 vertex. Two drawings D_1 and D_2 of the same graph are *combinatorially equivalent* if the same pairs of edges cross in D_1 as in D_2, moreover in the same order of the crossings on each edge, and their planarizations are equivalent plane graphs (i.e., with the same collection of faces).

Lemma 10. *Consider a graph G with a vertex cover X of size $k = |X|$. Then every abstract topological clustering of G has size at most singly exponential in k, and the number of combinatorially non-equivalent abstract topological clusterings of G is bounded from above by a doubly exponential function of k.*

Proof. A topological cluster in a drawing of G is uniquely determined by one of 2^k possible neighbourhood subsets in X, and one of up to $(k-1)!$ cyclic orders of the neighbours. Hence an abstract topological clustering C of G has at most $k + 2^k(k-1)! \leq k^{\mathcal{O}(k)}$ vertices. Hence the number of edges and of pairs of edges of C is bounded from above by $k^{\mathcal{O}(k)}$, which also implies $\mathrm{cr}(C) \leq k^{\mathcal{O}(k)}$. Hence the planarization of C has at most $k^{\mathcal{O}(k)}$ vertices, and there are altogether at most $2^{k^{\mathcal{O}(k)}}$ such possible nonequivalent planarizations of abstract topological clusterings of G. □

A consequence of Lemma 10 is that we can, in FPT time, process all possible abstract topological clusterings of any graph G with a small vertex cover. Therefore, from now on, we may just fix one abstract topological clustering C_X of G and discuss how to optimize the crossing number over all such drawings of G whose topological clustering comes from C_X. The latter problem will be reduced to a bounded instance of IQP, similarly as the special case of complete bipartite graphs has been handled in aforementioned [8].

4.1 IQP Formulation for Crossings

In regard of Definition 9 and the coming arguments, it will be useful to consider the following "compressed" representation of a graph G with a small vertex cover X. Let G_X denote the subgraph of G induced by X, and consider the function $h : 2^X \to \mathbb{N}_0$ such that, for any $Y \subseteq X$, $h(Y)$ is the number of vertices of G outside of X whose neighbourhood in G is exactly Y. Clearly, G_X and h determine G up to an isomorphism (and the size of this description can be only logarithmic compared to the size of G).

For given G and X, let us fix any abstract topological clustering C_X of G with respect to X. For $Y \subseteq X$, let $S(Y)$ be the set of vertices of $V(C_X) \setminus X$ whose neighborhood in X is exactly Y. Note that $h(Y)$ is non-zero iff $S(Y)$ is non-empty. Let $Y_1, \ldots, Y_l$ be an enumeration of all subsets of X which map to a non-zero value under h; then $\bigcup_{i=1}^l S(Y_i) = V(C_X) \setminus X$. For $i \in \{1, \ldots, l\}$, let $g(i) = |S(Y_i)|$ and let $S(Y_i) = \{v_{(i,1)}, \ldots, v_{(i,g(i))}\}$.

For an illustration, in Fig. 2 (on the right, but ignoring the weights since we are considering an abstract clustering) we have got $l = 2$ ($Y_1 = X$ and Y_2 are the two bottom vertices of X), and $g(1) = 2$ (blue and red clusters) and $g(2) = 1$ (yellow cluster). Altogether, $V(C_X) \setminus X$ has three vertices there.

Let I be an index set defined as $I := \{(i,j) \mid 1 \leq i \leq l, 1 \leq j \leq g(i)\}$. Similarly as in [8], we define the following *crossing vector* $\boldsymbol{p} = (p_\alpha \mid \alpha \in I)$ and the *crossing matrix* $\boldsymbol{Q} = (q_{\alpha,\beta} \mid \alpha, \beta \in I)$, such that the intended use of $\boldsymbol{p}$ is to count the crossings between the edges of G_X and the edges incident to each topological cluster corresponding to a vertex of $V(C_X) \setminus X$, and the intended use of Q is to count the cluster crossings of each one of the topological clusters (the diagonal entries) and the non-cluster crossings between pairs of the clusters (the other entries):

- The crossing vector p: Let $e_1, \ldots, e_r$ be an enumeration of the edges in G_X. For $\alpha \in I$ and $i \in \{1, \ldots, r\}$, let p_α^i be the number of edges incident to v_α that cross e_i in C_X. Then $p_\alpha = \sum_{i=1}^r p_\alpha^i$.
- The crossing matrix Q: Let $\alpha, \beta \in I$. If $\alpha \neq \beta$, define $q_{\alpha,\beta}$ as the number of crossings in C_X between the edges incident to v_α and the edges incident to v_β. If $\alpha = \beta = (i,j)$, then $q_{\alpha,\alpha} := Z(|Y_i|) = \left\lfloor \frac{|Y_i|}{2} \right\rfloor \cdot \left\lfloor \frac{|Y_i|-1}{2} \right\rfloor$.

($Z(\cdot)$ has been defined in Lemma 7 and Corollary 8.)

To recapitulate where we stand now; we have fixed an abstract topological clustering C_X of G, and in order to proceed to a drawing of G (underlied by C_X), we first need to assign suitable integer weights to the vertices of $V(C_X) \setminus X$. Our goal is to minimize the total number of crossings in the constructed drawing of G. However, we only have C_X and some assigned weights on $V(C_X) \setminus X$, which together define the topological clustering D_X of a drawing of G. The crossing number $\mathrm{cr}(D_X)$ is, via Lemma 6, related to the number of non-cluster crossings in a desired drawing of G (refer to the proof of Theorem 12 for a precise formulation). But it is not sufficient to minimize $\mathrm{cr}(D_X)$ since the cluster crossings in a drawing of G also play important role.

To complete the picture with cluster crossings, we define (cf. Corollary 8)

$$\mathrm{cl}(D_X) := \sum_{t \in V(D_X) \setminus X} \binom{c(t)}{2} \cdot Z\big(d(t)\big) = \sum_{t \in V(D_X) \setminus X} \binom{c(t)}{2} \cdot \left\lfloor \frac{d(t)}{2} \right\rfloor \cdot \left\lfloor \frac{d(t)-1}{2} \right\rfloor,$$

where c is the weight function of D_X and $d(t)$ denotes the degree of t (which is the same in D_X as in G). Again, we refer to the proof of Theorem 12 for further treatment of the relation of $\mathrm{cl}(D_X)$ to the cluster crossings in a drawing of G.

Lemma 11. *Let C_X be an abstract topological clustering of G with respect to a vertex cover X, and denote by $\mathcal{D}(C_X)$ the set of all topological clusterings of good drawings of G whose unweighted topological clustering is C_X. Let, furthermore, Y_i, g, I, p and Q be as above. Then the following IQP*

$$\begin{aligned}
\textit{Minimize} \quad & f(z) = z^T Q z + 2 \cdot p^T z & (2) \\
\textit{over all} \quad & z = \big(z_{(1,1)}, \ldots, z_{(1,g(1))}, \ldots, z_{(l,1)}, \ldots, z_{(l,g(l))}\big) \\
\textit{subject to} \quad & \sum_{j=1}^{g(i)} z_{(i,j)} = h(Y_i) \quad \textit{for } i \in \{1, \ldots, l\} \\
& z_{(i,j)} \geq 0 \quad \textit{for } (i,j) \in I \\
& z \in \mathbb{Z}^{|I|}
\end{aligned}$$

computes the minimum value of $2 \cdot (\mathrm{cr}(D) + \mathrm{cl}(D) - r)$ over all $D \in \mathcal{D}(C_X)$, where $r = \mathrm{cr}(C_X|_X)$ is the number of crossings in the subdrawing of C_X induced by the vertex set X.

Proof. First, note that for any $D \in \mathcal{D}(C_X)$ we have $\mathrm{cr}(D|_X) = r$ by definition. For a particular weight assignment $\boldsymbol{z}$, consider the corresponding topological clustering $D = D(C_X, \boldsymbol{z}) \in \mathcal{D}(C_X)$. We write $\mathrm{cr}(D) = r + r_1(D) + r_2(D)$ where $r_1(D)$ counts the (weighted) crossings in D which involve one edge with both ends in X, and $r_2(D)$ counts the crossings of which neither edge has both ends in X. From the definition of the crossing vector $\boldsymbol{p}$ we immediately have $r_1(D) = \boldsymbol{p}^T \boldsymbol{z}$. From the definition of the crossing matrix $\boldsymbol{Q}$ and that of $\mathrm{cl}(\cdot)$ we also get $r_2(D) + \mathrm{cl}(D) = \frac{1}{2} \cdot \boldsymbol{z}^T \boldsymbol{Q} \boldsymbol{z}$. Altogether, $\frac{1}{2} f(\boldsymbol{z}) = r_1(D) + r_2(D) + \mathrm{cl}(D) = \mathrm{cr}(D) + \mathrm{cl}(D) - r$. $\square$

We are now ready to prove the main result of this paper, which is as stated below.

Theorem 12 (refinement of Theorem 1). *Consider a simple graph G given on the input as follows: there is a set X (a vertex cover of G), a simple graph G_X (which is the subgraph of G induced by X), and a function $h : 2^X \to \mathbb{N}_0$ such that, for $Y \subseteq X$, $h(Y)$ is the number of vertices of G outside of X whose neighbourhood in G is exactly Y. The size of this input G equals the size of G_X plus the length of the bit-representation of function h.*

Then the problem to compute the crossing number of G and the corresponding topological clustering of an optimal drawing of G is fixed-parameter tractable with respect to the parameter $k = |X|$.

Note that, when the vertex cover size $k = |X|$ is fixed, the size of the input G described in Theorem 12 is logarithmic in the number of vertices of G. Although, in a typical use case, in which we do not get the input graph G in a parsed form as in Theorem 12, but rather as a list of vertices and edges, we can first compute, again in FPT [10], a vertex cover X of size $\leq k$ and the corresponding function h. Then, from the output topological clustering of an optimal drawing of G, we can easily in polynomial time construct the corresponding drawing of G. Hence Theorem 12 implies Theorem 1.

Proof Consider an optimal drawing D_0 of G, i.e., one for which $\mathrm{cr}(D_0) = \mathrm{cr}(G)$ holds. Then D_0 may be assumed a good drawing by folklore arguments. By Lemma 6, there is a topological clustering D_X of D_0 such that $\mathrm{cr}^n(D_0) \geq \mathrm{cr}(D_X)$. Recall that D_X is equipped with the weight function c, and that $\mathrm{cl}(D_X) = \sum_{t \in V(D_X) \setminus X} \binom{c(t)}{2} \cdot Z(d(t))$ where $d(t)$ denote the degree of t. By Corollary 8, the total number of cluster crossings in D_0 is at least $\mathrm{cl}(D_X)$.

Now, let C_X be the abstract topological clustering underlying D_X. Although we do not (yet) know D_X, we can "find" C_X by a brute force enumeration of all abstract topological clusterings of G, which is still in FPT by Lemma 10. Precisely, for every possible C_X (where "possible" is checked simply by brute force with respect to the parameter k), we compose an IQP as above (2). Then, using Theorem 3, we solve it to get an assignment $\boldsymbol{z}$ of weights to C_X, leading to a clustering D'_X, such that the objective value $\mathrm{cl}(D'_X) + \mathrm{cr}(D'_X)$ is minimized over all $D'_X \in \mathcal{D}(C_X)$ by Lemma 11. Let, furthermore, C_X° be an abstract topological clustering of G achieving the overall minimum value of the IQP solutions – this leads to a clustering D_X° with globally minimal $\mathrm{cl}(D_X^\circ) + \mathrm{cr}(D_X^\circ)$ for given G.

Consequently, counting separately the cluster and non-cluster crossings in D_0, and then considering the minimality of D_X°, we get

$$\mathrm{cr}(D_0) \geq \mathrm{cl}(D_X) + \mathrm{cr}^{\mathrm{n}}(D_0) \geq \mathrm{cl}(D_X) + \mathrm{cr}(D_X) \geq \mathrm{cl}(D_X^\circ) + \mathrm{cr}(D_X^\circ).$$

It is now enough to "lift" the clustering D_X° into a corresponding drawing D_1 of G with $\mathrm{cl}(D_X^\circ) + \mathrm{cr}(D_X^\circ)$ crossings, which follows straightforwardly in the same way as in [8], see Fig. 3. Hence $\mathrm{cl}(D_X^\circ) + \mathrm{cr}(D_X^\circ) \geq \mathrm{cr}(G) = \mathrm{cr}(D_0)$, and so $\mathrm{cr}(D_1) = \mathrm{cr}(D_0) = \mathrm{cr}(G)$.

It remains to address runtime of our procedure. In the IQP (2) we have $|I|$ bounded from above by the size of C_X, which is at most singly exponential in k by Lemma 10. The same asymptotic upper bound $k^{\mathcal{O}(k)}$ from the proof of Lemma 10 applies also to $\mathrm{cr}(C_X)$, and this clearly bounds all the entries of the matrix Q and the vector p. Let L be the length of the bit representation of h (from the input representation of G); then the length of the combined bit representation of the IQP (2) is at most $f_1(k) \cdot L^{\mathcal{O}(1)}$ for some computable (singly exponential) function f_1. Then from Theorem 3, (2) is solved by an algorithm in FPT time $f_2(k) \cdot L^{\mathcal{O}(1)}$ for some computable function f_2. This IQP step is repeated, by brute force and independently of L, at most $f_3(k)$ times where f_3 is a computable function (doubly exponential) coming from the bound on the number of abstract clusterings in Lemma 10.   □

5   Conclusions

In our work we have stressed simplicity of the considered graphs. A natural question is about what happens if we consider multigraphs with a vertex cover of size k. There is, unfortunately, no easy answer to this question since deep problems arise in two different places of our arguments. First, since the multiplicity of an edge may be unbounded in k, the entries of the crossing vector p and the crossing matrix Q would no longer be bounded in k. Second, when defining topological clusters, it would no longer be enough to consider a bounded number of neighbourhoods in X and a bounded number of cyclic orders, but also a potentially unbounded number of different multiplicities of the edges in a cluster. Each of these problems would completely ruin the runtime of our procedure.

Therefore, we leave the problem of computational complexity of the exact crossing number of multigraphs parameterized by a vertex cover size as open, for future research. On the other hand, in the special case of multigraphs with a vertex cover of size k and edge multiplicities bounded by a computable function of k, it is not difficult to extend our approach to obtain again an FPT algorithm (which we skip here due to space restrictions).

At last, we would like to very briefly mention the problem of minimizing the crossing number of a small perturbation of a given map of a graph, e.g. [13], which shares some common ground with our arguments. Although the objectives of the two problems are not easily comparable, we suggest that our approach can provide an efficient solution of the latter problem on graphs of small vertex cover.

References

1. Bhatt, S.N., Leighton, F.T.: A framework for solving VLSI graph layout problems. J. Comput. Syst. Sci. **28**(2), 300–343 (1984)
2. Biedl, T.C., Chimani, M., Derka, M., Mutzel, P.: Crossing number for graphs with bounded pathwidth. In: ISAAC 2017. LIPIcs, vol. 92, pp. 13:1–13:13. Schloss Dagstuhl (2017)
3. Cabello, S.: Hardness of approximation for crossing number. Discrete Comput. Geom. **49**(2), 348–358 (2013)
4. Cabello, S., Mohar, B.: Crossing and weighted crossing number of near-planar graphs. In: Tollis, I.G., Patrignani, M. (eds.) GD 2008. LNCS, vol. 5417, pp. 38–49. Springer, Heidelberg (2009). https://doi.org/10.1007/978-3-642-00219-9_5
5. Cabello, S., Mohar, B.: Adding one edge to planar graphs makes crossing number and 1-planarity hard. SIAM J. Comput. **42**(5), 1803–1829 (2013)
6. Chimani, M., Hliněný, P., Mutzel, P.: Vertex insertion approximates the crossing number for apex graphs. Eur. J. Comb. **33**, 326–335 (2012)
7. Chimani, M., Hliněný, P.: A tighter insertion-based approximation of the crossing number. J. Comb. Optim. **33**(4), 1183–1225 (2017)
8. Christian, R., Richter, R.B., Salazar, G.: Zarankiewicz's conjecture is finite for each fixed m. J. Comb. Theory, Ser. B **103**(2), 237–247 (2013)
9. Chuzhoy, J.: An algorithm for the graph crossing number problem. In: Proceedings of STOC 2011, pp. 303–312. ACM (2011)
10. Downey, R.G., Fellows, M.R.: Fixed-parameter tractability and completeness. Congr. Numer. **87**, 161–178 (1992)
11. Even, G., Guha, S., Schieber, B.: Improved approximations of crossings in graph drawings and VLSI layout areas. SIAM J. Comput. **32**(1), 231–252 (2002)
12. Fox, J., Pach, J., Suk, A.: Approximating the rectilinear crossing number. Comput. Geom. **81**, 45–53 (2019)
13. Fulek, R., Tóth, C.D.: Crossing minimization in perturbed drawings. In: Biedl, T., Kerren, A. (eds.) GD 2018. LNCS, vol. 11282, pp. 229–241. Springer, Cham (2018). https://doi.org/10.1007/978-3-030-04414-5_16
14. Garey, M.R., Johnson, D.S.: Crossing number is NP-complete. SIAM J. Alg. Discr. Meth. **4**, 312–316 (1983)
15. Gitler, I., Hliněný, P., Leanos, J., Salazar, G.: The crossing number of a projective graph is quadratic in the face-width. Electron. Notes Discrete Math. **29**, 219–223 (2007)
16. Grohe, M.: Computing crossing numbers in quadratic time. J. Comput. Syst. Sci. **68**(2), 285–302 (2004)
17. Hliněný, P., Chimani, M.: Approximating the crossing number of graphs embeddable in any orientable surface. In: Proceedings of SODA 2010, pp. 918–927. ACM (2010)
18. Hliněný, P., Salazar, G.: On the crossing number of almost planar graphs. In: Kaufmann, M., Wagner, D. (eds.) GD 2006. LNCS, vol. 4372, pp. 162–173. Springer, Heidelberg (2007). https://doi.org/10.1007/978-3-540-70904-6_17
19. Kawarabayashi, K., Reed, B.A.: Computing crossing number in linear time. In: Proceedings of STOC 2007, pp. 382–390. ACM (2007)
20. Lokshtanov, D.: Parameterized integer quadratic programming: Variables and coefficients. CoRR abs/1511.00310 (2015)

Morphing and Planarity

Maximizing Ink in Partial Edge Drawings of *k*-plane Graphs

Matthias Hummel⬤, Fabian Klute⬤, Soeren Nickel⬤,
and Martin Nöllenburg$^{(\boxtimes)}$⬤

Algorithms and Complexity Group, TU Wien, Vienna, Austria
matthiashummel@ymail.com, {fklute,noellenburg}@ac.tuwien.ac.at,
soeren.nickel@tuwien.ac.at

Abstract. Partial edge drawing (PED) is a drawing style for non-planar graphs, in which edges are drawn only partially as pairs of opposing stubs on the respective end-vertices. In a PED, by erasing the central parts of edges, all edge crossings and the resulting visual clutter are hidden in the undrawn parts of the edges. In symmetric partial edge drawings (SPEDs), the two stubs of each edge are required to have the same length. It is known that maximizing the ink (or the total stub length) when transforming a straight-line graph drawing with crossings into a SPED is tractable for 2-plane input drawings, but NP-hard for unrestricted inputs. We show that the problem remains NP-hard even for 3-plane input drawings and establish NP-hardness of ink maximization for PEDs of 4-plane graphs. Yet, for *k*-plane input drawings whose edge intersection graph forms a collection of trees or, more generally, whose intersection graph has bounded treewidth, we present efficient algorithms for computing maximum-ink PEDs and SPEDs. We implemented the treewidth-based algorithms and show a brief experimental evaluation.

1 Introduction

Visualizing non-planar graphs as node-link diagrams is challenging due to the visual clutter caused by edge crossings. The layout readability deteriorates as the edge density and thus the number of crossings increases [19]. Therefore alternative layout styles are necessary for non-planar graphs. A radical approach first used in applied network visualization work by Becker et al. [2] is to start with a traditional straight-line graph drawing and simply drop a large central part of each edge and with it many of the edge crossings. This idea relies on the closure and continuation principles in Gestalt psychology [17], which imply that humans can still see a full line segment based only on the remaining edge stubs by filling in the missing information. User studies have confirmed that such drawings remain readable while reducing clutter significantly [9,12] and Burch et al. [11] presented an interactive graph visualization tool using partially drawn edges combined with fully drawn edges.

The authors thank Michael Höller and Birgit Schreiber for the fruitful discussions during the "Seminar in Algorithms: Graphs and Geometry" held in 2017 at TU Wien. A preliminary abstract of this paper has been presented at EuroCG 2018.

D. Archambault and C. D. Tóth (Eds.): GD 2019, LNCS 11904, pp. 323–336, 2019.
https://doi.org/10.1007/978-3-030-35802-0_25

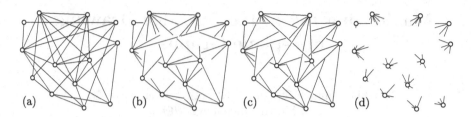

Fig. 1. Drawings of the same graph. (a) A straight-line drawing, (b) a maximum-ink SPED, (c) a maximum-ink PED, and (d) a maximum-ink SHPED.

The idea of drawing edges only partially has subsequently been formalized in graph drawing as follows [7]. A *partial edge drawing (PED)* is a graph drawing that maps vertices to points and edges to pairs of crossing-free edge stubs of positive length pointing towards each other. These edge stubs are obtained by erasing one contiguous central piece of the straight-line segment connecting the two endpoints of each edge. In other words each straight-line edge is divided into three parts, of which only the two outer ones are drawn (see Fig. 1). More restricted and better readable [4] variations of PEDs are *symmetric* PEDs, in which both stubs of an edge must have the same length (see Fig. 1(b)), and *homogeneous* PEDs, in which the ratio of the stub length to the total edge length is the same constant for all edges. Symmetric stubs facilitate finding adjacent vertices due to the identical stub lengths at both vertices, and symmetric homogeneous stubs (see Fig. 1(d)) additionally indicate the distance at which to find a neighboring vertex. Clearly, for very short stubs it is easy to hide all edge crossings, but reading adjacency information gets very difficult [12]. Therefore, the natural optimization problem in this formal setting is *ink maximization*, i.e., maximizing the total stub length, so that as much information as possible is given in the drawing while all crossings disappear in the negative background space.

We study the ink maximization problem for partial edge drawings (PEDs) and symmetric partial edge drawings (SPEDs) with a given geometric input drawing. These problems are known as MaxPED and MaxSPED, respectively [6,7]. Note that with a given input drawing, the ink maximization problem for symmetric homogeneous PEDs (SHPEDs) is trivial, as we can simply iterate over all crossings, choose the larger of the two stub ratios resolving the crossing and take the minimum of all these stub ratios, which yields the best solution.

Related Work. As a first result, Bruckdorfer and Kaufmann [7] presented an integer linear program for solving MaxSPED on general input drawings. Later, Bruckdorfer et al. [6] gave an $O(n \log n)$-time algorithm for MaxSPED on the class of 2-plane input drawings (no edge has more than two crossings), where n is the number of vertices. They also described an efficient 2-approximation algorithm for the dual problem of minimizing the amount of erased ink for arbitrary input drawings. The PhD thesis of Bruckdorfer [5] presents a sketch of an NP-hardness proof for MaxSPED, but left the complexity of MaxPED as an open problem, as well as the design of algorithms for MaxPED.

There are a number of additional results for PEDs without a given input drawing, i.e., having the additional freedom of placing the vertices in the plane. For example, the existence or non-existence of SHPEDs with a specified stub ratio δ for certain graph classes such as complete graphs, complete bipartite graphs, or graphs of bounded bandwidth has been investigated [6,7]. From a practical perspective, Bruckdorfer et al. [8] presented a force-directed layout algorithm to compute SHPEDs for stubs of 1/4 of the total edge length, but without a guarantee that all crossings are eliminated. Moreover, the idea of partial edge drawings has also been extended to orthogonal graph layouts [10].

Contribution. We extend the results of Bruckdorfer et al. [6] on 2-plane geometric graph drawings to k-plane graph drawings for $k > 2$, where a given graph drawing is k-*plane*, if every edge has at most k crossings. In particular, we strengthen the NP-hardness of MaxSPED [5] to the case of 3-plane input drawings without three (or more) mutually crossing edges. For MaxPED we show NP-hardness, even for 4-plane input drawings, which settles a conjecture of Bruckdorfer [5]. On the positive side, we give polynomial-time dynamic programming algorithms for both MaxSPED and MaxPED of k-plane graph drawings whose edge intersection graphs are collections of trees. More generally, we extend the algorithmic idea and obtain FPT algorithms if the edge intersection graph has bounded treewidth and also provide a proof-of-concept implementation. We evaluate the implementation using non-planar drawings from two classical layout algorithms, namely a force-based and a circular layout algorithm.

2 Preliminaries

Let G be a *simple graph* with edge set $E(G) = S = \{s_1, \ldots, s_m\}$ and Γ a straight-line drawing of G in the plane. We call Γ k-*plane* if every edge $s_i \in S$ is crossed by at most k other edges from S in Γ. We often use edge in S and segment in Γ interchangeably. Hence S can be seen as a set of line segments.

The *intersection graph* C of Γ is the graph containing a vertex v_i in $V(C)$ for every $s_i \in S$ and an edge $v_i v_j \in E(C)$ between vertices $v_i, v_j \in V(C)$ if the corresponding edges $s_i, s_j \in S$ intersect in Γ. We also denote the segment in S corresponding to a vertex $v \in V(C)$ by $s(v)$. Observe that the intersection graph C of a k-plane drawing Γ has maximum degree k. Using a standard sweep-line algorithm, computing the intersection graph C of a set of m line segments takes $O(m \log m + |E(C)|)$ time [3], where $|E(C)|$ is the number of intersections.

A *partial edge drawing* (PED) D of Γ draws a fraction $0 < f_s \leq 1$ of each edge $s = uv \in S$ by drawing edge stubs of length $f_u|s|$ at u and $f_v|s|$ at v, s.t., $f_u + f_v = f_s$. The *ink* or *ink value* $I(D)$ of a PED D is the total stub length $I(D) = \sum_{s \in S} f_s|s|$. In the problem MaxPED, the task is to find for a given drawing Γ a PED D^* such that $I(D^*)$ is maximum over all PEDs. A *symmetric partial edge drawing* (SPED) D of Γ is a PED, s.t., $f_u = f_v = f_s/2$ for every edge $s = uv \in S$. Then the MaxSPED problem is defined analogously to MaxPED.

Treewidth. A *tree decomposition* [20] for a graph G is a pair $(T, \mathcal{X})$ with T being a tree and $\mathcal{X}$ a collection of subsets $X_i \subseteq V(G)$. For every edge $uv \in E(G)$ we find $t \in V(T)$ such that $\{u, v\} \subseteq X_t$ and for every vertex $v \in V(G)$ we get $T[\{t \mid v \in X_t\}]$ is a connected and non-empty subtree of T. To differentiate the vertices of G and T we call the vertices of T *nodes* and a set $X_i \in \mathcal{X}$ a *bag*. Now the *width* of a tree decomposition $(T, \mathcal{X})$ is defined as $\max\{|X_t| - 1 \mid t \in V(T)\}$. For a graph G we say it has *treewidth* ω, if the tree decomposition with minimum width has width ω. For a node $t \in T$ we denote with $V_t \subseteq V(G)$ the union of all bags $X_{t'} \in \mathcal{X}$ such that t' is either t or a descendent of t in T.

In our algorithms we are using the well known *nice tree decomposition* [14]. For a graph G a nice tree decomposition $(T, \mathcal{X})$ is a special tree decomposition, where T is a rooted tree and we require that every node in T has at most two children. In case $t \in V(T)$ has two children $t_1, t_2 \in T$, then $X_t = X_{t_1} = X_{t_2}$. Such a node is called *join node*. For a node $t \in T$ with a single child $t_1 \in T$ we find either $|X_t| = |X_{t_1}| + 1$, $X_{t_1} \subset X_t$ or $|X_t| = |X_{t_1}| - 1$, $X_t \subset X_{t_1}$. The former we call *insert node* and the latter *forget node*. A leaf $t \in T$ is called a *leaf node* and its bag contains a single vertex. Finally let $r \in T$ be the root of T, then $X_r = \emptyset$. It is known that a tree decomposition can be transformed into a nice tree decomposition of the same width ω and with $O(\omega|V(G)|)$ tree nodes in time linear in the size of the graph G [14].

3   Complexity

We first investigate the complexity of MAXSPED and MAXPED, and prove both problems to be NP-hard for 3-plane and 4-plane input drawings, respectively.

3.1   Hardness of MaxSPED for $k \geq 3$

We close the gap between the known NP-hardness of MAXSPED [5] for general input drawings and the polynomial-time algorithm for 2-plane drawings [6].

Theorem 1. MAXSPED *is* NP-*hard even for 3-plane graph drawings.*

Proof. We reduce from the NP-hard problem PLANAR 3-SAT [18] using similar ideas as in Bruckdorfer's sketch of the hardness proof for general MAXSPED [5]. Here we specify precisely the maximum ink contributions of all gadgets needed for a satisfying variable assignment. Our variable gadgets are cycles of edge pairs that admit exactly two maximum-ink states. We construct clause gadgets consisting of three pairwise intersecting edges so that all crossings are between two edges only, while Bruckdorfer's gadgets have multiple edges intersecting in the same point. Let ϕ be a planar 3-SAT formula with n variables $\{x_1, \ldots, x_n\}$ and m clauses $\{c_1, \ldots, c_m\}$, each consisting of three literals. We can assume that ϕ comes with a planar drawing of its variable-clause graph H_ϕ, which has a vertex for each variable x_i and a vertex for each clause c_j. Each clause vertex is connected to the three variables appearing in the clause. In the drawing of H_ϕ all variable vertices are placed on a horizontal line and the clause vertices connect to the adjacent variable vertices either from above or from below the horizontal line. In our reduction (see Fig. 2) we mimic the drawing of H_ϕ by creating a 3-plane drawing Γ_ϕ as a set of line segments of two distinct lengths and define a value L such that Γ_ϕ has a SPED with ink at least L if and only if ϕ is satisfiable. The whole construction will be drawn onto a triangular grid of polynomial size.

All segments in the clause or variable gadgets are of length 8. The segments used for the connections are of length 4. We use pairs of intersecting segments, alternatingly colored red and green. The intersection point of each red-green segment pair is at distance 1 from an endpoint. Thus, the maximum amount of ink contributed by such a pair is 10 or 6, respectively (one full segment of length 8 or 4, respectively, and two stubs of length 1 each).

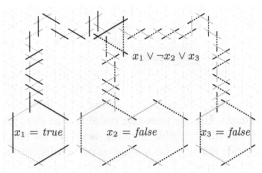

Fig. 2. Three variables and a satisfied clause gadget. Dotted parts do not belong to the SPED. (Color figure online)

Each variable gadget is a cycle of segment pairs, with (at least) one pair for each occurrence of the variable in ϕ, see Fig. 2. This cycle has exactly two ink-maximal SPEDs: either all red edges are full segments and all green edges are length-1 stubs or vice versa. We associate the configuration with green stubs and full red segments with the value *true* and the configuration with red stubs and full green segments with *false*.

For each clause we construct a triple of mutually intersecting segments, see the gadget on yellow background in the upper part of Fig. 2. Again, their intersection points are at distance 1 from the endpoints. It is clear that in such a clause triangle at most one of the three segments can be fully drawn, while the stubs of the other two can have length at most 1. Hence, the maximum amount of ink in a SPED contributed by a clause gadget is 12.

Finally, we connect variable and clause gadgets, such that a clause gadget can contribute its maximum ink value of 12 if and only if the clause is satisfied by the selected truth assignment to the variables. For a positive (negative) literal, we create a path of even length between a green (red) edge of the variable gadget and one of the three edges of the clause gadget as shown in Fig. 2. The first edge s of this path intersects the corresponding variable segment s' such that s' is split into a piece of length 3 and a piece of length 5, whereas s is split into a piece of length 1 and a piece of length 3. The last edge of the path intersects the corresponding clause edge again with a length ratio of 3 to 5. The path consists of a chain of red-green segment pairs, each contributing an ink value of at most 6.

It remains to argue that the resulting drawing has polynomial size and is a correct reduction. All segments are drawn on the underlying triangular grid and have integer lengths; all intersection points are grid points, too. Since the drawing of H_ϕ has polynomial size, so do the constructed gadgets. Additionally, no segment intersects more than three other segments, so the drawing is 3-plane.

For the correctness of the reduction, let L be the ink value obtained by counting 10 for each red-green segment pair in a variable, 6 for each red-green segment pair in a wire, and 12 for each clause gadget. First assume that ϕ has a satisfying truth assignment and put each variable gadget in its corresponding state. For each clause, select exactly one literal with value *true* in the satisfying truth assignment. We draw the clause segment that connects to the selected literal as a full segment and the other two as length-1 stubs. Recall that the literal paths are oriented from the variable gadget to the clause gadget. Since the last segment of the selected literal path must be drawn as two length-1 stubs, the only way of having a maximum contribution of that path is by alternating stubs and full segments. Hence, the first segment of the path must be a full segment. But because the variable is in the state that sets the literal to *true*, the intersecting variable segment is drawn as two stubs and the path configuration is valid. For the two non-selected literals, we can draw the last segments of their paths as full segments, as well as every segment at an even position, while the segments at odd positions are drawn as stubs. This is compatible with any of the two variable configurations and proves that we can indeed achieve ink value L.

Conversely, assume that we have a SPED with ink value L. By construction, every red-green segment pair and every clause gadget must contribute its respective maximum ink value. In particular, each variable gadget is either in state *true* or *false*. By design of the gadgets it is straight-forward to verify that the corresponding truth assignment satisfies ϕ.   □

3.2   Hardness of MaxPED for $k \geq 4$

We adapt our NP-hardness proof for MaxSPED to show that MaxPED is NP-hard for k-plane drawings with $k \geq 4$; see [16] for the full proof.

Theorem 2. MaxPED *is* NP-*hard even for 4-plane graph drawings.*

Proof (Sketch). As in the proof of Theorem 1, we show the result via a reduction from PLANAR 3-SAT. The key change for MaxPED comes from the fact, that the two stubs are now independent from each other. Take two crossing edges as an example. We now can draw the two segments with almost full ink value by just excluding an ε-sized gap in one of the two segments for some small $\varepsilon > 0$. We will use this placement of a gap in the variable and wire gadgets, to create two possible states. As before we use an underlying triangular grid, which we omit in the figures of this section for ease of presentation.

Let ϕ be a planar 3-SAT formula. For a variable x of ϕ we construct a variable gadget consisting of a cycle of p line segments $t_1, \ldots, t_p$, see Fig. 3a. Such a cycle has exactly two maximum-ink drawings. One, where for each segment t_i the gap is placed at its intersection with $t_{i+1} \pmod p$ and another drawing, in which the gap is placed at its intersection with $t_{i-1} \pmod p$. Figure 3a shows a gadget in its true state. The length of each segment t_i is $\alpha + 2\beta$, where α is the distance between the two intersection points and β is the length of each stub sticking out.

A clause gadget is a cycle of three pairwise intersecting segments r_1, r_2, r_3, which we call triangle segments. All segments are elongated at one end, such that the total length of a segment r_i, $i \in \{1, 2, 3\}$, is $4\alpha + 2\beta$. Since the stubs are independent we could draw all three triangle segments. To avoid this we attach a big 4-cycle to each r_i. Then r_i intersects the 4-cycle at a segment r_w, see Fig. 3b. If we place the gap of r_w at its intersection with r_i, we lose more units of ink than we gain by drawing every triangle segment r_i completely. Hence it is never possible to draw more than one full triangle segment in an ink-maximal PED.

Finally, a wire is a chain of segments $s_1, \ldots, s_z$ for each variable occurrence in a clause c in ϕ. We place the wire such that s_1 intersects the corresponding variable gadget at some segment t_j, and s_z intersects the clause gadget of c at one if its triangle segments r_i. For the variable we place this intersection point at distance β to its intersection with s_{i+1}, if it occurs positively, or with s_{i-1}, if it occurs negated. At the clause gadget we place the intersection of s_z with the corresponding r_i at distance β from the intersection between r_i and its successor r_{i+1}, see the small squares in Fig. 3b. Each segment s_i has length $\alpha/2 + 2\beta$.

Correctness follows similarly to the proof of Theorem 1. Let Γ_ϕ be the set of line segments constructed as above for a planar 3-SAT formula ϕ. We determine an ink value L, s.t., Γ_ϕ has a PED D with $I(D) \geq L$ if and only if ϕ has a satisfying variable assignment. The key property is that for each clause we find one wire such that its last segment is forced to place its gap

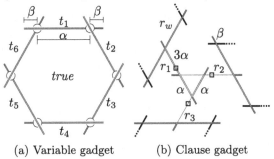

(a) Variable gadget   (b) Clause gadget

Fig. 3. Gadgets of our reduction. Squares mark connection points for wires.

at the intersection with the clause gadget in an ink-maximal PED. Then for each other segment s_i, $i \in \{1, \ldots, z-1\}$, we must place its gap at the intersection with s_{i+1}. Otherwise we would have to remove either $\alpha/2$ units of ink in the middle part of some s_i, or remove $\alpha - \beta$ units of ink at the variable gadget intersected by s_1. Both can be avoided if and only if ϕ has a satisfying assignment.   □

Corollary 1. MaxPED *and* MaxSPED *for k-plane drawings are not fixed-parameter tractable, when parameterized solely by k.*

4 Algorithms

Sections 3.1 and 3.2 showed that MaxSPED and MaxPED are generally NP-hard for $k \geq 3$ and $k \geq 4$ respectively. Now we consider the special case that the intersection graph of the k-plane input drawing is a tree or has bounded treewidth. In both cases we present polynomial-time dynamic programming algorithms for MaxSPED (Sects. 4.1 and 4.2) and MaxPED (Sect. 4.3).

Fig. 4. A segment $s(u)$ with five intersecting segments and the induced stub lengths. The boxed stub lengths are considered in $short(u)$ and do not affect $p(u)$.

Let C be the intersection graph of a given drawing Γ of a graph G as defined in Sect. 2. Let $u \in V(C)$ and $\delta_u = \deg(u)$. Then for the corresponding segment $s(u) \in S$ there are $\delta_u + 1$ relevant stub pairs including the whole segment, see Fig. 4. Let $\ell_1(u), \ldots, \ell_{\delta_u}(u) \in \mathbb{R}_+$ be the stub lengths induced by the intersection points of $s(u)$ with the segments of the neighbors of u, sorted from shorter to longer stubs. We define $\ell_0(u)$ as the length of the whole segment $s(u)$.

4.1 Trees

Here we assume that C is a rooted tree of maximum degree k. We give a bottom-up dynamic programming algorithm for solving MAXSPED on C. For each vertex $u \in V(C)$ we compute and store the maximum ink values $W_i(u)$ for $i = 0, \ldots, \delta_u$ for the subtree rooted at u such that $s(u)$ is drawn as a pair of stubs of length $\ell_i(u)$. For $u \in V(C)$ let $p(u)$ denote the parent of u in C and let $c(u)$ denote the set of its children. For $u \in V(C)$ let i_p be the index of the stub length $\ell_{i_p}(u)$ induced by the intersection point of $s(u)$ and $s(p(u))$. We define the following two values, which allow us to categorize the stub lengths into those not affecting the stubs of the parent and those that do affect the parent:

$$short(u) = \max\{W_1(u), \ldots, W_{i_p}(u)\} \quad long(u) = \max\{W_0(u), \ldots, W_{\delta_u}(u)\}.$$

Figure 4 highlights the stub lengths that are considered in $short(u)$. We recursively define

$$W_i(u) = \ell_i(u) + \sum_{v \in c(u)} \begin{cases} short(v) & \text{if } s(u) \text{ with length } \ell_i(u) \text{ intersects } s(v) \\ long(v) & \text{otherwise.} \end{cases} \quad (1)$$

The correctness of Recurrence (1) follows by induction. For a leaf u in C the set $c(u)$ is empty and the correctness of $W_i(u)$ is immediate. Further, $short(u) = W_1(u)$ and $long(u) = W_0(u)$ are set correctly for the parent $p(u)$. For an inner vertex u with degree δ_u we can assume by the induction hypothesis that the values $short(v)$ and $long(v)$ are computed correctly for all children $v \in c(u)$. Each value $W_i(u)$ for $0 \leq i \leq \delta_u$ is then the stub length $\ell_i(u)$ plus the sum of the maximum ink we can achieve among the children subject to the stubs of u being drawn with length $\ell_i(u)$. Setting $long(u)$ and $short(u)$ as above yields the two maximum ink values that are relevant for $p(u)$.

Recurrence (1) can be solved naively in $O(mk^2)$ time, where $m = |V(C)|$. Using the order on the stub lengths we can improve this to $O(mk)$ time by computing all $W_i(u)$ for one $u \in V(C)$ in $O(k)$ time. Let $u \in V(C)$ be a vertex with degree δ_u. The values $W_0(u) = \ell_0(u) + \sum_{v \in c(u)} short(v)$ and $W_1(u) = \ell_1(u) + \sum_{v \in c(u)} long(v)$ for the whole segment $s(u)$ and the shortest stubs can be computed in $O(k)$ time each. Now $W_{j+1}(u)$ can be computed from $W_j(u)$ in $O(1)$ time as follows. Let v_j be the neighbor of u that induces stub length $\ell_j(u)$ and assume $v_j \neq p(u)$. In $W_j(u)$ we could still count the value $long(v_j)$, but in $W_{j+1}(u)$ the stub length of u implies that v_j can contribute only to $short(v_j)$. Then $W_{j+1}(u) = W_j(u) - long(v_j) + short(v_j)$. If $v_j = p(u)$ the two values $W_j(u)$ and $W_{j+1}(u)$ are equal, as the corresponding change in stub length has no effect on the children of u. Then computing $short(u)$ and $long(u)$ takes $O(k)$ time.

Using standard backtracking we are able to find an optimal solution to the MAXSPED problem on G with drawing Γ by solving Recurrence (1) in $O(mk)$ time. Furthermore, the intersection graph C with m edges can be computed in $O(m \log m)$ time. We obtain the following theorem.

Theorem 3. *Let G be a simple graph with m edges and Γ a straight-line drawing of G. If the intersection graph C of Γ is a tree with maximum degree $k \in \mathbb{N}$, then* MAXSPED *can be solved in $O(mk + m \log m)$ time and space.*

4.2   Bounded Treewidth

Now we extend the case of a simple tree to the case that the intersection graph C has treewidth at most ω; see [16] for the omitted proofs. Our algorithm and proof of correctness follow a similar approach as the weighted independent set algorithm presented by Cygan et al. [13]. Let $(T, \mathcal{X})$ be a nice tree decomposition of C and k the maximum degree in C. For $V' \subseteq V(C)$ we define the *stub set* $S(V') := \{(u, \ell_i(u)) \mid u \in V' \text{ and } i = 0, \ldots, \delta_u\}$. For $(u, \ell_u), (v, \ell_v) \in S(V')$, $u \neq v$, we say (u, ℓ_u) *intersects* (v, ℓ_v) if $s(u)$ drawn with stub length ℓ_u intersects $s(v)$ drawn with length ℓ_v. Further we call $S(V')$ *valid* if $S(V')$ contains exactly one pair (u, ℓ) for each $u \in V'$ and no two pairs in $S(V')$ intersect, i.e., the stub lengths in $S(V')$ imply a SPED in the input drawing Γ. Further we define the ink of a stub set $S(V')$ as $I(S(V')) := \sum_{(u,\ell) \in S(V')} \ell$.

Lemma 1. *Let G be a simple graph, Γ a straight line drawing of G, C the intersection graph of Γ, and $(T, \mathcal{X})$ a tree decomposition of C. For any fixed $S \subseteq S(X_t), t \in T$, any two valid stub sets $S_1, S_2 \subseteq S(V(C))$ with maximum ink and $S_1 \cap S(X_t) = S_2 \cap S(X_t) = S$ have equal ink value $I(S_1 \cap S(V_t)) = I(S_2 \cap S(V_t))$.*

As a consequence of Lemma 1 we get that it suffices to store for every set of vertices V_t and a node $t \in T$ the value of a maximum-ink valid stub set for the choices of vertices in $S(X_t)$. Let $t \in T$ and $S \subseteq S(X_t)$ a stub set, then we define

$$W(t, S) = \max\{I(\hat{S}) \mid \hat{S} \text{ is a valid stub set, } S \subseteq \hat{S} \subseteq S(V_t), \text{ and } \hat{S} \cap S(X_t) = S\}.$$

If no such $\hat{S}$ exists, we set $W(t, S) = -\infty$. In other words, $W(t, S)$ is the maximum ink value achievable by any valid stub set in $S(V_t)$ choosing S as the stub set for the vertices in X_t. If the values $W(t,S)$ are computed correctly for every $t \in T$ we find the ink-value of a maximum-ink SPED by evaluating $W(r, \emptyset)$ with r being the root of T. Applying standard backtracking we can also reconstruct the stubs themselves. We now describe how to compute $W(t, S)$ for every node type $t \in T$ of a nice tree decomposition of C. All the recursion formulas are stated here. We provide the correctness proof for the introduce nodes; see [16] for the forget and join nodes.

Leaf Node. Let $t \in T$ be a leaf node and $v \in X_t$ the vertex contained in its bag, then we store $W(t, \{(v, \ell_i(v))\})$ for each pair $(v, \ell_i(v)) \in S(X_t)$ with $i \in [0, \delta_v]$.

Introduce Node. Suppose next $t \in T$ is an introduce node and t' its only child. Let $v \in X_t$ be the vertex introduced by t, $S \subseteq S(X_t)$, and $i \in [0, \delta_v]$, s.t., $(v, \ell_i(v)) \in S$. If S is not a valid stub set we set $W(t, S) = -\infty$, else

$$W(t, S) = W(t', S \setminus \{(v, \ell_i(v))\}) + \ell_i(v).$$

Correctness follows by considering a valid stub set $\hat{S}$ whose maximum is attained in the definition of $W(t, S)$. Then for the set $\hat{S} \setminus \{(v, \ell_i(v))\}$ it follows that it is considered in the definition of $W(t', S \setminus \{(v, \ell_i(v))\})$ and hence we get

$$W(t', S \setminus \{(v, \ell_i(v))\}) \geq I(\hat{S} \setminus \{(v, \ell_i(v))\}) = I(\hat{S}) - \ell_i(v) = W(t, S) - \ell_i(v)$$
$$W(t, S) \leq W(t', S \setminus \{(v, \ell_i(v))\}) + \ell_i(v).$$

On the other hand consider a valid stub set $\hat{S}'$ for which the maximum is attained in the definition of $W(t', S \setminus \{(v, \ell_i(v))\})$. We need to argue that $\hat{S}' \cup \{(v, \ell_i(v))\}$ is again a valid stub set. First, by assumption that S is a valid stub set, we immediately get that $(v, \ell_i(v))$ does not intersect any $(u, \ell_u) \in S \setminus \{(v, \ell_i(v))\} = \hat{S}' \cap X_{t'}$. Additionally, by the properties of the nice tree decomposition, we get that v has no neighbors in $V_{t'} \setminus X_{t'}$ and with $\hat{S}' \setminus X_{t'} \subseteq V_{t'} \setminus X_{t'}$ we can conclude that $\hat{S}' \cup \{(v, \ell_i(v))\}$ is a valid stub set. Furthermore it is considered in the definition of $W(t, S)$ and we have that

$$W(t, S) \geq I(\hat{S}' \cup \{(v, \ell_i(v))\}) = I(\hat{S}') + \ell_i(v) = W(t', S \setminus \{(v, \ell_i(v))\}) + \ell_i(v).$$

Forget Node. Suppose t is a forget node and t' its only child such that $X_t = X_{t'} \setminus \{v\}$ for some $v \in X_{t'}$. Let S be any subset of $S(X_t)$, if S is not a valid stub set we set $W(t, S) = -\infty$, else $W(t, S) = \max\{W(t', S \cup \{(v, \ell_i(v))\}) \mid i = 0, \ldots, \delta_v\}$.

Join Node. Suppose that t is a join node and t_1, t_2 its two children with $X_t = X_{t_1} = X_{t_2}$. Again let S be any subset of $S(X_t)$. If S is not a valid stub set we set $W(t, S) = -\infty$, else $W(t, S) = W(t_1, S) + W(t_2, S) - I(S)$.

It remains to argue about the running time. Let $m = |V(C)|$. We know there are $O(\omega m)$ many nodes in the tree T of the nice tree decomposition [14].

For each bag $t \in T$ we know it has at most $\omega + 1$ many elements and each element has at most $k + 1$ many possible stubs, hence we have to compute at most $(k + 1)^{\omega+1}$ values $W(t, S)$ per node $t \in T$. At each forget node we additionally need to compute the maximum of up to k entries. Consequently we perform $O((k + 1)^{\omega+2})$ many operations per node $t \in T$. All operations can be implemented in $O(k\omega)$ time. The only problematic one is to test a stub set for validity. We use a modified version of the data structure used in the independent set algorithm by Cygan et al. [13]. See [16] for the implementation details.

Theorem 4. *Let G be a simple graph with m edges and Γ a straight-line drawing of G. If the intersection graph C of Γ has treewidth at most $\omega \in \mathbb{N}$ and maximum degree $k \in \mathbb{N}$, MaxSPED can be solved in $O(m(k + 1)^{\omega+2}\omega^2 + m \log m)$ time and space.*

We remark that Theorem 3 shows a better running time in the case of C being a tree, than would follow from Theorem 4 with $\omega = 1$. Furthermore, since Theorem 4 is exponential only in the treewidth ω of C, it implies that MaxSPED is in the class XP[1] when parametrized by ω.

4.3 Algorithms for MaxPED

Let C be the intersection graph in a MaxPED problem. In contrast to MaxSPED we need to consider more combinations of stub lengths since they are not necessarily symmetric anymore. In fact there are $O(k^2)$ possible combinations of left and right stub lengths $\ell_i(u), \ell_j(u)$ for a vertex $u \in V(C)$. For the case of C being a tree our whole argumentation was based solely on the fact that we can subdivide the stub pairs into sets affecting the parent segment or not. This can also be done with the quadratic size sets of all stub pairs and we only get an additional factor of k in the running time.

Corollary 2. *Let G be a simple graph with m edges and Γ a straight-line drawing of G. If the intersection graph C of Γ is a tree with maximum degree $k \in \mathbb{N}$, then MaxPED can be solved in $O(mk^2 + m \log m)$ time and space.*

In case of C having bounded treewidth we again did never depend on the symmetry of the stubs, but only on them forming a finite set for each vertex. Consequently we can again just use these quadratic size sets of stub pairs, adding a factor of $k + 1$ compared to MaxSPED, and obtain the following.

Corollary 3. *Let G be a simple graph with m edges and Γ a straight-line drawing of G. If the intersection graph C of Γ has treewidth at most $\omega \in \mathbb{N}$ and maximum degree $k \in \mathbb{N}$, MaxPED can be solved in $O(m(k+1)^{\omega+3}\omega^2 + m \log m)$ time and space.*

---

[1] The class XP contains problems that can be solved in time $O(n^{f(k)})$, where n is the input size, k is a parameter, and f is a computable function.

5   Experiments

We implemented and tested the tree decomposition based algorithms.[2] To compute the nice tree decomposition we used the "htd" library [1] version 1.2, compiled with gcc version 8.3. The algorithm itself was implemented in Python 3.7, using the libraries[3] NetworkX 2.3 and Shapely 1.6. To run the experiments we used a cluster, each node equipped with an Intel Xeon E5-2640 v4 processor clocked at 2.4 GHz, 160 GB of Ram, and operating Ubuntu 16.04. Each run had a memory limit of at most 80 GB of RAM.

We generated random graphs using the NetworkX gnm algorithm. The graphs have $n = 40$ vertices and between $m = 40$ and 75 edges in increments of 5. This makes a total of 800 graphs, 100 for each $m \in \{40, 45, \ldots, 75\}$. For the layouts we used the NetworkX implementation of the spring embedder by Fruchterman and Reingold [15] and the graphviz[4] implementation "circo" of a circular layout, version 2.40.1. We could successfully run MAXSPED for all but four of the spring layouts and for all circle layouts with up to 60 edges.

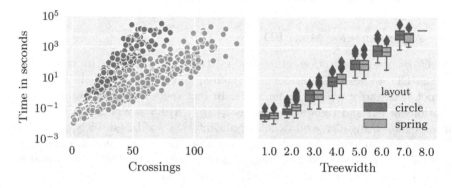

Fig. 5. Experimental results for the MAXSPED algorithm.

Since the time complexity of the algorithm depends exponentially on the treewidth of the intersection graph, we evaluated the running time relative to treewidth and number of crossings, see Fig. 5. The results are as expected, with the runtime quickly increasing to about 16 min (1,000 sec) for 80 crossings in case of the spring and 50 crossings in case of the circle layouts – or for a treewidth of 6 for both layouts. On the other hand we can handle up to 50 crossings and a treewidth of 4 for the spring layouts in about 10 seconds. The discrepancy in the runtime relative to the number of crossings between spring and circle layouts likely comes from different numbers of crossings per edge. To confirm this we took for each intersection graph its maximum degree k divided by the

---

[2] https://www.ac.tuwien.ac.at/partial-edge-drawing/.
[3] https://networkx.github.io/ and https://github.com/Toblerity/Shapely.
[4] https://www.graphviz.org/.

total number of input crossings. For the spring layouts this resulted in a ratio of 0.24 and for the circle layouts 0.33. Recall that the running time is dominated by $O((k+1)^{\omega+2})$. Hence an increase by a of factor 1.5 in the aforementioned value also results in an additional factor of $1.5^{\omega+2}$ in the asymptotic running time. Concerning ink, for the circle layouts an average of 84% ($\sigma = 0.09$) and for spring layouts an average of 90% ($\sigma = 0.06$) of the ink could be preserved.

For MAXPED we conducted the same experiments. In general one can say that the additional factor of $(k+1)^{\omega}$ makes a big difference. For details see [16].

In summary, our experiment confirmed the predicted running time behavior and showed that the amount of removed ink was moderate. Moreover, the "htd" library [1] performed very well for computing a nice tree decomposition so that we could focus on implementing the dynamic programming algorithm itself.

6   Conclusion

We extended the work by Bruckdorfer et al. [6] and showed NP-hardness for the MAXPED and MAXSPED problems, as well as polynomial-time algorithms for the case of the intersection graph of the input drawing being a tree or having bounded treewidth. For the latter, our proof-of-concept implementation worked reasonably well for small to medium-size instances.

An interesting open problem is to close the gap for MAXPED. While we showed it to be NP-hard for $k \geq 4$ and it is easy to solve for $k \leq 2$ [7], the case of $k = 3$ remains open. Another direction is to consider the existential question for homogeneous (but non-symmetric) PEDs with a fixed ratio δ, for which we can freely distribute the δ fraction of the ink to both stubs. We expect that our algorithms for trees and intersection graphs of bounded treewidth extend to that case, but we could not resolve if the problem is NP-hard or not.

References

1. Abseher, M., Musliu, N., Woltran, S.: htd - a free, open-source framework for (customized) tree decompositions and beyond. In: Salvagnin, D., Lombardi, M. (eds.) CPAIOR 2017. LNCS, vol. 10335, pp. 376–386. Springer, Heidelberg (2017). https://doi.org/10.1007/978-3-319-59776-8_30

2. Becker, R.A., Eick, S.G., Wilks, A.R.: Visualizing network data. IEEE Trans. Vis. Comput. Graph. 1(1), 16–28 (1995). https://doi.org/10.1109/2945.468391

3. de Berg, M., Cheong, O., van Kreveld, M., Overmars, M.: Computational Geometry: Algorithms and Applications, 3rd edn. Springer, Heidelberg (2008). https://doi.org/10.1007/978-3-540-77974-2

4. Binucci, C., Liotta, G., Montecchiani, F., Tappini, A.: Partial edge drawing: homogeneity is more important than crossings and ink. In: Information, Intelligence, Systems Applications (IISA 2016). IEEE (2016). https://doi.org/10.1109/IISA.2016.7785427

5. Bruckdorfer, T.: Schematics of Graphs and Hypergraphs. Ph.D. thesis, Universität Tübingen (2015). http://dx.doi.org/10.15496/publikation-8904

6. Bruckdorfer, T., Cornelsen, S., Gutwenger, C., Kaufmann, M., Montecchiani, F., Nöllenburg, M., Wolff, A.: Progress on partial edge drawings. J. Graph Algorithms Appl. **21**(4), 757–786 (2017). https://doi.org/10.7155/jgaa.00438
7. Bruckdorfer, T., Kaufmann, M.: Mad at edge crossings? Break the edges! In: Kranakis, E., Krizanc, D., Luccio, F. (eds.) FUN 2012. LNCS, vol. 7288, pp. 40–50. Springer, Heidelberg (2012). https://doi.org/10.1007/978-3-642-30347-0_7
8. Bruckdorfer, T., Kaufmann, M., Lauer, A.: A practical approach for 1/4-SHPEDs. In: Information, Intelligence, Systems and Applications (IISA 2015). IEEE (2015). https://doi.org/10.1109/IISA.2015.7387994
9. Bruckdorfer, T., Kaufmann, M., Leibßle, S.: PED user study. In: Di Giacomo, E., Lubiw, A. (eds.) GD 2015. LNCS, vol. 9411, pp. 551–553. Springer, Heidelberg (2015). https://doi.org/10.1007/978-3-319-27261-0_47
10. Bruckdorfer, T., Kaufmann, M., Montecchiani, F.: 1-bend orthogonal partial edge drawings. J. Graph Algorithms Appl. **18**(1), 111–131 (2014). https://doi.org/10.7155/jgaa.00316
11. Burch, M., Schmauder, H., Panagiotidis, A., Weiskopf, D.: Partial link drawings for nodes, links, and regions of interest. In: Information Visualisation (IV 2014), pp. 53–58 (2014). https://doi.org/10.1109/IV.2014.45
12. Burch, M., Vehlow, C., Konevtsova, N., Weiskopf, D.: Evaluating partially drawn links for directed graph edges. In: van Kreveld, M., Speckmann, B. (eds.) GD 2011). LNCS, vol. 7034, pp. 226–237. Springer, Heidelberg (2012).https://doi.org/10.1007/978-3-642-25878-7_22
13. Cygan, M., Fomin, F.V., Kowalik, L., Lokshtanov, D., Marx, D., Pilipczuk, M., Pilipczuk, M., Saurabh, S.: Parameterized Algorithms, vol. 3. Springer, Heidelberg (2015). https://doi.org/10.1007/978-3-319-21275-3
14. Downey, R.G., Fellows, M.R.: Parameterized Complexity. Springer, Heidelberg (2012)
15. Fruchterman, T.M.J., Reingold, E.M.: Graph drawing by force-directed placement. Softw. Pract. Exper. **21**(11), 1129–1164 (1991). https://doi.org/10.1002/spe.4380211102
16. Hummel, M., Klute, F., Nickel, S., Nöllenburg, M.: Maximizing ink inpartial edge drawings of k-plane graphs. CoRR **abs/1908.08905** (2019). http://arxiv.org/abs/1908.08905
17. Koffka, K.: Principles of Gestalt Psychology. Routledge, Abingdon (1935)
18. Lichtenstein, D.: Planar formulae and their uses. SIAM J. Comput. **11**(2), 329–343 (1982). https://doi.org/10.1137/0211025
19. Purchase, H.: Which aesthetic has the greatest effect on human understanding? In: Di Battista, G. (ed.) GD 1997. LNCS, vol. 1353, pp. 248–261. Springer, Heidelberg (1997). https://doi.org/10.1007/3-540-63938-1_67
20. Robertson, N., Seymour, P.D.: Graph minors. III. Planar tree-width. J. Comb. Theory Ser. B **36**(1), 49–64 (1984). https://doi.org/10.1016/0095-8956(84)90013-3

Graph Drawing with Morphing Partial Edges

Kazuo Misue$^{(\boxtimes)}$ ⓘ and Katsuya Akasaka

University of Tsukuba, Tsukuba, Japan
misue@cs.tsukuba.ac.jp, akasaka@vislab.cs.tsukuba.ac.jp

Abstract. A partial edge drawing (PED) of a graph is a variation of a node-link diagram. PED draws a link, which is a partial visual representation of an edge, and reduces visual clutter of the node-link diagram. However, more time is required to read a PED to infer undrawn parts. The authors propose a morphing edge drawing (MED), which is a PED that changes with time. In MED, links morph between partial and complete drawings; thus, a reduced load for estimation of undrawn parts in a PED is expected. Herein, a formalization of MED is shown based on a formalization of PED. Then, requirements for the scheduling of morphing are specified. The requirements inhibit morphing from crossing and shorten the overall time for morphing the edges. Moreover, an algorithm for a scheduling method implemented by the authors is illustrated and the effectiveness of PED from a reading time viewpoint is shown through an experimental evaluation.

Keywords: Graph drawing · Partial edge drawing · Morphing edge

1 Introduction

The partial edge drawing (PED) of a graph is a variation of a node-link diagram that is a visual representation of a graph. In PED, a link is drawn, which is a partial visual representation of an edge; that is, a part of the link is omitted, and then intersections of links are eliminated. Therefore, PED can reduce visual clutter of node-link diagrams. An experimental evaluation by Bruckdorfer et al. shows that PEDs reduces errors and provides higher accuracy when reading graphs than traditional node-link diagrams; however, longer reading time is required [6].

We propose a morphing edge drawing (MED), which is a PED that changes with time. In MED, links are morphed between partial and complete drawings. Therefore, reduced loads are expected to infer undrawn parts in a PED. However, the effect depends on the scheduling of morphing edges. We designed a scheduling algorithm that did not unnecessarily cause links to cross. Then, we performed a user study to evaluate the effectiveness of MEDs by implementation. The contributions herein are as follows: Proposal and formalization of MED. Setting scheduling requirements. Proposal of algorithm for scheduling morphing. Evaluation of MED via user study.

ⓒ Springer Nature Switzerland AG 2019
D. Archambault and C. D. Tóth (Eds.): GD 2019, LNCS 11904, pp. 337–349, 2019.
https://doi.org/10.1007/978-3-030-35802-0_26

2   Partial Edge Drawing

Let $G = (V, E)$ be a simple undirected graph and let $\Gamma(G) = (\Gamma(V), \Gamma(E))$ be a drawing of G, where $\Gamma(V) = \{\Gamma(v)|v \in V\}$ and $\Gamma(E) = \{\Gamma(e)|e \in E\}$. Let $\Gamma(G)$ be a traditional straight-line drawing. Let the drawing $\Gamma(v)$ of a node $v \in V$ be a small disk placed at a position p_v and let $\Gamma(e)$ of an edge $e \in E$ be a straight-line segment between two nodes (disks) incident to the edge. That is, $\Gamma(e) = \{s \cdot p_w + (1 - s) \cdot p_v|s \in [0, 1]\}$ when $e = (v, w)$. We call $\Gamma(G)$ a *complete edge drawing (CED)* because it draws every straight-line representing an edge completely.

We express the partial drawing of an edge $e = (v, w)$ as a function $\gamma_e :$ $[0, 1]^2 \rightarrow 2^{\Gamma(e)}$ shown in Exp. (1).

$$\gamma_e(\alpha, \beta) = \begin{cases} \{s \cdot p_w + (1 - s) \cdot p_v|s \in [0, \alpha] \cup [\beta, 1]\} & \text{for } \alpha < \beta \\ \Gamma(e) & \text{for } \alpha \geq \beta \end{cases} \quad (1)$$

That is, $\gamma_e(\alpha, \beta)$ of edge e comprise the parts that remain after removing the corresponding parts (α, β) from $\Gamma(e)$ when the entire $\Gamma(e)$ corresponds to the interval $[0, 1]$. Each of the remaining continuous parts is called a *stub*. The parameters α and β, which determine the stub lengths, are *partial edge parameters*. When $0 < \alpha$ and $\beta < 1$ for γ_e, the part to be deleted is not the end of $\Gamma(e)$; two stubs remain at the two nodes incident to the edge e. These are called a *pair of stubs*.

Drawing $\Gamma_{PED}(G) = (\Gamma(V), \Gamma_{PED}(E))$ is a *partial edge drawing (PED)* if for all edges $e \in E$, α_e and β_e are given, and at least an edge $e_1 \in E$ exists with $\alpha_{e_1} < \beta_{e_1}$, where $\Gamma_{PED}(E) = \{\gamma_e(\alpha_e, \beta_e)|e \in E\}$. When $\alpha_e = 1 - \beta_e$, i.e., the lengths of a pair of stubs are the same, the drawing is a *symmetric PED (SPED)*. The smaller parameter α_e is the *stub-edge ratio*. If the stub-ratios for all edges are the same δ, the drawing is a δ-*symmetric homogeneous PED (δ-SHPED)*.

Herein, we assume that $\Gamma(G)$ is given in advance and stubs may have intersections.

3   Related Work

Becker et al. conceived a drawing concept in which only half the links are used to reduce the visual clutter during the development of a tool called SeeNet [2]. Parallel Tagcloud, developed by Collins et al., adopts a method similar to PED [9]. Although Parallel Tagcloud is an extension of Tag Cloud, it can be regarded as a hierarchical layout of directed graphs; thus, it is useful against visual clutter caused by crossing of links. They can avoid drawing intersections by representing links as straight lines without drawing in the middle of the links.

Bruckdorfer et al. formalized the PED in their study [5]; our formalization of PED in Sect. 2 is a modified version of their formalization. They added continuity of the omitted parts of links as a condition, we have incorporated this into the formalization herein. Moreover, although they focused on a layout without crossing stubs in PED, this study allows stub crossings. Burch et al. applied PED to directed graphs using tapered links [8]. Schmauder et al. applied PED to weighted graphs by representing weights with edge colors [14].

Bruckdorfer et al. performed a comparison between CED[1] and 1/4-SHPED on reading performance of graphs [6]. Although the statistical significance was not shown, from their chart that visualizes the experimental results, in the task of reading graphs (for adjacency check of two nodes or search for adjacent nodes), we can guess that 1/4-SHPED is slightly more accurate than CED; however, the response time of 1/4-SHPED is longer. Binucci et al. conducted a more detailed evaluation to reveal that SHPED has high accuracy in reading graphs within SPED [3]. Burch examined the effect of stub orientation and length on graph reading accuracy [7].

Blass et al. avoided using arrows to facilitate grasping high-dimensional transitions in the state transition diagram and proposed moving the dashed pattern with animation [4]. Holten et al. compared the recognition accuracy of graphs in various edge drawing methods, such as tapered links and curved links, including animation [11]. They showed that the recognition accuracy of the graph is high by representation using animation. Romat et al. attempted to extend the design space using animation of edge textures [13]. The proposal herein can be considered as an application of animation to graph drawing, especially to drawing edges. However, the purpose is not to express the orientation, but to improve the reading accuracy and efficiency for graphs.

4   Morphing Edge Drawing

Let T be a set of times. Then, function $\mu_e : T \rightarrow 2^{\Gamma(e)}$, which determines a partial drawing of edge e for time $t \in T$, a *morphing function*. A dynamic drawing $\Gamma_{MED}(G) = (\Gamma(V), \Gamma_{MED}(E))$ of graph G with morphing functions is a *morphing edge drawing (MED)*, where $\Gamma_{MED}(E) = \{\mu_e | e \in E\}$ is a set of morphing functions.

Then, a function $\rho_e : T \rightarrow [0,1]^2$, which determines the partial edge parameters for a time $t \in T$, is a *ratio function*. The morphing function μ_e can be constructed as $\mu_e(t) = \gamma_e(\rho_e(t))$ using the ratio function.

4.1   Symmetric MED

When all ρ_e for all $e \in E$ satisfies $\rho_e(t) = (\delta_t, 1 - \delta_t)$ $(0 \le \delta_t \le 1/2)$ for all $t \in T$, we get SPED at any time. Thus, such a ratio function is a *symmetric ratio function*; furthermore, if a MED is composed of symmetric ratio functions, it is referred to as a *symmetric MED (SMED)*. As the two values obtained by a symmetric ratio function depend on each other, we can define the function as $\rho_e : T \rightarrow [0, 1/2]$ without ambiguity.

Morphing of edge e extending from stub-edge ratio δ_e to η_e and then shrinking to δ_e is expressed by a symmetric ratio function ρ_e, expressed as Exp. (2), where

---

[1] They call it the traditional straight-line model (TRA).

t_0 is the start time of the morphing, l is the length of $\Gamma(e)$, and s is the speed of the tips of the stubs (*morphing speed*). Here, let the morphing speed be constant. Figure 1 shows the graph of the function.

$$\rho_e(t) = \begin{cases} \delta_e & \text{for } t \leq t_0 \text{ or } t_2 < t \\ \delta_e + (t - t_0)s/l & \text{for } t_0 < t \leq t_1 \\ \eta_e - (t - t_1)s/l & \text{for } t_1 < t \leq t_2, \end{cases} \tag{2}$$

where t_1 is the time when the stub-edge ratio becomes η_e, and t_2 is the time when the stub-edge ratio returns to δ_e. Using one-way travel time $d_1 = (\eta_e - \delta_e)l/s$, $t_1 = t_0 + d_1$ and $t_2 = t_0 + 2d_1$.

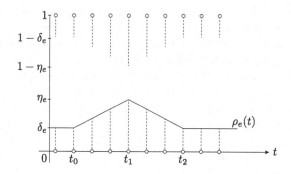

Fig. 1. Graph of ρ_e. Each pair (top and bottom) of dashed lines represent a pair of stubs expanding and contracting.

If the ratio functions have the same δ and η for all edges, the drawing is a (δ, η)-*symmetric homogeneous MED* ((δ, η)-*SHMED*). Note, this homogeneity does not always mean synchronicity. Drawings by (δ, η)-SHMED may not be SHPED at any time.

When $\eta = 1/2$, we omit η, like δ-*symmetric homogeneous MED* (δ-*SHMED*). In the drawing by δ-SHMED, a pair of stubs of edge e becomes $\Gamma(e)$ at a certain moment. Intuitively, the drawing by δ-SHMED changes between δ-SHPED and CED. However, a moment when it becomes CED does not always exist.

5   Scheduling of Morphing

We have set requirements to design the scheduling of morphing of all edges as follows:

R1: Morphing Does Not Make Crossings. To maintain the reading accuracy, visual clutter should be minimized. Therefore, morphing should not result in new crossings among stubs. However, if another edge exists that crosses a stub with a stub-edge ratio δ, then the crossing is inevitable. As mentioned earlier, rearrangements to avoid such crossings are beyond the scope of this paper. The requirement is to avoid crossings in the center area undrawn by δ-SHPED, (we refer to these areas as *blank areas*).

R2: Shorten Morphing Time for All Edges. The time taken for a viewer to focus on a stub should be minimized before morphing of the stub. We do not know in advance which stubs the viewer will focus on. Therefore, it is necessary to repeat morphing of all edges, and it is necessary to shorten the total morphing time of all edges.

5.1 Morphing Group

First, two non-crossing edges do not generate new crossings of stubs at any timing when morphing, i.e., they can morph simultaneously and independently. However, two edges that do not intersect may not be able to morph independently, depending on the relationship with other edges (see Fig. 2).

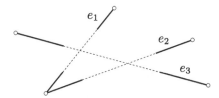

Fig. 2. Dependency between edges. The edges e_1 and e_2 do not intersect, but both intersect e_3, so they cannot be morphed independently. Dotted lines represent the omitted parts (blank areas).

A set of edges where the timing of morphing may affect each other is a *morphing group*. Different morphing groups can be scheduled independently. To determine the morphing groups, another graph is generated from the graph to be drawn. To avoid ambiguity, we will express the newly generated graph and its components as ⟨graph⟩, ⟨node⟩, and ⟨edge⟩. Let each edge be a ⟨node⟩. Suppose that there is an ⟨edge⟩ between ⟨nodes⟩ (i.e., edges) that intersect each other in a blank area. ⟨Nodes⟩ (i.e., edges) included in each connected component of the ⟨graph⟩ generated in this manner constitute a morphing group.

As two edges belonging to different morphing groups do not intersect, it is possible to schedule them independently in units of morphing groups. Hereafter, we describe scheduling of morphing of edges included in a morphing group.

5.2 Sequential Morphing

To prevent new stub crossings from being generated by morphing of two edges intersecting in blank areas, entry into a blank area should be exclusive. That is, the safest scheduling, satisfying requirement R1 (morphing does not result in crossings), is to perform edge morphing sequentially. However, with such simple scheduling, R2 cannot be satisfied.

5.3   Packing Morphing Intervals

Assume there are two intersecting edges e_1 and e_2, as shown in Fig. 3. If the scheduling is such that when a stub of edge e_1 has been stretched and contracted to the crossing point, a stub of e_2 extends to the crossing point, then no intersection will occur. Let the time it takes for the stub of e_1 to contract to the crossing point after it starts morphing be t_1 and the time it takes for the stub of e_2 to start morphing and then extend to the crossing point be t_2. If the stub of edge e_2 starts morphing $t_1 - t_2$ after the stub of e_1 starts morphing, no crossing occurs. In addition, when morphing is repeated alternately, e_1 will start morphing again next to e_2.

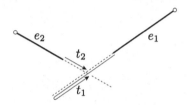

Fig. 3. Packing morphing intervals.

5.4   Parallel Morphing

Even if edges belong to the same morphing group, morphing of two non-intersecting edges may be performed simultaneously. By appropriately morphing parallelly, it is possible to shorten the total morphing time of all edges. For example, as edges e_1 and e_2 in Fig. 2 are not intersecting, their morphing can be parallelized if we can find adequate timing to avoid crossing with e_3. This can reduce the overall time.

5.5   Algorithm for Finding Morphing Start Time

In scheduling morphing, we decided to determine the start time from longer edges. Assuming that every stub has the same morphing speed, the longer the edge, the longer one cycle of morphing takes. By determining the start time from longer edges, while a long edge is morphing, the morphing of short edges that do not intersect with it can be embedded in the same time range.

Given a morphing group E (a set of edges), Algorithm 1 determines the morphing start time $t_s(e)$ for all edges $e \in E$. The algorithm determines the morphing start time in descending order of edge length. It checks the timing of every morphing stub of all edges in $C(e)$ that intersect with edge e, and allows the morphing start time of edge e to be the earliest time that does not result in intersection with the morphing stubs.

The $r_1(e, c)$ and $r_2(e, c)$ appearing in the algorithm represents the first and last time of the time range, respectively, when the start of the morphing of edge e is prohibited to avoid crossing with edge c. They are described as Exp. (3) and Exp. (4).

$$r_1(e, c) = t_s(c) + t_p(c, e) - t_r(e, c) \tag{3}$$
$$r_2(e, c) = t_s(c) + t_r(c, e) - t_p(e, c), \tag{4}$$

where $t_p(e, c)$ is the time it takes from the start of morphing of the stub of edge e to the first passing (passing while stretching) at the crossing point with edge c (cf. $t_p(e_2, e_1) = t_2$ in Fig. 3), and $t_r(e, c)$ is the time it takes from the start of morphing to the second passing (passing while shrinking) at the crossing point (cf. $t_r(e_1, e_2) = t_1$ in Fig. 3).

Algorithm 1. Determining the start time of morphing

Input: E — Set of edges included in a morphing group
Output: Start time is determined for all edges of E
1: **function** FINDSTARTTIME(E)
2:     **for** e in $sortByLength(E)$ **do**
3:         $I \leftarrow \{(r_1(e, c), r_2(e, c)) | c \in C(e) \wedge t_s(c) \text{ is defined.}\}$
4:         $t_s(e) \leftarrow earliestSpace(I)$
5:     **end for**
6: **end function**

Function $earliestSpace(I)$ yields the smallest value not included in the time ranges (intervals) in a given set I. If each pair (r_1, r_2) included in the set I is regarded as an interval $[r_1, r_2)$ of real numbers, function $earliestSpace(I)$ is defined as Exp. (5). We calculate $earliestSpace(I)$ using Algorithm 2. Note that T is a nonnegative real number in Algorithm 2.

$$earliestSpace(I) = \min\left\{ \left(\bigcup_{r \in I} r \right)^c \right\} \tag{5}$$

6   Evaluation Experiment

To investigate the effectiveness of MED, we conducted a comparative experiment with three types of visual representations: CED, 1/4-SHPED, and 1/4-SHMED.

Algorithm 2. Finding time when morphing can start

Input: I — Set of time ranges (pairs of times) during which morphing should not
    start
Output: Earliest time morphing can start
  1: **function** EARLIESTSPACE(I)
  2:    $t \leftarrow 0$
  3:    **for** (r_1, r_2) in $sortByStartTime(I)$ **do**
  4:        **if** $r_2 < t$ **then**
  5:            continue
  6:        **else if** $t < r_1$ **then**
  7:            **return** t
  8:        **else**
  9:            $t \leftarrow r_2$
 10:        **end if**
 11:    **end for**
 12:    **return** t
 13: **end function**

6.1   Hypothesis

We made the following hypothesis.

H1 1/4-SHMED requires less time to read a graph than 1/4-SHPED
H2 1/4-SHMED is more accurate at reading graphs than CED.

6.2   Tasks

We designed the following tasks to test the above hypotheses.

T1 Check if the two highlighted nodes are adjacent (connected by an edge).
T2 Select all the nodes to which the highlighted node is adjacent.

For T1, as shown in Fig. 4, a graph in which two nodes are highlighted is
displayed. Participants respond by pressing "Y" or "N" on the keyboard. When
creating sample graphs, the number of crossings of the edges connecting two
nodes were set to 8 or 16 when the nodes were adjacent.

For T2, as shown in Fig. 5, a graph in which one node is highlighted is displayed. Node selection is performed using a trackpad. When clicked, the pointed
node is selected and turns orange. Participants can also cancel the selection by
clicking again. Answers are confirmed by pressing the Enter key. When creating
sample graphs, we selected nodes to be highlighted such that the number of
adjacent nodes to it were 3, 6, and 9. Furthermore, we set the average number
of intersections of the edges of interest to be within 7.9–8.1 and the average of
lengths of the edges to be in the range of 3.3–3.7 cm on the screen to ensure
that the task difficulty was not excessively low or high. In this experiment, we
assumed that the distance between the participant's eyes and the screen was
40 cm.

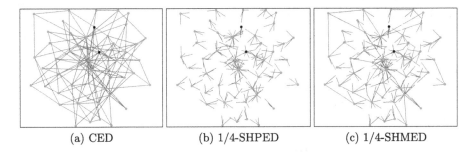

(a) CED   (b) 1/4-SHPED   (c) 1/4-SHMED

Fig. 4. Examples of visual representations used in T1

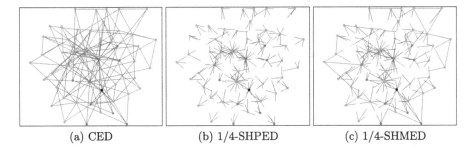

(a) CED   (b) 1/4-SHPED   (c) 1/4-SHMED

Fig. 5. Examples of visual representations used in T2

6.3 Graphs Used for the Experiment

We used the Barabási-Albert model [1] as a guideline to create a graph with
50 nodes and 144 edges. We used the Fruchterman-Reingold algorithm [10] to
determine the layout of the graph.

6.4 Morphing Speed

We set the morphing speed of each stub as $10°/s$. If morphing is too fast, the
human eye cannot track it. Conversely, if it is too slow, reading efficiency is
reduced. We derived the speed based on Robinson's experiment [12] such that
it is human eye-trackable while being as fast as possible. However, we set a
minimum one-way travel time 300 ms to make capturing morphing stubs easy.

6.5 Experimental Settings

We used a MacBook Pro 2017 (screen size 13.3 inches, screen resolution 1440×900) for the experiment. We set the display area to 1000×800 so the graph can
be viewed without scrolling.

The participants in this experiment were 12 students (4 university students
and 8 graduate students).

6.6   Experimental Procedure

The following procedure was used to conduct the experiment:

1 Overall explanation
2 Visual representation #1
    2-1 T1 practice (one question)
    2-2 T1 actual (nine questions)
    2-3 T2 practice (one question)
    2-4 T2 actual (nine questions)
    2-5 Questionnaire for visual representation #1
3 Visual representation #2 (flow similar to visual representation #1)
4 Visual representation #3 (flow similar to visual representation #1)
5 Questionnaire for whole experiment

We varied the order of presenting visual representations from each participant to eliminate the effects of order. Therefore, visual representations #1, #2, and #3 differ depending on the participant. We assigned two participants for each of the six (= 3!) orders.

6.7   Response Time

Figure 6 shows the distribution (boxplots) of response time (in millisecond) for each task and representation method. In both tasks, the average response time was the lowest for CED and highest for 1/4-SHPED. As the 1/4-SHMED is located in the middle, an improvement in the reading time for 1/4-SHPED can be expected. From the Shapiro-Wilk test ($\alpha = 0.05$), the time taken either task did not follow a normal distribution. Therefore, we performed multiple tests using the Friedman and Holm methods. Tables 1 and 2 show the test results for the response time for T1 and T2, respectively. As shown in Table 1, a significant difference was observed between the representation methods, i.e., 1/4-SHMED can shorten the time taken to confirm the adjacency between nodes, compared to 1/4-SHPED (H1). In contrast, no significant difference was found between the representation methods with respect to the response time of T2.

Table 1. Test result of response time of T1

Comparison	Test result (p value)	Significance level
CED vs 1/4-SHPED	2.035e−7	<0.0167
CED vs 1/4-SHMED	0.0343	<0.0500
1/4-SHPED vs 1/4-SHMED	0.0011	<0.0250

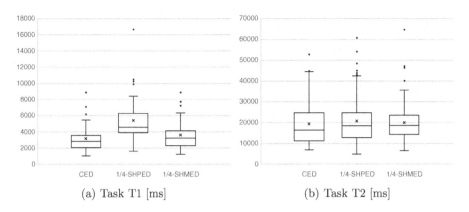

(a) Task T1 [ms]   (b) Task T2 [ms]

Fig. 6. Distribution of response time

Table 2. Test result of response time of Task T2

Comparison	Test result (p value)	Significance level
CED vs 1/4-SHPED	0.0543	>0.0167
CED vs 1/4-SHMED	0.1237	—
1/4-SHPED vs 1/4-SHMED	0.8474	—

6.8 Answer Accuracy

Figure 7(a) shows the number of correct answers and the number of incorrect answers for T1 in a stacked bar chart. The correct answer rate of 1/4-SHMED is the highest. However, independence between the representation methods was not recognized from the chi-square test. We defined the score for T2 as the Jaccard coefficient between the set of adjacent nodes and the set of answered nodes, i.e., it is 1 when the two sets completely match, and 0 when there is no common element. Figure 7(b) shows the distribution of scores according to each representation method for T2 in a boxplot. From the Shapiro-Wilk test ($\alpha = 0.05$), T2 did not followed a normal distribution. Therefore, we performed multiple tests using the Friedman and Holm methods. Table 3 shows the test results for the scores T2. As seen above, regarding the accuracy of answers, no significant difference was observed between the representation methods. Therefore, H2 is not supported.

Table 3. Test result of score of T2

Comparison	Test result (p value)	Significance level
CED vs 1/4-SHPED	0.5862	—
CED vs 1/4-SHMED	0.1489	—
1/4-SHPED vs 1/4-SHMED	0.07817	>0.0167

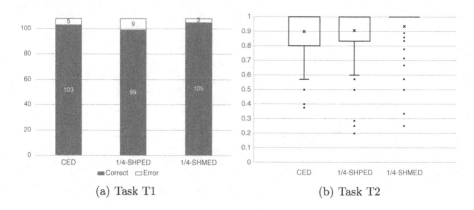

(a) Task T1   (b) Task T2

Fig. 7. Correct answer rate

6.9   Qualitative Feedback

We asked the participants for opinions on visual representations using questionnaires. The following comments were obtained on 1/4-SHMED.

- Positive opinions
  - Morphing made it easy to confirm the exact adjacency.
  - It can be judged whether two nodes are adjacent by observing the morphing of two stabs works simultaneously.
- Negative opinions
  - It is messy and difficult to see. My eyes are strained.
  - The stubs change too fast. The time for stubs to connect is too short.

Positive opinions indicate that morphing contributes to reading graphs. In contrast, from the negative opinions, it appears that visual clutter was not always resolved. The following can be considered as the main reasons. The first is that the morphing speed is too fast. In the implementation used for the experiment, to shorten the overall morphing time, the morphing speed was determined based on the tracking speed of human eyes; however, this appears to be too fast. The second is that there were a large number of stubs applying morphing. In the graph used in the experiment, out of the 144 edges, the average number of non-morphing edges is 24.5. Given that approximately 120 edges repeated morphing, the entire graph is considered to have caused visual clutter.

7   Concluding Remarks

We proposed morphing edge drawing (MED) which is time-varying partial edge drawing (PED) and showed the formalization of MED. We also developed a scheduling scheme for morphing such that dynamic stubs do not cause new crossings. We compared three visual representations, CED, 1/4-SHPED, and 1/4-SHMED, via a user study, and showed that 1/4-SHMED is better than 1/4-SHPED in terms of graph reading time. Thus, MED can function as a countermeasure against the time to read a graph by PED. In the future, it is important to investigate eye-friendly morphing that causes less strain and has improved scheduling.

References

1. Barabási, A.L., Albert, R.: Emergence of scaling in random networks. Science **286**(5439), 509–512 (1999). https://doi.org/10.1126/science.286.5439.509
2. Becker, R.A., Eick, S.G., Wilks, A.R.: Visualizing network data. IEEE Trans. Vis. Comput. Graph. **1**(1), 16–28 (1995). https://doi.org/10.1109/2945.468391
3. Binucci, C., Liotta, G., Montecchiani, F., Tappini, A.: Partial edge drawing: homogeneity is more important than crossings and ink. In: 2016 7th International Conference on Information, Intelligence, Systems Applications (IISA), pp. 1–6, July 2016. https://doi.org/10.1109/IISA.2016.7785427
4. Blaas, J., Botha, C., Grundy, E., Jones, M., Laramee, R., Post, F.: Smooth graphs for visual exploration of higher-order state transitions. IEEE Trans. Vis. Comput. Graph. **15**(6), 969–976 (2009). https://doi.org/10.1109/TVCG.2009.181
5. Bruckdorfer, T., Kaufmann, M.: Mad at edge crossings? Break the edges!. In: Kranakis, E., Krizanc, D., Luccio, F. (eds.) FUN 2012. LNCS, vol. 7288, pp. 40–50. Springer, Heidelberg (2012). https://doi.org/10.1007/978-3-642-30347-0_7
6. Bruckdorfer, T., Kaufmann, M., Leibßle, S.: PED user study. In: Di Giacomo, E., Lubiw, A. (eds.) GD 2015. LNCS, vol. 9411, pp. 551–553. Springer, Cham (2015). https://doi.org/10.1007/978-3-319-27261-0_47
7. Burch, M.: A user study on judging the target node in partial link drawings. In: 2017 21st International Conference Information Visualisation (IV), pp. 199–204, July 2017. https://doi.org/10.1109/iV.2017.43
8. Burch, M., Vehlow, C., Konevtsova, N., Weiskopf, D.: Evaluating partially drawn links for directed graph edges. In: van Kreveld, M., Speckmann, B. (eds.) GD 2011. LNCS, vol. 7034, pp. 226–237. Springer, Heidelberg (2012). https://doi.org/10.1007/978-3-642-25878-7_22
9. Collins, C., Viégas, F.B., Wattenberg, M.: Parallel tag clouds to explore and analyze faceted text corpora. In: 2009 IEEE Symposium on Visual Analytics Science and Technology, pp. 91–98, October 2009. https://doi.org/10.1109/VAST.2009.5333443
10. Fruchterman, T.M.J., Reingold, E.M.: Graph drawing by force-directed placement. Softw.: Pract. Exp. **21**(11), 1129–1164 (1991). https://doi.org/10.1002/spe.4380211102
11. Holten, D., Isenberg, P., van Wijk, J.J., Fekete, J.: An extended evaluation of the readability of tapered, animated, and textured directed-edge representations in node-link graphs. In: 2011 IEEE Pacific Visualization Symposium, pp. 195–202, March 2011. https://doi.org/10.1109/PACIFICVIS.2011.5742390
12. Robinson, D.A.: The mechanics of human smooth pursuit eye movement. J. Physiol. **180**(3), 569–591 (1965). https://doi.org/10.1113/jphysiol.1965.sp007718
13. Romat, H., Appert, C., Bach, B., Henry-Riche, N., Pietriga, E.: Animated edge textures in node-link diagrams: a design space and initial evaluation. In: Proceedings of the 2018 CHI Conference on Human Factors in Computing Systems, CHI 2018, pp. 187:1–187:13. ACM, New York (2018). https://doi.org/10.1145/3173574.3173761
14. Schmauder, H., Burch, M., Weiskopf, D.: Visualizing dynamic weighted digraphs with partial links. In: Proceedings of 6th International Conference on Information Visualization Theory and Applications (IVAPP), pp. 123–130 (2015)

A Note on Universal Point Sets
for Planar Graphs

Manfred Scheucher$^{(\boxtimes)}$, Hendrik Schrezenmaier, and Raphael Steiner

Institut für Mathematik, Technische Universität Berlin, Berlin, Germany
{scheucher,schrezen,steiner}@math.tu-berlin.de

Abstract. We investigate which planar point sets allow simultaneous straight-line embeddings of all planar graphs on a fixed number of vertices. We first show that at least $(1.293 - o(1))n$ points are required to find a straight-line drawing of each n-vertex planar graph (vertices are drawn as the given points); this improves the previous best constant 1.235 by Kurowski (2004).

Our second main result is based on exhaustive computer search: We show that no set of 11 points exists, on which all planar 11-vertex graphs can be simultaneously drawn plane straight-line. This strengthens the result by Cardinal, Hoffmann, and Kusters (2015), that all planar graphs on $n \leq 10$ vertices can be simultaneously drawn on particular n-universal sets of n points while there are no n-universal sets of size n for $n \geq 15$. We also provide 49 planar 11-vertex graphs which cannot be simultaneously drawn on any set of 11 points. This, in fact, is another step towards a (negative) answer of the question, whether every two planar graphs can be drawn simultaneously – a question raised by Brass, Cenek, Duncan, Efrat, Erten, Ismailescu, Kobourov, Lubiw, and Mitchell (2007).

Keywords: Simultaneously embedded · Stacked triangulation · Order type · Boolean satisfiability (SAT) · Integer programming (IP)

1 Introduction

A point set S in the Euclidean plane is called n-*universal for* a family $\mathcal{G}$ of planar n-vertex graphs if every graph G from $\mathcal{G}$ admits a plane straight-line embedding such that the vertices are drawn as points from S. A point set, which is n-universal for the family of all planar graphs, is simply called n-*universal*. We denote by $f_p(n)$ the size of a minimal n-universal set (for planar graphs), and by $f_s(n)$ the size of a minimal n-universal set for stacked triangulations, where stacked triangulations (a.k.a. planar 3-trees) are defined as follows:

M. Scheucher supported by DFG Grant FE 340/12-1. H. Schrezenmaier supported by DFG Grant FE-340/11-1. R. Steiner supported by DFG-GRK 2434.

Earlier versions of this paper (EuroCG 2019; arXiv versions 1 and 2) contained a flaw, which has been corrected. For more details see the full paper [27].

© Springer Nature Switzerland AG 2019
D. Archambault and C. D. Tóth (Eds.): GD 2019, LNCS 11904, pp. 350–362, 2019.
https://doi.org/10.1007/978-3-030-35802-0_27

Definition 1 (Stacked Triangulations). *Starting from a triangle, one may obtain any stacked triangulation by repeatedly inserting a new vertex inside a face (including the outer face) of the current triangulation and making it adjacent to all the three vertices contained in the face.*

An example of a stacked triangulation is shown in Fig. 1.

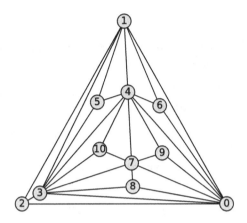

Fig. 1. A (labeled) stacked triangulation on 11 vertices in which every face is incident to a degree-3-vertex.

De Fraysseix, Pach, and Pollack [15] showed that every planar n-vertex graph admits a straight-line embedding on a $(2n - 4) \times (n - 2)$ grid – even if the combinatorial embedding (including the choice of the outer face) is prescribed. Moreover, the graphs are only embedded on a triangular subset of the grid. Hence, $f_p(n) \leq n^2 - O(n)$. This bound was further improved to the currently best known bound $f_p(n) \leq \frac{n^2}{4} - O(n)$ [7] (see also [8,28]). Also various subclasses of planar graphs have been studied intensively: Any stacked triangulation on n vertices (with a fixed outer face) can be drawn on a particular set of $f_s(n) \leq O(n^{3/2} \log n)$ points [19]. For outerplanar graphs, it is known that any set of n points in general position is n-universal [13,24]. For 2-outerplanar graphs and for simply nested graphs an upper bound of $O(n \log n)$ is known [5].

Concerning the lower bound on $f_p(n)$ and $f_s(n)$, respectively, the obvious relation $n \leq f_s(n) \leq f_p(n)$ holds for any $n \in \mathbb{N}$. The first non-trivial lower bound on the size of n-universal sets was also given by de Fraysseix, Pach, and Pollack [15], who showed a lower bound of $f_p(n) \geq n + (1 - o(1))\sqrt{n}$. Chrobak and Karloff [14] further improved the lower bound to $(1.098 - o(1))n$, and the multiplicative constant was later on improved to 1.235 by Kurowski [22]. In fact, Kurowski's lower bound even applies to $f_s(n)$.

Cardinal, Hoffmann, and Kusters [12] showed that n-universal sets of size n ·exist for every $n \leq 10$, whereas for $n \geq 15$ no such set exists – not even for stacked triangulations. Hence $f_p(n) = f_s(n) = n$ for $n \leq 10$ and $f_p(n) \geq f_s(n) > n$ for

$n \geq 15$. Moreover, they found a collection of 7,393 planar graphs on $n = 35$ vertices which cannot be simultaneously drawn straight-line on a common set of 35 points. We call such a collection of graphs a *conflict collection*. This was a first big step towards an answer to the question by Brass and others [9]:

Question 1 ([9]). Is there a conflict collection of size 2?

2   Outline

Our first result is the following theorem, which further improves the lower bound on $f_s(n)$. We present its proof in Sect. 3.

Theorem 1. *It holds that $f_s(n) \geq (\alpha - o(1))n$, where $\alpha = 1.293\ldots$ is the unique real-valued solution of the equation $\alpha^\alpha \cdot (\alpha - 1)^{1-\alpha} = 2$.*

In Sect. 4 we present our second result, which is another step towards a (negative) answer of Question 1 and strengthens the results from [12]. Its proof is based on exhaustive computer search.

Theorem 2 (Computer-assisted). *There is a conflict collection consisting of 49 stacked triangulations on 11 vertices. Furthermore, there is no conflict collection consisting of 36 triangulations on 11 vertices.*

Corollary 1. *There is no 11-universal set of size 11 – even for stacked triangulations. Hence, $f_p(11) \geq f_s(11) \geq 12$.*

Last but not least, since all known proofs for lower bounds make use of separating triangles, we also started the investigation of 4-connected triangulations. In Sect. 5 we present some n-universal sets of size n for 4-connected planar graphs for all $n \leq 17$.

3   Proof of Theorem 1

To prove the theorem, we use a refined counting argument based on a construction of a set of labeled stacked triangulations that was already introduced in [12]. There it was used to disprove the existence of n-universal sets of $n \geq 15$ points for the family of stacked triangulations.

Definition 2 (Labeled Stacked Triangulations, cf. [12, Sect. 3]). *For every integer $n \geq 4$, we define the family $\mathcal{T}_n$ of labeled stacked triangulations on the set of vertices $V_n := \{v_1, ..., v_n\}$ inductively as follows:*

- *$\mathcal{T}_4$ consists only of the complete graph K_4 with labels $v_1, \ldots, v_4$.*
- *If T is a labeled graph in $\mathcal{T}_{n-1}$ with $n \geq 5$, and $v_i v_j v_k$ defines a face of T, then the graph obtained from T by stacking the new vertex v_n to $v_i v_j v_k$ (i.e., connecting it to v_i, v_j, and v_k) is a member of $\mathcal{T}_n$.*

It is important to notice that, when speaking of $\mathcal{T}_n$, we distinguish between elements if they are distinct as *labeled graphs*, even if their underlying graphs are isomorphic. The essential ingredient we will need from [12] is the following.

Lemma 1 (cf. [12, Lemmas 1 and 2])

(i) For any $n \geq 4$, the family $\mathcal{T}_n$ contains exactly $2^{n-4}(n-3)!$ labeled stacked triangulations.

(ii) Let $P_n = \{p_1, \ldots, p_n\}$ be a set of $n \geq 4$ labeled points in the plane. Then for any bijection $\pi : V_n \to P_n$, there is at most one $T \in \mathcal{T}_n$ such that the embedding of T, which maps each vertex v_i to point $\pi(v_i)$, defines a straight-line-embedding of T.

Figure 1 illustrates the idea of item (ii) of Lemma 1.

We need the following simple consequence of the above:

Corollary 2. *Let $P = \{p_1, \ldots, p_m\}$ be a set of $m \geq n \geq 4$ labeled points in the plane. Then for any injection $\pi : V_n \to P$, there is at most one $T \in \mathcal{T}_n$ such that the embedding of T, which maps each vertex v_i to point $\pi(v_i)$, defines a straight-line-embedding of T.*

Proof. Let $T_1, T_2 \in \mathcal{T}_n$ be two stacked triangulations such that π describes a plane straight-line embedding of both. Since π is an injection, this means that π defines a straight-line embedding of both T_1, T_2 on the sub-point set $Q := \pi(V_n)$ of P of size n. Applying Lemma 1(ii) to the bijection $\pi : V_n \to Q$ and T_1, T_2, we deduce $T_1 = T_2$. This proves the claim. $\qquad\square$

We are now ready to prove Theorem 1.

Proof (Proof of Theorem 1). Let $n \geq 4$ be arbitrary and $m := f_s(n) \geq n$. There exists an n-universal point set $P = \{p_1, \ldots, p_m\}$ for all stacked triangulations, hence for every $T \in \mathcal{T}_n$ there exists a straight-line embedding of T on P, with (injective) vertex-mapping $\pi : V_n \to P$. By Corollary 2, we know that no two stacked triangulations from $\mathcal{T}_n$ (each of which has the same vertex set) yield the same injection π. Consequently, by Lemma 1(i), we have

$$2^{n-4}(n-3)! = |\mathcal{T}_n| \leq \frac{m!}{(m-n)!},$$

which means

$$\frac{1}{16n(n-1)(n-2)}2^n \leq \binom{m}{n} = \binom{f_s(n)}{n}.$$

Let $\beta(n) := \frac{f_s(n)}{n}$. Using the fact that (Stirling-approximation)

$$\binom{f_s(n)}{n} \sim \underbrace{\sqrt{\frac{f_s(n)}{2\pi n(f_s(n)-n)}}}_{\leq 1} \frac{f_s(n)^{f_s(n)}}{n^n(f_s(n)-n)^{f_s(n)-n}} \leq \left(\frac{\beta(n)^{\beta(n)}}{(\beta(n)-1)^{\beta(n)-1}}\right)^n,$$

we deduce (taking logarithms) that:

$$(1 - o(1))n \leq \log_2 \left(\frac{\beta(n)^{\beta(n)}}{(\beta(n) - 1)^{\beta(n)-1}} \right) n \iff 2 - o(1) \leq \frac{\beta(n)^{\beta(n)}}{(\beta(n) - 1)^{\beta(n)-1}}.$$

Consequently, $\beta(n) \geq (1 - o(1))\alpha$, where α is the unique solution to $\frac{\alpha^\alpha}{(\alpha-1)^{\alpha-1}} = 2$. This proves $f_s(n) = n \cdot \beta(n) \geq (1 - o(1))\alpha n$, which is the claim.   □

4   Proof of Theorem 2 and Corollary 1

In the following, we outline the strategy which we have used to find a conflict collection of 49 stacked 11-vertex triangulations. We refer the reader who is mainly interested in verifying our computational results directly to Sect. 4.5.

One fundamental observation is the following: if an n-universal point set has collinear points, then by perturbation one can obtain another n-universal point set *in general position*, i.e., with no collinear points. Hence, in the following we only consider point sets in general position. Also it is not hard too see that, if two point sets are *combinatorially equivalent*, i.e., there is a bijection such that the corresponding triples of points induce the same orientations, then both sets allow precisely the same straight-line drawings. Hence, in the following we further restrict our considerations to *(non-degenerated) order types*, i.e., the set of equivalence classes of point sets (in general position).

4.1   Enumeration of Order Types

The database of all order types of up to $n = 11$ points was developed by Aurenhammer, Aichholzer, and Krasser [3,4] (see also Krasser's dissertation [21]). The file for all order types of up to $n = 10$ points (each represented by a point set) is available online, while the file for $n = 11$ requires almost 100 GB of storage and is available on demand [2]. Their algorithm starts with an *abstract order type* on $k - 1$ points (which only encodes the triple orientations of a point set), computes its dual pseudoline arrangement, and inserts a k-th pseudoline in all possible ways. Due to geometrical constraints, there are in fact abstract order types enumerated which do not have a realization as a point set. However, since every order type is in fact also an abstract order type, it is sufficient for our purposes to test all abstract order types – independent from realizability.

For means of redundancy and to provide a fully checkable and autonomous proof, we have implemented an alternative algorithm to enumerate all abstract order types based on the following idea: Given a set of points $s_1, \ldots, s_n$ with $s_i = (x_i, y_i)$ sorted left to right[1], and let

$$\chi_{ijk} := \operatorname{sgn} \det \begin{pmatrix} 1 & 1 & 1 \\ x_i & x_j & x_k \\ y_i & y_j & y_k \end{pmatrix} \in \{-1, 0, +1\}$$

---

[1] In the dual line arrangement the lines are sorted by increasing slope.

denote the induced triple orientations, then the *signotope axioms* assert that, for every 4-tuple s_i, s_j, s_k, s_l with $i < j < k < l$, the sequence

$$\chi_{ijk}, \ \chi_{ijl}, \ \chi_{ikl}, \ \chi_{jkl}$$

(index-triples in lexicographic order) changes its sign at most once. For more information on the signotope axioms we refer to Felsner and Weil [18] (see also [6]).

Given an abstract order type on $k - 1$ points, we insert a k-th point in all possible ways, such that the signotope axioms are preserved. With our C++ implementation, we managed to verify the numbers of abstract order types from [3,4,21]. In fact, the enumeration of all 2,343,203,071 abstract order types of up to $n = 11$ points (cf. OEIS/A6247) can be done within about 20 CPU hours.

4.2   Enumeration of Planar Graphs

To enumerate all non-isomorphic maximal planar graphs on 11 vertices (i.e, triangulations), we have used the plantri graph generator (version 4.5) [10]. It is worth to note that also the nauty graph generator [23] can be used for the enumeration because the number of all (not necessarily planar) graphs on 11 vertices is not too large and the database can be filtered for planar graphs in reasonable time – negligible compared to the CPU time which we have used for later computations. For various computations on graphs, such as filtering stacked triangulations or to produce graphs for this paper, we have used SageMath [29][2].

4.3   Deciding Universality Using a SAT Solver

For a given point set S and a planar graph $G = (V, E)$ we model a propositional formula in conjunctive normal form (CNF) which has a solution if and only if G can be embedded on S – in fact, the variables encode a straight-line drawing.[3]

To model the CNF, we have used the variables M_{vp} to describe whether vertex v is mapped to point p, and the variables A_{pq} to describe whether the straight-line segment pq between the two points p and q is "active" in a drawing.

It is not hard to use a CNF to assert that such a vertex-to-point mapping is bijective. Also it is easy to assert that, if two adjacent vertices u and v are mapped to points p and q, then the straight-line segment pq is active. For each pair of crossing straight-line segments pq and rs (dependent on the order type of the point set) at least one of the two segments is not allowed to be active.

Implementation Detail: We have implemented a C++ routine which, given a point set and a graph as input, creates an instance of the above described model and then uses the solver MiniSat 2.2.0 [16] to decide whether the graph admits a straight-line embedding.

---

[2] We recommend the Sage Reference Manual on Graph Theory [30] and its collection of excellent examples.

[3] Cabello [11] showed that deciding embeddability is NP-complete in general. His reduction, however, constructs a 2-connected graph, and therefore the hardness remains unknown for 3-connected planar graphs.

4.4   Finding Conflict Collections – A Quantitive Approach

Before we actually tested whether a set of 11 points is 11-universal or not, we discovered a few necessary criteria for the point set, which can be checked much more efficiently. These considerations allowed a significant reduction of the total computation times.

Phase 1: There are various properties that a universal point set has to fulfill:

   Property 1: The planar graph depicted in Fig. 2 asserts an 11-universal set S – if one exists – to have a certain structure. If the embedding is as on the left of Fig. 2, then one of the two degree 3 vertices is drawn as extremal point of S, i.e., lies on the boundary of the convex hull $\text{conv}(S)$ of S. After the removal of this particular point, the remaining 10 points have 4 convex layers of sizes 3, 3, 3, and 1, respectively. If the embedding is as on the right of Fig. 2, then either one or two points of the blue triangle are drawn as extremal points of S (recall the triangular convex hull of S). And again, the points inside the blue triangle and outside the blue triangle have convex layers of sizes 3, 3, 1, and 3, 1, respectively.

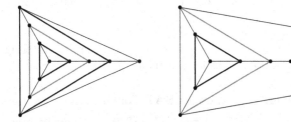

Fig. 2. The two embeddings of a graph, which forces the point set to have a certain structure. Each of the vertices of the blue triangle connects to one of the vertices of the two copies of K_4.

   Property 2: There exist a stacked triangulations on 11 points in which every face is incident to a degree-3-vertex; see for example Fig. 1. Independent from the embedding of this graph, there is a degree-3-vertex on the outer face, and hence all inner points lie inside a triangle spanned by an interior point and two extremal points. In particular, such a point set must have a triangular convex hull.

   Altogether, only 262,386,428 of the 2,343,203,071 abstract order types on 11 points fulfill Properties 1 and 2. (The computation time was about 10 CPU hours.)

Phase 2: For each of the 262,386,428 abstract order types on 11 points which fulfill the conditions above, we have tested the embeddability of all maximal planar graphs on n vertices separately using a SAT-solver based approach. In fact, as soon as one graph was not embeddable, the remaining graphs needed not to be checked. To speed up the computations we have used a priority queue:

a graph which does not admit an embedding gets increased priority for other point sets to be tested first.

To keep the conflict collection as small as possible, we first filtered out all point sets which do not allow a simultaneous embedding of all planar graphs on 11 vertices with maximum degree 10. Only 287,871 of the 262,386,428 abstract order types remained (about 100 CPU days). It is worth to note that there are 82 maximal planar graphs on 11 vertices with maximum degree 10 (cf. OEIS/A207), and that each of these graphs is a stacked triangulation.

At this point one can check with only a few CPU hours that the remaining 287,871 abstract order types are not 11-universal. Moreover, since some stacked triangulations on 11 vertices (e.g., G_{12} in [25]) contain the graph from Fig. 2 as a subgraph, the statement even applies to stacked triangulations. Consequently, the family of all 434 stacked triangulations on 11 vertices (cf. OEIS/A27610) is a conflict collection, and Corollary 1 follows directly.

Phase 3: To find a smaller conflict collection, we tested for each of the 434 stacked triangulations and each of the 287,871 remaining abstract order types, whether an embedding is possible (additional 35 CPU days). We used this binary information to formulate an integer program searching for a minimal set of triangulations without simultaneous embedding. Using the Gurobi solver (version 8.0.0) [20], we managed to find a collection $\mathcal{G}$ of 27 stacked triangulations which cannot be embedded simultaneously; see [25].

Since we asserted in Phases 1 and 2 that

(1) the graph in Fig. 2,
(2) a triangulation where every face is incident to a vertex of degree 3, and
(3) all 82 triangulations with maximum degree 10.

occur in the conflict collection, this yields a conflict collection of size 111 = $1 + 1 + 82 + 27$. In fact, since this subset of 27 stacked triangulations contains triangulations fulfilling properties (1) and (2) (see, e.g., graphs G_{12} und G_{10} in [25]), we indeed have a conflict collection of size 109.

We have also ran the computations for the collection of all 1,249 triangulations (cf. OEIS/A109), and the Gurobi solver showed that any conflict collection of (arbitrary) 11-vertex triangulations has size at least 26.

Phase 4: Recall that a minimal conflict collection not necessarily needs to fulfill the properties (1)–(3). Hence we again repeat the strategy from Phase 2, except that we test for the embeddability of the 27 stacked triangulations from the collection $\mathcal{G}$ obtained in Phase 3 instead of the 82 maximal planar graphs on 11 vertices with maximum degree 10.

After another 230 days of CPU time, our program had filtered out 2,194 of the 262,386,428 abstract order types (obtained in Phase 1) which allow a simultaneous embedding of the 27 stacked triangulations from $\mathcal{G}$.

Phase 5: As the reader might already guess, we proceed as in Phase 3: we tested for each of the 434 stacked triangulations and each of the 2,194 order types from Phase 4, whether an embedding is possible (only 2 CPU days). Using the Gurobi solver, we managed to find a collection $\mathcal{H}$ of 22 stacked triangulations, which cannot be simultaneously embedded on those order types; see [25].

Together with the 27 stacked triangulations from $\mathcal{G}$ we obtain a conflict collection of size 49, and the first part of Theorem 2 follows.

Phase 6: To further improve the lower bound, we have repeated our computations for the union of the two sets of point sets obtained in Phase 3 and Phase 5, respectively. Using Gurobi, we obtained that

- any conflict collection of stacked triangulations must have size at least 40, and
- any conflict collection of (arbitrary) triangulations must have size at least 37.

For means of redundancy, we have verified all obtained lower bounds also using CPLEX (version 12.8.0.0) [1], which performed similar to Gurobi.

This completes the proof of the second part of Theorem 2.

4.5   How to Verify Our Results?

To verify the computational results which are essential for the proof of the first part of Theorem 2, one can enumerate all order types on 11 points and test the conflict collection of 49 triangulations (`data/triangulations/n11_conflicting49.txt`). Starting with the unique order type on 3 points (`data/order_types/n3_order_types.bin`), it takes about 1 CPU day to enumerate all order types on 11 points. By falsifying simultaneous embeddability of the 49 graphs (about 200 CPU days, but can be run parallelized), the first part of Theorem 2 is then verified.

For the second part of the Theorem 2, one can filter the order types, which allow a simultaneous embedding of the triangulations from Phase 2 and 4, and then – using CPLEX or Gurobi – compute the minimum size of a conflicting collection among all 11-vertex triangulations and 11-vertex stacked triangulations, respectively. To save some computation time, we provide the filtered list in `data/triangulations/n11_after_phase2.bin.zip` and `n11_after_phase4.bin.zip`. The list of all (stacked) triangulations is provided in `data/triangulations/n11_all_triangulations.txt` and `n11_all_stacked_triangulations.txt`.

A more detailed description is provided in the full version [27]. The source codes of our programs and relevant data are available on the companion website [25].

5   Universal Sets for 4-Connected Graphs

For $n \leq 10$, examples of n-universal sets of n points for planar n-vertex graphs were already given in [12]. To provide n-universal sets for 4-connected planar graphs for $n = 11, \ldots, 17$, we slightly adapted our framework. Again, we enumerated 4-connected planar triangulations using the plantri graph generator, and using our C++ implementation, tested for universality. Our idea to find the proposed point sets for $n = 11, \ldots, 17$ was to start with an $(n - 1)$-universal set of $n - 1$ points and insert an n-th point in all possible ways (cf. Sect. 4.1). The abstract order types obtained in this way – if they turned out to be universal – were then realized as point sets using our framework *pyotlib*[4]. The obtained sets are given in Listing 1.1.

```
[(612,666),(754,635),(415,709),(884,597),(596,695),(890,977),
 (384,716),(834,609),(424,707),(974,10),(890,962),(306,805),
 (301,810),(4,736),(0,735),(975,6),(980,0)]
```

Listing 1.1. A set $\{p_1, \ldots, p_{17}\}$ of 17 points such that $\{p_1, \ldots, p_k\}$ is universal for 4-connected planar k-vertex graphs for all $k \in \{11, \ldots, 17\}$.

It is also worth to note that the numbers of 4-connected triangulations for $n = 11, \ldots, 20$ are 25; 87; 313; 1,357; 6,244; 30,926; 158,428; 836,749; 4,504,607; 24,649,284 (cf. OEIS/A7021). Hence, even if a universal point set is known, it is getting more and more time consuming to verify n-universality as n gets larger (at least using our SAT solver approach).

6   Discussion

In Sect. 3, we provided an improved lower bound for $f_p(n)$ and $f_s(n)$. However, the best known general upper bounds remain far from linear.

In Sect. 4, we have applied the ideas from Phases 2 and 3 twice (cf. Phases 4 and 5) to reduce the size of a conflict collection. One could further proceed with this strategy to find even smaller conflict collections (if such exist). Also one could simply test whether all elements from the conflict collection are indeed necessary, or whether certain elements can be removed. Note that, to compute a minimal conflict collection for $n = 11$, one could theoretically check which graphs admit an embedding on which point set and then find a minimal set cover as described in Phase 3 (Sect. 4). In practice, however, formulating such a minimal set cover instance (as integer program) is not reasonable because testing the embeddability of every graph in every point set would be an extremely time consuming task. (Recall that we used a priority queue to speed up our computation, so only a few pairs were actually tested. Also recall that, to generate the set cover instances, we only looked at a comparably small number of order types.) And even if such an instance was formulated, due to its size, the IP/set cover might not be solvable optimally in reasonable time.

---

[4] The "**python order type library**" was initiated during the Bachelor's studies of the first author [26] and provides many features to work with (abstract) order types such as local search techniques, realization or proving non-realizability of abstract order types, coordinate minimization and "beautification" for nicer visualizations. For more information, please consult the author.

Besides the computations for $n = 11$ points, we also adapted our program to find all n-universal order types on n points for every $n \leq 10$, and hence could verify the results from [12, Table 1]. To be precise, we found 5,956 9-universal abstract order types on $n = 9$ points, whereas only 5,955 are realizable as point sets. It is worth to note that there is exactly one non-realizable abstract order type on 9 points in the projective plane, which is dual to the simple non-Pappus arrangement, and that all abstract order types on $n \leq 8$ points are realizable. Besides the already known 2,072 realizable order types on 10 points, no further non-realizable 10-universal abstract order types were found. For more details on realizability see for example [21] or [17].

Unfortunately, we do not have an argument for subsets or supersets of n-universal point sets, and thus the question for $n = 12, 13, 14$ remains open. However, based on computational evidence (see also [12, Table 1]), we strongly conjecture that no n-universal set of n points exists for $n \geq 11$. It is also worth to note that 11-universal sets of 12 points exist (cf. Listing 1.2).

```
[(214,0),(0,13),(2,16),(9,26),(124,12),(133,11),
 (148,9),(213,1),(211,4),(210,6),(116,179),(122,197)]
```

Listing 1.2. An 11-universal set of 12 points.

As mentioned in the introduction of this paper, various graph classes have been studied for this problem. Even though our contribution on 4-connected planar graphs in Sect. 5 is rather small, it gives some evidence that comparably less points are needed to embed 4-connected planar graphs. In fact, we would not be surprised if n-universal sets of n points exist for 4-connected planar graphs.

References

1. IBM ILOG CPLEX Optimization Studio (2018). http://www.ibm.com/products/ilog-cplex-optimization-studio/
2. Aichholzer, O.: Enumerating Order Types for Small Point Sets with Applications. http://www.ist.tugraz.at/aichholzer/research/rp/triangulations/ordertypes/
3. Aichholzer, O., Aurenhammer, F., Krasser, H.: Enumerating order types for small point sets with applications. Order **19**(3), 265–281 (2002). https://doi.org/10.1023/A:1021231927255
4. Aichholzer, O., Krasser, H.: Abstract order type extension and new results on the rectilinear crossing number. Comput. Geom.: Theory Appl. **36**(1), 2–15 (2006). https://doi.org/10.1016/j.comgeo.2005.07.005
5. Angelini, P., Bruckdorfer, T., Di Battista, G., Kaufmann, M., Mchedlidze, T., Roselli, V., Squarcella, C.: Small universal point sets for k-outerplanar graphs. Discret. Comput. Geom. 1–41 (2018). https://doi.org/10.1007/s00454-018-0009-x
6. Balko, M., Fulek, R., Kynčl, J.: Crossing numbers and combinatorial characterization of monotone drawings of K_n. Discret. Comput. Geom. **53**(1), 107–143 (2015). https://doi.org/10.1007/s00454-014-9644-z
7. Bannister, M.J., Cheng, Z., Devanny, W.E., Eppstein, D.: Superpatterns and universal point sets. J. Graph Algorithms Appl. **18**(2), 177–209 (2014). https://doi.org/10.7155/jgaa.00318

8. Brandenburg, F.J.: Drawing planar graphs on $\frac{8}{9}n^2$ area. Electron. Notes Discret. Math. **31**, 37–40 (2008). https://doi.org/10.1016/j.endm.2008.06.005

9. Brass, P., Cenek, E., Duncan, C.A., Efrat, A., Erten, C., Ismailescu, D.P., Kobourov, S.G., Lubiw, A., Mitchell, J.S.: On simultaneous planar graph embeddings. Comput. Geom. **36**(2), 117–130 (2007). https://doi.org/10.1016/j.comgeo.2006.05.006

10. Brinkmann, G., McKay, B.D.: Fast generation of some classes of planar graphs. Electron. Notes Discret. Math. **3**, 28–31 (1999). https://doi.org/10.1016/S1571-0653(05)80016-2

11. Cabello, S.: Planar embeddability of the vertices of a graph using a fixed point set is NP-hard. J. Graph Algorithms Appl. **10**(2), 353–363 (2006). https://doi.org/10.7155/jgaa.00132

12. Cardinal, J., Hoffmann, M., Kusters, V.: On universal point sets for planar graphs. J. Graph Algorithms Appl. **19**(1), 529–547 (2015). https://doi.org/10.7155/jgaa.00374

13. Castañeda, N., Urrutia, J.: Straight line embeddings of planar graphs on point sets. In: Proceedings of the 8th Canadian Conference on Computational Geometry (CCCG 1996), pp. 312–318 (1996). http://www.cccg.ca/proceedings/1996/cccg1996_0052.pdf

14. Chrobak, M., Karloff, H.J.: A lower bound on the size of universal sets for planar graphs. ACM SIGACT News **20**(4), 83–86 (1989). https://doi.org/10.1145/74074.74088

15. De Fraysseix, H., Pach, J., Pollack, R.: How to draw a planar graph on a grid. Combinatorica **10**(1), 41–51 (1990). https://doi.org/10.1007/BF02122694

16. Eén, N., Sörensson, N.: An extensible SAT-solver. In: Giunchiglia, E., Tacchella, A. (eds.) SAT 2003. LNCS, vol. 2919, pp. 502–518. Springer, Heidelberg (2004). https://doi.org/10.1007/978-3-540-24605-3_37

17. Felsner, S., Goodman, J.E.: Pseudoline arrangements. In: Toth, C.D., O'Rourke, J., Goodman, J.E. (eds.) Handbook of Discrete and Computational Geometry, 3rd edn. CRC Press (2018). https://doi.org/10.1201/9781315119601

18. Felsner, S., Weil, H.: Sweeps, arrangements and signotopes. Discret. Appl. Math. **109**(1), 67–94 (2001). https://doi.org/10.1016/S0166-218X(00)00232-8

19. Fulek, R., Tóth, C.D.: Universal point sets for planar three-trees. J. Discret. Algorithms **30**, 101–112 (2015). https://doi.org/10.1016/j.jda.2014.12.005

20. Gurobi Optimization, LLC: Gurobi Optimizer (2018). http://www.gurobi.com

21. Krasser, H.: Order Types of Point Sets in the Plane. Ph.D. thesis, Institute for Theoretical Computer Science, Graz University of Technology, Austria (2003)

22. Kurowski, M.: A $1.235n$ lower bound on the number of points needed to draw all n-vertex planar graphs. Inf. Process. Lett. **92**(2), 95–98 (2004). https://doi.org/10.1016/j.ipl.2004.06.009

23. McKay, B.D., Piperno, A.: Practical graph isomorphism, II. J. Symb. Comput. **60**, 94–112 (2014). https://doi.org/10.1016/j.jsc.2013.09.003

24. Pach, J., Gritzmann, P., Mohar, B., Pollack, R.: Embedding a planar triangulation with vertices at specified points. Am. Math. Mon. **98**, 165–166 (1991). https://doi.org/10.2307/2323956

25. Scheucher, M.: Webpage: Source Codes and Data for Universal Point Sets. http://page.math.tu-berlin.de/~scheuch/supplemental/universal_sets

26. Scheucher, M.: On Order Types, Projective Classes, and Realizations. Bachelor's thesis, Graz University of Technology, Austria (2014). http://www.math.tu-berlin.de/~scheuch/publ/bachelors_thesis_tm_2014.pdf

27. Scheucher, M., Schrezenmaier, H., Steiner, R.: A Note On Universal Point Sets for Planar Graphs. arXiv:1811.06482 (2018)
28. Schnyder, W.: Embedding planar graphs on the grid. In: Proceedings of the First Annual ACM-SIAM Symposium on Discrete Algorithms, pp. 138–148. Society for Industrial and Applied Mathematics (1990)
29. Stein, W.A., et al.: Sage Mathematics Software (Version 8.1). The Sage Development Team (2018). http://www.sagemath.org
30. Stein, W.A., et al.: Sage Reference Manual: Graph Theory (Release 8.1) (2018). http://doc.sagemath.org/pdf/en/reference/number_fields/number_fields.pdf

Parameterized Complexity

Parameterized Algorithms for Book Embedding Problems

Sujoy Bhore[1] , Robert Ganian[1] , Fabrizio Montecchiani[2] ,
and Martin Nöllenburg[1]([✉])

[1] Algorithms and Complexity Group, TU Wien, Vienna, Austria
{sujoy,rganian,noellenburg}@ac.tuwien.ac.at
[2] Engineering Department, University of Perugia, Perugia, Italy
fabrizio.montecchiani@unipg.it

Abstract. A *k-page book embedding* of a graph G draws the vertices of G on a line and the edges on k half-planes (called *pages*) bounded by this line, such that no two edges on the same page cross. We study the problem of determining whether G admits a k-page book embedding both when the linear order of the vertices is fixed, called FIXED-ORDER BOOK THICKNESS, or not fixed, called BOOK THICKNESS. Both problems are known to be NP-complete in general. We show that FIXED-ORDER BOOK THICKNESS and BOOK THICKNESS are fixed-parameter tractable parameterized by the vertex cover number of the graph and that FIXED-ORDER BOOK THICKNESS is fixed-parameter tractable parameterized by the pathwidth of the vertex order.

1 Introduction

A *k-page book embedding* of a graph G is a drawing that maps the vertices of G to distinct points on a line, called *spine*, and each edge to a simple curve drawn inside one of k half-planes bounded by the spine, called *pages*, such that no two edges on the same page cross [21,26]; see Fig. 1 for an illustration. This kind of layout can be alternatively defined in combinatorial terms as follows. A k-page book embedding of G is a linear order $\prec$ of its vertices and a coloring of its edges which guarantee that no two edges uv, wx of the same color have their vertices ordered as $u \prec w \prec v \prec x$. The minimum k such that G admits a k-page book embedding is the *book thickness* of G, denoted by $bt(G)$, also known as the *stack number* of G. Book embeddings have been extensively studied in the literature, among others due to their applications in bioinformatics, VLSI, and parallel computing (see, e.g., [8,20] and refer also to [12] for a survey). A famous result by Yannakakis [30] states that every planar graph has book thickness at

Research of Fabrizio Montecchiani supported in part by MIUR under Grant 20174LF3T8 AHeAD: efficient Algorithms for HArnessing networked Data. Robert Ganian acknowledges support by the FWF (Project P 31336, "NFPC") and is also affiliated with FI MUNI, Brno, Czech Republic. Research of Sujoy Bhore and Martin Nöllenburg is supported by the Austrian Science Fund (FWF) grant P 31119.

© Springer Nature Switzerland AG 2019
D. Archambault and C. D. Tóth (Eds.): GD 2019, LNCS 11904, pp. 365–378, 2019.
https://doi.org/10.1007/978-3-030-35802-0_28

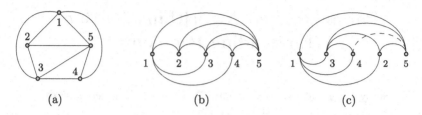

Fig. 1. (a) A planar graph G with book thickness two. (b) A 2-page book embedding of G. (c) A linear order of G such that its fixed-order book thickness is three (and the corresponding 3-page book embedding).

most four. Several other bounds are known for special graph families, for instance planar graphs with vertex degree at most four have book thickness two [3], while graphs of treewidth $w > 2$ have book thickness $w + 1$ [13,18].

Given a graph G and a positive integer k, the problem of determining whether $\mathrm{bt}(G) \leq k$, called BOOK THICKNESS, is known to be NP-complete. Namely, Bernhart and Kainen [4] proved that $\mathrm{bt}(G) \leq 2$ if and only if G is subhamiltonian, i.e., G is a subgraph of a planar Hamiltonian graph. Since deciding whether a graph is subhamiltonian is an NP-complete problem, BOOK THICKNESS is also NP-complete in general [8]. BOOK THICKNESS has been studied also when the linear order $\prec$ of the vertices is fixed, indeed, this is one of the original formulations of the problem, which arises in the context of sorting with parallel stacks [8]. We call this problem FIXED-ORDER BOOK THICKNESS and we denote by fo-bt$(G, \prec)$ the *fixed-order book thickness* of a graph G. Obviously, we have fo-bt$(G, \prec) \geq \mathrm{bt}(G)$, see Fig. 1. Deciding whether fo-bt$(G, \prec) \leq 2$ corresponds to testing the bipartiteness of a suitable conflict graph, and thus it can be solved in linear time. On the other hand, deciding if fo-bt$(G, \prec) \leq 4$ is equivalent to finding a 4-coloring of a circle graph and hence is an NP-complete problem [29].

Our Results. In this paper we study the parameterized complexity of BOOK THICKNESS and FIXED-ORDER BOOK THICKNESS. For both problems, when the answer is positive, we naturally also expect to be able to compute a corresponding k-page book embedding as a witness. While both problems are NP-complete already for small fixed values of k on general graphs, it is natural to ask which structural properties of the input (formalized in terms of structural parameters) allow us to solve these problems efficiently. Indeed, already Dujmovic and Wood [14] asked whether BOOK THICKNESS can be solved in polynomial time when the input graph has bounded treewidth [28]—a question which has turned out to be surprisingly resilient to existing algorithmic techniques and remains open to this day. Bannister and Eppstein [2] made partial progress towards answering Dujmovic and Wood's question by showing that BOOK THICKNESS is fixed-parameter tractable parameterized by the treewidth of G when $k = 2$.

We provide the first fixed-parameter algorithms for FIXED-ORDER BOOK THICKNESS and also the first such algorithm for BOOK THICKNESS that can be used when $k > 2$. In particular, we provide fixed-parameter algorithms for:

1. FIXED-ORDER BOOK THICKNESS parameterized by the vertex cover number of the graph;
2. FIXED-ORDER BOOK THICKNESS parameterized by the pathwidth of the graph and the vertex order; and
3. BOOK THICKNESS parameterized by the vertex cover number of the graph.

Results 1 and 2 are obtained by combining dynamic programming techniques with insights about the structure of an optimal book embedding. Result 3 then applies a kernelization technique to obtain an equivalent instance of bounded size (which can then be solved, e.g., by brute force). All three of our algorithms can also output a corresponding k-page book embedding as a witness (if it exists).

The remainder of this paper is organized as follows. Section 2 contains preliminaries and basic definitions. Results 1 and 2 on FIXED-ORDER BOOK THICKNESS are presented in Sect. 3, while Result 3 on BOOK THICKNESS is described in Sect. 4. Conclusions and open problems are found in Sect. 5. Some proofs are omitted due to space constraints; they are included in the full version [5].

2 Preliminaries

We use standard terminology from graph theory [10]. For $r \in \mathbb{N}$, we write $[r]$ as shorthand for the set $\{1, \ldots, r\}$. Parameterized complexity [9,11] focuses on the study of problem complexity not only with respect to the input size n but also a parameter $k \in \mathbb{N}$. The most desirable complexity class in this setting is FPT (*fixed-parameter tractable*), which contains all problems that can be solved by an algorithm running in time $f(k) \cdot n^{\mathcal{O}(1)}$, where f is a computable function. Algorithms running in this time are called *fixed-parameter algorithms*.

A k-page book embedding of a graph $G = (V, E)$ will be denoted by a pair $\langle \prec, \sigma \rangle$, where $\prec$ is a linear order of V, and $\sigma \colon E \to [k]$ is a function that maps each edge of E to one of k pages $[k] = \{1, 2, \ldots, k\}$. In a k-page book embedding $\langle \prec, \sigma \rangle$ it is required that for no pair of edges $uv, wx \in E$ with $\sigma(uv) = \sigma(wx)$ the vertices are ordered as $u \prec w \prec v \prec x$, i.e., each page is crossing-free.

We consider two graph parameters for our algorithms. A *vertex cover* C of a graph $G = (V, E)$ is a subset $C \subseteq V$ such that each edge in E has at least one end-vertex in C. The *vertex cover number* of G, denoted by $\tau(G)$, is the size of a minimum vertex cover of G. The second parameter is *pathwidth*, a classical graph parameter [27] which admits several equivalent definitions. The definition that will be most useful here is the one tied to linear orders [22]; see also [23,24] for recent works using this formulation. Given an n-vertex graph $G = (V, E)$ with a linear order $\prec$ of V such that $v_1 \prec v_2 \prec \cdots \prec v_n$, the *pathwidth* of $(G, \prec)$ is the minimum number κ such that for each vertex v_i ($i \in [n]$), there are at most κ vertices left of v_i that are adjacent to v_i or a vertex right of v_i. Formally, for each v_i we call the set $P_i = \{v_j \mid j < i, \exists q \geq i \text{ such that } v_j v_q \in E\}$ the *guard set* for v_i, and the pathwidth of $(G, \prec)$ is simply $\max_{i \in [n]} |P_i|$. The elements of the guard sets are called the *guards* (for v_i). We remark that the pathwidth of G is equal to the minimum pathwidth over all linear orders $\prec$.

3   Algorithms for Fixed-Order Book Thickness

Recall that in FIXED-ORDER BOOK THICKNESS the input consists of a graph $G = (V, E)$, a linear order $\prec$ of V, and a positive integer k. We assume that $V = \{v_1, v_2, \ldots, v_n\}$ is indexed such that $i < j \Leftrightarrow v_i \prec v_j$. The task is to decide if there is a page assignment $\sigma \colon E \rightarrow [k]$ such that $\langle \prec, \sigma \rangle$ is a k-page book embedding of G, i.e., whether fo-bt$(G, \prec) \leq k$. If the answer is 'YES' we shall return a corresponding k-page book embedding as a witness. In fact, our algorithms will return a book embedding with the minimum number of pages.

3.1   Parameterization by the Vertex Cover Number

As our first result, we will show that FIXED-ORDER BOOK THICKNESS is fixed-parameter tractable when parameterized by the *vertex cover number*. We note that the vertex cover number is a graph parameter which, while restricting the structure of the graph in a fairly strong way, has been used to obtain fixed-parameter algorithms for numerous difficult problems [1,15,16].

Let C be a minimum vertex cover of size $\tau = \tau(G)$; we remark that such a vertex cover C can be computed in time $\mathcal{O}(2^\tau + \tau \cdot n)$ [7]. Moreover, let $U = V \setminus C$. Our first observation shows that the problem becomes trivial if $\tau \leq k$.

Observation 1. *Every n-vertex graph G with a vertex cover C of size k admits a k-page book embedding with any vertex order $\prec$. Moreover, if G and C are given as input, such a book embedding can be computed in $\mathcal{O}(n + k \cdot n)$ time.*

Proof. Let $C = \{c_1, \ldots, c_k\}$ be a vertex cover of size k and let σ be a page assignment on k pages defined as follows. For each $i \in [k]$ all edges uc_i with $u \in U \cup \{c_1, \ldots, c_{i-1}\}$ are assigned to page i. Now, consider the edges assigned to any page $i \in [k]$. By construction, they are all incident to vertex c_i, and thus no two of them cross each other. Therefore, the pair $\langle \prec, \sigma \rangle$ is a k-page book embedding of G and can be computed in $\mathcal{O}(n + k \cdot n)$ time.   □

We note that the bound given in Observation 1 is tight, since it is known that complete bipartite graphs with bipartitions of size k and $h > k(k-1)$ have book thickness k [4] and vertex cover number k.

We now proceed to a description of our algorithm. For ease of presentation, we will add to G an additional vertex of degree 0, add it to U, and place it at the end of $\prec$ (observe that this does not change the solution to the instance).

If $\tau \leq k$ then we are done by Observation 1. Otherwise, let S be the set of all possible non-crossing page assignments of the edges whose both endpoints lie in C, and note that $|S| < \tau^{\tau^2}$ and S can be constructed in time $\mathcal{O}(\tau^{\tau^2})$ (recall that $k < \tau$ by assumption). As its first step, the algorithm branches over each choice of $s \in S$, where no pair of edges assigned to the same page crosses.

For each such non-crossing assignment s, the algorithm performs a dynamic programming procedure that runs on the vertices of the input graph in sequential (left-to-right) order. We will define a record set that the algorithm is going to compute for each individual vertex in left-to-right order. Let $c_1 \prec \ldots \prec c_\tau$ be

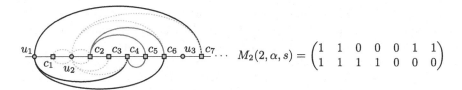

Fig. 2. A partial 2-page book embedding of a graph G with a vertex cover C of size 7. The visibilities of vertices in C (squares) from u_2 are marked by dashed edges (left). Corresponding visibility matrix $M_2(2, \alpha, s)$ (right).

the ordering of vertices of C, and let $u_1 \prec \ldots \prec u_{n-\tau}$ be the ordering of vertices of U.

In order to formalize our records, we need the notion of visibility. Let $i \in [n - \tau]$ and let $E_i = \{u_j c \in E \mid j < i, c \in C\}$ be the set of all edges with one endpoint outside of C that lies to the left of u_i. We call $\alpha \colon E_i \to [k]$ a *valid partial page assignment* if $\alpha \cup s$ maps edges to pages in a non-crossing fashion. Now, consider a valid partial page assignment $\alpha \colon E_i \to [k]$. We say that a vertex $c \in C$ is (α, s)-*visible* to u_t (for $t \in [n - \tau]$) on page p if it is possible to draw an edge from u_t to c on page p without crossing any other edge mapped to page p by $\alpha \cup s$. Figure 2 shows the visibilities of a vertex in two pages.

Based on this notion of visibility, for an index $a \in [n - \tau]$ we can define a $k \times \tau$ *visibility matrix* $M_i(a, \alpha, s)$, where an entry (p, b) of $M_i(a, \alpha, s)$ is 1 if c_b is (α, s)-visible to u_a on page p and 0 otherwise (see Fig. 2). Intuitively, this visibility matrix captures information about the reachability via crossing-free edges (i.e., *visibility*) to the vertices in C from u_a on individual pages given a particular assignment α of edges in E_i. Note that for a given tuple (i, a, α, s), it is straightforward to compute $M_i(a, \alpha, s)$ in polynomial time.

Observe that while the number of possible choices of valid partial page assignments $\alpha \colon E_i \to [k]$ (for some $i \in [n - \tau]$) is not bounded by a function of τ, for each $i, a \in [n - \tau]$ the number of possible visibility matrices is upper-bounded by 2^{τ^2}. On a high level, the core idea in the algorithm is to dynamically process the vertices in U in a left-to-right fashion and compute, for each such vertex, a bounded-size "snapshot" of its visibility matrices—whereas for each such snapshot we will store only one (arbitrarily chosen) valid partial page assignment. We will later (in Lemma 1) show that all valid partial page assignments leading to the same visibility matrices are "interchangeable".

With this basic intuition, we can proceed to formally defining our records. Let $X = \{x \in [n - \tau] \mid \exists c \in C : u_x$ is the immediate successor of c in $\prec\}$ be the set of indices of vertices in U which occur immediately after a cover vertex; we will denote the integers in X as $x_1, \ldots, x_z$ (in ascending order), and we note that $z \leq \tau$. For a vertex $u_i \in U$, we define our record set as follows: $\mathcal{R}_i(s) = \{\big(M_i(i, \alpha, s), M_i(x_1, \alpha, s), M_i(x_2, \alpha, s), \ldots, M_i(x_z, \alpha, s)\big) \mid \exists$ valid partial page assignment $\alpha \colon E_i \to [k]\}$. Note that each entry in $\mathcal{R}_i(s)$ captures one possible set (a "snapshot") of at most $\tau + 1$ visibility matrices: the visibility matrix for u_i itself, and the visibility matrices for the z non-cover vertices which follow immediately after the vertices in C. The intuition behind these latter visibility matrices is that they allow us to update our visibility matrix when our left-to-right dynamic programming algorithm reaches a vertex in C (in particular, as we will see later, for $i \in X$ it is not possible to update the visibility matrix $M_i(i, \alpha, s)$ only based on $M_{i-1}(i-1, \alpha, s)$). Along with $\mathcal{R}_i(s)$, we also store a mapping Λ_i^s from $\mathcal{R}_i(s)$ to valid partial page assignments of E_i which maps $(M_0, \ldots, M_z) \in \mathcal{R}_i(s)$ to some α such that $(M_0, \ldots, M_z) = (M_i(i, \alpha, s), M_i(x_1, \alpha, s), M_i(x_2, \alpha, s), \ldots, M_i(x_z, \alpha, s))$.

Let us make some observations about our records $\mathcal{R}_i(s)$. First, $|\mathcal{R}_i(s)| \leq 2^{\tau^3 + \tau^2}$. Second, if $\mathcal{R}_{n-\tau}(s) \neq \emptyset$ for some s, since $u_{n-\tau}$ is a dummy vertex of degree 0, then there is a valid partial page assignment $\alpha \colon E_{n-\tau} \to [k]$ such that $s \cup \alpha$ is a non-crossing page assignment of *all* edges in G. Hence we can output a k-page book embedding by invoking $\Lambda_{n-\tau}^s$ on any entry in $\mathcal{R}_{n-\tau}(s)$. Third (see the full version [5] for the proof):

Observation 2. *If for all $s \in S$ it holds that $\mathcal{R}_{n-\tau}(s) = \emptyset$, then $(G, \prec, k)$ is a NO-instance of* FIXED-ORDER BOOK THICKNESS.

The above implies that in order to solve our instance, it suffices to compute $\mathcal{R}_{n-\tau}(s)$ for each $s \in S$. As mentioned earlier, we do this dynamically, with the first step consisting of the computation of $\mathcal{R}_1(s)$. Since $E_1 = \emptyset$, the visibility matrices $M_1(1, \emptyset, s), M_1(x_1, \emptyset, s), \ldots, M_1(x_z, \emptyset, s)$ required to populate $\mathcal{R}_1(s)$ depend only on s and are easy to compute in polynomial time.

Finally, we proceed to the dynamic step. Assume we have computed $\mathcal{R}_{i-1}(s)$. We branch over each possible page assignment β of the (at most τ) edges incident to u_{i-1}, and each tuple $\rho \in \mathcal{R}_{i-1}(s)$. For each such β and $\gamma = \Lambda_{i-1}^s(\rho)$, we check whether $\beta \cup \gamma$ is a valid partial page assignment (i.e., whether $\beta \cup \gamma \cup s$ is non-crossing); if this is not the case, we discard this pair of (β, ρ). Otherwise we compute the visibility matrices $M_i(i, \beta \cup \gamma, s), M_i(x_1, \beta \cup \gamma, s), \ldots, M_i(x_z, \beta \cup \gamma, s)$, add the corresponding tuple into $\mathcal{R}_i(s)$, and set Λ_i^s to map this tuple to $\beta \cup \gamma$. We remark that here the use of $\Lambda_{i-1}^s(\rho)$ allows us not to distinguish between $i \in X$ and $i \notin X$—in both cases, the partial page assignment γ will correctly capture the visibility matrix for u_i.

Lemma 1. *The above procedure correctly computes $\mathcal{R}_i(s)$ from $\mathcal{R}_{i-1}(s)$.*

Proof. Consider an entry $(M_0, \ldots, M_z)$ computed by the above procedure from some $\beta \cup \gamma$. Since we explicitly checked that $\beta \cup \gamma$ is a valid partial page assignment, this implies that $(M_0, \ldots, M_z) \in \mathcal{R}_i(s)$, as desired.

For the opposite direction, consider a tuple $(M_0, \ldots, M_z) \in \mathcal{R}_i(s)$. By definition, there exists some valid partial page assignment α of E_i such that $M_0 = M_i(i, \alpha, s)$, $M_1 = M_i(x_1, \alpha, s)$, $\ldots$, $M_z = M_i(x_z, \alpha, s)$. Now let β be the restriction of α to the edges incident to u_{i-1}, and let γ' be the restriction of α to all other edges (i.e., all those not incident to u_{i-1}). Since $\gamma' \cup s$ is non-crossing and in particular γ' is a valid partial page assignment for E_{i-1}, $\mathcal{R}_{i-1}(s)$ must contain an entry $\omega = (M_{i-1}(i-1, \gamma', s), \ldots, (M_{i-1}(x_z, \gamma', s))$—let $\gamma = \Lambda_{i-1}^s(\omega)$.

To conclude the proof, it suffices to show that (1) $\beta \cup \gamma$ is a valid partial page assignment, and (2) $(M_i(i, \beta \cup \gamma', s), \ldots, M_i(x_z, \beta \cup \gamma', s))$, which is the original tuple corresponding to the hypothetical α, is *equal to* $(M_i(i, \beta \cup \gamma, s), \ldots, M_i(x_z, \beta \cup \gamma, s))$, which is the entry our algorithm computes from β and γ. Point (1) follows from the fact that $M_{i-1}(i-1, \gamma', s) = M_{i-1}(i-1, \gamma, s)$ in conjunction with the fact that u_{i-1} is adjacent only to vertices in C. Point (2) then follows by the same argument, but applied to each visibility matrix in the respective tuples: for each $x \in X$ we have $M_{i-1}(x, \gamma', s) = M_{i-1}(x, \gamma, s)$— meaning that the visibilities of u_x were identical before considering the edges incident to u_{i-1}—and so assigning these edges to pages as prescribed by β leads to an identical outcome in terms of visibility.   □

This proves the correctness of our algorithm. The runtime is upper-bounded by the product of $|S| < \tau^{\tau^2}$ (the initial branching factor), n (the number of times we compute a new record set $\mathcal{R}_i(s)$), and $2^{\tau^3 + \tau^2} \cdot \tau^\tau$ (to consider all combinations of γ and β so to compute a new record set from the previous one). A minimum-page book embedding can be computed by trying all possible choices for $k \in [\tau]$. We summarize Result 1 below.

Theorem 1. *There is an algorithm which takes as input an n-vertex graph G with a vertex order $\prec$, runs in time $2^{\mathcal{O}(\tau^3)} \cdot n$ where τ is the vertex cover number of G, and computes a page assignment σ such that $(\prec, \sigma)$ is a (fo-bt$(G, \prec)$)-page book embedding of G.*

3.2 Parameterization by the Pathwidth of the Vertex Ordering

As our second result, we show that FIXED-ORDER BOOK THICKNESS is fixed-parameter tractable parameterized by the pathwidth of $(G, \prec)$. We note that while the pathwidth of G is always upper-bounded by the vertex cover number, this does not hold when we consider a fixed ordering $\prec$, and hence this result is incomparable to Theorem 1. For instance, if G is a path, it has arbitrarily large vertex cover number while $(G, \prec)$ may have a pathwidth of 1, while on the other hand if G is a star, it has a vertex cover number of 1 while $(G, \prec)$ may have arbitrarily large pathwidth. To begin, we can show that the pathwidth of $(G, \prec)$ provides an upper bound on the number of pages required for an embedding (see the full version [5] for the proof).

Lemma 2. *Every n-vertex graph $G = (V, E)$ with a linear order $\prec$ of V such that $(G, \prec)$ has pathwidth k admits a k-page book embedding $\langle \prec, \sigma \rangle$, which can be computed in $\mathcal{O}(n + k \cdot n)$ time.*

We note that the bound given in Lemma 2 is also tight for the same reason as for Observation 1: complete bipartite graphs with bipartitions of size k and $h > k(k-1)$ have book thickness k [4], but admit an ordering $\prec$ with pathwidth k.

We now proceed to a description of our algorithm. Our input consists of the graph G, the vertex ordering $\prec$, and an integer k that upper-bounds the desired number of pages in a book embedding. Let κ be our parameter, i.e., the

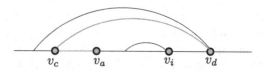

Fig. 3. An assignment of the edges of S_i to a page p, where the edge $v_c v_d$ is the (α, i, p)-important edge of v_a. Any vertex w with $v_c \prec w \prec v_a$ is visible to v_a, and any vertex $w' \prec v_c$ is not visible to v_a.

pathwidth of $(G, \prec)$; observe that due to Lemma 2, we may assume that $k \leq \kappa$. The algorithm performs a dynamic programming procedure on the vertices $v_1, v_2, \ldots, v_n$ of the input graph G in right-to-left order along $\prec$. For technical reasons, we initially add a vertex v_0 of degree 0 to G and place it to the left of v_1 in $\prec$; note that this does not increase the pathwidth of G.

We now adapt the concept of *visibility* introduced in Sect. 3.1 for use in this algorithm. First, let us expand our notion of guard set (see Sect. 2) as follows: for a vertex v_i, let $P^*_{v_i} = \{g^i_1, \ldots, g^i_m\}$ where for each $j \in [m-1]$, g^i_j is the j-th guard of v_i in reverse order of $\prec$ (i.e., g^i_1 is the guard that is nearest to v_i in $\prec$), and $g^i_m = v_0$. For a vertex v_i, let $E_i = \{v_a v_b \mid v_a v_b \in E, b > i\}$ be the set of all edges with at least one endpoint to the right of v_i and let $S_i = \{g^i_j v_b \mid g^i_j \in P^*_{v_i}, g^i_j v_b \in E_i\}$ be the restriction of E_i to edges between a vertex to the right of v_i and a guard in $P^*_{v_i}$. An assignment $\alpha \colon E_i \to [k]$ is called a *valid partial page assignment* if α maps the edges in E_i to pages in a non-crossing manner. Given a valid partial page assignment $\alpha \colon E_i \to [k]$ and a vertex v_a with $a \leq i$, we say a vertex v_x ($x < a$) is α-*visible* to v_a on a page p if it is possible to draw the edge $v_a v_x$ in page p without crossing any other edge mapped to p by α.

Before we proceed to describing our algorithm, we will show that the visibilities of vertices w.r.t. valid partial page assignments exhibit a certain regularity property. Given $a \leq i \leq n$, $p \in [k]$, and a valid partial page assignment α of E_i, let the (α, i, p)-*important* edge of v_a be the edge $v_c v_d \in S_i$ with the following properties: (1) $\alpha(v_c v_d) = p$, (2) $c < a$, and (3) $|a - c|$ is minimum among all such edges in S_i. If multiple edges with these properties exist, we choose the edge with minimum $|d - c|$. Intuitively, the (α, i, p)-important edge of v_a is simply the shortest edge of S_i which encloses v_a on page p; note that it may happen that v_a has no (α, i, p)-important edge. Observe that, if the edge exists, its left endpoint is $v_c \in P^*_{v_i}$, and we call v_c the (α, i, p)-*important* guard of v_a. The next observation easily follows from the definition of (α, i, p)-important edge, see also Fig. 3.

Observation 3. *If v_a has no (α, i, p)-important edge, then every vertex v_x with $x < a$ is α-visible to v_a. If the (α, i, p)-important guard of v_a is v_c, then v_x ($x < a$) is α-visible to v_a if and only if $x \geq c$.*

Observation 3 not only provides us with a way of handling vertex visibilities in the pathwidth setting, but also allows us to store all the information we require about vertex visibilities in a more concise way than via the matrices used in Sect. 3.1. For an index $i \in [n]$, a vertex v_a where $a \leq i$ and a valid partial page assignment α, we define the *visibility vector* $U_i(v_a, \alpha)$ as follows: the p-th component of $U_i(v_a, \alpha)$ is the (α, i, p)-important guard of v_a, and $\diamond$ if v_a has no (α, i, p)-important guard. Observe that since the number of pages is upper-bounded by κ by assumption and the cardinality of $P^*_{v_i}$ is at most $\kappa + 1$, there are at most $(\kappa + 2)^\kappa$ possible distinct visibility vectors for any fixed i.

Observe that thanks to Observation 3 the visibility vector $U_i(v_i, \alpha)$ provides us with complete information about the visibility of vertices v_b ($b < i$) from v_i—notably, v_b is not α-visible to v_i on page p if and only if v_b lies to the left of the (α, i, p)-important guard $U_i(v_i, \alpha)[p]$ (and, in particular, if $U_i(v_i, \alpha)[p] = \diamond$ then every such v_b is α-visible to v_i on page p). On a high level, the algorithm will traverse vertices in right-to-left order along $\prec$ and store the set of all possible visibility vectors at each vertex. To this end, it will use the following observation to update its visibility vectors.

Observation 4. *Let α be a valid partial page assignment of E_i and p be a page. If $v_{i-1} \notin P^*_{v_i}$, then a vertex v_b ($b < i - 1$) is α-visible to v_{i-1} on page p if and only if v_b is α-visible to v_i on page p.*

Proof. By definition v_{i-1} and v_i are consecutive in $\prec$. Let v_b (for $b < i - 1$) be a vertex that is α-visible to v_{i-1} on page p. If v_b is not α-visible to v_i on p, then there must be a vertex w between v_{i-1} and v_i that is incident to an edge in E_i separating v_{i-1} and v_i on page p. But this contradicts that v_{i-1} and v_i are consecutive in $\prec$. The other direction follows by the same argument. □

There is, however, a caveat: Observation 4 does not (and in fact cannot) allow us to compute the new visibility vector if $v_{i-1} \in P^*_i$. To circumvent this issue, our algorithm will not only store the visibility vector $U_i(v_i, \alpha)$ but also the visibility vectors for each guard of v_i. We now prove that we can compute the visibility vector for any vertex from the visibility vectors of the guards—this is important when updating our records, since we will need to obtain the visibility records for new guards that are introduced at some step of the algorithm.

Lemma 3. *Let $v_a \prec v_i$, α be a valid partial page assignment of E_i, $p \in [k]$ be a page, and assume $v_a \notin P^*_i$. Let $v_b \in P^*_i \cup \{v_i\}$ be such that $b > a$ and $|b - a|$ is minimized, i.e., v_b is the first guard to the right of v_a. Then $U_i(v_a, \alpha) = U_i(v_b, \alpha)$.*

Proof. Let v_x for $x < a$ be any vertex that is α-visible to v_a in page p and assume v_x is not α-visible to v_b. Then there must be an edge $wz \in E_i$ separating v_a from v_b in page p, i.e., $v_a \prec w \prec v_b$. But in that case w is a guard in P^*_i closer to v_a contradicting the choice of v_b. Conversely, let v_x for $x < a$ be a vertex that is not α-visible to v_a in page p. Then there must be an edge $wz \in E_i$ separating v_x from v_a on page p. Then edge wz also separates v_x from v_b and v_x is not α-visible to v_b. Therefore, the visibility vectors $U_i(v_a, \alpha)$ and $U_i(v_b, \alpha)$ corresponding to the vertices v_a and v_b, respectively, are equal. □

We can now formally define our record set as $Q_i = \{((U_i(v_i, \alpha), U_i(g_1^i, \alpha), \ldots, U_i(g_{m-1}^i, \alpha)) \mid \exists$ valid partial page assignment $\alpha \colon E_i \to [k]\}$, where each individual element (record) in Q_i can be seen as a queue starting with the visibility vector for v_i and then storing the visibility vectors for individual guards (note that there is no reason to store an "empty" visibility vector for g_m^i). To facilitate the construction of a solution, we will also store a function Λ_i from Q_i to valid partial page assignments of E_i which maps each tuple $\omega \in Q_i$ to some α such that $\omega = (U_i(v_i, \alpha), U_i(g_1^i, \alpha), \ldots, U_i(g_{m-1}^i, \alpha))$.

Let us make some observations about our records Q_i. First of all, since there are at most $(\kappa + 2)^\kappa$ many visibility vectors, $|Q_i| \leq (\kappa + 2)^{\kappa^2}$. Second, if $|Q_0| > 0$ then, since $E_0 = E$, the mapping $\Lambda_0(\omega)$ will produce a valid page assignment of E for any $\omega \in Q_0$. On the other hand, if G admits a k-page book embedding α with order $\prec$, then α witnesses the fact that Q_0 cannot be empty. Hence, the algorithm can return one, once it correctly computes Q_0 and Λ_0.

The computation is carried out dynamically and starts by setting $Q_n = \{\omega\}$, where $\omega = (\diamond)$, and $\Lambda_n(\omega) = \emptyset$. For the inductive step, assume that we have correctly computed Q_i and Λ_i, and the aim is to compute Q_{i-1} and Λ_{i-1}. For each $\omega = (\omega_1, \ldots, \omega_m) \in Q_i$, we compute an intermediate record ω' which represents the visibility vector of v_{i-1} w.r.t. $\alpha = \Lambda_i(\omega)$ as follows:

- if $v_{i-1} \in P_i^*$, then $\omega' = (\omega_2, \ldots, \omega_m)$, and
- if $v_{i-1} \notin P_i^*$, then $\omega' = (\omega_1, \ldots, \omega_m)$ (Recall Observation 4).

We now need to update our intermediate record ω' to take into account the new guards. In particular, we expand ω' by adding, for each new guard $g_j^{i-1} \in P_{i-1}^* \setminus P_i^*$, an intermediate visibility vector $U_{i-1}(g_j^{i-1}, \alpha)$ at the appropriate position in ω' (i.e., mirroring the ordering of guards in P_{i-1}^*). Recalling Lemma 3, we compute this new intermediate visibility vector $U_{i-1}(g_j^{i-1}, \alpha)$ by copying the visibility vector that immediately succeeds it in ω'.

Next, let $F_{i-1} = E_{i-1} \setminus E_i$ be the at most κ new edges that we need to account for, and let us branch over all assignments $\beta \colon F_{i-1} \to [k]$. For each such β, we check whether $\alpha \cup \beta$ is a valid partial page assignment of E_{i-1}, i.e., whether the new edges in F_{i-1} do not cross with each other or other edges in E_i when following the chosen assignment β and the assignment α obtained from Λ_i. As expected, we discard any β such that $\alpha \cup \beta$ is not valid.

Our final task is now to update the intermediate visibility vectors $U_{i-1}(*, \alpha)$ (with $*$ being a placeholder) to $U_{i-1}(*, \alpha \cup \beta)$. This can be done in a straightforward way by, e.g., looping over each edge $e \in F_{i-1}$, obtaining the page $p = \beta(e)$ that e is mapped to, reading $U_{i-1}(*, \alpha)[p]$ and replacing that value by the guard g incident to e if g occurs to the right of $U_{i-1}(*, \alpha)[p]$ and to the left of $*$. Finally, we enter the resulting record ω' into Q_{i-1}.

Lemma 4. *The above procedure correctly computes Q_{i-1} from Q_i.*

Proof. Consider an entry ω' computed by the above procedure from some $\alpha \cup \beta$ and ω. Since we explicitly checked that $\alpha \cup \beta$ is a valid partial page assignment for E_{i-1}, there must exist a record $(U_{i-1}(v_{i-1}, \alpha \cup \beta), U_{i-1}(g_1^{i-1}, \alpha \cup$

$\beta), \ldots, U_{i-1}(g_{m-1}^{i-1})) \in Q_{i-1}$, and by recalling Observation 3, Lemma 3 and Observation 4 it can be straightforwardly verified that this record is equal to ω'.

For the opposite direction, consider a tuple $\omega_0 \in Q_{i-1}$ that arises from the valid partial page assignment γ of E_{i-1}, and let β, α be the restrictions of γ to F_{i-1} and E_i, respectively. Since α is a valid partial page assignment of E_i, there must exist a tuple $\omega \in Q_i$ that arises from α. Let $\alpha' = \Lambda_i(\omega)$. To conclude the proof, it suffices to note that during the branching stage the algorithm will compute a record from a combination of α' (due to ω being in Q_i) and β, and the record computed in this way will be precisely ω_0.   □

This proves the correctness of the algorithm. The runtime is upper bounded by $\mathcal{O}(n \cdot (\kappa + 2)^{\kappa^2} \cdot \kappa^\kappa)$ (the product of the number of times we compute a new record, the number of records and the branching factor for β). A minimum-page book embedding can be obtained by trying all possible choices for $k \in [\kappa]$). We summarize Result 2 below.

Theorem 2. *There is an algorithm which takes as input an n-vertex graph $G = (V, E)$ with a vertex ordering $\prec$ and computes a page assignment σ of E such that $(\prec, \sigma)$ is a $(\text{fo-bt}(G, \prec))$-page book embedding of G. The algorithm runs in $n \cdot \kappa^{\mathcal{O}(\kappa^2)}$ time where κ is the pathwidth of $(G, \prec)$.*

4   Algorithms for Book Thickness

We now turn our attention to the general definition of book thickness (without a fixed vertex order). We show that, given a graph G, in polynomial time we can construct an equivalent instance G^* whose size is upper-bounded by a function of $\tau(G)$. Such an algorithm is called a *kernelization* and directly implies the fixed-parameter tractability of the problem with this parameterization [9,11].

Theorem 3. *There is an algorithm which takes as input an n-vertex graph $G = (V, E)$ and a positive integer k, runs in time $\mathcal{O}(\tau^{\tau^{\mathcal{O}(\tau)}} + 2^\tau \cdot n)$ where $\tau = \tau(G)$ is the vertex cover number of G, and decides whether $\text{bt}(G) \leq k$. If the answer is positive, it can also output a k-page book embedding of G.*

Proof. If $k > \tau$, by Observation 1 we can immediately conclude that G admits a k-page book embedding. Hence we shall assume that $k \leq \tau$. We will also compute a vertex cover C of size τ in time $\mathcal{O}(2^\tau \cdot n)$ using well-known results [7].

For any subset $U \subseteq C$ we say that a vertex of $V \setminus C$ is of *type U* if its set of neighbors is equal to U. This defines an equivalence relation on $V \setminus C$ and partitions $V \setminus C$ into at most $\sum_{i=0}^{\tau} \binom{\tau}{i} = 2^\tau$ distinct types. In what follows, we denote by V_U the set of vertices of type U. We claim the following.

Claim. Let $v \in V_U$ such that $|V_U| \geq 2 \cdot k^\tau + 2$. Then G admits a k-page book embedding if and only if $G' = G \setminus \{v\}$ does. Moreover, a k-page book embedding of G' can be extended to such an embedding for G in linear time.

Proof. (of the Claim). One direction is trivial, since removing a vertex from a book embedding preserves the property of being a book embedding of the resulting graph. So let $\langle \prec, \sigma \rangle$ be a k-page book embedding of G'. We prove that a k-page book embedding of G can be easily constructed by inserting v right next to a suitable vertex u in V_U and by assigning the edges of v to the same pages as the corresponding edges of u. We say that two vertices $u_1, u_2 \in V_U$ are *page equivalent*, if for each vertex $w \in U$, the edges $u_1 w$ and $u_2 w$ are both assigned to the same page according to σ. Each vertex in V_U has degree exactly $|U|$, hence this relation partitions the vertices of V_U into at most $k^{|U|} \leq k^\tau$ sets. Since $|V_U| \setminus \{v\} \geq 2 \cdot k^\tau + 1$, at least three vertices of this set, which we denote by u_1, u_2, and u_3, are page equivalent. Consider now the graph induced by the edges of these three vertices that are assigned to a particular page. By the above argument, such a graph is a $K_{h,3}$, for some $h > 0$. However, since already $K_{2,3}$ does not admit a 1-page book embedding, we have $h \leq 1$, that is, each u_i has at most one edge on each page. Then we can extend $\prec$ by introducing v right next to u_1 and assign each edge vw to the same page as $u_1 w$. Since each such edge vw runs arbitrarily close to the corresponding crossing-free edge $u_1 w$, this results in a k-page book embedding of G and concludes the proof of the claim. $\square$

We now construct a kernel G^* from G of size $\mathcal{O}(k^\tau)$ as follows. We first classify each vertex of G based on its type. We then remove an arbitrary subset of vertices from each set V_U with $|V_U| > 2 \cdot k^\tau + 1$ until $|V_U| = 2 \cdot k^\tau + 1$. Thus, constructing G^* can be done in $\mathcal{O}(2^\tau + \tau \cdot n)$ time, where 2^τ is the number of types and $\tau \cdot n$ is the maximum number of edges of G. From our claim above we can conclude that G^* admits a k-page book embedding if and only if G does. Determining the book thickness of G^* can be done by guessing all possible linear orders and page assignments in $O(k^\tau! \cdot k^{k^\tau}) = O(\tau^{\tau^{O(\tau)}})$ time. A k-page book embedding of G^* (if any) can be extended to one of G by iteratively applying the constructive procedure from the proof of the above claim, in $O(\tau \cdot n)$ time. $\square$

The next corollary easily follows from Theorem 3, by applying a binary search on the number of pages $k \leq \tau$ and by observing that a vertex cover of minimum size τ can be computed in $2^{\mathcal{O}(\tau)} + \tau \cdot n$ time [7].

Corollary 1. *Let G be a graph with n vertices and vertex cover number τ. A book embedding of G with minimum number of pages can be computed in $\mathcal{O}(\tau^{\tau^{O(\tau)}} + \tau \log \tau \cdot n)$ time.*

5   Conclusions and Open Problems

We investigated the parameterized complexity of BOOK THICKNESS and FIXED-ORDER BOOK THICKNESS. We proved that both problems can be parameterized by the vertex cover number of the graph, and that the second problem can be parameterized by the pathwidth of the fixed linear order. The algorithm for BOOK THICKNESS is the first fixed-parameter algorithm that works for general values of k, while, to the best of our knowledge, no such algorithms were known for FIXED-ORDER BOOK THICKNESS.

We believe that our techniques can be extended to the setting in which we allow edges on the same page to cross, with a given budget of at most c crossings over all pages. This problem has been studied by Bannister and Eppstein [2] with the number of pages k restricted to be either 1 or 2. It would also be interesting to investigate the setting where an upper bound on the maximum number of crossings *per edge* is given as part of the input, which is studied in [6].

The main question that remains open is whether BOOK THICKNESS (and FIXED-ORDER BOOK THICKNESS) can be solved in polynomial time (and even fixed-parameter time) for graphs of bounded treewidth, which was asked by Dujmović and Wood [14]. As an intermediate step towards solving this problem, we ask whether the two problems can be solved efficiently when parameterized by the treedepth [25] of the graph. Treedepth restricts the graph structure in a stronger way than treewidth, and has been used to obtain algorithms for several problems which have proven resistant to parameterization by treewidth [17,19].

References

1. Bannister, M.J., Cabello, S., Eppstein, D.: Parameterized complexity of 1-planarity. J. Graph Algorithms Appl. **22**(1), 23–49 (2018). https://doi.org/10.7155/jgaa.00457

2. Bannister, M.J., Eppstein, D.: Crossing minimization for 1-page and 2-page drawings of graphs with bounded treewidth. J. Graph Algorithms Appl. **22**(4), 577–606 (2018). https://doi.org/10.7155/jgaa.00479

3. Bekos, M.A., Gronemann, M., Raftopoulou, C.N.: Two-page book embeddings of 4-planar graphs. Algorithmica **75**(1), 158–185 (2016). https://doi.org/10.1007/s00453-015-0016-8

4. Bernhart, F., Kainen, P.C.: The book thickness of a graph. J. Comb. Theory Ser. B **27**(3), 320–331 (1979). https://doi.org/10.1016/0095-8956(79)90021-2

5. Bhore, S., Ganian, R., Montecchiani, F., Nöllenburg, M.: Parameterized algorithms for book embedding problems. CoRR abs/1908.08911 (2019). http://arxiv.org/abs/1908.08911

6. Binucci, C., Di Giacomo, E., Hossain, M.I., Liotta, G.: 1-page and 2-page drawings with bounded number of crossings per edge. Eur. J. Comb. **68**, 24–37 (2018). https://doi.org/10.1016/j.ejc.2017.07.009

7. Chen, J., Kanj, I.A., Xia, G.: Improved upper bounds for vertex cover. Theor. Comput. Sci. **411**(40–42), 3736–3756 (2010). https://doi.org/10.1016/j.tcs.2010.06.026

8. Chung, F., Leighton, F., Rosenberg, A.: Embedding graphs in books: a layout problem with applications to VLSI design. SIAM J. Algebraic Discret. Methods **8**(1), 33–58 (1987). https://doi.org/10.1137/0608002

9. Cygan, M., Fomin, F.V., Kowalik, L., Lokshtanov, D., Marx, D., Pilipczuk, M., Pilipczuk, M., Saurabh, S.: Parameterized Algorithms. Springer, Cham (2015). https://doi.org/10.1007/978-3-319-21275-3

10. Diestel, R.: Graph Theory. Graduate Texts in Mathematics, vol. 173. Springer, Heidelberg (2012)

11. Downey, R.G., Fellows, M.R.: Fundamentals of Parameterized Complexity. TCS. Springer, London (2013). https://doi.org/10.1007/978-1-4471-5559-1

12. Dujmović, V., Wood, D.R.: On linear layouts of graphs. Discrete Math. Theor. Comput. Sci. **6**(2), 339–358 (2004)
13. Dujmovic, V., Wood, D.R.: Graph treewidth and geometric thickness parameters. Discrete Comput. Geom. **37**(4), 641–670 (2007). https://doi.org/10.1007/s00454-007-1318-7
14. Dujmović, V., Wood, D.R.: On the book thickness of k-trees. Discrete Math. Theor. Comput. Sci. **13**(3), 39–44 (2011)
15. Fellows, M.R., Lokshtanov, D., Misra, N., Rosamond, F.A., Saurabh, S.: Graph layout problems parameterized by vertex cover. In: Algorithms and Computation (ISAAC 2008), pp. 294–305 (2008). https://doi.org/10.1007/978-3-540-92182-0_28
16. Ganian, R.: Improving vertex cover as a graph parameter. Discrete Math. Theor. Comput. Sci. **17**(2), 77–100 (2015)
17. Ganian, R., Ordyniak, S.: The complexity landscape of decompositional parameters for ILP. Artif. Intell. **257**, 61–71 (2018). https://doi.org/10.1016/j.artint.2017.12.006
18. Ganley, J.L., Heath, L.S.: The pagenumber of k-trees is $O(k)$. Discrete Appl. Math. **109**(3), 215–221 (2001). https://doi.org/10.1016/S0166-218X(00)00178-5
19. Gutin, G.Z., Jones, M., Wahlström, M.: The mixed Chinese postman problem parameterized by pathwidth and treedepth. SIAM J. Discrete Math. **30**(4), 2177–2205 (2016). https://doi.org/10.1137/15M1034337
20. Haslinger, C., Stadler, P.F.: RNA structures with pseudo-knots: graph-theoretical, combinatorial, and statistical properties. Bull. Math. Biol. **61**(3), 437–467 (1999). https://doi.org/10.1006/bulm.1998.0085
21. Kainen, P.C.: Some recent results in topological graph theory. In: Bari, R.A., Harary, F. (eds.) Graphs and Combinatorics, pp. 76–108. Springer, Berlin (1974). https://doi.org/10.1007/BFb0066436
22. Kinnersley, N.G.: The vertex separation number of a graph equals its pathwidth. Inf. Process. Lett. **42**(6), 345–350 (1992). https://doi.org/10.1016/0020-0190(92)90234-M
23. Lodha, N., Ordyniak, S., Szeider, S.: SAT-encodings for special treewidth and pathwidth. In: Gaspers, S., Walsh, T. (eds.) SAT 2017. LNCS, vol. 10491, pp. 429–445. Springer, Cham (2017). https://doi.org/10.1007/978-3-319-66263-3_27
24. Mallach, S.: Linear ordering based MIP formulations for the vertex separation or pathwidth problem. In: Brankovic, L., Ryan, J., Smyth, W.F. (eds.) IWOCA 2017. LNCS, vol. 10765, pp. 327–340. Springer, Cham (2018). https://doi.org/10.1007/978-3-319-78825-8_27
25. Nešetřil, J., Ossona de Mendez, P.: Sparsity. AC, vol. 28. Springer, Heidelberg (2012). https://doi.org/10.1007/978-3-642-27875-4
26. Ollmann, L.T.: On the book thicknesses of various graphs. In: 4th Southeastern Conference on Combinatorics, Graph Theory and Computing, vol. 8, p. 459 (1973)
27. Robertson, N., Seymour, P.D.: Graph minors. I. Excluding a forest. J. Comb. Theory Ser. B **35**(1), 39–61 (1983). https://doi.org/10.1016/0095-8956(83)90079-5
28. Robertson, N., Seymour, P.D.: Graph minors. II. Algorithmic aspects of tree-width. J. Algorithms **7**(3), 309–322 (1986). https://doi.org/10.1016/0196-6774(86)90023-4
29. Unger, W.: The complexity of colouring circle graphs. In: Finkel, A., Jantzen, M. (eds.) STACS 1992. LNCS, vol. 577, pp. 389–400. Springer, Heidelberg (1992). https://doi.org/10.1007/3-540-55210-3_199. (extended abstract)
30. Yannakakis, M.: Embedding planar graphs in four pages. J. Comput. Syst. Sci. **38**(1), 36–67 (1989). https://doi.org/10.1016/0022-0000(89)90032-9

Sketched Representations and Orthogonal Planarity of Bounded Treewidth Graphs

Emilio Di Giacomo⬤, Giuseppe Liotta⬤, and Fabrizio Montecchiani$^{(\boxtimes)}$⬤

Department of Engineering, University of Perugia, Perugia, Italy
{emilio.digiacomo,giuseppe.liotta,fabrizio.montecchiani}@unipg.it

Abstract. Given a planar graph G and an integer b, ORTHOGONALPLA-
NARITY is the problem of deciding whether G admits an orthogonal drawing with at most b bends in total. We show that ORTHOGONALPLANARITY
can be solved in polynomial time if G has bounded treewidth. Our proof
is based on an FPT algorithm whose parameters are the number of bends,
the treewidth and the number of degree-2 vertices of G. This result is
based on the concept of sketched orthogonal representation that synthet-
ically describes a family of equivalent orthogonal representations. Our
approach can be extended to related problems such as HV-PLANARITY
and FLEXDRAW. In particular, both ORTHOGONALPLANARITY and HV-
PLANARITY can be decided in $O(n^3 \log n)$ time for series-parallel graphs,
which improves over the previously known $O(n^4)$ bounds.

1 Introduction

An *orthogonal drawing* of a planar graph G is a planar drawing where each edge
is drawn as a chain of horizontal and vertical segments; see Fig. 1a. Orthogonal
drawings are among the most investigated research subjects in graph drawing,
see, e.g., [3–5,11,13,17,25,28,30,31] for a limited list of references, and also [12,
23] for surveys. The ORTHOGONALPLANARITY problem asks whether G admits
an orthogonal drawing with at most b bends in total, for a given $b \in \mathbb{N}$.

In a seminal paper, Garg and Tamassia [25] proved that ORTHOGONALPLA-
NARITY is NP-complete when $b = 0$, which implies that minimizing the number
of bends is also NP-hard. In fact, it is even NP-hard to approximate the minimum number of bends in an orthogonal drawing with an $O(n^{1-\varepsilon})$ error for any
$\varepsilon > 0$ [25]. On the positive side, Tamassia [30] showed that ORTHOGONALPLA-
NARITY can be decided in polynomial time if the input graph is *plane*, i.e., it has
a fixed embedding in the plane. When a planar embedding is not given as part
of the input, polynomial-time algorithms exist for some restricted cases, namely
subcubic planar graphs and series-parallel graphs, see, e.g., [11,13,17,28,31].

Research partially supported by: (i) MIUR, grant 20174LF3T8 AHeAD: efficient Algorithms for HArnessing networked Data.; (ii) Engineering Dep. - University of Perugia,
grants RICBASE2017WD and RICBA18WD: "Algoritmi e sistemi di analisi visuale di
reti complesse e di grandi dimensioni".

ⓒ Springer Nature Switzerland AG 2019
D. Archambault and C. D. Tóth (Eds.): GD 2019, LNCS 11904, pp. 379–392, 2019.
https://doi.org/10.1007/978-3-030-35802-0_29

Given the hardness result for ORTHOGONALPLANARITY, a natural research direction is to investigate its parameterized complexity [19]. Despite the rich literature about orthogonal drawings, this direction has been surprisingly disregarded. The only exception is a result by Didimo and Liotta [16], who described an algorithm for biconnected planar graphs that runs in $O(6^r n^4 \log n)$ time, where r is the number of degree-4 vertices. We recall that FPT algorithms have been proposed for other graph drawing problems, such as upward planarity [10,15,26], layered drawings [20], linear layouts [2,21,22], and 1-planarity [1].

Contribution. We describe an FPT algorithm for ORTHOGONALPLANARITY whose parameters are the number of bends, the treewidth and the number of degree-2 vertices of the input graph. We recall that the notion of treewidth [29] is commonly used as a parameter in the parameterized complexity analysis (see also Sect. 2). The algorithm works for planar graphs of degree four with no restriction on the connectivity. Our main contribution is summarized as follows.

Theorem 1. *Let G be an n-vertex planar graph with σ degree-2 vertices and let $b \in \mathbb{N}$. Given a tree-decomposition of G of width k, there is an algorithm that decides* ORTHOGONALPLANARITY *in $f(k,\sigma,b) \cdot n$ time, where $f(k,\sigma,b) = k^{O(k)}(\sigma+b)^k \log(\sigma+b)$. The algorithm computes a drawing of G, if one exists.*

For an n-vertex graph G of treewidth k, a tree-decomposition of G of width k can be found in $k^{O(k^3)} n$ time [7], while a tree-decomposition of width $O(k)$ can be computed in $2^{O(k)} n$ time [9]. The function $f(k,\sigma,b)$ depends exponentially on neither σ nor b. Since both σ and b are $O(n)$ [3], ORTHOGONALPLANARITY can be solved in time $n^{g(k)}$ for some polynomial function $g(k)$, and thus it belongs to the XP class when parameterized by treewidth [19]. Moreover, since the number of bends in a bend-minimum orthogonal drawing is $O(n)$, the next result follows from Theorem 1, performing a binary search on b.

Corollary 1. *Let G be an n-vertex planar graph. Given a tree-decomposition of G of width k, there is an algorithm that decides* ORTHOGONALPLANARITY *in $k^{O(k)} n^{k+1} \log n$ time. Also, a bend-minimum orthogonal drawing of G can be computed in $k^{O(k)} n^{k+1} \log^2 n$ time.*

By Corollary 1 ORTHOGONALPLANARITY can be decided in $O(n^3 \log n)$ time for graphs of treewidth two, and hence bend-minimum orthogonal drawings can be computed in $O(n^3 \log^2 n)$ time. We remark that the best previous result for these graph, dating back to twenty years ago, is an $O(n^4)$ algorithm by Di Battista et al. [13] which however is restricted to biconnected graphs (whereas ours is not).

Our FPT approach can be applied to related problems, namely to HV-PLANARITY and FLEXDRAW. HV-PLANARITY takes as input a planar graph G whose edges are each labeled H (horizontal) or V (vertical) and it asks whether G admits an orthogonal drawing with no bends and in which the direction of each edge is consistent with its label. As a corollary of our results, we can decide HV-PLANARITY in $O(n^3 \log n)$ time for series-parallel graphs, which improves a recent $O(n^4)$ bound by Didimo et al. [18] and addresses one of the open problems in that paper. FLEXDRAW takes as input a planar graph G whose edges have integer weights and it asks whether G admits an orthogonal drawing where each edge has a number of bends that is at most its weight [5,6].

Proof Strategy and Paper Organization. The first ingredient of our approach is a well-known combinatorial characterization of orthogonal drawings (see [12,30]) that transforms ORTHOGONALPLANARITY to the problem of testing the existence of a planar embedding along with a suitable angle assignment to each vertex-face and edge-face incidence (see Sect. 2). The second ingredient is the definition of a suitable data structure, called *orthogonal sketches*, that encodes sufficient information about any such combinatorial representation, and in particular it makes it possible to decide whether the representation can be extended with further vertices incident to a given vertex cutset of the graph (see Sect. 3). The proposed algorithm (see Sect. 4) traverses a tree-decomposition of the input graph and stores a limited number of orthogonal sketches for each node of the tree, rather than all its possible orthogonal drawings. This number depends on the width of the tree-decomposition, on the number of bends, and on the number of degree-2 vertices. The key observation is that a vertex of degree greater than two may correspond to a right turn when walking clockwise along the boundary of a face but not to a left turn, while a degree-2 vertex may correspond to both a left or a right turn. Thus, the number of degree-2 vertices, as well as the number of bends, have an impact in how much a face can "roll-up" in the drawing, which in our approach translates in possible weights that can be assigned to the edges of an orthogonal sketch. The extensions of our approach can be found in Sect. 5, while conclusions and open problems are in Sect. 6. For reasons of space, some proofs have been omitted and can be found in [14] (the corresponding statements are marked with an asterisk (*)).

2 Preliminaries

Embeddings. We assume familiarity with basic notions about graph drawings. A planar drawing of a planar graph G subdivides the plane into topologically connected regions, called *faces*. The infinite region is the *outer face*. A *planar embedding* of G is an equivalence class of planar drawings that define the same set of faces and with the same outer face. A planar embedding of a connected graph can be uniquely identified by specifying its *rotation system*, i.e., the clockwise circular order of the edges around each vertex, and the outer face. A *plane graph* G is a planar graph with a given planar embedding. The number of vertices encountered in a closed walk along the boundary of a face f of G is the *degree* of f, denoted as $\delta(f)$. If G is not 2-connected, a vertex may be encountered more than once, thus contributing more than one unit to the degree of the face.

Orthogonal Representations. Let $G = (V, E)$ be a planar graph with vertex degree at most four. A planar drawing Γ of G is *orthogonal* if each edge is a polygonal chain consisting of horizontal and vertical segments. A *bend* of an edge e in Γ is a point shared by two consecutive segments of e. An angle formed by two consecutive segments incident to the same vertex (resp. bend) is a *vertex-angle* (resp. *bend-angle*). An orthogonal representation of G can be derived from Γ and it specifies the values of all vertex- and bend-angles (see [12,30]). More formally, let $\mathcal{E}$ be a planar embedding of G, and let $e = (u, v)$ be an edge that belongs to the boundary of a face f of $\mathcal{E}$. The two possible orientations (u, v) and (v, u) of e are called *darts*. A dart is *counterclockwise with respect to f*, if f is on the left side when walking along the dart following its orientation. Let $D(u)$ be the set of darts exiting from u and let $D(f)$ be the set of counterclockwise darts with respect to f.

Definition 1. *Let $G = (V, E)$ be a planar graph with vertex degree at most four. An orthogonal representation H of G is a planar embedding $\mathcal{E}$ of G and an assignment to each dart (u, v) of two values $\alpha(u, v) = c_\alpha \cdot \frac{\pi}{2}$ and $\beta(u, v) = c_\beta \cdot \frac{\pi}{2}$, where $c_\alpha \in \{1, 2, 3, 4\}$ and $c_\beta \in \mathbb{N}$, that satisfies the following conditions.*

C1. *For each vertex u:* $\displaystyle\sum_{(u,v)\in D(u)} \alpha(u, v) = 2\pi$;

C2. *For each internal face f:* $\displaystyle\sum_{(u,v)\in D(f)} (\alpha(u, v) + \beta(v, u) - \beta(u, v)) = \pi(\delta(f) - 2)$;

C3. *For the outer face f_o:* $\displaystyle\sum_{(u,v)\in D(f_o)} (\alpha(u, v) + \beta(v, u) - \beta(u, v)) = \pi(\delta(f_o) + 2)$.

Let f be the face counterclockwise with respect to dart (u, v). The value $\alpha(u, v)$ represents the vertex-angle that dart (u, v) forms with the dart following it in the circular counterclockwise order around u; we say that $\alpha(u, v)$ is a *vertex-angle of u in f*. The value $\beta(u, v)$ represents the sum of the $\frac{\pi}{2}$ bend-angles that dart (u, v) forms in f. Condition **C1** guarantees that the sum of angles around each vertex is valid, while **C2** (respectively, **C3**) guarantees that the sum of the angles at the vertices and at the bends of an internal face (respectively, outer face) is also valid. Given an orthogonal representation of an n-vertex graph G, a corresponding orthogonal drawing can be computed in $O(n)$ time [30].

Tree-Decompositions. Let $(\mathcal{X}, T)$ be a pair such that $\mathcal{X} = \{X_1, X_2, \ldots, X_\ell\}$ is a collection of subsets of vertices of a graph G called *bags*, and T is a tree whose nodes are in a one-to-one mapping with the elements of $\mathcal{X}$. With a slight abuse of notation, X_i will denote both a bag of $\mathcal{X}$ and the node of T whose corresponding bag is X_i. The pair $(\mathcal{X}, T)$ is a *tree-decomposition* of G if : (i) For every edge (u, v) of G, there exists a bag X_i that contains both u and v, and (ii) For every vertex v of G, the set of nodes of T whose bags contain v induces a non-empty (connected) subtree of T. The *width* of a tree-decomposition $(\mathcal{X}, T)$ of G is $\max_{i=1}^{\ell} |X_i| - 1$, and the *treewidth* of G is the minimum width of any tree-decomposition of G. We use a particular tree-decomposition (which always exists [27]) that limits the number of possible transitions between bags.

Definition 2 [27]. *A tree-decomposition $(\mathcal{X}, T)$ of G is nice if T is a rooted tree and: (a) Every node of T has at most two children, (b) If a node X_i of T has two children whose bags are X_j and $X_{j'}$, then $X_i = X_j = X_{j'}$, (c) If a node X_i of T has only one child X_j, then there exists a vertex $v \in G$ such that either $X_i = X_j \cup \{v\}$ or $X_i \cup \{v\} = X_j$. In the former case of (c) we say that X_i introduces v, while in the latter case X_i forgets v.*

3   Orthogonal Sketches

Recall that an orthogonal representation of a planar graph G corresponds to a planar embedding of G and to an assignment of vertex- and bend-angles in each face of G. A fundamental observation for our approach is that the conditions that make an assignment of such angles a valid orthogonal representation of G can be verified for each vertex and for each face independently. In what follows we define two equivalence relations on the set of orthogonal representations of G that yields a set of equivalence classes whose size is bounded by some function of the width of T, of the number of degree-2 vertices of G, and of the number of bends.

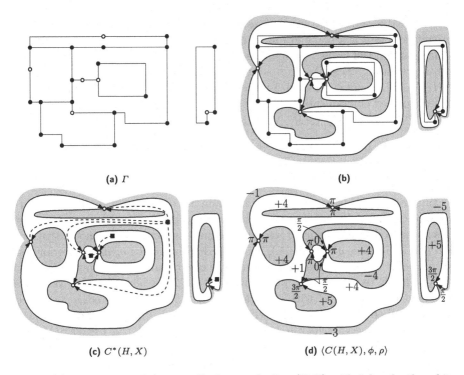

Fig. 1. (a) An orthogonal drawing Γ of a graph $G = (V, E)$ with 8 bends; the white vertices define a set $X \subseteq V$. (b) The representing cycles of the active faces of H with respect to X, where H denotes the orthogonal representation of Γ. (c) The connected sketched embedding $C^*(H, G)$. (d) The orthogonal sketch $\langle C(H, X), \phi, \rho \rangle$.

Sketched Embeddings. Let H be an orthogonal representation of a planar graph $G = (V, E)$ and let $X \subseteq V$; see for example Fig. 1a. The vertices in X are called *active*. A face f of H is *active* if it contains at least one active vertex. A *representing cycle* C_f of an active face f is an oriented cycle such that: (i) It contains all and only the active vertices of f in the order they appear in a closed walk along the boundary of f. (ii) C_f is counterclockwise with respect to f, that is, C_f is oriented coherently with the counterclockwise darts of face f. Notice that C_f may be non-simple because a cut-vertex may appear multiple times when walking along C_f. Also, if H contains distinct components, the outer face of each component is considered independently. See Fig. 1b for an illustration.

Let H be an orthogonal representation of a planar graph G. We may conveniently focus on an orthogonal drawing Γ that falls in the equivalence class of drawings having H as an orthogonal representation. Assume first that H is connected. The *sketched embedding* of H with respect to X is the plane graph $C(H, X)$ constructed as follows. For each active face f we draw in Γ its representing cycle C_f by identifying the vertices of C_f with the corresponding vertices of f and by drawing the edges of C_f inside f without creating crossings. Graph $C(H, X)$ is the embedded graph formed by the edges that we drew inside the active faces. This is a plane graph by construction, it may be disconnected, and it may contain self-loops and multiple edges. Graph $C(H, X)$ has a face f' for each representing cycle C_f of an active face f of H; we call f' an *active* face of $C(H, X)$. If H is not connected, a sketched embedding $C(H_i, X)$ is computed for each connected component H_i of H ($i = 1, \ldots, h$) and the sketched embedding of H is $C(H, X) = \bigcup_{i=1}^{h} C(H_i, X)$. See for example Fig. 1b.

Definition 3. *Let $G = (V, E)$ be a planar graph and let $X \subseteq V$. Let H_1 and H_2 be two orthogonal representations of G. H_1 and H_2 are X-equivalent if they have the same sketched embedding.*

Suppose that H is connected. We now aim at computing a connected supergraph of $C(H, X)$. By construction, the active faces of $C(H, X)$ may share vertices but not edges and hence $C(H, X)$ also contains faces that are not active. For each non-active face g of $C(H, X)$, we add a dummy vertex v_g in its interior and we connect it to all vertices on the boundary of g by adding dummy edges. This turns $C(H, X)$ to a connected plane graph $C^*(H, X)$, which we call a *connected sketched embedding* of H with respect to X. If H is not connected, a connected sketched embedding $C^*(H_i, X)$ is computed for each $C(H_i, X)$ independently, and the *connected* sketched embedding of H is $C^*(H, X) = \bigcup_{i=1}^{h} C^*(H_i, X)$. Figure 1c shows a connected sketched embedding obtained from Fig. 1b. Observe that it may be possible to construct different connected sketched embeddings of the same sketched embedding. However, any connected sketched embedding encodes the information about the global structure of H that is sufficient for the purposes of our algorithm. $C^*(H, X)$ (and hence $C(H, X)$) has a number of vertices that is $O(|X|)$ and a number of edges that is also $O(|X|)$ because $C^*(H, X)$ is planar and the multiplicity of an edge in $C^*(H, X)$ is at most four.

Lemma 1 (*). *Let $G = (V, E)$ be a planar graph and let $X \subseteq V$. Let $\mathcal{H}$ be the set of all possible orthogonal representations of G. The X-equivalent relation partitions $\mathcal{H}$ in at most $w^{O(w)}$ equivalence classes, where $w = |X|$.*

Orthogonal Sketches. Let H be an orthogonal representation of a plane graph $G = (V, E)$ and let $X \subseteq V$. Let $C(H, X)$ be a sketched embedding of H with respect to X. Recall that H is defined by two functions, α and β, that assign the vertex- and bend-angles made by darts inside their faces. The *shape* of $C(H, X)$ consists of two functions ϕ and ρ defined as follows. Let (u, v) be a dart of $C(H, X)$, which corresponds to a path Π_{uv} in H. Let z be the vertex of Π_{uv} adjacent to u (possibly $z = v$). We set $\phi(u, v) = \alpha(u, z)$; the value $\phi(u, v)$ still represents the vertex-angle that u makes in the face on the left of (u, v). Function ρ assigns to each dart (u, v) of $C(H, X)$ a number that describes the shape of Π_{uv} in H. More precisely, for each representing cycle C_f and for each dart (u, v) of C_f, $\rho(u, v, f) = n_{\frac{\pi}{2}}(u, v) - n_{\frac{3\pi}{2}}(u, v) - 2n_{2\pi}(u, v)$, where $n_a(u, v)$ ($a \in \{\frac{\pi}{2}, \frac{3\pi}{2}, 2\pi\}$) is the number of vertex- and bend-angles between u and v whose value is a. For example, Fig. 1d shows a sketched embedding together with its shape. We call $\rho(u, v, f)$ the *roll-up number*[1] of (u, v) in f. If $\phi(u, v) > \frac{\pi}{2}$ and f is the counterclockwise face with respect to dart (u, v), we say that u is *attachable* in f. Two X-equivalent sketched embeddings have the *same shape*, if they have the same values of ϕ and ρ. A sketched embedding $C(H, X)$, together with its shape $\langle\phi, \rho\rangle$, is called an *orthogonal sketch* and it is denoted by $\langle C(H, X), \phi, \rho\rangle$.

Definition 4. *Let $G = (V, E)$ be a planar graph and let $X \subseteq V$. Let H_1 and H_2 be two orthogonal representations of G. H_1 and H_2 are* shape-equivalent *if they are X-equivalent and their orthogonal sketches have the same shape.*

Lemma 2 (*). *Let $\langle C(H, X), \phi, \rho\rangle$ be an orthogonal sketch. Let C_f be a representing cycle of $C(H, X)$ and consider a closed walk along its boundary. Let ρ^* be the sum of the roll-up numbers over all the traversed edges, and let n_a be the number of encountered vertex-angles with value $a \in \{\frac{\pi}{2}, \frac{3\pi}{2}, 2\pi\}$. Then $\rho^* + n_{\frac{\pi}{2}} - n_{\frac{3\pi}{2}} - 2n_{2\pi} = c$, with $c = 4$ ($c = -4$) if f is an inner (the outer) face.*

Lemma 3 (*). *Let $G = (V, E)$ be a planar graph with σ vertices of degree two. Let $\mathcal{H}$ be the set of all possible orthogonal representations of G with at most b bends in total. The shape-equivalent relation partitions $\mathcal{H}$ in at most $w^{O(w)} \cdot (\sigma + b)^{w-1}$ equivalence classes, where $w = |X|$.*

Proof sketch. We shall prove that $n_X \cdot n_S \leq w^{O(w)} \cdot (\sigma + b)^{w-1}$, where n_X is the number of X-equivalent classes and n_S is the number of possible shapes for each X-equivalent class. By Lemma 1, $n_X \leq w^{O(w)}$; we can show that $n_S \leq w^{O(w)}(\sigma + b)^{w-1}$. For a fixed sketched embedding $C(H, X)$, a shape is defined by assigning to each dart (u, v) of $C(H, X)$ the two values $\phi(u, v)$ and $\rho(u, v, f)$. The number of choices for the values $\phi(u, v)$ is at most $4^{4w} \leq w^{O(w)}$. As for the possible choices for $\rho(u, v, f)$, we claim that $-(\sigma + b) \leq \rho(u, v, f) \leq \sigma + b + 4$ based on two observations. (1) The number of vertices and bends forming an angle of $\frac{3\pi}{2}$ inside a face cannot be greater than $b + \sigma$. (2) For each vertex forming an angle of 2π inside a face there are two vertex-angles of $\frac{\pi}{2}$ inside the same face. Finally, once the vertex-angles are fixed, the number of darts for which the roll-up number can be fixed independently is at most $w - 1$. Thus, we have $w - 1$ values to choose and $2(\sigma + b) + 5$ choices for each of them. □

---

[1] It may be worth observing that other papers used conceptually similar definitions, called *rotation* (see, e.g., [5]) and *spirality* (see, e.g., [13]).

4   The Parameterized Algorithm

Overview. We describe an algorithm, called ORTHOPLANTESTER, that decides whether a planar graph G admits an orthogonal drawing with at most b bends in total, by using a dynamic programming approach on a nice tree-decomposition T of G. The algorithm traverses T bottom-up and decides whether the subgraph associated with each subtree admits an orthogonal drawing with at most b bends. For each bag X, it stores all possible orthogonal sketches and, for each of them, the minimum number of bends of any orthogonal representation encoded by that orthogonal sketch. To generate this record, ORTHOPLANTESTER executes one of three possible procedures based on the type of transition with respect to the children of X in T. If the execution of the procedure results in at least one orthogonal sketch, then the algorithm proceeds, otherwise it halts and returns a negative answer. If the root bag contains at least one orthogonal sketch, then the algorithm returns a positive answer. In the positive case, the information corresponding to the embedding of the graph and to the vertex- and bend-angles can be reconstructed through a top-down traversal of T so to obtain an orthogonal representation of G, and consequently an orthogonal drawing [30].

The Algorithm. Let G be an n-vertex planar graph with vertex degree at most four and with σ vertices of degree two, and let $(\mathcal{X}, T)$ be a nice tree-decomposition of G of width k. Following a bottom-up traversal of T, let X_i be the next bag to be visited and let B_i the set of all orthogonal sketches of X_i. Let $w = k + 1$ and recall that $|X_i| \leq w$. Let T_i be the subtree of T rooted at X_i. Let G_i be the subgraph of G induced by all the vertices that belong to the bags in T_i. We distinguish the following four cases.

X_i is a Leaf. Without loss of generality, we can assume that X_i contains only one vertex v (as otherwise we can root in X_i a chain of bags that introduce the vertices of X_i one by one). Thus, G_i contains only v and it admits exactly one orthogonal representation with no bends. In particular, there is a unique sketched embedding consisting of a single representing cycle C_f having v and no edges on its boundary. Also, the functions ϕ and ρ are undefined.

X_i Forgets a Vertex. Let v be the vertex forgotten by X_i. Let X_j be the child of X_i in T. In this case $G_i = G_j$ and we generate the orthogonal sketches for B_i by suitably updating those in B_j. For each orthogonal sketch $\langle C(H, X_j), \phi, \rho \rangle$ in B_j and for each representing cycle C_f of $C(H, X_j)$ containing v, we apply the following operation. If v is the only vertex of C_f, we remove C_f from $C(H, X)$. Otherwise there are at most eight edges of C_f incident to v, based on whether v appears one or more times in a closed walk along C_f. We first remove all self-loops incident to v, if any. Let (u_1, v), (v, u_2) be any two edges of C_f incident to v that appear consecutively in a counterclockwise walk along C_f. For any such pair of edges we apply the following procedure. We remove the edges (u_1, v), (v, u_2) from C_f and we add an edge (u_1, u_2). We assign to the dart (u_1, u_2) roll-up number equal to the sum of the roll-up numbers of darts (u_1, v) and (v, u_2) plus a constant c defined as follows. If $\phi(v, u_2) = \pi$, then $c = 0$; if $\phi(v, u_2) = \frac{\pi}{2}$, then $c = 1$; if $\phi(v, u_2) = \frac{3\pi}{2}$, then $c = -1$; if $\phi(v, u_2) = 2\pi$, then $c = -2$. Once all consecutive pairs of edges incident to v have been processed, v is removed from C_f. It is immediate to verify that Lemma 2 holds for C_f after applying this operation. See Fig. 2a and b for an illustration.

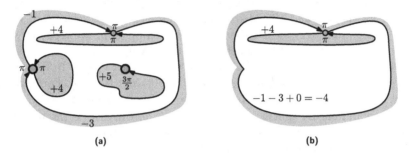

Fig. 2. A portion of a orthogonal sketch before and after removing the bigger vertices.

The above operation does not change the number of bends associated with the resulting orthogonal sketches, but it may create duplicated orthogonal sketches for B_i, which we delete. When deleting the duplicates, we shall pay attention on pairs of orthogonal sketches that are the same but with a different number of bends. To see this, let (u, v) be a dart of an orthogonal sketch $\langle C(H, X_j), \phi, \rho \rangle$, which corresponds to a path Π_{uv} in H, and let f be the face on the left of this path in H. An angle along Π_{uv} in f may be both a vertex-angle or a bend-angle. Hence, removing v from different orthogonal sketches (with different numbers of bends) may result in a set of orthogonal sketches that differ only in the number of bends. In this case, the algorithm stores the one with fewer bends, because in every step of the algorithm (see also the next two cases), the information about the total number of bends of an orthogonal sketch is only used to verify that it does not exceed the given parameter b.

X_i **Introduces a Vertex.** Let v be the vertex introduced in X_i. Let X_j be the child of X_i in T. If v does not have neighbors in X_i, then B_i is the union of B_j and the (unique) orthogonal sketch of the graph with the single vertex v (see the leaf case). Otherwise, let $u_1, \ldots, u_h$ be the neighbors of v in X_i, with $h \leq w$. We generate B_i from B_j by applying the following procedure. At a high level, we first update each sketched embedding that can be extracted from an orthogonal sketch in B_j by adding v, we then generate all shapes for the resulting sketched embeddings, and we finally discard those shapes that are not valid.

Let $C(H, X_j)$ be a sketched embedding for which there is at least one orthogonal sketch in B_j. Suppose first that $u_1, \ldots, u_h$ all belong to the same component of $C(H, X_j)$. By planarity, an orthogonal representation of G_j, whose sketched embedding is $C(H, X_j)$, can be extended with v only if it contains at least one face having all of v's neighbors on its boundary. This corresponds to verifying first the existence of a representing cycle C_f in $C(H, X_j)$ with all of v's neighbors on its boundary. We thus identify the representing cycles in which vertex v can be inserted and connected to its neighbors. We consider each possible choice independently; for each choice we duplicate $C(H, X_j)$ and insert v accordingly. For each of the resulting sketched embeddings, we generate all possible shapes. Namely, for each representing cycle, we generate all possible vertex-angle assignments for its vertices and all possible roll-up numbers for its edges that satisfy Lemma 2. Next, for every such assignment, denoted by $\langle \overline{\phi}, \overline{\rho} \rangle$, we verify its validity. Let S_j be the set of orthogonal sketches $\langle C(H, X_j), \phi, \rho \rangle$ of B_j such that the restriction of $\langle \overline{\phi}, \overline{\rho} \rangle$ to the edges of $C(H, X_j)$ corresponds to $\langle \phi, \rho \rangle$. If S_j is empty, $\langle \overline{\phi}, \overline{\rho} \rangle$ is discarded as it would not be possible to obtain it from B_j. Furthermore, observe that $\overline{\rho}(v, u_i)$ corresponds to the number of bends along the edge (v, u_i). Thus, we should ensure that $b^* + \sum_{i=1}^{h} \overline{\rho}(v, u_i) \leq b$, where b^* is the number of bends of $\langle C(H, X_j), \phi, \rho \rangle$. If this is not the case, again the shape is discarded. Finally, among the putative orthogonal sketches generated, we store in B_i only those for which Lemma 2 holds for each of its representing cycles.

X_i Has Two Children. Let X_j and $X_{j'}$ be the children of X_i in T. Recall that these three bags are all the same, although G_j and $G_{j'}$ differ. The only orthogonal representations of G_i are those that can be obtained by merging at the common vertices of X_i an orthogonal representation of G_j with an orthogonal representation of $G_{j'}$ in such a way that the resulting representation has a planar embedding, it has at most b bends in total, and it satisfies Definition 1. At a high level, this can be done by merging two connected sketched embeddings (one in B_j and one in $B_{j'}$) and then by verifying that there is a planar embedding for the merged graph such that Lemma 2 is verified for each representing cycle, and the overall number of bends is at most b. We split this procedure in two phases.

Let $C(H, X_j)$ be a sketched embedding for which there is at least one orthogonal sketch in B_j and let $C(H', X_{j'})$ be a sketched embedding for which there is at least one orthogonal sketch in $B_{j'}$. We first compute a connected sketched embedding $C^*(H, X_j)$ and a connected sketched embedding $C^*(H', X_{j'})$. Let $\overline{C}$ be the union of these two graphs disregarding the rotation system and the choice of the outer face. For each connected component of $\overline{C}$, we generate all possible planar embeddings. (The embeddings of $\overline{C}$ that are not planar are discarded because they correspond to non-planar embeddings of G_i.) For each planar embedding of $\overline{C}$, we verify that the planar embedding of $\overline{C}$ restricted to the edges of $C(H, X_j)$ corresponds to the planar embedding of $C(H, X_j)$ and the same holds for the edges of $C(H', X_{j'})$. This condition ensures that the embedding of $\overline{C}$ can be obtained from those of $C(H, X_j)$ and $C(H', X_{j'})$. We then remove the dummy vertices and the dummy edges from $\overline{C}$ and we analyze each face of the resulting plane graph to verify whether the orientation of its edges is consistent. Namely, a face of a sketched embedding contains only edges that are either all counterclockwise or clockwise with respect to it. If this condition is not satisfied, the candidate sketched embedding is discarded.

In the second phase, for each generated sketched embedding, we compute all of its possible shapes and test the validity of each of them. Let $\overline{C}$ be a sketched embedding. For each representing cycle of $\overline{C}$, we generate all possible vertex-angle assignments for its vertices and roll-up numbers for its edges, keeping only those that satisfy Lemma 2. For every such assignment $\langle \overline{\phi}, \overline{\rho} \rangle$, let S_j be the set of orthogonal sketches $\langle C(H, X_j), \phi, \rho \rangle$ of B_j such that the restriction of $\langle \overline{\phi}, \overline{\rho} \rangle$ to the edges of $C(H, X_j)$ corresponds to $\langle \phi, \rho \rangle$. Similarly, let $S_{j'}$ be the set of orthogonal sketches $\langle C(H', X_{j'}), \phi', \rho' \rangle$ of $B_{j'}$ such that the restriction of $\langle \overline{\phi}, \overline{\rho} \rangle$ to the edges of $C(H', X_{j'})$ corresponds to $\langle \phi', \rho' \rangle$. If any of S_j and $S_{j'}$ is empty, $\langle \overline{\phi}, \overline{\rho} \rangle$ is discarded as it would not be possible to obtain it from B_j and $B_{j'}$. Finally, let b_j^* and $b_{j'}^*$ be the minimum number of bends among the orthogonal sketches of S_j and $S_{j'}$, respectively. The set E_i of edges shared by $C(H, X_j)$ and $C(H', X_{j'})$ contains edges (if any) that connect pairs of vertices of X_i and that belong to G. In particular, for each edge in E_i, the absolute value of its roll-up number corresponds to the number of bends along it. Hence, we verify that $b_j^* + b_{j'}^* - \sum_{(u,v) \in E_i} |\rho(u,v)| \le b$, otherwise we discard $\langle \overline{\phi}, \overline{\rho} \rangle$. We conclude:

Lemma 4. *Graph G admits an orthogonal drawing with at most b bends if and only if algorithm* ORTHOPLANTESTER *returns a positive answer.*

Lemma 5 (*). *Algorithm* ORTHOPLANTESTER *runs in $k^{O(k)}(b + \sigma)^k \log(b + \sigma) \cdot n$ time, where k is the treewidth of G, σ is the number of degree-two vertices of G, and b is the maximum number of bends.*

Proof sketch. Let T' be a tree-decomposition of G of width k and with $O(n)$ nodes. We compute, in $O(k \cdot n)$ time, a nice tree-decomposition T of G of width k and $O(n)$ nodes [8,27]. In what follows, we prove that ORTHOPLANTESTER spends $k^{O(k)}(b+\sigma)^k \log(b+\sigma)$ time for each bag X_i of T. The claim trivially follows if X_i is a leaf of T. If X_i forgets a vertex v, ORTHOPLANTESTER considers each orthogonal sketch of the child bag X_j, which are $k^{O(k)} \cdot (b+\sigma)^k$ by Lemma 3 (a bag of T has at most $k + 1$ vertices). For each orthogonal sketch, ORTHO-PLANTESTER removes v from each of its $O(1)$ representing cycles. Clearly, this can be done in $O(1)$ time. Also, ORTHOPLANTESTER removes possible duplicates in B_i. Note that, before removing the duplicates, the elements in B_i are at most as many as those in B_j. To efficiently remove the duplicates in B_i, we represent each orthogonal sketch as a concatenation of three arrays, encoding its sketched embedding (i.e., the rotation system and the outer face), its function ϕ, and its function ρ. Thus we have a set of $N = k^{O(k)}(b + \sigma)^k$ arrays each of size $O(k)$. Sorting the elements of this set, and hence deleting all duplicates, takes $O(k) \cdot N \log N$ time, which, with some manipulations, can be rewritten as $k^{O(k)}(b+\sigma)^k \log(b+\sigma)$. If X_i introduces a vertex v, ORTHOPLANTESTER considers each sketched embedding that can be extracted from B_j. Each of them, is then extended with v in all possible ways, which are $k^{O(k)}$ (observe that $|X_j| \le k-1$). Next ORTHOPLANTESTER generates at most $k^{O(k)}(\sigma + b)^k$ orthogonal sketches. We remark that, as explained in the proof of Lemma 3, for each cycle of length $w \le k + 1$ it suffices to generate the roll-up numbers of $w - 1$ edges. Moreover, for the edges incident to v the roll-up number is restricted to the range $[-b, +b]$ and it is subject to the additional constraint that the total number of bends should not exceed b. For each generated shape, ORTHOPLANTESTER checks whether the corresponding subsets of the values of ϕ and ρ exist in the orthogonal sketches of B_j having the fixed sketched embedding. This can be done by encoding the values of ϕ and ρ as two concatenated arrays, each of size $O(k)$, by sorting the set of arrays, and by searching among this set. Since the number of orthogonal sketches is $N = k^{O(k)}(\sigma + b)^k$, this can be done in $O(k) \cdot N \log N = O(k) \cdot k^{O(k)}(\sigma + b)^k \, O(k) \log(k(\sigma + b)) = k^{O(k)}(\sigma + b)^k \log(\sigma + b)$ for a fixed sketched embedding, and in $k^{O(k)}(\sigma + b)^k \log(\sigma + b)$ time in total. Finally, it remains to check Lemma 2 for each representing cycle, which takes $O(k^2)$ time for each of the $k^{O(k)} \cdot (\sigma + b)^k$ orthogonal sketches in B_i.   □

Lemmas 4 and 5 imply Theorem 1 and Corollary 1.

5 Applications

HV Planarity. Let G be a graph such that each edge is labeled H (horizontal) or V (vertical). HV-PLANARITY asks whether G has a planar drawing such that each edge is drawn as a horizontal or vertical segment according to its label, called a *HV-drawing* (see, e.g., [18,24]). This problem is NP-complete [18]. The next theorem follows from our approach.

Theorem 2 (*). *Let G be an n-vertex planar graph with σ vertices of degree two. Given a tree-decomposition of G of width k, there is an algorithm that solves HV-PLANARITY in $k^{O(k)}\sigma^k \log \sigma \cdot n$ time.*

The next corollary improves the $O(n^4)$ bound in [18].

Corollary 2. *Let G be an n-vertex series-parallel graph. There is a $O(n^3 \log n)$-time algorithm that solves HV-PLANARITY.*

Flexible Drawings. Let $G = (V, E)$ be a planar graph with vertex degree at most four, and let $\psi : E \to \mathbb{N}$. The FLEXDRAW problem [5,6] asks whether G admits an orthogonal drawing such that for each edge $e \in E$ the number of bends of e is $b(e) \le \psi(e)$. The problem becomes tractable when $\psi(e) \ge 1$ [5] for all edges, while it can be parameterized by the number of edges e such that $\psi(e) = 0$ [6]. By subdividing $\psi(e)$ times each edge e, we can conclude the following.

Theorem 3 (*). *Let $G = (V, E)$ be an n-vertex planar graph, and let $\psi : E \to \mathbb{N}$. Given a tree-decomposition of G of width k, there is an algorithm that solves FLEXDRAW in $k^{O(k)}(n \cdot b^*)^{k+1} \log(n \cdot b^*)$ time, where $b^* = \max_{e \in E} \psi(e)$.*

6 Open Problems

The results in this paper suggest some interesting questions. First, we ask whether ORTHOGONALPLANARITY is FPT when parameterized by the number of bends and by treewidth. Improving the time complexity of Corollary 1 is also an interesting problem on its own. Since HV-PLANARITY is NP-complete even for graphs with vertex degree at most three [18], another research direction is to devise new FPT approaches for HV-PLANARITY on subcubic planar graphs.

References

1. Bannister, M.J., Cabello, S., Eppstein, D.: Parameterized complexity of 1-planarity. J. Graph Algorithms Appl. **22**(1), 23–49 (2018). https://doi.org/10.7155/jgaa. 00457
2. Bannister, M.J., Eppstein, D.: Crossing minimization for 1-page and 2-page drawings of graphs with bounded treewidth. J. Graph Algorithms Appl. **22**(4), 577–606 (2018). https://doi.org/10.7155/jgaa.00479
3. Biedl, T.C., Kant, G.: A better heuristic for orthogonal graph drawings. Comput. Geom. **9**(3), 159–180 (1998)
4. Biedl, T.C., Lubiw, A., Petrick, M., Spriggs, M.J.: Morphing orthogonal planar graph drawings. ACM Trans. Algorithms **9**(4), 29:1–29:24 (2013)
5. Bläsius, T., Krug, M., Rutter, I., Wagner, D.: Orthogonal graph drawing with flexibility constraints. Algorithmica **68**(4), 859–885 (2014)
6. Bläsius, T., Lehmann, S., Rutter, I.: Orthogonal graph drawing with inflexible edges. Comput. Geom. **55**, 26–40 (2016)
7. Bodlaender, H.L.: A linear-time algorithm for finding tree-decompositions of small treewidth. SIAM J. Comput. **25**(6), 1305–1317 (1996)
8. Bodlaender, H.L., Bonsma, P., Lokshtanov, D.: The fine details of fast dynamic programming over tree decompositions. In: Gutin, G., Szeider, S. (eds.) IPEC 2013. LNCS, vol. 8246, pp. 41–53. Springer, Cham (2013). https://doi.org/10.1007/978-3-319-03898-8_5
9. Bodlaender, H.L., Drange, P.G., Dregi, M.S., Fomin, F.V., Lokshtanov, D., Pilipczuk, M.: A c^k n 5-approximation algorithm for treewidth. SIAM J. Comput. **45**(2), 317–378 (2016)
10. Chan, H.: A parameterized algorithm for upward planarity testing. In: Albers, S., Radzik, T. (eds.) ESA 2004. LNCS, vol. 3221, pp. 157–168. Springer, Heidelberg (2004). https://doi.org/10.1007/978-3-540-30140-0_16
11. Chang, Y., Yen, H.: On bend-minimized orthogonal drawings of planar 3-graphs. In: SOCG 2017. LIPIcs, vol. 77, pp. 29:1–29:15. Schloss Dagstuhl - Leibniz-Zentrum fuer Informatik (2017)
12. Di Battista, G., Eades, P., Tamassia, R., Tollis, I.G.: Graph Drawing. Prentice-Hall, Upper Saddle River (1999)
13. Di Battista, G., Liotta, G., Vargiu, F.: Spirality and optimal orthogonal drawings. SIAM J. Comput. **27**(6), 1764–1811 (1998)
14. Di Giacomo, E., Liotta, G., Montecchiani, F.: Sketched representations and orthogonal planarity of bounded treewidth graphs. CoRR abs/1908.05015 (2019). http://arxiv.org/abs/1908.05015
15. Didimo, W., Giordano, F., Liotta, G.: Upward spirality and upward planarity testing. SIAM J. Discrete Math. **23**(4), 1842–1899 (2009). https://doi.org/10.1137/070696854
16. Didimo, W., Liotta, G.: Computing orthogonal drawings in a variable embedding setting. In: Chwa, K.-Y., Ibarra, O.H. (eds.) ISAAC 1998. LNCS, vol. 1533, pp. 80–89. Springer, Heidelberg (1998). https://doi.org/10.1007/3-540-49381-6_10
17. Didimo, W., Liotta, G., Patrignani, M.: Bend-minimum orthogonal drawings in quadratic time. In: Biedl, T., Kerren, A. (eds.) GD 2018. LNCS, vol. 11282, pp. 481–494. Springer, Cham (2018). https://doi.org/10.1007/978-3-030-04414-5_34
18. Didimo, W., Liotta, G., Patrignani, M.: HV-planarity: algorithms and complexity. J. Comput. Syst. Sci. **99**, 72–90 (2019)

19. Downey, R.G., Fellows, M.R.: Parameterized Complexity. Monographs in Computer Science. Springer, Heidelberg (1999)
20. Dujmović, V., Fellows, M.R., Kitching, M., Liotta, G., McCartin, C., Nishimura, N., Ragde, P., Rosamond, F.A., Whitesides, S., Wood, D.R.: On the parameterized complexity of layered graph drawing. Algorithmica 52(2), 267–292 (2008)
21. Dujmović, V., Fernau, H., Kaufmann, M.: Fixed parameter algorithms for one-sided crossing minimization revisited. J. Discrete Algorithms 6(2), 313–323 (2008)
22. Dujmović, V., Whitesides, S.: An efficient fixed parameter tractable algorithm for 1-sided crossing minimization. Algorithmica 40(1), 15–31 (2004)
23. Duncan, C.A., Goodrich, M.T.: Planar orthogonal and polyline drawing algorithms. In: Handbook of Graph Drawing and Visualization, pp. 223–246. Chapman and Hall/CRC (2013)
24. Durocher, S., Felsner, S., Mehrabi, S., Mondal, D.: Drawing HV-restricted planar graphs. In: Pardo, A., Viola, A. (eds.) LATIN 2014. LNCS, vol. 8392, pp. 156–167. Springer, Heidelberg (2014). https://doi.org/10.1007/978-3-642-54423-1_14
25. Garg, A., Tamassia, R.: On the computational complexity of upward and rectilinear planarity testing. SIAM J. Comput. 31(2), 601–625 (2001)
26. Healy, P., Lynch, K.: Two fixed-parameter tractable algorithms for testing upward planarity. Int. J. Found. Comput. Sci. 17(5), 1095–1114 (2006). https://doi.org/10.1142/S0129054106004285
27. Kloks, T.: Treewidth, Computations and Approximations. LNCS, vol. 842. Springer, Heidelberg (1994)
28. Rahman, M.S., Egi, N., Nishizeki, T.: No-bend orthogonal drawings of subdivisions of planar triconnected cubic graphs. IEICE Trans. 88–D(1), 23–30 (2005)
29. Robertson, N., Seymour, P.D.: Graph minors. II. Algorithmic aspects of tree-width. J. Algorithms 7(3), 309–322 (1986)
30. Tamassia, R.: On embedding a graph in the grid with the minimum number of bends. SIAM J. Comput. 16(3), 421–444 (1987)
31. Zhou, X., Nishizeki, T.: Orthogonal drawings of series-parallel graphs with minimum bends. SIAM J. Discrete Math. 22(4), 1570–1604 (2008)

Collinearities

4-Connected Triangulations on Few Lines

Stefan Felsner[✉][iD]

Institut für Mathematik, Technische Universität Berlin, Berlin, Germany
felsner@math.tu-berlin.de

Abstract. We show that 4-connected plane triangulations can be redrawn such that edges are represented by straight segments and the vertices are covered by a set of at most $\sqrt{2n}$ lines each of them horizontal or vertical. The same holds for all subgraphs of such triangulations.

The proof is based on a corresponding result for diagrams of planar lattices which makes use of orthogonal chain and antichain families.

1 Introduction

Given a planar graph G we denote by $\pi(G)$ the minimum number ℓ such that G has a plane straight-line drawing in which the vertices can be covered by a collection of ℓ lines. Clearly $\pi(G) = 1$ if and only if G is a forest of paths. The set of graphs with $\pi(G) = 2$, however, is already surprisingly rich, it contains trees, outerplanar graphs and subgraphs of grids, see [1,7].

The parameter $\pi(G)$ has received some attention in recent years, here is a list of known results:

- It is NP-complete to decide whether $\pi(G) = 2$ (Biedl et al. [2]).
- For a stacked triangulation G, a.k.a. planar 3-tree or Apollonian network, let d_G be the stacking depth (e.g. K_4 has stacking depth 1). On this class lower and upper bounds on $\pi(G)$ are $d_G + 1$ and $d_G + 2$ respectively, see Biedl et al. [2] and for the lower bound also Eppstein [6, Thm. 16.13].
- Eppstein [7] constructed a planar, cubic, 3-connected, bipartite graph G_ℓ on $O(\ell^3)$ vertices with $\pi(G_\ell) \geq \ell$.

Related parameters have been studied by Chaplick et al. [3,4].

The main result of this paper is the following theorem.

Theorem 1. *If G is a 4-connected plane triangulation on n vertices, then* $\pi(G) \leq \sqrt{2n}$.

The result is not far from optimal since, using a constant number of additional vertices and many additional edges, the graph G_ℓ mentioned above can be transformed into a 4-connected plane triangulation, i.e., in the class we have graphs with $\pi(G) \in \Omega(n^{1/3})$.

The full version of the paper with complete proofs is available at http://page.math.tu-berlin.de/~felsner/Paper/grid-lines.pdf and on arXiv:1908.04524 as version 2.
Partially supported by DFG grant FE-340/11-1.
Figures in this paper use colors which can be seen in the online versions.

© Springer Nature Switzerland AG 2019
D. Archambault and C. D. Tóth (Eds.): GD 2019, LNCS 11904, pp. 395–408, 2019.
https://doi.org/10.1007/978-3-030-35802-0_30

The proof of the theorem makes use of transversal structures, these are special colorings of the edges of a 4-connected inner triangulation of a 4-gon with colors red and blue.

In Sect. 2.1 we survey transversal structures. The red subgraph of a transversal structure can be interpreted as the diagram of a planar lattice. Background on posets and lattices is given in Sect. 2.2. Dimension of posets and the connection with planarity are covered in Sect. 2.3. In Sect. 2.4 we survey orthogonal partitions of posets. The theory implies that every poset on n elements can be covered by at most $\sqrt{2n} - 1$ subsets such that each of the subsets is a chain or an antichain.

In Sect. 3 we prove that the diagram of a planar lattice on n elements has a straight-line drawing with vertices placed on a set of $\sqrt{2n} - 1$ lines. All the lines used for the construction are either horizontal or vertical.

Finally in Sect. 4 we prove the main result: transversal structures can be drawn on at most $\sqrt{2n} - 1$ lines. In fact, the red subgraph of the transversal structure has such a drawing by the result of the previous section. It is rather easy to add the blue edges to this drawing. Theorem 1 is obtained as a corollary.

2   Preliminaries

2.1   Transversal Structures

Let G be an internally 4-connected inner triangulation of a 4-gon, in other words G is a plane graph with quadrangular outer face, triangular inner faces, and no separating triangle. Let s, a, t, b be the outer vertices of G in clockwise order. A *transversal structure* for G is an orientation and 2-coloring of the inner edges of G such that

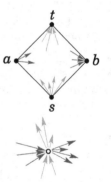

Fig. 1. The two local properties.

(1) All edges incident to s, a, t and b are red outgoing, blue outgoing, red incoming, and blue incoming, respectively.
(2) The edges incident to an inner vertex v come in clockwise order in four non-empty blocks consisting solely of red outgoing, blue outgoing, red incoming, blue incoming edges, respectively.

Figure 1 illustrates the properties and Fig. 2 shows an example. Transversal structures have been studied in [17], [12], and [13]. In particular it has been shown that every internally 4-connected inner triangulation of a 4-gon admits a transversal structure. Fusy [13] used transversal structures to prove the existence of straight-line drawings with vertices being placed on integer points (x, y) with $0 \le x \le W$, $0 \le y \le H$, and $H + W \le n - 1$.

An orientation of a graph G is said to be acyclic if it has no directed cycle. Given an acyclic orientation of G, a vertex having no incoming edge is called a *source*, and a vertex having no outgoing edge is called a *sink*. A *bipolar orientation* is an acyclic orientation with a unique source s and a unique sink t, cf. [11]. Bipolar orientations of plane graphs are also required to have s and t incident to the outer face.

A bipolar orientation of a plane graph has the property that at each vertex v the outgoing edges form a contiguous block and the incoming edges form a contiguous block. Moreover, each face f of G has two special vertices s_f and t_f such that the boundary of f consists of two non-empty oriented paths from s_f to t_f.

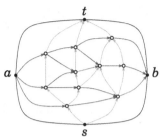

Fig. 2. An example of a transversal structure.

Let $G = (V, E)$ be an internally 4-connected inner triangulation of a 4-gon with outer vertices s, a, t, b in clockwise order, and let E_R and E_B respectively be the red and blue oriented edges of a transversal structure on G. We define $E_R^+ = E_R \cup \{(s,a),(s,b),(a,t),(b,t)\}$ and $E_B^+ = E_B \cup \{(a,s),(a,t),(s,b),(t,b)\}$, i.e., we think of the outer edges as having both, a red direction and a blue direction. The following has been shown in [17] and in [12].

Proposition 1. *The red graph $G_R = (V, E_R^+)$ and the blue graph $G_B = (V, E_B^+)$ are both bipolar orientations. G_R has source s and sink t, and G_B has source a and sink b.*

The following two properties are easy consequences of the previous discussion.

(R) The red and the blue graph are both transitively reduced, i.e., if (v, v') is an edge, then there is no directed path $v, u_1, \ldots, u_k, v'$ with $k \geq 1$.

(F) For every blue edge e there is a face f in the red graph such that e has one endpoint on each of the two oriented s_f to t_f paths on the boundary of f.

2.2 Posets

We assume basic familiarity with concepts and terminology for posets, referring the reader to the monograph [20] and survey article [21] for additional background material. In this paper we consider a poset $P = (X, <)$ as being equipped with a *strict* partial order.

A *cover relation* of P is a pair (x, y) with $x < y$ such that there is no z with $x < z < y$, we write $x \prec y$ to denote a cover relation of the two elements. A *diagram* (a.k.a. Hasse diagram) of a poset is an upward drawing of its transitive reduction. That is, X is represented by a set of points in the plane and a cover relation $x \prec y$ is represented by a y-monotone curve going upwards from x to y. In general these curves (edges) may cross each other but must not touch any vertices other than their endpoints. A diagram uniquely describes a poset, therefore, we usually show diagrams in our figures. A poset is said to be *planar* if it has a planar diagram.

It is well known that in discussions of graph planarity, we can restrict our attention to straight-line drawings. In fact, using for example a result of Schnyder [19], if a planar graph has n vertices, then it admits a planar straight-line drawing with vertices on an $(n-2) \times (n-2)$ grid. Discussions of planarity for posets can also be restricted to straight-line drawings; however, this may come at some cost in visual clarity. Di Battista et al. [5] have shown that an exponentially large grid may be required for upward planar drawings of directed acyclic planar graphs with straight lines. In the next subsection we will see that for certain planar posets the situation is more favorable.

2.3   Dimension of Planar Posets

Let $P = (X, <)$ be a poset. A *realizer* of P is a collection $L_1, L_2, \ldots, L_t$ of linear extensions of P such that $P = L_1 \cap L_2 \cap \cdots \cap L_t$. The *dimension* of $P = (X, <)$, denoted $\dim(P)$, is the least positive integer t such that P has a realizer of size t. Obviously, a poset P has dimension 1 if and only if it is a chain (total order). Also, there is an elementary characterization of posets of dimension at most 2 that we shall use.

Proposition 2. *A poset* $\mathbf{P} = (X, P)$ *has dimension as most* 2 *if and only if its incomparability graph is also a comparability graph.*

There are a number of results concerning the dimension of posets with planar order diagrams. Recall that an element is called a *zero* of a poset P when it is the unique minimal element. Dually, a *one* is a unique maximal element. A finite lattice, i.e., a poset which has well defined meet and join operations, always has a zero and a one.

The following result may be considered part of the folklore of the subject.

Theorem 2. *Let* P *be a finite lattice. Then* P *is planar if and only if it has dimension at most* 2.

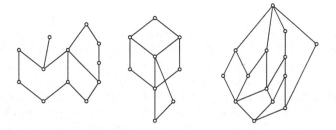

Fig. 3. Diagrams of a planar poset of dimension 3 (left), a non-planar lattice (middle), and a planar lattice (right).

For the reverse direction in the theorem, let P be a lattice of dimension at most 2. Let L_1 and L_2 be linear orders on X so that $P = L_1 \cap L_2$. For each

$x \in X$, and each $i = 1, 2$, let x_i denote the height of x in L_i. Then a planar diagram of P is obtained by locating each $x \in X$ at the point in the plane with integer coordinates (x_1, x_2) and joining points x and y with a straight segment when one of x and y covers the other in P. A pair of crossing edges in this drawing would violate the lattice property, indeed if $x \prec y$ and $x' \prec y'$ are two covers whose edges cross, then $x \leq y'$ and $x' \leq y$ whence x and x' have no unique least upper bound.

A planar digraph D with a unique sink and source, both of them on the outer face, and no transitive edges is the digraph of a planar lattice. Hence, the above discussion directly implies the following classical result.

Proposition 3. *A planar digraph D on n vertices with a unique sink and source on the outer face and no transitive edges has an upward planar drawing on an $(n-1) \times (n-1)$ grid.*

In this paper we will, henceforth, use the terms 2-dimensional poset and planar lattice respectively to refer to a poset $P = (X, <)$ together with a fixed ordered realizer $[L_1, L_2]$. In the case of the lattice, fixing the realizer can be interpreted as fixing a plane drawing of the diagram. By fixing the realizer of P we also have a well-defined *primary conjugate*, this is the poset Q on X with realizer $[L_1, \overline{L_2}]$, where $\overline{L_2}$ is the reverse of L_2. Define the *left of relation* on X such that x is left of y if and only of $x = y$ or x and y are incomparable in P and $x < y$ in Q.

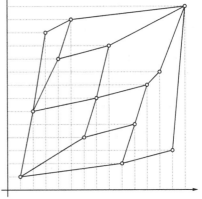

Fig. 4. The planar lattice from Fig. 3 with a realizer L_1, L_2.

2.4 Orthogonal Partitions of Posets

Let P be a finite poset, Dilworth's theorem states that *the maximum size of an antichain equals the minimum number of chains partitioning the elements of P.*

Greene and Kleitman [16] found a nice generalization of Dilworth's result. Define a *k-antichain* to be a family of k pairwise disjoint antichains.

Theorem 3. *For any partially ordered set P and any positive integer k*

$$\max \sum_{A \in \mathcal{A}} |A| = \min \sum_{C \in \mathcal{C}} \min(|C|, k)$$

where the maximum is taken over all k-antichains $\mathcal{A}$ and the minimum over all chain partitions $\mathcal{C}$ of P.

Greene [15] stated the dual of this theorem. Let an *ℓ-chain* be a family of ℓ pairwise disjoint chains.

Theorem 4. *For any partially ordered set P and any positive integer ℓ*

$$\max_{C \in \mathcal{C}} \sum |C| = \min_{A \in \mathcal{A}} \sum \min(|A|, \ell)$$

where the maximum is taken over all ℓ-chains C and the minimum over all antichain partitions A of P.

A further theorem of Greene [15] can be interpreted as a generalization of the Robinson-Schensted correspondence and its interpretation given by Greene [14]. With a partially ordered set P with n elements there is an associated partition λ of n, such that for the Ferrer's diagram $G(P)$ corresponding to λ we get:

Theorem 5. *The number of squares in the ℓ longest columns of G(P) equals the maximal number of elements covered by an ℓ-chain of P and the number of squares in the k longest rows of G(P) equals the maximal number of elements covered by a k-antichain.*

Figure 5 shows an example, in this case the Ferrer's diagram $G(P)$ corresponds to the partition $6 + 3 + 3 + 1 + 1 \models 14$. Several proofs of Greene's results are known, e.g. [8],[10], and [18]. For a not so recent, but at its time comprehensive survey we recommend [22].

The approach taken by Frank [10] is particularly elegant. Following Frank we call a chain family $\mathcal{C}$ and an antichain family $\mathcal{A}$ an *orthogonal pair* iff

1. $P = \left(\bigcup_{A \in \mathcal{A}} A \right) \cup \left(\bigcup_{C \in \mathcal{C}} C \right)$, and

2. $|A \cap C| = 1$ for all $A \in \mathcal{A}$, $C \in \mathcal{C}$.

If $\mathcal{C}$ is orthogonal to a k-antichain $\mathcal{A}$ and $\mathcal{C}^+$ is obtained from $\mathcal{C}$ by adding the rest of P as singletons, then

$$\sum_{A \in \mathcal{A}} |A| = \sum_{C \in \mathcal{C}^+} \sum_{A \in \mathcal{A}} |A \cap C| = \sum_{C \in \mathcal{C}^+} \min(|C|, k).$$

Thus $\mathcal{C}^+$ is a k optimal chain partition in the sense of Theorem 3. Similarly an ℓ optimal antichain partition in the sense of Theorem 4 can be obtained from an orthogonal pair $\mathcal{A}, \mathcal{C}$ where $\mathcal{C}$ is an ℓ-chain.

Using the minimum cost flow algorithm of Ford and Fulkerson [9], Frank proved the existence of a sequence of orthogonal chain and antichain families. This sequence is rich enough to allow the derivation of the whole theory. The sequence consists of an orthogonal pair for every point from the boundary of $G(P)$. With the point (k, ℓ) from the boundary of $G(P)$ we get an orthogonal pair $\mathcal{A}, \mathcal{C}$ such that $\mathcal{A}$ is a k-antichain and $\mathcal{C}$ an ℓ-chain, see Fig. 5. Since $G(P)$ is the Ferrer's diagram of a partition of n we can find a point (k, ℓ) on the boundary of $G(P)$ with $k + \ell \leq \sqrt{2n} - 1$ (This is because every Ferrer's shape of a partition of m which contains no point (x, y) with $x + y \leq s$ on the boundary contains the shape of the partition $(1, 2, \ldots, s + 1)$. From $m \geq \binom{s+2}{2}$ we get $s + 1 < \sqrt{2m}$).

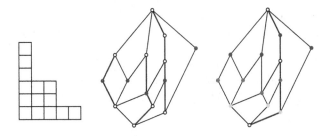

Fig. 5. The Ferrer's shape of the lattice L from Fig. 4 together with two orthogonal pairs of L corresponding to the boundary points $(1,3)$ and $(3,1)$ of $G(L)$; chains of $\mathcal{C}$ are blue, antichains of $\mathcal{A}$ are red, green, and yellow.

We will use the following corollary of the theory:

Corollary 1. *Let* $P = (X, <)$ *be a partial order on* n *elements, then there is an orthogonal pair* $\mathcal{A}, \mathcal{C}$ *where* $\mathcal{A}$ *is a* k*-antichain and* $\mathcal{C}$ *an* ℓ*-chain and* $k + \ell \leq \sqrt{2n} - 1$.

For our application we will need some additional structure on the antichains and chains of an orthogonal pair $\mathcal{A}, \mathcal{C}$.

The *canonical antichain partition* of a poset $P = (X, <)$ is constructed by recursively removing all minimal elements from P and make them one of the antichains of the partition. More explicitely $A_1 = \mathrm{Min}(X)$ and $A_j = \mathrm{Min}(X \setminus \bigcup\{A_i : 1 \leq i < j\})$ for $j > 1$. Note that by definition for each element $y \in A_j$ with $j > 1$ there is some $x \in A_{j-1}$ with $x < y$. Due to this property there is a chain of h elements in P if the canonical antichain partition consists of h non-empty antichains. This in essence is the dual of Dilworth's theorem, i.e., the statement: *the maximal size of a chain equals the minimal number of antichains partitioning the elements of* P.

Lemma 1. *Let* $\mathcal{A}, \mathcal{C}$ *be an orthogonal pair of* $P = (X, <)$ *and let* $P_{\mathcal{A}}$ *be the order induced by* P *on the set* $X_{\mathcal{A}} = \bigcup\{A : A \in \mathcal{A}\}$. *If* $\mathcal{A}'$ *is the canonical antichain partition of* $P_{\mathcal{A}}$, *then* $\mathcal{A}', \mathcal{C}$ *is again an orthogonal pair of* P

Proof. Let $\mathcal{A}$ be the family $A_1, \ldots, A_k$. Starting with this family we will change the antichains in the family while maintaining the invariant that the family of antichains together with $\mathcal{C}$ forms an orthogonal pair. At the end of the process the family of antichains will be the canonical antichain partition of $P_{\mathcal{A}}$.

The first phase of changes is the uncrossing phase. We iteratively choose two antichains A_i, A_j with $i < j$ from the present family and let $B_i = \{y \in A_i :$ there is an $x \in A_j$ with $x < y\}$ and $B_j = \{x \in A_j :$ there is a $y \in A_i$ with $x < y\}$. Define $A'_i = A_i - B_i + B_j$ and $A'_j = A_j - B_j + B_i$. It is easy to see that A'_i and A'_j are antichains and that the family obtained by replacing A_i, A_j by A'_i, A'_j is orthogonal to $\mathcal{C}$. This results in a family of k antichains such that if $i < j$ and $x \in A_i$ and $y \in A_j$ are comparable, then $x < y$.

The second phase is the push-down phase. We iteratively choose $i \in [k-1]$ and let $B = \{y \in A_{i+1} :$ there is no $x \in A_i$ with $x < y\}$ and define $A'_{i+1} = A_{i+1} - B$ and $A'_i = A_i + B$. It is again easy to see that A'_i and A'_{i+1} are antichains and that the family obtained by replacing A_i, A_{i+1} by A'_i, A'_{i+1} is orthogonal to $\mathcal{C}$. This results in a family of k antichains such that if $y \in A_{i+1}$, then there is an $x \in A_i$ with $x < y$. This implies that $A_j = \text{Min}(X_\mathcal{A} \setminus \bigcup\{A_i : 1 \leq i < j\})$, whence the family is the canonical antichain partition.   □

Let $P = (X, <)$ be a 2-dimensional poset with realizer $[L_1, L_2]$ and recall that the primary conjugate has realizer $[L_1, \overline{L_2}]$. The order Q corresponds to a transitive relation on the complement of the comparability graph of P, in particular chains of P and antichains of Q are in bijection.

The canonical antichain partition of Q yields the *canonical chain partition* of P. The canonical chain partition $C_1, C_2, \ldots, C_w$ of P can be characterized by the property that for each $1 \leq i < j \leq w$ and each element $y \in C_j$ there is some $x \in C_i$ with $x \parallel y$ and in L_1 element x comes before y. In particular C_1 is a maximal chain of P.

Let $\mathcal{A}, \mathcal{C}$ be an orthogonal pair of the 2-dimensional $P = (X, <)$. Applying the proof of Lemma 1 to the orthogonal pair $\mathcal{C}, \mathcal{A}$ of Q we obtain:

Lemma 2. *Let $\mathcal{A}, \mathcal{C}$ be an orthogonal pair of $P = (X, <)$ and let $P_\mathcal{C}$ be the order induced by P on the set $X_\mathcal{C} = \bigcup\{C : C \in \mathcal{C}\}$. If $\mathcal{C}'$ is the canonical chain partition of $P_\mathcal{C}$, then $\mathcal{C}', \mathcal{A}$ is again an orthogonal pair of P*

In a context where edges of the diagram are of interest, it is convenient to work with maximal chains. The canonical chain partition $C_1, C_2, \ldots, C_w$ of a 2-dimensional P induces a *canonical chain cover* of P which consists of maximal chains. With chain C_i associate a chain C_i^+ which is obtained by successively adding to C_i all compatible elements of $C_{i-1}, C_{i-2}, \ldots$ in this order. Alternatively C_i^+ can be described by looking at the conjugate of P with realizer $[\overline{L_1}, L_2]$ (this is the dual of the primary conjugate Q), and defining C_i^+ as the first chain in the canonical chain partition of the order induced by $\bigcup\{C_j : 1 \leq j \leq i\}$. The maximality of C_i^+ follows from the characterization of the canonical chain partition given above.

3   Drawing Planar Lattices on Few Lines

In this section we prove that planar lattices with n elements have a straight-line diagram with all vertices on a set of $\sqrt{2n} - 1$ horizontal and vertical lines. The following proposition covers the case where the lattice has an antichain partition of small size. We assume that a planar lattice is given with a realizer $[L_1, L_2]$ and, hence, with a fixed plane drawing of its diagram.

Proposition 4. *For any planar lattice $L = (X, <)$ with an extension $h : X \to \mathbb{R}$ of L there is a plane straight-line drawing Γ of the diagram D_L of L such that each element $x \in X$ is represented by a point with y-coordinate $h(x)$. Additionally all elements of the left boundary chain of D_L are aligned vertically in the drawing.*

The proof of this proposition can be found in the full version. The idea is to extend the drawing of a chain on a vertical line by ears. Such ear-extensions are also used in the proof of the next theorem.

Theorem 6. *For every planar lattice $L = (X, <)$ with $|X| = n$, there is a plane straight-line drawing of the diagram such that the elements are represented by points on a set of at most $\sqrt{2n} - 1$ lines. Additionally*

- *each of the lines is either horizontal or vertical,*
- *each crossing of a horizontal and a vertical line hosts an element of X.*

Proof. Let $\mathcal{A}, \mathcal{C}$ be an orthogonal pair of L such that $\mathcal{A}$ is a k-antichain, $\mathcal{C}$ an ℓ-chain, and $k + \ell \leq \sqrt{2n} - 1$. It follows from Corollary 1 that such a pair exists.

Since L has a fixed ordered realizer $[L_1, L_2]$, we can apply Lemma 1 to $\mathcal{A}$ and Lemma 2 to $\mathcal{C}$ to get an orthogonal pair $(A_1, \ldots, A_k), (C_1, \ldots, C_\ell)$ where the antichain family and the chain family are both canonical. Fix an extension $h : X \to \mathbb{R}$ of L with the property that $h(x) = i$ for all $x \in A_i$.

In the following we will construct a drawing Γ of D_L such that each element $x \in X$ is represented by a point with y-coordinate $h(x)$, and in addition all elements of chain C_i lie on a common vertical line $\mathbf{g}_i$ for $1 \leq i \leq \ell$. By Property 1 of orthogonal pairs, for each $x \in X$ there is an i such that $x \in A_i$ or a j such that $x \in C_j$ or both. Therefore, Γ will be a drawing such that the k horizontal lines $y = i$ with $i = 1, \ldots, k$ together with the ℓ vertical lines $\mathbf{g}_j$ with $j = 1, \ldots, \ell$ cover all the elements of X. Property 2 of orthogonal pairs implies the second extra property mentioned in the theorem.

If the number ℓ of chains is zero, then we get a drawing Γ with all the necessary properties from Proposition 4. Now let $\ell > 0$.

The chain family $C_1, \ldots, C_\ell$ is the canonical chain partition of the order induced on $X_C = \bigcup \{C_i : i = 1 \ldots \ell\}$. Let $C_1^+, \ldots, C_\ell^+$ be the corresponding canonical chain covering of X_C.

Let X_i for $1 \leq i \leq \ell$ be the set of all elements which are left of some element of C_i^+ in L, and let $X_{\ell+1} = X$. Define S_i as the suborder of L induced by X_i. Also let $Y_i = X_{i+1} - X_i + C_i^+$ and let T_i be the suborder of L induced by Y_i. Note the following properties of these sets and orders:

- $X_i \cap C_j = \emptyset$ for $1 \leq i < j \leq \ell$.
- Each S_i is a planar sublattice of L, its right boundary chain is C_i^+.
- T_i is a planar sublattice of L.

A drawing Γ_1 of S_1 with the right boundary chain being aligned vertically is obtained by applying Proposition 4 to the vertical reflection of the diagram $D_L[X_1]$ and reflecting the resulting drawing vertically.

We construct the drawing Γ of D_L in phases. In phase i we aim for a drawing Γ_{i+1} of S_{i+1} extending the given drawing Γ_i of S_i, i.e., we need to construct a drawing Λ_i of T_i such that

(1) The left boundary chain of Λ_i matches the right boundary chain of Γ_i.
(2) In Λ_i all elements of C_{i+1} are on a common vertical line $\mathbf{g}_{i+1}$.

The construction of Λ_i is done in three stages. First we extend C_i^+ to the right by adding 'ears'. Then we extend C_{i+1}^+ to the left by adding 'ears'. Finally we show that the left and the right part obtained from the first two stages can be combined to yield the drawing Λ_i.

To avoid extensive use of indices let $Y = Y_i$, $T = T_i$, $C^+ = C_i^+$, and let γ be a copy of the y-monotone polygonal right boundary of Γ_i, i.e., γ is a drawing of C. We initialize $\Lambda' = \gamma$.

A *left ear* of T is a face F in the diagram $D_L[Y]$ of T such that the left boundary of F is a subchain of the left boundary chain C^+ of $D_L[Y]$. The ear is *feasible* if the right boundary chain contains no element of C_{i+1}. Given a feasible ear we use the method from the proof of Proposition 4 to add F to γ. We represent the right boundary $z_0 < z_1 < \ldots < z_l$ excluding z_0 and z_l of F on a vertical line **g** by points $q_1, \ldots, q_{l-1}$ with y-coordinates as prescribed by h. The points q_0 and q_l representing z_0 and z_l respectively are already represented on γ. Then we place **g** at some distance β to the right of γ. The value of β has to be chosen large enough to ensure that edges q_0, q_1 and q_{l-1}, q_l are drawn such that they do not interfere with γ. Let Λ' be the drawing augmented by the polygonal path $q_0, q_1, \ldots, q_{l-1}, q_l$ and let C^+ again refer to the right boundary chain γ of Λ'. Delete all elements of the left boundary of F except z_0 and z_l from Y and T. This *shelling of a left ear* from T is iterated until there remains no left feasible ear. Upon stopping we have a drawing Λ' which can be glued to the right side of Γ_i. Let γ' be the right boundary chain of Λ'.

In the second stage the procedure is quite similar; starting from $C = C_{i+1}$ on a line **g** right ears taken from T are added to the drawing until there remains no feasible right ear. This yields a drawing Λ'' with left boundary γ''.

In the final stage we have to combine the drawings Λ', Λ'' into a single drawing. This is done by drawing the edges and chains which remain in T between the two boundary chains as straight segments between γ' and γ''. This will be possible because we can shift γ' and γ'' as far apart horizontally as necessary.

First we draw all the edges connecting the two chains. If we place γ' and γ'' at sufficient horizontal distance, then the edges of E can be drawn such that they do not interfere (introduce crossings) with γ' and γ''. Let Λ be the drawing consisting of γ', γ'', and the deges between them. An important feature of Λ is that if we move the two chains γ' and γ'' even further apart the drawing keeps the needed properties, i.e., the height of elements remains unaltered, vertices of a chain which should be vertically aligned remain vertically aligned because they are vertically aligned in either γ' or γ'' and the drawing is crossing-free.

Now assume that T contains elements which are not represented in Λ. Let B be a connected component of such elements where connectivity is with respect to D_L. All the elements of B have to be placed in a face F_B of Λ. Let δ' and δ'' be the left and right boundary of F_B.

In the following we will repeat the choice of a component B and a chain C from B which is to be drawn in the corresponding face F_B of Λ such that the minimum and the maximum of C have connecting edges to the two sides of the boundary of F_B. Let us consider the case that in D_L the maximum of C has an outgoing edge to an element which is represented by a point $p \in \delta'$ and the minimum of C has an incoming edge from an element represented by $q \in \delta''$. We represent the elements of C as points on the prescribed heights on a line segment ζ with endpoints p and q. It may become necessary to stretch the face horizontally to be able to place C. In this case we stretch the whole drawing between γ' and γ'' with a uniform stretch factor. There may be additional edges between elements of C and elements on δ' and δ''. They can also be drawn without crossing when the distance of δ' and δ'' exceeds some value b.

Stretching the whole drawing between γ' and γ'' allows us to draw the segment ζ and additional edges inside of F_B because of the following invariant. *For each face F of the drawing Λ and two points x and y from the boundary of F it holds that: if the segment x, y is not in the interior of F, then the parts of the boundary obstructing the segment x, y belong to γ' or γ''.*

When including a chain C in the drawing Λ, we place the elements of C at the prescribed heights on a common line segment ζ. This ensures that each new element contributes convex corners in all incident faces. Hence, new elements can not obstruct a visibility within a face. Therefore, obstructing corners correspond to elements of γ' or γ'' and the invariant holds.

Now consider the case where maximum and minimum of the chain C connect to two elements p and q on the same side of F. Since γ' and γ'' do not admit ear extensions we know that not both of p and q belong to one of γ' and γ''. If the segment from p to q is obstructed, then the invariant ensures that with sufficient horizontal stretch the segment ζ connecting p and q will be inside F. Hence, chain C can be drawn and Λ can be extended.

When there remains no component B containing a chain C which can be included in the drawing using the above strategy, then either all elements of Y are drawn or we have the following: every component B only connects to elements of a line segment ζ_B.

In this situation B is kind of a big ear over ζ_B. The details of how to draw B are given in the full version. But note, that in this situation we will not maintain or need the invariant.

Glueing the drawings Λ' with Λ at the polygonal path γ' and Λ with Λ'' at γ'' (a y-monotone collection of paths) yields a drawing Λ_i of T_i. The drawing Λ_i can be glued to Γ_i to form a drawing Γ_{i+1} of S_{i+1}. Eventually the drawing Γ_ℓ will be constructed. From there the drawing $\Gamma = \Gamma_{\ell+1}$ is obtained by adding some left ears. □

4 Transversal Structures on Few Lines

Theorem 7. *For every internally 4-connected inner triangulation of a 4-gon $G = (V, E)$ with n vertices there is a planar straight-line drawing such that the vertices are represented by points on a set of at most $\sqrt{2n} - 1$ lines. Additionally*

- *each of the lines is either horizontal or vertical,*
- *each crossing point of a horizontal and a vertical line hosts a vertex.*

Due to Theorem 6 it is possible to draw the red subgraph G_R of a transversal structure on $\sqrt{2n} - 1$ lines. It would be nice if we could include the blue edges of the transversal structure in the drawing. This, however, may yield crossings. Therefore we have to go through the proof of Theorem 6 and take care of blue edges while constructing the drawing of the red graph. The proof can be found in the full version. An example for an intermediate stage of the construction is shown in Fig. 6.

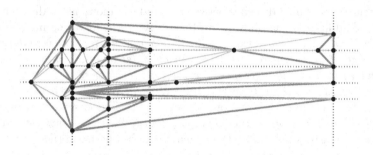

Fig. 6. A partially drawn transversal structure. The figure shows a drawing of Γ_4, these are the vertices left of some element in C_4^+ together with the induced edges.

It remains to see how Theorem 1 follows from Theorem 7. Let G be a 4-connected triangulation and let G' be obtained from G by deleting one of the outer edges. Now G' is an internally 4-connected inner triangulation of a 4-gon. Label the outer vertices of G' such that the deleted edge is the edge s, t. Slightly stretching Theorem 7 we prescribe $h(s) = -\infty$ and $h(t) = \infty$, this yields a planar straight-line drawing Γ of G' such that the vertices except s and t are represented by points on a set of at most $\sqrt{2n} - 1$ lines and the edges connecting to s and t are vertical rays. Moreover with every edge v, s or v, t there is an open cone K containing the vertical ray, such that the point representing v is the apex of K and this is the only vertex contained in K. Now let $\mathbf{g}$ be a vertical line which is disjoint from Γ. On $\mathbf{g}$ we find a point p_s which is contained in all the upward cones and a point p_t contained in all the downward cones. Taking p_s and p_t as representatives for s and t we can tilt the rays and make them finite edges ending in p_s and p_t respectively, and in addition draw the edge p_s, p_t.

We conclude with a remark and two open problems.

- Our results are constructive and can be complemented with algorithms running in polynomial time.
- Is $\pi(G) \in O(\sqrt{n})$ for every planar graph G on n vertices?
- What size of a grid is needed for drawings of 4-connected plane graphs on $O(\sqrt{n})$ lines?

Acknowledgments. Work on this problem began at the 2018 Bertinoro Workshop of Graph Drawing. I thank the organizers of the event for making this possible. Special thanks go to Pavel Valtr, Alex Pilz and Torsten Ueckerdt for helpful discussions.

References

1. Bannister, M.J., Devanny, W.E., Dujmovic, V., Eppstein, D., Wood, D.R.: Track layouts, layered path decompositions, and leveled planarity. Algorithmica **81**, 1561–1583 (2019)
2. Biedl, T., Felsner, S., Meijer, H., Wolff, A.: Line and plane cover numbers revisited. In: Archambault, D., Tóth, C.D. (eds.) GD 2019, LNCS 11904, pp. 409–415. Springer, Heidelberg (2019)
3. Chaplick, S., Fleszar, K., Lipp, F., Ravsky, A., Verbitsky, O., Wolff, A.: Drawing graphs on few lines and few planes. In: Hu, Y., Nöllenburg, M. (eds.) GD 2016. LNCS, vol. 9801, pp. 166–180. Springer, Cham (2016). https://doi.org/10.1007/978-3-319-50106-2_14
4. Chaplick, S., Fleszar, K., Lipp, F., Ravsky, A., Verbitsky, O., Wolff, A.: The complexity of drawing graphs on few lines and few planes. In: Ellen, F., Kolokolova, A., Sack, J.R. (eds.) WADS 2017. LNCS, vol. 10389, pp. 265–276. Springer, Cham (2017). https://doi.org/10.1007/978-3-319-62127-2_23
5. Di Battista, G., Tamassia, R., Tollis, I.G.: Area requirement and symmetry display of planar upward drawings. Discrete Comput. Geom. **7**, 381–401 (1992)
6. Eppstein, D.: Forbidden Configurations in Discrete Geometry. Cambridge University Press, Cambridge (2018)
7. Eppstein, D.: Cubic planar graphs that cannot be drawn on few lines. In: Proceedings of SoCG 2019. LIPIcs (2019)
8. Fomin, S.V.: Finite partially ordered sets and Young tableaux. Soviet Math. Dokl. **19**, 1510–1514 (1978)
9. Ford Jr., L.R., Fulkerson, D.R.: Flows in Networks. Princeton University Press, Princeton (1962)
10. Frank, A.: On chain and antichain families of partially ordered sets. J. Combin. Theory Ser. B **29**, 176–184 (1980)
11. de Fraysseix, H., de Mendez, P.O., Rosenstiehl, P.: Bipolar orientations revisited. Discrete Appl. Math. **56**(2–3), 157–179 (1995)
12. Fusy, E.: Combinatoire des cartes planaires et applications algorithmiques. Ph.D. thesis, LIX Polytechnique (2007). www.lix.polytechnique.fr/~fusy/Articles/these_eric_fusy.pdf
13. Fusy, E.: Transversal structures on triangulations: a combinatorial study and straight-line drawings. Discr. Math. **309**(7), 1870–1894 (2009)
14. Greene, C.: An extension of Schensted's theorem. Adv. Math. **14**, 254–265 (1974)
15. Greene, C.: Some partitions associated with a partially ordered set. J. Combin. Theory Ser. A **20**, 69–79 (1976)
16. Greene, C., Kleitman, D.J.: The structure of Sperner k-families. J. Combin. Theory Ser. A **20**, 41–68 (1976)
17. Kant, G., He, X.: Regular edge labeling of 4-connected plane graphs and its applications in graph drawing problems. Theor. Comput. Sci. **172**, 175–193 (1997)
18. Saks, M.: A short proof of the existence of k-saturated partitions of partially ordered sets. Adv. Math. **33**, 207–211 (1979)
19. Schnyder, W.: Embedding planar graphs on the grid. In: Proceedings of SODA 1990, pp. 138–148. ACM-SIAM (1990)
20. Trotter, W.T.: Combinatorics and Partially Ordered Sets: Dimension Theory. Johns Hopkins Series in the Mathematical Sciences. Johns Hopkins University Press, Baltimore (1992)

21. Trotter, W.T.: Partially ordered sets. In: Graham, R.L., Grötschel, M., Lovás, L. (eds.) Handbook of Combinatorics, North-Holland, vol. I, pp. 433–480 (1995)
22. West, D.B.: Parameters of partial orders and graphs: packing, covering and representation. In: Rival, I. (ed.) Graphs and Orders, pp. 267–350. D. Reidel, Dordrecht (1985)

Line and Plane Cover Numbers Revisited

Therese Biedl[1] , Stefan Felsner[2] , Henk Meijer[3], and Alexander Wolff[4]([⊠])

[1] University of Waterloo, Waterloo, Canada
[2] TU Berlin, Berlin, Germany
[3] University College Roosevelt, Middelburg, The Netherlands
[4] Universität Würzburg, Würzburg, Germany
usetheemailaddressonmyhomepage@gmail.com
http://www1.informatik.uni-wuerzburg.de/en/staff/wolff-alexander

Abstract. A measure for the visual complexity of a straight-line crossing-free drawing of a graph is the minimum number of lines needed to cover all vertices. For a given graph G, the minimum such number (over all drawings in dimension $d \in \{2,3\}$) is called the *d-dimensional weak line cover number* and denoted by $\pi_d^1(G)$. In 3D, the minimum number of *planes* needed to cover all vertices of G is denoted by $\pi_3^2(G)$. When edges are also required to be covered, the corresponding numbers $\rho_d^1(G)$ and $\rho_3^2(G)$ are called the *(strong) line cover number* and the *(strong) plane cover number*.

Computing any of these cover numbers—except $\pi_2^1(G)$—is known to be NP-hard. The complexity of computing $\pi_2^1(G)$ was posed as an open problem by Chaplick et al. [WADS 2017]. We show that it is NP-hard to decide, for a given planar graph G, whether $\pi_2^1(G) = 2$. We further show that the universal stacked triangulation of depth d, G_d, has $\pi_2^1(G_d) = d+1$. Concerning 3D, we show that any n-vertex graph G with $\rho_3^2(G) = 2$ has at most $5n - 19$ edges, which is tight.

1 Introduction

Recently, there has been considerable interest in representing graphs with as few objects as possible. The idea behind this objective is to keep the visual complexity of a drawing low for the observer. The types of objects that have been used are straight-line segments [5,8,14,15] and circular arcs [14,16].

Chaplick et al. [3] considered *covering* straight-line drawings of graphs by lines or planes and defined the following new graph parameters. Let $1 \leq l < d$, and let G be a graph. The *l-dimensional affine cover* number of G in $\mathbb{R}^d$, denoted by $\rho_d^l(G)$, is defined as the minimum number of l-dimensional planes in $\mathbb{R}^d$ such that G has a crossing-free straight-line drawing that is contained in the union of these planes. The *weak l-dimensional affine cover number* of G in $\mathbb{R}^d$, denoted by $\pi_d^l(G)$, is defined similarly to $\rho_d^l(G)$, but under the weaker restriction that only the vertices are contained in the union of the planes. Clearly, $\pi_d^l(G) \leq \rho_d^l(G)$,

The full version of this article is available at Arxiv [2]. S.F. was supported by DFG grant FE 340/11-1, A.W. by DFG grant WO 758/9-1, and T.B. by NSERC.

© Springer Nature Switzerland AG 2019
D. Archambault and C. D. Tóth (Eds.): GD 2019, LNCS 11904, pp. 409–415, 2019.
https://doi.org/10.1007/978-3-030-35802-0_31

and if $l' \leq l$ and $d' \leq d$ then $\pi_d^l(G) \leq \pi_{d'}^{l'}(G)$ and $\rho_d^l(G) \leq \rho_{d'}^{l'}(G)$. It turns out that it suffices to study the parameters $\rho_2^1, \rho_3^1, \rho_3^2,$ and $\pi_2^1, \pi_3^1, \pi_3^2$:

Theorem 1 (Collapse of the Affine Hierarchy [3]). *For any integers $1 \leq l < 3 \leq d$ and for any graph G, it holds that $\pi_d^l(G) = \pi_3^l(G)$ and $\rho_d^l(G) = \rho_3^l(G)$.*

Disproving a conjecture of Firman et al. [12], Eppstein [10] constructed planar, cubic, 3-connected, bipartite graphs on n vertices with $\pi_2^1(G) \geq n^{1/3}$. Answering a question of Chaplick et al. [3] he also constructed a family of subcubic series-parallel graphs with unbounded π_2^1-value. Felsner [11] proved that, for every 4-connected plane triangulation G on n vertices, it holds that $\pi_2^1(G) \leq \sqrt{2n}$. Chaplick et al. [4] also investigated the complexity of computing the affine cover numbers. Among others, they showed that in 3D, for $l \in \{1, 2\}$, it is NP-complete to decide whether $\pi_3^l(G) \leq 2$ for a given graph G. In 2D, the question has still been open, but a related question was raised by Dujmović et al. [7] already in 2004. They investigated so-called *track layouts* which are defined as follows. A graph admits a k-track layout if its vertices can be partitioned into k ordered independent subsets such that any pair of subsets induces a plane graph (w.r.t. the order of the subsets). The track number of a graph G, $\text{tn}(G)$, is the smallest k such that G admits a k-track layout. See also [6] for some recent developments. Note that in general $\pi_2^1(G) \neq \text{tn}(G)$; for example, $\pi_2^1(K_4) = 2$, whereas $\text{tn}(K_4) = 4$. Note further that a 3-track layout is necessarily plane (which is not the case for k-track layouts with $k > 3$). Dujmović posed the computational complexity of k-track layout as an open question.

While it is easy to decide efficiently whether a graph admits a 2-track layout, Bannister et al. [1] answered the open question of Dujmović et al. already for 3-track layouts in the affirmative. They first showed that a graph has a leveled planar drawing if and only if it is bipartite and has a 3-track layout. Combining this results with the NP-hardness of level planarity, proven by Heath and Rosenberg [13], immediately showed that it is NP-hard to decide whether a given graph has a 3-track layout. For $k > 3$, deciding the existence of a k-track layout is NP-hard, too, since it suffices to add to the given graph $k - 3$ new vertices each of which is incident to all original vertices of the graph [1].

Our contribution. We investigate several problems concerning the *weak line cover number* $\pi_2^1(G)$ and the *strong plane cover number* $\rho_3^2(G)$:

- We settle the open question of Chaplick et al. [4, p. 268] by showing that it is NP-hard to test whether, for a given planar graph G, $\pi_2^1(G) = 2$; see Sect. 2.
- We show that G_d, the universal stacked triangulation of depth d, (which has treewidth 3) has $\pi_2^1(G_d) = d + 1 = \log_3(2n_d - 5) + 1$, where n_d is the number of vertices of G_d; see Sect. 3.
- Eppstein has identified classes of treewidth-2 graphs with unbounded π_2^1-value. We give an easy direct argument showing that some 2-tree H_d with n_d' vertices has $\pi_2^1(H_d) \in \Omega(\log n_d')$; see the full version [2].
- Concerning 3D, we show that any n-vertex graph G with $\rho_3^2(G) = 2$ has at most $5n - 19$ edges; see Sect. 4. This bound is tight.

2 Complexity of Computing Weak Line Covers in 2D

In this section we investigate the computational complexity of deciding whether a graph can be drawn on two lines.

Theorem 2. *It is NP-hard to decide whether a given plane (or planar) graph G admits a drawing with $\pi_2^1(G) = 2$.*

Proof. Our proof is by reduction from the problem LEVEL PLANARITY, which Heath and Rosenberg [13] proved to be NP-hard. The problem is defined as follows. A planar graph G is *leveled-planar* if its vertex set can be partitioned into sets $V_1, \ldots, V_m$ such that G has a planar straight-line drawing where, for every $i \in \{1, \ldots, m\}$, vertices in V_i lie on the vertical line $\ell_i: y = i$ and each edge $v_j v_k$ of G connects two vertices on consecutive lines (that is, $|j - k| = 1$).

Chaplick et al. [3] have shown that every leveled-planar graph can be drawn on two lines. The converse, however, is not true. For example, K_4 is not leveled-planar, but $\pi_2^1(K_4) = 2$. Therefore, we modify the given graph in three ways. (a) We replace each edge of G by a $K_{2,4}$-gadget where the two nodes in one set of the bipartition replace the endpoints of the former edge; see Fig. 1a. (b) We add to the graph G' that resulted from the previous step a new subgraph G_0 (two copies of K_4 sharing exactly two vertices), which we connect by a path to a vertex on the outer face of G. (If the outer face is not fixed, we can try each vertex.) In Fig. 1b, G_0 is yellow and the path is red. The length L of the path is any upper bound on the number of levels of G', e.g., the diameter of G' (plus 1). (c) We attach to G_0 a triangulated spiral S (dark green in Fig. 1b). The spiral makes $L+2$ right turns; its final vertex is identified with the outermost vertex of the previous turn. Hence, apart from its many triangular faces, the graph $S + G_0$ has a large inner face F of degree $2(L+2)$ and a quadrangular outer face. Let G'' be the resulting graph. It remains to show that G is leveled-planar if and only if $\pi_2^1(G'') = 2$.

"$\Rightarrow$": Fix a leveled-planar drawing of G. By doubling the layers and using the new layers to place the large sides of $K_{2,4}$'s, one easily sees that G' is also leveled-planar, see Fig. 1a. As shown in Fig. 1b, the large inner face F of $S + G_0$ can be drawn so that it partitions the halflines emanating from the origin into L levels. (It is no problem that consecutive levels are turned by $90°$.) Since we chose L large enough (in particular $L \geq 2m - 1$), we can easily draw G' inside F. Note that the red path attached to G_0 is long enough to reach any vertex on the outer face of G'. Hence, $\pi_2^1(G'') = 2$.

"$\Leftarrow$": Fix a drawing of G'' on two lines. The two lines cannot be parallel since G'' contains $K_{2,4}$ and is not outer-planar; so after translation and/or skew we may assume that these two lines are the two coordinate axes. It is not hard to verify that G_0 must be drawn such that the origin is in its interior, at the common edge of the two K_4's. Furthermore, given this drawing of G_0, the 3-connected spiral S must be drawn as in Fig. 1b. Due to planarity and the fact that G' is connected to G_0 via the red path, G' can only be drawn in the interior of F. The drawing of $S + G_0$ partitions the halflines emanating from the origin

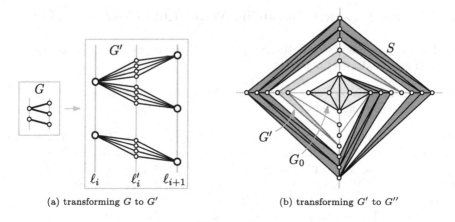

(a) transforming G to G' (b) transforming G' to G''

Fig. 1. Our reduction from LEVEL PLANARITY

into levels, which we number $1, 2, \ldots$ starting from the innermost level that contains a vertex of G'. Inside this face, the only way to draw the $K_{2,4}$-gadgets is as in Fig. 1a, spanning three consecutive levels. This forces all vertices of G to be placed on the odd-numbered levels and the vertices in $G' - G$ on the even-numbered levels. Now we can get a level assignment for G by reverting the transformation in Fig. 1a. Hence, G is leveled-planar.

This shows that our reduction is correct. It runs in polynomial time.   □

3   Weak Line Covers of Planar 3-Trees in 2D

In this section we consider the weak line cover number π_2^1 for planar graphs, i.e., we are interested in crossing-free straight-line drawings with vertices located on a small collection of lines. Clearly $\pi_2^1(G) = 1$ if and only if G is a forest of paths. The set of graphs with $\pi_2^1(G) = 2$, however, is already surprisingly rich, it contains all trees, outerplanar graphs and subgraphs of grids [1,10].

Stacked triangulations, a.k.a. planar 3-trees or Apollonian networks, are obtained from a triangle by repeatedly selecting a triangular face T and adding a new vertex (the *vertex stacked inside T*) inside T with edges to the vertices of T. This subdivides T into three smaller triangles, the *children* of T.

For $d \geq 0$ let G_d be the *universal stacked triangulation of depth d*, defined as follows. The graph G_0 is a triangle T_0, and G_d (for $d \geq 1$) is obtained from G_{d-1} by adding a stack vertex in each bounded face of G_{d-1}. Graph G_d has $n_d = \frac{1}{2}(3^d + 5)$ vertices and 3^d bounded faces. We show that its weak line cover number is $d + 1 = \log_3(2n_d - 5) + 1 \in \Theta(\log n_d)$. (A lower bound of d can also be found in Eppstein's recent book [9, Thm. 16.13].)

Theorem 3. *For $d \geq 1$ it holds that $\pi(G_d) = d + 1$.*

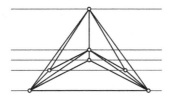

Fig. 2. A drawing of G_2 that can be extended to a drawing of G_3 on 5 parallel lines.

Proof. Here we prove only the lower bound; the construction for the upper bound is illustrated in Fig. 2 and given in the full version. Let $\mathcal{L}$ be a family of lines covering the vertices of a drawing of G_d. Let a, b, and c be the vertices of T_0. We first argue that at least d lines are needed to cover $V \setminus T_0$. Let x_1 be stacked into T_0. There is a line $L_1 \in \mathcal{L}$ covering x_1. Note that L_1 can intersect only two of the three child triangles of T_0 (where "intersect" here means "in the interior"). Let T_1 be a child triangle avoided by L_1, and let x_2 be the vertex stacked into T_1. There is a line $L_2 \in \mathcal{L}$ covering x_2. Let T_2 be a child triangle of T_1 avoided by L_2. Iterating this yields d pairwise distinct lines in $\mathcal{L}$.

To find one additional line in $\mathcal{L}$, we distinguish some cases. If a line $L \in \mathcal{L}$ covers two vertices of T_0, then it covers no inner vertex, and we are done.

Assume some line $L_a \in \mathcal{L}$ intersects x_1 and one vertex of T_0, say a. Let L_b and L_c be the lines intersecting b and c. The lines L_a, L_b, and L_c are pairwise different, else we are in the previous case. Of the three child triangles of T_0, at most one is intersected by L_a and at most two each are intersected by L_b and L_c. Therefore, some child triangle T_1 of T_0 is intersected by at most one of L_a, L_b, or L_c. The graph G_{d-1} inside T_1 requires at least $d-1$ lines for its interior points, and at most one of those lines is L_a, L_b, or L_c, so in total at least $d+1$ lines are needed.

The argument is similar if no line covers two of a, b, c, and x_1. The four distinct lines supporting a, b, c, and x_1 then intersect at most two child triangles each. So one child triangle T_1 is intersected by at most two of these lines. Combining the $d-1$ lines needed for the interior of T_1 with the two lines that do not intersect it, shows that $d+1$ lines are needed.   □

4   Maximal Graphs on Two Planes in 3D

We now switch to dimension $d = 3$ and the strong cover number. Obviously any graph G with a drawing that is covered by two planes has at most $6n - 12$ edges since it is the union of two planar graphs. Using maximality arguments and counting, we show that in fact G has at most $5n - 19$ edges if $n \geq 7$. (The restriction $n \geq 7$ is required since for $n = 3, 4, 5, 6$ we can have $3, 6, 9, 12$ edges.)

We argue first that our bound is tight. The *spine* is the intersection of two planes A and B. Put a path with $n - 4$ vertices on the spine. Add one vertex in each of the four halfplanes and connect each of these vertices to all vertices on the spine and to the vertex on the opposite halfplane. (We provide a figure in the full version.) This yields $n - 5$ edges on the path and $2(n - 4) + 1$ edges in each of the two planes, so $5n - 19$ edges in total.

Theorem 4. *Any graph G with $\rho_3^2(G) = 2$ and $n \geq 7$ vertices has at most $5n - 19$ edges.*

Proof. Fix a drawing of G on planes A and B, inducing planar graphs G_A and G_B within those planes. Let G_A^+ and G_B^+ be the graphs obtained from G_A and G_B by adding any edge that can be inserted without crossing, within the same plane, and with at most one bend on the spine. Clearly it suffices to argue that G_A^+ and G_B^+ together have at most $5n - 19$ edges. Let s be the number of vertices on the spine, let a be the number of vertices of G_A^+ not on the spine, and let b be the number of vertices of G_B^+ not on the spine. Clearly, $a + b + s = n$. We may assume $a \leq b$. We also assume that $1 \leq s \leq n - 4$ and that at least one edge of G_A^+ crosses the spine (so $2 \leq a \leq b$); see the full version.

Let t be the number of edges drawn along the spine. These are the only edges that belong to G_A^+ and G_B^+. Since G_A^+ and G_B^+ have at least three vertices each, we can bound the number of edges of G, $m(G)$, as follows:

$$m(G) \leq m(G_A^+) + m(G_B^+) - t \leq 3(s + a) - 6 + 3(s + b) - 6 - t \qquad (1)$$
$$= 3n - 12 + 3s - t \leq 4n - 16 + 2s - t.$$

So we must show that $2s - t \leq n - 3$. Let an *internal gap* be a line segment connecting two consecutive, non-adjacent vertices on the spine. There are $s - t - 1$ internal gaps. Let the *external gap* be the two infinite parts of the spine. Note that at least one edge of G_A^+ must cross the external gap, because G_A^+ has at least one vertex on each side of the gap, and we could connect the extreme such vertices (or re-route an existing edge) to cross the external gap, perhaps using a bend on the spine. We may further assume that even after such re-routing every internal gap is crossed by at least one edge of G_A^+. Otherwise we could delete all edges of G_B^+ passing through the gap, insert the edge between the spine vertices, and re-triangulate the drawing of G_B^+ where we removed edges. This would remove an internal gap, but would not decrease the number of edges. Since no edge can cross two gaps, at least $s - t$ edges of G_A^+ cross gaps. These edges form a planar bipartite graph with at most a vertices; therefore $s - t \leq 2a - 3$. [1] This yields $2s - t \leq s + 2a - 3 \leq s + a + b - 3 = n - 3$ as desired. ☐

We conjecture that the following more general statement holds:

Any n-vertex graph G with $\rho_3^2(G) = k$ has at most $(2k+1)(n-2k)+k-1$ edges, for all large enough n.

Acknowledgments. This research started at the Bertinoro Workshop on Graph Drawing 2017. We thank the organizers and other participants, in particular Will Evans, Sylvain Lazard, Pavel Valtr, Sue Whitesides, and Steve Wismath. We also thank Alex Pilz and Piotr Micek for enlightening conversations.

---

[1] One might be tempted to write a bound of $2a - 4$ here, but we must allow for the possibility of $a = 2$, in case of which the planar bipartite graph may have $1 = 2a - 3$ edges.

References

1. Bannister, M.J., Devanny, W.E., Dujmović, V., Eppstein, D., Wood, D.R.: Track layouts, layered path decompositions, and leveled planarity. Algorithmica **81**(4), 1561–1583 (2019). https://doi.org/10.1007/s00453-018-0487-5
2. Biedl, T., Felsner, S., Meijer, H., Wolff, A.: Line and plane cover numbers revisited. Arxiv report (2019). http://arxiv.org/abs/1908.07647
3. Chaplick, S., Fleszar, K., Lipp, F., Ravsky, A., Verbitsky, O., Wolff, A.: Drawing graphs on few lines and few planes. In: Hu, Y., Nöllenburg, M. (eds.) GD 2016. LNCS, vol. 9801, pp. 166–180. Springer, Cham (2016). https://doi.org/10.1007/978-3-319-50106-2_14. https://arxiv.org/abs/1607.01196
4. Chaplick, S., Fleszar, K., Lipp, F., Ravsky, A., Verbitsky, O., Wolff, A.: The complexity of drawing graphs on few lines and few planes. Algorithms and Data Structures. LNCS, vol. 10389, pp. 265–276. Springer, Cham (2017). https://doi.org/10.1007/978-3-319-62127-2_23. https://arxiv.org/abs/1607.06444
5. Dujmović, V., Eppstein, D., Suderman, M., Wood, D.R.: Drawings of planar graphs with few slopes and segments. Comput. Geom. Theory Appl. **38**(3), 194–212 (2007). https://doi.org/10.1016/j.comgeo.2006.09.002
6. Dujmovic, V., Joret, G., Micek, P., Morin, P., Ueckerdt, T., Wood, D.R.: Planar graphs have bounded queue-number. Arxiv report (2019). http://arxiv.org/abs/1904.04791
7. Dujmović, V., Pór, A., Wood, D.R.: Track layouts of graphs. Discrete Math. Theor. Comput. Sci. **6**(2), 497–522 (2004). https://hal.inria.fr/hal-00959023
8. Durocher, S., Mondal, D.: Drawing plane triangulations with few segments. In: Proceedings of 26th Canadian Confernce on Computational Geometry (CCCG 2014), pp. 40–45 (2014). http://cccg.ca/proceedings/2014/papers/paper06.pdf
9. Eppstein, D.: Forbidden Configurations in Discrete Geometry. Cambridge University Press, Cambridge (2018)
10. Eppstein, D.: Cubic planar graphs that cannot be drawn on few lines. In: Proceedings of 35th International Symposium on Computational Geometry (SoCG 2019). LIPIcs, vol. 129, pp. 32:1–32:15 (2019). https://doi.org/10.4230/LIPIcs.SoCG.2019.32. https://arxiv.org/abs/1903.05256
11. Felsner, S.: 4-connected triangulations on few lines. In: Archambault, D., Tóth, C.D. (eds.) Proceedings of 27th International Symposium on Graph Drawing & Network Visualization (GD 2019). LNCS, vol. 11904, pp. 395–408. Springer (2019). https://arxiv.org/abs/1908.04524
12. Firman, O., Lipp, F., Straube, L., Wolff, A.: Examining weak line covers with two lines in the plane. In: Biedl, T., Kerren, A. (eds.) Proceedings of International Symposium on Graph Drawing Network Visualization (GD 2018). LNCS, vol. 11282, pp. 643–645 (2018). https://link.springer.com/content/pdf/bbm:978-3-030-04414-5/1.pdf (poster)
13. Heath, L.S., Rosenberg, A.L.: Laying out graphs using queues. SIAM J. Comput. **21**(5), 927–958 (1992). https://doi.org/10.1137/0221055
14. Hültenschmidt, G., Kindermann, P., Meulemans, W., Schulz, A.: Drawing planar graphs with few geometric primitives. J. Graph Alg. Appl. **22**(2), 357–387 (2018). https://doi.org/10.7155/jgaa.00473
15. Kindermann, P., Meulemans, W., Schulz, A.: Experimental analysis of the accessibility of drawings with few segments. J. Graph Alg. Appl. **22**(3), 501–518 (2018). https://doi.org/10.7155/jgaa.00474
16. Schulz, A.: Drawing graphs with few arcs. J. Graph Alg. Appl. **19**(1), 393–412 (2015). https://doi.org/10.7155/jgaa.00366

Drawing Planar Graphs with Few Segments on a Polynomial Grid

Philipp Kindermann[1]([✉]) [iD], Tamara Mchedlidze[2], Thomas Schneck[3],
and Antonios Symvonis[4]

[1] Universität Würzburg, Würzburg, Germany
philipp.kindermann@uni-wuerzburg.de
[2] Karlsruhe Institute of Technology (KIT), Karlsruhe, Germany
mched@iti.uka.de
[3] Università Tübingen, Tübingen, Germany
thomas.schneck@uni-tuebingen.de
[4] National Technical University of Athens, Athens, Greece
symvonis@math.ntua.gr

Abstract. The visual complexity of a graph drawing can be measured by the number of geometric objects used for the representation of its elements. In this paper, we study planar graph drawings where edges are represented by few segments. In such a drawing, one segment may represent multiple edges forming a path. Drawings of planar graphs with few segments were intensively studied in the past years. However, the area requirements were only considered for limited subclasses of planar graphs. In this paper, we show that trees have drawings with $3n/4 - 1$ segments and n^2 area, improving the previous result of $O(n^{3.58})$. We also show that 3-connected planar graphs and biconnected outerplanar graphs have a drawing with $8n/3 - O(1)$ and $3n/2 - O(1)$ segments, respectively, and $O(n^3)$ area.

1 Introduction

The quality of a graph drawing can be assessed in a variety of ways: area, crossing number, bends, angular resolution, and many more. All these measures have their justification, but in general it is challenging to optimize all of them in a single drawing. Recently, the *visual complexity* was suggested as another quality measure for drawings [22]. The visual complexity denotes the number of simple geometric entities used in the drawing.

The visual complexity of a straight-line graph drawing can be formalized as the number of segments formed by its edges, which we refer to as *segment complexity*. Notice that edges constituting a single segment form a path in the graph. The idea of representing graphs with fewer segments complies with the Gestalt principles of perception, which are rules for the organization of perceptual scenes

This work was initiated at the Workshop on Graph and Network Visualization 2017.
The work of P. Kindermann was partially supported by DFG grant SCHU 2458/4-1.

D. Archambault and C. D. Tóth (Eds.): GD 2019, LNCS 11904, pp. 416–429, 2019.
https://doi.org/10.1007/978-3-030-35802-0_32

introduced in the area of psychology in the 19th century [16]. According to the law of continuation, the edges forming a segment may be easier grouped by our perception into a single entity. Therefore, drawing graphs with fewer segments may ease their perceptual processing. A recent user study [15] suggests that lowering the segment complexity may positively influence aesthetics, depending on the background of the observer, as long as it does not introduce unnecessarily sharp corners. From the theoretical perspective, it is natural to ask for a drawing of a graph with the smallest segment complexity. It is not surprising that it is NP-hard to determine whether a graph has a drawing with segment complexity k [9]. However, we can still expect to prove bounds for certain graph classes.

Dujmović et al. [7] were the first to study drawings with few segments and provided upper and lower bounds for several planar graph classes. Since then, several new results have been provided ([8,12,13,19,20], refer also to Table 1). These results shed only a little light on the area requirements of the drawings. In particular, in his thesis, Mondal [19] gives an algorithm for triangulations that produces drawings with $8n/3 - O(1)$ segments on a grid of size $2^{O(n \log n)}$ in general and $2^{O(n)}$ for triangulations of bounded degree. Even with this large grid, the algorithm uses substantially more segments than the best-known algorithm for triangulations without the grid requirement by Durocher and Mondal [8], which uses $7n/3 - O(1)$ segments. Recently, Hültenschmidt et al. [12] presented algorithms that produce drawings with $3n/4$ segments and $O(n^{3.58})$ area for trees, and $3n/2$ and $8n/3$ segments for outerplanar graphs and 3-trees, respectively, and $O(n^3)$ area. Igamberdiev et al. [13] have provided an algorithm to construct drawings of planar cubic 3-connected graphs with $n/2$ segments and $O(n^2)$ area.

Our Contribution. In this paper, we concentrate on finding drawings with low segment complexity on a small grid. Our contribution is summarized in Table 1. In Sect. 2, we show that every tree has a drawing with at most $3n/4 - 1$ segments on the $n \times n$ grid, improving the area bound by Hültenschmidt et al. [12]. We then focus on drawing 3-connected planar graphs in Sect. 3. Using a combination of Schnyder realizers and orderly spanning trees, we show that every 3-connected planar graph can be drawn with $m - (n - 4)/3 \le (8n - 14)/3$ segments on an $O(n) \times O(n^2)$ grid. Finally, in Sect. 4, we use this result to draw on an $O(n) \times O(n^2)$ grid maximal 4-connected graphs with $5n/2 - 4$ segments, biconnected outerplanar graphs with $(3n-3)/2$ segments, connected outerplanar graphs with $(7n - 9)/4$ segments, and connected planar graphs with $(17n - 38)/6$ segments. All our proofs are constructive and yield algorithms to obtain such drawings in $O(n)$ time. As a side result, we also prove that the total number of leaves in every Schnyder realizer of a 3-connected planar graph is at most $2n + 1$, which was only known for maximal planar graphs [2,18]. For the results on biconnected outerplanar 3- and 4-connected graphs, we use techniques that have been used to construct monotone drawings; thus, as a side result, these drawings are also monotone[1].

---

[1] A path P in a straight-line drawing of a graph is *monotone* if there exists a line l such that the orthogonal projections of the vertices of P on l appear along l in the order induced by P. A drawing is monotone if there is a monotone path between every pair of vertices.

Table 1. Upper and lower bounds on the visual complexity of segment drawings. Here, n is the number of vertices, m is the number of edges, ϑ is the number of odd-degree vertices, and b is the number of maximal biconnected components. Constant-term additions or subtractions have been omitted. Entries marked by a * are monotone drawings. FV corresponds to the full version [14].

Class	Segments		Segments on the grid		
	Lower b.	Upper b.	Segments	Grid	Ref.
tree	$\vartheta/2$ [7]	$\vartheta/2$ [7]	$3n/4$	$O(n^2) \times O(n^{1.58})$	[12]
			$3n/4$	$n \times n$	Th. 1
			$\vartheta/2$	quasipoly.	[12]
max. outerplanar	n [7]	n [7]	$3n/2$	$O(n) \times O(n^2)$	[12]
2-conn. outerplanar	n [7]		* $m - n/2$	$O(n) \times O(n^2)$	FV
			* $3n/2$	$O(n) \times O(n^2)$	FV
outerplanar	n [7]		$3n/2 + b$	$O(n) \times O(n^2)$	FV
			$7n/4$	$O(n) \times O(n^2)$	FV
2-tree	$3n/2$ [7]	$3n/2$ [7]			
planar 3-tree	$2n$ [7]	$2n$ [7]	$8n/3$	$O(n) \times O(n^2)$	[12]
2-conn. planar	$2n$ [7]	$8n/3$ [8]	$\rightarrow$ **planar**		
3-conn. planar	$2n$ [7]	$5n/2$ [7]	* $m - n/3$	$O(n) \times O(n^2)$	Th. 2
			* $8n/3$	$O(n) \times O(n^2)$	Cor. 1
cubic 3-conn. planar	$n/2$ [20]	$n/2$ [13]	$n/2$	$O(n) \times O(n)$	[13]
triangulation	$2n$ [8]	$7n/3$ [8]	* $8n/3$	$O(n) \times O(n^2)$	Cor. 1
4-conn. planar	$2n$ [8]	$21n/8$ [8]	* $\rightarrow$ **3-conn.**		
4-conn. triang.	$2n$ [8]	$9n/4$ [8]	* $5n/2$	$O(n) \times O(n^2)$	FV
planar	$2n$ [8]	$8n/3$ [8]	$17n/3 - m$	$O(n) \times O(n^2)$	FV
			$17n/6$	$O(n) \times O(n^2)$	FV

We note that there are three trivial lower bounds for the segment complexity of a general graph $G = (V, E)$ with n vertices and m edges: (i) $\vartheta/2$, where ϑ is the number of odd-degree vertices, (ii) $\max_{v \in V} \lceil \deg(v)/2 \rceil$, and (iii) $\lceil m/(n-1) \rceil$. These trivial lower bounds are the same as for the slope number of graphs [23], that is, the minimum number of slopes required to draw all edges, and the slope number is upper bounded by the number of segments required.

Relevant to segment complexity are the studies by Chaplick et al. [3,4] who consider drawings where all edges are to be covered by few lines (or planes); the difference to our problem is that collinear segments are counted only once in their model. In the same fashion, Kryven et al. [17] aim to cover all edges by few circles (or spheres).

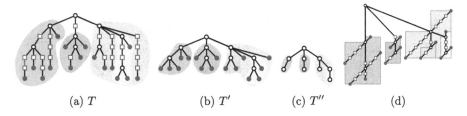

Fig. 1. (a) A tree T. Degree-2 vertices are squared, leaves are filled. (b) The tree T' obtained from T by contracting the degree-2 vertices. (c) The tree T'' obtained from T' by removing all leaves. (d) The drawing of our algorithm.

2   Trees

Let $T = (V, E)$ be a tree with n vertices. In this section, we describe an algorithm to draw T with at most $3n/4 - 1$ segments on an $n \times n$ grid in $O(n)$ time.

If T consist only of vertices of degree 1 and 2, then it is a path and we can draw it with 1 segment and $n \times 1$ area. So, we will assume that there is at least one vertex with higher degree. We choose such a vertex as the root of T. Denote the number of degree-2 vertices by β and the number of leaves by α. In the first step, we create another tree T' with $n - \beta$ vertices by contracting all edges incident to a degree-2 vertex. We say that a degree-2 vertex u *belongs* to a vertex v if v is the first descendent of u in T that has degree greater than 2. Note that T' has the same number of leaves as T. In the next step, we remove all leaves from T' and obtain a tree T'' with $n - \beta - \alpha$ vertices; see Fig. 1.

The main idea of our algorithm is as follows. We draw T'' with $n - \beta - \alpha - 1$ segments. Then, we add the α leaves in such a way that they either extend the segment of an edge, or that two of them share a segment, which results in at most $\alpha/2$ new segments. Finally, we place the β degree-2 vertices onto the segments without increasing the number of segments. This way, we get a drawing with at most $n - \beta - \alpha/2$ segments. Since T' has no degree-2 vertices, more than half of its vertices are leaves, so $\alpha > (n - \beta)/2$. Hence, the drawing has at most $3(n - \beta)/4 < 3n/4$ segments. Unfortunately, there are a few more details we have to take care of to achieve this bound.

Let v be a vertex in T'', and let $T[v]$ be the subtree of T rooted at v. Let n_v denote the number of vertices in $T[v]$. Let $v_1, \ldots, v_k$ be the children of v in T''. As induction hypotheses, we assume that each $T[v_i]$ is drawn inside a polygon B_i of *dimensions* (edge lengths) $\ell_i, r_i, t_i, b_i, w_i, h_i$ as indicated in Fig. 2a such that

($\mathcal{I}_1$) no vertex of $T[v_i]$ lies to the top-left of v_i, and
($\mathcal{I}_2$) B_i has area $n_i \times n_i$.

Using three steps, we describe how to draw $T[v]$ inside a polygon B_v of dimensions $\ell_v, r_v, t_v, b_v, w_v, h_v$ such that v lies at coordinate $(0, 0)$. First, we place $T[v_1], \ldots, T[v_k]$. Second, we add the degree-2 vertices that belong to $v_1, \ldots, v_k$. Finally, we add the leaf-children of v and the degree-2 vertices belonging to them.

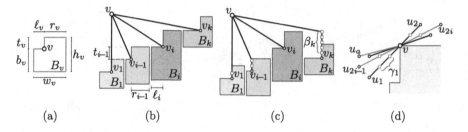

Fig. 2. Drawing of $T[v]$ with $k = 4$. (a) B_v; (b) the children of v in T''; (c) the degree-2 vertices belonging to these children; and (d) the remaining vertices of $T[v]$ which form C_v.

Step 1. We aim at placing v_1 directly below v, and each polygon $B_i, i \geq 2$, to the right of polygon B_{i-1}, aligning v_i with the top boundary of B_{i-1}; see Fig. 2b. We place v_1 at coordinate $(0, -1 - \sum_{i=1}^{k} t_i)$, and each v_i at coordinate $(x(v_{i-1}) + r_{i-1} + \ell_i + 1, y(v_{i-1}) + t_{i-1})$, where $x(v)$ and $y(v)$ are the x- and y-coordinates of v, respectively. By invariant ($\mathcal{I}_2$), the total width and height of the drawings of $B_1, \ldots, B_k$ are both at most $\sum_{i=1}^{k} n_{v_i}$.

Step 2. Let β_i be the number of degree-2 vertices that belong to v_i. We move each polygon B_i downwards by β_i, and place the degree-2 vertices above v_i; see Fig. 2c. This does not change the placement of any edge of v, the polygons are only moved downwards and are still disjoint, so the drawing remains planar. The height of the drawing increases by at most $\max_{i=1}^{k} \beta_i \leq \sum_{i=1}^{k} \beta_i$ to $\sum_{i=1}^{k}(n_{v_i} + \beta_i)$, while the width remains $\sum_{i=1}^{k} n_{v_i}$.

Step 3. Let C_v the subtree of $T[v]$ that consists of v, its leaf-children in T', and the degree-2 vertices belonging to them. Let $u_1, \ldots, u_a$ be the leaves of C_v and let $\gamma_1, \ldots, \gamma_a$ be the number of degree-2 vertices that belong to them. Without lost of generality, assume that $\gamma_1 \geq \ldots \geq \gamma_a$. We first consider the case where a is even. We place the leaves alternatively to the bottom-left and to the top-right of v with as many rows between them and v as degree-2 vertices belong to them; we draw each u_{2i-1} and u_{2i} on a segment through v with slope $1/i$. To this end, we place u_{2i-1} at coordinate $(-(\gamma_{2i-1} + 1) \cdot i, -\gamma_{2i-1} - 1)$ and u_{2i} at coordinate $((\gamma_{2i} + 1) \cdot i, \gamma_{2i} + 1)$ (recall that u is placed at $(0, 0)$). We are able to place the degree-2 vertices that belong to these leaves between them and v; see Fig. 2d.

If a is odd, then we apply the procedure described above for $u_1, \ldots, u_{a-1}$. Vertex u_a is placed as follows. If v is a leaf in T'', then we place u_a below v at coordinate $(0, -\gamma_a - 1)$. If v is not a leaf in T'', and no degree-2 vertex belongs to v, and v is not the first child of its parent in T'' (that is, there will be no edge that leaves v vertically above), then we place u_a above v at coordinate $(0, \gamma_a + 1)$ such that it shares a segment with (v, v_1). Otherwise, we place u_a as every other vertex u_i with odd index at coordinate $(-(\gamma_a + 1) \cdot i, -\gamma_a - 1)$.

By construction, the segments through v drawn at step 3 cannot intersect $B_2, \ldots, B_k$, but there might be an intersection between the segment from u_1 to v and B_1. In this case, we move B_1 downwards until the crossing disappears, which makes the drawing planar again. We call this action **Step 4**. Thus, we have created a drawing of $T[v]$ inside the polygon B_v that complies with invariant $(\mathcal{I}_1)$. In the following, we show that B_v satisfies invariant $(\mathcal{I}_2)$.

We analyze the width and height of the part of the drawing of C_v. Let $\gamma^L = \sum_{i=1}^{\lceil a/2 \rceil} \gamma_{2i-1}$ and $\gamma^R = \sum_{i=1}^{\lfloor a/2 \rfloor} \gamma_{2i}$ be the number of degree-2 vertices drawn to the left and right of v, respectively, and let $\gamma = \gamma^L + \gamma^R$.

Recall that $\gamma_1 \geq \ldots \geq \gamma_a$ and leaf u_i was placed at y-coordinate $\pm(\gamma_i + 1)$. Hence, the vertices with the lowest and highest y-coordinate are u_1 at $y(u_1) = -\gamma_1 - 1$ and u_2 at $y(u_2) = \gamma_2 + 1$, respectively. Thus, the height of the drawing of C_v is $1 = 1 + a + \gamma$ if $a = 0$; $2 + \gamma_1 = 1 + a + \gamma$ if $a = 1$; and $3 + \gamma_1 + \gamma_2 \leq 1 + a + \gamma$ if $a \geq 2$, so at most $1 + a + \gamma$ in total.

For analyzing the width of the drawing of C_v, we first consider those vertices that are drawn to the right of v. Let r be such that u_{2r} is the rightmost vertex at x-coordinate $(\gamma_{2r} + 1) \cdot r$. Since $\gamma_1 \geq \ldots \geq \gamma_a$, we have that

$$\gamma^R = \sum_{i=1}^{\lfloor a/2 \rfloor} \gamma_{2i} \geq \sum_{i=1}^{r} \gamma_{2i} \geq r \cdot \gamma_{2r}.$$

Symmetrically, let ℓ be such that $u_{2\ell-1}$ is the leftmost vertex at x-coordinate $-(\gamma_{2\ell-1} + 1) \cdot \ell$. We have that

$$\gamma^L = \sum_{i=1}^{\lceil a/2 \rceil} \gamma_{2i-1} \geq \sum_{i=1}^{\ell} \gamma_{2i-1} \geq \ell \cdot \gamma_{2\ell-1}.$$

Hence, the total width of this part of the drawing is at most

$$1 + (\gamma_{2r} + 1) \cdot r + (\gamma_{2\ell-1} + 1) \cdot \ell \leq 1 + \ell + r + \gamma^L + \gamma^R \leq 1 + a + \gamma.$$

Recall that before step 3 the width of the drawing of $T[v]$ was $\sum_{i=1}^{k} n_{v_i}$ and the height was at most $\sum_{i=1}^{k} (n_{v_i} + \beta_i)$. In step 3, the width increases by at most $1 + a + \gamma$. In step 4, we move the drawing of $T[v_1]$ downwards if it is crossed by the segment between u_1 and v until this crossing is resolved. There cannot be a crossing if $y(u_1) > y(v_1)$, so we move it by at most $|y(u_1)|$ downwards, which is exactly the height of the part of the drawing of C_v that lies below v. Hence, the height in Steps 3 and 4 increases by at most the height of the drawing of C_v, which is $1 + a + \gamma$. Since $n_v = 1 + \sum_{i=1}^{k} (n_{v_i} + \beta_i) + a + \gamma$, the width and the height of B_v is at most n_v. With this we complete the proof of invariant $(\mathcal{I}_2)$.

We will now discuss the number of segments in T. Let r be the root of T, and let $v \in T'' \setminus \{r\}$. We need a few definitions; see Fig. 3. Let p_v be the parent of v in T''. Let P_v be the path between v and p_v in T; let $T^+[v] = T[v] \cup P_v$; let n_v^+ be the number of vertices in $T^+[v] \setminus \{p_v\}$; let e_v be the edge of P_v incident to p_v; and let s_v be the number of segments used in the drawing of $T^+[v]$.

Lemma 1. *For any vertex $v \neq r$ of T'', if e_v is drawn vertical, then $s_v \leq (3n_v^+ - 1)/4$, otherwise $s_v \leq 3n_v^+/4$.*

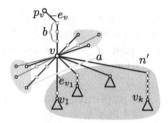

Fig. 3. Illustration of $T^+[v]$ in the proof of Lemma 1.

Proof. We prove the lemma by induction on the height of T'', so we can assume that the bound holds for all children of v in T''. Recall that $u_1, \ldots, u_a$ are the leaf-children of v in T', $v_1, \ldots, v_k$ are the children of v in T'', and v_1 is connected to v by a vertical segment. Let b be the number of degree-2 vertices that belong to v. Let $n' = \sum_{i=1}^{k} n_{v_i}^+$; then, $n_v^+ \geq n' + a + b + 1$. (There might be degree-2 vertices between v and its leaf-children in T' which we do not count.) By induction, $s_{v_1} \leq 3(n_{v_1}^+ - 1)/4$ and $s_{v_i} \leq 3n_{v_i}^+/4$ for $2 \leq i \leq k$, so $\sum_{i=1}^{k} s_{v_i} \leq (3n' - 1)/4$. It remains to analyze the number of segments for C_v and for the path P_v.

Case 1. v is a leaf in T'' and $b = 0$. Then, $n_v^+ \geq a + 1$. Since v is a leaf in T'', it has at least two children in T, so $a \geq 2$.

Case 1.1. a is even. We use $a/2$ for C_v plus one for the edge e_v. Thus, $s_v \leq a/2 + 1 \leq (n_v^+ - 1)/2 + 1 = (3n_v^+ - n_v^+ + 2)/4 \leq (3n_v^+ - 1)/4$ since $n_v^+ \geq 3$.

Case 1.2. a is odd and e_v is vertical, so $a \geq 3$ and $n_v^+ \geq 4$. We use $(a-1)/2$ segments for $u_1, \ldots, u_{a-1}$ and one segment for u_a and e_v. Thus, $s_v \leq (a-1)/2 + 1 \leq n_v^+/2 \leq 3n_v^+/4 - 1$.

Case 1.3. a is odd and e_v is not vertical. We use one more segment than in Case 1.2, so $s_v \leq 3n_v^+/4$.

Case 2. v is a leaf in T'' and $b > 0$. Then, $a \geq 2$ and $n_v^+ \geq a+b+1 \geq a+2 \geq 4$

Case 2.1. a is even and e_v is vertical. We use $a/2$ segments for $u_1, \ldots, u_2$, and the degree-2 vertices that belong to v lie on a vertical segment with e_v. Hence, we have $s_v \leq a/2 + 1 \leq n_v^+/2 \leq 3n_v^+/4 - 1$.

Case 2.2. a is even and e_v is not vertical. We again have $n_v^+ \geq 4$. The degree-2 vertices that belong to v now lie on a different segment than e_v, so we have one more segment than in Case 2.1, so $s_v \leq 3n/4$.

Case 2.3. a is odd. We have $a \geq 3$ and thus $n_v^+ \geq 5$. We have drawn $u_1, \ldots, u_{a-1}$ paired up. We have drawn u_a on a vertical segment with the degree-2 vertices that belong to v, and we have possibly one more segment for e_v. Hence, we have $s_v \leq (a-1)/2 + 2 \leq (n_v^+ + 1)/2 = (3n_v^+ - n_v^+ + 2)/4 \leq (3n_v^+ - 3)/4$.

Case 3. v is not a leaf in T'' and $b = 0$. We have $n_v^+ \geq n' + a + 1$, so $n' \leq n_v^+ - a - 1$.

Case 3.1. $a = 0$ and e_v is vertical. Then, $n_v^+ = n' + 1$ and e_v lies on a vertical segment with the edge e_{v_1}. Hence, $s_v \le (3n' - 1)/4 = (3n_v^+ - 4)/4$.

Case 3.2. $a = 0$ and e_v is not vertical. Again, $n_v^+ = n' + 1$. We use one segment for e_v, so we have $s_v \le (3n' - 1)/4 + 1 = 3n_v^+/4$.

Case 3.3. $a \ge 2$ is even. We use $a/2$ segments for C_v and one more for e_v. Hence, $s_v \le (3n' - 1)/4 + a/2 + 1 = (3n' + 2a + 3)/4 \le (3n_v^+ - a)/4 \le (3n_v^+ - 2)/4$.

Case 3.4. a is odd and e_v is vertical. We use $(a+1)/2$ segments for $u_1, \ldots, u_a$, but e_v shares its vertical segment with e_{v_1}. Hence, $s_v \le (3n'-1)/4 + (a+1)/2 = (3n' + 2a + 1)/4) \le (3n_v^+ - a - 2)/4 \le (3n_v^+ - 3)/4$.

Case 3.5. a is odd and e_v is not vertical. In this case, we place u_a above v such that it lies on a segment with e_{v_1}. We use $(a - 1)/2$ segments for $u_1, \ldots, u_{a-1}$ and one segment for e_v, so we have the same number of segments as in Case 3.4.

Case 4. v is not a leaf in T'' and $b > 0$. We have $n_v^+ \ge n' + a + b + 1 \ge n' + a + 2$.

Case 4.1. a is even. We use $a/2$ segments for $u_1, \ldots, u_a$. The edges of the path P_v share a vertical segment with e_{v_1}. We use at most one more segment for e_v, so $s_v \le (3n' - 1)/4 + a/2 + 1 = (3n' + 2a + 3)/4 \le (3n_v^+ - a - 3)/4 \le (3n_v^+ - 3)/4$.

Case 4.2. a is odd and e_v is vertical. We use the exact same number of segments as in Case 3.4, so $s_v \le (3n_v^+ - 3)/4$.

Case 4.3. a is odd and e_v is not vertical. We use $(a+1)/2$ segments for $u_1, \ldots, u_a$. The edges of the path P_v share a vertical segment with e_{v_1}, and we need one more segment for e_v. Hence, $s_v \le (3n' - 1)/4 + (a+1)/2 + 1 = (3n' + 2a + 5)/4 \le (3n_v^+ - a - 1)/4 \le (3n_v^+ - 2)/4$. □

Now we can bound the total number of segments in the drawing of T.

Lemma 2. *Our algorithm draws T with at most $3n/4 - 1$ segments if $n \ge 3$.*

Proof. If T is a path with $n \ge 3$, then the bound trivially holds. If T is a subdivision of a star, then the bound also clearly holds. Otherwise, T'' consists of more than one vertex. Let $v_1, \ldots, v_k$ be the children of the root r of T'' such that v_1 is connected by a vertical edge. Recall that $n' = \sum_{i=1}^{k} n_{v_i}^+$. By Lemma 1, the subtrees $T[v_i]^+$, $i = 1, \ldots, k$ contribute at most $(3n' - 1)/4$ segments to the drawing of T. Let a be the number of leaf children of r in T'. If a is even, then we use $a/2$ segments to draw them. If a is odd, then we align one of them with the vertical segment of v_1, and draw the remaining with $(a - 1)/2$ segments. Since $n \ge n' + a + 1$, the total number of segments is at most $(3n' - 1)/4 + a/2 \le 3n/4 - a/4 - 1 \le 3n/4 - 1$. □

All steps of the algorithm work in linear time. Sorting the leaf-children by the number of degree-2 vertices belonging to them can also be done in linear time with, e.g., CountingSort, as the numbers are bounded by n. Thus, Theorem 1 follows. Figure 1d shows the result of our algorithm for the tree of Fig. 1a.

Theorem 1. *Any tree with $n \ge 3$ vertices can be drawn planar on an $n \times n$ grid with $3n/4 - 1$ segments in $O(n)$ time.*

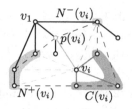

Fig. 4. Edges in a Schnyder realizer.

Fig. 5. Definition of orderly spanning tree (bold).

Fig. 6. Definition of slope-disjointness.

3   3-Connected Planar Graphs

In this section, we present an algorithm to compute planar drawings with at most $(8n - 14)/3$ segments for 3-connected planar graphs.

Let G be a triangulation. Let v_1, v_2, v_3 be the vertices of the outer face. We decompose the interior edges into three *Schnyder trees* T_1, T_2, and T_3 rooted at v_1, v_2, and v_3, respectively. The edges of the trees are oriented towards their roots. For $k \in \{1, 2, 3\}$, we call each edge in T_k a *k-edge* and the parent of a vertex in T_k its *k-parent*. The decomposition is a *Schnyder realizer* [21] if at every interior vertex the edges are counter-clockwise ordered as: outgoing 1-edge, incoming 3-edges, outgoing 2-edge, incoming 1-edges, outgoing 3-edge, and incoming 2-edges; see Fig. 4. A Schnyder tree T_k also contains the exterior edges of v_k, so each exterior edges lies in two Schnyder trees and each v_k is a leaf in the other two Schnyder trees; hence, each Schnyder tree is a spanning tree.

For 3-connected planar graphs, Schnyder realizers also exist [6,10], but the interior edges can be *bidirected*: an edge (u, v) is bidirected if it is an outgoing i-edge at u and an outgoing j-edge at v with $i \neq j$. All other edges are *unidirected*, that is, they are an outgoing i-edge at u and an incoming i-edge at v (or vice-versa). The restriction on the cyclic ordering around each vertex remains the same, but now the Schnyder trees are not necessarily edge-disjoint.

Chiang et al. [5] have introduced the notion of *orderly spanning trees*. Recently, orderly spanning trees were redefined by Hossain and Rahman [11] as *good spanning trees*. We will use the definition by Chiang et al., but note that these two definitions are equivalent. Two vertices in a rooted spanning tree are *unrelated* if neither of them is an ancestor of the other one. A tree is *ordered* if the circular order of the edges around each vertex is fixed. Let $G = (V, E)$ be a plane graph and let $r \in V$ lie on the outer face. Let T be an ordered spanning tree of G rooted at r that respects the embedding of G. Let $v_1, \ldots, v_n$ be the vertices of T as encountered in a counter-clockwise pre-order traversal. For any vertex v_i, let $p(v_i)$ be its parent in T, let $C(v_i)$ be the children of v in T, let $N(v_i)$ be the neighbors of v_i in G that are unrelated to v_i; see Fig. 5. Further, let $N^-(v_i) = \{v_j \in N(v_i) \mid j < i\}$ and $N^+(v_i) = \{v_j \in N(v_i) \mid j > i\}$. Then, T is called *orderly* if the neighbors around every vertex v_i are in counter-clockwise order $p(v_i)$, $N^-(v_i)$, $C(v_i)$, $N^+(v_i)$. In particular, this means that there is no edge in G between v_i and an ancestor in T that is not its parent and there is no edge in G between v_i and a descendent in T that is not its child. This fact is crucial, as it allows us to draw a path in an orderly spanning tree on a single segment without introducing overlapping edges.

Angelini et al. [1] have introduced the notion of a *slope-disjoint* drawing of a rooted tree T, which is defined as follows; see Fig. 6.

(S1) For every vertex u in T, there exist two slopes $\alpha_1(u)$ and $\alpha_2(u)$ with $0 < \alpha_1(u) < \alpha_2(u) < \pi$, such that, for every edge e that is either $(p(u), u)$ or lies in $T[u]$, it holds that $\alpha_1(u) < \text{slope}(e) < \alpha_2(u)$;

(S2) for every directed edge (v, u) in T, it holds that $\alpha_1(u) < \alpha_1(v) < \alpha_2(v) < \alpha_2(u)$ (recall that edges are directed towards the root); and

(S3) for every two vertices u, v in T with $p(u) = p(v)$, it holds that either $\alpha_1(u) < \alpha_2(u) < \alpha_1(v) < \alpha_2(v)$ or $\alpha_1(v) < \alpha_2(v) < \alpha_1(u) < \alpha_2(u)$.

Lemma 3 ([1]). *Every slope-disjoint drawing of a tree is planar and monotone.*

We will now create a special slope-disjoint drawing for rooted orderly trees.

Lemma 4. *Let $T = (V, E)$ be an ordered tree rooted at a vertex r with λ leaves. Then, T admits a slope-disjoint drawing with λ segments on an $O(n) \times O(n^2)$ grid such that all slopes are integer. Such a drawing can be found in $O(n)$ time.*

Proof sketch. Let $v_1, \ldots, v_n = r$ be the vertices of T as encountered in a counter-clockwise post-order traversal. Let $e_i = (v_i, p(v_i)), 1 \le i < n$. We assign the slopes to the edges of T in the order $e_1, \ldots, e_{n-1}$. We start with assigning slope $s_1 = 1$ to e_1. For any other edge $e_i, 1 < i < n$, if v_i is a leaf in T, then we assign the slope $s_i = s_{i-1} + 1$ to e_i. Otherwise, since we traverse the vertices in a post-order, $p(v_{i-1}) = v_i$ and we assign the slope $s_i = s_{i-1}$ to e_i.

We create a drawing Γ of T as follows. We place $r = v_n$ at coordinate $(0, 0)$. For every other vertex v with parent p that is drawn at coordinate (x, y), we place v at coordinate $(x + 1, y + \text{slope}(v))$.

We now analyze the number of segments used in Γ; slope-disjointness, area, and running time are proven in the full version [14]. The root r is an endpoint of $\deg(r)$ segments and every leaf is an endpoint of exactly 1 segment. For every other vertex v, its incoming edge and one of its outgoing edges lie on the same segment, so it is an endpoint of $\deg(v) - 2$ segments. Since every segment has two endpoints, the total number of segments is

$$\frac{1}{2} \left(\deg(r) + \sum_{v \text{ not leaf}, v \ne r} (\deg(v) - 2) + \sum_{v \text{ leaf}} \deg(v) \right)$$

$$= \frac{1}{2} \left(\sum_v \deg(v) - 2(n - \lambda - 1) \right) = \frac{1}{2} (2n - 2 - 2n + 2\lambda + 2) = \lambda.$$

□

Lemma 5. *Let $G = (V, E)$ be a planar graph and let T be an orderly spanning tree of G with λ leaves. Then, G admits a planar monotone drawing with at most $m - n + 1 + \lambda$ segments on an $O(n) \times O(n^2)$ grid in $O(n)$ time.*

Proof. We first create a drawing of T according to Lemma 4. Now, we will plug this tree drawing into the algorithm by Hossain and Rahman [11].

This algorithm takes a slope-disjoint drawing of an orderly spanning tree T of G and stretches the edges of T such that the remaining edges of G can be inserted without crossings. In this stretching operation, the slopes of the edges of T are not changed. Further, the total width of the drawing only increases by a constant factor. Since T is drawn slope-disjoint, this produces a planar monotone drawing of G on an $O(n) \times O(n^2)$ grid. The algorithm runs in $O(n)$ time.

To count the number of segments, assume that every edge of G that does not lie on T is drawn with its own segment. We have drawn T with λ segments and the slopes of the edges of T. Hence, our algorithm draws G with λ segments for T and with $m - n + 1$ segments for the remaining edges.   □

Both Chiang et al. [5] and Hossain and Rahman [11] have shown that every planar graph has an embedding that admits an orderly spanning tree. However, we do not know anything about the number of leaves in an orderly spanning tree. Miura et al. [18] have shown that Schnyder trees are orderly spanning trees, and it is known that every 3-connected planar graph has a Schnyder realizer.

Lemma 6 ([18]). *Let $G = (V, E)$ be a 3-connected planar graph and let T_1, T_2, and T_3 be the Schnyder trees of a Schnyder realizer of G. Then, T_1, T_2, and T_3 are orderly spanning trees of G.*

Bonichon et al. [2] showed that there is a Schnyder realizer for every triangulated graph such that the total number of leaves in T_1, T_2, and T_3 is at most $2n+1$, which already gives us a good bound on the number of segments for triangulations. We will now show that the same holds for every Schnyder realizer of a 3-connected graph. Let v be a leaf in one of the Schnyder trees T_k, $k \in \{1, 2, 3\}$, that is not the root of a Schnyder tree, so v has no incoming k-edge. Hence, the outgoing $(k + 1)$-edge (v, u) and the outgoing $(k - 1)$-edge (v, w) are consecutive in the cyclical ordering around v, so they lie on a common face f. We assign the pair (v, k) to f. We first show two lemmas.

Lemma 7. *Let $u_1, \ldots, u_p$ be the vertices on an interior face f in ccw order. If (u_1, k) and (u_2, i) are assigned to f for some $i, k \in \{1, 2, 3\}$, then $i = k$.*

Proof. Refer to Fig. 7. By definition, (u_1, u_2) is an outgoing $(k + 1)$-edge at u_1. Since u_1 is a leaf in T_k, (u_1, u_2) cannot be an outgoing k-edge at u_2. Hence, (u_1, u_2) is either an incoming $(k + 1)$-edge at u_2 (if it is unidirected), or an outgoing $(k - 1)$-edge at u_2 (if it is bidirected); it cannot be an outgoing $(k + 1)$-edge since bidirected edges have to belong to two different Schnyder trees. For (u_2, i) to be assigned to f, u_2 must have two outgoing edges at f, so we are in the latter case. Hence, (u_2, u_3) is outgoing at u_2, and by the cyclical ordering of the edges around u_2, it is an outgoing $(k + 1)$-edge. Thus, u_2 has an outgoing $(k + 1)$-edge and an outgoing $(k - 1)$-edge at f, so $i = k$.   □

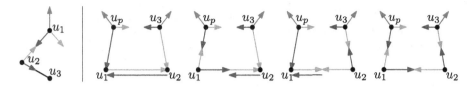

Fig. 7. (Left) Proof of Lemma 7 and (right) proof of Lemma 8.

Lemma 8. *Let $u_1, u_2, \ldots, u_p$ be vertices on an interior face f in counter-clockwise order. If $u_3, \ldots, u_p$ are assigned to f, then neither u_1 nor u_2 are.*

Proof sketch. From Lemma 7, it follows that $(u_3, k), \ldots, (u_p, k)$ are assigned to f for some $k \in \{1, 2, 3\}$, so (u_1, u_p) is an outgoing $(k+1)$-edge at u_p and (u_2, u_3) is an outgoing $(k-1)$-edge at u_3; since u_1 and u_p are leaves in T_k, (u_1, u_p) is either an incoming $(k+1)$-edge or an outgoing $(k-1)$-edge at u_1 and (u_2, u_3) is either an incoming $(k-1)$-edge or an outgoing $(k+1)$-edge at u_2. However, each of the four possible configurations violates the properties of a Schnyder realizer, as illustrated in Fig. 7. The full proof is given in the full version [14]. □

Now we prove the bound on the number of leaves in a Schnyder realizer.

Lemma 9. *Let T_1, T_2, T_3 be a Schnyder realizer of a 3-connected planar graph $G = (V, E)$. Then, there are at most $2n + 1$ leaves in total in T_1, T_2, and T_3.*

Proof. Consider any interior face f of G. By definition of the assignment, no vertex can be assigned to f twice. By Lemma 8, at least two vertices on f are not assigned to f, so we assign at most $\deg(f) - 2$ leaves to f. At the outer face f^*, every vertex that is not the root of a Schnyder tree can be assigned as a leaf at most once. However, the root of each of the Schnyder trees has no outgoing edges, but it can be a leaf in both the other two Schnyder trees. Hence, we assign at most $\deg(f^*) + 3$ leaves to the outer face. Let F be the faces in G. Since, for every Schnyder tree, each of its leaves gets assigned to exactly one face, the total number of leaves in T_1, T_2, and T_3 is at most

$$\sum_{f \in F} (\deg(f) - 2) + 5 = 2m - 2|F| + 5 = 2m + 2n - 2m - 4 + 5 = 2n + 1.$$

□

Now we have the tools to prove the main result of this section.

Theorem 2. *Any 3-connected planar graph can be drawn planar monotone on an $O(n) \times O(n^2)$ grid with $m - (n - 4)/3$ segments in $O(n)$ time.*

Proof. Let $G = (V, E)$ be a 3-connected planar graph. We compute a Schnyder realizer of G, which is possible in $O(n)$ time. By Lemma 9, the Schnyder trees have at most $2n + 1$ leaves in total, so one of them, say T_1, has at most $(2n+1)/3$

leaves. By Lemma 6, T_1 is an orderly spanning tree, so we can use Lemma 5 to obtain a planar monotone drawing of G on an $O(n) \times O(n^2)$ grid with at most $m - n + 1 + (2n + 1)/3 = m - n/3 + 4/3$ segments in $O(n)$ time.   □

Since a planar graph has at most $m \leq 3n - 6$ edges, we have the following.

Corollary 1. *Any 3-connected planar graph can be drawn planar monotone on an $O(n) \times O(n^2)$ grid with $(8n - 14)/3$ segments in $O(n)$ time.*

4   Other Planar Graph Classes

We can use the results of Sect. 3 to obtain grid drawings with few segments for other planar graph classes on an $O(n) \times O(n^2)$ grid in $O(n)$ time. In particular, we can draw (i) 4-connected triangulations with $5n/2 - 4$ segments; (ii) biconnected outerplanar graphs with $m - (n - 3)/2 \leq (3n - 3)/2$ segments; (iii) outerplanar graphs with $(7n - 9)/4$ segments, or with $(3n - 5)/2 + b$ segments, where b is its number of maximal biconnected components; and (iv) planar graphs with $(17n - 38)/3 - m$ or $(17n - 38)/6$ segments. Details are given in the full version [14].

Acknowledgements. We thank Roman Prutkin for the initial discussion of the problem and Therese Biedl for helpful comments.

References

1. Angelini, P., Colasante, E., Battista, G.D., Frati, F., Patrignani, M.: Monotone drawings of graphs. J. Graph Algorithms Appl. **16**(1), 5–35 (2012). https://doi.org/10.7155/jgaa.00249
2. Bonichon, N., Le Saëc, B., Mosbah, M.: Wagner's theorem on realizers. In: Widmayer, P., Eidenbenz, S., Triguero, F., Morales, R., Conejo, R., Hennessy, M. (eds.) ICALP 2002. LNCS, vol. 2380, pp. 1043–1053. Springer, Heidelberg (2002). https://doi.org/10.1007/3-540-45465-9_89
3. Chaplick, S., Fleszar, K., Lipp, F., Ravsky, A., Verbitsky, O., Wolff, A.: Drawing graphs on few lines and few planes. In: Hu, Y., Nöllenburg, M. (eds.) GD 2016. LNCS, vol. 9801, pp. 166–180. Springer, Cham (2016). https://doi.org/10.1007/978-3-319-50106-2_14
4. Chaplick, S., Fleszar, K., Lipp, F., Ravsky, A., Verbitsky, O., Wolff, A.: The complexity of drawing graphs on few lines and few planes. In: Ellen, F., Kolokolova, A., Sack, J.R. (eds.) WADS 2017. LNCS, vol. 10389, pp. 265–276. Springer, Cham (2017). https://doi.org/10.1007/978-3-319-62127-2_23
5. Chiang, Y.T., Lin, C.C., Lu, H.I.: Orderly spanning trees with applications. SIAM J. Comput. **34**(4), 924–945 (2005). https://doi.org/10.1137/s0097539702411381
6. Di Battista, G., Tamassia, R., Vismara, L.: Output-sensitive reporting of disjoint paths. Algorithmica **23**(4), 302–340 (1999). https://doi.org/10.1007/PL00009264
7. Dujmović, V., Eppstein, D., Suderman, M., Wood, D.R.: Drawings of planar graphs with few slopes and segments. Comput. Geom. Theory Appl. **38**(3), 194–212 (2007). https://doi.org/10.1016/j.comgeo.2006.09.002
8. Durocher, S., Mondal, D.: Drawing plane triangulations with few segments. Comput. Geom. **77**, 27–39 (2019). https://doi.org/10.1016/j.comgeo.2018.02.003

9. Durocher, S., Mondal, D., Nishat, R.I., Whitesides, S.: A note on minimum-segment drawings of planar graphs. J. Graph Algorithms Appl. **17**(3), 301–328 (2013). https://doi.org/10.7155/jgaa.00295

10. Felsner, S., Zickfeld, F.: Schnyder woods and orthogonal surfaces. Discret. Comput. Geom. **40**(1), 103–126 (2008). https://doi.org/10.1007/s00454-007-9027-9

11. Hossain, M.I., Rahman, M.S.: Good spanning trees in graph drawing. Theor. Comput. Sci. **607**, 149–165 (2015). https://doi.org/10.1016/j.tcs.2015.09.004

12. Hültenschmidt, G., Kindermann, P., Meulemans, W., Schulz, A.: Drawing planar graphs with few geometric primitives. J. Graph Algorithms Appl. **22**(2), 357–387 (2018). https://doi.org/10.7155/jgaa.00473

13. Igamberdiev, A., Meulemans, W., Schulz, A.: Drawing planar cubic 3-connected graphs with few segments: algorithms & experiments. J. Graph Algorithms Appl. **21**(4), 561–588 (2017). https://doi.org/10.7155/jgaa.00430

14. Kindermann, P., Mchedlidze, T., Schneck, T., Symvonis, A.: Drawing planar graphs with few segments on a polynomial grid. arXiv report (2019). http://arxiv.org/abs/1903.08496

15. Kindermann, P., Meulemans, W., Schulz, A.: Experimental analysis of the accessibility of drawings with few segments. J. Graph Algorithms Appl. **22**(3), 501–518 (2018). https://doi.org/10.7155/jgaa.00474

16. Kobourov, S.G., Mchedlidze, T., Vonessen, L.: Gestalt principles in graph drawing. In: Di Giacomo, E., Lubiw, A. (eds.) GD 2015. LNCS, vol. 9411, pp. 558–560. Springer, Cham (2015). https://doi.org/10.1007/978-3-319-27261-0_50

17. Kryven, M., Ravsky, A., Wolff, A.: Drawing graphs on few circles and few spheres. In: Panda, B.S., Goswami, P.P. (eds.) CALDAM 2018. LNCS, vol. 10743, pp. 164–178. Springer, Cham (2018). https://doi.org/10.1007/978-3-319-74180-2_14

18. Miura, K., Azuma, M., Nishizeki, T.: Canonical decomposition, realizer, schnyder labeling and orderly spanning trees of plane graphs. Int. J. Found. Comput. Sci. **16**(01), 117–141 (2005). https://doi.org/10.1142/s0129054105002905

19. Mondal, D.: Visualizing graphs: optimization and trade-offs. Ph.d. thesis, University of Manitoba (2016). http://hdl.handle.net/1993/31673

20. Mondal, D., Nishat, R.I., Biswas, S., Rahman, M.S.: Minimum-segment convex drawings of 3-connected cubic plane graphs. J. Comb. Optim. **25**(3), 460–480 (2013). https://doi.org/10.1007/s10878-011-9390-6

21. Schnyder, W.: Embedding planar graphs on the grid. In: Johnson, D.S. (ed.) SODA 1990, pp. 138–148. SIAM (1990). http://dl.acm.org/citation.cfm?id=320191

22. Schulz, A.: Drawing graphs with few arcs. J. Graph Algorithms Appl. **19**(1), 393–412 (2015). https://doi.org/10.7155/jgaa.00366

23. Wade, G.A., Chu, J.: Drawability of complete graphs using a minimal slope set. Comput. J. **37**(2), 139–142 (1994). https://doi.org/10.1093/comjnl/37.2.139

Variants of the Segment Number
of a Graph

Yoshio Okamoto[1,2]📵, Alexander Ravsky[3], and Alexander Wolff[4](✉)📵

[1] University of Electro-Communications, Chōfu, Japan
[2] RIKEN Center for Advanced Intelligence Project, Tokyo, Japan
[3] Pidstryhach Institute for Applied Problems of Mechanics and Mathematics,
National Academy of Sciences of Ukraine, Lviv, Ukraine
alexander.ravsky@uni-wuerzburg.de
[4] Universität Würzburg, Würzburg, Germany
usetheemailaddressonmyhomepage@gmail.com
http://www1.informatik.uni-wuerzburg.de/en/staff/wolff-alexander

Abstract. The *segment number* of a planar graph is the smallest number of line segments whose union represents a crossing-free straight-line drawing of the given graph in the plane. The segment number is a measure for the visual complexity of a drawing; it has been studied extensively.

In this paper, we study three variants of the segment number: for planar graphs, we consider crossing-free polyline drawings in 2D; for arbitrary graphs, we consider crossing-free straight-line drawings in 3D and straight-line drawings with crossings in 2D. We first construct an infinite family of planar graphs where the classical segment number is asymptotically twice as large as each of the new variants of the segment number. Then we establish the $\exists \mathbb{R}$-completeness (which implies the NP-hardness) of all variants. Finally, for cubic graphs, we prove lower and upper bounds on the new variants of the segment number, depending on the connectivity of the given graph.

1 Introduction

When drawing a graph, a way to keep the visual complexity low is to use few geometric objects for drawing the edges. This idea is captured by the *segment number* of a (planar) graph, that is, the smallest number of crossing-free line segments that together constitute a straight-line drawing of the given graph. The *arc number* of a graph is defined analogously with respect to circular-arc drawings. So far, both numbers have only been studied for planar graphs. Two obvious lower bounds for the segment number are known [5]: (i) $\eta(G)/2$, where $\eta(G)$ is the number of odd-degree vertices of G, and (ii) the planar slope number of G, that is, the smallest number k such that G admits a crossing-free straight-line drawing whose edges have k different slopes.

A.W. acknowledges support from DFG grant WO 758/9-1.

D. Archambault and C. D. Tóth (Eds.): GD 2019, LNCS 11904, pp. 430–443, 2019.
https://doi.org/10.1007/978-3-030-35802-0_33

Dujmović et al. [5], who introduced segment number and planar slope number, showed among others that trees can be drawn without crossings such that the optimum segment number and the optimum planar slope number are achieved simultaneously. In fact, any tree T admits a drawing with $\eta(T)/2$ segments and $\Delta(T)/2$ slopes, where $\Delta(T)$ is the maximum degree of T. Unfortunately, these drawings need exponential area. Therefore, Schulz [19] suggested to study the arc number of planar graphs. Among other things, he showed that any n-vertex tree can be drawn on a polynomial-size grid ($O(n^{1.81}) \times n$) using at most $3n/4$ arcs.

Another measure for the visual complexity of a drawing of a graph is the minimum number of *lines* whose union contains a straight-line crossing-free drawing of the given graph. This parameter is called the *line cover number* of a graph G and denoted by $\rho_2^1(G)$ for 2D (where G must be planar) and $\rho_3^1(G)$ for 3D. Together with the plane cover number $\rho_3^2(G)$ and other variants, these parameters have been introduced by Chaplick et al. [2]. They also showed that both line cover numbers are $\exists\mathbb{R}$-hard to compute [3]. (For background on $\exists\mathbb{R}$, see Schaefer's work [18]).

Upper bounds for the segment number and the arc number (in terms of the number of vertices, n, ignoring constant additive terms) are known for series-parallel graphs ($3n/2$ vs. n), planar 3-trees ($2n$ vs. $11n/6$), and triconnected planar graphs ($5n/2$ vs. $2n$) [5,19]. The upper bound on the segment number for triconnected planar graphs has been improved for the special cases of triangulations and 4-connected triangulations (from $5n/2$ to $7n/3$ and $9n/4$, respectively) by Durocher and Mondal [6]. For the special case of triconnected cubic graphs, Dujmović et al. [5] showed that the segment number is upperbounded by $n + 2$. (A cubic graph with n vertices has $3n/2$ edges.) The result of Dujmović et al. was improved by Mondal et al. [16] who gave two linear-time algorithms based on cannonical decompositions; one that uses at most $n/2 + 3$ segments for $n \geq 6$ and one that uses $n/2 + 4$ segments but places all vertices on a grid of size $n \times n$. Both algorithms use at most six different slopes. Note that $n/2 + 3$ segments are optimal for cubic planar graphs since in every vertex at least one segment must end and in the at least three vertices on the convext hull all three incident segments must end. Igamberdiev et al. [12] fixed a bug in the algorithm of Mondal et al., presented two conceptually different (but slower) algorithms that meet the lower bound and compared them experimentally in terms of common metrics such as angular resolution.

Hültenschmidt et al. [11] provided bounds for segment and arc number under the additional constraint that vertices must lie on a polynomial-size grid. They also showed that n-vertex triangulations can be drawn with at most $5n/3$ arcs, which is better than the lower bound of $2n$ for the segment number on this class of graphs. For 4-connected triangulations, they need at most $3n/2$ arcs. Kindermann et al. [13] recently strengthened some of these results by showing that many classes of planar graphs admit nontrivial bounds on the segment number even when restricting vertices to a grid of size $O(n) \times O(n^2)$. For drawing n-vertex trees with at most $3n/4$ segments, they reduced the grid size to

$n \times n$. Among other things, Durocher et al. [7] showed that the segment number is NP-hard to compute *with respect to a fixed embedding*, even in the special case of arrangement graphs. They also showed that the following partial representation extension problem is NP-hard: given an outerplanar graph G, an integer k, and a straight-line drawing δ of a subgraph of G, is there a k-segment drawing that contains δ? It is still open, however, whether the segment number is fixed-parameter tractable.

In this paper, we consider several variants of the planar segment number seg_2 that has been studied extensively. In particular, we study the *3D segment number* seg_3, which is the most obvious generalization of the planar segment number. It is the smallest number of straight-line segments needed for a crossing-free straight-line drawing of a given graph in 3D. We also study the *crossing segment number* seg_x in 3D, where edges are allowed to cross, but they are not allowed to overlap or to contain vertices in their interiors. In this case, by Lemma 1, the minimum number of segments constituting a drawing of a given graph can be achieved by a plane drawing. Finally, for planar graphs, we study the *bend segment number* $seg_\angle$ in 2D, which is the smallest number of straight-line segments needed for a crossing-free polyline drawing of a given graph in 2D.

Durocher et al. [7] were also interested in the 3D segment number. They stated that their proof of the NP-hardness of the above-mentioned partial representation problem can be adjusted to 3D. They suspected that the 3D segment number remains NP-hard to compute even if the given graph is subcubic. Instead, they showed that a variant of the 3D segment number is NP-hard where one is given a 3D drawing and additional co-planarity constraints that must be fulfilled in the final drawing.

Our Contribution. First, we establish some relationships between the variants of the segment number; see Sect. 2. Then we turn to the complexity of computing the new variants of the segment number; see Sect. 3. By re-using ideas from the $\exists \mathbb{R}$-completeness proof of Chaplick et al. [3] regarding the computation of the line cover numbers ρ_2^1 and ρ_3^1, we establish the $\exists \mathbb{R}$-completeness (and hence the NP-hardness) of all variants of the segment number – seg_2, seg_3, seg_x, and $seg_\angle$ – even for graphs of maximum degree 4. Thus, we nearly answer the open problem of Durocher et al. [7] concerning the computational complexity of the 3D segment number for subcubic graphs. Note that Hoffmann [10] recently established the $\exists \mathbb{R}$-hardness of computing the slope number $slope(G)$ of a planar graph G.

Our main contribution consists in algorithms and lower-bound constructions for connected ($\gamma = 1$), biconnected ($\gamma = 2$), and triconnected ($\gamma = 3$) cubic graphs; see Table 1. To put these results into perspective, recall that any cubic graph with n vertices needs at least $n/2 + 3$ and at most $3n/2$ segments to be drawn, regardless of the drawing style. (In contrast, four slopes suffice for cubic graphs [17]). We prove our bounds in Sect. 4. Note that for cubic graphs, vertex- and edge-connectivity are the same [4, Thm. 2.17].

Before we start, we introduce the following notation. For a given polyline drawing δ of a graph in 2D or 3D, we denote by $seg(\delta)$ the number of (inclusion-wise maximal) straight-line segments of which the drawing δ consists.

Table 1. Overview over existing and new bounds on variants of the segment number of cubic graphs. The upper bounds hold for all n-vertex graphs of a certain vertex connectivity γ. The lower bounds are existential; there exist graphs for which they hold. Note that seg_2 and $\text{seg}_\angle$ are defined only for planar graphs. We skip more specialized known results (e.g., concerning grid size [11] or triangulations [6]).

γ	$\text{seg}_2(G)$		$\text{seg}_3(G)$		$\text{seg}_\angle(G)$		$\text{seg}_\times(G)$	
1	$\geq 5n/6$	[Prp. 2]	$\geq 5n/6$	[Prp. 2]	$\geq 5n/6$	[Prp. 2]	$\geq 5n/6$	[Prp. 2]
2			$\leq n+2$	[Th. 5]	$\leq n+1$	[Th. 4]	$\leq n+2$	[Th. 5]
	$\geq 3n/4$	[Prp. 4]	$\geq 5n/6$	[Prp. 3]	$\geq 3n/4$	[Prp. 4]	$\geq 3n/4$	[Prp. 4]
3	$= n/2+3$	[12,16]	$\leq n+2$	[Th. 5]			$\leq n+2$	[Th. 5]
	(except for $G = K_4$)		$\geq 7n/10$	[Prp. 5]	$\text{seg}_\angle \equiv \text{seg}_2$			

2 Relationships Between Segment Number Variants

Lemma 1. *Given a graph G and a straight-line drawing δ of G in 3D with the property that no two edges overlap and no edge contains a vertex in its interior, then there exists a plane drawing δ' of G with $\text{seg}(\delta') \leq \text{seg}(\delta)$ and with the same property as δ. (Note that both in δ and δ' edges may cross).*

Proof. For each triplet u, v, w of points in δ that correspond to three distinct vertices of G, let $P(u, v, w)$ be the plane or line spanned by the vectors $\overrightarrow{uv}$ and $\overrightarrow{wv}$, and let $\mathcal{P}$ be the set of all such planes or lines. Choose a point A in $\mathbb{R}^3 \setminus \bigcup \mathcal{P}$ that does not lie in the xy-plane. Let δ' be the drawing that results from projecting δ parallel to the vector OA onto the xy-plane. Due to the choice of the projection, δ' may contain crossings, but no edge contains a vertex to which it is not incident, and no two edges overlap. By construction, $\text{seg}(\delta') \leq \text{seg}(\delta)$. $\qquad\square$

Corollary 1. *For any graph G it holds that $\text{seg}_\times(G) \leq \text{seg}_3(G)$.*

Proposition 1. *There is an infinite family of planar graphs $(\mathcal{S}_i)_{i \geq 3}$ such that $\mathcal{S}_i$ has $n_i = i^3 - i + 6$ vertices and the ratios $\text{seg}_2(\mathcal{S}_i)/\text{seg}_3(\mathcal{S}_i)$, $\text{seg}_2(\mathcal{S}_i)/\text{seg}_\angle(\mathcal{S}_i)$, and $\text{seg}_2(\mathcal{S}_i)/\text{seg}_\times(\mathcal{S}_i)$ all converge to 2 with increasing i.*

Proof. We construct, for $i \geq 3$, a triangulation $\mathcal{T}_i$ with maximum degree 6 and $t_i = i^2 - 2i + 3$ vertices (and, hence, $3t_i - 6$ edges and $2t_i - 4$ faces), as follows. Take two triangular grids of side length $i - 1$ (a single triangle is a grid of side length 1) and glue their boundaries, identifying corresponding vertices and edges. Clearly, the result is a (planar) triangulation. Let $s_i = \text{seg}_2(\mathcal{T}_i)$. Then, by the result of Dujmović et al. [5], $s_i \leq 5t_i/2$.

We assume that i is even. To each vertex v of the triangulation, we attach an i-*fan*, that is, a path of length i each of whose vertices is connected to v. Let $\mathcal{S}_i$ be the resulting graph, which has $n_i = t_i(i + 2)$ vertices.

In 2D, no matter how the triangulation is drawn, only three vertices lie on the outer face. Consider an i-fan incident to one of the $t_i - 3$ inner vertices; see

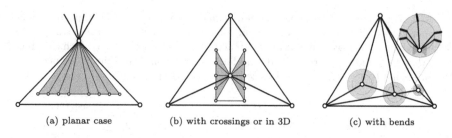

(a) planar case   (b) with crossings or in 3D   (c) with bends

Fig. 1. Attaching a fan (thin edges) to a vertex of a triangulation (thick edges) of maximum degree 6

Fig. 1a. Each such i-fan must be placed into a triangular face and needs at least $i - 3$ segments that are disjoint from the drawing of the triangulation. (Here we use that every vertex has degree at most 6.) Hence, $\operatorname{seg}_2(\mathcal{S}_i) \geq (t_i - 3) \cdot (i - 3) = i^3 - O(i^2)$.

In 3D on the other hand, we can draw every fan in a plane different from the triangulation such that the fan's path lies on three segments and the remaining edges are paired such that each pair shares a segment; see Fig. 1b. Hence, $\operatorname{seg}_3(\mathcal{S}_i) \leq t_i \cdot (i/2 + 3) + s_i = i^3/2 + O(i^2)$. Due to Corollary 1, $\operatorname{seg}_\times(\mathcal{S}_i) \leq \operatorname{seg}_3(\mathcal{S}_i)$.

To bound $\operatorname{seg}_\angle(\mathcal{S}_i)$, observe that we can modify the layout of the triangulation as in Fig. 1c such that every vertex is incident to an angle greater than π without any incoming edges. This can be achieved as follows. On each inner vertex v, place a disk D_v whose radius is (slightly smaller than) the minimum over the lengths of the incident edges divided by 2 and over the distances to all non-incident edges. The resulting disks have positive radii and are pairwise disjoint. Now we go through all vertices. Let v be the current vertex and let ∂D_v be the boundary of D_v. We bend all edges incident to v at ∂D_v and place v on some unused point on ∂D_v. As a result, every vertex is incident to an angle greater than π without any incoming edges. In this area (marked red in Fig. 1c), we can place the corresponding fan. The modification introduces at most two bends in every edge of the triangulation. Hence, $\operatorname{seg}_\angle(\mathcal{S}_i) \leq t_i \cdot (i/2 + 3) + 3 \cdot (3t_i - 6) = i^3/2 + O(i^2)$.   □

Open Problem 1. *What are upper bounds for the ratios* $\operatorname{seg}_2(G)/\operatorname{seg}_3(G)$, $\operatorname{seg}_2(G)/\operatorname{seg}_\angle(G)$, *and* $\operatorname{seg}_2(G)/\operatorname{seg}_\times(G)$ *with* G *ranging over all planar graphs?*

3   Computational Complexity

Chaplick et al. [3, Theorem 1] showed that it is ∃ℝ-hard to decide for a planar graph G and an integer k whether $\rho_2^1(G) \leq k$ and whether $\rho_3^1(G) \leq k$. We follow their approach to show the hardness of all variants of the segment number that we study in this paper.

A *simple line arrangement* is a set $\mathcal{L}$ of k lines in $\mathbb{R}^2$ such that each pair of lines has one intersection point and no three lines share a common point. We define the *arrangement graph* for a set of lines as follows [1]: The vertices correspond to the intersection points of lines and two vertices are adjacent in the graph if and only if they lie on the same line and no other vertex lies between them. The ARRANGEMENT GRAPH RECOGNITION problem is to decide whether a given graph is the arrangement graph of some set of lines.

Bose et al. [1] showed that this problem is NP-hard by reduction from a version of PSEUDOLINE STRETCHABILITY for the Euclidean plane, whose NP-hardness was proved by Shor [20]. It turns out that ARRANGEMENT GRAPH RECOGNITION is actually an $\exists\mathbb{R}$-complete problem [8, p. 212]. This stronger statement follows from the fact that the Euclidean PSEUDOLINE STRETCHABILITY is $\exists\mathbb{R}$-hard as well as the original projective version [15,18].

Theorem 1. *Given a planar graph G of maximum degree 4 and an integer k, it is $\exists\mathbb{R}$-hard to decide whether $\mathrm{seg}_2(G) \leq k$, whether $\mathrm{seg}_\angle(G) \leq k$, and whether $\mathrm{seg}_\times(G) \leq k$.*

Proof. Similarly to Chaplick et al. [3, proof of Theorem 1], we first observe that if G is an arrangement graph, there must be an integer ℓ such that G has $\ell(\ell - 1)/2$ vertices (of degree $d \in \{2,3,4\}$) and $\ell(\ell - 2)$ edges. This uniquely determines ℓ. We set the parameter k from the statement of our theorem to this value of ℓ. Again, as Chaplick et al., we construct a graph G' from G by appending a tail (i.e., a degree-1 vertex) to each degree-3 vertex of G and two tails to each degree-2 vertex of G.

We claim that the following five conditions are equivalent: (i) G is an arrangement graph on k lines, (ii) $\rho_2^1(G') \leq k$, (iii) $\mathrm{seg}_2(G') \leq k$, (iv) $\mathrm{seg}_\angle(G') \leq k$, and (v) $\mathrm{seg}_\times(G') \leq k$. Once the equivalence is established, the $\exists\mathbb{R}$-hardness of deciding (i) implies the $\exists\mathbb{R}$-hardness of deciding any of the other statements.

Indeed, according to Chaplick et al. [3, proof of Theorem 1], G is an arrangement graph if and only if $\rho_2^1(G') \leq k$, that is, (i) and (ii) are equivalent.

Assume (i). If G corresponds to a line arrangement of k lines, all edges of G lie on these k lines and the tails of G' can be added without increasing the number of lines. This arrangement shows that $\mathrm{seg}_2(G') \leq k$, that is, (i) implies (iii).

Assume (iii), i.e., $\mathrm{seg}_2(G') \leq k$. Then $\mathrm{seg}_\angle(G') \leq k$ (iv) and $\mathrm{seg}_\times(G') \leq k$ (v).

Assume (iv), i.e., $\mathrm{seg}_\angle(G') \leq k$. Let Γ' be a polyline drawing of G' on $\mathrm{seg}_\angle(G')$ segments. The graph G' contains $\binom{k}{2}$ degree-4 vertices. As each of these vertices lies on the intersection of two segments in Γ', we need k segments to get enough intersections, that is, $\mathrm{seg}_\angle(G') \geq k$. Thus $\mathrm{seg}_\angle(G') = k$ and each intersection of the segments of Γ' (in particular, each bend) is a vertex of G'. Therefore edges in Γ' do not bend in interior points and Γ' witnesses that $\mathrm{seg}_2(G) \leq k$. Thus (iv) implies (ii).

Finally, assume (v), i.e., $\mathrm{seg}_\times(G') \leq k$. Let Γ be a straight-line drawing with possible crossings on $\mathrm{seg}_\times(G')$ segments. Again, we need k segments to get enough intersections, that is, $\mathrm{seg}_\times(G') \geq k$. Thus $\mathrm{seg}_\times(G') = k$ and each intersection of the segments of Γ' is a vertex of G'. Therefore edges in Γ' do not cross and Γ' witnesses that $\mathrm{seg}_2(G) \leq k$. Thus (v) implies (ii).

Summing up, (iii) implies (iv) and (v), which both imply (ii), which implies (i), which implies (iii). Hence, all statements are equivalent. $\square$

Theorem 2. *Given a graph G of maximum degree 4 and an integer k, it is $\exists\mathbb{R}$-hard to decide whether $\text{seg}_3(G) \leq k$.*

Proof. Chaplick et al. [3, proof of Theorem 1] argued that for the graph G' constructed in the proof of Theorem 1 above, it holds that $\rho_2^1(G') = \rho_3^1(G')$. Then, by the proof of Theorem 1, we have $\rho_3^1(G') = \text{seg}_\times(G')$.

By definition, we immediately obtain $\text{seg}_3(G') \leq \rho_3^1(G')$. By Corollary 1, we have that $\text{seg}_\times(G') \leq \text{seg}_3(G')$. Therefore, $\text{seg}_\times(G') = \text{seg}_3(G')$. Together with the arguments in the proof of Theorem 1, this implies the theorem.   □

Theorem 3. *Given a planar graph G and an integer k, it is $\exists\mathbb{R}$-complete to decide whether $\text{seg}_2(G) \leq k$, whether $\text{seg}_3(G) \leq k$, whether $\text{seg}_\angle(G) \leq k$, and whether $\text{seg}_\times(G) \leq k$.*

Proof. Given the hardness results in Theorems 1 and 2, it remains to show that each of the four problems lies in $\exists\mathbb{R}$. Chaplick et al. [3] [ArXiv version, Sect. 2] have shown that deciding whether $\rho_1^2(G) \leq k$ and $\rho_1^3(G) \leq k$ both lie in $\exists\mathbb{R}$. To this end, they showed that these questions can be formulated as first-order existential expressions over the reals. We now show how to extend their expression for deciding whether $\rho_1^2(G) \leq k$ to an expression for deciding whether $\text{seg}_2(G) \leq k$. The expressions for the other variants can be extended in a similar way.

Their existential statement over the reals starts with the quantifier prefix $\exists v_1 \ldots \exists v_n \exists p_1 \exists q_1 \ldots \exists p_k \exists q_k$, where quantification $\exists a$ over a point $a = (x, y)$ means the quantifier block $\exists x \exists y$, the points $v_1, \ldots, v_n$ are the points to which the vertices of G, $\{1, \ldots, n\}$, are mapped, and the pairs $(p_1, q_1) \ldots, (p_k, q_k)$ define the k lines that cover the drawing of G. The expression Π over which they quantify uses a subexpression that takes as input three points in $\mathbb{R}^2$; for a, b, and c, they define the expression $B(a, b, c)$ such that it is true if and only if a lies on the line segment $\overline{bc}$.

To the expression Π we simply add a term that ensures that, for each pair of consecutive points v_i and v_j on the same line, vertices i and j are adjacent in G:

$$\bigwedge_{l \in \{1, \ldots, k\}, i, j, k \in V} B(v_i, p_l, q_l) \wedge B(v_j, p_l, q_l) \wedge \neg B(v_k, v_i, v_j) \Rightarrow \{i, j\} \in E,$$

where V is the vertex set and E is the edge set of the graph G.   □

4   Algorithms and Lower Bounds for Cubic Graphs

Consider a polyline drawing δ of a cubic graph (in 2D or 3D). Note that there are two types of vertices; those where exactly one segment ends and those where three segments end. We call these vertices *flat vertices* and *tripods*, respectively. Let $f(\delta)$ be the number of flat vertices, $t(\delta)$ the number of tripods, and $b(\delta)$ the number of bends in δ.

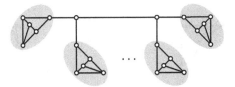

Fig. 2. The graph G_k (here $k = 4$) is a caterpillar with $k - 2$ inner vertices of degree 3 where each leaf has been replaced by a copy of the 5-vertex graph K_4' (shaded gray).

Lemma 2. *For any straight-line drawing δ of a cubic graph with n vertices,* $\operatorname{seg}(\delta) = 3n/2 - f(\delta) + b(\delta) = n/2 + t(\delta) + b(\delta)$.

Proof. Clearly, $n = f(\delta) + t(\delta)$. The number of "segment ends" is $3t(\delta) + f(\delta) + 2b(\delta) = 3n - 2f(\delta) + 2b(\delta) = n + 2t(\delta) + 2b(\delta)$. The claim follows since every segment has two ends. ◻

4.1 Singly-Connected Cubic Graphs

Proposition 2. *There is an infinite family $(G_k)_{k \geq 1}$ of connected cubic graphs such that G_k has $n_k = 6k - 2$ vertices and $\operatorname{seg}_2(G_k) = \operatorname{seg}_3(G_k) = \operatorname{seg}_\angle(G_k) = \operatorname{seg}_\times(G_k) = 5k - 1 = 5n_k/6 + 2/3$.*

Proof. Let K_4' be the graph K_4 with a subdivided edge. Consider the graph G_k depicted in Fig. 2 (for $k = 4$). It consists of a caterpillar with $k - 2$ inner vertices (of degree 3) where each of the k leaf nodes is replaced by a copy of K_4'. The convex hull of every polyline drawing of K_4' has at least three extreme points. One of these points may connect K_4' to $G_k - K_4'$, but each of the remaining two must be a tripod or a bend. This holds for every copy of K_4'. Hence, for any drawing δ of G, $t(\delta) + b(\delta) \geq 2k$. Now Lemma 2 yields that $\operatorname{seg}(\delta) \geq 5k - 1$. For the drawing in Fig. 2, the bound is tight. ◻

4.2 Biconnected Cubic Graphs

Proposition 3. *There is an infinite family of Hamiltonian (and hence biconnected) cubic graphs $(H_k)_{k \geq 3}$ such that H_k has $n_k = 6k$ vertices, $\operatorname{seg}_3(H_k) = 5k = 5n_k/6$, and $\operatorname{seg}_\times(H_k) = 4k = 2n_k/3$.*

Proof. Consider the graph H_k depicted in Fig. 3 (for $k = 4$). It is a k-cycle where each vertex is replaced by a copy of a 6-vertex graph K ($K_{3,3}$ minus an edge). The graph H_k has $n_k = 6k$ vertices and is not planar.

In any 2D drawing of the subgraph K, at least three vertices lie on the convex hull of the drawing of K. Two of these vertices may connect K to $H_k - K$, but at least one of the convex-hull vertices is a tripod. This holds for every copy of K. Hence, for any (3D) drawing δ of H_k, $t(\delta) \geq k$. Now Lemma 2 yields that $\operatorname{seg}(\delta) \geq n_k/2 + k = 2n_k/3$. The same bound holds for $\operatorname{seg}_\times(H_k)$.

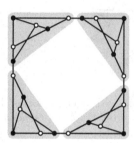

Fig. 3. The cubic graph H_k (here $k = 4$) is a k-cycle whose vertices are replaced by $K_{3,3}$ minus an edge (shaded).

Fig. 4. The planar cubic graph I_k (here $k = 9$) is a k-cycle whose vertices are replaced by K_4 minus an edge (shaded).

In order to bound $\mathrm{seg}_3(H_k)$ we consider two possibilities for the drawing of the subgraph K; either it lies in a plane or it doesn't. In the planar case, the two vertices that connect K to $H_k - K$ cannot lie in the same face of the planar embedding of K (otherwise we could connect these two vertices without crossings, contradicting the fact that $K_{3,3}$ is not planar). Hence, at least two vertices on the convex hull of K must be tripods. In the non-planar case, the convex hull consists of four vertices. Two of these may connect K to $H_k - K$, but again at least two must be tripods. In both cases we hence have $t(\delta) \geq 2k$ for any 3D drawing δ of H_k. Now Lemma 2 yields $\mathrm{seg}(\delta) \geq n_k/2 + 2k = 5n_k/6$. The same bound holds for $\mathrm{seg}_3(H_k)$.

For the drawing in Fig. 3, the bound for $\mathrm{seg}_\times$ is tight. Lifting the k white vertices that do not lie on the outer face from the xy-plane ($z = 0$) to the plane $z = 1$, yields a crossing-free 3D drawing where the bound for seg_3 is tight. $\quad\square$

Proposition 4. *There is an infinite family of planar cubic Hamiltonian (and hence biconnected) graphs $(I_k)_{k\geq 3}$ such that I_k has $n_k = 4k$ vertices and $\mathrm{seg}_2(I_k) = \mathrm{seg}_3(I_k) = \mathrm{seg}_\angle(I_k) = \mathrm{seg}_\times(I_k) = 3k = 3n_k/4$.*

Proof. Consider the graph I_k depicted in Fig. 4 (for $k = 9$). It is a k-cycle where each vertex is replaced by a copy of the graph K', which is K_4 minus an edge. Therefore, I_k has $4k$ vertices. The depicted drawing consists of $3k$ segments. This yields the upper bounds.

Concerning the lower bounds, note that, in any drawing style, each subgraph K' has an extreme point not connected to $I_k - V(K')$. This point must be a tripod or a bend. Hence, in any drawing δ of I_k, $t(\delta) + b(\delta) \geq k$ and, by Lemma 2, $\mathrm{seg}_2(I_k) = \mathrm{seg}_3(I_k) = \mathrm{seg}_\angle(I_k) = \mathrm{seg}_\times(I_k) \geq 2k + t(\delta) + b(\delta) \geq 3k$. $\quad\square$

Theorem 4. *For any biconnected planar cubic graph G with n vertices, it holds that $\mathrm{seg}_\angle(G) \leq n + 1$. A corresponding drawing can be found in linear time.*

Proof. We draw G using the algorithm of Liu et al. [14] that draws any planar biconnected cubic graph except the tetrahedron orthogonally with at most one bend per edge and at most $n/2+1$ bends in total. It remains to count the number of segments in this drawing. In any vertex exactly one segment ends; in any bend exactly two segments end. In total, this yields at most $n + 2 \cdot (n/2 + 1) = 2n + 2$ segment ends and at most $n + 1$ segments.

Concerning the special case of the tetrahedron (K_4), note that it can be drawn with five segments when bending one of its six edges. $\Box$

Open Problem 2. *What about 4-regular graphs? They have $2n$ edges. If we bend every edge once, we already need $2n$ segments – and not all 4-regular graphs can be drawn with at most one bend per edge.*

Every biconnected graph G admits an *st-numbering*, that is, an ordering $\langle v_1, \ldots, v_n \rangle$ of the vertex set $\{v_1, \ldots, v_n\}$ of G such that for every $j \in \{2, \ldots, n-1\}$ vertex v_j has at least one predecessor (that is, a neighbor v_i with $i < j$) and at least one successor (that is, a neighbor v_k with $k > j$). Such a numbering can be computed in linear time [9]. Given a cubic graph with an st-numbering $\langle v_1, \ldots, v_n \rangle$, we call a vertex v_j with $j \in \{1, \ldots, n\}$ a *p-vertex* if it has p predecessors; $p \in \{0, 1, 2, 3\}$.

Lemma 3. *Given a biconnected cubic graph with an st-numbering $\langle v_1, \ldots, v_n \rangle$, there is one 0-vertex and one 3-vertex and there are $(n-2)/2$ 1-vertices and $(n-2)/2$ 2-vertices.*

Proof. Direct every edge from the vertex with smaller index to the vertex with higher index. In the resulting directed graph, the sum of the indegrees equals the sum of the outdegrees. Hence, the number of 1-vertices (with indegree 1 and outdegree 2) and the number of 2-vertices (with indegree 2 and outdegree 1) must be equal. It is obvious that there is one 0- and 3-vertex each. $\Box$

Theorem 5. *For any biconnected cubic graph G with n vertices, $\mathrm{seg}_3(G) \le n+2$ and $\mathrm{seg}_\times(G) \le n + 2$[1].*

Proof. We show that $\mathrm{seg}_3(G) \le n+2$. Then Corollary 1 yields $\mathrm{seg}_\times(G) \le n+2$. For two different points x and y in $\mathbb{R}^3$, we denote the line that goes through x and y by xy.

Let $\langle v_1, \ldots, v_n \rangle$ be an st-numbering of G. We construct a drawing δ of G, going through the vertices according to the st-numbering and using x-coordinate $j \pm \varepsilon$ for vertex v_j, where $0 < \varepsilon \ll 1$. We place v_1 at $(1, 1, 1)$. At every step $j = 2, \ldots, n$, we maintain a set $\mathcal{L}$ of lines that are directed to the right such that any two lines in $\mathcal{L}$ are either skew (that is, they don't lie in the same plane) or they intersect and their unique intersection point is the location of a vertex v_k with $k \le j$ (that is, the intersection point is v_j or it lies to the left of v_j). Initially, $\mathcal{L}$ is empty.

---

[1] After submitting this article, we realized that our proof is incomplete. The correct statement of the theorem and its proof can be found in the full version https://arxiv.org/abs/1908.08871.

If v_j is a 1-vertex, we differentiate two cases depending on the unique predecessor v_i of v_j.

Case I: If v_i is the last vertex on a line ℓ in $\mathcal{L}$, we place v_j on the intersection point of ℓ with the plane $x = j$. In this case, the set $\mathcal{L}$ doesn't change.

Case II: Otherwise, we place v_j in the plane $x = j$ such that the line $v_i v_j$ is skew with respect to all lines in $\mathcal{L}$ except for the line ℓ that contains v_i and the unique predecessor of v_i. (Note that the predecessor of v_i and the line ℓ don't exist if $i = 1$.) Clearly, $v_i v_j$ and ℓ intersect in v_i and $i < j$. Hence, we can add the line $v_i v_j$ to the set $\mathcal{L}$.

If v_j is a 2-vertex, let v_i and $v_{i'}$ be the two predecessors of v_j. Again, we consider two cases.

Case I': At least one of v_i or $v_{i'}$ is flat (that is, it lies on an inner point of the segment created by its incident edges that have already been drawn) or one of them is the vertex v_1.

In this case, we treat v_j similarly as in Case II above; we make sure that the lines $v_i v_j$ and $v_{i'} v_j$ are skew with respect to all lines in $\mathcal{L}$ except that $v_i v_j$ won't be skew with respect to the at most two lines that connect v_i to its predecessors and $v_{i'} v_j$ won't be skew with respect to the at most two lines that connect $v_{i'}$ to its predecessors. Note that $v_i v_j$ intersects any line through v_i and its neighbors in v_i, and it holds that $i < j$. Similarly, $v_{i'} v_j$ intersects any line through $v_{i'}$ and its neighbors in $v_{i'}$, and it holds that $i' < j$. The lines $v_i v_j$ and $v_{i'} v_j$ intersect in v_j. Hence, we can add the lines $v_i v_j$ and $v_{i'} v_j$ to the set $\mathcal{L}$.

Case II': Both v_i and $v_{i'}$ are the last vertices on their lines ℓ and ℓ', respectively.

If one of them, say v_i, has a successor v_k with $k > j$, we extend the line ℓ of v_i and put v_j on the intersection of ℓ and the plane $x = j$.

Otherwise v_i has a successor v_k with $k < j$ and $v_{i'}$ has a successor $v_{k'}$ with $k' < j$, which both don't lie on the lines ℓ and ℓ'. In this case, we put v_j on one of ℓ and ℓ', say ℓ, and add the line $v_{i'} v_j$ to the set $\mathcal{L}$. Now we pick some $0 < \varepsilon \ll 1$ such that we can place v_j at the intersection of ℓ and $x = j + \varepsilon$. We must avoid to place v_j on a plane spanned by any two non-skew lines in $\mathcal{L}$ (intersecting to the left of $x = j$). With this trick, the invariant for $\mathcal{L}$ still holds since the new line in $\mathcal{L}$, $v_{i'} v_j$, intersects only ℓ' (in $v_{i'}$, hence to the left).

Finally, we place v_n (which is a 3-vertex) at a point in the plane $x = n$ that does not lie on any of the lines spanned by pairs and planes spanned by triples of previously placed vertices.

This finishes the description of the drawing δ of G. Due to our invariant regarding the set $\mathcal{L}$, no two edges of G intersect in δ.

To bound the number of segments in δ, we use a simple charging argument. Each non-first and non-last vertex v has a predecessor which is a flat vertex or v_1. To this predecessor v pays a coin. On the other hand, v_1 receives at most three coins and every flat vertex receives at most two coins. Hence, $f(\delta) \geq (n - 5)/2$. Since n is even, $f(\delta) \geq n/2 - 2$. Now, Lemma 2 yields the claim. $\qquad \square$

4.3   Triconnected Cubic Graphs

Proposition 5. *There is an infinite family of triconnected cubic graphs* $(F_k)_{k\geq 4}$ *such that* F_k *has* $n_k = 5k$ *vertices and* $\mathrm{seg}_3(F_k) = 3.5k = 7n_k/10$.

Proof. Let G_k be an arbitrary triconnected cubic graph with k vertices (k even). By Steinitz's theorem, there exists a drawing of the graph G_k as a 1-skeleton of a 3D convex polyhedron. Replace each vertex v of G_k by a copy of $K_{2,3}$ as shown in Fig. 5, where v is the central (orange) vertex—a tripod—, all other vertices of the copy are flat, and the three arrows correspond to the three edges of G_k. The resulting geometric graph F_k has $n_k = 5k$ vertices and is not planar. Since F_k has k tripod vertices, by Lemma 2, $\mathrm{seg}_3(F_k) \leq n_k/2 + k = 3.5k = 7n_k/10$.

In order to bound $\mathrm{seg}_3(F_k)$ from below, we consider two possibilities for the drawing of each subgraph $K_{2,3}$; either it lies in a plane or it doesn't. In the planar case, the convex hull of the drawing has at least three extreme points. If none of them was a tripod then there would be exactly three extreme points, each a black vertex. Thus we could place an additional white vertex in the exterior of the convex hull and connect it to all black vertices, obtaining an impossible plane drawing of $K_{3,3}$. In the

Fig. 5. Gadget for the proof of Proposition 5 (Color figure online)

non-planar case, the convex hull consists of at least four vertices. Three of these may connect $K_{2,3}$ to $F_k - V(K_{2,3})$, but again at least one must be a tripod.

In both cases we hence have $t(\delta) \geq k$ for any 3D drawing δ of F_k. Now Lemma 2 yields $\mathrm{seg}(\delta) = n_k/2 + t(\delta) \geq 3.5k$.   □

5   Open Problems

Apart from improving our bounds, we have the following open problem.

Open Problem 3. *Can we produce drawings in 3D (or with bends or crossings in 2D) that fit on grids of small size?*

Acknowledgments. We thank the organizers and participants of the 2019 Dagstuhl seminar "Beyond-planar graphs: Combinatorics, Models and Algorithms". In particular, we thank Günter Rote and Martin Gronemann for suggestions that led to some of this research. We also thank Carlos Alegría. We thank our reviewers for an idea that improved the bound in Proposition 5, for suggesting the statement of Lemma 1, and for many other helpful comments.

References

1. Bose, P., Everett, H., Wismath, S.K.: Properties of arrangement graphs. Int. J. Comput. Geom. Appl. **13**(6), 447–462 (2003). https://doi.org/10.1142/S0218195903001281

2. Chaplick, S., Fleszar, K., Lipp, F., Ravsky, A., Verbitsky, O., Wolff, A.: Drawing graphs on few lines and few planes. In: Hu, Y., Nöllenburg, M. (eds.) GD 2016. LNCS, vol. 9801, pp. 166–180. Springer, Cham (2016). https://doi.org/10.1007/978-3-319-50106-2_14

3. Chaplick, S., Fleszar, K., Lipp, F., Ravsky, A., Verbitsky, O., Wolff, A.: The complexity of drawing graphs on few lines and few planes. In: Ellen, F., Kolokolova, A., Sack, J.R. (eds.) Algorithms and Data Structures. LNCS, vol. 10389, pp. 265–276. Springer, Cham (2017). https://doi.org/10.1007/978-3-319-62127-2_23. arxiv.org/1607.06444

4. Chartrand, G., Zhang, P.: Chromatic Graph Theory, 1st edn. Chapman & Hall/CRC, Boca Raton (2008)

5. Dujmović, V., Eppstein, D., Suderman, M., Wood, D.R.: Drawings of planar graphs with few slopes and segments. Comput. Geom. Theory Appl. 38(3), 194–212 (2007). https://doi.org/10.1016/j.comgeo.2006.09.002

6. Durocher, S., Mondal, D.: Drawing plane triangulations with few segments. In: Proceedings Canadian Conference on Computational Geometry (CCCG 2014), pp. 40–45 (2014). http://cccg.ca/proceedings/2014/papers/paper06.pdf

7. Durocher, S., Mondal, D., Nishat, R., Whitesides, S.: A note on minimum-segment drawings of planar graphs. J. Graph Algorithms Appl. 17(3), 301–328 (2013). https://doi.org/10.7155/jgaa.00295

8. Eppstein, D.: Drawing arrangement graphs in small grids, or how to play planarity. J. Graph Algorithms Appl. 18(2), 211–231 (2014). https://doi.org/10.7155/jgaa.00319

9. Even, S., Tarjan, R.E.: Computing an st-numbering. Theoret. Comput. Sci. 2(3), 339–344 (1976). https://doi.org/10.1016/0304-3975(76)90086-4

10. Hoffmann, U.: On the complexity of the planar slope number problem. J. Graph Algorithms Appl. 21(2), 183–193 (2017). https://doi.org/10.7155/jgaa.00411

11. Hültenschmidt, G., Kindermann, P., Meulemans, W., Schulz, A.: Drawing planar graphs with few geometric primitives. In: Bodlaender, H.L., Woeginger, G.J. (eds.) WG 2017. LNCS, vol. 10520, pp. 316–329. Springer, Cham (2017). https://doi.org/10.1007/978-3-319-68705-6_24

12. Igamberdiev, A., Meulemans, W., Schulz, A.: Drawing planar cubic 3-connected graphs with few segments: algorithms and experiments. J. Graph Algorithms Appl. 21(4), 561–588 (2017). https://doi.org/10.7155/jgaa.00430

13. Kindermann, P., Mchedlidze, T., Schneck, T., Symvonis, A.: Drawing planar graphs with few segments on a polynomial grid. In: Archambault, D., Tóth, C.D. (eds.) GD 2019. LNCS, vol. 11904, pp. 416–429. Springer. Cham (2019). https://arxiv.org/abs/1903.08496

14. Liu, Y., Marchioro, P., Petreschi, R.: At most single-bend embeddings of cubic graphs. Appl. Math. 9(2), 127–142 (1994). https://doi.org/10.1007/BF02662066

15. Matoušek, J.: Intersection graphs of segments and $\exists\mathbb{R}$. ArXiv report (2014). http://arxiv.org/abs/1406.2636

16. Mondal, D., Nishat, R.I., Biswas, S., Rahman, M.S.: Minimum-segment convex drawings of 3-connected cubic plane graphs. J. Comb. Optim. 25(3), 460–480 (2013). https://doi.org/10.1007/s10878-011-9390-6

17. Mukkamala, P., Szegedy, M.: Geometric representation of cubic graphs with four directions. Comput. Geom. Theory Appl. 42(9), 842–851 (2009). https://doi.org/10.1016/j.comgeo.2009.01.005

18. Schaefer, M.: Complexity of some geometric and topological problems. In: Eppstein, D., Gansner, E.R. (eds.) GD 2009. LNCS, vol. 5849, pp. 334–344. Springer, Heidelberg (2010). https://doi.org/10.1007/978-3-642-11805-0_32

19. Schulz, A.: Drawing graphs with few arcs. J. Graph Algorithms Appl. **19**(1), 393–412 (2015). https://doi.org/10.7155/jgaa.00366
20. Shor, P.W.: Stretchability of pseudolines is NP-hard. In: Gritzmann, P., Sturmfels, B. (eds.) Applied Geometry and Discrete Mathematics-The Victor Klee Festschrift. DIMACS Series in Mathematics and Theoretical Computer Science, vol. 4, pp. 531–554. American Mathematical Society (1991)

Topological Graph Theory

Local and Union Page Numbers

Laura Merker and Torsten Ueckerdt[(✉)]

Karlsruhe Institute of Technology (KIT), Institute of Theoretical Informatics,
Karlsruhe, Germany
laura.merker@student.kit.edu, torsten.ueckerdt@kit.edu

Abstract. We introduce the novel concepts of local and union book embeddings, and, as the corresponding graph parameters, the local page number $\mathrm{pn}_\ell(G)$ and the union page number $\mathrm{pn}_u(G)$. Both parameters are relaxations of the classical page number $\mathrm{pn}(G)$, and for every graph G we have $\mathrm{pn}_\ell(G) \leqslant \mathrm{pn}_u(G) \leqslant \mathrm{pn}(G)$. While for $\mathrm{pn}(G)$ one minimizes the total number of pages in a book embedding of G, for $\mathrm{pn}_\ell(G)$ we instead minimize the number of pages incident to any one vertex, and for $\mathrm{pn}_u(G)$ we instead minimize the size of a partition of G with each part being a vertex-disjoint union of crossing-free subgraphs. While $\mathrm{pn}_\ell(G)$ and $\mathrm{pn}_u(G)$ are always within a multiplicative factor of 4, there is no bound on the classical page number $\mathrm{pn}(G)$ in terms of $\mathrm{pn}_\ell(G)$ or $\mathrm{pn}_u(G)$. We show that local and union page numbers are closer related to the graph's density, while for the classical page number the graph's global structure can play a much more decisive role. We introduce tools to investigate local and union book embeddings in exemplary considerations of the class of all planar graphs and the class of graphs of tree-width k. As an incentive to pursue research in this new direction, we offer a list of intriguing open problems.

Keywords: Book embedding · Page number · Stack number · Local covering number · Planar graph · Tree-width

1 Introduction

A *linear embedding* of a graph $G = (V, E)$ is a tuple $(\prec, \mathcal{P})$ where $\prec$ is a total ordering[1] of the vertex set V and $\mathcal{P} = \{P_1, \ldots, P_k\}$ is a partition of the edge set E. The ordering $\prec$ is sometimes called the *spine ordering*, and each part P_i of $\mathcal{P}$ is called a *page*. For a given spine ordering $\prec$, two edges $uv, xy \in E$ with $u \prec v$ and $u \prec x \prec y$ are said to be *crossing* if $u \prec x \prec v \prec y$. A linear embedding $(\prec, \mathcal{P})$ is a *book embedding* if for any two edges uv and xy in E we have

$$\text{if } u \prec x \prec v \prec y \text{ and } uv \in P_i, xy \in P_j \text{ then } i \neq j. \tag{1}$$

---

[1] We define $\prec$ as a linear ordering. However, in a few places we shall think of $\prec$ as a cyclic ordering. This is legitimate as we are interested in crossing edges only, and these are preserved under cyclic shifts.

© Springer Nature Switzerland AG 2019
D. Archambault and C. D. Tóth (Eds.): GD 2019, LNCS 11904, pp. 447–459, 2019.
https://doi.org/10.1007/978-3-030-35802-0_34

So Eq. (1) simply states that no two edges in the same page are crossing, or equivalently, any two crossing edges are assigned to distinct pages in $\mathcal{P}$.

Book embeddings were introduced by Ollmann [24] as well as Bernhart and Kainen [3], see also [17]. Besides their apparent applications in real-world problems (see e.g. [8,26] and the numerous references in [9]), book embeddings enjoy steady popularity in graph theory; see for example [11,16,21,27,30,33], just to name a few. In most cases (also including the generalizations for directed graphs [4] or pages with limited crossings [5]), one seeks to find a book embedding with as few pages as possible for given graph G. In particular, $(\prec, \mathcal{P})$ is a *k-page book embedding* if $|\mathcal{P}| = k$, and the *page number* of G, denoted by $\mathrm{pn}(G)$, is the smallest k for which we can find a k-page book embedding of G. (We remark that $\mathrm{pn}(G)$ is sometimes also called the book thickness [3] or stack number [9] of G.)

As the main contribution of the present paper, we propose two relaxations of the page number parameter: The local page number $\mathrm{pn}_\ell(G)$ and the union page number $\mathrm{pn}_u(G)$. We initiate the study of these parameters by comparing $\mathrm{pn}_\ell(G)$, $\mathrm{pn}_u(G)$, and $\mathrm{pn}(G)$ for graphs G in some natural graph classes, such as planar graphs (c.f. Sect. 3), graphs of bounded density (c.f. Sect. 2), and graphs of bounded tree-width (c.f. Sect. 4). Besides these bounds, a (perhaps not surprising) result showing computational hardness (c.f. Theorem 4), and a few structural observations, we also give some intriguing open problems at the end of the paper in Sect. 5.

Before listing our specific results in Sect. 1.1 below, let us define and motivate the novel parameters local and union page numbers.

Local Page Numbers. For a book embedding $(\prec, \mathcal{P})$ of graph $G = (V, E)$ and a vertex $v \in V$, let us denote by $\mathcal{P}_v$ the subset of pages that contain at least one edge incident to v. Then we define:

- A book embedding is *k-local* if $|\mathcal{P}_v| \leqslant k$ for each $v \in V$, i.e., each vertex has incident edges on at most k pages.
- The *local page number*, denoted by $\mathrm{pn}_\ell(G)$, is the smallest k for which we can find a k-local book embedding of G.

Thus, we seek to find a book embedding with any number of pages (possibly more than $\mathrm{pn}(G)$), but with no vertex having incident edges on more than k of these pages. As each k-page book embedding is a k-local book embedding,

$$\text{for any graph } G \text{ we have} \qquad \mathrm{pn}_\ell(G) \leqslant \mathrm{pn}(G). \qquad (2)$$

However, $\mathrm{pn}_\ell(G)$ can be strictly smaller than $\mathrm{pn}(G)$. For example, K_5 and $K_{3,3}$ both have page number 3 and local page number 2. As illustrated in Fig. 1, K_5 admits a 2-local 3-page book embedding, i.e., this book embedding simultaneously certifies $\mathrm{pn}(K_5) \leqslant 3$ and $\mathrm{pn}_\ell(K_5) \leqslant 2$. In the left of Fig. 1 we have a 2-local 4-page book embedding of $K_{3,3}$ (when the three orange/thick edges are put into three separate pages). So here, the introduction of "extra" pages, additionally to the necessary $\mathrm{pn}(K_{3,3}) = 3$ pages in every book embedding of $K_{3,3}$, allowed us to actually reduce the maximum number of pages incident to any one vertex from 3 to $\text{pn}_\ell(K_{3,3}) = 2$. And for some graphs G with $\text{pn}_\ell(G) = k$, in fact all k-local book embeddings have more than $\text{pn}(G)$ pages.

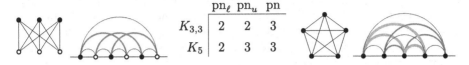

	pn_ℓ	pn_u	pn
$K_{3,3}$	2	2	3
K_5	2	3	3

Fig. 1. Comparison of local, union, and classical page numbers on the examples of $K_{3,3}$ and K_5. Left: 2-page union embedding of $K_{3,3}$. Right: 2-local book embedding of K_5.

Union Page Numbers. For a linear embedding $(\prec, \mathcal{P})$ (so not necessarily a book embedding) of graph $G = (V, E)$ and a page $P \in \mathcal{P}$, let us denote by G_P the subgraph of G on all edges in P and all vertices with some incident edge in P. Then we define:

- A linear embedding $(\prec, \mathcal{P})$ is a *union embedding* if $(\prec, \{E(C)\})$ is a (1-page) book embedding for each connected component C of G_P and each $P \in \mathcal{P}$, i.e., each connected component of each page is crossing-free.
- The *union page number*, denoted by $\text{pn}_u(G)$, is the smallest k for which we can find a k-page union embedding of G.

In other words, in a union embedding, each page is the vertex-disjoint union of crossing-free graphs; hence the name "union page number". So we allow crossing edges on a single page P, as long as these are contained in different components of G_P. For the union page number $\text{pn}_u(G)$ we minimize the number of pages, just like for the classical page number $\text{pn}(G)$.

Again, each k-page book embedding is also a k-page union embedding, giving $\text{pn}_u(G) \leqslant \text{pn}(G)$. Moreover, each k-page union embedding can be transformed into a k-local book embedding by putting each component of each page onto a separate page, giving $\text{pn}_\ell(G) \leqslant \text{pn}_u(G)$. Summarizing,

$$\text{for any graph } G \text{ we have} \qquad \text{pn}_\ell(G) \leqslant \text{pn}_u(G) \leqslant \text{pn}(G). \qquad (3)$$

Consider again the linear embedding of $K_{3,3}$ in the left of Fig. 1, but this time put all three orange/thick edges on the same page P. These edges are pairwise crossing, so this is not a book embedding. However these edges lie in separate connected components of G_P, so this is a union embedding. As we found a 2-page union embedding of $K_{3,3}$, we see $\text{pn}_u(K_{3,3}) \leqslant 2 < 3 = \text{pn}(K_{3,3})$.

Comparing union and local page numbers, we have that $\text{pn}_\ell(G)$ can be strictly smaller than $\text{pn}_u(G)$. For example, we have already seen in Fig. 1 that $\text{pn}_\ell(K_5) \leqslant 2$, and we claim that $\text{pn}_u(K_5) > 2$. Indeed, for the cyclic spine ordering $v_1 \prec \cdots \prec v_5$ and pages P_1, P_2 we may assume by symmetry that $v_1 v_3, v_1 v_4, v_2 v_5 \in P_1$ and $v_2 v_4, v_3 v_5 \in P_2$. As each connected component of G_{P_1} and G_{P_2} is crossing-free, v_2 and v_3 are in distinct components in both page P_1 and page P_2, leaving no way to assign the edge $v_2 v_3$.

Motivation. Local and union page numbers are motivated by local and union covering numbers as introduced by Knauer and the second author [19]. In order to give a brief summary of the covering number framework, consider a graph H and a graph class $\mathcal{G}$. An *injective $\mathcal{G}$-cover of H* is a set $S = \{G_1, \ldots, G_m\}$ of subgraphs[2] of H such that $H = G_1 \cup \cdots \cup G_m$ and $G_i \in \mathcal{G}$ for $i = 1, \ldots, m$. In other words, H is covered by (is the union of) some m (possibly isomorphic) graphs from $\mathcal{G}$. Moreover, let $\overline{\mathcal{G}}$ denote the class of all finite vertex-disjoint unions of graphs in $\mathcal{G}$, meaning that $G \in \overline{\mathcal{G}}$ if and only if G is the vertex-disjoint union of some number of graphs in $\mathcal{G}$.

The *global $\mathcal{G}$-covering number* of H, denoted by $\mathrm{cn}_g^{\mathcal{G}}(H)$, is the smallest m such that there exists an injective $\mathcal{G}$-cover of H of size m, i.e., using m graphs in $\mathcal{G}$. The *union $\mathcal{G}$-covering number* of H, denoted by $\mathrm{cn}_u^{\mathcal{G}}(H)$, is the smallest m such that there exists an injective $\overline{\mathcal{G}}$-cover of H of size m, i.e., using m vertex-disjoint unions of graphs in $\mathcal{G}$. The *local $\mathcal{G}$-covering number* of H, denoted by $\mathrm{cn}_\ell^{\mathcal{G}}(H)$, is the smallest k such that there exists an injective $\mathcal{G}$-cover of H in which every vertex of H is contained in at most k graphs of the cover, i.e., using any number of graphs from $\mathcal{G}$ but with no vertex of H being contained in more than k of these.[3]

Many graph parameters (including arboricities, thickness parameters, variants of chromatic numbers, several Ramsey numbers, and interval representations) are $\mathcal{G}$-covering numbers of a certain type and for a certain graph class $\mathcal{G}$. Moreover, recently the global-union-local framework was extended to settings that do not directly concern graph covers, such as the local and union boxicity [6], and the local dimension of posets [29], which has stimulated research drastically [2,7,12,18,20,28]. Our proposed local and union page numbers naturally arise from the covering number framework by using ordered graphs and ordered subgraphs in the above definitions and taking $\mathcal{G}$ to be the class of all crossing-free ordered graphs.

Particularly the local page number might be very useful in applications. For example, oftentimes the spine ordering $\prec$ of G is already given from the problem formulation (by time stamps, geographic positions or a genetic sequence). Then the edges of G model some kind of connections and classical book embeddings are used to distribute the connections to machines that can process sets of connections that satisfy the LIFO (last-in-first-out) property. Local book embeddings could be used to model situations in which the total number of machines is not the scarce resource but rather the number of machines working on the same element, i.e., vertex. Imagine for example limited capacity at each element in terms of computing power (as for cell phones) or simply spatial restrictions (as for genes). This kind of task is precisely modeled by local book embeddings and the local page numbers.

---

[2] In a general $\mathcal{G}$-cover one considers graph homomorphisms from graphs in $\mathcal{G}$ into H. However, we consider here only injective $\mathcal{G}$-covers, which is equivalent to considering subgraphs of H.

[3] The covering number framework includes a fourth covering number, the folded $\mathcal{G}$-covering number of H, which we omit here, so as not to congest the discussion.

1.1 Our Contribution

First, we show that the new parameters $\mathrm{pn}_\ell(G)$ and $\mathrm{pn}_u(G)$ can be arbitrarily smaller than the classical page number $\mathrm{pn}(G)$, while local and union page number are always at most a multiplicative factor of 4 apart.

Theorem 1. *For any $k \geqslant 3$ and infinitely many values of n, there exist n-vertex graphs G with*

$$\mathrm{pn}_\ell(G) \leqslant \mathrm{pn}_u(G) \leqslant k+2 \quad and \quad \mathrm{pn}(G) = \Omega\left(\sqrt{k}\, n^{1/2 - 1/k}\right).$$

In contrast, for every graph G we have $\mathrm{pn}_u(G) \leqslant 4\,\mathrm{pn}_\ell(G) + 2$.

While for every planar graph G we have $\mathrm{pn}(G) \leqslant 4$ [33], it is not known whether there is a planar graph G with $\mathrm{pn}(G) = 4$. The best known lower bound was given by Bernhart and Kainen [3], who presented a planar graph G with $\mathrm{pn}(G) = 3$. That very graph satisfies $\mathrm{pn}_\ell(G) = 2$, but we can augment it to a planar graph with local page number 3.

Theorem 2. *There is a planar graph G with $\mathrm{pn}_\ell(G) = 3$.*

For graphs G with tree-width k, it is known that $\mathrm{pn}(G) \leqslant k$ if $k \in \{1, 2\}$ [25] and $\mathrm{pn}(G) \leqslant k+1$ if $k \geqslant 3$ [13], and both bounds are best possible [10, 30]. For the local and union page number we get a lower bound of k.

Theorem 3. *For every $k \geqslant 1$ there is a graph G of tree-width k with $\mathrm{pn}_u(G) \geqslant \mathrm{pn}_\ell(G) \geqslant k$.*

Finally, it is known that $\mathrm{pn}(G) \leqslant 2$ if and only if G is a subgraph of a planar Hamiltonian graph [3]. Hence, it follows from [31] that deciding $\mathrm{pn}(G) \leqslant 2$ is NP-complete, which easily generalizes to $\mathrm{pn}(G) \leqslant k$ for each $k \geqslant 2$. (Since $\mathrm{pn}(G) = 1$ is equivalent to G being outerplanar, this can be efficiently tested.) If the spine ordering $\prec$ is already given, the problem of finding an edge partition into k crossing-free pages is equivalent to that of properly k-coloring circle graphs and hence determining the smallest such k is NP-complete [14]. While properly k-coloring circle graphs is polynomial-time solvable for $k = 2$, it is open whether the problem becomes NP-hard for fixed $k \geqslant 3$. For the local page number we have NP-completeness for fixed spine ordering $\prec$ and each fixed $k \geqslant 3$.

Theorem 4. *For any $k \geqslant 3$, it is NP-complete to decide for a given graph G and given spine ordering $\prec$, whether there exists an edge partition $\mathcal{P}$ such that $(\prec, \mathcal{P})$ is a k-local book embedding.*

For a proof of Theorem 4 we refer the interested reader to the Bachelor's thesis of the first author [22].

2   Bounds in Terms of Density

Though not a fixed mathematical concept, the density of a graph $G = (V, E)$ quantifies the number $|E|$ of edges in terms of the number $|V|$ of vertices. An important specification of density is the *maximum average degree* of G defined by

$$\mathrm{mad}(G) = \max \left\{ \frac{2|E(H)|}{|V(H)|} \mid H \subseteq G, H \neq \emptyset \right\}.$$

Recall that for a linear embedding $(\prec, \mathcal{P})$ of $G = (V, E)$ and a page $P \in \mathcal{P}$ we denote by $G_P = (V_P, P)$ the subgraph of G on all edges in P and all vertices of G with at least one incident edge in P. Clearly, if P is crossing-free, then G_P is outerplanar and thus $|P| \leqslant 2|V_P| - 3$. As $\bigcup_{P \in \mathcal{P}} P = E$ and $V_P \subseteq V$ for each page P, we immediately get an upper bound on the density of any graph with a k-local book embedding.

Lemma 5. *For any graph $G = (V, E)$ we have*

$$\mathrm{pn}_\ell(G) \geqslant \max \left\{ \frac{|E(H)|}{2|V(H)| - 3} \mid H \subseteq G, H \neq \emptyset \right\}.$$

Proof. Let H be any non-empty subgraph of a graph G of local page number $\mathrm{pn}_\ell(G) = k$. Then there is a k-local book embedding $(\prec, \mathcal{P})$ of H, each page P of which describes an outerplanar graph $H_P = (V_P, P)$. Thus

$$|E(H)| \leqslant \sum_{P \in \mathcal{P}} (2|V_P| - 3) \leqslant 2k|V(H)| - 3|\mathcal{P}| \leqslant \mathrm{pn}_\ell(G) \cdot (2|V(H)| - 3).$$

$\square$

From Lemma 5 and Eq. (3) we conclude for every graph G that

$$\mathrm{pn}(G) \geqslant \mathrm{pn}_u(G) \geqslant \mathrm{pn}_\ell(G) \geqslant \mathrm{mad}(G)/4. \tag{4}$$

In other words, the graph's density gives a lower bound on all three kinds of page numbers. Perhaps surprisingly, there is also an *upper* bound on the union and local page numbers in terms of the graph's density.

Nash-Williams [23] proved that any graph G edge-partitions into k forests if and only if

$$k \geqslant \max \left\{ \frac{|E(H)|}{|V(H)| - 1} \mid H \subseteq G, |V(H)| \geqslant 2 \right\}.$$

The smallest such k, the *arboricity* $\mathrm{a}(G)$ of G, thus satisfies $\frac{1}{2}\mathrm{mad}(G) < \mathrm{a}(G) \leqslant \frac{1}{2}\mathrm{mad}(G) + 1$. The *star arboricity* $\mathrm{sa}(G)$ of G is the minimum k such that G edge-partitions into k star forests. Thus $\mathrm{sa}(G)$ is the union $\mathcal{G}$-covering number of G with respect to the class $\mathcal{G} = \{K_{1,n} \mid n \in \mathbb{N}\}$ of all stars. Using the covering number framework, Knauer and the second author [19] introduced the corresponding local $\mathcal{G}$-covering number, the *local star arboricity* $\mathrm{sa}_\ell(G)$, as the minimum k such that G edge-partitions into some number of stars, but with each vertex having an incident edge in at most k of these stars. It is known [1,19] that $\mathrm{sa}(G)$ and $\mathrm{sa}_\ell(G)$ can be bound in terms of $\mathrm{a}(G)$ as

$$\mathrm{a}(G) \leqslant \mathrm{sa}(G) \leqslant 2\,\mathrm{a}(G) \qquad \text{and} \qquad \mathrm{a}(G) \leqslant \mathrm{sa}_\ell(G) \leqslant \mathrm{a}(G) + 1.$$

Theorem 6. *For any graph G we have $\mathrm{pn}_\ell(G) \leqslant \mathrm{sa}_\ell(G)$ and $\mathrm{pn}_u(G) \leqslant \mathrm{sa}(G)$. In particular, we have*

$$\frac{\mathrm{mad}(G)}{4} \leqslant \mathrm{pn}_\ell(G) \leqslant \frac{\mathrm{mad}(G)}{2} + 2 \quad \text{and} \quad \frac{\mathrm{mad}(G)}{4} \leqslant \mathrm{pn}_u(G) \leqslant \mathrm{mad}(G) + 2.$$

Proof. Take an arbitrary spine ordering $\prec$ and an edge-partition $\mathcal{P}$ into stars. Then each page is crossing-free, which shows $\mathrm{pn}_\ell(G) \leqslant \mathrm{sa}_\ell(G)$. Now take an arbitrary spine ordering $\prec$ and an edge-partition $\mathcal{P}$ into star forests. Then each connected component on each page is a star and thus crossing-free, which shows $\mathrm{pn}_u(G) \leqslant \mathrm{sa}(G)$. □

Though Theorem 6 is merely an observation, it has a number of interesting consequences. First of all, the local and union page number are not too far apart: $\mathrm{pn}_\ell(G) \leqslant \mathrm{pn}_u(G) \leqslant 4\,\mathrm{pn}_\ell(G) + 2$. However, the local and union page numbers can be very far from the classical page number. For example, we have $\mathrm{sa}_\ell(G) \leqslant k = \mathrm{mad}(G)$ for every k-regular graph G, and hence $\mathrm{pn}_\ell(G) \leqslant k$ and $\mathrm{pn}_u(G) \leqslant k + 2$ whenever G is k-regular. On the other hand, Malitz [21] proved that for every $k \geqslant 3$ there are n-vertex k-regular graphs G with page number $\mathrm{pn}(G) = \Omega(\sqrt{k}n^{1/2-1/k})$. Together this proves Theorem 1.

For planar G we have $\mathrm{a}(G) \leqslant 3$ [23], hence $\mathrm{sa}_\ell(G) \leqslant 4$ [19], as well as $\mathrm{sa}(G) \leqslant 5$ [15]. Hence, Theorem 6 immediately gives the following (without relying on Yannakakis' result [33]).

Corollary 7. *For every planar graph G we have $\mathrm{pn}_\ell(G) \leqslant 4$ and $\mathrm{pn}_u(G) \leqslant 5$.*

3   Planar Graphs

In this section we consider planar graphs. In particular, we prove Theorem 2 stating the existence of a planar graph with local page number 3. Our planar graph will be a large enough stacked triangulation (also known as planar 3-trees, chordal triangulations, or Apollonian networks). For this let $T_0 \cong K_3$ and for $n \geqslant 1$ define T_n as obtained from T_{n-1} by placing a new vertex v_Δ in each facial triangle Δ of T_{n-1}, and connecting v_Δ by edges to each of the three vertices of Δ. Thus, for $n \geqslant 0$ we have $|V(T_n)| = 3^n + 2$.

Suppose for the sake of contradiction there is a 2-local book embedding $(\prec, \mathcal{P})$ of T_9. We consider the subgraphs $T_0 \subseteq T_1 \subseteq \cdots \subseteq T_9$ of T_9.

Claim. There exists an edge vw in T_1 with $\mathcal{P}_v = \mathcal{P}_w$ and $|\mathcal{P}_v| = |\mathcal{P}_w| = 2$.

Indeed, consider the four vertices v_1, v_2, v_3, v_4 of one of the two K_4 subgraphs in T_1. Without loss of generality assume that $|\mathcal{P}_{v_1}| = \cdots = |\mathcal{P}_{v_4}| = 2$. As $\mathcal{P}_{v_i} \cap \mathcal{P}_{v_j} \neq \emptyset$ for any $i,j \in \{1,\ldots,4\}$, we can see $\mathcal{P}_{v_1},\ldots,\mathcal{P}_{v_4}$ as four pairwise

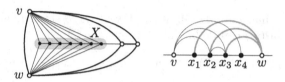

Fig. 2. Part of the planar graph with local page number 3 (left) and part of the hypothetical 2-local book embedding (right).

incident edges in a multigraph I on vertex set $\mathcal{P}$, where two vertices of I are connected by an edge if there is common vertex of G on the two respective pages. Thus, if $\mathcal{P}_{v_1}, \ldots, \mathcal{P}_{v_4}$ were pairwise distinct, they would form a star, i.e., all pairwise intersections would be the same page $P \in \mathcal{P}$. But then the whole K_4 subgraph on $v_1, \ldots, v_4$ would be embedded on page P, which is impossible as K_4 is not outerplanar.

So let vw be an edge in T_1 with $\mathcal{P}_v = \mathcal{P}_w = \{P_1, P_2\}$. By the inductive construction of stacked triangulations, there is a set $X = \{x_1, \ldots, x_7\}$ of seven vertices in $T_8 - T_1$ that are incident to v and w and induce a path in T_9; see Fig. 2. By pigeon-hole principle and cyclic shifts of $\prec$, we may assume that $v \prec x_1 \prec x_2 \prec x_3 \prec x_4 \prec w$, where $x_1, \ldots, x_4$ are consecutive in $\prec$ when restricted to X. Each of vx_i and wx_i, $i = 1, \ldots, 4$, lies on P_1 or P_2; say $vx_4 \in P_1$. Then $wx_1, wx_2, wx_3 \in P_2$, and thus $vx_2, vx_3 \in P_1$. In particular, we have $\mathcal{P}_{x_2} = \mathcal{P}_{x_3} = \{P_1, P_2\}$.

Now observe that x_2 cannot be adjacent to any vertex y with $x_3 \prec y$ and $y \neq w$. Indeed, such an edge $x_2 y$ would cross the edge $vx_3 \in P_1$ and one of $wx_1, wx_3 \in P_2$. Symmetrically, x_3 cannot be adjacent to any vertex y with $y \prec x_2$ and $y \neq v$. As X induces a path in T_9 and no vertex of X lies between x_2 and x_3 in $\prec$, it follows that $x_2 x_3$ is an edge of the path. By symmetry assume $x_2 x_3 \in P_1$. This implies that v cannot be adjacent to any vertex y with $x_2 \prec y \prec x_3$, as such an edge vy would cross the edges $x_2 x_3 \in P_1$ and $wx_2 \in P_2$.

But then v, x_2, x_3 form a facial triangle Δ of T_8 with all three edges on page P_1. However, there is no possible placement for the vertex $y = v_\Delta$ in T_9 that is adjacent to each of v, x_2, x_3. Thus, the planar graph T_9 admits no 2-local book embedding, which proves Theorem 2.

4   Graphs with Bounded Tree-Width

In this section we investigate the largest union page number and the largest local page number among all graphs of tree-width k. Clearly it suffices to consider edge-maximal graphs of tree-width k, the so-called k-trees, which are inductively defined as follows: A graph G is a k-tree if and only if $G \cong K_{k+1}$ or G is obtained from a smaller k-tree G' by adding one new vertex v whose neighborhood in G' is a clique of order k.

As our main tool in this section, let us define a linear embedding $(\prec, \mathcal{P})$ to be a *forest embedding* if the edges on each page $P \in \mathcal{P}$ form a forest. For a graph G, we say that a book embedding $(\prec, \mathcal{P})$ of some other graph $\bar{G} = (\bar{V}, \bar{E})$ *contains a forest embedding* of G if there exists a set $X \subseteq \bar{V}$ such that $G \cong \bar{G}[X]$ and $(\prec, \mathcal{P})$ restricted to $\bar{G}[X]$ is a forest embedding of G.

Lemma 8. *For every $\ell \in \mathbb{N}$ and every k-tree G there exists a k-tree $\bar{G}$ such that every ℓ-local book embedding of $\bar{G}$ contains a forest embedding of G.*

Proof. We find $\bar{G}$ based on $G = (V, E)$ by induction on $|V|$ as follows.

In the base case we have $G \cong K_{k+1}$ and we find $\bar{G}$ by induction on k. In the base case of this inner induction we have $k = 1$ and it suffices (for any ℓ) to take $\bar{G} = G \cong K_2$. For $k > 1$, we get from induction a $(k-1)$-tree $\bar{G}_{k-1}$ all of whose ℓ-local book embeddings contain a forest embedding of K_k. Starting with $\bar{G}_{k-1}$, add for each k-clique C in $\bar{G}_{k-1}$ an independent set I_C of $3k^2\ell$ vertices, together with all possible edges between C and I_C. The resulting graph has tree-width k and hence can be augmented to a k-tree $\bar{G}$. Consider any book embedding $(\prec, \mathcal{P})$ of $\bar{G}$. The inherited book embedding of $\bar{G}_{k-1} \subseteq \bar{G}$ contains a forest embedding of K_k, i.e., we have a forest embedding of some k-clique C in $\bar{G}_{k-1}$. If one vertex v in I_C has its k incident edges on k pairwise different pages, then we have a forest embedding of $C \cup v \cong K_{k+1}$, as desired. Otherwise, each vertex v in I_C has two incident edges on the same page in $\mathcal{P}$ joining v with two vertices in C. By pigeon-hole principle, for a set I' of at least $|I_C|/k^2 = 3\ell$ vertices of I_C these are the same two vertices c, c' of C. Since each of c, c' has incident edges on at most ℓ pages, again by pigeon-hole principle, one page in $\mathcal{P}$ contains the edges between c, c' and at least $|I'|/\ell = 3$ vertices in I'. However this is a contradiction as $K_{2,3}$ is not outerplanar.

Now for the induction step of the outer induction, assume that G is a k-tree with $|V| > k + 1$ vertices. Then G is obtained from a k-tree G' by adding one vertex v whose neighborhood in G' is a clique of order k. From induction we get a k-tree $\bar{G}'$ all of whose ℓ-local book embeddings contain a forest embedding of G'. Now we can do the same argument as before: Obtain $\bar{G}$ from $\bar{G}'$ by adding for each k-clique C in $\bar{G}'$ an independent set I_C of size $3k^2\ell$, together with all possible edges between C and I_C. Then any ℓ-local book embedding of $\bar{G}$ induces an ℓ-local book embedding of $\bar{G}'$, which hence contains a forest embedding of G'. Let C be the k-clique in G' that forms the neighborhood of v in G. The same argumentation as above then shows that at least one vertex in I_C has its k incident edges to C on k distinct pages, giving the desired forest embedding of G. (Essentially, the only difference to the base case is that adding the independent sets to $\bar{G}'$ gives a full k-tree, since $\bar{G}'$ is already a k-tree.) $\square$

Having Lemma 8, Theorem 3 (the existence of a k-tree with local page number k) follows with two simple edge counts.

If $G = (V, E)$ admits a ℓ-local forest embedding $(\prec, \mathcal{P})$, then

$$|E| \leqslant \sum_{P \in \mathcal{P}} (|V_P| - 1) \leqslant \ell|V| - |\mathcal{P}| \leqslant \ell(|V| - 1). \tag{5}$$

If $G = (V, E)$ is a k-tree, then

$$|E| = k|V| - \binom{k+1}{2}. \tag{6}$$

To prove Theorem 3, we shall find for each $k \geqslant 1$ a k-tree whose local page number is at least k. For $k = 1$ there is nothing to show. For $k \geqslant 2$, let $G_0 = (V, E)$ be any k-tree with $|V| > \binom{k+1}{2} - (k-1)$ (Note that this is a vertex count!) and let $G = \bar{G}_0$ be the corresponding k-tree given by Lemma 8 for $\ell = k - 1$. Assuming for the sake of contradiction that $\mathrm{pn}_\ell(G) \leqslant k - 1$, we obtain a $(k-1)$-local forest embedding $(\prec, \mathcal{P})$ of G_0. Then

$$|E| \overset{5}{\leqslant} (k-1)(|V|-1) = k|V| - (|V| + (k-1)) < k|V| - \binom{k+1}{2} \overset{6}{=} |E|,$$

a contradiction. Hence $\mathrm{pn}_\ell(G) \geqslant k$, as desired.

To end this section, let us also discuss some further implications of Lemma 8. We leave it open whether every k-tree has local page number at most k, i.e., whether the lower bound in Theorem 3 is tight. By Lemma 8 this is equivalent to every k-tree admitting a k-local forest book embedding. By putting each tree in each forest on a separate page, we even get a k-local forest embedding $(\prec, \mathcal{P})$ with a tree on each page. Moreover, by Eqs. (5) and (6) we would have $|\mathcal{P}| \leqslant \binom{k+1}{2}$, i.e., no more than $\binom{k+1}{2}$ trees in total, while at most k at any one vertex.

And we get a similar statement for the maximum union page number of k-trees. Suppose $(\prec, \mathcal{P})$ is an ℓ-union embedding of some graph, and that on all pages in $\mathcal{P}$ together we have m connected components. Putting each connected component on a separate (new) page, we obtain an ℓ-local book embedding $(\prec, \tilde{\mathcal{P}})$ with $|\tilde{\mathcal{P}}| = m$ pages. Now if $\mathrm{pn}_u(G) \leqslant k$ for all k-trees, then Lemma 8 implies that every k-tree even admits a k-union forest embedding. Moreover, by Eqs. (5) and (6) we get a forest embedding with $m \leqslant \binom{k+1}{2}$ trees in total, while having at most k at any one vertex.

Specifically, in order to prove $\mathrm{pn}_\ell(G) \leqslant k$ for every k-tree G, our task is to find a partition $\mathcal{P}$ of the edges in G into at most $\binom{k+1}{2}$ trees, such that every vertex is contained in no more than k of these trees, as well as a spine ordering $\prec$ for which each of the trees is non-crossing. The first part has a very natural solution: Every k-tree G has chromatic number $k+1$ and admits a unique[4] proper $(k+1)$-vertex coloring ϕ. Moreover, there are exactly $\binom{k+1}{2}$ pairs of colors in ϕ, any pair of color classes induces a tree in G, and each vertex of G is contained in exactly k of these trees. Hence every k-tree G edge-partitions into $\binom{k+1}{2}$ trees with each vertex being contained in k of these trees. Note that in this cover, every $(k+1)$-clique in G has all $\binom{k+1}{2}$ edges in pairwise distinct trees.

We have however not been able to prove (or disprove) the existence of a spine ordering $\prec$ under which no pair of color classes induces a crossing. If such exists, it would show $\mathrm{pn}_\ell(G) \leqslant k$ for all $k \geqslant 1$ and $\mathrm{pn}_u(G) \leqslant k$ for k odd and

---

[4] Up to relabeling of color classes.

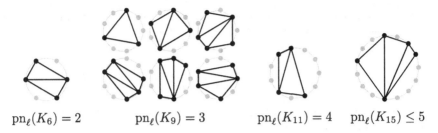

$$\text{pn}_\ell(K_6) = 2 \qquad \text{pn}_\ell(K_9) = 3 \qquad \text{pn}_\ell(K_{11}) = 4 \qquad \text{pn}_\ell(K_{15}) \le 5$$

Fig. 3. Illustrations of some k-local book embeddings of K_n for $n = 6, 9, 11, 15$. The shown page for $n = 6$ ($n = 11$, $n = 15$) is repeated 3 times (11 times, 15 times), each shifted cyclically by one position.

$\text{pn}_u(G) \le k + 1$ for k even. Note that for the union page number we also need to group the $\binom{k+1}{2}$ trees into as few forests of vertex-disjoint trees as possible. Due to the nature of our coloring, this is equivalent to properly edge-coloring K_{k+1}; hence the distinction on the parity of k.

5   Conclusions and Open Problems

In this paper we presented two novel graph parameters: the local page number $\text{pn}_\ell(G)$ and the union page number $\text{pn}_u(G)$. Both parameters are weakenings of the classical page number $\text{pn}(G)$ and we have $\text{pn}_\ell(G) \le \text{pn}_u(G) \le \text{pn}(G)$. Hence, one might be able to strengthen existing lower bounds of the form $\text{pn}(G) \ge X$ by showing $\text{pn}_u(G) \ge X$ or even $\text{pn}_\ell(G) \ge X$. On the other hand, one might be able to support conjectured upper bounds of the form $\text{pn}(G) \le X$ by showing the weaker bounds $\text{pn}_\ell(G) \le X$ or even $\text{pn}_u(G) \le X$.

In this paper we started to pursue this direction of research. Let us list some concrete cases that are still open:

– For the complete graph K_n it is known [3] that $\text{pn}(K_n) = \lceil n/2 \rceil$. On the other hand, the density of K_n implies that $\text{pn}_\ell(K_n) \ge \lceil (n-1)/4 \rceil$ (Lemma 5). In Fig. 3 we indicate some k-local book embeddings of K_n for some small values of n. According to this $\text{pn}_\ell(K_6) \le 2$, $\text{pn}_\ell(K_9) \le 3$, $\text{pn}_\ell(K_{11}) \le 4$, and $\text{pn}_\ell(K_{15}) \le 5$. Using the inequality $|E(G)| \le 2\,\text{pn}_\ell(G)|V(G)| - 3\,\text{pn}(G)$ from the proof of Lemma 5, we see that $\text{pn}_\ell(K_7) \ge 3$. (And with one further trick we get $\text{pn}_\ell(K_{10}) \ge 4$.) We refer to [22] for more details, and state it is an open problem to improve the following general bounds:

$$\left\lceil \frac{n-1}{4} \right\rceil \le \text{pn}_\ell(K_n) \le \text{pn}_u(K_n) \le \text{pn}(K_n) = \left\lceil \frac{n}{2} \right\rceil$$

– In 1989, Yannakakis [33] proved that for any planar graph G we have $\text{pn}(G) \le 4$, while removing an earlier claim [32] that there would be some planar graph G with $\text{pn}(G) \ge 4$. Ganley and Heath [13] observed that stacked triangulation T_2 (using our notation from Sect. 3, but also known as the Goldner-Harary graph) is a planar graph with $pn(T_2) = 3$, which remains until today the best known lower bound. While $pn_\ell(T_2) = 2$, we show in Sect. 3 that $pn_\ell(T_9) = 3$, while we leave it as an open problem to improve on the bounds

$$3 \leqslant \max_{G \text{ planar}} pn_\ell(G) \leqslant \max_{G \text{ planar}} pn_u(G) \leqslant \max_{G \text{ planar}} pn(G) \leqslant 4.$$

- We have a similar open problem for k-trees, where we refer to the detailed discussion at the end of Sect. 4.

$$k \leqslant \max_{G \text{ } k\text{-tree}} pn_\ell(G) \leqslant \max_{G \text{ } k\text{-tree}} pn_u(G) \leqslant \max_{G \text{ } k\text{-tree}} pn(G) = \begin{cases} k & \text{if } k \leqslant 2 \\ k+1 & \text{if } k \geqslant 3 \end{cases}$$

Besides determining the local and union page numbers for other graph classes (like for example regular graphs), it is also interesting to further analyze the relation between $pn_\ell(G), pn_u(G), a(G)$ and $sa(G)$. For example, what is the maximum of $pn_u(G)/pn_\ell(G)$ over all graphs G?

Finally, let us mention that changing the non-crossing condition Eq. (1) underlying the notion of book embeddings to for example a non-nesting condition, we get local and union versions of queue numbers. Interestingly, the proof of Theorem 6 remains valid and so does Corollary 7, giving that every planar graph has local queue number at most 4 and union queue number at most 5.

References

1. Alon, N., McDiarmid, C., Reed, B.: Star arboricity. Combinatorica **12**(4), 375–380 (1992)
2. Barrera-Cruz, F., Prag, T., Smith, H., Taylor, L., Trotter, W.T.. Comparing Dushnik-Miller dimension, Boolean dimension and local dimension. CoRR (2017)
3. Bernhart, F., Kainen, P.C.: The book thickness of a graph. J. Comb. Theory Ser. B **27**(3), 320–331 (1979)
4. Binucci, C., Da Lozzo, G., Di Giacomo, E., Didimo, W., Mchedlidze, T., Patrignani, M.: Upward book embeddings of st-graphs. CoRR, abs/1903.07966 (2019)
5. Binucci, C., Di Giacomo, E., Hossain, M.I., Liotta, G.: 1-page and 2-page drawings with bounded number of crossings per edge. Eur. J. Comb. **68**, 24–37 (2018)
6. Bläsius, T., Stumpf, P., Ueckerdt, T.: Local and union boxicity. Discrete Math. **341**(5), 1307–1315 (2018)
7. Bosek, B., Grytczuk, J., Trotter, W.T.: Local dimension is unbounded for planar posets. CoRR (2017)
8. Chung, F.R., Leighton, F.T., Rosenberg, A.L.: Embedding graphs in books: a layout problem with applications to VLSI design. SIAM J. Algebr. Discrete Methods **8**(1), 33–58 (1987)
9. Dujmovic, V., Wood, D.R.: On linear layouts of graphs. Discrete Math. Theor. Comput. Sci. **6**(2) (2004)
10. Dujmović, V., Wood, D.R.: Graph treewidth and geometric thickness parameters. Discrete Comput. Geom.d **37**(4), 641–670 (2007)
11. Enomoto, H., Nakamigawa, T., Ota, K.: On the pagenumber of complete bipartite graphs. J. Comb. Theory Ser. B **71**(1), 111–120 (1997)

12. Felsner, S., Ueckerdt, T.: A note on covering Young diagrams with applications to local dimension of posets. CoRR (2019). to appear at EUROCOMB 2019
13. Ganley, J.L., Heath, L.S.: The pagenumber of k-trees is $O(k)$. Discrete Appl. Math. **109**(3), 215–221 (2001)
14. Garey, M.R., Johnson, D.S., Miller, G.L., Papadimitriou, C.H.: The complexity of coloring circular arcs and chords. SIAM J. Algebr. Discrete Methods **1**(2), 216–227 (1980)
15. Hakimi, S.L., Mitchem, J., Schmeichel, E.: Star arboricity of graphs. Discrete Math. **149**(1), 93–98 (1996)
16. Heath, L.S., Istrail, S.: The pagenumber of genus g graphs is $o(g)$. J. ACM **39**(3), 479–501 (1992)
17. Kainen, P.C.: Some recent results in topological graph theory. In: Bari, R.A., Harary, F. (eds.) Graphs and Combinatorics, pp. 76–108. Springer, Heidelberg (1974). https://doi.org/10.1007/BFb0066436
18. Kim, J., Martin, R.R., Masařík, T., Shull, W., Smith, H.C., Uzzell, A., Wang, Z.: On difference graphs and the local dimension of posets. CoRR abs/1812.00832 (2018)
19. Knauer, K., Ueckerdt, T.: Three ways to cover a graph. Discrete Math. **339**(2), 745–758 (2016)
20. Majumder, A., Mathew, R.: Local boxicity, local dimension, and maximum degree. CoRR (2018)
21. Malitz, S.M.: Graphs with e edges have pagenumber $o(\sqrt{e})$. J. Algorithms **17**(1), 71–84 (1994)
22. Merker, L.: Local page numbers. Bachelor's thesis, Karlsruhe Institute of Technology, Germany (2018)
23. Nash-Williams, C.S.J.A.: Decomposition of finite graphs into forests. J. Lond. Math. Soc. **39**(1), 12 (1964)
24. Ollmann, L.T.: On the book thicknesses of various graphs. In: Proceeedings 4th Southeastern Conference on Combinatorics, Graph Theory and Computing, vol. 8, p. 459 (1973)
25. Rengarajan, S., Veni Madhavan, C.E.: Stack and queue number of 2-trees. In: Du, D.Z., Li, M. (eds.) Computing and Combinatorics, pp. 203–212. Springer, Heidelberg (1995). https://doi.org/10.1007/BFb0030834
26. Rosenberg, A.L.: The Diogenes approach to testable fault-tolerant arrays of processors. IEEE Trans. Comput. C-32(10), 902–910 (1983)
27. Togasaki, M., Yamazaki, K.: Pagenumber of pathwidth-k graphs and strong pathwidth-k graphs. Discrete Math. **259**(1), 361–368 (2002)
28. Trotter, W.T., Walczak, B.: Boolean dimension and local dimension. Electron. Notes Discrete Math. **61**, 1047–1053 (2017). The European Conference on Combinatorics, Graph Theory and Applications (EUROCOMB'17)
29. Ueckerdt, T.: Order & geometry workshop (2016). Gułtowy
30. Vandenbussche, J., West, D.B., Gexin, Y.: On the pagenumber of k-trees. SIAM J. Discrete Math. **23**(3), 1455–1464 (2009)
31. Wigderson, A.: The complexity of the Hamiltonian circuit problem for maximal planar graphs. Technical report, Technical Report EECS 198, Princeton University, USA (1982)
32. Yannakakis, M.: Four pages are necessary and sufficient for planar graphs. In: Proceedings of the Eighteenth Annual ACM Symposium on Theory of Computing, STOC 1986, pp. 104–108. ACM, New York (1986)
33. Yannakakis, M.: Embedding planar graphs in four pages. J. Comput. Syst. Sci. **38**(1), 36–67 (1989)

Mixed Linear Layouts:
Complexity, Heuristics, and Experiments

Philipp de Col[ID], Fabian Klute[ID], and Martin Nöllenburg[✉][ID]

Algorithms and Complexity Group, TU Wien, Vienna, Austria
philipp.decol@gmx.at, {fklute,noellenburg}@ac.tuwien.ac.at

Abstract. A k-page linear graph layout of a graph $G = (V, E)$ draws
all vertices along a line ℓ and each edge in one of k disjoint halfplanes
called *pages*, which are bounded by ℓ. We consider two types of pages.
In a *stack page* no two edges should cross and in a *queue page* no edge
should be nested by another edge. A crossing (nesting) in a stack (queue)
page is called a *conflict*. The algorithmic problem is twofold and requires
to compute (i) a vertex ordering and (ii) a page assignment of the edges
such that the resulting layout is either conflict-free or conflict-minimal.
While linear layouts with only stack or only queue pages are well-studied,
mixed s-stack q-queue layouts for $s, q \geq 1$ have received less attention.
We show NP-completeness results on the recognition problem of certain
mixed linear layouts and present a new heuristic for minimizing conflicts.
In a computational experiment for the case $s, q = 1$ we show that the new
heuristic is an improvement over previous heuristics for linear layouts.

1 Introduction

Linear graph layouts, in particular book embeddings [1,12] (also known as stack
layouts) and queue layouts [9,10], form a classic research topic in graph drawing
with many applications beyond graph visualization as surveyed by Dujmović and
Wood [6]. A k-page linear layout $\Gamma = (\prec, \mathcal{P})$ of a graph $G = (V, E)$ consists of an
order $\prec$ on the vertex set V and a partition of E into k subsets $\mathcal{P} = \{P_1, \ldots, P_k\}$
called *pages*. Visually, we may represent Γ by mapping all vertices of V in the
order $\prec$ onto a line ℓ. Each page can be represented by mapping all edges to
semi-circles connecting their endpoints in a halfplane bounded by ℓ. If a page
P is a *stack page*, then no two edges in P may cross, or at least the number
of crossings should be minimized. More precisely, two edges uv, wx in P cross
(assuming $u \prec v$, $w \prec x$, and $u \prec w$) if and only if their vertices are ordered as
$u \prec w \prec v \prec x$. Conversely, if a page P is a *queue page*, then no two edges in P
may be nested, or at least the number of nestings should be minimized. Here,
an edge wx is nested by an edge uv if and only if their vertices are ordered as
$u \prec w \prec x \prec v$ (under the same assumptions as above).

Stack and queue layouts have mostly been studied for planar graphs with
a focus on investigating the stack number (also called book thickness) and the
queue number of graphs, which correspond to the minimum integer k, for which

© Springer Nature Switzerland AG 2019
D. Archambault and C. D. Tóth (Eds.): GD 2019, LNCS 11904, pp. 460–467, 2019.
https://doi.org/10.1007/978-3-030-35802-0_35

a graph admits a k-stack or k-queue layout. It is known that recognizing graphs with queue number 1 or with stack number 2 is NP-complete [1,10]. Further, it is known that every planar graph admits a 4-stack layout [16], but it is open whether the stack number of planar graphs is actually 3. Due to their practical relevance, book drawings, i.e., stack layouts in which crossings are allowed, have also been investigated from a practical point of view. Klawitter et al. [11] surveyed the literature and performed an experimental study on several state-of-the-art book drawing algorithms aiming to minimize the number of crossings in layouts on a fixed number of stack pages. Conversely, for queue layouts it was a longstanding open question whether planar graphs have bounded queue number [9]; this was recently answered positively by Dujmović et al. [5].

Mixed layouts, which combine $s \geq 1$ stack pages and $q \geq 1$ queue pages, are studied less. For an s-stack q-queue layout $\Gamma = (\prec, \mathcal{P})$, the set of pages $\mathcal{P}$ is itself partitioned into the stack pages $\mathcal{S} = \{S_1, \ldots, S_s\}$ and the queue pages $\mathcal{Q} = \{Q_1, \ldots, Q_q\}$. Heath and Rosenberg [10] conjectured that every planar graph admits a 1-stack 1-queue layout, but this has been disproved recently by Pupyrev [13], who conjectured that instead every bipartite planar graph has a 1-stack 1-queue layout. Pupyrev further provides a SAT-based online tool for testing the existence of an s-stack q-queue layout[1].

Contributions. We first show two NP-completeness results in Sect. 2. The first one shows that testing the existence of a 2-stack 1-queue layout is NP-complete, and the other proves that an NP-complete mixed layout recognition problem with fixed vertex order remains NP-complete under addition of stack or queue pages. Next, we focus our attention on 1-stack 1-queue layouts and propose, to the best of our knowledge, the first heuristic targeted at minimizing conflicts in 1-stack 1-queue layouts, see Sect. 3. In a computational experiment in Sect. 4 we show that our heuristic achieves fewer conflicts compared to previous heuristics for stack layouts with a straightforward adaptation to mixed layouts.

Due to space constraints, proofs of statements marked with $\star$, as well as some additional plots are available only in the full version of this paper [4].

2   Complexity

In this section we give new complexity results regarding mixed linear layouts. For Theorem 1 we first make some useful observations, see the full paper [4] for details. Let K_8 be the complete graph on eight vertices and $\Gamma = (\prec, \{S_1, S_2\}, \{Q\})$ a 2-stack 1-queue layout of a K_8. Using exhaustive search[2] we verified that in such a 2-stack 1-queue layout the three longest edges are in $S_1 \cup S_2$ and the edges between the first and third, and the sixth and eighth vertex in $\prec$ are in Q. Finally, for two K_8's only the last and first vertex from each K_8 can interleave.

---

[1] http://be.cs.arizona.edu.
[2] Source code available at https://github.com/pdecol/mixed-linear-layouts.

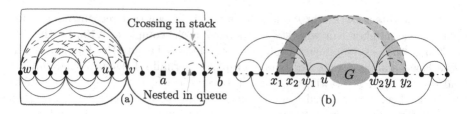

Fig. 1. (a) A 2-stack 1-queue layout of a double-K_8. Only the left K_8 is drawn fully, dashed edges are in the queue page. (b) Sketch of the gadget for Theorem 1.

Let $G_1 = (V_1, E_1)$ and $G_2 = (V_2, E_2)$ be two distinct K_8's. A *double-K_8* G is formed, by identifying two so-called *shared* vertices u, v of G_1 and G_2 with each other and adding one more edge wz between so-called *outer* vertices $w \in V_1$, $z \in V_2$ such that neither w nor z is one of the shared vertices, see Fig. 1(a). For any 2-stack, 1-queue layout $\Gamma = (\prec, \{S_1, S_2\}, \{Q\})$ of a double-K_8 we verified with exhaustive search that the shared vertices have to be the two middle vertices in $\prec$ and that the outer vertices are always the first and last vertices in the ordering of such a layout.

Lemma 1 ($\star$). *Let $G = (V, E)$ be a double-K_8 with outer vertices w, z and two additional vertices $a, b \in V$ with an edge $ab \in E$. In every 2-stack 1-queue layout $\Gamma = (\prec, \{S_1, S_2\}, \{Q\})$ of G, with $w \prec a \prec z$ and $b \prec w$ or $z \prec b$, it holds that $ab \in Q$ and a is between the first or last three vertices in the double-K_8.*

Corollary 1 ($\star$). *Let $G_1 = (V_1, E_1)$ and $G_2 = (V_2, E_2)$ be two double-K_8's. In a 2-stack 1-queue layout Γ of $G_1 \cup G_2$ with linear order $\prec$, either $u \prec v$ or $v \prec u$ for all $u \in V_1$, $v \in V_2$.*

Let $G = (V, E)$ be a graph consisting of two double-K_8's $G_1 = (V_1, E_1)$ and $G_2 = (V_2, E_2)$. Let $w_1 \in V_1$ and $w_2 \in V_2$ be two outer vertices. Further, let $x_1, x_2 \in V_1$ and $y_1, y_2 \in V_2$ be four vertices such that x_1, x_2 are in the same K_8 as w_1, w_2 respectively for y_1, y_2, and none of them is a shared vertex. Finally, add a vertex u to V and the edges $x_1 y_1$, $x_2 y_2$, $w_1 u$, and $w_2 u$ to E, see Fig. 1(b).

Lemma 2 ($\star$). *Let $G = (V, E)$ be the graph constructed as above. Then in any 2-stack 1-queue layout $\Gamma = (\prec, \{S_1, S_2\}, \{Q\})$ of G we find, w.l.o.g., $w_1 \prec u \prec w_2$ and $w_1 u, w_2 u \in S_1 \cup S_2$.*

Theorem 1. *Let $G = (V, E)$ be a simple undirected graph. It is NP-complete to decide if G admits a 2-stack 1-queue layout.*

Proof. The problem is clearly in NP. We show the result by a reduction from the problem of deciding the existence of a 2-stack layout, which is NP-complete and equivalent to decide whether a graph is subhamiltonian [1,2]. Let $G' = (V', E')$ be a graph constructed as in Lemma 2. Identify the special vertex u in V' with any vertex in G and add the rest of G to G'. Clearly, if G has a 2-stack layout, we can construct a 2-stack 1-queue layout of G' as sketched in Fig. 1(b). Conversely, let $\Gamma = (\prec, \{S_1, S_2\}, \{Q\})$ be a 2-stack 1-queue layout of G'. As for u in Lemma 2, we find that $w_1 \prec v \prec w_2$ for every neighbor $v \in V$ of u. By induction we find for all $v' \in V$ that $w_1 \prec v' \prec w_2$. Hence all edges in G are nested by $x_1 y_1$ and $x_2 y_2$, which are both in Q. It follows that G has a 2-stack layout. $\qquad\square$

Our second complexity result shows that adding stack or queue pages to the specification of an already NP-complete mixed linear layout problem with given vertex order $\prec$ remains NP-complete. Note that for $s = 4$ and $q = 0$ the problem of deciding if the edges can be assigned to the pages even when the vertex order is fixed is known to be NP-complete [15].

Theorem 2. ($\star$). *Let $G = (V, E)$ be a simple undirected graph and $\prec$ a fixed order of V. If it is NP-complete to decide if G admits an s-stack q-queue layout respecting $\prec$, then it is also NP-complete to decide if G admits (i) an s-stack $(q + 1)$-queue layout or (ii) an $(s + 1)$-stack q-queue layout respecting $\prec$, respectively.*

3   Heuristic Algorithm

Most heuristics for minimizing crossings in book drawings work in two steps [11]. First compute a vertex order and then a page assignment. We propose a new page assignment heuristic, specifically tailored to mixed linear layouts. To the best of our knowledge, this is the first heuristic for minimizing crossings and nestings in mixed linear layouts. It uses stack and queue data structures for keeping track of conflicts and estimating possible future conflicts. The design allows us to consider the assigned and unassigned edges at the same time while efficiently processing them to run in $O(m^2)$ time for a graph with m edges.

In the following, we describe how the algorithm works for a 1-stack 1-queue layout. We note that it is straight forward to adapt our approach to s-stack, q-queue layouts for arbitrary s and q. A stack (queue) can be used to validate that a given page has no crossings (nestings) by visiting the vertices in their linear order and inserting (removing) each edge when the left (right) end-vertex is visited, respectively [9]. We can use a similar strategy for our heuristic. Here it is allowed to remove edges even if they are not on top of the stack or in front of the queue, but of course this might produce conflicts. Let S be a stack and Q a queue. We additionally keep two counters for each edge e, the so called *crossing counter* $c(e)$ and the *nesting counter* $n(e)$. The vertices are processed from left to right in a given vertex order. For the current vertex u we insert all edges $e = uv$ into S and Q. If there are multiple edges uv, we add them according to their length to S and Q. For S we sort them from long to short, for Q from short to long. Once the second vertex v of an edge $e = uv$ is visited, we remove e from S and Q, and decide to which page we assign it. Let s_e be the number of edges on top of e in S and q_e the number of edges in front of e in Q. Then we assign e to the stack page if $c(e) + 0.5s_e \leq n(e) + 0.5q_e$ and update $c(e')$ for each edge e' on top of e in S. Otherwise we assign e to the queue page and update $n(e')$ for each edge e' in front of e in Q. Intuitively we estimate for each edge e how many conflicts this edge produces for edges e' to be processed later. The advantage of this estimation is that we potentially assign the edge to a page that adds more conflicts now, but might create fewer conflicts in the future.

4 Experiments

We denote our algorithm as *stack-queue* heuristic and compare it with the two page assignment heuristics *eLen* [3] and *ceilFloor* [14] that are commonly used with book drawings [11] (For the source code and benchmark instances see footnote 2). Both can be adapted to mixed layouts, while other book drawing heuristics try to explicitly partition the edges into planar sets, which is obviously not suitable for queue pages in mixed layouts. Both process the edges by decreasing length and greedily assign each edge to the page where it causes fewer conflicts at the time of insertion. In case of ties, a stack page is preferred over a queue page. The difference is that *eLen* computes the length based on the linear vertex order and *ceilFloor* based on the corresponding cyclic vertex order as follows. Given a vertex order $1 \prec 2 \prec \ldots \prec n$, *eLen* considers the edge $(1, n)$ first and the edges $(i, i+1)$ last. In *ceilFloor* the length of an edge uv is defined as $min(|u - v|, n - |u - v|)$. All three heuristics run in $O(m^2)$.

The goal of our experiment is to explore the performance differences in terms of the number of conflicts per edge of the adapted book drawing algorithms compared to our new heuristic. We thus measure the resulting number of conflicts per edge for all three algorithms, as well as record for each instance the algorithm with the fewest number of conflicts. We first tested the algorithms on the complete graphs with up to 50 vertices. Furthermore, we generated 500 random graphs for each number of vertices in $\{25, 50, \ldots, 400\}$ from different *sparse* graph classes, see Fig. 2, since it is known that both, stack and queue page, can contain at most $2n - 3$ conflict-free edges [1,10]. All experiments ran on a Linux cluster (Ubuntu 16.04.6 LTS), where each node has two Intel Xeon E5540 (2.53 GHz Quad Core) processors and 24 GB RAM. The running time of one run of each algorithm was relatively low, taking less than one second on average and at most 2.5 s for the denser random graphs with 400 vertices.

For the complete graphs *stack-queue* was the best page assignment heuristic producing about 2/3 of the conflicts of *eLen* and *ceilFloor*. For other graphs, the vertex order has a strong effect on the results. In our experiments we first compared three state-of-the-art vertex order heuristics (breadth-first search (*rbfs* [14]), depth-first search (*AVSDF* [8]) and connectivity (*conGreedy* [11])) before applying the page assignment heuristics. It turned out that for all of them the same vertex order heuristic performed best on the same graph class, so that all experiments could be run without bias on exactly the same input order.

Benchmark Graphs. We first generated random (not necessarily planar) graphs of n vertices and either $m = 3n$ or $m = 6n$ edges. The graphs were created by drawing uniformly at random the required number of edges, discarding disconnected graphs. The best vertex order heuristic was *conGreedy*.

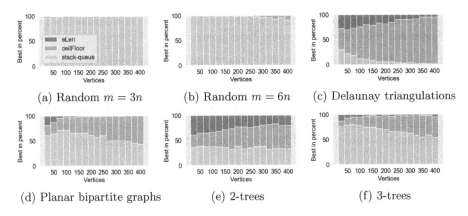

(a) Random $m = 3n$ (b) Random $m = 6n$ (c) Delaunay triangulations

(d) Planar bipartite graphs   (e) 2-trees   (f) 3-trees

Fig. 2. How many times each algorithm obtained the fewest conflicts in percent.

Since 1-stack 1-queue layouts are especially interesting for planar and planar bipartite graphs [13], we generated random planar and maximal planar bipartite graphs. The planar graphs are generated as Delaunay triangulations of n random points in the plane. Since every planar bipartite graph has a 2-stack embedding [7], we randomly generated a vertex order of alternating vertices from both vertex sets to ensure that a Hamiltonian path exists. We then randomly selected two vertices of the two sets and added the edge to the graph if it was possible to do so without a crossing. We repeated the process of randomly selecting the vertices until the maximum number of $2n - 4$ edges had been reached. The vertex order for the Delaunay triangulations was computed by *rbfs* and for maximal planar bipartite graphs by *AVSDF*. As it turned out that *stack-queue* did not perform as well as *ceilFloor* and *eLen* on the Delaunay triangulations, which is in contrast to the random and planar bipartite graphs, we wondered whether the presence of many triangles might be the reason. Hence, we considered two graph classes with many triangles and the same maximal edge densities as planar and planar bipartite graphs, respectively, namely planar 3- and 2-trees. For 2-trees the best vertex order was computed by *AVSDF* and for 3-trees by *conGreedy*.

Results. Aggregated results of our experiments are shown in Fig. 2. Additional plots are provided in the full version [4]. For random graphs *stack-queue* performs best among the three heuristics for almost all instances, even though the difference to *ceilFloor* in conflicts per edge is small. For Delaunay triangulations, *stack-queue* performs best for small graphs ($n \leq 25$), but for the larger instances *ceilFloor* computes better solutions for the majority of instances. In terms of conflicts per edge, however, all three algorithms are quite close together. In the case of planar bipartite graphs, *stack-queue* is the best algorithm for up to 300 vertices. Afterwards *ceilFloor* performs slightly better, but in both cases, again, the difference in the number of conflicts is small. For 2-trees, the results are more or less evenly split among all three algorithms. Yet, for 3-trees, *stack-queue* computes the best solutions for 70–80% of the instances with up to 100 vertices. The differences in the number of conflicts per edge is also more noticeable. For larger instances *ceilFloor* catches up with *stack-queue*.

Discussion. The results of our experiments showed that the proposed *stack-queue* heuristic beats or competes with previously existing and suitably adapted page assignment heuristics for book drawings on most of the tested benchmark graph classes with the exception of Delaunay triangulations, where *ceilFloor* performed best. Since the running time of all three algorithms is $O(m^2)$ this does make *stack-queue* a suitable method for computing 1-stack 1-queue layouts.

5   Conclusion

We believe it is possible to adapt our technique from Theorem 1 for $s > 2$ and $q = 1$. The biggest obstacle is to find such rigid structures as the double-K_8. In the algorithmic direction it could be interesting to investigate specialized heuristics for finding vertex orders in the queue- and mixed layout case. Whether every planar bipartite graph admits a 1-stack 1-queue layout remains open.

References

1. Bernhart, F., Kainen, P.C.: The book thickness of a graph. J. Comb. Theory Ser. B **27**(3), 320–331 (1979). https://doi.org/10.1016/0095-8956(79)90021-2
2. Chung, F., Leighton, F., Rosenberg, A.: Embedding graphs in books: a layout problem with applications to VLSI design. SIAM J. Alg. Discret. Meth. **8**(1), 33–58 (1987). https://doi.org/10.1137/0608002
3. Cimikowski, R.J.: Algorithms for the fixed linear crossing number problem. Discret. Appl. Math. **122**(1–3), 93–115 (2002). https://doi.org/10.1016/S0166-218X(01)00314-6
4. de Col, P., Klute, F., Nöllenburg, M.: Mixed linear layouts: complexity, heuristics, and experiments. CoRR abs/1908.08938 (2019). http://arxiv.org/abs/1908.08938
5. Dujmovic, V., Joret, G., Micek, P., Morin, P., Ueckerdt, T., Wood, D.R.: Planar graphs have bounded queue-number. CoRR abs/1904.04791 (2019)
6. Dujmović, V., Wood, D.R.: On linear layouts of graphs. Discret. Math. Theor. Comput. Sci. **6**(2), 339–358 (2004)
7. de Fraysseix, H., de Mendez, P.O., Pach, J.: A left-first search algorithm for planar graphs. Discret. Comput. Geom. **13**, 459–468 (1995). https://doi.org/10.1007/BF02574056
8. He, H., Sýkora, O.: New circular drawing algorithms. In: Workshop on Information Technologies - Applications and Theory (ITAT) (2004)
9. Heath, L.S., Leighton, F.T., Rosenberg, A.L.: Comparing queues and stacks as mechanisms for laying out graphs. SIAM J. Discret. Math. **5**(3), 398–412 (1992). https://doi.org/10.1137/0405031
10. Heath, L.S., Rosenberg, A.L.: Laying out graphs using queues. SIAM J. Comput. **21**(5), 927–958 (1992). https://doi.org/10.1137/0221055
11. Klawitter, J., Mchedlidze, T., Nöllenburg, M.: Experimental evaluation of book drawing algorithms. In: Frati, F., Ma, K.L. (eds.) GD 2017. LNCS, vol. 10692, pp. 224–238. Springer, Cham (2018). https://doi.org/10.1007/978-3-319-73915-1_19

12. Ollmann, L.T.: On the book thicknesses of various graphs. In: 4th Southeastern Conference on Combinatorics, Graph Theory and Computing, vol. 8, p. 459 (1973)
13. Pupyrev, S.: Mixed linear layouts of planar graphs. In: Frati, F., Ma, K.L. (eds.) GD 2017. LNCS, vol. 10692, pp. 197–209. Springer, Cham (2018). https://doi.org/10.1007/978-3-319-73915-1_17
14. Satsangi, D., Srivastava, K., Srivastava, G.: K-page crossing number minimization problem: an evaluation of heuristics and its solution using GESAKP. Memet. Comput. 5(4), 255–274 (2013). https://doi.org/10.1007/s12293-013-0115-5
15. Unger, W.: On the k-colouring of circle-graphs. In: Cori, R., Wirsing, M. (eds.) STACS 1988. LNCS, vol. 294, pp. 61–72. Springer, Heidelberg (1988). https://doi.org/10.1007/BFb0035832
16. Yannakakis, M.: Embedding planar graphs in four pages. J. Comput. Syst. Sci. 38(1), 36–67 (1989). https://doi.org/10.1016/0022-0000(89)90032-9

Homotopy Height, Grid-Major Height and Graph-Drawing Height

Therese Biedl[1], Erin Wolf Chambers[2], David Eppstein[3(✉)],
Arnaud De Mesmay[4], and Tim Ophelders[5]

[1] David R. Cheriton School of Computer Science, University of Waterloo,
Waterloo, Canada
biedl@uwaterloo.ca
[2] Department of Computer Science, Saint Louis University, St. Louis, USA
erin.chambers@gmail.com
[3] Computer Science Department, University of California, Irvine, Irvine, USA
eppstein@uci.edu
[4] Univ. Grenoble Alpes, CNRS, Grenoble INP, GIPSA-lab, 38000 Grenoble, France
ademesmay@gmail.com
[5] Department of Computational Mathematics, Science and Engineering,
Michigan State University, East Lansing, USA
tim.ophelders@gmail.com

Abstract. It is well-known that both the pathwidth and the outer-planarity of a graph can be used to obtain lower bounds on the height of a planar straight-line drawing of a graph. But both bounds fall short for some graphs. In this paper, we consider two other parameters, the (simple) homotopy height and the (simple) grid-minor height. We discuss the relationship between them and to the other parameters, and argue that they give lower bounds on the straight-line drawing height that are never worse than the ones obtained from pathwidth and outer-planarity.

1 Introduction

Straight-line drawings of planar graphs are one of the oldest and most intensely studied problems in graph drawing [1–6]. It has been known since the 1990s that every planar graph has a straight-line drawing of height $n-1$ [5,6] and that some planar graphs require height $\frac{2}{3}n$ if the outer-face must be respected [7,8]. Nevertheless many problems surrounding the height of planar straight-line drawings remain open; it is not even known whether minimizing height is NP-hard (although the problem is NP-hard when edges may only connect adjacent rows [9] and it is fixed-parameter tractable in the output height [10]).

Erin Chambers was supported in part by NSF grants CCF-1614562 and DBI-1759807. David Eppstein was supported in part by NSF grants CCF-1618301 and CCF-1616248. Arnaud de Mesmay was supported in part by grants ANR-18-CE40-0004-01 (FOCAL) and ANR-16-CE40-0009-01 (GATO). This work began at the Fifth Annual Workshop on Geometry and Graphs, at the Bellairs Research Institute of McGill University.

© Springer Nature Switzerland AG 2019
D. Archambault and C. D. Tóth (Eds.): GD 2019, LNCS 11904, pp. 468–481, 2019.
https://doi.org/10.1007/978-3-030-35802-0_36

One of the chief obstacles is that very few tools are known for arguing that a planar graph requires a certain height in all planar straight-line drawings. Two graph parameters are commonly used for this: the pathwidth (as the height is at least $pw(G)$ [10,11]) and the outer-planarity (as the height is at least twice the outer-planarity minus 1 [7,8]); for detailed definitions see Sect. 2. However, both parameters may be constant in graphs that require linear height [12].

In this paper, we study two other graph parameters, the *homotopy height* $Hh(G)$ and the *grid-major height* $GMh(G)$ and their simple variants $sHh(G)$ and $sGMh(G)$. Roughly speaking, the homotopy height is defined as the minimum k such that a sequence of paths of length at most k sweep the graph[1], while the grid-major height is the minimum height of a grid of which the graph is a minor. Figure 1 illustrates this and graph parameters used in the paper. Our simple variants add simplicity constraints to the paths involved in the sweeping or the columns of the grid-major representation. We show that despite their apparent differences, homotopy height and grid-major height are equal, and that both the normal and the simple variants are lower bounds on the graph drawing height. More precisely, any planar triangulated graph G has

$$pw(G) \overset{(*)}{\leq} Hh(G) = GMh(G) \overset{(*)}{\leq} sHh(G) = sGMh(G) \overset{(*)}{\leq} VRh(G) = SLh(G),$$
$$(1)$$

where $VRh(G)$ and $SLh(G)$ are the minimum height of a visibility representation and straight-line drawing of G. As we will show, the inequalities marked with $(*)$ are strict for some planar graphs. More strongly, the parameters separated by these inequalities can differ by non-constant factors from each other.

In particular, the (simple) grid-major height and homotopy-height can both serve as lower bounds on the height of a straight-line drawing. For some graphs (e.g. the one in Fig. 4(b)) this gives a better lower bound than can be achieved via pathwidth, though not a better lower bound than what was known [12]. We should mention that the outer-planarity op is also related to these parameters via

$$2op(G) - 1 \overset{(*)}{\leq} GMh(G),$$
$$(2)$$

so the homotopy-height and grid-major height can also can replace outer-planarity as lower-bound tool for graph-drawing height, and in fact, provide a convenient vehicle for unifying both tools. While these results have not yet led us to new lower bound results for straight-line drawings, they provide new tools which had not been considered previously, and suggest a promising new line of inquiry.

Our results naturally raise the question of the complexity of computing these parameters. Computing the optimal height of homotopies is conjectured but not

---

[1] We note that there are *many* possible variants of homotopy height, all quantifying in slightly different ways the optimal way to sweep a planar graph with a curve. We have chosen here one particular variant that seems to be most suitable for graph drawing purposes, and we only study it for triangulated graphs. We refer the reader to other recent works on this parameter [13–15] for further discussion.

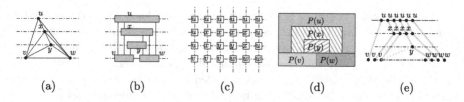

(a)	(b)	(c)	(d)	(e)

Fig. 1. The same graph with (a) a straight-line drawing, (b) a flat visibility representation, (c) a simple grid-major representation, (d) the corresponding contact-representation, (e) a simple homotopy. The height is always four. The moves of the homotopy are: a face-flip at $\{u, v, x\}$, a boundary-move, an edge-slide at (v, y), a face-flip at $\{x, y, w\}$ and face-flip at $\{u, x, w\}$.

known to be NP-hard [13]; even arguing that it is in NP is non-trivial [15], although it has a logarithmic approximation [14,15]. Our equalities imply that computing the homotopy-height $k = Hh(G)$ is (non-uniform) fixed-parameter tractable in k. Indeed, it equals grid-major height, which is closed under taking minors. Minor testing can be expressed in second-order logic, and the graphs of bounded grid-major height have bounded pathwidth, so it follows from graph minor theory and Courcelle's theorem [16] that for any k, the graphs with grid-major height k can be recognized in linear time. However, this method uses the (unknown!) forbidden minors for grid-major height; finding them remains an open problem of independent interest. We can also show more directly using Courcelle's theorem that simple grid-major height is fixed-parameter tractable.

All our results are only for *triangulated planar graphs*, planar graphs where all faces (including the outer-face) are triangles. This is not a big restriction for graph drawing height, as any planar graph G is a subgraph of a triangulated planar graph G' that has a straight-line drawing of the same height, up to a small additive term. (Obtain G' by triangulating the convex hull of a drawing of G and adding three vertices that surround the drawing.) Most of our parameters naturally carry over to non-triangulated planar graphs, but some parameters would be much more cumbersome to define and work with for non-triangle faces.

For space reasons we defer our algorithmic results and many proof details to the full version of this paper.

2   Definitions

All graphs in this paper are *planar*: they can be drawn in the plane without crossings. Their *faces* are maximal connected regions that remain when removing the drawing); we call the unbounded face the *outer-face*. Unless otherwise stated, we study only *simple* graphs that have no loops and at most one edge between any two vertices, and we almost always study *triangulated graphs*, where all faces (including the outer-face) are bounded by a simple cycle of length 3. Such a graph is *maximal planar*: no edge can be added without violating simplicity or planarity. Its planar embedding is unique up to the choice of outer-face.

Let G be a triangulated graph with fixed outer-face f. We define *outer-planarity* $op(G)$ via a removal process as follows: In a first step, remove all vertices on the outer-face. In each subsequent step, remove all vertices on the outer-face of the remaining graph. Then $op(G, f)$ is the number of steps until no vertices remain, and $op(G)$ is the minimum of $op(G, f)$ over all choices of face f.

Graph-Drawing Parameters: The $W \times H$-*grid* has vertices at the *grid-points* $\{1, \ldots, W\} \times \{1, \ldots, H\}$ and an edge between any two grid-points of distance one. A *straight-line drawing* of G consists of a mapping of G to grid-points such that if all edges are drawn as straight-line segments between their endpoints, no two edges cross and no edge overlaps a non-incident vertex. Every planar graph has such a drawing [1–3] whose supporting grid has height at most $n - 1$ [5,6]. We use $SLh(G)$ to denote the smallest height h of a straight-line drawing of G.

A *flat visibility-representation* of G consists of an assignment of a horizontal segment (*bar*) to every vertex of G such that for any edge (v, w) there exists a *line of visibility*, i.e., a line segment connecting bars of v, w that intersects no other bar. In the original definition lines of visibility had to be horizontal; for us it will be more convenient to allow both horizontal and vertical lines of visibility, as long as they do not cross. Every planar graph has a flat visibility representation [17–19]. We use $VRh(G)$ to denote the smallest height h of such a representation, presuming all bars reside at positive integral y-coordinates.

Width Parameters: A *path decomposition* of a graph G is a collection $X_1, \ldots, X_L$ of vertex-sets (*bags*) that satisfies the following: each vertex v appears in at least one bag, the bags containing v are consecutive, and for each edge (v, w) at least one bag contains both v and w. The *width* of a path decomposition of G is the largest bag-size minus 1, and the *pathwidth* $pw(G)$ of a graph is the smallest possible width of a path decomposition.

We introduce another width parameter which is quite natural, but to our knowledge has not been studied before. A *grid-representation* of a graph G consists of a $W \times H$-grid where each gridpoint is labelled with one vertex of G in such a way that (1) every vertex appears at least once as a label, (2) for any vertex v the grid-points that are labelled v induce a connected subgraph of the grid, and (3) for any edge (v, w) of G there exists a grid-edge where the ends are labelled v and w. In particular, if G has a grid-representation then it is a minor of the $W \times H$-grid. Let $GMh(G)$ be the *grid-major height*, i.e., the smallest h such that G has a grid-representation where the grid has height h.

We say that a grid-major representation of G is *simple* if in every column c of the grid and for any vertex v of G, the nodes labeled v in c form a path. The *simple grid-major height* of G, denoted $sGMh(G)$, is the smallest h such that G has a simple grid-major representation of height h.

A grid-major representation of height h can be viewed, equivalently, as a *contact-representation* with integral orthogonal polygons as follows: Assign to every vertex v the polygon $P(v)$ that we obtain if we replace every grid-point labelled v with a unit square centered at that grid-point and take their union. Since the grid-representation uses integral points, the coordinates of sides of

$P(v)$ are halfway between integers. See Fig. 1(d). We get a set of interior-disjoint orthogonal polygons with integer edge-lengths whose union is a rectangle of height h, where (v, w) is an edge of G if and only if $P(v)$ and $P(w)$ share at least one unit-length segment on their boundaries. Conversely any contact-representation with integral orthogonal polygons that uses all points inside a bounding rectangle can be viewed as a grid-major representation. A simple grid-major representation becomes a contact representation with *x-monotone polygons* (every vertical line intersects the polygon in an interval) and vice versa. Contact-representations of graphs have been studied extensively (see e.g. [20] and the references therein), but to our knowledge the question of the required height of such representations has not previously been considered.

Homotopy Parameters: A *(discrete) simple homotopy* is defined for a planar triangulated graph G with a fixed outer-face $\{u, v, w\}$, and it consists of a sequence $h_0, \dots, h_W$ of walks in G (we call these *curves*) such that:

1. h_0 and h_W are trivial curves at two distinct vertices of the outer-face, say u and v.
2. The vertices u and v partition the outer-face into two subpaths $s(uv)$ and $t(uv)$. For $0 \le i \le W$, the curve h_i starts on $s(uv)$ and ends on $t(uv)$.
3. For all $0 \le i < W$ we can obtain h_{i+1} from h_i with a face-flip, edge-slide, a boundary-move or a boundary-edge-slide.
4. Each curve h_i is a simple path and for any $0 \le i < j \le W$, if vertex v belongs to h_i and h_j then it also belongs to all curves in between.

Here a *face-flip* consists of picking an inner face $\{x, y, z\}$ such that the subsequence x-y is in h_i, and replacing the sub-path x-y by x-z-y to obtain h_{i+1}. The reverse move, going from x-z-y to x-y, is also allowed. An *edge-slide*[2] consists of picking an edge $e = (x, y)$ adjacent to two inner faces $\{x, y, z\}$ and $\{x, y, t\}$, such that the subsequence z-x-t is in h_i. Then replace the subpath z-x-t in h_i by z-y-t to obtain h_{i+1}. A *boundary-move* consists of picking an edge $e = (x, y)$ on the outer face, and, if $e \in s(uv)$, and x is the start of h_i, it appends y so that it becomes the new starting point (thus replacing x by the subsequence y-x). If $e \in t(uv)$ and x is the end of h_i, it appends y at the end. The reverse operations are also allowed. A *boundary-edge-slide* consists of picking an edge $e = (x, y)$ on the outer face adjacent to an inner face $\{x, y, z\}$, and, if $e \in s(uv)$ and h_i starts with x-z, we flip $\{x, y, z\}$ and remove e, i.e., we replace the starting subsequence x-z by y-x. The symmetric operation for edges on $t(uv)$ is also allowed. (Observe that this boundary-edge-slide is the same as flipping a face and removing the boundary edge with a boundary move.) see Fig. 1(e).

The *height* of a simple homotopy is the length of the longest path h_i, counting as path-length the number of vertices. Let $sHh(G, f)$ be the minimum height of a simple homotopy of G that uses f as outer-face, and set $sHh(G)$ (the *simple*

---

[2] Edge-slides are typically not allowed in discrete homotopies, but the result of one edge-slide is the same as flipping two inner faces consecutively. Thus, allowing edge-slides only results in an additive difference of at most one for the homotopy height.

homotopy height) to be the minimum of $sHh(G, f)$ over all choices of outer-faces f. (Since we only study triangulated graphs the rotation scheme is unique and so this covers all possible planar embeddings.)

The definition of a (non-simple) homotopy is obtained by removing the simplicity assumption on the curves h_i, and allowing two other kinds of moves (spikes and unspikes) leveraging the non-simplicity of the curves. For technical reasons and to obtain a maximal generality, we will also relax the conditions on the endpoints and the starting and ending curves. Since the precise definition is somewhat technical, we postpone it to Sect. 3.2 and the full version.

Some Simple Results: We briefly review some relationships that are well-known, or easily derived.

- $pw(G) \leq GMh(G)$ since a $W \times H$-grid has pathwidth at most H and pathwidth is closed under taking minors.
- Obviously $GMh(G) \leq sGMh(G)$.
- $sGMh(G) \leq VRh(G)$ since a flat visibility representation can easily be converted into a simple grid-major representation by assigning label v to all grid-points of the bar of v as well as all grid-points that this bar can see downward or rightward without intersecting other bars or non-incident edges. See Fig. 1(b–c).
- $VRh(G) = SLh(G)$ since flat visibility representations can be transformed into straight-line drawings of the same height, and vice versa ([21]. using [22,23]).
- Finally we have $2op(G) - 1 \leq GMh(G)$. To this end, assume that we have a grid-major representation Γ of G of height h. Observe that the grid-graph Γ has outer-planarity $\lceil h/2 \rceil$. Since outer-planarity does not increase when taking minors it follows that $op(G) \leq \lceil h/2 \rceil \leq \frac{h+1}{2}$.

3   Homotopy-Height and Grid-Major Height

The above inequalities fill in most of the chain in Eq. 1, but one key new part is missing: how does the (simple) homotopy height relate to the (simple) grid-major height?

3.1   Simple Grid-Major Height and Simple Homotopy Height

Lemma 1. *For any triangulated planar graph G we have $sGMh(G) \leq sHh(G)$.*

Proof. (Sketch) Let $h_0, \ldots, h_W$ be a simple homotopy of height $k = sHh(G)$. The rough idea is to label a $W \times k$-grid by giving the gridpoints in the ith column the labels of the vertices in h_i, in top-to-bottom-order, and adding some duplicate copies of vertices in h_i to fill the column. However, we must insert more columns in between to ensure the properties of a simple grid-major representation.

It will be easier to describe this by giving a contact-representation of G where all polygons are x-monotone and the height is k. Curve h_0 is a vertex u; we initialize $P(u)$ as a $1 \times k$ rectangle. Now assume that for some $i \geq 0$, we

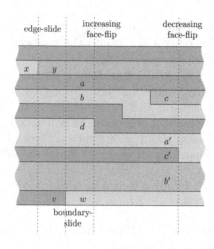

Fig. 2. Converting a discrete simple homotopy into a contact-representation.

have built a contact representation Γ_i of the graph that was swept by $h_0, \ldots, h_i$. Furthermore, the right boundary of the bounding box of Γ_i contains sides of the vertices in h_i, in order. Figure 2 sketches how to expand Γ_i rightwards, depending on the next move used for the homotopy; full details are in the full version.

In all cases the polygons remain connected and are x-monotone. Furthermore we realized exactly those incidences that were added to the graph when sweeping to h_{i+1}, and the right boundary contains exactly the polygons of vertices of h_{i+1}, in order. Therefore, repeating gives a contact-representation of height k that uses x-monotone polygons, and thus the desired simple grid-major representation. □

Lemma 2. *For any triangulated planar graph G we have $sHh(G) \leq sGMh(G)$.*

Proof. (Sketch) Fix a simple grid representation Γ of G of height $sGMh(G)$. For this proof it will be easier to interpret Γ as a contact representation. So for any vertex z, let $P(z)$ be the orthogonal polygon obtained by taking the unions of all unit squares whose centerpoint is a grid point labelled z. Since the grid-representation uses integral points, the coordinates of sides of $P(v)$ are halfway between integers.

A *junction* is a point that belongs to at least three sides of polygons; we call it *interior/exterior* depending on whether it lies on the boundary of the rectangle $\mathcal{R}$ that encloses the contact representation. No junction can belong to four sides since G is maximal, so it can be classified as *horizontal* or *vertical* depending on the majority among its incident sides. A *corner* is a point that belongs to exactly two sides of polygons. It is not possible for both ends of a side to be exterior junctions, or else the corresponding edge of G would be a bridge of the graph, contradicting the 3-connectivity of the triangulated graph G.

We show in the full version that with suitable local changes to the contact representation, we can ensure the following while maintaining the same height:

(1) Every interior junction is horizontal, (2) no two interior vertical sides have the same x-coordinate, (3) exactly one vertex u touches the left boundary, exactly one vertex v touches the right boundary, and $u \neq v$, and (4) exactly three vertices u, v, w touch the boundary. Therefore $\{u, v, w\}$ forms a face f; declare f to be the outer-face. As in the definition of homotopy, let $s(uv)$ and $t(uv)$ be the subpaths of G between u and v on f. By definition, they consist of the vertices occupying the top and bottom boundaries of $\mathcal{R}$.

Let the contact representation now have x-range $[-\frac{1}{2}, W + \frac{1}{2}]$. For $i = 0, \ldots, W$, define h_i to be the vertices whose polygons intersect the vertical line $\{x = i\}$, enumerated from top to bottom. Clearly $h_0 = \langle u \rangle$, $h_W = \langle v \rangle$, and any h_i begins on $s(uv)$ and ends on $t(uv)$. It remains to show that for $0 \leq i < W$ going from h_i to h_{i+1} is one of the permitted moves. Consider some vertical side e that has x-coordinate $i + \frac{1}{2}$ (if there is none then $h_i = h_{i+1}$ and we are done). Note that the change from h_i to h_{i+1} affects *only* vertices that are incident to e or participate in junctions at the ends of e, because no other vertical side has x-coordinate $i + \frac{1}{2}$ (by $0 \leq i < W$ and assumption) and so there is no difference between the curves elsewhere. Figure 3 shows (up to symmetry) all possibilities for what the ends of e are. One observe that this results in the following situations: (a) $h_i = h_{i+1}$ and no move is needed, (b) this is impossible if polygons are x-monotone, (c) a face-flip, (d) a boundary-move, (e) an edge-slide and (f) a boundary-edge-slide. Therefore we only use allowed moves and have found a homotopy. It is simple since polygons are x-monotone. The height equals the maximum number of intersected polygons, which is no more than the height of the contact representation, hence the height of the grid representation. □

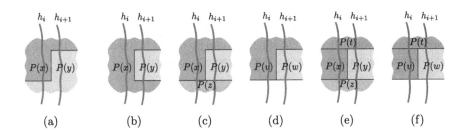

Fig. 3. The cases of how curves change.

Putting the two results together, the homotopy-height and the simple grid-major-height are exactly the same value.

3.2 Homotopy Height and Grid-Major Height

One can reasonably argue that the notion of grid-major height is more natural than simple grid-major height. Furthermore, it is trivially a minor-closed quantity, which is advantageous from a structural and algorithmic point of view. In this subsection we show that grid-major height can also be interpreted as the height of a more general notion of homotopy than the one defined in the preliminaries. Compared to the case of simple homotopies, in a (non-simple) homotopy, we remove the hypothesis that the curves are simple and we allow two new moves, spikes and unspikes, leveraging this non-simplicity. Figure 3(b) illustrates a spike. Furthermore, the conditions are slightly relaxed: the endpoints are allowed to move along an edge instead of a face, and the starting and ending vertices are allowed to be the same. This could allow "trivial" homotopies (for example an empty one, idling on a single vertex), and thus we add a new condition on topological non-triviality to disallow those. The precise definition of a discrete homotopy can be found in the full version.

Lemma 3. *For any triangulated planar graph we have $gmh(G) \leq Hh(G)$.*

Lemma 4. *For any triangulated graph we have $Hh(G) \leq gmh(G)$.*

The proofs are in spirit very similar to those in the previous subsection (we may now have spikes or unspikes as moves, but these simply correspond to Fig. 3b). However, numerous details need attention, in particular it is not at all obvious why the polygons created from a homotopy would be connected if they are not x-monotone, and some of the steps in the proof of Lemma 2 do not seem to hold in the non-simple setting anymore (this is why we relaxed the conditions on the outer-face and the distinctness of the start and the end of the homotopy). The full version gives the (somewhat lengthy) details.

Since grid-major height is trivially minor-closed, testing whether a graph has grid-major height at most k can be decided in time $O(f(k)|G|^3)$ by testing the (unknown!) forbidden minors, which are in finite number by Robertson-Seymour theory. Because minor testing can be expressed in second-order logic, and the graphs of bounded grid-major height have bounded pathwidth, it follows from Courcelle's theorem [16] that these minors can be tested in linear time. Therefore, the two previous lemmas give us the following corollary.

Corollary 1. *We can decide whether a triangulated planar graph has homotopy height $Hh(G)$ at most k in time $O(f(k)poly(|G|))$ for some computable function $f(k)$. In particular, the problem of computing the homotopy height is FPT when parameterized by the output.*

4   Strictness Examples

We have now given all the inequalities needed for Eqs. 1 and 2. In this section, we argue that many of these inequalities are strict by exhibiting suitable planar triangulations.

Lemma 5. *There exists a planar triangulated graph G with $pw(G) = 3$ and $GMh(G) \in \Omega(n)$.*

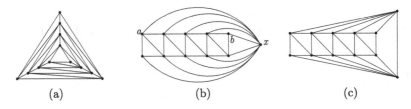

(a)   (b)   (c)

Fig. 4. (a) Nested triangles. (b) The graph from [12]. (c) A graph of which it is a minor.

Proof. Graph G is the "nested triangles graph" from [7,8], consisting of $n/3$ triangles that are stacked inside each other and connected in such a way that the result is triangulated and has pathwidth 3. See Fig. 4(a). For any choice of outer-face there are at least $n/6$ triangles that remain stacked inside each other. Therefore $op(G) \geq n/6$ and $GMh(G) \geq n/3 - 1$.   □

Lemma 6. *There exists a planar triangulated graph that has grid-major height at most 4, but simple grid-major-height $\Omega(n)$.*

Proof. Consider graph G in Fig. 4(b), which is taken from [12]. It is a minor of the graph G' in Fig. 4(c), which has a straight-line drawing of height 4. Therefore $sGMh(G') \leq SLh(G') \leq 4$, which implies $GMh(G) \leq 4$ since G is a minor of G'.

We claim that $sGMh(G) \in \Omega(n)$, and prove this by arguing that $sHh(G) \in \Omega(n)$; the two parameters are the same. Crucial to our argument is that for many vertex-pairs any path connecting them *without* using x has length $\Omega(n)$; we will find such a path from the curves in a homotopy.

So consider a simple discrete homotopy of height k, and let f be the face it uses as the outer-face (it need not be the outer-face used in Fig. 4(b)). Graph $G \setminus x$ is connected, but $d_{G \setminus x}(a, b) = (n - 3)/2$. Define d_a to be the minimum distance in $G \setminus x$ from a to some vertex on face f, and similarly define d_b. Since f is a triangle, we can combine two such shortest paths to obtain a path from a to b in $G \setminus x$ of length at most $d_a + d_b + 1$, therefore (up to renaming) $d_a \geq (n-5)/4$.

In particular, for $n \geq 7$ vertex a is not on f. Let h_i be a curve of the homotopy that contains a and note that it begins and ends on f. Split h_i into two paths π_1 and π_2 at vertex a. These paths are vertex-disjoint except for a since the homotopy is simple. At most one of these paths contains x. Say π_1 does not contain x and hence connects f to a without visiting x. Therefore $|\pi_1| \geq d_a \geq (n - 5)/4$, and the height of the homotopy is $\Omega(n)$.   □

In particular, Lemma 6 provides a different (and in our opinion more accessible) proof that the graph in Fig. 4(b) requires $\Omega(n)$ height in any straight-line drawing [12].

We now want to show that the inequality $sGMh(G) \leq SLh(G)$ can be strict, and for this, need a definition. A graph G is called a *series-parallel graph (with terminals s and t)* if it either is an edge (s, t), or if it was obtained via a combination in series or in parallel. Here, a *combination in series* takes two such graphs

G_i with terminals s_i, t_i for $i = 1, 2$, and identifies t_1 with s_2. A *combination in parallel* also takes two such graphs and identifies s_1 with s_2 and t_1 with t_2. It is well-known that such graphs are planar.

Lemma 7. *Any series-parallel graph has a simple grid-major representation of height $O(\log n)$.*

Proof. Roughly speaking, we "bend" some of the bars in the visibility representations of series-parallel graphs from [12] to guarantee logarithmic height. Formally we proceed by induction on m, and prove that if G has m edges, then it has a simple grid-major representation of height $2\lceil \log m \rceil + 2$ where the top-right corner is labelled s and the bottom-right corner is labelled t. Furthermore, any column that contains s and/or t also has its topmost/bottommost grid point labelled with s/t. In the base case G is an edge (s, t) and we can simply label a 1×2 grid with s and t.

Assume first that G was obtained by parallel combinations of G_1 and G_2. Consider Fig. 5. After renaming we may assume $m(G_2) \leq m(G_1)$, so $m(G_2) \leq m/2$. Recursively obtain a grid-major representation Γ_1 of G_1, and pad it with duplicate rows (if needed) so that it has height $2\lceil \log m \rceil + 2$. Recursively obtain a grid-major representation Γ_2 of G_2 of height at most $2\lceil \log(m(G_2)) \rceil + 2 \leq 2\lceil \log m \rceil$. Place Γ_2 to the right of Γ_1, leaving the top and bottom row unused. Label the points above Γ_2 with s and the points below Γ_2 with t and verify all conditions.

Now assume that G was obtained by a series combination of two graphs G_1, G_2 where G_1 had terminals s, x and G_2 had terminals x, t. We assume $m(G_2) \leq m(G_1)$, the other case is symmetric. Recursively obtain grid-major representations Γ_1 and Γ_2 of G_1 and G_2 of height $2\lceil \log m \rceil + 2$ and $2\lceil \log m \rceil$ as before. Place Γ_2 to the right of Γ_1, leaving the top two rows unused, and leaving one column between the representations unused. All grid-points in this column, as well as in the row above Γ_2, are labelled x. (In particular, the grid-points labelled x form an "S-shape" as if we had bent a bar in the middle.) The second row above Γ_2 is labelled s so that again the top-right corner has s. One easily verifies that this is a simple grid-major representation of G with height $2\lceil \log m \rceil + 2 \in O(\log n)$. $\square$

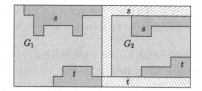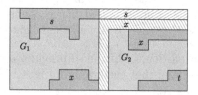

Fig. 5. Grid-Major representations of series-parallel graphs.

Fig. 6. A graph with $op(G) = 2$ but $GMh(G) \in \Omega(\log n)$ (Lemma 8): a complete binary tree (thick black edges), augmented to become maximal outer-planar (dashed blue edges), with a new vertex added in the outer face (thin red edges). (Color figure online)

Theorem 1. *There exists a planar triangulated graph G for which $sGMh(G) \in O(\log n)$ but $SLh(G) \in \Omega(2^{\sqrt{\log n}})$.*

Proof. (Sketch) We know from Frati [24] that for any N, there exists a series-parallel graph G_N with $n \geq N$ vertices for which any planar straight-line drawing has height $\Omega(2^{\sqrt{\log n}})$. Also $sGMh(G_N) \in O(\log n)$ by Lemma 7. A suitable supergraph of G_N (see the full version) is triangulated and satisfies all properties. □

Lemma 8. *There exists a planar triangulated graph G with $op(G) = 2$ but $GMh(G) \in \Omega(\log n)$.*

Proof. Take any tree T that has pathwidth $\Omega(\log n)$, for example a complete binary tree. This is an outer-planar graph; add edges to the graph while maintaining outer-planarity until the graph is maximal outer-planar, hence 2-connected and all faces except the outer-face are triangles. Insert a new vertex in the outer-face and make it adjacent to all other vertices; the result (see Fig. 6) is a triangulated planar graph G with outer-planarity 2 and $GMh(G) \geq pw(G) \geq pw(T) \in \Omega(\log n)$. □

5 Outlook

In this paper, we studied two parameters of planar triangulated graphs, the homotopy height (well-known in computational geometry but not previously used for graph drawing) and the grid-major height (related to contact-representations, but not explicitly expressed as a graph parameter before). We argue that these two seemingly unrelated parameters are actually equal, and that they, as well as their variations that require simplicity in some sense, can serve as lower bounds for the height of straight-line drawings of planar graphs. Their equality also implies that testing whether homotopy height is at most k is fixed-parameter tractable in k. We leave many open problems:

- What is the complexity of computing these various graph parameters? In particular, while it is strongly believed that computing the minimum height of a planar drawing is NP-hard, we are not aware of any proof of this. Similarly, it is not known whether computing the homotopy height, or equivalently the grid-major height, is NP-hard or polynomial. The same goes for the simple variants. On the other hand, computing the pathwidth is NP-hard even for planar graphs [25], while computing the outerplanarity is polynomial [26].
- The trivial minor-closedness of grid-major height proves the existence of an FPT algorithm to compute it when parameterized by the output. However, this algorithm relies on finding the forbidden minors, which are unknown. Finding an explicit algorithm for this problem is still open.
- We focused on straight-line drawings, but *poly-line drawings* of G (i.e., straight-line drawings of some subdivision G' of G) are also of interest. Letting $PLh(G)$ be the smallest height of such drawings, one sees that $GMh(G) \leq sGMh(G') \leq SLh(G') \leq PLh(G)$, but is it true that $sGMh(G) \leq PLh(G)$?

References

1. Wagner, K.: Bemerkungen zum Vierfarbenproblem. Jahresber. Dtsch. Math. Ver. **46**, 26–32 (1936)
2. Fáry, I.: On straight line representation of planar graphs. Acta Sci. Math. (Szeged) **11**(4), 229–233 (1948)
3. Stein, S.K.: Convex maps. Proc. Am. Math. Soc. **2**, 464–466 (1951)
4. Tutte, W.T.: How to draw a graph. Proc. London Math. Soc. **13**, 743–767 (1963)
5. Fraysseix, H.D., Pach, J., Pollack, R.: How to draw a planar graph on a grid. Combinatorica **10**, 41–51 (1990)
6. Schnyder, W.: Embedding planar graphs on the grid. In: Symposium on Discrete Algorithms (SODA 1990), pp. 138–148. SIAM (1990)
7. Dolev, D., Leighton, T., Trickey, H.: Planar embedding of planar graphs. Adv. Comput. Res. **2**, 147–161 (1984)
8. Fraysseix, H.D., Pach, J., Pollack, R.: Small sets supporting Fáry embeddings of planar graphs. In: Symposium Theory of Computing (STOC 1988), pp. 426–433. ACM (1988)
9. Heath, L.S., Rosenberg, A.L.: Laying out graphs using queues. SIAM J. Comput. **21**(5), 927–958 (1992)
10. Dujmović, V., Fellows, M.R., Kitching, M., Liotta, G., McCartin, C., Nishimura, N., Ragde, P., Rosamond, F., Whitesides, S., Wood, D.R.: On the parameterized complexity of layered graph drawing. Algorithmica **52**(2), 267–292 (2008)
11. Felsner, S., Liotta, G., Wismath, S.: Straight-line drawings on restricted integer grids in two and three dimensions. J. Graph Algorithms Appl. **7**(4), 335–362 (2003)
12. Biedl, T.: Small drawings of outerplanar graphs, series-parallel graphs, and other planar graphs. Discret. Comput. Geom. **45**(1), 141–160 (2011)
13. Chambers, E.W., Letscher, D.: On the height of a homotopy. In: Canadian Conference on Computational Geometry (CCCG 2009), pp. 103–106 (2009)
14. Har-Peled, S., Nayyeri, A., Salavatipour, M., Sidiropoulos, A.: How to walk your dog in the mountains with no magic leash. Discret. Comput. Geom. **55**(1), 39–73 (2016)

15. Chambers, E.W., de Mesmay, A., Ophelders, T.: On the complexity of optimal homotopies. In: Symposium on Discrete Algorithms (SODA 2018), pp. 1121–1134. SIAM (2018)
16. Courcelle, B.: The monadic second-order logic of graphs. I. Recognizable sets of finite graphs. Inf. Comput. **85**(1), 12–75 (1990)
17. Wismath, S.K.: Characterizing bar line-of-sight graphs. In: Symposium on Computational Geometry (SoCG 1985), pp. 147–152. ACM (1985)
18. Tamassia, R., Tollis, I.: A unified approach to visibility representations of planar graphs. Discret. Comput. Geom. **1**, 321–341 (1986)
19. Rosenstiehl, P., Tarjan, R.E.: Rectilinear planar layouts and bipolar orientation of planar graphs. Discret. Comput. Geom. **1**, 343–353 (1986)
20. Alam, M.J., Biedl, T., Felsner, S., Kaufmann, M., Kobourov, S.G., Ueckerdt, T.: Computing cartograms with optimal complexity. Discret. Comput. Geom. **50**(3), 784–810 (2013)
21. Biedl, T.: Height-preserving transformations of planar graph drawings. In: Duncan, C., Symvonis, A. (eds.) GD 2014. LNCS, vol. 8871, pp. 380–391. Springer, Heidelberg (2014). https://doi.org/10.1007/978-3-662-45803-7_32
22. Pach, J., Tóth, G.: Monotone drawings of planar graphs. J. Graph Theory **46**(1), 39–47 (2004)
23. Eades, P., Feng, Q., Lin, X., Nagamochi, H.: Straight-line drawing algorithms for hierarchical graphs and clustered graphs. Algorithmica **44**(1), 1–32 (2006)
24. Frati, F.: Lower bounds on the area requirements of series-parallel graphs. Discret. Math. Theor. Comput. Sci. **12**(5), 139–174 (2010)
25. Gustedt, J.: On the pathwidth of chordal graphs. Discret. Appl. Math. **45**(3), 233–248 (1993)
26. Bienstock, D., Monma, C.L.: On the complexity of embedding planar graphs to minimize certain distance measures. Algorithmica **5**(1), 93–109 (1990)

On the Edge-Vertex Ratio
of Maximal Thrackles

Oswin Aichholzer[1], Linda Kleist[2(✉)], Boris Klemz[3], Felix Schröder[4],
and Birgit Vogtenhuber[1]

[1] Graz University of Technology, Graz, Austria
{oaich,bvogt}@ist.tugraz.at
[2] Technische Universität Braunschweig, Brunswick, Germany
kleist@ibr.cs.tu-bs.de
[3] Freie Universität Berlin, Berlin, Germany
klemz@inf.fu-berlin.de
[4] Technische Universität Berlin, Berlin, Germany
fschroed@math.tu-berlin.de

Abstract. A drawing of a graph in the plane is a *thrackle* if every pair
of edges intersects exactly once, either at a common vertex or at a proper
crossing. Conway's conjecture states that a thrackle has at most as many
edges as vertices. In this paper, we investigate the edge-vertex ratio of
maximal thrackles, that is, thrackles in which no edge between already
existing vertices can be inserted such that the resulting drawing remains
a thrackle. For maximal geometric and topological thrackles, we show
that the edge-vertex ratio can be arbitrarily small. When forbidding isolated vertices, the edge-vertex ratio of maximal geometric thrackles can
be arbitrarily close to the natural lower bound of $1/2$. For maximal topological thrackles without isolated vertices, we present an infinite family
with an edge-vertex ratio of $5/6$.

1 Introduction

A drawing of a graph in the plane is a *thrackle* if every pair of edges intersects exactly once, either at a common vertex or at a proper crossing. Conway's
conjecture from the 1960s states that a thrackle has at most as many edges
as vertices [7]. While it is known that the conjecture holds true for geometric
thrackles in which edges are drawn as straight-line segments [18], it is widely
open in general. In this paper, we investigate *maximal thrackles*. A thrackle is
maximal if no edge between already existing vertices can be inserted such that
the resulting drawing remains a thrackle. Our work is partially motivated by the
results of Hajnal et al. [11] on saturated k-simple graphs. A graph is k-simple if
every pair of edges has at most k common points, either proper crossings and/or
a common endpoint. A k-simple graph is *saturated* if no further edge can be
added while maintaining th k-simple property. In [11], simple graphs on n vertices with only $7n$ edges are constructed, as well as saturated 2-simple graphs
on n vertices with $14.5n$ edges.

© Springer Nature Switzerland AG 2019
D. Archambault and C. D. Tóth (Eds.): GD 2019, LNCS 11904, pp. 482–495, 2019.
https://doi.org/10.1007/978-3-030-35802-0_37

If true, Conway's conjecture implies that in every thrackle the ratio between the number of edges and the number of vertices is at most 1. We denote the edge-vertex ratio of a thrackle T by $\varepsilon(T)$. In this paper, we investigate the other extreme, namely maximal thrackles with a low edge-vertex ratio.

In Sect. 2, we consider geometric thrackles. We show that for this class the edge-vertex ratio can be arbitrarily small. This is done by a construction that allows to add isolated vertices while maintaining maximality. If we disallow isolated vertices, then a natural lower bound for the edge-vertex ratio is $\frac{1}{2}$. A similar construction can be used to get arbitrarily close to this bound.

Theorem 1. *For any $c > 0$, there exists*

(a) a maximal geometric thrackle T_a such that $\varepsilon(T_a) < c$, as well as
(b) a maximal geometric thrackle T_b without isolated vertices such that $\varepsilon(T_b) < \frac{1}{2} + c$.

We then consider topological thrackles in Sect. 3. Similar as before we show that the edge-vertex ratio can approach zero using isolated vertices.

Theorem 2. *For every $c > 0$, there is a maximal thrackle T' with $\varepsilon(T') < c$.*

Note that Theorem 2 is not just a trivial implication of Theorem 1, as a maximal geometric thrackle is not necessarily a maximal topological thrackle. As our main result, in Sect. 4, we show that there exists an infinite family of thrackles without isolated vertices which has an edge-vertex ratio of $\frac{5}{6}$.

Theorem 3. *There exists an infinite family of thrackles $\mathcal{F}$ without isolated vertices, such that for all $T \in \mathcal{F}$ it holds that $\varepsilon(T) = \frac{5}{6}$.*

Our construction is based on an example presented by Kynčl [12] in the context of simple drawings where he showed that not every simple drawing can be extended to a simple drawing of the complete graph. The example was also used in [13] for a related problem.

Due to space constraints, several proofs of this work are either sketched or completely omitted. They can be found in the arXiv version [1].

Related Work. In one of the first works on Conway's Thrackle Conjecture, Woodall [22] characterized all thrackles under the assumption that the conjecture is true. For example, he showed that a cycle C_n has a thrackle embedding with straight edges if and only if n is odd. It is not hard to come up with other graphs on n vertices with n edges that have a thrackle embedding, but adding an additional edge always seems to be impossible. Consequently, two lines of research emerged from Conway's conjecture. In the first, the goal is to prove the conjecture for special classes of drawings, while the second direction aims for upper bounds on the number of pairwise crossing or incident edges in any simple topological drawing with n vertices.

For straight line drawings of thrackles, so called *geometric* thrackles, already Erdős provided a proof for the conjecture, actually answering a question from 1934 by Hopf and Pannwitz on distances between points. Probably the most elegant argument is due to Perles and can be found in [18]. Extending geometric drawings, a drawing is called *x-monotone* if each curve representing an edge is intersected by every vertical line in at most one point. In the same paper, Pach and Sterling [18] show that the conjecture holds for *x*-monotone drawings by imposing a partial order on the edges.

A drawing of a graph is called *outerplanar* if its vertices lie on a circle and its edges are represented by continuous curves contained in the interior of this circle. In [5] several properties for outerplanar thrackles are shown, with the final result that outerplanar thrackles are another class where the conjecture is true. Misereh and Nikolayevsky [16] generalized this further to thrackle drawings where all vertices lie on the boundaries of $d \leq 3$ connected domains which are in the complement of the drawing. They characterize annular thrackles ($d = 2$) and pants thrackles ($d = 3$) and show that in all cases Conway's conjecture holds. Finally, Cairns, Koussas, and Nikolayevsky [2] prove that the conjecture holds for spherical thrackles, that is, thrackles drawn on the sphere such that the edges are arcs of great circles.

In a similar direction, several attempts show that some types of thrackles are *non-extensible*. A thrackle is called non-extensible if it cannot be a subthrackle of a counterexample to Conway's conjecture. Wehner [21] stated the hypothesis that a potential counterexample to Conway's conjecture would have certain graphtheoretic properties. Li, Daniels, and Rybnikov [14] support this hypothesis by reducing Conway's conjecture to the problem of proving that thrackles from a special class (which they call 1-2-3 group) are non-extensible. Actually, already Woodall [22] had shown that if the conjecture is false, then there exists a counterexample consisting of two even cycles that share a vertex.

On the negative side, we mention tangled- and generalized thrackles. A tangled-thrackle is a thrackle where two edges can have a common point of tangency instead of a proper crossing. Besides the fact that tangled-thrackles with at least $\lfloor 7n/6 \rfloor$ edges are known [17] – and therefore Conway's conjecture can not be extended to tangled-thrackles – Ruiz-Vargas, Suk, and Tóth [20] show that the number of edges for tangled-thrackles is $O(n)$. A *generalized* thrackle is a drawing where any pair of edges shares an odd number of points. Lovász, Pach, and Szegedy [15] showed that a bipartite graph can be drawn as a generalized thrackle if and only if it is planar. As planar bipartite graphs can have up to $2n - 4$ edges, this implies that generalized thrackles exist with a edge-vertex ratio close to 2. A tight upper bound of $2n - 2$ edges for generalized thrackles was later provided by Cairns and Nikolayevsky [3].

The race for an upper bound on the number m of edges of a thrackle was started by the two just mentioned papers. Lovász, Pach, and Szegedy [15] provided the first linear bound of $m \leq 2n - 3$ and Cairns and Nikolayevsky [3] improved this to $m \leq \frac{3}{2}(n - 1)$. They also consider more general drawings of thrackles on closed orientable surfaces; see also [4].

By exploiting certain properties of the structure of possible counterexamples, Fulek and Pach [8] gave an algorithm that, for any $c > 0$, decides whether the number of edges are at most $(1 + c)n$ for all thrackles with $n \geq 3$. As the running time of this algorithm is exponential in $1/c$, the possible improvement by the algorithm is limited, but the authors managed to show an upper bound of $m \leq \frac{167}{117}n \approx 1.428n$. Combining several previous results in a clever way, Goddyn and Xu [10] slightly improved this bound to $m \leq 1.4n - 1.4$. Among other observations they also used the fact that it was known that Conway's conjecture holds for $n \leq 11$. This has been improved to $n \leq 12$ in the course of enumerating all path-thrackles for n up to 12 in [19]. The currently best known upper bound of $m \leq 1.3984n$ is again provided by Fulek and Pach [9]. They also show that for *quasi*-thrackles Conway's conjecture does not hold. A quasi-thrackle is a thrackle where two edges that do not share a vertex are allowed to cross an odd number of times. For this class they provide an upper bound of $m \leq \frac{3}{2}(n-1)$ and show that this bound is tight for infinitely many values of n.

2   Geometric Thrackles

For maximal geometric thrackles, the edge-vertex ratio can be arbitrarily small. Even if we forbid isolated vertices, it may be arbitrarily close to the natural lower bound of $\frac{1}{2}$, which is implied by the handshaking lemma.

Theorem 1. *For any $c > 0$, there exists*

(a) a maximal geometric thrackle T_a such that $\varepsilon(T_a) < c$, as well as
(b) a maximal geometric thrackle T_b without isolated vertices such that
   $\varepsilon(T_b) < \frac{1}{2} + c$.

Proof sketch. Consider the thrackle T formed by the seven dark, thick edges in Fig. 1, which we call the *butterfly*. The butterfly is a maximal thrackle: Any segment between the *bottom* three vertices b_1, b_2, b_3 or between the *top* seven

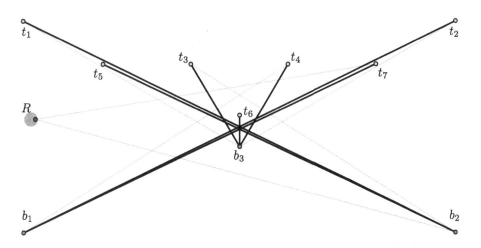

Fig. 1. The butterfly T (thick, dark edges).

vertices $t_1, \ldots, t_7$ is disjoint from the *central* edge $b_3 t_6$ or from one of the *long* edges $b_1 t_2$, $b_1 t_7$, $b_2 t_1$, and $b_2 b_5$. Moreover, aside from $b_1 t_6$ and $b_2 t_6$, all segments with one bottom and one top vertex as an endpoint are disjoint from the central edge or one of the long edges. Finally, the two remaining segments $b_1 t_6$ and $b_2 t_6$ are disjoint from $b_3 t_4$ or $b_3 t_3$, respectively.

To prove the theorem, we extend the butterfly in two different ways.

(a) To obtain T_a from T, we insert a sufficient number of isolated vertices in a small circular region R (indicated in Fig. 1) that is placed to the left of t_6 such that the lower tangent of R that passes through t_6 is below all top vertices other than t_6, and the upper tangent of R that passes through b_3 is above all bottom vertices except for b_3. These properties imply each segment between R and a vertex of T is disjoint from the central edge or one of the long edges. Hence, T_a is indeed a maximal thrackle.

(b) To obtain T_b from T, we add a sufficient number of segments $u_i v_i$ with $i = 1, 2, \ldots, m$ as indicated in Fig. 2. All these segments pass through a common point along the central edge. All upper endpoints u_i are placed on the line through t_1 and t_2, and all lower endpoints v_i are placed on the line through b_1 and b_2. For each index i, the slope $s(u_i v_i)$ is negative. Moreover, we have $s(u_i v_i) < s(u_j v_j)$ for $i < j$.

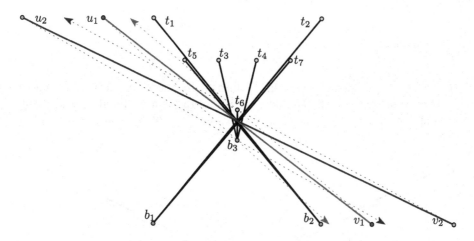

Fig. 2. The thrackle T_b is obtained by adding several segments $u_i v_i$.

Suppose that the first $i-1$ segments have already been created for some $i \geq 1$. Then we choose the slope of $u_i v_i$ such that the vertices

- $V_i^+ = \{v_1, v_2, \ldots, v_{i-1}\} \cup \{b_1, b_2\}$ are below the line $u_i b_3$; and
- $V_i^- = \{u_1, u_2, \ldots, u_{i-1}\} \cup \{t_1, t_2, t_3, t_4, t_5, t_7\}$ are above the line $v_i t_6$.

This choice implies that all non-edge segments between vertices of T_b are disjoint from the central edge or one of the long edges. Hence, T_b is maximal.   □

3 Topological Thrackles of Arbitrarily Small Edge-Vertex Ratio

In this section, we show that the edge-vertex ratio of a maximal thrackle in the topological setting may be arbitrarily small, unless isolated vertices are forbidden.

Theorem 2. *For every $c > 0$, there is a maximal thrackle T' with $\varepsilon(T') < c$.*

Proof sketch. Consider the thrackle T of a simple cycle on six vertices depicted in Fig. 3. Adding a sufficiently large number of isolated vertices into the central triangular face f_0 of T yields a thrackle T' with $\varepsilon(T') < c$. It remains to show that T' is maximal. Towards a contradiction, assume that it is possible to insert an edge uv into T' such that the resulting drawing remains a thrackle. Our plan is to show that uv is self-intersecting or intersects one of the edges of T twice, which yields the desired contradiction. To this end, we explore the drawing of e, going from u to v. We distinguish three cases, depending on how many of the vertices u, v are isolated vertices of T'.

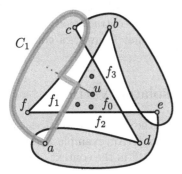
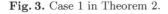
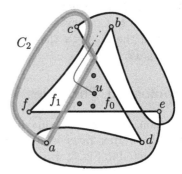

Fig. 3. Case 1 in Theorem 2.   **Fig. 4.** Case 2 in Theorem 2.

Case 1: Both u and v are isolated vertices of T'. To begin with, the edge uv has to leave f_0 and, by symmetry, we may assume that it does so by intersecting ab. The thereby entered face f_1 has degree four. Consequently, there are three options for uv to proceed. First, assume that uv leaves f_1 by intersecting the edge af, as depicted in Fig. 3. By planarity, in order to reach v, the edge uv has to intersect the closed curve C_1 formed by parts of ab and af, and the part of uv that intersects f_1. This implies that uv intersects itself, or it intersects ab or af at least twice, which yields the desired contradiction. It follows that uv leaves f_1 via cd or ef. This implies that leaving f_0 via f_1 already requires crossings with two of the three segments ab, cd, and ef that bound f_0. However, traversing e in reverse, that is, going from v to u, requires us to leave f_0 via one of the other adjacent faces f_2 and f_3. By symmetry, this requires two additional crossings with the segments ab, cd, and ef. Consequently, one of these segments is crossed at least twice, which again yields a contradiction.

Case 2: Precisely one of u and v is isolated in T'. Without loss of generality, we may assume that u is the isolated endpoint of uv. As in the previous case, we may assume that uv leaves f_0 via ab and enters f_1. Given that uv has to intersect the edge de (among others), it has to leave f_1 (by passing through af, ef, or cd).

The case that f_1 is left via af can be excluded using similar arguments as in Case 1. It remains to consider the cases that uv leaves f_1 via cd or ef, respectively. First, consider the former case, for an illustration refer to Fig. 4. Given that uv has already intersected ab and cd, it follows that $v \in \{e, f\}$. By planarity, it is not possible that $v = f$, since this would imply that uv has to intersect the closed curve C_2, which is composed of parts of the already intersected edges ab and cd and the edge af, which is incident to f. It follows that $v = e$. At some point, the edge uv intersects the edge af in its interior and, thereby, enters the region interior to C_2 that does not contain e. However, the edges bounding C_2 have now all been intersected and, hence, it is no longer possible to reach e. It follows that uv does actually not leave f_1 via cd. It remains to consider the case that f_1 is left via ef. While not symmetric, this case can be handled similarly to the previous one.

Case 3: Both u and v belong to T. Note that this implies that $T + uv$ is a counterexample to Conways's conjecture. We obtain a contradiction, as it was established in the master's thesis by Pammer [19] that Conways's conjecture holds for $n \leq 12$. $\qquad\square$

4 Topological Thrackles Without Isolated Vertices

In this section, we investigate maximal thrackles without isolated vertices, such that the edge-vertex ratio is strictly smaller than 1. An example of such a thrackle, depicted in Fig. 5, was presented by Kynčl [12] in the context of simple drawings, i.e., drawings in which every two edges intersect at most once.

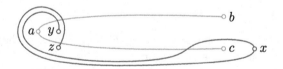

Fig. 5. Kynčl's example K.

Proposition 1. *Kynčl's example K is a maximal thrackle.*

Note that the edge-vertex ratio of Kynčl's example is $\frac{4}{6} = \frac{2}{3}$. To date, we know of no maximal thrackle without isolated vertices that has a lower edge-vertex ratio, with the exception of the trackle consisting of one edge, namely $K_{1,1}$. Now, we present an infinite family of thrackles with a low edge-vertex ratio.

Theorem 3. *There exists an infinite family of thrackles $\mathcal{F}$ without isolated vertices, such that for all $T \in \mathcal{F}$ it holds that $\varepsilon(T) = \frac{5}{6}$.*

We start with a high-level overview of the proof strategy. We start our construction with a geometric star-shaped thrackle T of the cycle C_{2n+1}, for some $n \geq 2$, as depicted in Fig. 6 for $n = 4$. In the first step, we duplicate every vertex and edge of T. This results in a thrackle drawing T_1 of the cycle C_{4n+2}. Then we apply another vertex/edge duplication step that consists of adding a copy of Kynčl's example to each edge. This yields a thrackle T_2. We show that if T_2 was not maximal, we can assume that the additional edge starts from vertices of T_1. Therefore, the maximality of T_1 implies the maximality of T_2.

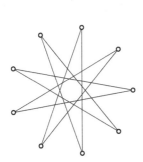

Fig. 6. C_{2n+1} as a star trackle.

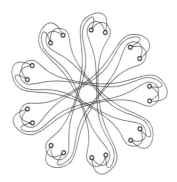

Fig. 7. C_{4n+2} as a blown up star trackle.

Now, we define T_1 precisely. To this end, we choose an orientation of C_{2n+1} and consider three consecutive vertices u, v, and w of C_{2n+1}. We replace every vertex v of T by two vertices v_1 and v_2 very close to v. Every directed edge uv of T is replaced by the edges $u_2 v_1$ and $u_1 v_2$, which are routed in a thin *tunnel* around uv in the following way: The edge starting at u_1 goes along uv without crossing it, surrounds v_1, and then crosses the edge vw of T to connect to v_2. Analogously, the edge starting at u_2 goes along uv, surrounds v_2, and then crosses the edge vw of T as well as $u_1 v_2$ to connect to v_1; see Fig. 8 for an illustration. The edges emanating from v_1 and v_2 are drawn analogously and hence intersect the edges $u_1 v_2$ and $u_2 v_1$, respectively.

Fig. 8. Step 1: Duplicating the vertices and edges. The tunnel of uv is depicted by the gray region.

The result T_1 is a drawing of the cycle C_{4n+2}; a drawing for $n = 4$ is depicted in Fig. 7. It is not hard to see that every pair of edges of T_1 intersects and, hence, T_1 is a thrackle.

Lemma 1. T_1 *is a thrackle.*

Moreover, T_1 is maximal.

Proposition 2. *The thrackle T_1 of C_{4n+2} is maximal.*

For the next step, we introduce the *Kynčl belt construction*, which is applied to T_1 in order to obtain a drawing T_2. We will show that T_2 is a maximal thrackle with edge-vertex-ratio of $\frac{5}{6}$.

The Kynčl belt construction creates a copy of Kynčl's example for each edge of T_1. The edges of T_1 are preserved and the Kynčl copy K_e created for an edge e of T_1 is drawn very close to e and interlaced with e and its incident edges, in order to ensure that the edges of K_e intersect with all edges of T_1 (and T_2). For an illustration consider Fig. 9.

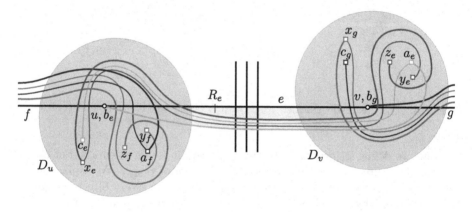

Fig. 9. Kynčl belt construction, the original edges (thick) are preserved

More precisely, the construction works as follows: for each vertex v of T_1 there exists a small disk D_v containing v such that the intersection of D_v with T_1 is a simple curve consisting of parts of the two edges incident to v. In particular, the disk D_v is disjoint from all edges that are not incident to v. We refer to D_v as the *vicinity* of v. We may assume without loss of generality that the vertex vicinities are pairwise disjoint. As in the previous step, we consider the edges of T_1 to be directed. Consider a directed edge $e = uv$ of T_1 and let f and g denote the edges that precede and succeed e along T_1, respectively. The vertices of the Kynčl copy K_e that is created for e are denoted by a_e, b_e, c_e and x_e, y_e, z_e, where i_e corresponds to its pendant $i \in \{a, b, c, x, y, z\}$ of Kynčl's example illustrated in Fig. 5. We may assume that the small triangular faces incident to e are to the right side of e at u and to the left side of e at v; note that this property holds for every second edge of T_1; see again Fig. 7. The vertices a_e, y_e, and z_e are placed in D_v, to the left side of the directed path eg. On the other hand, the vertices c_e and x_e are placed in D_u, to the right side of the directed path fe. Finally, the vertex b_e is identified with u.

All intersections between the edges of K_e are placed inside D_v as illustrated in Fig. 9. All edges of K_e cross g in D_v and then follow the edge e closely in order to reach D_u. In particular, we draw the edges close enough to e such that they are disjoint from all vertex vicinities except for D_v and D_u. Note that in this way, the edges pass through all edges of $E(T_1) \setminus \{f, e, g\}$. Finally, inside D_u, the edges of K_e that are non-incident to b_e cross e and then f.

This construction is repeated for every second edge of T_1; recall that T_1 is a cycle of even length. For the remaining edges of T_1, we proceed analogously, except that we use a reflected version of Kynčl's example and we exchange the roles of the two sides of the directed paths eg and fe inside the disks D_u and D_v, as illustrated in Fig. 9, by this ensuring that all additional vertices are located in the small triangular cells. Note that this ensures that each edge e' of K_e crosses each edge of K_f (and K_g) precisely once. Additionally, the edges of the remaining Kynčl copies are intersected by the part of e' that is disjoint from D_u and D_v. This shows that T_2 is indeed a thrackle. Moreover, for each edge of the cycle T_1, we have added four new edges and five new vertices, which results in the claimed edge-vertex-ratio of $\frac{5}{6}$. We will refer to $B_e := E(K_e) \cup \{e\}$ as the *edge bundle of e*. Note that these are exactly the edges that run in parallel close to each other, when outside of D_u or D_v. The *region R_e* of this bundle is the region of $T_2 \setminus (D_u \cup D_v)$ that is enclosed by its outer edges e and $a_e b_e$ (see Fig. 9).

It remains to prove that T_2 is a maximal thrackle. Therefore, we assume by contradiction that there exists a new edge s that can be introduced into T_2 such that $T_2 \cup s$ is a thrackle. To arrive at contradiction, we show the following properties of s.

Property 1. *For every vertex u and edge $e = uv$ of T_1 it holds that a new edge s does not enter D_u within a bundle, i.e., $s \cap R_e \cap \partial D_u = \emptyset$.*

Property 2. *Let e and f be two edges of T_1 sharing an endpoint u. If s has one of its endpoints v in $D_u \setminus \{u\}$, it intersects all edges of $B_e \cup B_f$ inside D_u. Moreover, $v \in \{a_f, y_f, z_f\}$.*

Property 3. *If there exists a new edge s with vertices in T_2 such that $T_2 \cup s$ is a thrackle, then there exists an edge s' such that $T_2 \cup s'$ is a thrackle, the vertices of s' belong to T_1, and the vertices of s' do not share an edge in T_1.*

Proof sketch. Let $UV := s$. If both U, V are vertices of T_1, then the claim is proved. Therefore, we may assume that U does not belong to T_1. Let u denote the vertex of T_1 such that U is contained in D_u; likewise, let v denote the vertex of T_1 such that V is contained in D_v. When constructing T_2 from T_1, we ensure to place all new vertices in the small triangular faces incident to each vertex of T_1, see Fig. 7. Due to this placement, it may be derived from Property 2 that $u \neq v$.

We now show that u and v do not share an edge in T_1. Suppose for a contradiction, that $e := uv$ is an edge of T_1. If $U \neq u$ and $V \neq v$, then by Property 2, s intersects all edges of B_e in both D_u and D_v; a contradiction. Similarly, if $U \neq u$ and $V = v$, then s intersects all edges of B_e in D_u and $e = Uv$ in D_v ; a contradiction. Consequently, u and v do not share an edge.

Now we use the fact that s intersects all edges present in D_u (by Property 2) to reroute s inside D_u. As before, let the sections of e and f inside D_u partition D_u in its top and bottom half.

Let $w_1, w_2, \ldots, w_k$ denote the sequence of intersections of s with ∂D_u. Since the vertex U of s is inside D_u, k is an odd integer. Moreover, by Properties 1 and 2, no section $w_{2i-1}w_{2i}$ connects the top and bottom half. Consequently, $w_1 w_2, \ldots, w_{k-2}w_{k-1}$ form pairs contained in the top or bottom part that are additionally nested since s has no self-intersections. We replace the sections $w_{2i-1}w_{2i}$ of s by curves close to the boundary of D_U such that no edge of D_u is intersected.

The last part $w_k U$ we reroute as follows, see also Fig. 10: If w_k is contained in the top half of D_u, we replace the part of s inside D_u by a straight line segment that connects u and $\partial D_u \cap s$; note that this segment intersects all edges in D_u. If w_k is contained in the bottom half of D_u, we replace $w_k U$ inside D_u with a curve from u to $\partial D_u \cap s$ as illustrated; note that this curve intersects all edges of D_u.

After this replacement, the new edge s' intersects the same set of edges as s. Therefore, $T_2 + s'$ is a thrackle. Moreover, the vertex U of s is replaced by the vertex u of s' where u is in T_1. If $V \neq v$, we apply the same rerouting for the other vertex V of s.   ◁

Property 3 implies that if T_1 is maximal, then T_2 is maximal. Therefore, Proposition 2 implies that T_2 is a maximal thrackle with $\varepsilon(T_2) = \frac{5}{6}$. This completes the proof of Theorem 3.

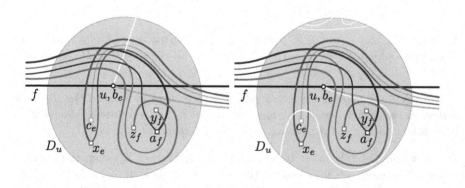

Fig. 10. Illustration of Property 3.

5   Ongoing Work and Open Problems

We believe that by *repeating the Kynčl belt construction*, one obtains a class of maximal trackles such that for every c, there exists maximal thrackle T with $\varepsilon(T) < \frac{4}{5} + c$. The idea is as follows: Since the original edges of T_1 are preserved in T_2, we can apply the Kynčl belt construction to T_2 by using only the edges of T_1. This results in a thrackle T_3. To do this, we find new, smaller vicinities around every vertex of T_1 which are free of other vertices and non-incident edges. For an illustration, consider Fig. 11. By repeating the procedure k times, we obtain a trackle T_k with

$$\varepsilon(T_k) = \frac{2n + 1 + 4k}{2n + 1 + 5k} = \frac{4}{5} + \frac{2n + 1}{10n + 5 + 25k} < \frac{4}{5} + c \Leftrightarrow k > \frac{(1 - 5c)(2n + 1)}{25c}.$$

Showing that T_k is (potentially) maximal is more involved and ongoing work, in which we are done with proving most appearing cases.

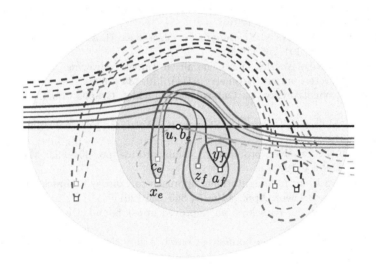

Fig. 11. Applying the Kynčl belt construction multiple times.

We conclude with a list of interesting open problems:

- What is the minimal number of edges that a maximal thrackle without isolated vertices can have? Can such a maximal thrackle T have $\varepsilon(T) < \frac{4}{5}$?
- Is it true that for every maximal thrackle T it holds that $\varepsilon(T) > \frac{1}{2}$ or do maximal matching thrackles (other than $K_{1,1}$) exist? It has been very recently shown [6] that geometric matching thrackles are not maximal. The question remains open for topological thrackles.
- Does Conway's conjecture hold?

Acknowledgements. This research was initiated during the 15th European Research Week on Geometric Graphs (GGWeek 2018) at Haus Tornow am See (Märkische Schweiz, Germany) and Freie Universität Berlin. The workshop was supported by the Deutsche Forschungsgemeinschaft (DFG) through the Research Training Network *Facets of Complexity* and the collaborative DACH project *Arrangements and Drawings*. We thank the organizers and all participants for the stimulating atmosphere. In particular, we thank André Schulz for proposing the study of maximal thrackles as a research question, and Viola Mészáros and Stefan Felsner for joining some of our discussions and contributing valuable ideas.

Within the collaborative DACH project *Arrangements and Drawings*, O.A. and B.V. were partially supported by the Austrian Science Fund (FWF) under grant I 3340-N35 and F.S. was partially supported by the DFG under grant FE 340/12-1.

References

1. Aichholzer, O., Kleist, L., Klemz, B., Schröder, F., Vogtenhuber, B.: On the edge-vertex ratio of maximal thrackles. https://arxiv.org/abs/1908.08857v2
2. Cairns, G., Koussas, T., Nikolayevsky, Y.: Great-circle spherical thrackles. Discrete Math. **338**(12), 2507–2513 (2015)
3. Cairns, G., Nikolayevsky, Y.: Bounds for generalized thrackles. Discrete Comput. Geometry **23**(2), 191–206 (2000)
4. Cairns, G., Nikolayevsky, Y.: Generalized thrackle drawings of non-bipartite graphs. Discrete Comput. Geometry **41**(1), 119–134 (2009)
5. Cairns, G., Nikolayevsky, Y.: Outerplanar thrackles. Graphs Comb. **28**(1), 85–96 (2012)
6. Cleve, J., Mulzer, W., Perz, D., Steiner, R., Welzl, E.: Personal communication, August 2019
7. Conway, J.H.: Unsolved problems in combinatorics, pp. 351–363. Mathematical Institute, Oxford (1972)
8. Fulek, R., Pach, J.: A computational approach to Conway's Thrackle Conjecture. Comput. Geom. Theor. Appl. **44**(6–7), 345–355 (2011)
9. Fulek, R., Pach, J.: Thrackles: an improved upper bound. Discrete Appl. Math. **259**, 226–231 (2019)
10. Goddyn, L., Xu, Y.: On the bounds of Conway's thrackles. Discrete Comput. Geom. **58**(2), 410–416 (2017)
11. Hajnal, P., Igamberdiev, A., Rote, G., Schulz, A.: Saturated simple and 2-simple topological graphs with few edges. J. Graph Algorithms Appl. **22**(1), 117–138 (2018)
12. Kynčl, J.: Improved enumeration of simple topological graphs. Discrete Comput. Geom. **50**, 727 (2013). https://doi.org/10.1007/s00454-013-9535-8
13. Kynčl, J., Pach, J., Radoičić, R., Tóth, G.: Saturated simple and k-simple topological graphs. Comput. Geom. **48**(4), 295–310 (2015). https://doi.org/10.1016/j.comgeo.2014.10.008
14. Li, W., Daniels, K., Rybnikov, K.: A study of Conway's Thrackle Conjecture. Vertex **2**(4), 1 (2006)
15. Lovász, L., Pach, J., Szegedy, M.: On Conway's thrackle conjecture. Discrete Comput. Geom. **18**(4), 369–376 (1997)
16. Misereh, G., Nikolayevsky, Y.: Annular and pants thrackles. Discrete Math. Theor. Comput. Sci. **20**(1) (2018). https://doi.org/10.23638/DMTCS-20-1-16

17. Pach, J., Radoičić, R., Tóth, G.: Tangled thrackles. In: Márquez, A., Ramos, P., Urrutia, J. (eds.) EGC 2011. LNCS, vol. 7579, pp. 45–53. Springer, Heidelberg (2012). https://doi.org/10.1007/978-3-642-34191-5_4
18. Pach, J., Sterling, E.: Conway's conjecture for monotone thrackles. Am. Math. Mon. **118**(6), 544–548 (2011)
19. Pammer, J.: Rotation Systems and Good Drawings, pp. 1–83. TUGraz (2014)
20. Ruiz-Vargas, A.J., Suk, A., Tóth, C.D.: Disjoint edges in topological graphs and the tangled-thrackle conjecture. Eur. J. Comb. **51**, 398–406 (2016)
21. Wehner, S.: On the thrackle problem (2013). http://www.thrackle.org/thrackle.html
22. Woodall, D.: Thrackles and deadlock. Comb. Math. Appl. **348**, 335–347 (1969)

Best Paper in Track 2

Symmetry Detection and Classification
in Drawings of Graphs

Felice De Luca[ID], Md. Iqbal Hossain[(✉)][ID], and Stephen Kobourov[ID]

Department of Computer Science, University of Arizona, Tucson, USA
{felicedeluca,hossain,kobourov}@cs.arizona.edu

Abstract. Symmetry is a key feature observed in nature (from flowers and leaves, to butterflies and birds) and in human-made objects (from paintings and sculptures, to manufactured objects and architectural design). Rotational, translational, and especially reflectional symmetries, are also important in drawings of graphs. Detecting and classifying symmetries can be very useful in algorithms that aim to create symmetric graph drawings and in this paper we present a machine learning approach for these tasks. Specifically, we show that deep neural networks can be used to detect reflectional symmetries with 92% accuracy. We also build a multi-class classifier to distinguish between reflectional horizontal, reflectional vertical, rotational, and translational symmetries. Finally, we make available a collection of images of graph drawings with specific symmetric features that can be used in machine learning systems for training, testing and validation purposes. Our datasets, best trained ML models, source code are available online.

1 Introduction

The surrounding world contains symmetric patterns in objects, animals, plants and celestial bodies. A symmetric feature is defined by the repetition of a pattern along one of more axes, called *axes of symmetry*. Depending on how the repetition occurs the symmetry is classified as *reflection* when the feature is reflected across the reflection axis, and *translation* when the pattern is shifted in the space. Special cases of reflection symmetries are horizontal (reflective) symmetry when the axis of symmetry is horizontal or a vertical (reflective) symmetry when such axis is vertical. Rotational symmetries occur when the translational axes of symmetry are radial.

Symmetry has been studied in many different fields such as psychology, art, computer vision, and even graph drawing. In psychology, for example, studies on the impact of symmetry on humans show that the vertical symmetry in objects is perceived pre-attentively. A similar study conducted in the context of graph drawing also shows that the vertical symmetry in drawings of graphs is best perceived among all others [8]. In this context, algorithms to measure symmetries in graph drawings have been proposed although it has been shown that these measures do not always agree with what humans perceive as symmetric [34].

© Springer Nature Switzerland AG 2019
D. Archambault and C. D. Tóth (Eds.): GD 2019, LNCS 11904, pp. 499–513, 2019.
https://doi.org/10.1007/978-3-030-35802-0_38

Convolutional Neural Networks (CNN) have become a standard image classification technique [18]. CNNs automatically extract features by using information about adjacent pixels to down-sample the image in the first layers, followed by a prediction layer at the end.

Led by the lack of a reliable way to identify a symmetric layout and eventually classify it by the symmetry it contains, in this paper we consider CNNs for the detection and classification of symmetries in graph drawing. Specifically we consider the following two problems: (i) Binary classification of symmetric and non-symmetric layout; and (ii) multi-class classification of symmetric layouts by their type: horizontal, vertical, rotational, translational. In particular, our contributions are as follows:

1. We describe a machine learning model that can be used to determine whether a given drawing of a graph has reflectional symmetry or is not-symmetric (binary classification). This model provides 92% accuracy on our test dataset.
2. We describe a multi-class classification model to determine whether a given drawing of a graph has vertical, horizontal, rotational, or translational symmetry. This model provides 99% accuracy on our test dataset.
3. We make available training datasets, as well as the algorithms to generate them.

The full version of this paper contains more details, figures and tables [7].

2   Related Work

Symmetry detection has applications in different areas such as computer vision, computer graphics, medical imaging, and robotics. Competitions for symmetry detection algorithms have taken place several times; for example, see Liu et al. [20]. For reflection and translation symmetries the problem can be interpreted as computing one or more axes of symmetry [17]. In the context of graph drawing, symmetry is one of the main aesthetic criteria [26].

Symmetry Detection and Computer Vision: Detection of symmetry is an important subject of study in computer vision [1,21,24]. The last decades have seen a growing interest in this area although the study of bilateral symmetries in shapes dates back to the 1930s [2]. The main focus is on the detection of symmetry in real-world 2D or 3D images. As Park et al. [25] point out, although symmetry detection in real-world images has been widely studied it still remains a challenging, unsolved problem in computer vision. The method proposed by Loy and Eklundh [22] performed best in a competition for symmetry detection [20] and is considered a state-of-the-art algorithm for computer vision symmetry detection [6,25]. Symmetries in 2D points set have also been studied and Highnam [11] proposes an algorithm for discovering mirror symmetries. More recently, Cicconet et al. [6] proposed a computer vision technique to detect the line of reflection (mirror) symmetry in 2D and the straight segment that divides the symmetric object into its mirror symmetric parts. Their technique outperforms the winner of the 2013 competition [20] on single symmetry detection.

Symmetry Detection and Graphs: In graph theory the symmetry of a graphs is known as automorphism [23] and testing whether a graph has any axial symmetry is an NP-complete problem [3]. A mathematical heuristic to detect symmetries in graphs is given in [9]. Klapaukh [15,16] and Purchase [26] describe algorithms for measuring the symmetry of a graph drawing. While the first measure analyzes the drawing to find reflection, rotation and translation symmetries, the latter considers only the reflection. Welsh and Kobourov [34] evaluate how well the measures of symmetry agree with human evaluation of symmetry. The results show that in cases where the Klapaukh and Purchase measures strongly disagreed on the scoring of symmetry, human judgment agrees more often with the Purchase metric.

Symmetry Detection and Machine Learning: Convolutional neural networks can be a powerful tool for the automatic detection of symmetries. Vasudevan *et al.* [33] use this approach for the detection of symmetries in atomically resolved imaging data. The authors train a deep convolutional neural network for symmetry classification using 4000 simulated images, 3 convolutional layers, a fully connected layer, and a final "softmax" output layer on this training dataset. After training over 30 epochs, the authors obtained an accuracy of 85% on the validation set. Tsogkas and Kokkinos [32] propose a learning-based approach to detect symmetry axes in natural images, where the symmetry axes are contours lying in the middle of elongated structures. To the best of our knowledge, there are no prior machine learning approaches for detecting or classifying symmetries in graph drawings.

Neural Networks for Image Classification and Detection: Convolutional Neural Networks (CNNs) are standard in image recognition and classification, object detection, and video analysis. The Mark I Perception machine was the first implementation of the perceptron algorithm in 1957 by Rosenblatt [27]. Widrow and Hoff proposed a multilayer perceptron [35]. Back-propagation was introduced by Rumelhart *et al.* [28]. LeNet-5 [19] was deployed for zip code and digit recognition. In 2012, Alex Krizhevsky [18] introduced CNNs with AlexNet. Szegedy *et al.* [14] introduced GoogLeNet and the Inception module. Other notable developments include VGGNet [30] and residual networks (ResNet) [10].

3 Background and Preliminaries

In this section we give a brief overview of machine learning in the context of our experiments. We also attempt to clarify some of the terminology we use throughout the paper, focusing in particular on *Deep Neural Networks* and *Convolutional Neural Networks*.

A deep neural network is made of several layers of neurons. Information flows through a neural network in two ways: via the *feedforward network* and via *backpropagation*. During the training phase, information is fed into the network via the input units, which trigger the layers of hidden units, and these in turn arrive at the output units. This common design is called a *feedforward network*. Not all units fire all the time. Each unit receives inputs from the units of the previous layer, and the inputs are multiplied by the weights of the connections they travel along. Every unit adds up all the inputs it receives in this way and if the sum exceeds a certain threshold value, the unit fires and triggers the units it is connected to in the next layer.

Importantly, there is a feedback process called *backpropagation* that can be used to improve the weights. This involves the comparison of the output the network produces with the output it was meant to produce, and using the difference between them to modify the weights of the connections between the units in the network, working from the output units, through the hidden units, and to the input units. Over time, backpropagation helps the network to "learn," reducing the difference between actual and intended outputs.

Convolutional neural network (CNN) are used mainly for image data classification where intermediate layers and computations are a bit different then fully connected neural networks. Each pixel of input image is mapped with a neuron of the input layer. Output neurons are mapped to target classes. Figure 1 shows a simple CNN architecture. Different types of layers in a typical CNN include:

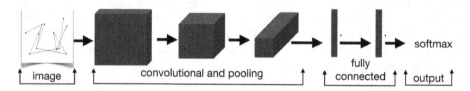

Fig. 1. A typical convolutional neural network.

- *convolution layer (convnet)*: in this layer a small filter (usually 3 × 3) is taken and moved over the image. Applying filters in the layer helps to detect low and high level features in the image so that spatial features are preserved in the layer. The convolutional layer helps to reduce the number of parameters compared to a fully connected layer. Keeping the same set of filters helps to share parameters and sparsity helps to further reduce the parameters. For example, in a 3 × 3 filter every node in the next layer is only connected to 9 nodes in the previous layer. This sparse connection helps to avoid over-fitting.
- *activation layer*: this layer applies an activation function from the previous layer. Example functions include ReLU, tanh and sigmoid.
- *pooling*: the pooling layer is used to reduce size of the convnet. Filter size f, stride s, padding p are used as parameters of the pooling layer. Average pooling or max pooling are the standard options. After applying the pooling to a given image shape $(N_h \times N_w \times N_c)$, it turns into $\lfloor \frac{N_h-f}{s}+1 \rfloor \times \lfloor \frac{N_w-f}{s}+1 \rfloor \times N_c$.

– *Fully Connected Layer* (FCL): a fully connected layer creates a complete bipartite graph with the previous layer. Adding a fully-connected layer is useful when learning combinations of non-linear features.

We now review some common machine learning terms. *Training loss* is the error on the training set of data, and *validation loss* is the error after running the validation set of data through the trained network. Ideally, train loss and validation loss should gradually decrease, and training and validation accuracy should increase over training epochs. The *training set* is the data used to adjust the weights on the neural network. The *validation set* is used to verify that increase in accuracy over the training data actually yields an increase in accuracy. If the accuracy over the training data set increases, but the accuracy over the validation data decreases, it is a sign of *overfitting*. The *testing set* is used only for testing the final solution in order to confirm the actual predictive power of the network. A *confusion matrix* is a table summarizing the performance in classification tasks. Each row of the matrix represents the instances in a predicted class while each column represents the instances in an actual class. The *precision* p represents how many selected item are relevant and *recall* r represents how many relevant items are selected. $F1$ *-score* is measured by the formula $2 * \frac{r*p}{p+r}$.

4   Datasets

In this section we describe how we generated datasets for our machine learning systems. To the best of our knowledge, there is no dataset of images suitable for training machine learning systems for symmetry detection in graph drawings. Our dataset contains images that feature different types of symmetries, including reflection, translation or rotation symmetries and variants thereof. An overview all types of layouts is given in Fig. 2.

We started with a dataset of simple symmetric images and inspected the results trying to identify which characteristic of the layout leads to its classification as symmetric or not symmetric. If we observed a characteristic in the symmetric layouts we generated non symmetric layouts that expose it and symmetric layouts without it. Then we fed them to the system for the classification. In case of inaccurate results we included the new layouts (that we call *breaking instances* of the dataset) in the training system and repeated the process until we could not identify any other specific feature.

In order to distinguish inputs of different sizes, we refer to layouts in our dataset as *small* or *large* based on the number of vertices, $|V|$. A small layout has $|V| \in [5, 8]$ while a large layout has $|V| \in [10, 20]$. The number of edges is a random integer $|E| \in [|V|, \lfloor 1.2 \rfloor * |V|]$. The layouts included in the global dataset used for all experiments can be summarized as follows:

– *SmallSym*: small reflective symmetric layout
– *SmallNonSym*: non symmetric generated from SmallSym with random node positions

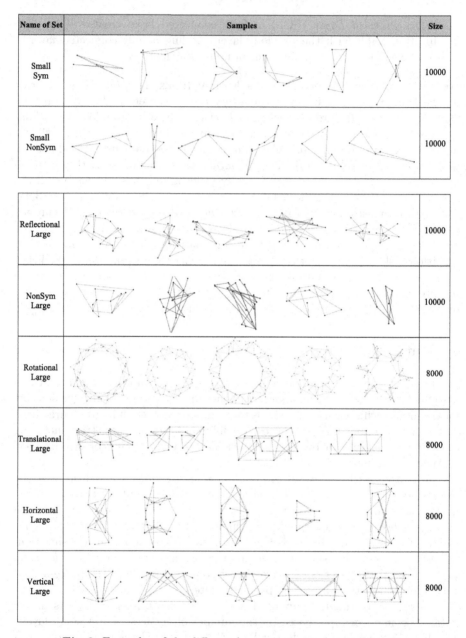

Name of Set	Samples	Size
Small Sym		10000
Small NonSym		10000
Reflectional Large		10000
NonSym Large		10000
Rotational Large		8000
Translational Large		8000
Horizontal Large		8000
Vertical Large		8000

Fig. 2. Examples of the different layout instances in our dataset.

- *ReflectionalLarge*: large reflective symmetric layouts with random axis of symmetry
- *NonSymLarge*: non symmetric generated from ReflectionalLarge layouts
- *HorizontalLarge*: large reflective symmetric layouts with a 0° axis of symmetry

- *VerticalLarge*: large reflective symmetric layouts with a 90° axis of symmetry
- *RotationalLarge*: rotational symmetric with random axes between 4 and 10
- *TranslationalLarge*: translational symmetric translated along x-axis

In the remainder of this section we discuss how we generated our layouts and the process that led to them.

4.1 Reflectional Layout Generation

A reflectional symmetric layout may expose different characteristics such as "parallel lines" orthogonal to the axis of symmetry and edge crossings on the axis of symmetry.

The generation procedure for symmetric graphs and layouts thereof differs slightly depending on the type of symmetry we attempt to capture.

We used the procedure for generating a graph and a reflectional symmetric layout with the "parallel lines" feature following the algorithm in [8] as follows. Given a graph with $\frac{n}{2}$ vertices, called a *component*, we assign to each vertex of the component positive random coordinates. Then we copy this component and replace the x-coordinates of each vertex with the negative value of the original. This results in a layout with two disjoint components that are then connected by a random number of edges in $[1, \lfloor |V|/3 \rfloor]$ selecting random vertices in one component and connecting them to their corresponding vertices in the other component. This results in layouts with vertical axis of symmetry; see Fig. 3(b). To create layouts with horizontal axis of symmetry we add a 90° rotation; see Fig. 3(a).

The procedure for generating a graph and a reflectional symmetric layout without the "parallel lines" feature is described in Algorithm *SymGG*. This algorithm gives an overview on how to create the symmetric versions with the different features. In the following we explain how we defined *SymGG* based on experimental improvements of our dataset. Given a symmetric graph with n vertices by Algorithm *SymGG*, we create a non-symmetric layout by assigning to each vertex of the input graph any random y-coordinate and a positive random x-coordinate to the vertices with identifier $< \frac{n}{2}$ and a negative random x-coordinate, otherwise.

To create reflectional symmetric layouts, instead, if a vertex with identifier $i < \frac{n}{2}$ gets coordinates (x_r, y_r) then the vertex with identifier $i_c = i + \frac{n}{2}$ gets assigned coordinates $(-x_r, y_r)$. If the graph has an odd number of vertices then the vertex with identifier $n - 1$ gets $x = 0$. Note that, by construction, the resulting layouts have a vertical axis of symmetry; see Fig. 3(e). To create layouts with horizontal axis of symmetry we add a 90° rotation; see Fig. 3(f).

4.2 Dataset Definition

Here we describe the process that led to us to the dataset of reflectional symmetric layouts.

To this aim we generated the SmallSym, SmallNonSym, NonSymLarge and ReflectionalLarge layouts.

Algorithm SymGG(n,m): Symmetric graph generation with n vertices and m edges

---

1: define $G = (V, E)$ where $|V| = n$ with id $[0, n - 1]$ and $|E| = 0$
2: add m edges to G selecting one or more edge types from [3-6] and continuing with steps [7-12]
3: for a random edge choose random integers u, v in $[0, n - 1]$ such as $(u, v) \notin E$;
4: for a random edge that does not cross the axis of reflection choose random integers u, v in $[0, \lfloor \rfloor * n/2 - 1]$ such as $(u, v) \notin E$;
5: for parallel edge feature choose random integer u in $[0, \lfloor \rfloor * n/2 - 1]$ and $v = u + \lfloor \rfloor * n/2$ such as $(u, v) \notin E$;
6: for crossing edge feature choose random integer u in $[0, \lfloor \rfloor * n/2 - 1]$ and v in $[n/2, n - 1]$ such as $(u, v) \notin E$;
7: Generate the symmetric edge (u_sym, v_sym) of (u, v)
8: $u\_sym = u \mp \lfloor \rfloor * n/2$ if $u \gtrless \lfloor \rfloor * n/2$
9: $v\_sym = v \mp \lfloor \rfloor * n/2$ if $v \gtrless \lfloor \rfloor * n/2$
10: $u\_sym = u$ if n is odd and $u = n - 1$
11: $v\_sym = v$ if n is odd and $v = n - 1$
12: add (u, v) and (u_sym, v_sym) to E

First Improvement: At first, we trained our system with the reflective symmetric layouts and random layouts generated using the approach in [8] as described above.

Observations: Using this simple dataset we observed that the system could always classify the layouts correctly for any of the used layouts.

Layouts Characteristic: Analyzing the used dataset we observed that the generation algorithm used gives symmetric layout for reflective symmetry with a clear symmetric feature that is 'parallel lines' orthogonal to the reflection axis. These lines separate two identical but reflected subcomponents, as Fig. 3(a-b) show.

Breaking Layout: After identifying the 'parallel lines' feature, we generated non-symmetric layouts with the same feature. These layouts were created starting from the symmetric layouts and then assigning random positions to the vertices not linked to the parallel edges; an example of random layout with parallel edges is shown in Fig. 4b. Without re-training the system, these layouts are misclassified as symmetric, breaking the previously built model.

Second Improvement: Here we added to our dataset the breaking instances of the previous model and new symmetric layouts that do not show the 'parallel lines' feature. The parallel lines of a symmetric layouts are given by vertices that are connected to their reflected copy (since they share either the x or y coordinate in the space). The new layouts we generated have the two subcomponents not only connected by edges between a vertex and his reflected copy but also by edges connecting a random vertex of one component to a random vertex of the

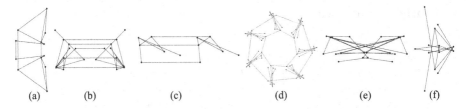

Fig. 3. Symmetric layouts in the dataset: (a) Horizontal, (b) Vertical, (c) Translational, (d) Rotational, (e) Vertical without parallel lines, (f) Horizontal without parallel lines.

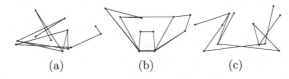

Fig. 4. Non symmetric layouts in the dataset: (a) Random, (b) with Parallel Lines, (c) with Crossings

other (and viceversa to keep the symmetry). These edges generate crossings on the axis of symmetry of the symmetric layout, instead of the parallel lines. Pseudocode for the symmetric graph generation Algorithm SymGG (with even number of edges as input) can be found above.

Analogously we generated some random layouts that show the same feature, starting from a symmetric layout with non-parallel edges and shuffling the position of the vertices not connected to such edges. Figure 3(e) illustrates and example of symmetric layout with crossings while Fig. 4(c) depicts a non symmetric layout with crossings.

Observations: Training the system with these new layouts we obtained good results on all layouts, including those misclassified in the previous setup.

Breaking Instance: Inspecting the current dataset we identified another characteristic of the current symmetric layouts: an even number of vertices. We then generates symmetric layouts with an odd number of vertices. The generation algorithm for these layouts is given in Algorithm SymGG. Again, without training, the the current system fails on such layouts misclassifying them as a non-symmetric. Further, we observed that rotating the symmetric layouts also makes our machinery fail.

Final Improvement: Here we added to our dataset instances with odd number of vertices for both symmetric and non symmetric layouts. We also added instances rotated by a random angle between 0 and 360. Since we could not find further breaking instance for this dataset, we used it for our experiments.

4.3   Other Symmetric Layouts

In addition to the instances above we generated the translational layouts and rotational layouts using the algorithm in [8], as follows.

To create translational symmetric layouts we use the same process of generation of reflectional symmetric layout with parallel edges above but instead of taking the negative value of the x-coordinate of the copied component we shift each component by a predefined value δ. If a vertex in the given component gets coordinates (x, y) then the vertex in the copied component is assigned coordinates $(x - \delta, y)$; see Fig. 3(c).

The generation process for rotational symmetric layouts is different, since the number of vertices depends on the number of symmetric axes. To generate such layouts we start from a given graph component with n vertices and then we select a random number of radial symmetric axes in the range $[4, 10]$. After assigning a random position to the vertices of the component we copy and shift it over the reflection axes. Then we choose two random vertices in the component and use them to connect pairs of rotationally consecutive components; see Fig. 3(d).

5   Experimental Setup

Our images are in black and white with a size of 200×200 pixels. We use 1 pixel for the edge width and 3×3 pixels for a vertex. We configured our system with the following settings: 1 grayscale channel, with resealing by $1/255$, batch size 16 and number of epochs 20. In all of our experiment we use 80% of our data as training set, 10% as validation set, and 10% as test set. Test sets are never used in during training, those are reserved for computing the final accuracy. During training, in every epoch we check the validation accuracy and save the best trained model as checkpoint. The best trained model is used on the final test set.

Since images of graph drawings have different features than that of real-world images (e.g., textures and shapes), we tested different popular CNN architectures with same parameter settings.

We use CNN architectures from the Keras implementation; Keras is a high-level API of Tensorflow that supports training with multiple CPUs[1]. For our experiments, we used the High Performance Computing system at the University of Arizona. Specifically, training was done on 28 CPUs, each with Intel Xeon 3.2 GHz processor and 6 GB of memory. Training time for the different models ranged from 6 to 29 h; see Table 1.

6   Detecting Reflectional Symmetry

Small Binary Classification (SPBC) **Experiment:** In this experiment we test how accurately we can distinguish between drawings of graphs with

---

[1] https://github.com/keras-team/keras/tree/master/keras/applications.

Table 1. Overview of the CNN models used in the experiment.

Name	Parameters	Layers	References	Our training time (h)
ResNet50	23.59M	177	[10]	15.25
MobileNet	3.23M	93	[12]	6.22
MobileNetV2	2.26M	157	[29]	8.36
NASNetMobile	4.27M	771	[36]	5.79
NASNetLarge	84.93M	1041	[36]	10.21
VGG16	107.01M	23	[30]	24.24
VGG19	112.32M	26	[4]	25.32
Xception	20.87M	134	[5]	19.59
InceptionResNetV2	54.34M	782	[31]	15.18
DenseNet121	7.04M	429	[13]	20.11
DenseNet201	18.32M	709	[13]	28.49

reflectional symmetry and ones without. We use a binary classifier trained on *SmallSym* and *SmallNonSym* instances from our dataset; see Fig. 2. We use the *InceptionResNet* CNN model with 12000 images for training, 2000 images for validation, and 2000 image for testing. The model achieves 92% accuracy. We evaluated several different models before settling on *InceptionResNet*; see the full paper for more details [7].

We cross-validate our results with two earlier metrics specifically designed to evaluate the symmetry in drawing of graphs, namely the Purchase metric [26] and the Klapaukh metric [15]. These two metrics were not designed for binary classification, but given a graph layout they provide a score in the range $[0, 1]$. We interpret a score of ≥ 0.5 as a vote for "symmetry" and a score of < 0.5 as a vote of "no symmetry". We can now compare the performance of our CNN model against those of the Purchase metric and the Klapaukh metric on the same set of 2000 test images. We report accuracy, precision, recall and F1-score in Table 2. We can see that while the two older metrics perform well, the CNN is better in all aspects (except recall, where the Purchase metric is .01% better).

Table 2. Comparison between the CNN model and existing symmetry metrics.

Model	Accuracy	Precision	recall	F1-Score
Purchase [26]	82%	0.67	0.96	0.79
Klapaukh [15]	82%	0.80	0.86	0.83
InceptionResNet	92%	0.90	0.95	0.93

Training loss, validation loss, training accuracy and validation accuracy for our *E*xperiment SPBC are shown in the full version of the paper [7].

7 Detecting Different Types of Symmetries

Multi-class symmetric layouts classification (LHVRT) Experiement:
In this experiment we test how accurately we can distinguish between drawings of graphs with different types of symmetries. We use a multi-class classifier trained on several types of symmetries: Horizontal, Vertical, Rotational and
Translational. Recall that Horizontal and Vertical are special cases of reflection
symmetry, where the axis of reflection is horizontal or vertical, respectively.

We train the CNN with *HorizontalLarge, VerticalLarge, RotationalLarge*, and
TranslationalLarge instances from our dataset; see Fig. 2.

We use the *ResNet50* CNN model with 16000 images for training, 2000 images
for validation, and 4480 images for testing. The model achieves 99% accuracy.
Table 3 shows the corresponding confusion matrix. We evaluated several different models before settling on *ResNet50*. Training loss, validation loss, training
accuracy and validation accuracy for our *E*xperiment LHVRT are shown in the
full version of the paper, where more results and discussion thereof can also be
found [7].

Table 3. Confusion matrix from *ResNet50*. Each row of the matrix represents the
instances in a predicted class while each column represents the instances in an actual
class.

	HorizontalLarge	RotationalLarge	TranslationalLarge	VerticalLarge
HorizontalLarge	1280	0	0	0
RotationalLarge	0	800	0	0
TranslationalLarge	0	0	798	2
VerticalLarge	0	0	1	1599

8 Conclusions

In the experiments above we achieved high accuracy for both detection and
classification. Compared to existing evaluation metrics for symmetric layout we
observed that our machinery outperforms the mathematical formulae proposed
when used as classifiers.

Note, however, that there are many limitations to consider. First of all, we
generated all the datasets and have not tested the models on layouts obtained
from other layout algorithms. Further, the graphs we used are small and we have
not confirmed how well humans agree with the decisions of the machine learning
system. Finally, the two tasks we performed are limited in power, and we do
not yet have a model that can accurately predict whether a graph drawing is
symmetric or not, or which of two drawings of the same graph is more symmetric.

Nevertheless, we believe our dataset can be useful for future experiments and
our initial results on limited tasks indicate that a machine learning framework

can be useful for symmetry detection and classification. Our dataset, models, results details can be found in https://github.com/enggiqbal/mlsymmetric.

Acknowledgement. This work is supported in part by NSF grants CCF-1740858, CCF-1712119, DMS-1839274, and DMS-1839307. This experiment uses High Performance Computing resources supported by the University of Arizona TRIF, UITS, and RDI and maintained by the University of Arizona Research Technologies department.

References

1. Atallah, M.J.: On symmetry detection. IEEE Trans. Comput. **C−34**(7), 663–666 (1985). https://doi.org/10.1109/TC.1985.1676605
2. Birkhoff, G.D.: Aesthetic Measure. Cambridge (1932)
3. Manning, J.B.: Geometric symmetry in graphs. ETD Collection for Purdue University (1990)
4. Chen, L., Zhang, H., Xiao, J., Nie, L., Shao, J., Liu, W., Chua, T.S.: SCA-CNN: spatial and channel-wise attention in convolutional networks for image captioning. In: Proceedings - 30th IEEE Conference on Computer Vision and Pattern Recognition, CVPR 2017, vol. 2017, pp. 6298–6306 (2017). https://doi.org/10.1109/CVPR.2017.667
5. Chollet, F.: Xception: deep learning with depthwise separable convolutions. In: Proceedings - 30th IEEE Conference on Computer Vision and Pattern Recognition, CVPR 2017. vol. 2017, pp. 1800–1807 (2017). https://doi.org/10.1109/CVPR.2017.195
6. Cicconet, M., Birodkar, V., Lund, M., Werman, M., Geiger, D.: A convolutional approach to reflection symmetry. Pattern Recognit. Lett. **95**, 44–50 (2016). https://doi.org/10.1016/j.patrec.2017.03.022
7. De Luca, F., Hossain, M.I., Kobourov, S.: Symmetry detection and classification in drawings of graphs. arXiv preprint arXiv:1907.01004 (2019)
8. De Luca, F., Kobourov, S., Purchase, H.: Perception of symmetries in drawings of graphs. In: Biedl, T., Kerren, A. (eds.) GD 2018. LNCS, vol. 11282, pp. 433–446. Springer, Cham (2018). https://doi.org/10.1007/978-3-030-04414-5_31
9. de Fraysseix, H.: An heuristic for graph symmetry detection. In: Kratochvíyl, J. (ed.) GD 1999. LNCS, vol. 1731, pp. 276–285. Springer, Heidelberg (1999). https://doi.org/10.1007/3-540-46648-7_29
10. He, K., Zhang, X., Ren, S., Sun, J.: Deep residual learning for image recognition. In: Proceedings of the IEEE Computer Society Conference on Computer Vision and Pattern Recognition, vol. 2016, pp. 770–778 (2016). https://doi.org/10.1109/CVPR.2016.90
11. Highnam, P.T.: Optimal algorithms for finding the symmetries of a planar point set. Technical report 5, Carnegie Mellon University, Pittsburgh, PA, August 1986. https://doi.org/10.1016/0020-0190(86)90097-9
12. Howard, A.G., Zhu, M., Chen, B., Kalenichenko, D., Wang, W., Weyand, T., Andreetto, M., Adam, H.: MobileNets: efficient convolutional neural networks for mobile vision applications. arXiv preprint arXiv:1704.04861 (2017)
13. Huang, G., Liu, Z., Van Der Maaten, L., Weinberger, K.Q.: Densely connected convolutional networks. In: Proceedings - 30th IEEE Conference on Computer Vision and Pattern Recognition, CVPR 2017. vol. 2017, pp. 2261–2269 (2017). https://doi.org/10.1109/CVPR.2017.243

14. Ioffe, S., Szegedy, C.: Batch normalization: accelerating deep network training by reducing internal covariate shift. In: 32nd International Conference on Machine Learning, ICML 2015, vol. 1, pp. 448–456 (2015)
15. Klapaukh, R.: An Empirical Evaluation of Force-Directed Graph Layout. Ph.D. thesis, Victoria University of Wellington (2014)
16. Klapaukh, R., Marshall, S., Pearce, D.: A symmetry metric for graphs and line diagrams. In: Chapman, P., Stapleton, G., Moktefi, A., Perez-Kriz, S., Bellucci, F. (eds.) Diagrams 2018. LNCS (LNAI), vol. 10871, pp. 739–742. Springer, Cham (2018). https://doi.org/10.1007/978-3-319-91376-6_71
17. Kokkinos, I., Maragos, P., Yuille, A.: Bottom-up amp;amp; top-down object detection using primal sketch features and graphical models. In: 2006 IEEE Computer Society Conference on Computer Vision and Pattern Recognition (CVPR 2006), vol. 2, pp. 1893–1900, June 2006. https://doi.org/10.1109/CVPR.2006.74
18. Krizhevsky, A., Sutskever, I., Hinton, G.E.: ImageNet classification with deep convolutional neural networks. In: Advances in Neural Information Processing Systems, vol. 2, pp. 1097–1105 (2012)
19. LeCun, Y., Bottou, L., Bengio, Y., Haffner, P.: Gradient-based learning applied to document recognition. Proc. IEEE 86(11), 2278–2323 (1998). https://doi.org/10.1109/5.726791
20. Liu, J., Slota, G., Zheng, G., Wu, Z., Park, M., Lee, S., Rauschert, I., Liu, Y.: Symmetry detection from realworld images competition 2013: summary and results. In: Proceedings of the IEEE Conference on Computer Vision and Pattern Recognition Workshops, pp. 200–205 (2013)
21. Liu, Y., Hel-Or, H., Kaplan, C.S., Van Gool, L.: Computational Symmetry in Computer Vision and Computer Graphics. Now, Delft (2010)
22. Loy, G., Eklundh, J.-O.: Detecting symmetry and symmetric constellations of features. In: Leonardis, A., Bischof, H., Pinz, A. (eds.) ECCV 2006. LNCS, vol. 3952, pp. 508–521. Springer, Heidelberg (2006). https://doi.org/10.1007/11744047_39
23. Lubiw, A.: Some NP-complete problems similar to graph isomorphism. SIAM J. Comput. 10(1), 11–21 (1981). https://doi.org/10.1137/0210002
24. Mitra, N.J., Pauly, M., Wand, M., Ceylan, D.: Symmetry in 3D geometry: extraction and applications. Comput. Graph. Forum 32(6), 1–23 (2013). https://doi.org/10.1111/cgf.12010
25. Park, M., Lee, S., Chen, P.C., Kashyap, S., Butt, A.A., Liu, Y.: Performance evaluation of state-of-the-art discrete symmetry detection algorithms. In: 26th IEEE Conference on Computer Vision and Pattern Recognition, CVPR, pp. 1–8, June 2008. https://doi.org/10.1109/CVPR.2008.4587824
26. Purchase, H.C.: Metrics for graph drawing aesthetics. J. Vis. Lang. Comput. 13(5), 501–516 (2002). https://doi.org/10.1016/S1045-926X(02)90232-6
27. Rosenblatt, F.: The perceptron: a probabilistic model for information storage and organization in the brain. Psychol. Rev. 65(6), 386–408 (1958). https://doi.org/10.1037/h0042519
28. Rumelhart, D.E., Hinton, G.E., Williams, R.J.: Learning representations by back-propagating errors. Nature 323(6088), 533–536 (1986). https://doi.org/10.1038/323533a0
29. Sandler, M., Howard, A., Zhu, M., Zhmoginov, A., Chen, L.C.: MobileNetV2: inverted residuals and linear bottlenecks. In: Proceedings of the IEEE Computer Society Conference on Computer Vision and Pattern Recognition, pp. 4510–4520 (2018). https://doi.org/10.1109/CVPR.2018.00474
30. Simonyan, K., Zisserman, A.: Very deep convolutional networks for large-scale image recognition. arXiv preprint arXiv:1409.1556 (2014)

31. Szegedy, C., Ioffe, S., Vanhoucke, V., Alemi, A.A.: Inception-v4, inception-ResNet and the impact of residual connections on learning. In: 31st AAAI Conference on Artificial Intelligence, AAAI 2017, pp. 4278–4284 (2017)
32. Tsogkas, S., Kokkinos, I.: Learning-based symmetry detection in natural images. In: Fitzgibbon, A., Lazebnik, S., Perona, P., Sato, Y., Schmid, C. (eds.) ECCV 2012. LNCS, vol. 7578, pp. 41–54. Springer, Heidelberg (2012). https://doi.org/10. 1007/978-3-642-33786-4_4
33. Vasudevan, R.K., Dyck, O., Ziatdinov, M., Jesse, S., Laanait, N., Kalinin, S.V.: Deep convolutional neural networks for symmetry detection. Microsc. Microanal. **24**(S1), 112–113 (2018). https://doi.org/10.1017/s1431927618001058
34. Welch, E., Kobourov, S.: Measuring symmetry in drawings of graphs. Comput. Graph. Forum **36**(3), 341–351 (2017). https://doi.org/10.1111/cgf.13192
35. Widrow, B., Lehr, M.A.: 30 years of adaptive neural networks: perceptron, madaline, and backpropagation. Proc. IEEE **78**(9), 1415–1442 (1990). https://doi.org/ 10.1109/5.58323
36. Zoph, B., Vasudevan, V., Shlens, J., Le, Q.V.: Learning transferable architectures for scalable image recognition. In: Proceedings of the IEEE Conference on Computer Vision and Pattern Recognition, pp. 8697–8710 (2018)

Level Planarity

An SPQR-Tree-Like Embedding Representation for Upward Planarity

Guido Brückner[1(✉)], Markus Himmel[1], and Ignaz Rutter[2]

[1] Karlsruhe Institute of Technology, Karlsruhe, Germany
brueckner@kit.edu, markus.himmel@studentkit.edu
[2] University of Passau, Passau, Germany
rutter@fim.uni-passau.de

Abstract. The SPQR-tree is a data structure that compactly represents all planar embeddings of a biconnected planar graph. It plays a key role in constrained planarity testing.

We develop a similar data structure, called the UP-tree, that compactly represents all upward planar embeddings of a biconnected single-source directed graph. We demonstrate the usefulness of the UP-tree by solving the upward planar embedding extension problem for biconnected single-source directed graphs.

1 Introduction

A natural extension of planarity to directed graphs (digraphs) is to consider planar drawings where each edge is drawn as a y-monotone curve. Such drawings are called *upward planar*, and a graph admitting an upward planar drawing is *upward planar*. A planar (combinatorial) embedding $\mathcal{E}$ of a graph G is an *upward planar embedding* if G has an upward planar drawing whose (combinatorial) embedding is $\mathcal{E}$. Whereas undirected graphes can be tested for planarity in linear time, upward planarity testing is NP-complete in general, though there are efficient algorithms for graphs with a single source [4, 24] and graphs with a fixed embedding [3]. In the special case of *st-graphs*, i.e., graphs with a single source s and a single sink t with s and t on the same face, planar embeddings are the same as the upward planar embeddings [28], and hence upward planarity and planarity are equivalent.

A related but different planarity notion for digraphs is *level planarity*, where the vertices of the graph have fixed levels that correspond to horizontal lines in the drawing. The task is to order the vertices on each level so that the drawing is planar. Level planarity can be tested in linear time [26] by a quite involved algorithm, or in quadratic time by several simpler algorithms [11, 20, 29].

In a constrained embedding problem, one seeks a planar embedding of a given graph that satisfies additional constraints. Typical examples are simultaneous

---

This work was partially supported by grant RU 1903/3-1 of the German Research Foundation (DFG).

D. Archambault and C. D. Tóth (Eds.): GD 2019, LNCS 11904, pp. 517–531, 2019.
https://doi.org/10.1007/978-3-030-35802-0_39

embeddings with fixed edges [5], cluster planarity [19], constraints on the face sizes [13,14], optimizing the depth of the embedding [2] and optimizing the bends in an orthogonal drawing [6,7,18]. One of the most prominent examples of the last years is the *partial drawing extension* problem, which asks whether a given drawing of a subgraph can be extended to a planar drawing of the whole graph. The *partial embedding extension* problem is strongly related, here the input is a planar embedding of a subgraph and the question is whether it can be extended to a planar embedding of the whole graph. For undirected planar graphs the two problems are equivalent and can be solved in linear time [1,25].

One of the key tools for all of these applications is the SPQR-tree, which compactly represents all planar embeddings of a biconnected planar graph G and breaks down the complicated task of choosing a planar embedding of G into simpler independent embedding choices of its triconnected components [15–17,23,27,30]. In fact, these embeddings are either uniquely determined up to reversal, or they consist in arbitrarily choosing a permutation of parallel edges between two pole vertices. The common approach for attacking the above-mentioned constrained embedding problems is to project the constraints on the global embedding to local constraints on the skeleton embeddings that can then be satisfied by consistent local choices. While the implementation details are often highly technical and non-trivial, the approach has proven to be extremely successful.

In comparison, relatively little is known about constrained planarity problems for planarity notions of digraphs. Brückner and Rutter [10] study the problem of extending a given partial drawing of a level graph and Da Lozzo et al. [12] study the same question for upward planarity. In general, extending a given partial upward planar drawing requires to determine an upward planar embedding that (i) extends the embedding of the partial drawing, and (ii) admits a drawing that extends the given drawing. Here step (i) requires solving the embedding extension problem but with additional constraints that ensure that a drawing extension is feasible. It is worth noting that for upward planarity the embedding extension problem and the drawing extension problem are distinct; Da Lozzo et al. show that, generally, even if an upward planar embedding of the whole graph is given, it is NP-complete to decide whether it can be drawn such that it extends a given partial drawing [12, Theorem 2]. On the positive side, they present tractability results for directed paths and cycles with a given upward planar embedding, and for st-graphs. The restriction to st-graphs allows a relatively simple characterization of the upward planar embeddings that extend the given partial drawings [12, Lemma 6], which yields an $O(n \log n)$-time algorithm for step (ii). For step (i), Da Lozzo et al. exploit the fact that for st-graphs, the choice of an upward planar embedding is equivalent to choosing a planar embedding, and hence the SPQR-tree allows to efficiently search for an upward planar embedding satisfying the additional constraints required by condition (ii).

In this paper, we seek to generalize the approach of Da Lozzo et al. to biconnected single-source graphs. The key difficulty in this case is that neither do we have access to all the upward planar embeddings of such graphs, nor is it known what the necessary and sufficient conditions are for an upward planar embedding to admit a drawing that extends a given subdrawing.

Contribution and Outline. We construct a novel SPQR-tree-like embedding representation, called the UP-tree, that represents exactly the upward planar embeddings of a biconnected single-source graph. As in SPQR-trees, the embedding choices in the UP-tree are broken down into independent embedding choices of skeleton graphs that are either unique up to reversal or allow to arbitrarily permute parallel edges between two poles. As such, UP-trees can take the role of SPQR-trees for constrained embedding problems in upward planarity, making them a powerful tool with a broad range of applications. We demonstrate this by giving an quadratic-time algorithm for the upward planar embedding extension problem for biconnected single-source graphs.

After introducing some preliminaries in Sect. 2, we review the results on decomposing upward planar single-source digraphs due to Hutton and Lubiw [24] in Sect. 3. We proceed to extend this idea from a single decomposition to decomposition trees. Proofs of statements marked with a star ($\star$) can be found in the full version [9]. In Sect. 4, we define the UP-tree and in Sect. 5 we use it to solve the partial upward embedding extension problem.

2   Preliminaries

Let $G = (V, E)$ be a connected simple undirected graph. A *cutvertex* of G is a vertex whose removal disconnects G. We say that G is *biconnected* if it has no cutvertex. We say that $\{u, v\}$ is a *cutpair* if there are connected subgraphs H_1, H_2 of G with $H_1 \cup H_2 = G$ and $H_1 \cap H_2 = \{u, v\}$. If a graph has no cutpair it is *triconnected*.

Decomposition Trees. Assume that G is biconnected. A *decomposition* along a cutpair $\{u, v\}$ of G is defined as follows. Let μ_1, μ_2 be two nodes connected by an undirected arc (μ_1, μ_2). Node μ_i is equipped with a multigraph $H_i \cup \{(u, v)\}$ called its *skeleton* denoted by $\text{skel}(\mu_i)$. The newly added edge (u, v) is a *virtual edge* and corresponds to μ_2 in μ_1 and to μ_1 in μ_2, respectively. This is formalized as functions $\text{corr}_{\mu_1} \colon (u, v) \mapsto \mu_2$ and $\text{corr}_{\mu_2} \colon (u, v) \mapsto \mu_1$. If there exists a cutpair $\{u', v'\}$ in $\text{skel}(\mu_i)$ and we may once again decompose along that cutpair. By repeating this process we obtain an unrooted *decomposition tree* $\mathcal{T}$.

Let $a = (\mu, \nu)$ be an arc of $\mathcal{T}$. Then $\text{skel}(\mu)$ and $\text{skel}(\nu)$ share two vertices u, v and the existence of a can be traced back to a decomposition along u, v. We then say that the *poles* of μ in ν are u and v. When ν is clear from the context we also simply refer to u and v as the poles of μ.

A decomposition can be reverted by contracting an arc (μ, ν) of $\mathcal{T}$ and merging the skeletons of μ and ν. To merge $\text{skel}(\mu)$ and $\text{skel}(\nu)$, remove from $\text{skel}(\mu)$ the virtual edge e with $\text{corr}_\mu(e) = \nu$ and from $\text{skel}(\nu)$ the virtual edge e' with $\text{corr}_\nu(e') = \mu$ and set the union of these two graphs as the skeleton of the node obtained by contracting (μ, ν) in $\mathcal{T}$. This is a *composition* along (μ, ν).

Consider an arc $a = (\mu, \nu)$ of $\mathcal{T}$. Removing a from $\mathcal{T}$ separates $\mathcal{T}$ into two subtrees $\mathcal{T}_\mu$ and $\mathcal{T}_\nu$ containing μ and ν, respectively. Define the *pertinent graph* of μ in ν as the skeleton of the single node obtained by contracting all arcs in $\mathcal{T}_\mu$. Again, when ν is clear from the context we simply refer to this graph as the pertinent graph of μ and denote it by $G(\mu)$.

Rooted Decomposition Trees and Planar Embeddings. Throughout this paper let an *embedding* of a graph denote a rotation system together with an outer face. Decomposition trees can be used to decompose not only a graph, but also an embedding of it. Consider a biconnected graph G together with a planar embedding $\mathcal{E}$. Let $e^\star$ be an edge of G incident to the outer face of $\mathcal{E}$. Further, let $\mathcal{T}$ be a decomposition tree of G rooted at a node whose skeleton contains $e^\star$. Equip the skeleton of each node μ of $\mathcal{T}$ with an embedding as follows. The embedding of $\mathrm{skel}(\mu)$ is obtained from $\mathcal{E}$ by contracting for each virtual edge (u, v) of $\mathrm{skel}(\mu)$ the pertinent graph $G(\mathrm{corr}_\mu(u, v))$ into a single edge. These embeddings of the skeletons of the nodes of $\mathcal{T}$ are referred to as a *configuration*. The fact that $e^\star$ is incident to the outer face gives two properties. First, the edge $e^\star$ lies on the outer face of the skeleton of the root node of $\mathcal{T}$. Second, every non-root node ν of $\mathcal{T}$ has some parent node μ and the virtual edge e with $\mathrm{corr}_\nu(e) = \mu$ lies on the outer face of $\mathrm{skel}(\nu)$. We extend our notion of a *configuration* to any set of embeddings of the skeletons of the nodes of $\mathcal{T}$ that fulfills these two properties.

Recall that decomposition trees allow for (graph) composition along arcs. We can also compose embeddings. When contracting an arc (μ, ν), we merge $\mathrm{skel}(\mu)$ and $\mathrm{skel}(\nu)$ as described above. Obtain the embedding of the merged skeleton by replacing the occurrences of the virtual edge in the rotation system around its poles by the appropriate rotation system in the embedding of the other skeleton. This means that G together with a planar embedding can be decomposed into a decomposition tree $\mathcal{T}$ together with a configuration. And symmetrically, $\mathcal{T}$ together with any configuration can be composed into a planar embedding of G.

SPQR-Trees. As described in the previous paragraph, decomposition trees separate independent choices in finding planar embeddings of a graph. We may either choose an embedding of the entire graph, which is generally very complex, or we may decompose the graph into smaller skeletons, independently choose embeddings of these skeletons and compose them into an embedding of the entire graph. In this sense decomposition trees implement a tradeoff between making few complex choices or many simple choices.

The *SPQR-tree* is a decomposition tree that makes this tradeoff in favor of many simple choices. SPQR-trees have four kinds of nodes, all of whose skeletons offer only few and well-structured embedding choices. (i) R-nodes are nodes whose skeleton is triconnected. Such skeletons have a unique planar embedding up to flipping. (ii) S-nodes are nodes whose skeleton is a simple cycle. Such skeletons offer no embedding choice (recall that the outer face is fixed by the rooting). Adjacent S-nodes are contracted into one larger S-node, i.e., an S-node whose skeleton is a larger simple cycle. This means that in SPQR-trees no two S-nodes are adjacent. (iii) P-nodes are nodes whose skeleton is a multigraph that

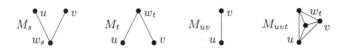

Fig. 1. The four markers used by Hutton and Lubiw. The markers are digraphs; in the figure, all edges are directed upward.

consists of two vertices connected by three or more edges. The order of these edges may be arbitrarily permuted. Again, adjacent P-nodes are contracted into one larger P-node, i.e., no two P-nodes are adjacent. (iv) Q-nodes are nodes whose skeleton consists of two vertices connected by two edges, namely one virtual edge and one non-virtual edge. They offer no embedding choice. Note that only the skeletons of Q-nodes contain non-virtual edges. See Fig. 3(a) and (b) for a graph and its SPQR-tree decomposition.

3   Decomposition Trees and Upward Planar Embeddings

Recall from the previous section that for biconnected graphs we can decompose any planar embedding into planar embeddings of the skeletons of a decomposition tree; and symmetrically, we can compose a planar embedding of the whole graph from planar embeddings of the skeletons. In this section we find a similar relationship between upward planar embeddings of a biconnected single-source digraph G and upward planar embeddings of the skeletons of a suitably-defined decomposition tree of G.

3.1   Decompositions and Upward Planar Embeddings

In this section we review the decomposition result of Hutton and Lubiw and formulate the interface between their result and our results.

Let G be a biconnected single-source digraph together with an upward planar embedding $\mathcal{E}$. Further, let $e^\star$ denote the edge around the source of G that is leftmost in $\mathcal{E}$. Now let H_1, H_2 be two subgraphs of G with (i) $H_1 \cup H_2 = G$, (ii) $H_1 \cap H_2 = \{u, v\}$, (iii) $e^\star \in H_1$ and (iv) $H_1 \setminus \{u, v\}$ or $H_2 \setminus \{u, v\}$ is connected.

Hutton and Lubiw construct two graphs H_1' and H_2' from H_1 and H_2 by including one of the *markers* shown in Fig. 1. Markers are simple digraphs with two vertices u, v that connect the marker to the remaining graph. The marker in H_1' is designed to represent H_2 and the marker in H_2' is designed to represent H_1. If there exists a directed path from u to v we say that u *dominates* v and write $u < v$ for short. Otherwise u and v are *incomparable*. The vertex v is a *source* if it has no incoming edges in G, a *sink* if it has no outgoing edges in G and an *internal* vertex if it has both incoming and outgoing edges in G. Markers are determined based on whether $u < v$ and whether v is a source, sink or internal vertex in H_1 and H_2: If u and v are incomparable in G, set $H_1' = H_1 \cup M_t$ and $H_2' = H_2 \cup M_s$. Otherwise, assume $u < v$. Define H_1' as follows. If v is a source in H_2 set $H_1' = H_1 \cup M_t$. If v is a sink in H_2 set $H_1' = H_1 \cup M_{uv}$. Otherwise v is an internal vertex in H_2 and we set $H_1' = H_1 \cup M_{uvt}$. Define H_2' as follows. If v is a source in H_1 set $H_2' = H_2 \cup M_t$, otherwise set $H_2' = H_2 \cup M_{uv}$. See Fig. 4 in the full version for example decompositions.

Recall that decomposition trees of planar graphs allow for (de-)composition of planar embeddings. Hutton and Lubiw provide a similar property for the graphs G, H_1' and H_2'.

Theorem 1 ($\star$, implicit in [24]). *Let $\mathcal{E}$ be an upward planar embedding of G with $e^\star$ as the leftmost edge around s. Then $\mathcal{E}$ induces upward planar embeddings $\mathcal{F}_1, \mathcal{F}_2$ of H_1', H_2', respectively with the following properties. In $\mathcal{F}_1$, $e^\star$ is the leftmost edge around the source of H_1'. In $\mathcal{F}_2$, the edges of the marker are leftmost around the source of H_2'. Conversely, if $\mathcal{F}_1$ and $\mathcal{F}_2$ are upward planar embeddings of H_1' and H_2' such that $e^\star$ is the leftmost edge around the source of H_1' and the edges of the marker are leftmost around the source of H_2', then the composition of these embeddings is upward planar.*

Hutton and Lubiw do not explicitly state Theorem 1. Instead, Theorems 6.5, 6.7, 6.8 and 6.9 in [24] discuss the same situation as Theorem 1, but are only concerned with upward planarity, not with the embeddings involved. See the full version for a detailed discussion.

3.2   Decomposition Trees and Upward Planar Embeddings

The approach of Hutton and Lubiw is to decompose a single-source digraph G into two smaller single-source digraphs G_1, G_2 and use Theorem 1 to translate upward-planarity testing of G to upward-planarity testing of two smaller instances H_1', H_2'. Observe that the markers and the replacement rules are defined so that both H_1' and H_2' are single-source digraphs. This means that H_1' and H_2' can be recursively decomposed. Note that in the context of connectivity markers are treated simply as edges, i.e., markers are not decomposed further. When a graph cannot be further decomposed it is triconnected and therefore has a unique planar embedding which can be tested for upward planarity in linear time using the algorithm of Bertolazzi et al. [4]. In the context of upward planarity testing the full marker graph is considered. Upward planar embeddings of H_1' and H_2' can then be composed to an upward planar embedding of G. In the context of embedding composition markers are again treated simply as edges. In particular, it does not matter whether the clockwise order of the edges incident to u in M_{uvt} is $(u, v), (u, x), (u, w_t)$ or $(u, w_t), (u, x), (u, v)$.

We use a different approach. Instead of testing H_1' and H_2' for upward-planarity separately, we manage them as the skeletons of two nodes in a decomposition tree $\mathcal{T}$. Note that Theorem 1 requires $H_1 \setminus \{u, v\}$ or $H_2 \setminus \{u, v\}$ to be connected. We call such a decomposition *maximal*. We then decompose these skeletons further, which grows the decomposition tree. A *maximal-decomposition tree* is a decomposition tree obtained by performing only maximal decompositions. A *configuration* equips the skeleton of each node in the tree with an upward planar embedding. In this embedding, $e^\star$ or the marker that represents the component that contains $e^\star$ must be incident to the outer face and leftmost around the source of the skeleton. See Fig. 3(c) for an example of a maximal-decomposition tree. Applying Theorem 1 at each decomposition step gives the following.

Theorem 2. *Let G be a biconnected graph with a single source s, let $e^\star$ be an edge of G incident to s and let T denote a maximal-decomposition tree of G. Then the upward-planar embeddings of G in which $e^\star$ is the leftmost edge around s correspond bijectively to the configurations of T.*

We could use Theorem 2 directly to represent all upward planar embeddings of a graph. But we also show that decomposition trees are uniquely defined by the decompositions that are executed, but not by the order of these decompositions. This means that just like we can talk about *the* SPQR-tree decomposition for a graph we will be able to talk about *the* UP-tree decomposition. The benefit of this is that we can use a UP-tree decomposition to determine that some constrained representation problem has no solution without having to consider other conceivable UP-tree decompositions.

To prove uniqueness, we show that the order of the decompositions is irrelevant. We then apply the decompositions as defined by the SPQR-tree decomposition, which is unique, and obtain the unique UP-tree decomposition. To this end, we prove that the marker replacement rules do not depend on the order of the decompositions. Recall that the marker replacement rules depend on vertex dominance and the local neighborhood of certain vertices. We prove Lemma 1, which states that decompositions preserve vertex dominance and Lemma 2, which states that decompositions preserve the local neighborhood of certain vertices.

Lemma 1 ($\star$). *Let G be a biconnected single-source digraph and let H_1', H_2' denote the result of decomposing along a cutpair $\{u, v\}$ of G. For $i = 1, 2$ and any two vertices x, y in H_i' it is $x < y$ in H_i' if and only if $x < y$ in G.*

Lemma 2 ($\star$). *Let G be a biconnected single-source digraph and let H_1', H_2' denote the result of decomposing along a cutpair $\{u, v\}$ of G. For $i = 1, 2$ let $\{x, y\}$ denote a cutpair of H_i' that separates H_i' into F_1 and F_2 and G into D_1 and D_2. Then y is a source in F_1 if and only if y is a source in D_1. Moreover, y is a source, sink or internal vertex in F_2 if and only if y is a source, sink or internal vertex in D_2, respectively.*

Lemmas 1 and 2 immediately give the following.

Lemma 3. *Let G be a biconnected graph with a single source s, let $e^\star$ be an edge of G incident to s and let T denote a decomposition tree of G. Then T relative to $e^\star$ is uniquely defined by the decompositions regardless of their order.*

A configuration of T can be computed as follows. Recall that all skeletons are single-source digraphs. We may therefore run the algorithm due to

Bertolazzi et al. [4] on each skeleton. Observe that in a configuration of $\mathcal{T}$ relative to e^* the skeleton of each node μ of $\mathcal{T}$ must be embedded so that e^* or the marker that corresponds to the component that contains e^* must appear leftmost around the source of $\text{skel}(\mu)$. We can enforce this by rooting the decomposition tree constructed by the algorithm of Bertolazzi et al. at the Q-node corresponding to e^* or an edge of the marker that corresponds to the component that contains e^*.

4   UP-Trees

We are ready to construct the UP-tree, a maximal-decomposition tree designed to mimic the SPQR-tree. Let G be a biconnected directed single-source graph. The base of the construction is the decomposition tree obtained by performing the same set of decompositions as in the construction of the SPQR-tree decomposition of the underlying undirected graph of G. We then perform two additional steps. The first step is to split P-nodes into chains of smaller nodes. The second step is to determine whether skeletons of R-nodes can be reversed and to contract some arcs of the decomposition tree. In both steps, we reason about upward planarity of fixed embeddings with the following lemma due to Bertolazzi et al. [4].

Let G be a biconnected single-source graph together with a planar embedding. The *face-sink graph* F of G has the vertices and faces of G as its vertices. It contains an undirected edge $\{f, v\}$ if f is a face of G and v is a vertex of G that is incident to f and both edges incident to v and f are directed towards v. The following lemma implies a linear-time algorithm that tests an embedding for upward planarity and outputs for each face whether it can be the outer face.

Lemma 4 ([4, Theorem 1]). *Let G be an embedded planar single-source digraph and let h be a face of G. Graph G has an upward planar drawing that preserves the embedding with outer face h if and only if all of the following is true: (i) graph F is a forest (ii) there is exactly one tree T of F with no internal vertices of G, while the remaining trees have exactly one internal vertex; (iii) h is in tree T; and (iv) the source of G is in the boundary of h.*

4.1   P-Node Splits

In SPQR-trees, the edges of P-nodes may be arbitrarily permuted. In decomposition trees for upward planar graphs there are stricter rules for the ordering of the markers in P-nodes. In this section, we determine these rules and find that by breaking up the P-nodes into chains of smaller nodes we obtain a decomposition tree for upward planarity whose P-nodes exhibit the same behavior as in SPQR-trees, i.e., their edges may be arbitrarily permuted. The idea is that certain kinds of markers must appear consecutively.

First, we argue that all M_{uv} markers must appear consecutively. To see this, note that if M_s appears between two M_{uv} markers then the outer face is not

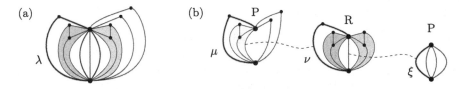

Fig. 2. Splitting a P-node λ obtained from the SPQR-tree (a) into a chain of smaller nodes μ, ν, ξ (b). The bold marker represents the component that contains the edge $e^\star$.

incident to the source of the skeleton, which is vertex w_s of M_s. If a marker M with $M = M_t$ or $M = M_{uvt}$ appears between two M_{uv} markers then the face incident to w_t of M and a marker M_{uv} is not connected to the outer face and not connected to an internal vertex. In all cases the conditions from Lemma 4 are violated.

Moreover, all M_{uv} and M_{uvt} markers must appear consecutively. To see this, note that if M_t appears between two markers M_{uv} or M_{uvt} the vertex w_t of M_t cannot be connected to an internal vertex or the outer face and apply Lemma 4.

These observations motivate the following restructuring of P-nodes. Let λ denote a P-node obtained from the SPQR-tree. The *parent marker* in skel(λ) is the marker that corresponds to the parent node of λ. If the parent marker in skel(λ) is M_s all other markers must be M_t. In this case these markers can already be arbitrarily permuted and nothing further needs to be shown. Otherwise the parent marker is M_t or M_u (recall that by definition of H_2' the parent marker is not M_{uvt}). See Fig. 2(a) where the parent marker is M_t (the case for M_u is similar). Because all M_{uv} and M_{uvt} markers must appear consecutively, we create a new P-node μ that contains the parent marker of skel(λ), all M_t markers of skel(λ) and a single M_{uvt} marker to represent all M_{uv} and M_{uvt} markers of skel(λ). This marker corresponds to a new node ν that contains all M_{uvt} markers of skel(λ) and—because all M_{uv} markers must appear consecutively— a single M_{uv} marker. This marker corresponds to a new node ξ that contains all M_{uv} markers of skel(λ). If skel(λ) contains no M_{uvt} marker we can include a M_{uv} marker instead of a M_{uvt} marker in skel(μ) and connect it directly to ξ, the node ν can then be omitted.

The new node μ has the property that its markers can be arbitrarily permuted, i.e., it is a P-node. Observe that there can be at most two M_{uvt} markers in skel(λ). This means that skel(ν) has at most four markers and its embedding is fixed up to reversal, i.e., it is an R-node. Finally, the new node ξ also has the property that its markers can be arbitrarily permuted, i.e., it is also a P-node. See Fig. 2(b) and Fig. 3(c) and (d) for a larger example. We conclude the following.

Lemma 5. *Let G be a biconnected digraph with a single source s and let $e^\star$ denote an edge incident to s. There exists a decomposition tree $\mathcal{T}$ that (i) represents all upward planar embeddings of G in which $e^\star$ is the leftmost edge around s, and (ii) the children of all P-nodes in $\mathcal{T}$ can be arbitrarily permuted.*

4.2   Arc Contractions

Recall that in SPQR-trees the skeletons of R-nodes are triconnected, i.e., their planar embedding is fixed up to reversal. So, every R-node offers one degree of freedom, namely, whether it has some reference embedding or the reversal thereof. In this section we alter our decomposition tree so that it has this same property.

By definition the marker corresponding to the parent node is leftmost in any embedding of a skeleton. Hence, this marker is incident to the outer face. Reversing the embedding of the skeleton is equivalent to choosing the other face incident to the marker as the outer face. Theorem 2 guarantees that any configuration of $\mathcal{T}$ can be composed to an upward planar embedding. This means that a skeleton can be reversed if and only if both faces incident to the parent marker can be chosen as the outer face. This can be checked with the upward planarity test for embedded single-source graphs due to Bertolazzi et al. [4], which also outputs the set of faces that can be chosen as the outer face. If both incident faces are candidates for the outer face this node does indeed offer a degree of freedom and we leave it unchanged. Otherwise, if only one incident face is a candidate for the outer face this node does not offer a degree of freedom. We then merge it with its parent node and contract the corresponding arc in the decomposition tree. This leads to an R-node with a larger skeleton.

See Fig. 5 (a) in the full version for an upward planar graph G and (b) a decomposition tree thereof. Parts of the face sink graphs of $\mathrm{skel}(\mu)$ and $\mathrm{skel}(\nu)$ are shown in red, namely the two quadratic vertices dual to the faces incident to the parent marker and the edges incident to those vertices. One criterion for a face to be a candidate for becoming the outer face due to Bertolazzi et al. is that there has to be a path from this face to the outer face in the face sink graph. This holds true for both faces incident to the parent marker in $\mathrm{skel}(\nu)$, but not in $\mathrm{skel}(\mu)$. Therefore the arc (μ, ν) is not contracted but the arc (λ, μ) is contracted. This leads to the decomposition tree shown in (c). See also Fig. 3(c) and (d) for a larger example.

Lemma 6. *Let G be a biconnected digraph with a single source s and let $e^\star$ denote an edge incident to s. There exists a decomposition tree $\mathcal{T}$ that (i) represents all upward planar embeddings of G in which $e^\star$ is the leftmost edge around s, and (ii) the children of all P-nodes in $\mathcal{T}$ can be arbitrarily permuted. (iii) the skeletons of all R-nodes in $\mathcal{T}$ can be reversed.*

We call the decomposition tree $\mathcal{T}$ the *UP-tree of G relative to $e^\star$.*

4.3   Computation in Linear Time

Let G be a biconnected digraph with a single source s and let $e^\star$ denote an edge incident to s. Recall that the construction of the UP-tree $\mathcal{T}$ of G relative to $e^\star$ consists of the following seven steps. 1. Construct the SPQR-tree $\mathcal{T}$ of G in linear time [22,23]. 2. For each pair of vertices u, v that are the poles of a marker in some skeleton of $\mathcal{T}$, we have to determine whether $u < v$ in G.

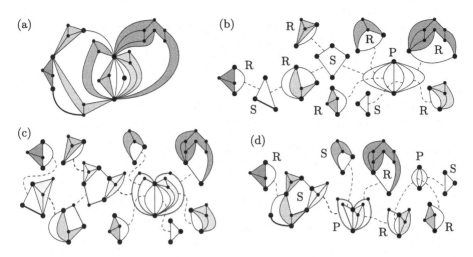

Fig. 3. Construction of the UP-tree. An upward planar biconnected single-source graph (a), the SPQR-tree of its underlying undirected graph (b) with the Q-nodes omitted, the result of replacing virtual edges with markers (c) and the UP-tree after splitting P-nodes and contracting arcs (d).

To compute this information for all pairs in linear time, we use a union-find-based technique described by Bläsius et al. [8]. Process all skeletons of T and for every pair of poles u, v that is encountered register v as a *candidate* at u and register u as a candidate at v. Next, initialize every vertex of G in its own singleton set. Then, process each vertex u in some reverse topological order of G. Unify the singleton set of u with the sets of its direct descendants in G. Now for any candidate v stored at u we can query in whether u and v belong to the same set, which is equivalent to $u < v$. Note that the operands to all unify operations are completely determined by the structure of G. We exploit this fact to run the linear-time union-find algorithm due to Gabow and Tarjan [21]. 3. For each arc $a = (\mu, \nu)$ of T, decide whether the poles of a are sources, sinks or internal vertices in $G(\mu)$ and $G(\nu)$. This information can be found using a simple bottom-up technique. We first compute the indegree and outdegree of every node of G. We then perform a depth-first traversal of T. We maintain a list of the number of incoming and outgoing edges for each node seen so far, which is updated when a Q-node is visited. Upon entering a subtree, we store these numbers for the poles of the arc leaving the subtree at the root of the subtree. Upon leaving a subtree, we can now calculate the differences between the current numbers and the stored numbers, which gives the in- and outdegree of the poles in the graph $G(\mu)$. Using the in- and outdegree of the poles in G computed earlier, we can also compute the in- and outdegree of the poles in $G(\nu)$. This step clearly takes linear time. 4. In each skeleton, replace all virtual edges with their respective markers. With the information that was computed in the previous step and the fact that all markers have constant size this step is feasible in linear time. 5. Construct a configuration of T by running the linear-time upward planar embedding algorithm of Bertolazzi et al. [4] on every skeleton. Because the size of all skeletons is linear in the size of G this step takes linear time. 6. Perform P-node splits. The running time spent on one P-node is clearly linear in the size of its skeleton. This gives linear running time overall. 7. Perform arc contractions. The upward planarity test for fixed embeddings due to Bertolazzi et al. runs in linear time. Contracting an arc is feasible in constant time. This gives linear running time overall.

Theorem 3. *Let G be a biconnected digraph with a single source s and let $e^\star$ denote an edge incident to s. The UP-tree $\mathcal{T}$ of G relative to $e^\star$ is a decomposition tree whose internal nodes are (i) S-nodes whose skeletons have a fixed embedding, (ii) R-nodes whose skeletons have a fixed embedding up to reversal, or (iii) P-nodes where the markers can be arbitrarily permuted in the skeleton and whose leaves are Q-nodes that offer no embedding choice. The configurations of $\mathcal{T}$ correspond bijectively to the upward planar embeddings of G where $e^\star$ appears leftmost around s. Moreover, $\mathcal{T}$ can be computed in linear time.*

5   Partial Upward Embedding

In this section we apply the UP-tree to solve the partial upward embedding problem in quadratic time. A *partially embedded graph* is a tuple $(G, H, \mathcal{H})$, where G is a planar graph, H is a subgraph of G and $\mathcal{H}$ is a planar embedding of H. An embedding $\mathcal{G}$ of G *extends* the partial embedding $\mathcal{H}$ if all edges e, f, g in H that share a common endpoint v appear in the same cyclic order around v in $\mathcal{G}$ and $\mathcal{H}$. The *partial embedding problem* asks whether there exists an embedding $\mathcal{G}$ of G that extends $\mathcal{H}$. Angelini et al. solve the partial embedding problem in linear time [1]. The algorithm considers every triple of edges (e, f, g) in H that share a common endpoint v and enforces the constraints imposed by these edges in the SPQR-tree $\mathcal{T}$. Note that e, f, g each correspond to a Q node in $\mathcal{T}$. Because $\mathcal{T}$ is a tree there is exactly one node μ in $\mathcal{T}$ so that the paths from μ to these Q nodes are disjoint. The relative order of e, f, g in the embedding represented by $\mathcal{T}$ is determined by the embedding of $\mathrm{skel}(\mu)$. If $\mathrm{skel}(\mu)$ offers no embedding choice (as in S nodes) determine whether the ordering of e, f, g given by $\mathcal{H}$ is the same as the one given by the unique embedding of $\mathrm{skel}(\mu)$. If not, reject the instance. If $\mathrm{skel}(\mu)$ has two possible embeddings (as in R nodes) the ordering of e, f, g given by $\mathcal{H}$ fixes one of the two embeddings of $\mathrm{skel}(\mu)$ as the only candidate. Finally, if μ is a P node the ordering of e, f, g given by $\mathcal{H}$ restricts the set of admissible permutations of the virtual edges in $\mathrm{skel}(\mu)$. The algorithm collects all these constraints and checks whether they can be fulfilled at the same time.

A *partially embedded upward graph* is defined as a tuple $(G, H, \mathcal{H})$, where G is an upward planar graph, H is a subgraph of H and $\mathcal{H}$ is an upward planar embedding of H. Note that UP-trees have all properties of SPQR-trees that are needed in the algorithm described above. In particular, the markers in P-nodes may be arbitrarily permuted, R-nodes may be reversed and all other nodes offer no embedding choice. Hence, we use the UP-tree as a drop-in replacement for the SPQR-tree in the algorithm of Angelini et al. to obtain an algorithm that solves the partial upward embedding problem. Note that the UP-tree is rooted at some edge that must be embedded as the leftmost edge around the source of the graph. We may have to try a linear number of candidate edges in the worst case. This gives the following.

Theorem 4. *The partial upward embedding problem can be solved in quadratic running time for biconnected single-source digraphs.*

6 Conclusion

We have developed the UP-tree, which is an SPQR-tree-like embedding representation for upward planarity. We expect that the UP-tree is a valuable tool that makes it possible to translate existing constrained planar embedding algorithms that use SPQR-trees to the upward planar setting. As an example, we have demonstrated how to use the UP-tree as a drop-in replacement for the SPQR-tree in the partial embedding extension problem, solving the previously open partial upward embedding extension problem for the biconnected single-source case.

References

1. Angelini, P., Di Battista, G., Frati, F., Jelínek, V., Kratochvíl, J., Patrignani, M., Rutter, I.: Testing planarity of partially embedded graphs. ACM Trans. Algorithms **11**(4), 32:1–32:42 (2015). https://doi.org/10.1145/2629341
2. Angelini, P., Di Battista, G., Patrignani, M.: Finding a minimum-depth embedding of a planar graph in $o(n^4)$ time. Algorithmica **60**(4), 890–937 (2011). https://doi.org/10.1007/s00453-009-9380-6
3. Bertolazzi, P., Di Battista, G., Liotta, G., Mannino, C.: Upward drawings of triconnected digraphs. Algorithmica **12**(6), 476–497 (1994)
4. Bertolazzi, P., Di Battista, G., Mannino, C., Tamassia, R.: Optimal upward planarity testing of single-source digraphs. SIAM J. Comput. **27**(1), 132–169 (1998)
5. Bläsius, T., Kobourov, S.G., Rutter, I.: Simultaneous embedding of planar graphs. In: Tamassia, R. (ed.) Handbook of Graph Drawing and Visualization, Discrete Mathematics and its Applications, pp. 349–373. CRC Press (2014)
6. Bläsius, T., Lehmann, S., Rutter, I.: Orthogonal graph drawing with inflexible edges. Comput. Geom. **55**, 26–40 (2016). https://doi.org/10.1016/j.comgeo.2016.03.001
7. Bläsius, T., Rutter, I., Wagner, D.: Optimal orthogonal graph drawing with convex bend costs. ACM Trans. Algorithms **12**(3), 33:1–33:32 (2016). https://doi.org/10.1145/2838736
8. Bläsius, T., Karrer, A., Rutter, I.: Simultaneous embedding: edge orderings, relative positions, cutvertices. Algorithmica **80**(4), 1214–1277 (2018)
9. Brückner, G., Himmel, M., Rutter, I.: An SPQR-tree-like embedding representation for upward planarity. CoRR abs/1908.00352v1 (2019). https://arxiv.org/abs/1908.00352v1

10. Brückner, G., Rutter, I.: Partial and constrained level planarity. In: Klein, P.N. (ed.) Proceedings of 28th Annual ACM-SIAM Symposium on Discrete Algorithms (SODA 2017), pp. 2000–2011. SIAM (2017)

11. Brückner, G., Rutter, I., Stumpf, P.: Level planarity: transitivity vs. even crossings. In: Biedl, T., Kerren, A. (eds.) GD 2018. LNCS, vol. 11282, pp. 39–52. Springer, Cham (2018). https://doi.org/10.1007/978-3-030-04414-5_3

12. Da Lozzo, G., Di Battista, G., Frati, F.: Extending upward planar graph drawings. CoRR abs/1902.06575 (2019)

13. Da Lozzo, G., Jelínek, V., Kratochvíl, J., Rutter, I.: Planar embeddings with small and uniform faces. In: Ahn, H.-K., Shin, C.-S. (eds.) ISAAC 2014. LNCS, vol. 8889, pp. 633–645. Springer, Cham (2014). https://doi.org/10.1007/978-3-319-13075-0_50

14. Da Lozzo, G., Rutter, I.: Approximation algorithms for facial cycles in planar embeddings. In: Hsu, W.L., Lee, D.T., Liao, C.S. (eds.) Proceedings of the 29th International Symposium on Algorithms and Computation (ISAAC 2018). LIPIcs, vol. 123, pp. 41:1–41:13. Schloss Dagstuhl - Leibniz-Zentrum fuer Informatik (2018). https://doi.org/10.4230/LIPIcs.ISAAC.2018.41

15. Di Battista, G., Tamassia, R.: Incremental planarity testing. In: Proceedings of the 30th Annual Symposium on Foundations of Computer Science, pp. 436–441, October 1989. https://doi.org/10.1109/SFCS.1989.63515

16. Di Battista, G., Tamassia, R.: On-line graph algorithms with SPQR-trees. In: Paterson, M.S. (ed.) ICALP 1990. LNCS, vol. 443, pp. 598–611. Springer, Heidelberg (1990). https://doi.org/10.1007/BFb0032061

17. Di Battista, G., Tamassia, R.: On-line maintenance of triconnected components with SPQR-trees. Algorithmica 15(4), 302–318 (1996). https://doi.org/10.1007/BF01961541

18. Didimo, W., Liotta, G., Patrignani, M.: Bend-minimum orthogonal drawings in quadratic time. In: Biedl, T., Kerren, A. (eds.) GD 2018. LNCS, vol. 11282, pp. 481–494. Springer, Cham (2018). https://doi.org/10.1007/978-3-030-04414-5_34

19. Feng, Q.-W., Cohen, R.F., Eades, P.: Planarity for clustered graphs. In: Spirakis, P. (ed.) ESA 1995. LNCS, vol. 979, pp. 213–226. Springer, Heidelberg (1995). https://doi.org/10.1007/3-540-60313-1_145

20. Fulek, R., Pelsmajer, M.J., Schaefer, M., Štefankovič, D.: Hanani-Tutte, monotone drawings, and level-planarity. In: Pach, J. (ed.) Thirty Essays on Geometric Graph Theory, pp. 263–287. Springer, New York (2013). https://doi.org/10.1007/978-1-4614-0110-0_14

21. Gabow, H.N., Tarjan, R.E.: A linear-time algorithm for a special case of disjoint set union. J. Comput. Syst. Sci. 30(2), 209–221 (1985)

22. Gutwenger, C., Mutzel, P.: A linear time implementation of SPQR-trees. In: Marks, J. (ed.) GD 2000. LNCS, vol. 1984, pp. 77–90. Springer, Heidelberg (2001). https://doi.org/10.1007/3-540-44541-2_8

23. Hopcroft, J.E., Tarjan, R.E.: Dividing a graph into triconnected components. SIAM J. Comput. 2(3), 135–158 (1973)

24. Hutton, M.D., Lubiw, A.: Upward planar drawing of single-source acyclic digraphs. SIAM J. Comput. 25(2), 291–311 (1996)

25. Jelínek, V., Kratochvíl, J., Rutter, I.: A Kuratowski-type theorem for planarity of partially embedded graphs. Comput. Geom.: Theory Appl. 46(4), 466–492 (2013)

26. Jünger, M., Leipert, S.: Level planar embedding in linear time. In: Kratochvíyl, J. (ed.) GD 1999. LNCS, vol. 1731, pp. 72–81. Springer, Heidelberg (1999). https://doi.org/10.1007/3-540-46648-7_7

27. Mac Lane, S.: A structural characterization of planar combinatorial graphs. Duke Math. J. **3**(3), 460–472 (1937). https://doi.org/10.1215/S0012-7094-37-00336-3
28. Platt, C.R.: Planar lattices and planar graphs. J. Comb. Theory Ser. B **21**(1), 30–39 (1976)
29. Randerath, B., Speckenmeyer, E., Boros, E., Hammer, P., Kogan, A., Makino, K., Simeone, B., Cepek, O.: A satisfiability formulation of problems on level graphs. Electron. Notes Discret. Math. **9**, 269–277 (2001). IICS 2001 Workshop on Theory and Applications of Satisfiability Testing (SAT 2001)
30. Tutte, W.T.: Connectivity in Graphs. University of Toronto Press, Toronto (1966)

A Natural Quadratic Approach to the Generalized Graph Layering Problem

Sven Mallach[✉]

Department of Mathematics and Computer Science,
University of Cologne, Cologne, Germany
mallach@informatik.uni-koeln.de

Abstract. We propose a new exact approach to the generalized graph
layering problem that is based on a particular quadratic assignment formulation. It expresses, in a natural way, the associated layout restrictions
and several possible objectives, such as a minimum total arc length, minimum number of reversed arcs, and minimum width, or the adaptation
to a specific drawing area. Our computational experiments show a competitive performance compared to prior exact models.

Keywords: Graph drawing · Layering · Integer programming

1 Introduction

Hierarchical graph drawing is an indispensable tool to automate the cleaned-
up illustration of e.g. flow diagrams or data dependency representations. Here,
the dominant methodological framework studied in research and implemented in
software is the one proposed by Sugiyama et al. [9]. It involves four successive and
interdependent steps for cycle removal, vertex layering, crossing minimization,
and, finally, horizontal coordinate assignment and arc routing.

Classically, the workflow is carried out by solving the feedback arc set problem, i.e., reversing (a minimum number of) arcs, in the first step such that all
arcs have a common direction during the others. Then, for the final drawing,
original directions are restored. As a result, the height of a graph's layering is
bounded from below by the total height or number of vertices on a longest path
after the first step. In particular, a poor aspect ratio of the final drawing may
thus be inevitable from the very beginning. Also, the number and placement of
dummy vertices, usually introduced if an arc spans a layer to facilitate the other
steps and to more accurately account for the width, is strongly affected.

This motivates the recently studied integration of the first two steps [4,6–8].
Here, the central idea is to identify (a small number of) suitable arcs to be drawn
reverse to the intended hierarchical direction such that this enables a layout that
is *two-dimensionally* compact, possibly meets other common aesthetic criteria
such as a minimum total arc length, or even adapts to a drawing area of a certain
aspect ratio. Figure 1 exemplifies the potential effects on aesthetics and readability. Moreover, previous experimental studies in the referenced articles show that

---

The original version of this chapter was revised: the final formula in section 4.1 was
corrected. The correction to this chapter is available at https://doi.org/10.1007/978-
3-030-35802-0_44

D. Archambault and C. D. Tóth (Eds.): GD 2019, LNCS 11904, pp. 532–544, 2019.
https://doi.org/10.1007/978-3-030-35802-0_40

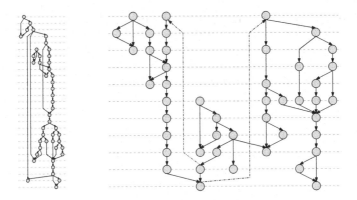

Fig. 1. Two layered drawings of the same directed acyclic graph. On the left a classic one, i.e., adhering to its longest path with all arcs pointing downwards, and on the right with a better aspect ratio achieved by reversing only two arcs (drawn dash-dotted).

significant improvements in terms of the drawing area or aspect ratio are frequent when optimizing the layering with respect to the corresponding objectives.

In this paper, we present a new exact approach to integrate vertex layering, the feedback arc set problem, and width or drawing area optimization. Its most appealing novelty is that it adheres to the quadratic nature of the problem rather than avoiding it. This quadratic nature does not only exist geometrically. Even when optimizing only for one direction (e.g. width), several aspects of a layered drawing depend on *conjunctive* decisions. For instance, the questions whether an arc is reversed, or whether it causes a dummy vertex on a particular layer, depend on the layer assignments of both of the arc's end vertices at the same time.

Our approach allows to express such conditions that depend on two simultaneous decisions, and thus all the present generalizations of the classical graph layering problem and their objective functions, in a natural and intuitive way. It is based on a quadratic assignment problem (QAP) formulated and solved as a mixed-integer program (MIP). As our computational results show, it can compete with the currently best known, but less intuitive MIP formulations. This is surprising since the QAP is considered to be among the hardest $\mathcal{NP}$-hard optimization problems. However, the graph layering problem poses a particularly well-suited special case: Our model does not require artificial constructions to model common drawing restrictions and objectives as linear expressions but profits from a sophisticated linearization technique, is compact in the number of constraints, and comparably insensitive to a graph's density. Ideally, our drawn links to the QAP inspire also new models for related layout styles or heuristic approaches to tackle larger instances or to support interactive user applications.

The paper is organized as follows. In Sect. 2, we give a quick survey of the state of the art generalizations of the classical graph layering problem. The major existing exact approaches to solve these are highlighted in Sect. 3. We then present our quadratic formulation of the most generalized graph layering problem variants in Sect. 4. Finally, we report in Sect. 5 on our computational evaluation, and close with a conclusion in Sect. 6.

2   A Landscape of Graph Layering Problems

A *layering* L of a directed graph $G = (V, A)$ with vertex set V and arc set A is a mapping $L\colon V \to \mathbb{N}^+$ that assigns each vertex $v \in V$ a unique *layer* index $L(v)$. Classically, and presuming that G is acyclic, a layering L is considered *feasible* if $L(v) - L(u) \geq 1$ holds for all $uv \in A$. In 1993, the following associated problem was introduced and shown to be polynomial time solvable by Gansner et al. [2]:

Problem 1. *Directed Layering Problem (DLP). Let $G = (V, A)$ be a directed acyclic graph. Find a feasible layering L of G minimizing the total arc length*

$$\sum_{uv \in A} \big(L(v) - L(u)\big).$$

Since, in a final layered drawing of non-acyclic graphs, the presence of reversed arcs is inevitable, and to overcome the limitations mentioned in the introduction (as well for acyclic graphs), a straightforward generalization of DLP discussed by Rüegg et al. [6,7], is to integrate arc reversals into the layering phase, and thus to consider a layering L feasible if $L(v) \neq L(u)$, i.e., $|L(v) - L(u)| \geq 1$, holds for all $uv \in A$. This gives rise to a second objective besides edge length minimization, namely the minimization of the number of reversed arcs. The trade off between both goals may be addressed by introducing respective weights ω_{len} and ω_{rev} into the objective function. The resulting optimization problem is then:

Problem 2. *Generalized Layering Problem (GLP). Let $G = (V, A)$ be a directed graph. Find a feasible layering L of G minimizing*

$$\omega_{len} \Big(\sum_{uv \in A} |L(v) - L(u)| \Big) + \omega_{rev} \, |\{uv \in A \mid L(v) < L(u)\}|.$$

As opposed to DLP, GLP is $\mathcal{NP}$-hard, and it remains so even if one of ω_{len} and ω_{rev} is zero [6]. Both problems have also been combined with width minimization which is worthwhile, even though the final drawing width is further influenced by the horizontal coordinate assignment and arc routing [7]. In this context, the *estimated width* $\mathcal{W}$ of a layering is given by the maximum number or maximum total width of original and dummy vertices in any of its layers. With an associated objective function weight ω_{wid}, we have the two according problems:

Problem 3. *Minimum Width Directed Layering Problem (DLP-W). Find a layering L of a directed acyclic graph $G = (V, A)$ feasible for DLP that minimizes*

$$\omega_{len} \Big(\sum_{uv \in A} (L(v) - L(u)) \Big) + \omega_{wid} \, \mathcal{W}.$$

Problem 4. *Minimum Width Generalized Layering Problem (GLP-W)[1]. Find a layering L of a directed graph $G = (V, A)$ feasible for GLP that minimizes*

---

[1] This problem is called *Compact Generalized Layering Problem (CGLP)* in [4,8] but renamed here to harmonize with the other variants.

$$\omega_{len} \left(\sum_{uv \in A} |L(v) - L(u)| \right) + \omega_{rev} |\{uv \in A \mid L(v) < L(u)\}| + \omega_{wid} \, \mathcal{W}.$$

Here, it should be mentioned that DLP-W is equivalent to the precedence-constrained multiprocessor scheduling problem (when ignoring dummy vertices), and thus as well $\mathcal{NP}$-hard [10].

Finally, Rüegg et al. propose to optimize a layering with respect to a target drawing area of width r_W and height r_H [8]. Informally, a 'best' such drawing is considered one that can be maximally scaled ('zoomed in') until it exhausts one of the two dimensions (cf. Figs. 1 and 2).

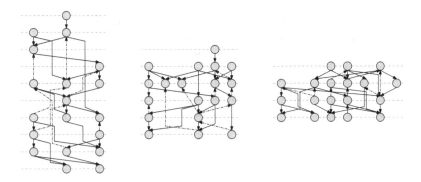

Fig. 2. A directed graph drawn based on a layering created by solving GLP-MS with the ratio $r_W : r_H$ set to $1:2$ (left), $1:1$ (middle), and $2:1$ (right). The obtained optimal $\mathcal{W} : \mathcal{H}$ combinations are respectively $5:10$, $6:6$, and $8:4$.

Formally, define $\mathcal{H} := \max_{v \in V} L(v)$ to be the height of a layering L. This definition is suitable as we may assume w.l.o.g. (and, if necessary, enforce a posteriori for any given and feasible layering) that vertices are assigned to consecutive layers starting from index one. The *scaling factor* $\mathcal{S}$ to be maximized is then the minimum of the ratios between the targeted and the actually used width and height, respectively, i.e., $\mathcal{S} = \min\{\frac{r_W}{\mathcal{W}}, \frac{r_H}{\mathcal{H}}\}$. Adding once more a corresponding weight ω_{scl}, the problem can be expressed as follows:

Problem 5. *Maximum Scale Generalized Layering Problem (GLP-MS). Given a drawing area of (normalized²) width r_W and (normalized) height r_H, find a feasible layering L of a directed graph $G = (V, A)$ that minimizes the expression*

$$\omega_{len} \left(\sum_{uv \in A} |L(v) - L(u)| \right) + \omega_{rev} |\{uv \in A \mid L(v) < L(u)\}| - \omega_{scl} \, \mathcal{S}.$$

---

² We remark that the parameters r_W and r_H used to characterize the target drawing area introduce undesired economies of scale since different values representing the same aspect ratio lead to different numeric maxima of $\mathcal{S}$. At the same time, it is not necessary to specify the dimensions of the drawing area in *absolute* values as the goal is a best effort layering for the *relative* aspect ratio of the targeted area. We thus propose to normalize r_W and r_H to $\frac{r_W}{\min\{r_W, r_H\}}$ and $\frac{r_H}{\min\{r_W, r_H\}}$, respectively.

A slight variation of GLP-MS as well proposed in [8] will be denoted GLP-MS*: Here, instead of minimizing $-\omega_{scl}\,\mathcal{S}$, one minimizes $\omega_{scl}\,\bar{\mathcal{S}}$ where $\bar{\mathcal{S}} := \frac{1}{\mathcal{S}}$.

3   Evolution of Exact Approaches to Graph Layering

To our best knowledge, all existing *exact* methods are based on integer programming, and can be coarsely classified based on three different types of variables involved to model the *layering* of the vertices of a directed graph $G = (V, A)$:

- Assignment-based: Binary variables $x_{v,k}$ taking on value one if $L(v) = k$, and taking on value zero otherwise.
- Ordering-based: This refers to binary variables $y_{k,v}$ taking on value one if $L(v) > k$, and taking on value zero otherwise (see, e.g., the appendix).
- Direct: $L(v)$ is directly modeled as a general *integer* variable.

Table 1 gives a quick overview of the formulations proposed so far. Some of them are named to ease their reference. Of course, GLP models can solve DLP (by setting $\omega_{rev} = \infty$), width minimization may be turned off (by setting $\omega_{wid} = 0$), and models to maximize the scaling factor can be used to minimize the width (by setting $r_W = 1$ and $r_H = \infty$).

Table 1. An overview of pre-existing MIP models for the different layering problems.

Problem	Assignment-based	Ordering-based	Direct
DLP			[2]
DLP-W	WHS [3]		
GLP			[7]
GLP-W	EXT [4]	CGL-W [4]	
GLP-MS*		CGL-MS* [8]	

The direct approach is a perfect fit for the DLP since the resulting problem can be solved in polynomial time by combinatorial algorithms as discussed by Gansner et al. [2]. However, this property (more precisely, the underlying structure) is lost when incorporating width constraints or arc reversals, and the direct method becomes inferior to models with binary variables in practice [4,7].

The first assignment-based formulation (WHS) was proposed by Healy and Nikolov in [3]. They considered width *constraints*, but their model may be easily altered to solve DLP-W. Here, the breakthrough to computational tractability for small and medium-sized graphs was to exploit the fixed direction of arcs.

Naturally, this could again not be preserved when it comes to GLP. Rather, modeling whether an arc is reversed or causes a dummy vertex comes then at the cost of additional variables and linearization constraints. Moreover, as Jabrayilov et al. [4] show, the corresponding assignment-based formulation (model EXT) for GLP-W rendered inferior to the first ordering-based one (called CGL-W). Consequently, in [8], Rüegg et al. used the latter as a basis to design a MIP model (that is, to the best of our knowledge, the only one so far) for GLP-MS* and that we thus consistently refer to as CGL-MS*.

However, as described in the following, a *quadratic* assignment-based approach re-enables the possibility to express conditions regarding arc reversals and dummy vertices intuitively, and without *artificial* linearization constraints.

4   A Natural Quadratic Graph Layering Framework

4.1   A Basic Quadratic Layer Assignment Model (QLA)

Let $G = (V, A)$ be a directed graph, and let Y be an upper bound on the number of layers such that assignment variables $x_{v,k}$ are to be introduced for all $v \in V$ and all $k \in \{1, \ldots, Y\}$. Consider as well the variables $p_{u,k,v,\ell}$, for all $uv \in A$ and all $k, \ell \in \{1, \ldots, Y\}$, that shall express the product $x_{u,k} \cdot x_{v,\ell}$. Then, a basic formulation of the layering constraints for any of the GLP problems (DLP as well, but this would permit to be more restrictive) can be expressed as follows:

$$\sum_{k=1}^{Y} x_{v,k} \qquad = 1 \qquad \text{for all } v \in V \tag{1}$$

$$\sum_{\ell=1}^{Y} p_{u,k,v,\ell} \qquad = x_{u,k} \qquad \text{for all } uv \in A, \; k \in \{1, \ldots, Y\} \tag{2}$$

$$\sum_{k=1}^{Y} p_{u,k,v,\ell} \qquad = x_{v,\ell} \qquad \text{for all } uv \in A, \; \ell \in \{1, \ldots, Y\} \tag{3}$$

$$p_{u,k,v,k} \qquad = 0 \qquad \text{for all } uv \in A, \; k \in \{1, \ldots, Y\} \tag{4}$$

$$x_{v,k} \qquad \in \{0,1\} \qquad \text{for all } v \in V, \; k \in \{1, \ldots, Y\}$$

$$p_{u,k,v,\ell} \qquad \in [0,1] \qquad \text{for all } uv \in A, \; k,\ell \in \{1, \ldots, Y\}$$

Equations (1) let each vertex be assigned a unique layer. Following the compact linearization approach [5], Eqs. (2) and (3) establish that variable $p_{u,k,v,\ell} = x_{u,k} \cdot x_{v,\ell}$ if the latter two take binary values[3]. As a nice feature of this model, the condition that two adjacent vertices cannot share a common layer can simply be expressed as the variable fixings (4), i.e., the variables may just be omitted in practice. Without accounting for these, the total number of constraints is $2|A| \cdot Y + |V|$, and the total number of variables is $|V| \cdot Y + |A| \cdot (Y-1)^2$.

---

[3] An intuitive interpretation for (2) is: If $x_{u,k}$ is zero, all products involving it must be zero as well. Conversely, if $x_{u,k}$ is one, then exactly one of the $p_{u,k,v,\ell}$ on the left hand side (which are *all* the products of $x_{u,k}$ with different $x_{v,\ell}$ for some fixed $v \in V$, $v \neq u$) need to be equal to one as well due to (1). Equations (3) imply the same for the second factor of any product variable $p_{u,k,v,\ell}$. Of course, these as well as (2) and (3) could be avoided under employment of an evolved *non-linear* solution method.

In terms of the latter, the ordering-based CGL models (a compacted reformulation of those in [4] and [8] is displayed in the appendix) are more economical. Even if auxiliary variables for arc reversals or dummy vertices are introduced, their total number is still only $(|V| + |A|) \cdot (Y - 1)$. However, the CGL models induce about twice the number of constraints compared to the model above (which is more critical to a MIP solver), and they are more sensitive to a graph's density.

For any arc $uv \in A$, there is exactly one pair of layers ℓ and k, $\ell \neq k$, such that $x_{u,\ell} \cdot x_{v,k} = 1$. All other products are zero. The length of $uv \in A$ thus equals

$$\sum_{\ell=2}^{Y}\sum_{k=1}^{\ell-1} ((\ell - k) \cdot (x_{u,k} \cdot x_{v,\ell} + x_{u,\ell} \cdot x_{v,k})) = \sum_{\ell=2}^{Y}\sum_{k=1}^{\ell-1} ((\ell - k) \cdot (p_{u,k,v,\ell} + p_{u,\ell,v,k})).$$

Similarly, the arc $uv \in A$ is reversed if and only if the expression

$$\sum_{\ell=2}^{Y} \left(x_{u,\ell} \cdot \sum_{k=1}^{\ell-1} x_{v,k} \right) = \sum_{\ell=2}^{Y}\sum_{k=1}^{\ell-1} p_{u,\ell,v,k}$$

evaluates to one. Otherwise, the expression will evaluate to zero.

Finally, an arc $uv \in A$ causes a dummy vertex on a layer $k \in \{2, \ldots, Y-1\}$ if and only if k is between the layers u and v are assigned to, i.e., if the expression

$$\sum_{\ell=1}^{k-1}\sum_{m=k+1}^{Y} (p_{u,\ell,v,m} + p_{u,m,v,\ell})$$

evaluates to one. Again, it will be zero otherwise.

4.2   Quadratic Layer Assignment for GLP-W (QLA-W)

To build model QLA-W from the above basis, it suffices to add a additional continuous variable $\mathcal{W} \in \mathbb{R}_{\geq 0}$ to capture the width together with the Y constraints:

$$\sum_{v \in V} x_{v,k} \qquad\qquad\qquad \leq \mathcal{W} \quad \text{for all } k \in \{1, Y\}$$

$$\sum_{uv \in A}\sum_{\ell=1}^{k-1}\sum_{m=k+1}^{Y}(p_{u,\ell,v,m} + p_{u,m,v,\ell}) + \sum_{v \in V} x_{v,k} \leq \mathcal{W} \quad \text{for all } k \in \{2, \ldots, Y-1\}$$

The objective function can be stated as

$$\text{minimize} \sum_{uv \in A} \left(\sum_{\ell=2}^{Y}\sum_{k=1}^{\ell-1}\omega_{len}(\ell - k)(p_{u,k,v,\ell} + p_{u,\ell,v,k}) + \omega_{rev}\, p_{u,\ell,v,k} \right) + \omega_{wid}\, \mathcal{W}.$$

4.3   Quadratic Layer Assignment for GLP-MS* (QLA-MS*)

Recalling the definitions from Sect. 2, GLP-MS* asks for the (weighted) minimization of the inverse scaling factor

$$\bar{S} = \frac{1}{S} = \frac{1}{\min\{\frac{r_W}{\mathcal{W}}, \frac{r_H}{\mathcal{H}}\}} = \max\left\{ \frac{\mathcal{W}}{r_W}, \frac{\mathcal{H}}{r_H} \right\}.$$

Thus, to obtain model QLA-MS*, the basic model is to be extended with an according variable $\bar{S} \in \mathbb{R}_{\geq 0}$, and with the $Y + |V|$ constraints:

$$\sum_{v \in V} x_{v,k} \leq r_W \, \bar{S} \quad \text{for all } k \in \{1, Y\}$$

$$\sum_{uv \in A} \sum_{\ell=1}^{k-1} \sum_{m=k+1}^{Y} (p_{u,\ell,v,m} + p_{u,m,v,\ell}) + \sum_{v \in V} x_{v,k} \leq r_W \, \bar{S} \quad \text{for all } k \in \{2, \ldots, Y-1\}$$

$$\sum_{k=1}^{Y} k \, x_{v,k} \leq r_H \, \bar{S} \quad \text{for all } v \in V$$

Finally, the objective function is

$$\text{minimize} \sum_{uv \in A} \left(\sum_{\ell=2}^{Y} \sum_{k=1}^{\ell-1} \omega_{len}(\ell - k)(p_{u,k,v,\ell} + p_{u,\ell,v,k}) + \omega_{rev} \, p_{u,\ell,v,k} \right) + \omega_{scl} \, \bar{S}.$$

5   Experimental Evaluation

The *aesthetic effects* when integrating GLP-MS and GLP-W into the hierarchical framework by Sugiyama et al. were extensively studied already in [4,6–8]. Here, we thus confine ourselves to evaluate (i) *the computational effects when targeting different aesthetic objectives and aspect ratios*, and (ii) *the competitiveness of QLA-W and QLA-MS* with respect to CGL-W and CGL-MS*.

To accomplish this, we employed the original models CGL-W from [4] and CGL-MS* from [8], except for leaving out constraints that enforce at least one vertex to be placed on the first layer[4]. Also, we employ the same two instance sets as in the mentioned prior studies: The first set ATTAR are the AT&T graphs from [1] whereof we extracted all non-tree instances with $20 \leq |V| \leq 60$, and $20 \leq |A| \leq 168$. Their density $\frac{|A|}{|V|}$ is within $[1, 4.72]$ (on average 1.47). The second set RANDOM consists of 180 randomly generated and also acyclic and non-tree graphs with 17 to 60 vertices, and 30 to 91 arcs. These were obtained as follows: First, a respective number of vertices was created. Then, for each vertex, a random number of outgoing arcs (with arbitrary random target) is added such that the total number of arcs is 1.5 times the number of vertices. Finally, isolated vertices were removed.

During our experiments, all MIPs were solved using Gurobi[5] (release version 8) single-threadedly on a Debian Linux system with an Intel Core i7-3770T CPU (2.5 GHz) and 8 GB RAM, and with a time limit set to half an hour.

---

[4] Any solution violating these constraints may be normalized a posteriori by simply ignoring empty layers. Moreover, in case of GLP-MS*, they are implied if the height imposes a stronger restriction on the minimization of $\bar{S}$ than the width (which depends on the adjacency structure of the graph as well as on the choice of r_W, r_H, and Y). In any other case, they do break some symmetries, but did not lead to better experimental results.

[5] A proprietary MIP solver, see https://www.gurobi.com.

5.1   GLP and GLP-W

Here, we consider two experiments. In experiment (1), we set $Y = \lceil 1.6 \cdot \sqrt{|V|} \rceil$, $\omega_{len} = 1$, $\omega_{rev} = Y\omega_{len}|A|$, and $\omega_{wid} = 1$ as in [4]. This can be seen as an almost pure GLP setting, since one saved unit of width has the same (small) effect in the objective function as one saved unit of (total) edge length. Moreover, arcs are reversed only if this is inevitable due to the choice of Y or because they are part of a cycle. In experiment (2), we instead give priority to width minimization by increasing ω_{wid} to $\omega_{rev}|A| + |A| \cdot Y + 1$.

The results are shown in Fig. 3. The most distinctive observation is that the layering problem is considerably harder to solve with *both* MIPs if emphasis is given to width minimization. In this case, the solution times are higher and several timeouts occur for the larger graphs, while the pure GLP setting can be solved routinely for all instances considered. With respect to experiment (2), the results obtained with QLA-W are slightly better on average than with CGL-W for the ATTar graphs, whereas the opposite is true for the Random graphs, especially due to the increased number of timeouts for the larger ones. Thus, in total, there is no clear superior or inferior model – on average both show a comparable or competitive performance with the MIP solver employed.

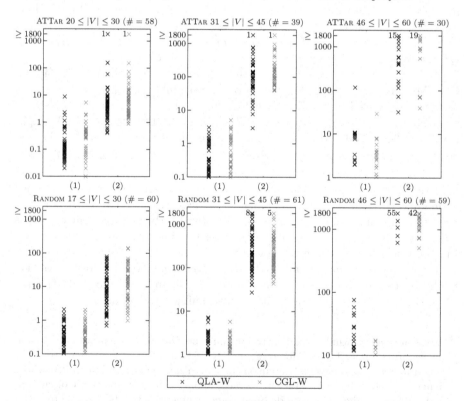

Fig. 3. The plots depict the solution times in seconds (a cross per graph) for GLP with neglected (1) and emphasized width minimization (2). Crosses at the top (1800 s) correspond to timeouts, their quantities are given to the left of the respective cross.

5.2   GLP-MS*

In this experiment, the parameters are set (almost) as in [8]: $Y = |V|$, $\omega_{len} = 1$, $\omega_{rev} = Y\omega_{len}|A|$, and $\omega_{scl} = \omega_{rev}|A| + |A| \cdot Y + 1$. Priorities are thus the same as in experiment (2) before, except now emphasizing on a maximum scaling factor instead of the width alone. The $r_H : r_W$ ratios considered are $1 : 2$, $1 : 1$, and $2 : 1$.

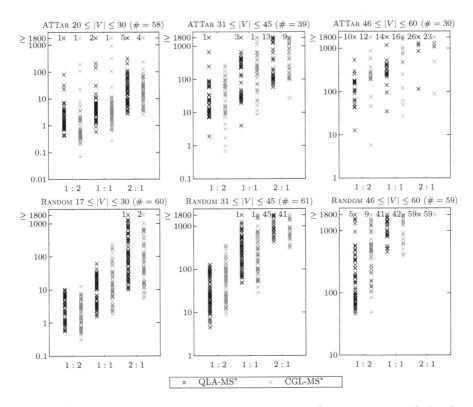

Fig. 4. The plots depict the solution times in seconds (a cross per graph) for the different $r_H : r_W$ combinations. Crosses at the top (1800 s) correspond to timeouts, their quantities are given to the left of the respective cross.

As Fig. 4 shows, the results are again diverse under fine-grained inspection, and comparable from a coarse perspective. For both QLA-MS* and CGL-MS* there are instance subsets and $r_H : r_W$ ratios where the use of either one leads to faster solutions and less timeouts.

Also, for both formulations, the 2 : 1-case, where the height is constrained to be *at most* twice the width, turns out to be the hardest - which is not surprising as e.g. the ATTAR graphs are *at least* twice as high as wide when drawn with the classic framework [4]. As opposed to that, in the 1 : 2-case, the condition that the height is restricted to be no more than half of the width is a strong one, as most of the considered graphs cannot be layered very widely. This entails better bounds on the objective function during the MIP solution process.

An interesting question is the relative tractability of GLP-W and (especially the 2 : 1-case of) the more general GLP-MS*, since (in a sense) both aim at a (relatively) small width, but the first with respect to a fixed (but larger) height limit, and the second additionally requiring to find the best $\mathcal{H} : \mathcal{W}$-pair under the given aspect ratio constraint. As could be expected, comparing the two results indicates that GLP-W appears to be considerably easier to solve to optimality, at least with the present modeling techniques.

6   Conclusion

In recent years, several extensions of the graph layering problem of increasing generality have been studied, and several algorithmic solution approaches have been proposed. Exact methods are based on integer programming where assignment and ordering variables have dominated the most successful models for the more general variants of the problem.

In this paper, we proposed a new *quadratic* assignment approach that can be adapted to each of these, allows a natural expression of the associated layout restrictions or aesthetic objectives, and turns out to be computationally competitive. However, as soon as the emphasis is on width minimization or even a particular drawing area is targeted, still neither of the present methods is able to *routinely* solve *arbitrary* instances beyond about 50 vertices. Moreover, as long as the proposed quadratic model is solved based on linearization, its problem size will become a limitation if the number of vertices is considerably increased. Nevertheless, depending on the adjacency structure of the graphs to be layered, this may be possible, and our results show that different settings (aspect ratios, emphasis or neglection of the width, restrictive or non-restrictive heights) have a strong impact on the tractability of the problem – which also means that some can be handled quite well.

Also, the recent generalizations of the layering problem are a clear step towards the requirements of real-world applications – and real-world displays. We hope that our drawn links to the quadratic assignment problem inspire also some novel heuristic solution approaches, or exact models for related graph drawing paradigms.

A   Ordering-Based Reference Models

As a reference, we display here slightly more compact reformulations of CGL-W and CGL-MS* compared to their respective original presentations in [4] and [8].

Let $G = (V, A)$ be a directed graph, and let Y be an upper bound on the height of the layering to be chosen a priori. A basic CGL model then involves binary variables $y_{k,v}$, for all $v \in V$ and all layer indices $k \in \{1, \ldots, Y-1\}$, being equal to 1 if $k < L(v)$ and being equal to zero otherwise. In particular $L(v) = 1$ if and only if $y_{1,v} = 0$, $L(v) = Y$ if and only if $y_{k,v} = 1$ for all $k \in \{1, \ldots, Y-1\}$, and $L(v) = k$ if and only if $y_{k-1,v} - y_{k,v} = 1$. The number of basic variables thus amounts to $|V| \cdot (Y - 1)$.

Moreover, there are only the following $|V| \cdot (Y - 2)$ basic constraints that establish transitivity in the sense that $L(v) > k$ implies $L(v) > k - 1$ for each $k \in \{2, \ldots, Y - 1\}$:

$$y_{k,v} - y_{k-1,v} \leq 0 \qquad \text{for all } v \in V,\ k \in \{2, \ldots, Y - 1\}$$

However, to model GLP, $|A|$ further auxiliary variables r_{uv} as well as $|A| \cdot Y$ constraints – ensuring that r_{uv} is equal to one if $uv \in A$ is reversed and equal to zero otherwise – are required.

These constraints are:

$$
\begin{aligned}
y_{1,u} - r_{uv} &\geq 0 && \text{for all } uv \in A \\
y_{1,v} + r_{uv} &\geq 1 && \text{for all } uv \in A \\
y_{k-1,u} - y_{k,v} - r_{uv} &\leq 0 && \text{for all } uv \in A,\ k \in \{2, \ldots, Y - 1\} \\
y_{k-1,v} - y_{k,u} + r_{uv} &\leq 1 && \text{for all } uv \in A,\ k \in \{2, \ldots, Y - 1\} \\
y_{Y-1,u} - r_{uv} &\leq 0 && \text{for all } uv \in A \\
y_{Y-1,v} + r_{uv} &\leq 1 && \text{for all } uv \in A
\end{aligned}
$$

Moreover, to model dummy vertices (and thus the width), $|A| \cdot (Y-2)$ further variables $d_{uv,k}$ for each arc $uv \in A$ and each layer $k \in \{2, \ldots, Y - 1\}$ are required. The associated constraints enforcing $d_{uv,k}$ to be one if $uv \in A$ causes a dummy vertex on layer k (otherwise, an optimum solution has $d_{uv,k} = 0$ whenever $\omega_{len} > 0$) are:

$$
\begin{aligned}
y_{k,u} - y_{k-1,v} - d_{uv,k} &\leq 0 && \text{for all } uv \in A,\ k \in \{2, \ldots, Y - 1\} \\
y_{k,v} - y_{k-1,u} - d_{uv,k} &\leq 0 && \text{for all } uv \in A,\ k \in \{2, \ldots, Y - 1\}
\end{aligned}
$$

To obtain CGL-W, one further adds a variable $\mathcal{W}$ and the Y constraints:

$$\sum_{v \in V} (1 - y_{1,v}) \leq \mathcal{W} \tag{5}$$

$$\sum_{v \in V} y_{Y-1,v} \leq \mathcal{W} \tag{6}$$

$$\sum_{v \in V} (y_{k-1,v} - y_{k,v}) + \sum_{uv \in A} d_{uv,k} \leq \mathcal{W} \qquad \text{for all } k \in \{2, \ldots, Y - 1\} \tag{7}$$

The total number of constraints is then $(4|A|+|V|+1) \cdot (Y-2)+4|A|+2$. One can now exploit that the length of an arc (i.e., the difference of the layer indices its endpoints are assigned to) is equivalent to the number of dummy vertices it causes plus one. Thus, the objective function for CGL-W can be expressed as:

$$\text{minimize} \sum_{uv \in A} \left(\omega_{rev}\, r_{uv} + \omega_{len} \left(1 + \sum_{k=2}^{Y-1} d_{uv,k} \right) \right) + \omega_{wid}\, \mathcal{W}$$

To rather obtain CGL-MS*, it suffices to introduce the variable $\bar{S}$ instead, replace $\mathcal{W}$ with $r_W\,\bar{S}$ in (5)–(7), and to add the $|V|$ additional constraints:

$$1 + \sum_{k\in\{1,\dots,Y-1\}} y_{k,v} \qquad \leq r_H\,\bar{S} \qquad \text{for all } v \in V$$

Then, the objective is

$$\text{minimize} \sum_{uv\in A}\left(\omega_{rev}\,r_{uv} + \omega_{len}(1 + \sum_{k=2}^{Y-1} d_{uv,k})\right) + \omega_{scl}\,\bar{S}.$$

and the total number of constraints amounts to $(4|A|+|V|+1)\cdot(Y-2)+4|A|+|V|+2$.

References

1. Di Battista, G., Garg, A., Liotta, G., Parise, A., Tamassia, R., Tassinari, E., Vargiu, F., Vismara, L.: Drawing directed acyclic graphs: an experimental study. In: North, S. (ed.) GD 1996. LNCS, vol. 1190, pp. 76–91. Springer, Heidelberg (1997). https://doi.org/10.1007/3-540-62495-3_39

2. Gansner, E.R., Koutsofios, E., North, S.C., Vo, K.P.: A technique for drawing directed graphs. Softw. Eng. 19(3), 214–230 (1993)

3. Healy, P., Nikolov, N.S.: A branch-and-cut approach to the directed acyclic graph layering problem. In: Goodrich, M.T., Kobourov, S.G. (eds.) GD 2002. LNCS, vol. 2528, pp. 98–109. Springer, Heidelberg (2002). https://doi.org/10.1007/3-540-36151-0_10

4. Jabrayilov, A., Mallach, S., Mutzel, P., Rüegg, U., von Hanxleden, R.: Compact layered drawings of general directed graphs. In: Hu, Y., Nöllenburg, M. (eds.) GD 2016. LNCS, vol. 9801, pp. 209–221. Springer, Cham (2016). https://doi.org/10.1007/978-3-319-50106-2_17

5. Mallach, S.: Compact linearization for binary quadratic problems subject to assignment constraints. 4OR 16(3), 295–309 (2018). https://doi.org/10.1007/s10288-017-0364-0

6. Rüegg, U., Ehlers, T., Spönemann, M., von Hanxleden, R.: A generalization of the directed graph layering problem. Technical Report 1501, Kiel University, Department of Computer Science (2015). ISSN 2192–6247

7. Rüegg, U., Ehlers, T., Spönemann, M., von Hanxleden, R.: A generalization of the directed graph layering problem. In: Hu, Y., Nöllenburg, M. (eds.) GD 2016. LNCS, vol. 9801, pp. 196–208. Springer, Cham (2016). https://doi.org/10.1007/978-3-319-50106-2_16

8. Rüegg, U., Ehlers, T., Spönemann, M., von Hanxleden, R.: Generalized layerings for arbitrary and fixed drawing areas. J. Graph Algorithms Appl. 21(5), 823–856 (2017). https://doi.org/10.7155/jgaa.00441

9. Sugiyama, K., Tagawa, S., Toda, M.: Methods for visual understanding of hierarchical system structures. IEEE Trans. Syst. Man Cybern. 11(2), 109–125 (1981)

10. Ullman, J.D.: NP-complete scheduling problems. J. Comput. Syst. Sci. 10(3), 384–393 (1975). https://doi.org/10.1016/S0022-0000(75)80008-0

Graph Stories in Small Area

Manuel Borrazzo, Giordano Da Lozzo$^{(\boxtimes)}$, Fabrizio Frati,
and Maurizio Patrignani

Roma Tre University, Rome, Italy
{borrazzo,dalozzo,frati,patrigna}@dia.uniroma3.it

Abstract. We study the problem of drawing a dynamic graph, where
each vertex appears in the graph at a certain time and remains in the
graph for a fixed amount of time, called the window size. This defines a
graph story, i.e., a sequence of subgraphs, each induced by the vertices
that are in the graph at the same time. The drawing of a graph story
is a sequence of drawings of such subgraphs. To support readability,
we require that each drawing is straight-line and planar and that each
vertex maintains its placement in all the drawings. Ideally, the area of
the drawing of each subgraph should be a function of the window size,
rather than a function of the size of the entire graph, which could be too
large. We show that the graph stories of paths and trees can be drawn on
a $2W \times 2W$ and on an $(8W + 1) \times (8W + 1)$ grid, respectively, where W
is the window size. These results yield linear-time algorithms. Further,
we show that there exist graph stories of planar graphs whose subgraphs
cannot be drawn within an area that is only a function of W.

1 Introduction

We consider a graph that changes over time. Its vertices enter the graph one after
the other and persist in the graph for a fixed amount of time, called the *window
size*. We call such a dynamic graph a *graph story*. More formally, let V be the
set of vertices of a graph G. Each vertex $v \in V$ is equipped with a label $\tau(v)$,
which specifies the time instant at which v appears in the graph. The labeling
$\tau : V \to \{1, 2, \ldots, |V|\}$ is a bijective function specifying a total ordering for V.
At any time t, the graph G_t is the subgraph of G induced by the set of vertices
$\{v \in V : t - W < \tau(v) \leq t\}$. We denote a graph story by $\langle G, \tau, W \rangle$.

We are interested in devising an algorithm for visualizing graph stories. The
input of the algorithm is an entire graph story and the output is what we call
a drawing story. A *drawing story* is a sequence of drawings of the graphs G_t.
The typical graph drawing conventions can be applied to a drawing story. E.g.,
a drawing story is planar, straight-line, or on the grid if all its drawings are
planar, straight-line, or on the grid, respectively.

This research was supported by MIUR Proj. "MODE" n° 20157EFM5C, by MIUR
Proj. "AHeAD" n° 20174LF3T8, by MIUR-DAAD JMP n° 34120, by H2020-MSCA-
RISE Proj. "CONNECT" n° 734922, and by Roma Tre University Proj. "GeoView".

D. Archambault and C. D. Tóth (Eds.): GD 2019, LNCS 11904, pp. 545–558, 2019.
https://doi.org/10.1007/978-3-030-35802-0_41

A trivial way for constructing a drawing story would be to first produce a drawing of G and then to obtain a drawing of G_t, for each time t, by filtering out vertices and edges that do not belong to G_t. However, if the number of vertices of G is much larger than W, this strategy might produce unnecessarily large drawings. Ideally, the area of the drawing of each graph G_t should be a (polynomial) function of W, rather than a function of the size of the entire graph.

In this paper we show that the graph stories of paths and trees can be drawn on a $2W \times 2W$ and on an $(8W + 1) \times (8W + 1)$ grid, respectively, so that all the drawings of the story are straight-line and planar, and so that vertices do not change their position during the drawing story. Further, we show that there exist graph stories of planar graphs that cannot be drawn within an area that is only a function of W, if planarity is required and vertices are not allowed to change their position during the drawing story.

The visualization of dynamic graphs is a classic research topic in graph drawing. In what follows we compare our model and results with the literature. We can broadly classify the different approaches in terms of the following features [2]. (i) The objects that appear and disappear over time can be vertices or edges. (ii) The lifetime of the objects may be fixed or variable. (iii) The story may be entirely known in advance (*off-line model*) or not (*on-line model*). In this paper, the considered objects are vertices, the lifetime is fixed and the model is off-line.

A considerable amount of the literature on the theoretical aspects of dynamic graphs focuses on trees. In [3], the objects are edges, the lifetime W is fixed, and the model is on-line; an algorithm is shown for drawing a tree in $O(W^3)$ area, under the assumption that the edges arrive in the order of a Eulerian tour of the tree. In [10], the objects are vertices, the lifetime W is fixed, the model is off-line (namely, the sequence of vertices is known in advance, up to a certain threshold k), and the vertices can move. A bound in terms of W and k is given on the total amount of movement of the vertices. In [22], each subgraph of the story is given (each subgraph is a tree, whereas the entire graph may be arbitrary), each object can have an arbitrary lifetime, the model is off-line, and the vertices can move. Aesthetic criteria as in the classical Reingold-Tilford algorithm [19] are pursued.

Other contributions consider more general types of graphs. In [15], the objects are edges, which enter the drawing and never leave it, the model is on-line, the considered graphs are outerplanar, and the vertices are allowed to move by a polylogarithmic distance. In [8], the objects are edges, the lifetime is fixed, and the model is off-line; NP-completeness is shown for the problem of computing planar topological drawings of the graphs in the story; other results for the topological setting are presented in [1,21]. Further related results appear in [7,11, 12,16,18,20]; in particular, geometric simultaneous embeddings [4,6] are closely related to the setting we consider in this paper. Contributions focused on the information-visualization aspects of dynamic graphs are surveyed in [2]; further, a survey on temporal graph problems appears in [17].

Missing proofs can be found in the full version of the paper [5].

2   Preliminaries

In this section, we present definitions and preliminaries.

Graphs and Drawings. We denote the set of vertices and edges of a graph G by $V(G)$ and $E(G)$, respectively. A *drawing* of G maps each vertex in $V(G)$ to a distinct point in the plane and each edge in E to Jordan arc between its end-points. A drawing is *straight-line* if each arc is a straight-line segment, it is *planar* if no two arcs intersect, except at a common endpoint, and it is *on the grid* (or, a *grid drawing*) if each vertex is mapped to a point with integer coordinates.

Rooted Ordered Forests and Their Drawings. A *rooted tree* T is a tree with one distinguished vertex, called *root* and denoted by $r(T)$. We denote by $T(u)$ the subtree of T rooted at a node u.

A *rooted ordered tree* T is a rooted tree such that, for each vertex $v \in V(T)$, a *left-to-right* (linear) order $u_1, \ldots, u_k$ of the children of v is specified. A sequence $\mathcal{F} = T_1, T_2, \ldots, T_k$ of rooted ordered trees is a *rooted ordered forest*.

A *strictly-upward* drawing of a rooted tree T is such that each edge is represented by a curve monotonically increasing in the y-direction from a vertex to its parent. A *strictly-upward* drawing Γ of a rooted forest $\mathcal{F}$ is such that the drawing of each tree $T_i \in \mathcal{F}$ in Γ is strictly-upward. A strictly-upward drawing of an ordered tree T is *order-preserving* if, for each vertex $v \in V(T)$, the left-to-right order of the edges from v to its children in the drawing is the same as the order associated with v in T. A strictly-upward drawing of a rooted ordered forest $\mathcal{F}$ is *order-preserving* if the drawing of each tree $T_i \in \mathcal{F}$ is order-preserving. The definitions of (*order-preserving*) *strictly-leftward*, *strictly-downward*, and *strictly-rightward* drawings of (ordered) rooted trees and forests are similar.

Geometry. Given two points p_1 and p_2 in $\mathbb{R}^2$, we denote by $[p_1, p_2]$ the closed straight-line segment connecting p_1 and p_2. The *width* and *height* of a grid $[a, b] \times [c, d]$ are $|[a, b] \cap \mathbb{Z}| = b - a + 1$ and $|[c, d] \cap \mathbb{Z}| = d - c + 1$, respectively. A grid drawing Γ of a graph *lies on a $W \times H$ grid* if Γ is enclosed by the bounding box of some grid of width W and height H, and *lies on the* grid $[a, b] \times [c, d]$ if Γ is enclosed by the bounding box of the grid $[a, b] \times [c, d]$.

Graph Stories. A graph story $\langle G, \tau, W \rangle$ is naturally *associated with* a sequence $\langle G_1, G_2, \ldots, G_{n+W-1} \rangle$; for any $t \in \{1, \ldots, n + W - 1\}$, the graph G_t is the subgraph of G induced by the set of vertices $\{v \in V : t - W < \tau(v) \le t\}$. Clearly, $|V(G_t)| \le W$. Note that $G_1, G_2, \ldots, G_{W-1}$ are subgraphs of G_W and $G_{n+1}, G_{n+2}, \ldots, G_{n+W-1}$ are subgraphs of G_n, while each of G_t and G_{t+1} has a vertex that the other graph does not have, for $t = W, \ldots, n - 1$. A graph story $\langle G, \tau, W \rangle$ in which G is a planar graph, a path, or a tree is a *planar graph story*, a *path story*, or a *tree story*, respectively.

A *drawing story* for $\langle G, \tau, W \rangle$ is a sequence $\langle \Gamma_1, \Gamma_2, \ldots, \Gamma_{n+W-1} \rangle$ of drawings such that, for every $t = 1, \ldots, n+W-1$: (i) Γ_t is a drawing of G_t, (ii) a vertex v is drawn at the same position in all the Γ_t's such that $v \in V(G_t)$, and (iii) an edge (u, v) is represented by the same curve in all the Γ_t's such that $(u, v) \in E(G_t)$.

The above definition implies that the drawings $\Gamma_1, \Gamma_2, \ldots, \Gamma_{W-1}$ are the restrictions of Γ_W to the vertices and edges of $G_1, G_2, \ldots, G_{W-1}$, respectively, and that the drawings $\Gamma_{n+1}, \Gamma_{n+2}, \ldots, \Gamma_{n+W-1}$ are the restrictions of Γ_n to the vertices and edges of $G_{n+1}, G_{n+2}, \ldots, G_{n+W-1}$, respectively. Hence, an algorithm that constructs a drawing story only has to specify the drawings $\Gamma_W, \Gamma_{W+1}, \ldots, \Gamma_n$.

We only consider drawing stories Γ that are planar, straight-line, and on the grid. Storing each drawing in Γ would require $\Omega(n \cdot W)$ space in total. However, since each vertex has the same position in all the drawings where it appears, since edges are straight-line segments, and since any two consecutive graphs in a graph story differ by $O(1)$ vertices, we can encode Γ in total $O(n)$ space.

Let Γ be a straight-line drawing story of a graph story $\langle G, \tau, W \rangle$ and let G' be a subgraph of G (possibly $G' = G$). The *drawing of G' induced by Γ* is the straight-line drawing of G' in which each vertex has the same position as in every drawing $\Gamma_t \in \Gamma$ where it appears. Note that the drawing of G' induced by Γ might have crossings even if Γ is planar. For a subset $B \subseteq V(G)$, let $G[B]$ be the subgraph of G induced by the vertices in B and let $\Gamma[B]$ be the straight-line drawing of $G[B]$ induced by Γ.

Given a graph story $\langle G, \tau, W \rangle$, we will often consider a partition of V into *buckets* $B_1, \ldots, B_h$, where $h = \lceil \frac{n}{W} \rceil$. For $i = 1, \ldots, h$, the bucket B_i is the set of vertices v such that $(i-1)W + 1 \leq \tau(v) \leq \min\{i \cdot W, n\}$. Note that all buckets have W vertices, except, possibly, for B_h. We have the following useful property.

Property 1. For any t, let $i = \lceil \frac{t}{W} \rceil$. Then G_t is a subgraph of $G[B_{i-1} \cup B_i]$.

3   Planar Graph Stories

In this section we prove a lower bound on the size of any drawing story of a planar graph story. Let $n \equiv 0 \mod 3$. An n-vertex *nested triangles graph* G contains the vertices and the edges of the 3-cycle $C_i = (v_{i-2}, v_{i-1}, v_i)$, for $i = 3, 6, \ldots, n$, plus the edges (v_i, v_{i+3}), for $i = 1, 2, \ldots, n-3$. The nested triangles graphs with $n \geq 6$ vertices are 3-connected and thus they have a unique combinatorial embedding (up to a flip) [23]. We have the following.

Theorem 1. Let $\langle G, \tau, 9 \rangle$ be a planar graph story such that G is an n-vertex nested triangles graph and $\tau(v_i) = i$. Then any drawing story of $\langle G, \tau, 9 \rangle$ lies on a $\Omega(n) \times \Omega(n)$ grid.

Proof. In order to prove the statement we show that, for any drawing story Γ of $\langle G, \tau, 9 \rangle$, the straight-line drawing of G induced by Γ is planar. Then the statement follows from well-known lower bounds in the literature [9,13,14].

Let $\Gamma = \Gamma_1, \Gamma_2, \ldots, \Gamma_{n+8}$. Note that, for any $m = 9, 12, \ldots, n$, the graph G_m contains the cycles $C_{m-6} = (v_{m-8}, v_{m-7}, v_{m-6})$, $C_{m-3} = (v_{m-5}, v_{m-4}, v_{m-3})$, and $C_m = (v_{m-2}, v_{m-1}, v_m)$. For any $m = 9, 12, \ldots, n$, let H_m and I_m be the subgraphs of G induced by $v_1, v_2, \ldots, v_m$ and by $v_{m-5}, v_{m-4}, \ldots, v_m$, respectively, and let Γ_m^H and Γ_m^I be the drawings of H_m and I_m induced by Γ, respectively.

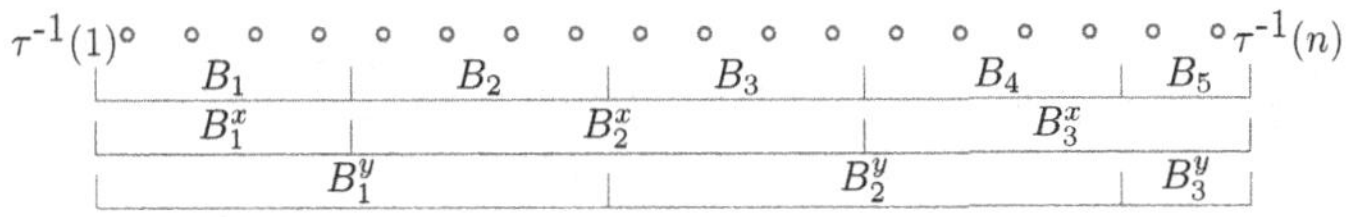

Fig. 1. Examples of buckets, x-buckets, and y-buckets when $W = 4$.

Since $G = H_n$, in order to show that the drawing of G induced by Γ is planar, it suffices to show that Γ_m^H is planar, for each $m = 9, 12, \ldots, n$. We prove this statement by induction on m. The induction also proves that the cycle C_m bounds a face of Γ_m^H and of Γ_m^I.

Suppose first that $m = 9$. Then $H_9 = G_9$ and the drawing $\Gamma_9^H = \Gamma_9$ of G_9 is planar since Γ is a planar straight-line drawing of $\langle G, \tau, 9\rangle$ and $\Gamma_9 \in \Gamma$. The 3-connectivity of H_9 and I_9 implies that C_9 bounds a face of Γ_9^H and Γ_9^I.

Suppose now that $m > 9$. We show that Γ_m^H is planar. By induction, Γ_{m-3}^H is planar. Thus, we only need to prove that no crossing is introduced by placing the vertices v_{m-2}, v_{m-1}, and v_m, which belong to H_m and not to H_{m-3}, in Γ_{m-3}^H as they are placed in Γ_m and by drawing their incident edges as straight-line segments. First, by induction, the cycle $C_{m-3} = (v_{m-5}, v_{m-4}, v_{m-3})$ bounds a face of Γ_{m-3}^H and of Γ_{m-3}^I. Second, the vertices v_{m-2}, v_{m-1}, and v_m, as well as their incident edges, lie inside such a face in Γ_m. Namely, v_{m-2}, v_{m-1}, and v_m lie inside the same face of Γ_{m-3}^I in Γ_m, as otherwise the cycle C_m would cross edges of I_{m-3} in Γ_m; further, the face of Γ_{m-3}^I in which v_{m-2}, v_{m-1}, and v_m lie in Γ_m is incident to all of v_{m-5}, v_{m-4}, and v_{m-3}, as otherwise the edges (v_{m-5}, v_{m-2}), (v_{m-4}, v_{m-1}), and (v_{m-3}, v_m) would cross edges of I_{m-3} in Γ_m; however, no face of Γ_{m-3}^I other than the one delimited by C_{m-3} is incident to all of v_{m-5}, v_{m-4}, and v_{m-3}. This proves the planarity of Γ_m^H. The induction and the proof of the theorem are completed by observing that C_m bounds a face of Γ_m^H and Γ_m^I. □

4   Path Stories

Let $\langle G, \tau, W\rangle$ be a path story, where $G = (v_1, v_2, \ldots, v_n)$. Note that the ordering of $V(G)$ given by the path is, in general, different from the ordering given by τ.

The x-*buckets* of G are the sets B_i^x, with $i = 1, \ldots, \lceil\frac{h+1}{2}\rceil$, such that $B_1^x = B_1$, and $B_i^x = B_{2i-2} \cup B_{2i-1}$, for $i = 2, \ldots, \lceil\frac{h+1}{2}\rceil$. Note that each x-bucket has $2W$ vertices, except for B_1 and, possibly, the last x-bucket; see Fig. 1. The y-*buckets* of G are the sets B_j^y, with $j = 1, \ldots, \lceil\frac{h}{2}\rceil$, such that $B_j^y = B_{2j-1} \cup B_{2j}$. Note that each y-bucket has $2W$ vertices, except possibly for the last y-bucket; see Fig. 1. Also, each vertex belongs to exactly one x-bucket and to exactly one y-bucket.

We now present the following theorem; its proof is similar in spirit to the proof that any two paths admit a simultaneous geometric embedding [6].

Theorem 2. For any path story $\langle G, \tau, W\rangle$, it is possible to compute in $O(n)$ time a drawing story that is planar, straight-line, and lies on a $2W \times 2W$ grid.

Proof. Let G be the path $(v_1, v_2, \ldots, v_n)$ and let $h = \lceil \frac{n}{W} \rceil$ be the number of buckets of $V(G)$. We now order the vertices in each x-bucket and in each y-bucket; this is done according to the ordering in which the vertices appear in the path G. Formally, for $i = 1, 2, \ldots, \lceil \frac{h+1}{2} \rceil$, let $x_i : B_i^x \to \{1, \ldots, |B_i^x|\}$ be a bijective function such that, for any $v_a, v_b \in B_i^x$, we have $x_i(v_a) < x_i(v_b)$ if and only if $a < b$. Similarly, for $i = 1, 2, \ldots, \lceil \frac{h}{2} \rceil$, let $y_i : B_i^y \to \{1, \ldots, |B_i^y|\}$ be a bijective function such that, for any $v_a, v_b \in B_i^y$, we have $y_i(v_a) < y_i(v_b)$ if and only if $a < b$. We assign the coordinates to the vertices of G in Γ as follows. For any vertex v of G, let B_i^x and B_j^y be the x-bucket and the y-bucket containing v, respectively. We place v at the point $(x_i(v), y_j(v))$ in all the drawings $\Gamma_t \in \Gamma$ such that v belongs to G_t. Also, we draw each edge as a straight-line segment.

We now prove that the constructed drawing story Γ satisfies the properties in the statement. Since x-buckets and y-buckets have size at most $2W$, we have that each drawing $\Gamma_t \in \Gamma$ lies on the $[1, 2W] \times [1, 2W]$ grid. We have the following.

Property 2. For $i = 1, \ldots, \lceil \frac{h+1}{2} \rceil$, the straight-line drawing $\Gamma[B_i^x]$ of $G[B_i^x]$ is planar. For $j = 1, \ldots, \lceil \frac{h}{2} \rceil$, the straight-line drawing $\Gamma[B_j^y]$ of $G[B_j^y]$ is planar.

Proof. By Property 1, there exists an x-bucket B_i^x or a y-bucket B_j^y that contains $V(G_t)$. Then Γ_t coincides with $\Gamma[B_i^x]$ or with $\Gamma[B_j^y]$, respectively, restricted to the vertices in $V(G_t)$. By Property 2, we have that Γ_t is planar.

Finally, the time needed to compute Γ coincides with the time needed to compute the functions x_i and y_j, for each $i \in \{1, 2, \ldots, \lceil \frac{h+1}{2} \rceil\}$ and $j \in \{1, 2, \ldots, \lceil \frac{h}{2} \rceil\}$. This can be done in total $O(n)$ time as follows. For $i = 1, \ldots, n$, traverse the path G from v_1 to v_n. When a vertex v_k is considered, the buckets B_i^x and B_j^y where v_k should be inserted are determined in $O(1)$ time; then v_k is inserted in each of these buckets as the currently last vertex. This process provides each x-bucket B_i^x and each y-bucket B_j^y with the desired orderings x_i and y_j, respectively. $\qquad\square$

5 Tree Stories

In this section we show how to draw a tree story $\langle T, \tau, W \rangle$. Our algorithm partitions $V(T)$ into buckets $B_1, \ldots, B_h$ and then partitions the subtrees of T induced by each bucket B_i into two rooted ordered forests. For odd values of i, the forests corresponding to B_i are drawn "close" to the y-axis, while for even values of i, the forests corresponding to B_i are drawn "close" to the x-axis. The drawings of these forests need to satisfy strong visibility properties, as edges of T might connect vertices in a bucket B_i with the roots of the forests corresponding to the bucket B_{i+1}, and vice versa. We now introduce a drawing standard for (static) rooted ordered forests that guarantees these visibility properties.

For a vertex v in a drawing Γ, denote by $\ell_\search(v)$ the half-line originating at v with slope -2. Also, consider a horizontal half-line originating at v and directed rightward; rotate such a line in clockwise direction around v until it overlaps with $\ell_\search(v)$; this rotation spans a closed wedge centered at v, which we call the $\triangledown$-*wedge* of v and denote by $S_\triangledown(v)$. We have the following definition.

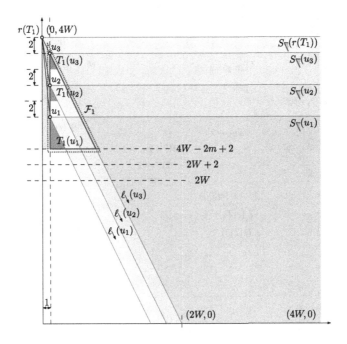

Fig. 2. Construction of a ⋏-drawing: the first inductive case, in which $k = 1$ and $m > 1$.

Definition 1. *Let $\mathcal{F} = T_1, T_2, \ldots, T_k$ be a rooted ordered forest, with a total of $m \leq W$ vertices. A ⋏-**drawing** Γ of $\mathcal{F}$ is a planar straight-line strictly-upward strictly-leftward order-preserving grid drawing of $\mathcal{F}$ such that:*

(i) *Γ lies on the $[0, m-1] \times [4W - 2m + 2, 4W]$ grid;*

(ii) *the roots $r(T_1), r(T_2), \ldots, r(T_k)$ lie along the segment $[(0, 2W+2), (0, 4W)]$, in this order from bottom to top, and $r(T_k)$ lies on $(0, 4W)$;*

(iii) *the vertices of T_i have y-coordinates strictly smaller than the vertices of T_{i+1}, for $i = 1, \ldots, k-1$;*

(iv) *for each tree T_i and for each vertex v of T_i, let $u_1, u_2, \ldots, u_\ell$ be the children of v in left-to-right order; then the vertices of $T_i(u_j)$ have y-coordinates strictly smaller than the vertices of $T_i(u_{j+1})$, for $j = 1, \ldots, \ell - 1$; and*

(v) *for each vertex v of $\mathcal{F}$, the wedge $S_{\mathrm{Y}}(v)$ does not intersect Γ other than along $\ell_{\mathrm{\diagdown}}(v)$.*

We are going to use the following property.

Property 3. *For each vertex v in a ⋏-drawing Γ of $\mathcal{F}$, the wedge $S_{\mathrm{Y}}(v)$ contains the segment $[(2W + 2, 0), (4W, 0)]$ in its interior.*

We can similarly define ⊁-, ⋎-, and ⋞-*drawings*; in particular, a drawing of $\mathcal{F}$ is a ⊁-, ⋎-, or ⋞- drawing if and only if it can be obtained from a ⋏-drawing

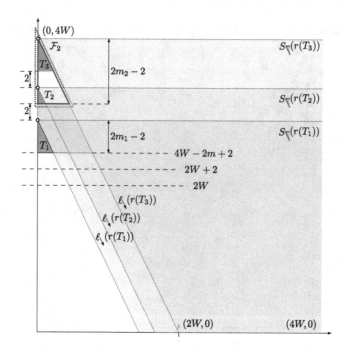

Fig. 3. Construction of a ⅄-drawing: the second inductive case, in which $k > 1$.

by a clockwise rotation around the origin of the Cartesian axes by 90°, 180°, or 270°, respectively. A property similar to Property 3 can be stated for such drawings.

We now present an algorithm, called ⅄-DRAWER, that constructs a ⅄-drawing Γ of $\mathcal{F} = T_1, \ldots, T_k$. The algorithm ⅄-DRAWER constructs Γ by induction, primarily, on k and, secondarily, on the number of vertices m of $\mathcal{F}$.

The base case of the algorithm ⅄-DRAWER happens when $m = 1$ (and thus $k = 1$); then we obtain Γ by placing $r(T_1)$ at $(0, 4W)$. In the first inductive case, we have $k = 1$ and $m > 1$; see Fig. 2. Let $\mathcal{F}_1$ be the rooted ordered forest $T_1(u_1), T_1(u_2), \ldots, T_1(u_{\ell_1})$, where $u_1, u_2, \ldots, u_{\ell_1}$ are the children of $r(T_1)$ in left-to-right order. Inductively construct a ⅄-drawing Γ_1 of $\mathcal{F}_1$. We obtain Γ by placing $r(T_1)$ at $(0, 4W)$ and by translating Γ_1 one unit to the right and two units down. In the second inductive case, we have $k > 1$; see Fig. 3. We inductively construct a ⅄-drawing Γ_1 of T_1 and a ⅄-drawing Γ_2 of the rooted ordered forest $\mathcal{F}_2 = T_2, T_3, \ldots, T_k$. Then, we obtain Γ from Γ_1 and Γ_2 by translating Γ_1 down so that $r(T_1)$ lies two units below the lowest vertex in Γ_2. We have the following.

Lemma 1. The algorithm ⅄-DRAWER constructs a ⅄-drawing of $\mathcal{F}$ in $O(m)$ time.

Algorithms ⌐-DRAWER, ⅄-DRAWER, and ⅃-DRAWER that construct a ⌐-drawing, a ⅄-drawing, and a ⅃-drawing of $\mathcal{F}$ can be defined analogously.

We now go back to the problem of drawing a tree story $\langle T, \tau, W \rangle$. Let $n = |V(T)|$. Recall that $V(T)$ is partitioned into buckets $B_1, \ldots, B_h$, where $h = \lceil \frac{n}{W} \rceil$. We now show how to partition the subtrees of T induced by each bucket into up to two rooted ordered forests, so that the algorithms $\wedge$-, $\succ$-, $\vee$-, and $\prec$-DRAWER can be exploited in order to draw such forests, thus obtaining a drawing story of $\langle T, \tau, W \rangle$. We proceed in several phases as follows.

PHASE 1: We label each vertex v of T belonging to a bucket B_i with the label $b(v) = i$ and we remove from T all the edges (u, v) such that $|b(u) - b(v)| > 1$. Observe that such edges are not visualized in a drawing story of $\langle T, \tau, W \rangle$.

PHASE 2: As T might be a forest because of the previous edge removal, we add dummy edges to T to turn it back into a tree, while ensuring that $|b(u) - b(v)| \leq 1$ for every edge (u, v) of T. This is possible due to the following.

Lemma 2. Dummy edges can be added to T in total $O(n)$ time so that T becomes a tree and every edge (u, v) of T is such that $|b(u) - b(v)| \leq 1$.

PHASE 3: We now root T at an arbitrary vertex $r(T)$ in B_1. A *pertinent component* of T is a maximal connected component of T composed of vertices in the same bucket. We assign a *label* $b(P) = i$ to a pertinent component P if every vertex of P belongs to B_i. We now construct the following sets $R_1, R_2, \ldots, R_k$ of pertinent components of T; see Fig. 4. The set R_1 only contains the pertinent component of T the vertex $r(T)$ belongs to. For $j > 1$, the set R_j contains every pertinent component P of T such that (i) P does not belong to $\bigcup_{i=1}^{j-1} R_i$ and (ii) P contains a vertex that is adjacent to a vertex belonging to a pertinent component in R_{j-1}.

By the construction of the R_j's, since $|b(u) - b(v)| \leq 1$ for every edge (u, v) of T, and by the rooting of T, we have the following simple property.

Property 4. For every vertex $v \in R_j$, each child of v belongs to either R_j or R_{j+1}.

PHASE 4: Next, we turn T into an ordered tree. Consider any vertex v and let R_j be the set v belongs to. Then, by Proposition 4, each child of v is either in R_j or in R_{j+1}. We set the left-to-right order of the children of v so that all those in R_j come first, in any order, and all those in R_{j+1} come next, in any order.

PHASE 5: We now define rooted ordered forests. For $i = 1, \ldots, h$ with i odd, let $\mathcal{F}_{i \cdot \wedge}$ (resp., $\mathcal{F}_{i \cdot \vee}$) be the forest containing all the pertinent components P such that $b(P) = i$ and such that $P \in R_j$ with $j \equiv 1 \mod 4$ (resp., $j \equiv 3 \mod 4$). Also, for $i = 1, \ldots, h$ with i even, let $\mathcal{F}_{i \cdot \succ}$ (resp., $\mathcal{F}_{i \cdot \prec}$) be the forest containing all the pertinent components P such that $b(P) = i$ and such that $P \in R_j$ with $j \equiv 2 \mod 4$ (resp., $j \equiv 0 \mod 4$). We have the following.

Observation 1. Let v be a vertex of T and u be its parent. Let R_i and R_j be the sets containing the pertinent components u and v belong to, respectively, where $j = i$ or $j = i + 1$, by Property 4. Then the following cases are possible.

1a If $j = i$, then u and v both belong to either $\mathcal{F}_{i\cdot\wedge}$, $\mathcal{F}_{i\cdot\backslash}$, $\mathcal{F}_{i\cdot\succ}$, or $\mathcal{F}_{i\cdot\angle}$.

1b If $j = i + 1$, then v is the root of a pertinent component in R_j. Also, either:

(i) i is odd, $u \in \mathcal{F}_{i\cdot\wedge}$, and $v \in \mathcal{F}_{i+1\cdot\succ}$;

(ii) i is even, $u \in \mathcal{F}_{i\cdot\succ}$, and $v \in \mathcal{F}_{i+1\cdot\backslash}$;

(iii) i is odd, $u \in \mathcal{F}_{i\cdot\backslash}$, and $v \in \mathcal{F}_{i+1\cdot\angle}$; or

(iv) i is even, $u \in \mathcal{F}_{i\cdot\angle}$, and $v \in \mathcal{F}_{i+1\cdot\wedge}$.

For each pertinent component P in any $\mathcal{F}_{i\cdot X}$, with $i \in \{1, \ldots, h\}$ and $X \in \{\wedge, \succ, \backslash, \angle\}$, let R_j be the set P belongs to. If $j = 1$, then the root of P is $r(T)$, otherwise the root of P is the vertex of P that is adjacent to a vertex in R_{j-1}. Further, the left-to-right order of the children of every vertex of P is the one inherited from T. Finally, the linear ordering of the pertinent components in $\mathcal{F}_{i\cdot X}$ is defined as follows. Let P_1 and P_2 be any two pertinent components in $\mathcal{F}_{i\cdot X}$ and let R_j and R_k be the sets containing P_1 and P_2, respectively. If $j < k$, then P_1 precedes P_2 in $\mathcal{F}_{i\cdot X}$. If $j > k$, then P_1 follows P_2 in $\mathcal{F}_{i\cdot X}$. Otherwise $j = k$; let x be the lowest common ancestor of the roots of P_1 and of P_2 in T. Also, let p_1 and p_2 be the paths connecting the roots of P_1 and of P_2 with x, respectively. Further, let x_1 and x_2 be the children of x belonging to p_1 and to p_2, respectively. Then, P_1 precedes P_2 in $\mathcal{F}_{i\cdot X}$ if and only if x_1 precedes x_2 in the left-to-right order of the children of x. We have the following.

Lemma 3. Given the sets $R_1, \ldots, R_k$, the rooted ordered forests $\mathcal{F}_{i\cdot X}$, with $i = 1, \ldots, h$ and $X \in \{\wedge, \succ, \backslash, \angle\}$, can be computed in total $O(n)$ time.

We are now ready to state the following main result.

Theorem 3. For any tree story $\langle T, \tau, W \rangle$ such that T has n vertices, it is possible to construct in $O(n)$ time a drawing story that is planar, straight-line, and lies on an $(8W + 1) \times (8W + 1)$ grid.

Proof. We construct a planar straight-line drawing story Γ of $\langle T, \tau, W \rangle$. We perform PHASES 1–5 in order to construct the rooted ordered forests $\mathcal{F}_{i\cdot X}$ with $i \in \{1, \ldots, h\}$ and $X \in \{\wedge, \succ, \backslash, \angle\}$. Note that PHASE 2 introduces in T some dummy edges, which are removed from the actual drawing story of $\langle T, \tau, W \rangle$ after the construction of Γ. Since Γ is a straight-line drawing, we only need to assign coordinates to the vertices of T; see Fig. 4. We apply the algorithm X-DRAWER to construct an X-drawing $\Gamma_{i\cdot X}$ of each rooted ordered forest $\mathcal{F}_{i\cdot X}$. We let the coordinates of each vertex v of $\mathcal{F}_{i\cdot X}$ in Γ coincide with its coordinates in $\Gamma_{i\cdot X}$.

We now prove that Γ satisfies the properties in the statement of the theorem. By Condition i of Definition 1, a $\wedge$-drawing lies on the $[0, 4W] \times [0, 4W]$ grid. Similarly, a $\succ$-, a $\backslash$-, and a $\angle$-drawing lies on the $[0, 4W] \times [-4W, 0]$ grid, on the $[-4W, 0] \times [-4W, 0]$ grid, and on the $[-4W, 0] \times [0, 4W]$ grid, respectively. Thus, Γ lies on the $[-4W, 4W] \times [-4W, 4W]$ grid.

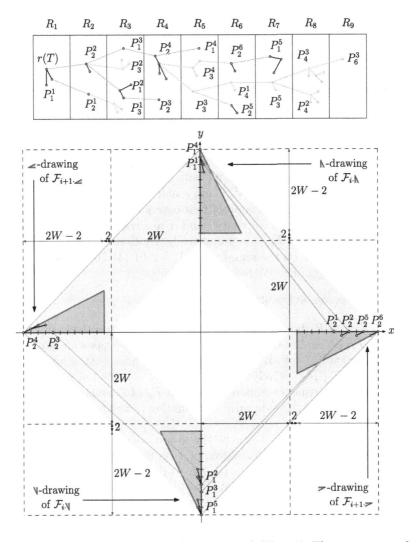

Fig. 4. Illustration for the proof of Theorem 3, with $W = 12$. The upper part of the figure shows the rooted ordered tree T; vertices and edges that are not in $T[B_{1,2}]$ are gray. A pertinent component P_i^j of T belongs to the bucket B_i; further, the index j represents the order of the components in the corresponding rooted forests. The lower part of the figure shows the drawing $\Gamma[B_{1,2}]$ of $T[B_{1,2}]$ constructed by the algorithm.

Concerning the running time for the construction of Γ, we have that the initial modification of T, which ensures that $|b(u) - b(v)| \leq 1$ for every edge (u, v) of T, can be done in $O(n)$ time, by Lemma 2. The labeling $b(u)$ of each vertex u of T is easily done in $O(n)$ time. The construction of the sets $R_1, \ldots, R_k$ can be accomplished by an $O(n)$-time traversal of T starting from $r(T)$. The construction of the rooted ordered forests $\mathcal{F}_{i \cdot X}$, with $i = 1, \ldots, h$ and with

$X \in \{\text{\\},\text{\\},\text{\\},\text{\\}\}$, can be performed in total $O(n)$ time by Lemma 3. Finally, by Lemma 1, the algorithm X-DRAWER runs in linear time in the size of its input $\mathcal{F}_{i \cdot X}$.

It remains to show that each drawing $\Gamma_t \in \Gamma$ is planar. We exploit the following lemma. Let $B_{i,i+1} = B_i \cup B_{i+1}$.

Lemma 4. For $i = 1, \ldots, h-1$, the drawing $\Gamma[B_{i,i+1}]$ of $T[B_{i,i+1}]$ is planar.

Proof. Suppose that i is odd; the case in which i is even can be treated analogously. Let (u,v) and (w,z) be any two edges of $T[B_{i,i+1}]$. We prove that (u,v) and (w,z) do not cross in $\Gamma[B_{i,i+1}]$. We only discuss the most interesting case, in which u and w both belong to $\mathcal{F}_{i \cdot \backslash}$ and v and z both belong to $\mathcal{F}_{i+1 \cdot \nearrow}$; see the full version of the paper for a proof that covers the other cases.

By Observation 1b, we have that v and z are the roots of two pertinent components P_v and P_z of $\mathcal{F}_{i+1 \cdot \nearrow}$, respectively. By Condition ii of Definition 1, the vertices v and z lie on the segment $[(2W+2,0),(4W,0)]$. Assume w.l.o.g. that z lies to the right of v. We have the following.

Claim 1. The vertex w lies above the vertex u in $\Gamma[B_{i,i+1}]$.

Proof. Let R_j and R_k be the sets containing v and z, respectively. By the construction of $\mathcal{F}_{i+1 \cdot \nearrow}$ and since z lies to the right of v, two cases are possible.

In the first case $j < k$. By Observation 1b, we have that u and w belong to R_{j-1} and R_{k-1}, respectively. Since $j-1 < k-1$, we have that u and w belong to distinct pertinent components P_u and P_w of T, respectively, where P_u precedes P_w in $\mathcal{F}_{i \cdot \backslash}$. By Condition iii of Definition 1, we have that w lies above u in $\Gamma[B_{i,i+1}]$.

In the second case $j = k$. Let x be the lowest common ancestor of v and z in T. Also, let v' and z' be the children of x on the paths from x to v and z, respectively. Then v' precedes z' in the left-to-right order of the children of x. Note that $j-1 = k-1$ and that x is also the lowest common ancestor of u and w in T, where u lies on the path between v and x and w lies on the path between z and x.

If u and w belong to distinct pertinent components P_u and P_w of T then, as in the first case, P_u precedes P_w in $\mathcal{F}_{i \cdot \backslash}$. By Condition iii of Definition 1 we have that w lies above u in $\Gamma[B_{i,i+1}]$.

If u and w belong to the same pertinent component of T and neither of them is x, then by Condition iv of Definition 1 we have that w lies above u in $\Gamma[B_{i,i+1}]$.

If u and w belong to the same pertinent component of T and $w = x$, then we have that w lies above u in $\Gamma[B_{i,i+1}]$ since $\Gamma_{i \cdot \backslash}$ is strictly-upward.

Finally, note that $u \neq x$. Indeed, suppose, for a contradiction, that u is the lowest common ancestor of v and z. By the choice of the left-to-right ordering of the children of u (given in PHASE 4), we have that z' precedes v in this ordering. Therefore, by the construction of the ordering of $\mathcal{F}_{i+1 \cdot \nearrow}$ (given in PHASE 5), we have that P_z precedes P_v in $\mathcal{F}_{i+1 \cdot \nearrow}$; by Condition ii of Definition 1, this contradicts the assumption that z lies to the right of v. □

By Claim 1 and by Condition i of Definition 1, we have that w lies above u, which lies above v and z; recall that these last two vertices lie on the segment $[(2W + 2, 0), (4W, 0)]$. Hence, the edge (u, v) crosses the edge (w, z) if and only if u lies to the right of the edge (w, z). By Property 3, the edge (w, z) lies in the wedge $S_{\nabla}(w)$, except at w. However, if u lies to the right of the edge (w, z), then $S_{\nabla}(w)$ contains u, which contradicts Condition v of Definition 1.   □

By Property 1 and Lemma 4, we have that Γ is planar.   □

6 Conclusions and Open Problems

We have shown how to draw dynamic trees with straight-line edges, using an area that only depends on the number of vertices that are simultaneously present in the tree, while maintaining planarity. This result is obtained by partitioning the vertices of the tree into buckets and by establishing topological and geometric properties for the forests induced by pairs of consecutive buckets. Further, we proved that this result cannot be generalized to arbitrary planar graphs.

Several interesting problems arise. **(OP1)** Which families of planar graphs admit a planar straight-line drawing story on a grid whose size is polynomial in W? How about outerplanar graphs? **(OP2)** Which bounds can be shown for a dynamic graph that is not a tree, while each graph of the story is a forest?

Acknowledgments. We are indebted to Giuseppe Di Battista for interesting conversations and suggestions that helped us improve and direct our investigation.

References

1. Angelini, P., Bekos, M.A.: Hierarchical partial planarity. Algorithmica **81**(6), 2196–2221 (2019). https://doi.org/10.1007/s00453-018-0530-6
2. Beck, F., Burch, M., Diehl, S., Weiskopf, D.: The state of the art in visualizing dynamic graphs. In: Borgo, R., Maciejewski, R., Viola, I. (eds.) Eurographics Conference on Visualization, EuroVis 2014 - State of the Art Reports, STARs, Swansea, UK, 9–13 June 2014. Eurographics Association (2014). https://doi.org/10.2312/eurovisstar.20141174
3. Binucci, C., Brandes, U., Di Battista, G., Didimo, W., Gaertler, M., Palladino, P., Patrignani, M., Symvonis, A., Zweig, K.A.: Drawing trees in a streaming model. Inf. Process. Lett. **112**(11), 418–422 (2012). https://doi.org/10.1016/j.ipl.2012.02.011
4. Bläsius, T., Kobourov, S.G., Rutter, I.: Simultaneous embedding of planar graphs. In: Tamassia, R. (ed.) Handbook on Graph Drawing and Visualization, pp. 349–381. Chapman and Hall/CRC (2013)
5. Borrazzo, M., Da Lozzo, G., Frati, F., Patrignani, M.: Graph stories in small area. CoRR abs/1908.09318 (2019)
6. Braß, P., Cenek, E., Duncan, C.A., Efrat, A., Erten, C., Ismailescu, D., Kobourov, S.G., Lubiw, A., Mitchell, J.S.B.: On simultaneous planar graph embeddings. Comput. Geom. **36**(2), 117–130 (2007). https://doi.org/10.1016/j.comgeo.2006.05.006

7. Cohen, R.F., Di Battista, G., Tamassia, R., Tollis, I.G.: Dynamic graph drawings: trees, series-parallel digraphs, and planar ST-digraphs. SIAM J. Comput. **24**(5), 970–1001 (1995). https://doi.org/10.1137/S0097539792235724

8. Da Lozzo, G., Rutter, I.: Planarity of streamed graphs. In: Paschos, V.T., Widmayer, P. (eds.) CIAC 2015. LNCS, vol. 9079, pp. 153–166. Springer, Cham (2015). https://doi.org/10.1007/978-3-319-18173-8_11

9. de Fraysseix, H., Pach, J., Pollack, R.: How to draw a planar graph on a grid. Combinatorica **10**(1), 41–51 (1990)

10. Demetrescu, C., Di Battista, G., Finocchi, I., Liotta, G., Patrignani, M., Pizzonia, M.: Infinite trees and the future. In: Kratochvíyl, J. (ed.) GD 1999. LNCS, vol. 1731, pp. 379–391. Springer, Heidelberg (1999). https://doi.org/10.1007/3-540-46648-7_39

11. Di Battista, G., Tamassia, R.: On-line planarity testing. SIAM J. Comput. **25**(5), 956–997 (1996). https://doi.org/10.1137/S0097539794280736

12. Di Battista, G., Tamassia, R., Vismara, L.: Incremental convex planarity testing. Inf. Comput. **169**(1), 94–126 (2001). https://doi.org/10.1006/inco.2001.3031

13. Dolev, D., Leighton, F., Trickey, H.: Planar embedding of planar graphs. Adv. Comput. Res. **2**, 147–161 (1984)

14. Frati, F., Patrignani, M.: A note on minimum-area straight-line drawings of planar graphs. In: Hong, S.-H., Nishizeki, T., Quan, W. (eds.) GD 2007. LNCS, vol. 4875, pp. 339–344. Springer, Heidelberg (2008). https://doi.org/10.1007/978-3-540-77537-9_33

15. Goodrich, M.T., Pszona, P.: Streamed graph drawing and the file maintenance problem. In: Wismath, S., Wolff, A. (eds.) GD 2013. LNCS, vol. 8242, pp. 256–267. Springer, Cham (2013). https://doi.org/10.1007/978-3-319-03841-4_23

16. Italiano, G.F.: Fully dynamic planarity testing. In: Kao, M.Y. (ed.) Encyclopedia of Algorithms, pp. 806–808. Springer, New York (2016). https://doi.org/10.1007/978-1-4939-2864-4_157

17. Michail, O.: An introduction to temporal graphs: an algorithmic perspective. Internet Math. **12**(4), 239–280 (2016)

18. Poutré, J.A.L.: Alpha-algorithms for incremental planarity testing (preliminary version). In: Leighton, F.T., Goodrich, M.T. (eds.) Proceedings of the Twenty-Sixth Annual ACM Symposium on Theory of Computing, Montréal, Québec, Canada, 23–25 May 1994, pp. 706–715. ACM (1994). https://doi.org/10.1145/195058.195439

19. Reingold, E.M., Tilford, J.S.: Tidier drawings of trees. IEEE Trans. Softw. Eng. **7**(2), 223–228 (1981). https://doi.org/10.1109/TSE.1981.234519

20. Rextin, A., Healy, P.: A fully dynamic algorithm to test the upward planarity of single-source embedded digraphs. In: Tollis, I.G., Patrignani, M. (eds.) GD 2008. LNCS, vol. 5417, pp. 254–265. Springer, Heidelberg (2009). https://doi.org/10.1007/978-3-642-00219-9_24

21. Schaefer, M.: Picking planar edges; or, drawing a graph with a planar subgraph. In: Duncan, C., Symvonis, A. (eds.) GD 2014. LNCS, vol. 8871, pp. 13–24. Springer, Heidelberg (2014). https://doi.org/10.1007/978-3-662-45803-7_2

22. Skambath, M., Tantau, T.: Offline drawing of dynamic trees: algorithmics and document integration. In: Hu, Y., Nöllenburg, M. (eds.) GD 2016. LNCS, vol. 9801, pp. 572–586. Springer, Cham (2016). https://doi.org/10.1007/978-3-319-50106-2_44

23. Whitney, H.: Congruent graphs and the connectivity of graphs. Am. J. Math. **54**(1), 150–168 (1932). http://www.jstor.org/stable/2371086

Level-Planar Drawings with Few Slopes

Guido Brückner$^{(\boxtimes)}$, Nadine Davina Krisam, and Tamara Mchedlidze

Karlsruhe Institute of Technology, Karlsruhe, Germany
brueckner@kit.edu, nadine.krisam@student.kit.edu, mched@iti.uka.de

Abstract. We introduce and study level-planar straight-line drawings with a fixed number λ of slopes. For proper level graphs, we give an $O(n \log^2 n / \log \log n)$-time algorithm that either finds such a drawing or determines that no such drawing exists. Moreover, we consider the partial drawing extension problem, where we seek to extend an immutable drawing of a subgraph to a drawing of the whole graph, and the simultaneous drawing problem, which asks about the existence of drawings of two graphs whose restrictions to their shared subgraph coincide. We present $O(n^{4/3} \log n)$-time and $O(\lambda n^{10/3} \log n)$-time algorithms for these respective problems on proper level-planar graphs.

We complement these positive results by showing that testing whether non-proper level graphs admit level-planar drawings with λ slopes is NP-hard even in restricted cases.

1 Introduction

Directed graphs explaining hierarchy naturally appear in multiple industrial and academic applications. Some examples include PERT diagrams, UML component diagrams, text edition networks [1], text variant graphs [19], philogenetic and neural networks. In these, and many other applications, it is essential to visualize the implied directed graph so that the viewer can perceive the hierarchical structure it contains. By far the most popular way to achieve this is to apply the *Sugyiama framework* – a generic network visualization algorithm that results in a drawing where each vertex lies on a horizontal line, called *layer*, and each edge is directed from a lower layer to a higher layer [14].

The Sugyiama framework consists of several steps: elimination of directed cycles in the initial graph, assignment of vertices to layers, vertex ordering and coordinate assignement. During each of these steps several criteria are optimized, by leading to more readable visualizations, see e.g. [14]. In this paper we concentrate on the last step of the framework, namely coordinate assignment. Thus, the subject of our study are *level graphs* defined as follows. Let $G = (V, E)$ be a directed graph. A *k-level assignment* of G is a function $\ell : V \to \{1, 2, \ldots, k\}$ that assigns each vertex of G to one of k levels. We refer to G together with ℓ as to a *(k-)level graph*. The *length* of an edge (u, v) is defined as $\ell(v) - \ell(u)$. We say that G is *proper* if all edges have length one. The level graph shown in Fig. 1(a) is proper, whereas the one shown in (b) is not. For a non-proper level graph G

© Springer Nature Switzerland AG 2019
D. Archambault and C. D. Tóth (Eds.): GD 2019, LNCS 11904, pp. 559–572, 2019.
https://doi.org/10.1007/978-3-030-35802-0_42

there exists a *proper subdivision* obtained by subdividing all edges with length greater than one which result in a proper graph.

A *level drawing* Γ of a level graph G maps each vertex $v \in V$ to a point on the horizontal line with y-coordinate $\ell(v)$ and a real x-coordinate $\Gamma(v)$, and each edge to a y-monotone curve between its endpoints. A level drawing is called *level-planar* if no two edges intersect except in common endpoints. It is *straight-line* if the edges are straight lines. A level drawing of a proper (subdivision of a) level graph G induces a total left-to-right order on the vertices of a level. We say that two drawings are *equivalent* if they induce the same order on every level. An equivalence class of this equivalence relation is an *embedding* of G. We refer to G together with an embedding to as *embedded level graph* $\mathcal{G}$. The third step of Sugyiama framework, vertex ordering, results in an embedded level graph.

The general goal of the coordinate assignment step is to produce a final visualization while further improving its readability. The criteria of readability that have been considered in the literature for this step include straightness and verticality of the edges [14]. Here we study the problem of coordinate assignment step with bounded number of slopes. The *slope* of an edge (u, v) in Γ is defined as $(\Gamma(v) - \Gamma(u))/(\ell(v) - \ell(u))$. For proper level graphs this simplifies to $\Gamma(v) - \Gamma(u)$. We restrict our study to drawings in which all slopes are non-negative; such drawings can be transformed into drawings with negative slopes by shearing; see Fig. 1. A level drawing Γ is a λ-*slope drawing* if all slopes in Γ appear in the set $\{0, 1, \ldots, \lambda - 1\}$. We study embedding-preserving straight-line level-planar λ-slope drawings, or λ-*drawings* for short and refer to the problem of finding these drawings as λ-DRAWABILITY. Since the possible edge slopes in a λ-drawing are integers all vertices lie on the integer grid.

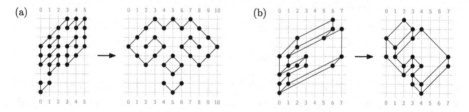

Fig. 1. Shearing drawings to change the slopes. In (a), the left drawing with slopes 0 and 1 is transformed into the right orthogonal drawing, i.e., one with slopes -1 and 1. In (b), the left drawing with slopes $0, 1$ and 2 is transformed into a drawing with slopes $-1, 0$ and 1.

Related Work. The number of slopes used for the edges in a graph drawing can be seen as an indication of the simplicity of the drawing. For instance, the measure *edge orthogonality*, which specifies how close a drawing is to an *orthogonal drawing*, where edges are polylines consisting of horizontal and vertical segments only, has been proposed as a measure of aesthetic quality of a graph drawing [31]. In a similar spirit, Kindermann et al. studied the effect reducing the segment complexity on the aesthetics preference of graph drawings and observed that in some cases people prefer drawings using lower segment complexity [22]. More generally, the use of few slopes for a graph drawing may contribute to the formation of "Prägnanz" ("good figure" in German) of the visualization, that accordingly to the Gestalt law of Prägnanz, or law of simplicity, contributes to the perceived quality of the visualizations. This is design principle often guides the visualization of metro maps. See [28] for a survey of the existing approaches, most of which generate octilinear layouts of metro maps, and [27] for a recent model for drawing more general k-linear metro maps.

Level-planar drawing with few slopes have not been considered in the literature but drawings of undirected graphs with few slopes have been extensively studied. The *(planar) slope number* of a (planar) graph G is the smallest number s so that G has a (planar) straight-line drawing with edges of at most s distinct slopes. Special attention has been given to proving bounds on the (planar) slope number of undirected graph classes [4, 9–12, 20, 21, 23, 24, 29]. Determining the planar slope number is hard in the existential theory of reals [16].

Several graph visualization problems have been considered in the partial and simultaneous settings. In the *partial drawing extension* problem, one is presented with a graph and an immutable drawing of some subgraph thereof. The task is to determine whether the given drawing of the subgraph can be completed to a drawing of the entire graph. The problem has been studied for the planar setting [25, 30] and also the level-planar setting [8]. In the *simultaneous drawing* problem, one is presented with two graphs that may share some subgraph. The task is to draw both graphs so that the restrictions of both drawings to the shared subgraph are identical. We refer the reader to [5] for an older literature overview. The problem has been considered for orthogonal drawings [2] and level-planar drawings [3]. Up to our knowledge, neither partial nor simultaneous drawings have been considered in the restricted slope setting.

Contribution. We introduce and study the λ-DRAWABILITY problem. To solve this problem for proper level graphs, we introduce two models. In Sect. 3 we describe the first model, which uses a classic integer-circulation-based approach. This model allows us to solve the λ-DRAWABILITY in $O(n \log^3 n)$ time and obtain a λ-drawing within the same running time if one exists. In Sect. 4, we describe the second distance-based model. It uses the duality between flows in the primal graph and distances in the dual graph and allows us to solve the λ-DRAWABILITY in $O(n \log^2 n / \log \log n)$ time. We also address the partial and simultaneous settings. The classic integer-circulation-based approach can be used to extend connected partial λ-drawings in $O(n \log^3 n)$ time. In Sect. 5, we build on the distance-based model to extend not-necessarily-connected partial λ-drawings in $O(n^{4/3} \log n)$ time, and to obtain simultaneous λ-drawings in $O(\lambda n^{10/3} \log n)$ time if they exist. We finish with some concluding remarks in Sect. 6 and refer to the full version [7] for a proof that 2-DRAWABILITY is NP-hard even for biconnected graphs where all edges have length one or two.

2   Preliminaries

Let Γ be a level-planar drawing of an embedded level-planar graph $\mathcal{G}$. The *width* of Γ is defined as $\max_{v \in V} \Gamma(v) - \min_{v \in V} \Gamma(v)$. An integer $\bar{x}$ is a *gap* in Γ if it is $\Gamma(v) \neq \bar{x}$ for all $v \in V$, $\Gamma(v_1) < \bar{x}$ and $\Gamma(v_2) > \bar{x}$ for some $v_1, v_2 \in V$, and $\Gamma(u) < \bar{x} < \Gamma(v)$ for no edge $(u, v) \in E$. A drawing Γ is *compact* if it has no gap. Note that a λ-drawing of a connected level-planar graph is inherently compact. While in a λ-drawing of a non-connected level-planar graph every gap can be eliminated by a shift. The fact that we only need to consider compact λ-drawings helps us to limit their width. Thus, any compact λ-drawing has a width of at most $(\lambda - 1)(n - 1)$.

Let u and v be two vertices on the same level i. With $[u, v]_{\mathcal{G}}$ (or $[u, v]$ when $\mathcal{G}$ is clear from the context) we denote the set of vertices that contains u, v and all vertices in between u and v on level i in $\mathcal{G}$. We say that two vertices u and v are *consecutive in* $\mathcal{G}$ when $[u, v] = \{u, v\}$. Two edges $e = (u, w), e' = (v, x)$ are *consecutive in* $\mathcal{G}$ when the only edges with one endpoint in $[u, v]_{\mathcal{G}}$ and the other endpoint in $[w, x]_{\mathcal{G}}$ are e and e'.

A *flow network* $F = (N, A)$ consists of a set of nodes N connected by a set of directed arcs A. Each arc has a *demand* specified by a function $d : A \to \mathbb{N}_0$ and a *capacity* specified by a function $c : A \to \mathbb{N} \cup \{\infty\}$ where ∞ encodes unlimited capacity. A *circulation* in F is a function $\varphi : A \to \mathbb{N}_0$ that assigns an integral flow to each arc of F and satisifies the two following conditions. First, the circulation has to respect the demands and capacities of the arcs, i.e., for each arc $a \in A$ it is $d(a) \leq \varphi(a) \leq c(a)$. Second, the circulation has to respect flow conservation, i.e., for each node $v \in N$ it is $\sum_{(u,v) \in A} \varphi(u, v) = \sum_{(v,u) \in A} \varphi(v, u)$. Depending on the flow network no circulation may exist.

3   Flow Model

In this section, we model the λ-DRAWABILITY as a problem of finding a circulation in a flow network. Let $\mathcal{G}$ be an embedded proper k-level graph together with a level-planar drawing Γ. As a first step, we add two directed paths p_{left} and p_{right} that consist of one vertex on each level from 1 to k to $\mathcal{G}$. Insert p_{left} and p_{right} into Γ to the left and right of all other vertices as the *left* and *right* boundary, respectively. See Fig. 2(a) and (c). From now on, we assume that $\mathcal{G}$ and Γ contain the left and right boundary.

The flow network $F_{\mathcal{G}}^{\lambda}$ consists of nodes and arcs and is similar to a directed dual of $\mathcal{G}$ with the difference that it takes the levels of $\mathcal{G}$ into account. In particular, for every edge e of $\mathcal{G}$, $F_{\mathcal{G}}^{\lambda}$ contains two nodes e_{left} and e_{right}, in the left and the right faces incident to e, and a dual *slope arc* $e^{\star} = (e_{\text{right}}, e_{\text{left}})$ with demand 0 and capacity $\lambda - 1$; see the blue arcs in Fig. 2(b) and (c). The flow across $e^{\star}$ determines the slope of e. Additionally, for every pair of consecutive vertices u, v we add two nodes $[u, v]_{\text{low}}$ and $[u, v]_{\text{high}}$ to $F_{\mathcal{G}}^{\lambda}$ and connect them by a *space arc* $[u, v]^{\star}$; see the red arcs in Fig. 2(b) and (c). The flow across $[u, v]^{\star}$ determines the space between u and v. The space between u and v needs to be

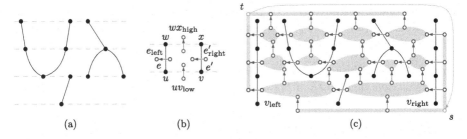

Fig. 2. (a) An embedded level graph $\mathcal{G}$. (b) The definition of the arcs of the flow network. (c) The graph $\mathcal{G}$ together with the paths p_{left} and p_{right} in black. The resulting flow network $F_{\mathcal{G}}^{\lambda}$ (c) consists of the blue slope arcs and the red space arcs, its nodes are formed by merging the nodes in the gray areas. The red space arcs have a demand of 1 and a capacity of $(\lambda - 1)(n - 1)$ and the blue slope arcs have a demand of zero and a capacity of $\lambda - 1$.

at least one to prevent u and v from colliding and can be at most $(\lambda - 1)(n - 1)$ due to the restriction to compact drawings. So, assign to $[u, v]^{\star}$ a demand of one and a capacity of $(\lambda - 1)(n - 1)$. To obtain the final flow network we merge certain nodes. Let $e = (u, w)$ and $e' = (v, x)$ be consecutive edges. Merge the nodes $e_{\text{right}}, e'_{\text{left}}$, the nodes $\{\{u', v'\}_{\text{high}} : \forall u', v' \text{ consecutive in } [u, v]\}$ and the nodes $\{\{w', x'\}_{\text{low}} : \forall w', x' \text{ consecutive in } [w, x]\}$ into a single node. Next, merge all remaining source and sink nodes into one source node s and one sink node t, respectively. See Fig. 2(c), where the gray areas touch nodes that are merged into a single node. Finally, insert an arc from t to s with unlimited capacity.

The network $F_{\mathcal{G}}^{\lambda}$ is designed in such a way that the circulations in $F_{\mathcal{G}}^{\lambda}$ correspond bijectively to the λ-drawings of $\mathcal{G}$. Let Γ be a drawing of $\mathcal{G}$ and let x be the function that assigns to each vertex of $\mathcal{G}$ its x-coordinate in Γ. We define a dual circulation $x^{\star}$ as follows. Recall that every arc a of $F_{\mathcal{G}}^{\lambda}$ is either dual to an edge of $\mathcal{G}$ or to two consecutive vertices in $\mathcal{G}$. Hence, the left and right incident faces f_{left} and f_{right} of a in $F_{\mathcal{G}}^{\lambda}$ contain vertices of $\mathcal{G}$. Define the circulation $x^{\star}$ by setting $x^{\star}(a) := x(f_{\text{right}}) - x(f_{\text{left}})$. We remark the following, although we defer the proof to the next section.

Lemma 1. *Let $\mathcal{G}$ be an embedded proper level-planar graph together with a λ-drawing Γ. The dual $x^{\star}$ of the function x that assigns to each vertex of $\mathcal{G}$ its x-coordinate in Γ is a circulation in $F_{\mathcal{G}}^{\lambda}$.*

In the reverse direction, given a circulation φ in $F_{\mathcal{G}}^{\lambda}$ we define a dual function $\varphi^{\star}$ that, when interpeted as assigning an x-coordinate to the vertices of $\mathcal{G}$, defines a λ-drawing of G. Refer to the level-1-vertex p_{right} as v_{right}. Start by setting $\varphi^{\star}(v_{\text{right}}) = 0$, i.e., the x-coordinate of v_{right} is 0. Process the remaining vertices of the right boundary in ascending order with respect to their levels. Let (u, v) be an edge of the right boundary so that u has already been processed and v has not been processed yet. Then set $\varphi^{\star}(v) = \varphi^{\star}(u) + \varphi((u, v)^{\star})$, where $(u, v)^{\star}$ is the slope arc dual to (u, v). Let w, x be a pair of consecutive vertices so that x has already been processed and w has not yet been processed yet. Then set $\varphi^\star(w) = \varphi^\star(x) + \varphi([w,x]^\star)$, where $[w,x]^\star$ is a space arc. It turns out that $\varphi^\star$ defines a λ-drawing of $\mathcal{G}$.

Lemma 2. *Let $\mathcal{G}$ be an embedded proper level-planar graph, let $\lambda \in \mathbb{N}$ and let φ be a circulation in $F_{\mathcal{G}}^\lambda$. Then the dual $\varphi^\star$, when interpeted as assigning an x-coordinate to the vertices of $\mathcal{G}$, defines a λ-drawing of G.*

While both Lemmas 1 and 2 can be proven directly, we defer their proofs to Sect. 4 where we introduce the distance model and prove Lemmas 3 and 4, the stronger versions of Lemmas 1 and 2, respectively. Combining Lemmas 1 and 2 we obtain the following.

Theorem 1. *Let $\mathcal{G}$ be an embedded proper level-planar graph and let $\lambda \in \mathbb{N}$. The circulations in $F_{\mathcal{G}}^\lambda$ correspond bijectively to the λ-drawings of $\mathcal{G}$.*

Theorem 1 implies that a λ-drawing can be found by applying existing flow algorithms to $F_{\mathcal{G}}^\lambda$. For that, we transform our flow network with arc demands to the standard maximum flow setting without demands by introducing new sources and sinks. We can then use the $O(n \log^3 n)$-time multiple-source multiple-sink maximum flow algorithm due to Borradaile et al. [6] to find a circulation in $F_{\mathcal{G}}^\lambda$ or to determine that no circulation exists.

Corollary 1. *Let $\mathcal{G}$ be an embedded proper level-planar graph and let $\lambda \in \mathbb{N}$. It can be tested in $O(n \log^3 n)$ time whether a λ-drawing of $\mathcal{G}$ exists, and if so, such a drawing can be found within the same running time.*

3.1   Connected Partial Drawings

Recall that a partial λ-drawing is a tuple $(\mathcal{G}, \mathcal{H}, \Pi)$, where $\mathcal{G}$ is an embedded level-planar graph, $\mathcal{H}$ is an embedded subgraph of $\mathcal{G}$ and Π is a λ-drawing of $\mathcal{H}$. We say that $(\mathcal{G}, \mathcal{H}, \Pi)$ is λ-*extendable* if $\mathcal{G}$ admits a λ-drawing Γ whose restriction to $\mathcal{H}$ is Π. Here Γ is referred to as a λ-*extension* of $(\mathcal{G}, \mathcal{H}, \Pi)$.

In this section we show that in case $\mathcal{H}$ is connected, we can use the flow model to decide whether $(\mathcal{G}, \mathcal{H}, \Pi)$ is λ-extendable. Observe that when $\mathcal{H}$ is connected Π is completely defined by the slopes of the edges in $\mathcal{H}$ up to horizontal translation. Let $F_{\mathcal{G}}^\lambda$ be the flow network corresponding to $\mathcal{G}$. In order to fix the slopes of an edge e of $\mathcal{H}$ to a value ℓ, we fix the flow across the dual slope arc $e^\star$ in $\mathcal{H}$ to ℓ. Checking whether a circulation in the resulting flow network exists can be reduced to a multiple-source multiple-sink maximum flow problem, which once again can be solved by the algorithm due to Borradaile et al. [6].

Corollary 2. *Let $(\mathcal{G}, \mathcal{H}, \Pi)$ be a partial λ-drawing where $\mathcal{H}$ is connected. It can be tested in $O(n \log^3 n)$ time whether $(\mathcal{G}, \mathcal{H}, \Pi)$ is λ-extendable, and if so, a corresponding λ-extension can be constructed within the same running time.*

4   Dual Distance Model

A minimum cut (and, equivalently, the value of the maximum flow) of an st-planar graph G can be determined by computing a shortest $(s^\star, t^\star)$-path in a dual of G [17,18]. Hassin showed that to construct a flow, it is sufficient to compute the distances from $s^\star$ to all other vertices in the dual graph [13]. To the best of our knowledge, this duality has been exploited only for flow networks with arc capacities, but not with arc demands. In this section, we extend this duality to arcs with demands. The resulting dual distance model improves the running time for the λ-DRAWABILITY, lets us test the existence of λ-extensions of partial λ-drawings for non-connected subgraphs, and allows us to develop an efficient algorithm for testing the existence of simultaneous λ-drawings.

We define $D_{\mathcal{G}}^{\lambda}$ to be the directed dual of $F_{\mathcal{G}}^{\lambda}$ as follows. Let $a = (u, v)$ be an arc of $F_{\mathcal{G}}^{\lambda}$ with demand $d(a)$ and capacity $c(a)$. Further, let f_{left} and f_{right} denote the left and the right faces of a in $F_{\mathcal{G}}^{\lambda}$, respectively. The dual $D_{\mathcal{G}}^{\lambda}$ contains f_{left} and f_{right} as vertices connected by one edge $(f_{\text{left}}, f_{\text{right}})$ with length $c(a)$ and another edge $(f_{\text{right}}, f_{\text{left}})$ with length $-d(a)$; see Fig. 3.

Observe that to obtain $F_{\mathcal{G}}^{\lambda}$ from $\mathcal{G}$ (with left and right paths p_{left} and p_{right}) we added dual arcs to edges of $\mathcal{G}$ and dual arcs to the space between two consecutive vertices on one level. Consider for a moment the graph $\mathcal{G}'$ obtained from $\mathcal{G}$ by adding edges (u, v) for all consecutive vertices u v, where u is to the right of v. Graph G' and $D_{\mathcal{G}}^{\lambda}$ are identical and therefore $D_{\mathcal{G}}^{\lambda}$ has the vertex set V of $\mathcal{G}$ and contains a subset of its edges. Recall that the dual slope arcs in $F_{\mathcal{G}}^{\lambda}$ have demand 0 and capacity $\lambda - 1$, therefore the edges of $D_{\mathcal{G}}^{\lambda}$ that connect vertices on different layers have non-negative length. While the edges of $D_{\mathcal{G}}^{\lambda}$ between consecutive vertices on the same level have length -1.

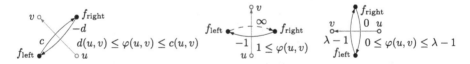

Fig. 3. Definition of the dual edges for a flow network arc $a = (u, v)$ with demand $d(a)$ and capacity $c(a)$. Let f_{left} and f_{right} denote the vertices corresponding to the faces to the left and right of a in $F_{\mathcal{G}}^{\lambda}$. Then add the edge $(f_{\text{left}}, f_{\text{right}})$ with length $c(a)$ and the reverse edge $(f_{\text{right}}, f_{\text{left}})$ with length $-d(a)$. Edges with infinite length are not created because they do not add constraints.

A *distance labeling* is a function $x : V \to \mathbb{Z}$ that for every edge (u, v) of $D_{\mathcal{G}}^{\lambda}$ with length l satisfies $x(v) \leq x(u) + l$. We also say that (u, v) *imposes the distance constraint* $x(v) \leq x(u) + l$. A distance labeling for $D_{\mathcal{G}}^{\lambda}$ is the x-coordinate assignment for a λ-drawing: For an edge (u, v) of $D_{\mathcal{G}}^{\lambda}$ where u, v are consecutive vertices in $\mathcal{G}$, the distance labeling guarantees $x(v) \leq x(u) - 1$, i.e., the consecutive vertices are in the correct order and do not overlap. If an edge (u, v) between layers has length $\lambda - 1$, then the distance labeling ensures $x(v) \leq x(u) + \lambda - 1$, i.e., (u, v)

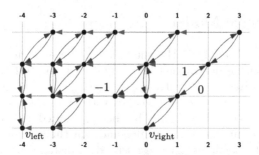

Fig. 4. The distance network $D_{\mathcal{G}}^2$ obtained from the flow network $F_{\mathcal{G}}^2$ shown in Fig. 2(c). The x-coordinate of every vertex is its distance from v_{right} in $D_{\mathcal{G}}^2$. All red arcs have length -1, all blue arcs pointing up have length 1 and all blue arcs pointing down have length 0. For every red arc there exists an arc in the reverse direction with length ∞. We omit these arcs because they do not impose any constraints on the shortest distance labeling.

has a slope in $\{0, \ldots, \lambda - 1\}$. Computing the shortest distances from v_{right} in $D_{\mathcal{G}}^{\lambda}$ to every vertex (if they are well-defined) gives a distance labeling that we refer to as the *shortest distance labeling*. A distance labeling of $D_{\mathcal{G}}^{\lambda}$ does not necessarily exist. This is the case when $D_{\mathcal{G}}^{\lambda}$ contains a negative cycle, e.g., when the in- or out-degree of a vertex in $\mathcal{G}$ is strictly larger than λ. For a distance labeling x of $D_{\mathcal{G}}^{\lambda}$ we define a dual circulation $x^{\star}$ by setting $x^{\star}(a) := x(f_{\text{right}}) - x(f_{\text{left}})$ for each arc a of $F_{\mathcal{G}}^{\lambda}$ with left and right incident faces f_{left} and f_{right} (Fig. 4).

Lemma 3. *Let $\mathcal{G}$ be an embedded level-planar graph and Γ be a λ-drawing of $\mathcal{G}$. The function x that assigns to each vertex of $\mathcal{G}$ its x-coordinate in Γ is a distance labeling of $D_{\mathcal{G}}^{\lambda}$ and its dual $x^{\star}$ is a circulation in $F_{\mathcal{G}}^{\lambda}$.*

Proof. Since Γ preserves the embedding of $\mathcal{G}$, for each consecutive vertices v, u, with v preceeding u in $\mathcal{G}$ it holds that $\Gamma(v) < \Gamma(u)$. Since Γ is a grid drawing $\Gamma(v) \leq \Gamma(u) - 1$, which implies $x(v) \leq x(u) + \ell$, where $\ell = -1$ is the length of (u, v). Since Γ is a λ-drawing, i.e. every edge (u, v) between the two levels has a slope in $\{0, \ldots \lambda - 1\}$, it holds that $\Gamma(u) \leq \Gamma(v) \leq \Gamma(u) + \lambda - 1$, which implies $x(u) \leq x(v) + 0$, for the edge (v, u) of $D_{\mathcal{G}}^{\lambda}$ with length zero and $x(v) < x(u) + \lambda - 1$ for the edge (u, v) of $D_{\mathcal{G}}^{\lambda}$ with length $\lambda - 1$. Hence, x is a distance labeling of $D_{\mathcal{G}}^{\lambda}$.

We now show that $x^{\star}$ is a circulation in $F_{\mathcal{G}}^{\lambda}$. Let $f_1, f_2, \ldots, f_t, f_{t+1} = f_1$ be the faces incident to some node v of $F_{\mathcal{G}}^{\lambda}$ in counter-clockwise order. Let a be the arc incident to v and dual to the edge between f_i and f_{i+1} with $1 \leq i \leq t$. If a is an incoming arc, it adds a flow of $x(f_{i+1}) - x(f_i)$ to v. If a is an outgoing arc, it removes a flow of $x(f_i) - x(f_{i+1})$ from v, or, equivalently, it adds a flow of $x(f_{i+1}) - x(f_i)$ to v. Therefore, the flow through v is $\sum_i (x(f_{i+1}) - x(f_i))$. This sum cancels to zero, i.e., the flow is preserved at v. Recall that the edge $(f_{\text{left}}, f_{\text{right}})$ with length $c(a)$ in $D_{\mathcal{G}}^{\lambda}$ ensures $x(f_{\text{right}}) \leq x(f_{\text{left}}) + c(a)$, which gives $x^{\star}(a) \leq c(a)$. So, no capacities are exceeded. Analogously, the edge $(f_{\text{right}}, f_{\text{left}})$ with length $-d(a)$ in $D_{\mathcal{G}}^{\lambda}$ ensures $x(f_{\text{left}}) \leq x(f_{\text{right}}) - d(a)$, which gives $x^\star(a) \geq d(a)$. Hence, all demands are fulfilled and $x^\star$ is indeed a circulation in $F_{\mathcal{G}}^\lambda$. □

Recall from Sect. 3 that for a circulation φ in $F_{\mathcal{G}}^\lambda$ we define a dual drawing $\varphi^\star$ by setting the x-coordinates of the vertices of $\mathcal{G}$ as follows. For the lowest vertex of the right boundary set $\varphi^\star(v_{\mathrm{right}}) = 0$. Process the remaining vertices of the right boundary in ascending order with respect to their levels. Let (u, v) be an edge of the right boundary so that u has already been processed and v has not been processed yet. Then set $\varphi^\star(v) = \varphi^\star(u) + \varphi((u, v)^\star)$, where $(u, v)^\star$ is the slope arc dual to (u, v). Let w, x be a pair of consecutive vertices so that x has already been processed and w has not yet been processed yet. Then set $\varphi^\star(w) = \varphi^\star(x) + \varphi([w, x]^\star)$, where $[w, x]^\star$ is a space arc. It turns out that $\varphi^\star$ is a distance labeling of $D_{\mathcal{G}}^\lambda$ and a λ-drawing of $\mathcal{G}$.

Lemma 4. *Let $\mathcal{G}$ be an embedded level-planar graph, let $\lambda \in \mathbb{N}$, and let φ be a circulation in $F_{\mathcal{G}}^\lambda$. The dual $\varphi^\star$ is a distance labeling of $D_{\mathcal{G}}^\lambda$ and the drawing induced by interpreting the distance label of a vertex as its x-coordinate is a λ-drawing of $\mathcal{G}$.*

Proof. We show that $\varphi^\star$ is a distance labeling in $D_{\mathcal{G}}^\lambda$. The algorithm described above assings a value to every vertex of $D_{\mathcal{G}}^\lambda$. We now show that $\varphi^\star$ is indeed a distance labeling by showing that every edge satisfies a distance constraint.

Observe that the distance constraints imposed by edges dual to the space arcs are satisfied by construction. To show that the distance constraints imposed by edges dual to the slope arcs are also satisfied, we prove that for every edge (u, v), it holds that $\varphi^\star(v) = \varphi^\star(u) + \varphi((u, v)^\star)$. We refer to this as *condition $\mathcal{C}$* for short. Since $\varphi((u, v)^\star) \leq \lambda - 1$ and the length ℓ of (u, v) is $\lambda - 1$ we obtain $\varphi^\star(v) = \varphi^\star(u) + \ell$, which implied that $\phi^\star$ is a distance labeling of $D_{\mathcal{G}}^\lambda$.

The proof is by induction based on the bottom to top and right to left order among the edges of $D_{\mathcal{G}}^\lambda$. We say that (a, b) *precedes* (c, d) if either $\ell(a) < \ell(c)$, or $\ell(a) = \ell(c)$ and a is to the right of c, or $\ell(a) = \ell(c)$ and b is to the right of d (in case $a = c$). For the base case observe that the edges with both end-vertices on the first level and the edges of p_{right} satisfy condition $\mathcal{C}$ by the definition of $\varphi^\star$. Now let (u, v) be an edge not addressed in the base case and assume that for every edge (u', v') preceding edge (u, v) condition $\mathcal{C}$ holds. For the inductive step we show that condition $\mathcal{C}$ also holds for (u, v). Let (u', v') denote the edge to the right of (u, v) so that (u, v) and (u', v') are consecutive; see Fig. 5. Because v is not the rightmost vertex on its level this edge exists. Let A denote the set of space arcs $v_1 v_2^\star$ in $F_{\mathcal{G}}^\lambda$ with $v_1, v_2 \in [v', v]$. Analogously, let B denote the set of space arcs $u_1 u_2^\star$ in $F_{\mathcal{G}}^\lambda$ with $u_1, u_2 \in [u', u]$. It is $\varphi^\star(v) = \varphi^\star(v') + \sum_{a \in A} \varphi(a)$ by definition of $\varphi^\star$. Further, by induction hypothesis and since (u', v') precedes (u, v) it holds that $\varphi^\star(v') = \varphi^\star(u') + \varphi((u', v')^\star)$. Inserting the latter into the former equation, we obtain $\varphi^\star(v) = \varphi^\star(u') + \varphi((u', v')^\star) + \sum_{a \in A} \varphi(a)$. Again, by definition of $\varphi^\star$, it is $\varphi^\star(u) = \varphi^\star(u') + \sum_{b \in B} \varphi(b)$. By subtracting $\varphi^\star(u)$ from $\varphi^\star(v)$ we obtain

$$\varphi^\star(v) = \varphi^\star(u) - \sum_{b \in B} \varphi(b) + \varphi((u', v')^\star) + \sum_{a \in A} \varphi(a) \tag{1}$$

Flow conservation on the vertex of $F_{\mathcal{G}}^{\lambda}$ to which edges of A and B are incident gives $\varphi((u,v)^\star) - \sum_{a \in A} \varphi(a) - \varphi((u',v')^\star) + \sum_{b \in B} \varphi(b) = 0$. Solving this equation for $\varphi(u,v)$ and inserting it into (1) yields $\varphi^\star(v) = \varphi^\star(u) + \varphi((u,v)^\star)$, i.e. the condition $\mathcal{C}$ holds for (u,v). Therefore $\varphi^\star$ is a distance labeling, which we have shown to define a λ-drawing of $\mathcal{G}$.   □

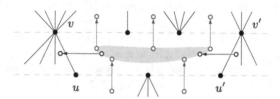

Fig. 5. Proof of Lemma 4. Sets A and B contain the outgoing and incoming red flow network arcs incident to the gray oval, respectively.

Because $D_{\mathcal{G}}^{\lambda}$ is planar we can use the $O(n \log^2 n / \log \log n)$-time shortest path algorithm due to Mozes and Wulff-Nilsen [26] to compute the shortest distance labeling. This improves our $O(n \log^3 n)$-time algorithm from Sect. 3.

Theorem 2. *Let $\mathcal{G}$ be an embedded proper level-planar graph. The distance labelings of $D_{\mathcal{G}}^{k}$ correspond bijectively to the λ-drawings of $\mathcal{G}$. If such a drawing exists, it can be found in $O(n \log^2 n / \log \log n)$ time.*

5   Partial and Simultaneous Drawings

In this section we use the distance model from Sect. 4 to construct partial and simultaneous λ-drawings. We start with introducing a useful kind of drawing. Let Γ be a λ-drawing of $\mathcal{G}$. We call Γ a λ-*rightmost* drawing when there exists no λ-drawing Γ' with $\Gamma(v) < \Gamma'(v)$ for some $v \in V$. In this definition, we assume $x(\Gamma(v_{\text{right}})) = x(\Gamma'(v_{\text{right}})) = 0$ to exclude trivial horizontal translations. Hence, a drawing is rightmost when every vertex is at its rightmost position across all level-planar λ-slope grid drawings of $\mathcal{G}$. It is not trivial that a λ-rightmost drawing exists, but it follows directly from the definition that if such a drawing exists, it is unique. The following lemma establishes the relationship between λ-rightmost drawings and shortest distance labelings of $D_{\mathcal{G}}^{\lambda}$.

Lemma 5. *Let $\mathcal{G}$ be an embedded proper level-planar graph. If $D_{\mathcal{G}}^{\lambda}$ has a shortest distance labeling it describes the λ-rightmost drawing of $\mathcal{G}$.*

Proof. The shortest distance labeling of $D_{\mathcal{G}}^{\lambda}$ is maximal in the sense that for any vertex v there exists a vertex u and an edge (u,v) with length l so that it is $x(v) = x(u) + l$. Recall that the definition of distance labelings only requires $x(v) \leq x(u) + l$. The claim then follows by induction over V in ascending order with respect to the shortest distance labeling.   □

5.1   Partial Drawings

Let $(\mathcal{G}, \mathcal{H}, \Pi)$ be a partial λ-drawing. In Sect. 3.1 we have shown that the flow model can be adapted to check whether $(\mathcal{G}, \mathcal{H}, \Pi)$ has a λ-extension, in case $\mathcal{H}$ is connected. In this section, we show how to adapt the distance model to extend partial λ-drawings, including the case $\mathcal{H}$ is disconnected. Recall that the distance label of a vertex v is its x-coordinate. A partial λ-drawing fixes the x-coordinates of the vertices of $\mathcal{H}$. The idea is to express this with additional constraints in $D_{\mathcal{G}}^{\lambda}$. Let v_{ref} be a vertex of $\mathcal{H}$. In a λ-extension of $(\mathcal{G}, \mathcal{H}, \Pi)$, the relative distance along the x-axis between a vertex v of $\mathcal{H}$ and vertex v_{ref} should be $d_v = \Pi(v_{\mathrm{ref}}) - \Pi(v)$. This can be achieved by adding an edge (v, v_{ref}) with length d_v and an edge (v_{ref}, v) with length $-d_v$. The first edge ensures that it is $x(v_{\mathrm{ref}}) \le x(v) + d_v$, i.e., $x(v) \ge x(v_{\mathrm{ref}}) - d_v$ and the second edge ensures $x(v) \le x(v_{\mathrm{ref}}) - d$. Together, this gives $x(v) = x(v_{\mathrm{ref}}) - d_v$. Let $D_{\mathcal{G}, \Pi}^{\lambda}$ be $D_{\mathcal{G}}^{\lambda}$ augmented by the edges $\{(v, v_{\mathrm{ref}}), (v_{\mathrm{ref}, v}) : \forall v \in \mathcal{H}\}$ with lengths as described above.

To decide existence of λ-extension and in affirmative construct the corresponding drawing we compute the shortest distance labeling in $D_{\mathcal{G}, \Pi}^{\lambda}$. Observe that this network can contain negative cycles and therefore no shortest distance labeling. Unfortunately, $D_{\mathcal{G}, \Pi}^{\lambda}$ is not planar, and thus we cannot use the embedding-based algorithm of Mozes and Wulff-Nilsen. However, since all newly introduced edges have v_{ref} as one endpoint, v_{ref} is an *apex* of $D_{\mathcal{G}}^{\lambda}$, i.e., removing v_{ref} from $D_{\mathcal{G}, \Pi}^{\lambda}$ makes it planar. Therefore $D_{\mathcal{G}, \Pi}^{\lambda}$ can be recursively separated by separators of size $O(\sqrt{n})$. We can therefore use the shortest-path algorithm due to Henzinger et al. to compute the shortest distance labeling of $D_{\mathcal{G}, \Pi}^{\lambda}$ in $O(n^{4/3} \log n)$ time [15].

Theorem 3. *Let $(\mathcal{G}, \mathcal{H}, \Pi)$ be a partial λ-drawing. In $O(n^{4/3} \log n)$ time it can be determined whether $(\mathcal{G}, \mathcal{H}, \Pi)$ has a λ-extension and in the affirmative the corresponding drawing can be computed within the same running time.*

5.2   Simultaneous Drawings

In the simultaneous λ-drawing problem, we are given a tuple $(\mathcal{G}_1, \mathcal{G}_2)$ of two embedded level-planar graphs that share a common subgraph $\mathcal{G}_{1\cap 2} = \mathcal{G}_1 \cap \mathcal{G}_2$. We assume w.l.o.g. that G_1 and G_2 share the same right boundary and that the embeddings of $\mathcal{G}_1$ and $\mathcal{G}_2$ coincide on $\mathcal{G}_{1\cap 2}$. The task is to determine whether there exist λ-drawings Γ_1, Γ_2 of $\mathcal{G}_1, \mathcal{G}_2$, respectively, so that Γ_1 and Γ_2 coincide on the shared graph $\mathcal{G}_{1\cap 2}$. The approach is the following. Start by computing the rightmost drawings of $\mathcal{G}_1$ and $\mathcal{G}_2$. Then, as long as these drawings do not coincide on $\mathcal{G}_{1\cap 2}$ add necessary constraints to $D_{\mathcal{G}_1}^{\lambda}$ and $D_{\mathcal{G}_2}^{\lambda}$. This process terminates after a polynomial number of iterations, either by finding a simultaneous λ-drawing, or by determining that no such drawing exist.

Finding the necessary constraints works as follows. Suppose that Γ_1, Γ_2 are the rightmost drawings of $\mathcal{G}_1, \mathcal{G}_2$, respectively. Because both $\mathcal{G}_1$ and $\mathcal{G}_2$ have the same right boundary they both contain vertex v_{right}. We define the coordinates in the distance labelings of $D_{\mathcal{G}_1}^{\lambda}$ and $D_{\mathcal{G}_2}^{\lambda}$ in terms of this reference vertex.

Now suppose that for some vertex v of $\mathcal{G}_{1\cap2}$ the x-coordinates in Γ_1 and Γ_2 differ, i.e., it is $\Gamma_1(v) \neq \Gamma_2(v)$. Assume $\Gamma_1(v) < \Gamma_2(v)$ without loss of generality. Because Γ_1 is a rightmost drawing, there exists no drawing of $\mathcal{G}_1$ where v has an x-coordinate greater than $\Gamma_1(v)$. In particular, there exist no simultaneous drawings where v has an x-coordinate greater than $\Gamma_1(v)$. Therefore, we must search for a simultaneous drawing where $\Gamma_2(v) \leq \Gamma_1(v)$. We can enforce this constraint by adding an edge (v_{right}, v) with length $\Gamma_1(v)$ into $D_{\mathcal{G}_2}^{\lambda}$. We then attempt to compute the drawing Γ_2 of $\mathcal{G}_2$ defined by the shortest distance labeling in $D_{\mathcal{G}_2}^{\lambda}$. This attempt produces one of two possible outcomes. The first possibility is that there now exists a negative cycle in $D_{\mathcal{G}_2}^{\lambda}$. This means that there exists no drawing Γ_2 of G_2 with $\Gamma_2(v) \leq \Gamma(v)$. Because Γ_1 is a rightmost drawing, this means that no simultaneous drawings of $\mathcal{G}_1$ and $\mathcal{G}_2$ exist. The algorithm then terminates and rejects this instance. The second possiblity is that we obtain a new drawing Γ_2. This drawing is rightmost among all drawings that satisfy the added constraint $\Gamma_2(v) \leq \Gamma_1(v)$. In this case there are again two possibilities. Either we have $\Gamma_1(v) = \Gamma_2(v)$ for each vertex v in $\mathcal{G}_{1\cap2}$. In this case Γ_1 and Γ_2 are simultaneous drawings and the algorithm terminates. Otherwise there exists at least one vertex w in $\mathcal{G}_{1\cap2}$ with $\Gamma_1(w) \neq \Gamma_2(w)$. We then repeat the procedure just described for adding a new constraint.

We repeat this procedure of adding other constraints. To bound the number of iterations, recall that we only consider compact drawings, i.e., drawings whose width is at most $(\lambda - 1)(n - 1)$. In each iteration the x-coordinate of at least one vertex is decreased by at least one. Therefore, each vertex is responsible for at most $(\lambda - 1)(n - 1)$ iterations. The total number of iterations is therefore bounded by $n(\lambda - 1)(n - 1) \in O(\lambda n^2)$.

Note that due to the added constraints $D_{\mathcal{G}_1}^{\lambda}$ and $D_{\mathcal{G}_2}^{\lambda}$ are generally not planar. We therefore apply the $O(n^{4/3} \log n)$-time shortest-path algorithm due to Henzinger et al. that relies not on planarity but on $O(\sqrt{n})$-sized separators to compute the shortest distance labellings. This gives the following.

Theorem 4. *Let $\mathcal{G}_1, \mathcal{G}_2$ be embedded level-planar graphs that share a common subgraph $\mathcal{G}_{1\cap2}$. In $O(\lambda n^{10/3} \log n)$ time it can be determined whether $\mathcal{G}_1, \mathcal{G}_2$ admit simultaneous λ-drawings and if so, such drawings can be computed within the same running time.*

6   Conclusion

In this paper we studied λ-drawings, i.e., level-planar drawings with λ slopes. We model λ-drawings of proper level-planar graphs as integer flow networks. This lets us find λ-drawings and extend connected partial λ-drawings in $O(n \log^3 n)$ time. We extend the duality between integer flows in a primal graph and shortest distances in its dual to obtain a more powerful distance model. This distance model allows us to find λ-drawings in $O(n \log^2 n / \log \log n)$ time, extend not-necessarily-connected partial λ-drawings in $O(n^{4/3} \log n)$ time and find simultaneous λ-drawings in $O(\lambda n^{10/3} \log n)$ time.

In the non proper case, testing the existence of a 2-drawing becomes NP-hard, even for biconnected graphs with maximum edge length two [7].

References

1. FAUSTEDITION. http://www.faustedition.net/macrogenesis/dag
2. Angelini, P., Chaplick, S., Cornelsen, S., Da Lozzo, G., Di Battista, G., Eades, P., Kindermann, P., Kratochvíl, J., Lipp, F., Rutter, I.: Simultaneous orthogonal planarity. In: Hu, Y., Nöllenburg, M. (eds.) GD 2016. LNCS, vol. 9801, pp. 532–545. Springer, Cham (2016). https://doi.org/10.1007/978-3-319-50106-2_41
3. Angelini, P., Da Lozzo, G., Di Battista, G., Frati, F., Patrignani, M., Rutter, I.: Beyond level planarity. In: Hu, Y., Nöllenburg, M. (eds.) GD 2016. LNCS, vol. 9801, pp. 482–495. Springer, Cham (2016). https://doi.org/10.1007/978-3-319-50106-2_37
4. Barát, J., Matouvsek, J., Wood, D.R.: Bounded-degree graphs have arbitrarily large geometric thickness. Electr. J. Comb. 13(1), 3 (2006)
5. Bläsius, T., Kobourov, S.G., Rutter, I.: Simultaneous embedding of planar graphs. In: Tamassia [32], pp. 349–381
6. Borradaile, G., Klein, P.N., Mozes, S., Nussbaum, Y., Wulff-Nilsen, C.: Multiple-source multiple-sink maximum flow in directed planar graphs in near-linear time. In: Proceedings of the 52nd Annual IEEE Symposium on Foundations of Computer Science, pp. 170–179. IEEE Press, New York (2011)
7. Brückner, G., Krisam, N.D., Mchedlidze, T.: Level-planar drawings with few slopes. CoRR, abs/1907.13558v1 (2019)
8. Brückner, G., Rutter, I.: Partial and constrained level planarity. In: Klein, P.N. (ed.) Proceedings of the 28th Annual ACM-SIAM Symposium on Discrete Algorithms, pp. 2000–2011. SIAM (2017)
9. Di Giacomo, E., Liotta, G., Montecchiani, F.: Drawing outer 1-planar graphs with few slopes. J. Graph Algorithms Appl. 19(2), 707–741 (2015)
10. Dujmović, V., Eppstein, D., Suderman, M., Wood, D.R.: Drawings of planar graphs with few slopes and segments. Comput. Geom. 38(3), 194–212 (2007)
11. Dujmović, V., Suderman, M., Wood, D.R.: Graph drawings with few slopes. Comput. Geom. 38(3), 181–193 (2007)
12. Di Giacomo, E., Liotta, G., Montecchiani, F.: Drawing subcubic planar graphs with four slopes and optimal angular resolution. Theor. Comput. Sci. 714, 51–73 (2018)
13. Hassin, R.: Maximum flow in (s, t) planar networks. Inform. Process. Lett. 13(3), 107 (1981)
14. Healy, P., Nikolov, N.S.: Hierarchical drawing algorithms. In: Tamassia [32], pp. 409–453
15. Henzinger, M.R., Klein, P.N., Rao, S., Subramanian, S.: Faster shortest-path algorithms for planar graphs. J. Comput. Syst. Sci. 55(1), 3–23 (1997)
16. Hoffmann, U.: On the complexity of the planar slope number problem. J. Graph Algorithms Appl. 21(2), 183–193 (2017)
17. Te Chiang, H.: Integer Programming and Network Flows. Addison-Wesley, Reading (1969)
18. Itai, A., Shiloach, Y.: Maximum flow in planar networks. SIAM J. Comput. 8(2), 135–150 (1979)

19. Jänicke, S., Geßner, A., Franzini, G., Terras, M., Mahony, S., Scheuermann, G.: Traviz: a visualization for variant graphs. DSH **30**(Suppl-1), i83–i99 (2015)
20. Keszegh, B., Pach, J., Pálvölgyi, D.: Drawing planar graphs of bounded degree with few slopes. SIAM J. Discrete Math. **27**(2), 1171–1183 (2013)
21. Keszegh, B., Pach, J., Pálvölgyi, D., Tóth, G.: Drawing cubic graphs with at most five slopes. Comput. Geom. **40**(2), 138–147 (2008)
22. Kindermann, P., Meulemans, W., Schulz, A.: Experimental analysis of the accessibility of drawings with few segments. J. Graph Algorithms Appl. **22**(3), 501–518 (2018)
23. Knauer, K.B., Micek, P., Walczak, B.: Outerplanar graph drawings with few slopes. Comput. Geom. **47**(5), 614–624 (2014)
24. Lenhart, W., Liotta, G., Mondal, D., Nishat, R.I.: Planar and plane slope number of partial 2-trees. In: Wismath, S., Wolff, A. (eds.) GD 2013. LNCS, vol. 8242, pp. 412–423. Springer, Cham (2013). https://doi.org/10.1007/978-3-319-03841-4_36
25. Mchedlidze, T., Nöllenburg, M., Rutter, I.: Extending convex partial drawings of graphs. Algorithmica **76**(1), 47–67 (2016)
26. Mozes, S., Wulff-Nilsen, C.: Shortest paths in planar graphs with real lengths in $O(n\log^2 n/\log\log n)$ time. In: de Berg, M., Meyer, U. (eds.) ESA 2010. LNCS, vol. 6347, pp. 206–217. Springer, Heidelberg (2010). https://doi.org/10.1007/978-3-642-15781-3_18
27. Nickel, S., Nöllenburg, M.: Drawing k-linear metro maps. CoRR, abs/1904.03039 (2019)
28. Nöllenburg, M.: A survey on automated metro map layout methods. In: Schematic Mapping Workshop. Essex, UK, April 2014
29. Pach, J., Pálvölgyi, D.: Bounded-degree graphs can have arbitrarily large slope numbers. Electr. J. Comb. **13**(1), N1 (2006)
30. Patrignani, M.: On extending a partial straight-line drawing. Int. J. Found. Comput. Sci. **17**(5), 1061–1070 (2006)
31. Purchase, H.C.: Metrics for graph drawing aesthetics. J. Vis. Lang. Comput. **13**(5), 501–516 (2002)
32. Tamassia, R. (ed.): Handbook on Graph Drawing and Visualization. Chapman and Hall/CRC, London/Boca Raton (2013)

Graph Drawing Contest Report

Graph Drawing Contest Report

Philipp Kindermann[1]([✉]), Tamara Mchedlidze[2], and Ignaz Rutter[3]

[1] Universität Würzburg, Würzburg, Germany
philipp.kindermann@uni-wuerzburg.de
[2] Karlsruhe Institute of Technology, Karlsruhe, Germany
mched@iti.uka.de
[3] Universität Passau, Passau, Germany
rutter@fim.uni-passau.de

Abstract. This report describes the 26th Annual Graph Drawing Contest, held in conjunction with the 27th International Symposium on Graph Drawing and Network Visualization (GD'19) in Průhonice/Prague, Czech Republic. The mission of the Graph Drawing Contest is to monitor and challenge the current state of the art in graph-drawing technology.

1 Introduction

Following the tradition of the past years, the Graph Drawing Contest was divided into two parts: the *creative topics* and the *live challenge*.

Creative topics were comprised by two data sets. The first data set described appearances of superheroes in the Marvel Cinematic Universe movie. The second data set was comprised by occurrences of ingredients in popular meals. The data sets were published a year in advance, and contestants submitted their visualizations before the conference started. Submissions were evaluated according to aesthetic appeal, domain-specific requirements, readability and clarity of the visualization and novelty of the visualization concept.

The live challenge took place during the conference in a format similar to a typical programming contest. Teams were presented with a collection of *challenge graphs* and had one hour to submit their highest scoring drawings. This year's topic was to minimize the number of crossings an upward straight-line drawing of a graph with vertex locations restricted to a grid.

Overall, we received 29 submissions: 10 submissions for the creative topics and 19 submissions for the live challenge.

2 Creative Topics

The general goal of the creative topics was to model each data set as a graph and visualize it with complete artistic freedom, and with the aim of communicating as information much as possible from the provided data in the most readable and clear way.

© Springer Nature Switzerland AG 2019
D. Archambault and C. D. Tóth (Eds.): GD 2019, LNCS 11904, pp. 575–583, 2019.
https://doi.org/10.1007/978-3-030-35802-0_43

We received 6 submissions for the first topic, and 4 for the second. For each topic, we selected the top three submissions before the conference, which were printed on large poster boards and presented at the Graph Drawing Symposium. During the conference dinner, we presented these submissions and announced the winners. We will now review the top three submissions for each topic (for a complete list of submissions, refer to http://www.graphdrawing.org/gdcontest/contest2019/results.html).

2.1   Marvel Cinematic Universe

The Marvel Cinematic Universe is a media franchise and shared universe that is centered on a series of superhero films, based on characters that appear in comic books published by Marvel. The data set describes a selection of 28 characters (heroes) and in which of the 24 movies released so far they appeared. The data was compiled from the Marvel Cinematic Universe Wiki[1].

Third Place: Velitchko Filipov, Alessio Arleo, Davice Ceneda, and Silvia Miksch (TU Vienna). The committee appreciated the extensive use of glyphs, the use of the non-provided information on when the movies were filmed, the clarity, the minimalistic style, and the aesthetics of the visualization.

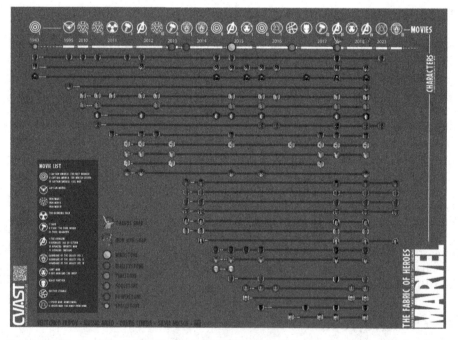

Second Place: Markus Wallinger (TU Vienna). The committee valued the choice of the metro-map visualization metaphor, the well-constructed layout, the use of glyphs. On the other hand the committee observed that the visualization

---

[1] https://marvelcinematicuniverse.fandom.com/wiki/Marvel_Cinematic_Universe_Wiki.

would gain in readability from a better choice of color palette and from the selective use of text labels on the lines.

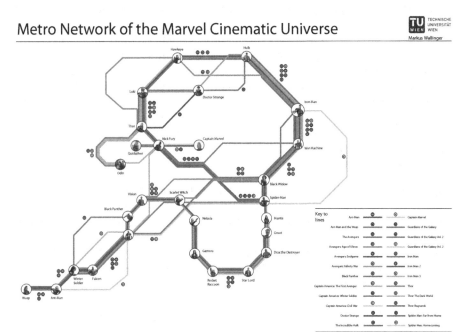

Winner: Evmorfia Argyriou, Christian Brunnermeier, Anne Eberle, and Johannes Rössel (yWorks). The committee especially valued the minimalism and the clarity of the presentation, the choice of the hierarchical layout resembling Sankey diagrams, the use of the non-provided information such as the timeline and the screen time, the choice of the color pallete and the glyphs. The committee was also impressed by the design choices of the overall poster. The visualization and an explanation of the drawing process is available online: http://yworks.com/marvel.

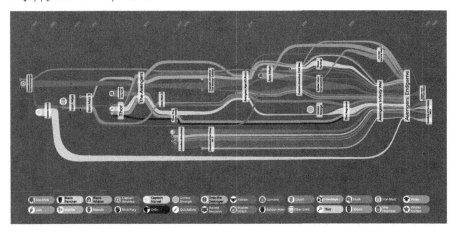

❝ In this visualization of the Marvel Cinematic Universe and its characters, we've arranged all movies in their in-universe timeline. Characters
are shown as edges that pass through the movies they appear in. The
thickness of edges models the relative screen time of characters within
a movie. The overall layout is automatically computed and uses yFiles'
support for Sugiyama-style graph drawing with several adjustments and
enhancements for parts of the graph. The accompanying interactive application (yworks.com/marvel) also allows to filter the graph by character or film series. ❞

Evmorfia Argyriou

2.2   Meal Ingredients

The data set describes 151 food recipes extracted from the TheMealDB
database[2]. TheMealDB was built in 2016 to provide a free data source API
for recipes online. It originated on the Kodi forums as a way to browse recipes
on a TV.

The provided data set consisted of three types. There were 297 food ingredients, e.g. "Beef", "Flour", "Red Wine". There were 11 areas (countries) that
are popular for their dishes around the globe: "American", "British", "Chinese", "French", "Greek", "Indian", "Italian", "Japanese", "Mexican", "Span-
ish", "Thai". Finally, there were 151 recipes that contain a list of ingredients
and belong to one area.

The task was to visualize the data either in a form of a graph or any other
form the authors prefer. The authors could decide whether they omit relatively
uninformative parts of the data.

**Third Place: James Wood, Marni Torkel, Ereina Gomez, Amyra Mei-
diana, Peter Eades, and Seok-Hee Hong (University of Sydney).** The
committee noticed that the constructed graph of meals, where the edges represent shared ingredients, and the graph of ingredients, where the edges represent
the number of meals using both ingredients, provide a interesting overview on
the data, by revealing the clusters of similar meals and relative ingredients.

---

[2] https://www.themealdb.com.

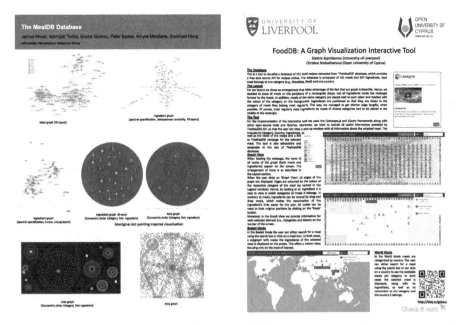

Second Place: Elektra Kypridemou (University of Liverpool) and Christos Rodosthenous (Open University of Cyprus). The committee appreciated the interactive design of the system which allowed to investigate the data set in details. The interactive tool is available online: http://cognition-srv1. ouc.ac.cy/food_db.

Winner: Guangping Li, Soeren Nickel, Martin Nöllenburg, Ivan Viola, and Hsiang-Yun Wu (TU Vienna). The committee valued the attempt to visualize both the details of the data set but also an overview, which was presented by showing the clusters of the meals by the country of origin. The committee also valued the idea of splitting the ingredient vertices to untangle the visualization.

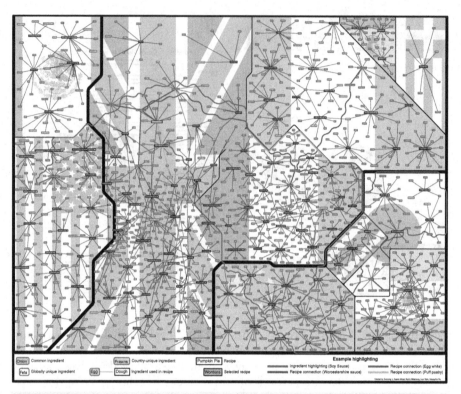

    In our World Map of Recipes, we used a multi-level force-based algorithm to partition screen space among the countries and to harmonize the label territory. The visibility management is achieved by duplicating high-frequency ingredient nodes coupled with a spanning-tree-based visual integration. The technique automatically grouped countries sharing common ingredients in their recipes close to each other, which happened to produce continental clusters, and it visually discriminates ingredient nodes with different levels of importance. It also allows us to highlight ingredients and recipes of interest using an occlusion-free curve routing scheme.

Martin Nöllenburg

3   Live Challenge

The live challenge took place during the conference and lasted exactly one hour. During this hour, local participants of the conference could take part in the manual category (in which they could attempt to draw the graphs using a supplied tool[3]), or in the automatic category (in which they could use their own software to draw the graphs). At the same time, remote participants could also take part in the automatic category.

---

[3]   http://graphdrawing.org/gdcontest/tool/.

The challenge focused on minimizing the number of crossings in an upward straight-line embedding of a given directed graph, with vertex locations restricted to a grid. The results were judged solely with respect to the number of crossings; other aesthetic criteria were not taken into account. This allows an objective way to evaluate each drawing.

3.1 The Graphs

In the manual category, participants were presented with six graphs. These were arranged from small to large and chosen to contain different types of graphs and graph structures. In the automatic category, participants had to draw the same six graphs as in the manual category, and in addition another six larger graphs. Again, the graphs were constructed to have different structure.

For illustration, we include the fifth graph in its initial state, the best manual solution we received (by team Dinosaurs), and the best automatic solution we received (by team Tübingen-Bus).

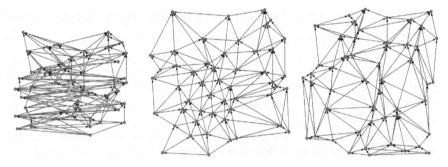

For the complete set of graphs and submissions, refer to the contest website at http://www.graphdrawing.org/gdcontest/contest2019/results.html. The graphs are still available for exploration and solving Graph Drawing Contest Submission System[4].

Similarly to the past years, the committee observed that manual (human) drawings of graphs often display a deeper understanding of the underlying graph structure than automatic and therefore gain in readability. The committee was also impressed by the fact that for all the small graphs the manual drawing were almost as good as the automatic drawings.

3.2 Results: Manual Category

We are happy to present the full list of scores for all teams. The numbers listed are the numbers of crossings in the drawings; the horizontal bars visualize the corresponding scores.

---

[4] https://graphdrawingcontest.appspot.com.

graph	1	2	3	4	5	6
balloon	2	14	7	27	-	424
UpperCut	2	21	9	5	32	323
UPGD19	0	13	21	8	32	242
1 device per team	4	47	70	25	77	380
Vertigo	0	20	5	9	32	394
Discrete geometers	6	18	5	5	32	251
36% Battery Remaining	2	4	7	7	32	278
PAF	0	9	6	10	38	396
scho+	2	17	7	12	35	196
Dinosaurs	4	16	7	12	32	270
#OnlyGoodDrawings	2	8	8	6	32	271
Complicated Graphs	2	24	147	26	35	371
IV	0	22	49	10	34	283
Now Austrians	2	14	5	8	32	308
Aaaaaah	0	8	6	6	215	267

Third place: **Discrete geometers**, consisting of Jan Kyncl and Birgit Vogten-huber.

Second place: **#OnlyGoodCrossings**, consisting of Fabrizio Montecchiani, Luca Castelli Aleardi, and Giacomo Ortali.

Winner: **36% Battery Remaining League**, consisting of Evmorvia Argyriou, Henry Förster, and Martin Gronemann.

66 For the manual contest, we followed first of all the basic rules that are independent of the problem to be solved. So at the beginning we submitted all instances right away with their initial layout to ensure that at the end we have a feasible entry for every instance. We started with the smaller instances in which we invested more time compared to the larger instances. For this particular problem, it was quite beneficial to try to figure out how these instances have been created. For example, while trying to untangle Instance 5, it became quickly evident that the underlying undirected graph is a 1-planar graph with a planar skeleton that resembles some kind of grid graph. This then led to a pattern that we used for the layout and we were quickly done with the second largest instance. The last and largest instance, however, was very difficult. After a "good" start trying to figure out the structure, we managed to increase the number of crossings from 448 to over 600 which forced us to start over. On the second try we simply followed a greedy approach as there were only 5 minutes and 40% of battery left. 99

Martin Gronemann

3.3 Results: Automatic Category

We are happy to present the full list of scores for all teams that participated in the automatic category. The numbers listed are the numbers of crossings in the drawings; the horizontal bars visualize the corresponding scores.

graph	1	2	3	4	5	6	7	8	9	10	11	12
JáMa	2	19	10	13	48	297						
InfitIntersect? Uni Kassel	2	17	6	7	40	354	267	2730	495	6405	14697	706388
UoA	14	108	176	39	32	437						
Tübingen-Bus	0	4	5	4	32	81	0	307	38	1568	1721	147628

Third place: **JáMa**, consisting of Tomáš Masařík and Jana Novotná.
Second place: **InfitIntersect? Uni Kassel**, consisting of Dominik Dürrschnabel, Jannik Raabe, Christoph Sandrock, and Joschka Wittich.
Winner: **Tübingen-Bus**, consisting of Solveig Klepper, Axel Kuckuk, Paul Palomero Bernardo, Maximilian Pfister, Patrizio Angelini, Michalis Bekos, and Michael Kaufmann.

> ❝ We adopted a variant of the probabilistic hill-climbing method that gave us the first place last year, which we adjusted and optimized to the given task. The performed optimization made it even faster than the Tübingen-Bus that brought us to Prague. We are now looking forward ❞ to participate next year and defend our title.
> *Maximilian Pfister*

Acknowledgments. The contest committee would like to thank the organizing committee of the conference for providing a room with hardware for the live challenge and medals for the winners; the generous sponsors of the symposium; and all the contestants for their participation. Further details including all submitted drawings and challenge graphs can be found at the contest website:

http://www.graphdrawing.org/gdcontest/contest2019/results.html

Correction to: A Natural Quadratic Approach to the Generalized Graph Layering Problem

Sven Mallach

Correction to:
Chapter "A Natural Quadratic Approach to the Generalized Graph Layering Problem" in: D. Archambault and C. D. Tóth (Eds.): *Graph Drawing and Network Visualization*, LNCS 11904, https://doi.org/10.1007/978-3-030-35802-0_40

The original version of this chapter was revised. The final formula in section 4.1 was corrected.

The updated version of this chapter can be found at
https://doi.org/10.1007/978-3-030-35802-0_40

© Springer Nature Switzerland AG 2019
D. Archambault and C. D. Tóth (Eds.): GD 2019, LNCS 11904, p. C1, 2019.
https://doi.org/10.1007/978-3-030-35802-0_44

Poster Abstracts

A 1-planarity Testing and Embedding Algorithm

Carla Binucci$^{(\boxtimes)}$ [ID], Walter Didimo [ID], and Fabrizio Montecchiani [ID]

Università degli Studi di Perugia, Perugia, Italy
{carla.binucci, walter.didimo, fabrizio.montecchiani}@unipg.it

Abstract. Recognizing whether a graph is 1-planar is NP-complete, even for restricted graph classes. We present a testing and embedding algorithm for general 1-planar graphs, based on backtracking. We implemented our approach and experimented it on two popular graph suites.

Introduction. A graph is *1-planar* if it can be drawn in the plane such that each edge is crossed at most once. The family of 1-planar graphs naturally extends that of planar graphs and it has received increasing attention in the last years [11, 15]. Recognizing whether a graph is 1-planar is an NP-complete problem [12, 16], in contrast with well-known efficient algorithms for testing planarity. The problem is NP-complete even for restricted graph classes, for instance graphs of bounded treewidth [5] (see also [4, 8]). The problem becomes fixed-parameter tractable when parameterized by vertex-cover number, cyclomatic number, or tree-depth [5]. Polynomial-time testing algorithms have been designed for some subfamilies of 1-planar graphs (see, e.g., [3, 6, 7, 14]). However, no practical algorithms for general graphs exist that can be effectively implemented and adopted in applications. This scenario naturally motivates our research, which goes in the direction of filling the gap between theory and practice, and which poses some interesting foundations for further advances. Our contribution is as follows: (1) We describe a testing and embedding algorithm for 1-planar graphs, based on a backtracking strategy and relatively easy to implement. Our algorithm can be also used as a preliminary step for those algorithms taking a 1-planar embedding as input. (2) We experiment our algorithm on two well-established graph suites, the ROME and NORTH graphs [1, 10]. We measure its running time and compare its number of crossings with respect to a state-of-the-art planarizer. The classified solved instances are publicly available [2].

Algorithm. Our 1-planarity testing and embedding algorithm, 1PlanarTester, works as follows. It processes each biconnected component C of the input graph G independently. First, it preliminary checks whether C can be immediately labeled as 1-planar or as not 1-planar based on simple criteria. Else, 1PlanarTester runs a backtracking procedure whose output is either a 1-planar embedding of C or a negative answer. At the end, 1PlanarTester will either output an embedding for each biconnected component of G, or return a component that is not 1-planar. The backtracking procedure takes as input a biconnected graph G with n vertices and m edges. Let $\overline{E}$ be the set of all pairs of independent

© Springer Nature Switzerland AG 2019
D. Archambault and C. D. Tóth (Eds.): GD 2019, LNCS 11904, pp. 587–589, 2019.
https://doi.org/10.1007/978-3-030-35802-0

Table 1. Summary of the experiments.

Graphs	# Instances	Solved (%)	1-planar (%)	Runtime (minutes)			Solved by Backtracking	Crossing Ratio
				AVG	SD	MAX		
Rome 10–20	91	91.2%	100.0%	0.07	0.60	5.50	14.5%	1.07
Rome 21–30	164	69.5%	100.0%	0.38	3.61	38.40	24.6%	1.12
Rome 31–40	388	43.8%	100.0%	1.34	12.06	115.38	21.2%	1.30
Rome 41–50	119	37.8%	100.0%	0.01	0.01	0.02	20.0%	1.09
North 10–20	121	73.6%	88.8%	1.86	10.10	92.69	39.3%	1.78
North 21–30	69	39.1%	77.8%	3.85	10.14	39.58	25.9%	1.64
North 31–40	55	38.2%	57.1%	3.78	11.65	39.80	9.5%	1.00
North 41–50	32	18.8%	83.3%	0.01	0.01	0.01	0.0%	1.00

edges of G, and let $k = |\overline{E}|$. We choose an ordering of $\overline{E}$, and we encode a *candidate solution* as a binary array y of length k where $y[i] = 0$ (resp. $y[i] = 1$) means that the i-th pair of edges of $\overline{E}$ do not cross (resp. cross) in a 1-planar embedding of G (if it exists). We say that y is TRUE if: (i) Each edge is crossed at most once, and (ii) by replacing each crossing with a dummy vertex, the resulting graph is planar; otherwise y is FALSE. We can prove that G is 1-planar if and only if the set of candidate solutions $\mathcal{C}$ contains a TRUE solution. The set $\mathcal{C}$ is generated incrementally, by computing a binary search tree T. Each node ν of T has an array y_ν of length $i_\nu < k$ that represents a partial candidate solution. When visiting a node ν of T (in a top-down order), we run a routine that outputs one of three values: SOL, if y_ν is (or can be extended to) a TRUE solution; CUT, if y_ν is a FALSE solution and hence the subtree at ν can be pruned; CNT, otherwise. The main idea behind such routine is to verify whether conditions (i) and (ii) apply to the graph induced by those edges that either cross in y_ν or will not cross in any extension of y_ν. If the conditions apply, the routine tries to complete the solution and returns either SOL or CNT, else it returns CUT.

Experiments. We implemented 1PlanarTester in the C# language, integrating the OGDF library [9]. We executed experiments to evaluate the running time and the size of the instances that our algorithm can handle in a reasonable time. For those instances classified as 1-planar, we compare the number of crossings produced by 1PlanarTester with respect to a planarizer [13] available in OGDF (which is not restricted to produce 1-planar drawings). We used the non-planar instances of two well-established suites: the Rome and the North graphs [1, 10]; see [2] for the resulting classification. Table 1 groups the instances by size and summarizes the experimental results. For each group it reports the number of instances, the % of solved instances (i.e., whose computations took less than 3 h), the % of 1-planar instances among the solved ones, the running time (average, std dev and max) took for the solved instances, the % of solved instances settled by using the backtracking procedure, the ratio of the number of crossings produced by 1PlanarTester over the OGDF planarizer. The majority of the solved instances are 1-planar, which corroborates the interest on 1-planar graphs from an application perspective. Overall, 1PlanarTester solved most of the ROME (resp. NORTH) graphs with up to 40 (resp. 20) vertices. Also, the average crossing ratio is always below 1.78, hence restricting the number of crossings per edge did not affect too much the total number of crossings.

References

1. http://www.graphdrawing.org/data.html. Accessed July 2019
2. http://mozart.diei.unipg.it/montecchiani/1planarity/labels.xlsx. Accessed July 2019
3. Auer, C., Bachmaier, C., Brandenburg, F.J., Gleiner, A., Hanauer, K., Neuwirth, D., Reislhuber, J.: Outer 1-planar graphs. Algorithmica **74**(4), 1293–1320 (2016)
4. Auer, C., Brandenburg, F.J., Gleißner, A., Reislhuber, J.: 1-planarity of graphs with a rotation system. J. Graph Algorithms Appl. **19**(1), 67–86 (2015)
5. Bannister, M.J., Cabello, S., Eppstein, D.: Parameterized complexity of 1-planarity. J. Graph Algorithms Appl. **22**(1), 23–49 (2018). https://doi.org/10.7155/jgaa.00457
6. Brandenburg, F.J.: Recognizing optimal 1-planar graphs in linear time. Algorithmica **80**(1), 1–28 (2018)
7. Brandenburg, F.J.: Characterizing and recognizing 4-map graphs. Algorithmica **81**(5), 1818–1843 (2019)
8. Cabello, S., Mohar, B.: Adding one edge to planar graphs makes crossing number and 1-planarity hard. SIAM J. Comput. **42**(5), 1803–1829 (2013)
9. Chimani, M., Gutwenger, C., Jünger, M., Klau, G.W., Klein, K., Mutzel, P.: The open graph drawing framework (OGDF). In: Handbook of Graph Drawing and Visualization, pp. 543–569. Chapman and Hall/CRC (2013)
10. Di Battista, G., Garg, A., Liotta, G., Tamassia, R., Tassinari, E., Vargiu, F.: An experimental comparison of four graph drawing algorithms. Comput. Geom. **7**, 303–325 (1997)
11. Didimo, W., Liotta, G., Montecchiani, F.: A survey on graph drawing beyond planarity. ACM Comput. Surv. **52**(1), 4:1–4:37 (2019)
12. Grigoriev, A., Bodlaender, H.L.: Algorithms for graphs embeddable with few crossings per edge. Algorithmica **49**(1), 1–11 (2007). https://doi.org/10.1007/s00453-007-0010-x
13. Gutwenger, C., Mutzel, P.: An experimental study of crossing minimization heuristics. In: Liotta, G. (ed.) GD 2003. LNCS, vol. 2912, pp. 13–24. Springer, Heidelberg (2004). https://doi.org/10.1007/978-3-540-24595-7_2
14. Hong, S., Eades, P., Katoh, N., Liotta, G., Schweitzer, P., Suzuki, Y.: A linear-time algorithm for testing outer-1-planarity. Algorithmica **72**(4), 1033–1054 (2015)
15. Kobourov, S.G., Liotta, G., Montecchiani, F.: An annotated bibliography on 1-planarity. Comput. Sci. Rev. **25**, 49–67 (2017). https://doi.org/10.1016/j.cosrev.2017.06.002
16. Korzhik, V.P., Mohar, B.: Minimal obstructions for 1-immersions and hardness of 1-planarity testing. J. Graph Theory **72**(1), 30–71 (2013)

Stretching Two Pseudolines in Planar Straight-Line Drawings

Tamara Mchedlidze[1], Marcel Radermacher[1(✉)], Ignaz Rutter[2],
and Peter Stumpf[2]

[1] Department of Informatics, Karlsruhe Institute of Technology (KIT),
Karlsruhe, Germany
mched@iti.uka.de, radermacher@kit.edu
[2] Faculty of Computer Science and Mathematics, University of Passau,
Passau, Germany
{rutter, stumpf}@fim.uni-passau.de

Every planar graph $G = (V, E)$ has a straight-line drawing [4, 6]. In a restricted setting one seeks a drawing of G that obeys given constraints, e.g., Biedl et al. [1, 2] studied whether a bipartite planar graph has a drawing where the two sets of the partition can be separated by a straight line. Da Lozzo et al. [3] generalized the previous result and characterized the planar graphs with a partition $L \cup R \cup S = V$ of the vertex set that have a planar straight-line drawing such that the vertices in L and R lie left and right of a common line l, respectively, and the vertices in S lie on l. In this case S is called *collinear*. In particular, they showed that S is collinear if and only if there is a drawing of G such that there is an open simple curve $\mathcal{P}$ that starts and ends in the outer face of G, separates L from R, collects all vertices in S, and that, for each edge e, either entirely contains e or intersects e at most once. We refer to $\mathcal{P}$ as a *pseudoline with respect to G*.

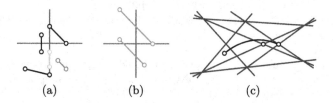

       (a)   (b)   (c)

Fig. 1. Throughout the paper, blue curve indicate pseudolines. (a) Allowed types of edges in aligned graphs of alignment complexity $(1, 0, 0)$. The green edge is aligned. The purple edge is free. (b) Aligned graph of alignment complexity $(2, 1, \perp)$. (c) Aligned graph of alignment complexity $(\perp, 3, \perp)$ that does not have an aligned drawing [5]. (Color figure online)

Mchedlidze et al. [5] generalized this concept to arrangements of pseudolines and introduced the notion of *aligned graphs*, i.e, a pair $(G, \mathcal{A})$ where G is a planar embedded graph and $\mathcal{A}$ is a set of pseudolines with respect to G that intersect

Work was partially supported by grants RU 1903/3-1 and WA 654/21-1 of the German Research Foundation(DFG).

D. Archambault and C. D. Tóth (Eds.): GD 2019, LNCS 11904, pp. 590–592, 2019.
https://doi.org/10.1007/978-3-030-35802-0

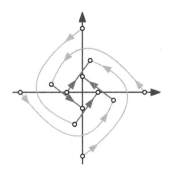

Fig. 2. 2-aligned graph that does not have an aligned drawing. (Color figure online)

pairwise at most once. Informally, a pair (Γ, A) of a straight-line drawing Γ of G and a line arrangement A, is an *aligned drawing* of (G, A) if (Γ, A) and (G, A) have the same combinatorial structure.

For $i = 0, 1, 2$, let E_i be the set of edges with i endpoints on distinct pseudolines. For an edge e, let l^e be the number of pseudolines in A such that their intersection with e lies in the interior of e. Then we define l_i as $\bot$ if E_i is empty and, otherwise, as the maximum value l^e over all edges $e \in E_i$. The complexity of an aligned graph can be described with the triple (l_0, l_1, l_2). Mchedlidze et al. showed that every aligned graph of alignment complexity $(1, 0, \bot)$, i.e., for each edge e there is at most one pseudoline $\mathcal{L} \in A$ that has a non-empty intersection with e, has an aligned drawing. They asked which combinations of number of pseudolines and alignment complexities always admit an aligned drawing. For example, The 8-aligned graph in Fig. 1c of alignment complexity $(\bot, 3, \bot)$ does not have an aligned drawing [5]. We provide an example of an aligned graph that does not have an aligned drawing with a *smaller* alignment complexity and that uses fewer pseudolines.

Theorem 1. *The 2-aligned graph in Fig. 2 has alignment complexity $(\bot, 1, \bot)$ and does not have an aligned drawing.*

The crux of the example in Fig. 2 is that the source vertices of the red (green) edges are free (aligned). We refer to aligned graphs without the red edges as *counterclockwise aligned graphs* or as *ccw aligned graphs*.

Theorem 2. *Every ccw-aligned graph $(G, \{\mathcal{X}, \mathcal{Y}\})$ has an aligned drawing.*

Open Questions. To fully answer the open question of Mchedlidze et al., the following questions are of particular interest.

- Does every aligned graph $(G, \{\mathcal{X}, \mathcal{Y}\})$ have an aligned drawing, if $\mathcal{X}$ and $\mathcal{Y}$ do not intersect?
- Does every counterclockwise-aligned graph (G, A) have an aligned drawing, if the pseudolines in A intersect in a single point?
- For general stretchable pseudoline arrangements A, does every aligned graph (G, A) of alignment complexity $(1, 0, 0)$ have an aligned drawing?

References

1. Biedl, T., Kaufmann, M., Mutzel, P.: Drawing planar partitions II: HH-drawings. In: Hromkovič, J., Sýkora, O. (eds.) WG 1998. LNCS, vol. 1517, pp. 124–136. Springer, Heidelberg (1998). https://doi.org/10.1007/10692760_11
2. Biedl, T.C.: Drawing planar partitions I: LL-drawings and LH-drawings. In: Proceedings of the 14th Annual Symposium on Computational Geometry (SoCG 1998), pp. 287–296. ACM (1998). https://doi.org/10.1145/276884.276917
3. Da Lozzo, G., Dujmovic, V., Frati, F., Mchedlidze, T., Roselli, V.: Drawing planar graphs with many collinear vertices. J. Comput. Geom. **9**(1), 94–130 (2018). https://doi.org/10.20382/jocg.v9i1a4
4. Fáry, I.: On straight line representation of planar graphs. Acta Universitatis Szegediensis. Sectio Scientiarum Mathematicarum **11**, 229–233 (1948)
5. Mchedlidze, T., Radermacher, M., Rutter, I.: Aligned drawings of planar graphs. J. Graph Algorithms Appl. **22**(3), 401–429 (2018). https://doi.org/10.7155/jgaa.00475
6. Tutte, W.T.: How to draw a graph. Proc. London Math. Soc. **s3–13**(1), 743–767 (1963)

Adventures in Abstraction: Reachability in Hierarchical Drawings

Panagiotis Lionakis[1,2], Giacomo Ortali[3(✉)], and Ioannis G. Tollis[1,2]

[1] Computer Science Department, University of Crete, Heraklion, Greece
{lionakis, tollis}@csd.uoc.gr
[2] Tom Sawyer Software, Inc., Berkeley, CA 94707, USA
[3] University of Perugia, Perugia, Italy
giacomo.ortali@gmail.com

We present algorithms and experiments for the visualization of directed graphs that focus on displaying their reachability information. Our algorithms are based on the concepts of the path and channel decomposition as proposed in the framework presented in [5]. They reduce the visual complexity of the resulting drawings by (a) drawing the vertices of the graph in some vertical lines, and (b) by progressively *abstracting* some transitive edges thus showing only a subset of the edge set in the output drawing. The process of progressively abstracting the edges gives different visualization results, but they all have the same transitive closure as the input graph. Notice that this type of abstraction has additional applications in storing the transitive closure of huge graphs, which is a significant problem in the area of graph databases and big data [2, 3, 6, 8, 9]. We also present experimental results that show a very interesting interplay between bends, crossings, clarity of the drawings, and the abstraction of edges. Our algorithms require at most $O(km)$ time, where k is the number of paths/channels and m is the number of edges. They produce progressively more abstract drawings of the input graph. No dummy vertices are introduced and the vertices of each path/channel are *vertically aligned.*

A *path* and a *channel* are both ordered sets of vertices. In a path every vertex is connected by a direct edge to its successor, while in a channel any vertex is connected to it by a directed path (which may be a single edge). Figure 1 shows an example of three different hierarchical drawings: Part (a) shows the drawing of a directed graph G computed by Tom Sawyer Perspectives [1] that (as almost all implementations) follows the Sugiyama framework [7]; Part (b) shows a hierarchical drawing computed by our first variant algorithm taking G as input; Part (c) shows an abstracted hierarchical drawing computed by our final variant that removes all path edges and selected transitive cross edges. The advantages of the last drawing are (i) clarity of the drawing due to the sparse representation, (ii) all path edges and transitive edges (within a path) are implied by the x and y coordinates, (iii) the drawn graph has the same transitive closure as G, (iv) it gives a compact data structure to store the transitive closure of G, and (v) a path between vertices that are on different (decomposition) paths can be obtained by traversing exactly one cross edge.

The algorithms presented here are variants of the path based algorithm presented in [5]. Namely, we present seven variants (v0−v6, including the original

© Springer Nature Switzerland AG 2019
D. Archambault and C. D. Tóth (Eds.): GD 2019, LNCS 11904, pp. 593–595, 2019.
https://doi.org/10.1007/978-3-030-35802-0

one) that progressively remove edges, crossings and bends. The full set of variants and results can be found in [4]. Each variant has its own advantages and disadvantages that can be exploited in various applications. Furthermore, due to its flexibility, new variants can be created based on the needs of specific applications. We present experimental results that demonstrate the power of edge abstraction and its impact on the number of bends, crossings, bundling, etc.

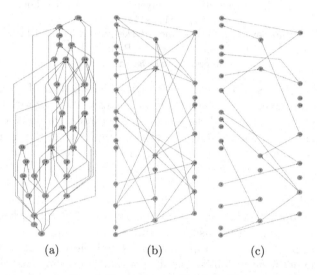

(a)   (b)   (c)

Fig. 1. (a) A drawing of a Graph G as computed by Tom Sawyer Perspectives following the Sugiyama framework; (b) a drawing based on G computed by our first variant; (c) an abstracted hierarchical drawing computed by our final variant.

All variants run very fast ($<<0.2$ s). Table 1 summarizes the results regarding the *number of bends and crossings* for each variant. An interesting observation is that our variants produce hierarchical drawings that can be suitable for large datasets since the reachability information is clearly visible while the running time is rather small. Observe that variants v4 and v6 give the most promising results since they outperform in the number of crossings, bends and drawn edges.

As expected, the number of crossings is influenced heavily by the number of edges drawn and the extent of edge bundling. Variant v1 is slightly better than v0. This can be explained by the fact that in v1 there are more bundles of edges and this naturally decreases the number of crossings, at the expense of the number of bends. All other variants have much better performance than v0 and v1 because the corresponding drawings contain significantly fewer edges, as indicated by column m3. Similar variants can be used using the channel based framework.

Table 1. Results on *number of crossings* (m1), *number of bends* (m2) and (c) *% number of edges drawn* (m3) for each variant over all DAGs.

Variant	DAG1			DAG2			DAG3			DAG4			DAG5		
	m1	m2	m3	m1	m2	m3	m1	m2	m3	m1	m2	m3	m1	m2	m3
v0	60	1	78%	20	2	93%	357	7	89%	1785	16	86%	11506	46	80%
v1	52	7	78%	22	6	93%	326	22	89%	1514	40	86%	8640	88	80%
v2	46	1	72%	20	2	93%	189	7	77%	1102	13	77%	3160	32	56%
v3	24	1	71%	20	2	93%	180	7	79%	989	14	75%	2733	38	54%
v4	17	1	65%	20	2	93%	92	7	67%	618	12	66%	866	25	30%
v5	53	1	37%	15	2	18%	331	7	47%	1722	16	47%	11353	46	60%
v6	12	1	24%	15	2	18%	73	7	26%	574	12	27%	801	25	10%

References

1. Tom Sawyer Software. www.tomsawyer.com
2. Jagadish, H.V.: A compression technique to materialize transitive closure. ACM Trans. Database Syst. **15**(4), 558–598 (1990). https://doi.org/10.1145/99935.99944
3. Jin, R., Ruan, N., Dey, S., Yu, J.X.: SCARAB: scaling reachability computation on large graphs. In: Proceedings of the ACM SIGMOD International Conference on Management of Data, SIGMOD 2012, Scottsdale, AZ, USA, 20–24 May 2012, pp. 169–180 (2012). https://doi.org/10.1145/2213836.2213856
4. Lionakis, P., Ortali, G., Tollis, I.G.: Adventures in abstraction: Reachability in hierarchical drawings. arXiv:1907.11662. https://arxiv.org/abs/1907.11662
5. Ortali, G., Tollis, I.G.: Algorithms and bounds for drawing directed graphs. In: Biedl, T., Kerren, A. (eds.) GD 2018. LNCS, vol. 11282, pp. 579–592. Springer, Cham (2018). https://doi.org/10.1007/978-3-030-04414-5_41
6. van Schaik, S.J., de Moor, O.: A memory efficient reachability data structure through bit vector compression. In: Proceedings of the ACM SIGMOD International Conference on Management of Data, SIGMOD 2011, Athens, Greece, 12–16 June 2011, pp. 913–924 (2011). https://doi.org/10.1145/1989323.1989419
7. Sugiyama, K., Tagawa, S., Toda, M.: Methods for visual understanding of hierarchical system structures. IEEE Trans. Syst. Man Cybern. **11**(2), 109–125 (1981)
8. Veloso, R.R., Cerf, L., Jr., W.M., Zaki, M.J.: Reachability queries in very large graphs: a fast refined online search approach. In: Proceedings of the 17th International Conference on Extending Database Technology, EDBT 2014, Athens, Greece, 24–28 March 2014, pp. 511–522 (2014). https://doi.org/10.5441/002/edbt.2014.46
9. Yildirim, H., Chaoji, V., Zaki, M.J.: GRAIL: a scalable index for reachability queries in very large graphs. VLDB J. **21**(4), 509–534 (2012). https://doi.org/10.1007/s00778-011-0256-4

On Topological Book Embedding
for k-plane Graphs

Michael Kaufmann and Axel Kuckuk[⊠]

Wilhelm-Schickhard-Institut für Informatik, Universität Tübingen,
Tübingen, Germany
mk@informatik.uni-tuebingen.de, axel.kuckuk@student.uni-tuebingen.de

Introduction. In topological book embedding (TBE) [6], the vertices of a given planar graph G are placed along a horizontal line (the spine) and the edges are realized as simple non-crossing arcs above or below the spine (top/bottom arcs) or as sequences of arcs on both sides of the spine. For planar graphs at most one spine crossing per edge is sufficient [5]. In [7], each edge is represented either by a top or bottom arc, or a monotone bottom-top biarc (bt-biarc), consisting of a bottom arc followed by a top arc, resp.

TBE has been heavily used for planar graphs [1, 5, 9], in particular in connection with point set embedding problems [3, 8, 11–13]. In [7], Everett et al. showed that every planar graph can be embedded on a so-called necklace point set with at most one bend per edge using an appropriate TBE.

We will generalize this result for k-plane graphs under some restrictions. k-plane graphs play an outstanding role in the area of graphs beyond planarity, which is an important recent research direction. The extension of the techniques for TBE and point set embedding to graphs beyond planarity is a challenging task [10]. The graphs we consider are embedded in the plane s.t. their crossing-free edges form a biconnected spanning subgraph and their crossing edges are being crossed $\leq k$ times, i.e. they are k-plane. Further we assume that adjacent edges do not cross while two non-adjacent edges might cross at most once.

We recall the concept of an open ear decomposition, a well-known characteristic of biconnected graphs. An open ear of graph G is a simple path p with distinct endpoints. An open ear decomposition is an edge partition into a sequence of ears s.t. only the two endpoints of each ear belong to earlier ears. Biconnected graphs have an open ear decomposition [14, 15]. For planar graphs it can be assumed that every subsequent path might close a face. We assume this property as well so that the faces are added one by one when adding the ear-defining paths [4].

The Algorithm. Let G be the k-plane graph with a biconnected spanning subgraph G_p of crossing-free edges. Let $p_1, p_2, \ldots, p_f$ be an open ear decomposition of G_p s.t. p_1 consists of an edge (v_1, v_2) on the outer face. We start placing v_1 and v_2 on the spine and connect them by the top arc (v_1, v_2).

Iteratively we add the paths from the open ear decomposition. We keep the invariant that the sequence of edges along the outer face from v_1 to v_2 consist of simple top arcs. Let $v_1 = u_1, u_2, \ldots, u_k = v_2$ be the sequence of vertices along the

© Springer Nature Switzerland AG 2019
D. Archambault and C. D. Tóth (Eds.): GD 2019, LNCS 11904, pp. 596–598, 2019.
https://doi.org/10.1007/978-3-030-35802-0

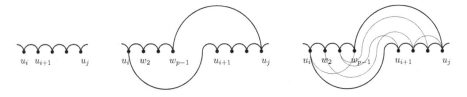

Fig. 1. Inserting a new path and filling the new face

outer face at step ℓ (Fig. 1 left) and $p_\ell = w_1, ..., w_p$ be the next path connecting the vertices $u_i = w_1$ and $u_j = w_p$ where $j > i$ w.l.o.g. in the ear decomposition.

We place the path $p'_\ell = w_2, ..., w_{p-1}$ between u_i and u_{i+1}, s.t. the path edges form top arcs maintaining the invariant. Note that the last edge (w_{p-1}, w_p) will separate all the vertices $u_{i+1}, ..., u_{k-1}$ from the outer face. If there are no vertices between u_i and u_{i+1} on the spine then the old edge (u_i, u_{i+1}) becomes a bottom arc, else it becomes a bt-biarc bridging all the vertices that are between u_i and u_{i+1}. In the latter case any other existing top arc with u_i as its left adjacent vertex will also become a bt-biarc in the same way (Fig. 1 middle).

Crossing edges inside the new face are either bottom arcs (if both endpoints are in $w_1, ..., w_{p-1}$), top arcs (if both endpoints are in $u_{i+1}, ..., u_k$) or bt-biarcs (otherwise). We route the bt-biarcs s.t. they cross the spine in the reverse order of their left adjacent vertices, avoiding unnecessary crossings (Fig. 1 right).

This concludes the embedding algorithm and we summarize by

Theorem 1. *For any k-plane graph with a biconnected spanning subgraph of crossing-free edges we can construct a TBE using only simple arcs and bt-biarcs.*

Corollary 1. *Any k-plane graph G with a biconnected spanning subgraph of crossing-free edges can be embedded appropriately on a necklace point set (integer points along a parabola, see [7]) using at most one bend on each edge.*

Proof. Idea: Apply the edge-routing algorithm of Everett et al. [7] using the TBE for G as described above.   □

Corollary 2. *Any 1-plane graph can be embedded appropriately on a necklace point set using at most one bend on each edge.*

Proof. Idea: Extend a given 1-plane graph by planar edges (kites) s.t. it has a biconnected spanning planar subgraph. We apply Corollary 1.   □

Corollary 3. *Optimal 2-plane and 3-plane graphs can be embedded appropriately on a necklace point set using at most one bend on each edge.*

Proof. Idea: Those graphs have spanning planar subgraphs [2]. We apply Corollary 1.   □

Remark 1. We were able to generalize the above technique for TBE to k-plane graphs with spanning subgraph of crossing-free edges, s.t. each edge consists of at most 5 arc-segments. Open questions are how to improve this bound and how to generalize the TBE to k-plane graphs without spanning planar subgraphs.

References

1. Badent, M., Di Giacomo, E., Liotta, G.: Drawing colored graphs on colored points. Theor. Comput. Sci. **408**(2–3), 129–142 (2008). https://doi.org/10.1016/j.tcs.2008.08.004

2. Bekos, M.A., Kaufmann, M., Raftopoulou, C.N.: On optimal 2- and 3-planar graphs. In: Aronov, B., Katz, M.J. (eds.) Symposium on Computational Geometry, LIPIcs, vol. 77, pp. 16:1–16:16. Schloss Dagstuhl - Leibniz-Zentrum fuer Informatik (2017). https://doi.org/10.4230/LIPIcs.SoCG.2017.16

3. Cabello, S.: Planar embeddability of the vertices of a graph using a fixed point set is np-hard. J. Graph Algorithms Appl. **10**(2), 353–363 (2006). http://jgaa.info/accepted/2006/Cabello2006.10.2.pdf

4. Di Battista, G., Eades, P., Tamassia, R., Tollis, I.G.: Graph Drawing: Algorithms for the Visualization of Graphs. Prentice-Hall (1999)

5. Di Giacomo, E., Didimo, W., Liotta, G., Wismath, S.K.: Curve-constrained drawings of planar graphs. Comput. Geom. **30**(1), 1–23 (2005). https://doi.org/10.1016/j.comgeo.2004.04.002

6. Enomoto, H., Miyauchi, M.S., Ota, K.: Lower bounds for the number of edge-crossings over the spine in a topological book embedding of a graph. Discrete Appl. Math. **92**(2–3), 149–155 (1999). https://doi.org/10.1016/S0166-218X(99)00044-X

7. Everett, H., Lazard, S., Liotta, G., Wismath, S.K.: Universal sets of n points for one-bend drawings of planar graphs with n vertices. Discrete Comput. Geom. **43**(2), 272–288 (2010). https://doi.org/10.1007/s00454-009-9149-3

8. Fulek, R., Tóth, C.D.: Universal point sets for planar three-trees. J. Discrete Algorithms **30**, 101–112 (2015). https://doi.org/10.1016/j.jda.2014.12.005

9. Giordano, F., Liotta, G., Mchedlidze, T., Symvonis, A.: Computing upward topological book embeddings of upward planar digraphs. In: Tokuyama, T. (ed.) ISAAC 2007. LNCS, vol. 4835, pp. 172–183. Springer, Heidelberg (2007). https://doi.org/10.1007/978-3-540-77120-3_17

10. Kaufmann, M.: On point set embeddings for k-planar graphs with few bends per edge. In: Catania, B., Královič, R., Nawrocki, J., Pighizzini, G. (eds.) SOFSEM 2019. LNCS, vol. 11376, pp. 260–271. Springer, Cham (2019). https://doi.org/10.1007/978-3-030-10801-4_21

11. Kaufmann, M., Wiese, R.: Embedding vertices at points: Few bends suffice for planar graphs. J. Graph Algorithms Appl. **6**(1), 115–129 (2002). http://www.cs.brown.edu/publications/jgaa/accepted/2002/KaufmannWiese2002.6.1.pdf

12. Kurowski, M.: A 1.235 lower bound on the number of points needed to draw all n-vertex planar graphs. Inf. Process. Lett. **92**(2), 95–98 (2004). https://doi.org/10.1016/j.ipl.2004.06.009

13. Pach, J., Gritzmann, P., Mohar, B., Pollack, R.: Embedding a planar triangulation with vertices at specified points. Am. Math. Monthly **98**, 165–166 (1991). Professor Pach's number: [065]

14. Schmidt, J.M.: A simple test on 2-vertex- and 2-edge-connectivity. Inf. Process. Lett. **113**(7), 241–244 (2013). https://doi.org/10.1016/j.ipl.2013.01.016

15. Whitney, H.: Non-separable and planar graphs. Trans. Am. Math. Soc. **34**, 339–362 (1932). https://doi.org/10.1090/S0002-9947-1932-1501641-2

On Compact RAC Drawings

Henry Förster[✉] and Michael Kaufmann

Wilhelm-Schickard-Institut für Informatik,
University of Tübingen, Tübingen, Germany
{foersth, mk}@informatik.uni-tuebingen.de

Since real-world graphs are often nonplanar, beyond planar graphs have been studied [10]. Two important parameters are the angles formed by edges at their intersections [12, 13] and the number of bends per edge [14, 15]. Unsurprisingly, one of the first papers on beyond planar graphs [9] introduced *RAC* (or *right-angle-crossing*) drawings, where angles formed by edges at their intersections are always 90°. Previously, the main questions for RAC drawings have been recognition [2, 3, 6], characterization [11] and relations to other graph classes [2, 5, 7].

We study the area for RAC drawings of dense graphs depending on the number of bends per edge. With two bends only $\mathcal{O}(n)$ edges can be drawn [1, 4, 9]. For denser graphs, the following bounds are known: (i) three bends per edge in $\mathcal{O}(n^4)$ area [9], and, (ii) four bends per edge in $\mathcal{O}(n^3)$ area [8]. Here, vertices and bends must be placed on a grid. We achieve the following new results:

Theorem 1 (Fig. 1). *Every simple graph on n vertices admits a RAC drawing with three bends per edge in $\mathcal{O}(n^3)$ area.*

Theorem 2. *There exists no RAC drawing of K_n with three bends per edge in $\mathcal{O}(n^2)$ area for sufficiently large n.*

Proof (Sketch). We prove by contradiction. We observe that $\Omega(n^2)$ segments require $\Omega(n)$ length. Further, $\Theta(n^4)$ intersections occur on segments of $\Omega(n)$ length with $\mathcal{O}(1)$ different slopes while there are many long *start segments* (i.e., incident to vertices) with few crossings forming obstacles. Hence, groups of start segments are almost parallel and vertices are close to other vertices. Vertices that are too close are not connected with segments with many intersections. □

Theorem 3 (Fig. 2). *Every simple graph on n vertices admits a RAC drawing with eight bends per edge in $\mathcal{O}(n^2)$ area.*

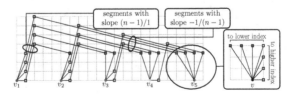

Fig. 1. In contrast to the previously known construction by Didimo et al. [9] for $\mathcal{O}(n^4)$ area, each vertex is incident to two different types of bends (green and red squares) which lead to vertices with larger and smaller (resp.) indices. For the segments between bends, we use slopes that are almost horizontal or vertical. (Color figure online)

© Springer Nature Switzerland AG 2019
D. Archambault and C. D. Tóth (Eds.): GD 2019, LNCS 11904, pp. 599–601, 2019.
https://doi.org/10.1007/978-3-030-35802-0

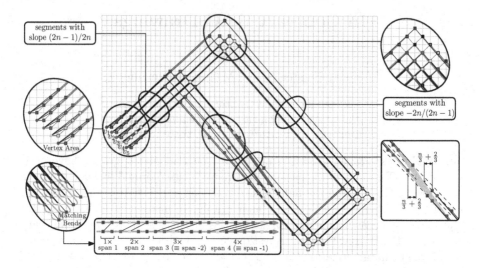

Fig. 2. Drawing of K_5 computed by our algorithm in the proof of Theorem 3.

Proof (Sketch). Edges are composed of two half edges routed from vertex area to matching bends. An appropriate matching realizes each edge: We connect the first bottom right matching bend with the second top right matching bend. We connect the next two bottom right matching bends with the following two top left matching bends. This pattern continues by increasing the number of matched pairs and the *span*. As spans i and $n - i$ are cyclically equivalent, all connections are realized once. As spans are at most $n - 1$, segments are straight-line.   □

Theorem 4 (Fig. 3). *Every simple k-partite graph on n vertices admits a RAC drawing with three bends per edge in $\mathcal{O}(k^2 n^2)$ area.*

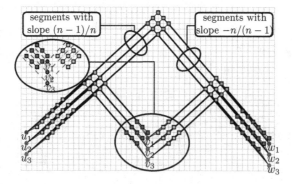

Fig. 3. Example of $K_{3,3,3}$. Vertices of the same partition are in connected regions. As in Theorem 1, we have two types of bends at vertices (green and red squares). (Color figure online)

References

1. Angelini, P., Bekos, M.A., Förster, H., Kaufmann, M.: On RAC drawings of graphs with one bend per edge. In: Biedl, T., Kerren, A. (eds.) GD 2018. LNCS, vol. 11282, pp. 123–136. Springer, Cham (2018). https://doi.org/10.1007/978-3-030-04414-5_9

2. Angelini, P., Cittadini, L., Didimo, W., Frati, F., Di Battista, G., Kaufmann, M., Symvonis, A.: On the perspectives opened by right angle crossing drawings. J. Graph Algorithms Appl. **15**(1), 53–78 (2011). https://doi.org/10.7155/jgaa.00217

3. Argyriou, E.N., Bekos, M.A., Symvonis, A.: The straight-line RAC drawing problem is NP-hard. J. Graph Algorithms Appl. **16**(2), 569–597 (2012). https://doi.org/10.7155/jgaa.00274

4. Arikushi, K., Fulek, R., Keszegh, B., Moric, F., Tóth, C.D.: Graphs that admit right angle crossing drawings. Comput. Geom. **45**(4), 169–177 (2012). https://doi.org/10.1016/j.comgeo.2011.11.008

5. Bachmaier, C., Brandenburg, F.J., Hanauer, K., Neuwirth, D., Reislhuber, J.: NIC-planar graphs. Discrete Appl. Math. **232**, 23–40 (2017). https://doi.org/10.1016/j.dam.2017.08.015

6. Bekos, M.A., Didimo, W., Liotta, G., Mehrabi, S., Montecchiani, F.: On RAC drawings of 1-planar graphs. Theor. Comput. Sci. **689**, 48–57 (2017). https://doi.org/10.1016/j.tcs.2017.05.039

7. Brandenburg, F.J., Didimo, W., Evans, W.S., Kindermann, P., Liotta, G., Montecchiani, F.: Recognizing and drawing ic-planar graphs. Theor. Comput. Sci. **636**, 1–16 (2016). https://doi.org/10.1016/j.tcs.2016.04.026

8. Di Giacomo, E., Didimo, W., Liotta, G., Meijer, H.: Area, curve complexity, and crossing resolution of non-planar graph drawings. Theory Comput. Syst. **49**(3), 565–575 (2011). https://doi.org/10.1007/s00224-010-9275-6

9. Didimo, W., Eades, P., Liotta, G.: Drawing graphs with right angle crossings. Theor. Comput. Sci. **412**(39), 5156–5166 (2011). https://doi.org/10.1016/j.tcs.2011.05.025

10. Didimo, W., Liotta, G., Montecchiani, F.: A survey on graph drawing beyond planarity. ACM Comput. Surv. **52**(1), 4:1–4:37 (2019). https://doi.org/10.1145/3301281

11. Eades, P., Liotta, G.: Right angle crossing graphs and 1-planarity. Discrete Appl. Math. **161**(7–8), 961–969 (2013). https://doi.org/10.1016/j.dam.2012.11.019

12. Huang, W.: Using eye tracking to investigate graph layout effects. In: Hong, S., Ma, K. (eds.) 6th International Asia-Pacific Symposium on Visualization 2007, APVIS 2007, pp. 97–100. IEEE Computer Society (2007). https://doi.org/10.1109/APVIS.2007.329282

13. Huang, W., Eades, P., Hong, S.: Larger crossing angles make graphs easier to read. J. Vis. Lang. Comput. **25**(4), 452–465 (2014). https://doi.org/10.1016/j.jvlc.2014.03.001

14. Purchase, H.C.: Effective information visualisation: a study of graph drawing aesthetics and algorithms. Interact. Comput. **13**(2), 147–162 (2000). https://doi.org/10.1016/S0953-5438(00)00032-1

15. Purchase, H.C., Carrington, D.A., Allder, J.: Empirical evaluation of aesthetics-based graph layout. Empir. Softw. Eng. **7**(3), 233–255 (2002)

FPQ-Choosable Planarity Testing

Giuseppe Liotta[1] ⓘ, Ignaz Rutter[2] ⓘ, and Alessandra Tappini[1(✉)] ⓘ

[1] Dipartimento di Ingegneria, Università degli Studi di Perugia, Perugia, Italy
giuseppe.liotta@unipg.it, alessandra.tappini@studenti.unipg.it
[2] Department of Computer Science and Mathematics, University of Passau,
Passau, Germany
rutter@fim.uni-passau.de

Abstract. Hierarchical embedding constraints define a set of allowed
cyclic orders for the edges incident to the vertices of a graph. These
constraints are expressed in terms of FPQ-trees. Let G be a graph such
that every vertex of G is equipped with a set of FPQ-trees encoding
hierarchical embedding constraints for its incident edges. We study the
problem of testing whether G admits a planar embedding such that, for
each vertex v of G, the cyclic order of the edges incident to v is described
by at least one of the FPQ-trees associated with v. We prove that the
problem is fixed-parameter tractable for biconnected graphs, where the
parameters are the treewidth of G and the number of FPQ-trees per
vertex, and we show that if one of these two parameters is dropped,
then the problem is not fixed-parameter tractable. We also apply our
techniques to the study of NodeTrix planarity testing of clustered graphs.

Introduction. Graph planarity with hierarchical embedding constraints
addresses the problem of testing whether a graph G admits a planar embedding
where the cyclic order of the edges incident to (some of) its vertices is totally or
partially fixed. The term "hierarchical" reflects the fact that these constraints
describe ordering relationships both between sets of edges incident to a same
vertex and, recursively, between edges within a same set. Hierarchical embedding constraints can be conveniently encoded by using FPQ-trees, a variant of
PQ-trees that includes F-nodes in addition to P- and to Q-nodes. An F-node
encodes a permutation that cannot be reversed. See Fig. 1 for an example.

In a seminal work, Gutwenger et al. [4] study the planarity testing problem
with hierarchical embedding constraints by allowing *at most one* FPQ-tree per
vertex, and they present a linear-time algorithm to solve the problem. In [5], we
generalize their study by allowing *more than one* FPQ-tree per vertex. To this
aim, we introduce and study a problem called FPQ-Choosable Planarity
Testing. An *FPQ-choosable graph* consists of a pair (G, D), where G is a
(multi-)graph and D is a mapping that associates each vertex v of G with a

Work partially supported by: MIUR, grant 20174LF3T8 AHeAD: efficient Algorithms for HArnessing networked Data; Dip. Ingegneria Univ. Perugia, grants
RICBASE2017WD-RICBA18WD: "Algoritmi e sistemi di analisi visuale di reti complesse e di grandi dimensioni"; German Science Found. (DFG), grant Ru 1903/3-1.

D. Archambault and C. D. Tóth (Eds.): GD 2019, LNCS 11904, pp. 602–604, 2019.
https://doi.org/10.1007/978-3-030-35802-0

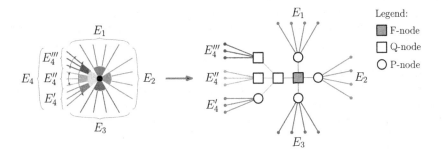

Fig. 1. A vertex v with hierarchical embedding constraints and the corresponding FPQ-tree. The sets of edges must appear in one of the following clockwise orders around v: $E_1, E_2, E_3, E_4', E_4'', E_4'''$; $E_1, E_2, E_3, E_4''', E_4'', E_4'$. The edges of E_1, E_2, E_3, and E_4' can be arbitrarily permuted, while the edges of E_4'' and E_4''' have only two possible cyclic orders that are the reverse of one another.

set $D(v)$ of FPQ-trees whose leaves represent the edges incident to v. An FPQ-choosable graph is *FPQ-choosable planar* if there exists a planar embedding of G such that, for each vertex v of G, the cyclic order of the edges incident to v is encoded by an FPQ-tree in $D(v)$. FPQ-CHOOSABLE PLANARITY TESTING asks whether a given FPQ-choosable graph (G, D) is FPQ-choosable planar.

Besides being interesting on its own right, this generalization can be used to model and study other graph planarity testing problems. As a proof of concept, we apply our results to the study of NodeTrix planarity with fixed sides of clustered graphs; see [1–3] for some references about NodeTrix planarity testing.

Main Results. In this section, we give a list of our main results. We first study the computational complexity of FPQ-CHOOSABLE PLANARITY TESTING, and we prove the following.

Theorem 1. FPQ-CHOOSABLE PLANARITY TESTING *with a bounded number of FPQ-trees per vertex is NP-complete. It remains NP-complete even when the FPQ-trees have only P-nodes.*

Theorem 2. FPQ-CHOOSABLE PLANARITY TESTING *parameterized by treewidth is W[1]-hard. It remains W[1]-hard even when the FPQ-trees have only P-nodes.*

The above results imply that FPQ-CHOOSABLE PLANARITY TESTING is not fixed-parameter tractable if parameterized by the treewidth only or by the number of FPQ-trees per vertex only. For a contrast, we show the following.

Theorem 3. *Let (G, D) be a biconnected FPQ-choosable (multi-)graph such that $G = (V, E)$ and $|V| = n$. Let $D(v)$ be the set of FPQ-trees associated with vertex $v \in V$. There exists an $O(\delta^{\frac{9}{4}t} \cdot n^2 + n^3)$-time algorithm to test whether (G, D) is FPQ-choosable planar, where t is the treewidth of G and $\delta = \max_{v \in V} |D(v)|$.*

We finally analyze the interplay between FPQ-CHOOSABLE PLANARITY TEST-
ING and NodeTrix planarity testing with fixed sides, which is known to be NP-
complete even when the size of the matrices is bounded by a constant [2, 3].

Theorem 4. *Let G be an n-vertex clustered graph whose clusters have size at most k. Let t be the treewidth of G. If the multi-graph obtained by collapsing each cluster of G into a vertex is biconnected, then there exists an $O(k!^{\frac{9}{4}t} \cdot n^2 + n^3)$-time algorithm to test whether G is NodeTrix planar with fixed sides.*

References

1. Besa Vial, J.J., Da Lozzo, G., Goodrich, M.T.: Computing k-modal embeddings of planar digraphs. CoRR abs/1907.01630 (2019). http://arxiv.org/abs/1907.01630
2. Da Lozzo, G., Di Battista, G., Frati, F., Patrignani, M.: Computing NodeTrix representations of clustered graphs. J. Graph Algorithms Appl. **22**(2), 139–176 (2018). https://doi.org/10.7155/jgaa.00461
3. Di Giacomo, E., Liotta, G., Patrignani, M., Rutter, I., Tappini, A.: Node-Trix planarity testing with small clusters. Algorithmica **81**(9), 3464–3493 (2019). https://doi.org/10.1007/s00453-019-00585-6
4. Gutwenger, C., Klein, K., Mutzel, P.: Planarity testing and optimal edge insertion with embedding constraints. J. Graph Algorithms Appl. **12**(1), 73–95 (2008). https://doi.org/10.7155/jgaa.00160
5. Liotta, G., Rutter, I., Tappini, A.: Graph planarity testing with hierarchical embedding constraints. CoRR abs/1904.12596 (2019). http://arxiv.org/abs/1904.12596

Packing Trees into 1-Planar Graphs

Felice De Luca[1] , Emilio Di Giacomo[2] , Seok-Hee Hong[3] ,
Stephen Kobourov[1] , William Lenhart[4] , Giuseppe Liotta[2] , Henk Meijer[5],
Alessandra Tappini[2(✉)] , and Stephen Wismath[6]

[1] Department of Computer Science, University of Arizona, Tucson, USA
felicedeluca@email.arizona.edu, kobourov@cs.arizona.edu
[2] Dipartimento di Ingegneria, Università degli Studi di Perugia, Perugia, Italy
{emilio.digiacomo, giuseppe.liotta}@unipg.it,
alessandra.tappini@studenti.unipg.it
[3] School of Computer Science, University of Sydney, Sydney, Australia
seokhee.hong@sydney.edu.au
[4] Department of Computer Science, Williams College, Williamstown, USA
wlenhart@williams.edu
[5] Department of Computer Science, University College Roosevelt, Middelburg,
The Netherlands
h.meijer@ucr.nl
[6] Department of Computer Science, University of Lethbridge, Lethbridge, Canada
wismath@uleth.ca

Introduction. In the *graph packing problem* we are given a collection of n-vertex graphs G_1, G_2, ..., G_k and we are requested to find an n-vertex graph G that contains the given graphs as edge-disjoint spanning subgraphs. Various settings of the problem can be defined depending on the type of graphs that have to be packed and on the restrictions put on the packing graph G (see, e.g., [1, 7, 8, 11–13]). García et al. [4] consider the *planar packing problem*, that is the case when the graph G is required to be planar. They conjecture that every pair of non-star trees can be packed into a planar graph. Notice that, when G is required to be planar, two is the maximum number of trees that can be packed (because three trees have more than $3n - 6$ edges). García et al. prove their conjecture for some restricted cases. After a sequence of other partial results [2, 3, 5, 9], the conjecture was finally proved true by Geyer et al. [6].

We initiate the study of the *1-planar packing problem*, i.e., the problem of packing a set of graphs into a 1-planar graph (a graph is 1-planar if it can be drawn with at most one crossing per edge). Since any two non-star trees admit a planar packing, a natural question is whether we can pack more than two trees into a 1-planar graph. On the other hand, since each 1-planar graph has at most $4n - 8$ edges [10], it is not possible to pack more than three trees into a 1-planar graph. Thus, our main question is whether any three trees with maximum vertex degree $n - 3$ admit a 1-planar packing. The restriction about the degree is

This work started at the Bertinoro Workshop on Graph Drawing 2019 and it is partially supported by: (*i*) MIUR, under grant 20174LF3T8 "AHeAD: efficient Algorithms for HArnessing networked Data", (*ii*) Dipartimento di Ingegneria - Università degli Studi di Perugia, under grants RICBASE2017WD and RICBA18WD.

D. Archambault and C. D. Tóth (Eds.): GD 2019, LNCS 11904, pp. 605–607, 2019.
https://doi.org/10.1007/978-3-030-35802-0

necessary because a vertex of degree larger than $n - 3$ in one tree cannot have degree at least one in the other two trees.

Results. We first prove that there exist triples of trees that do not admit a 1-planar packing, even for structurally simple trees.

Theorem 1. *For every $n \geq 10$, there exists a triple of caterpillars that does not admit a 1-planar packing.*

Theorem 2. *There exists a triple consisting of a path and two caterpillars with $n = 7$ vertices that does not admit a 1-planar packing.*

Motivated by the two theorems above we consider triples consisting of two paths P_1 and P_2 and a caterpillar T and we prove that every such triple in which T is 5-legged admits a 1-planar packing if and only if it has at least six vertices. A caterpillar is *h-legged* if all its non-leaf vertices have degree either 2 or at least $h + 2$. Our proof is constructive and at a very high-level can be described as follows. Let P be the backbone of T (the *backbone* of a caterpillar is a path obtained by removing all leaves except two) and let P'_1 and P'_2 be two paths with the same length as P. We first construct a 1-planar packing of P, P'_1 and P'_2. We then modify the computed packing to include the leaves of the caterpillar; this requires transforming some edges of P'_1 and P'_2 to sub-paths that pass through the added leaves. The resulting packing is a 1-planar packing of P_1, P_2 and T. See Fig. 1 for an example.

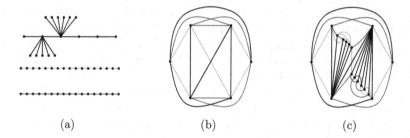

(a)   (b)   (c)

Fig. 1. (a) A 5-legged caterpillar T and two paths P_1 and P_2; (b) a 1-planar packing of the three paths P, P'_1, and P'_2; (c) a 1-planar packing of T, P_1 and P_2.

Theorem 3. *Two paths and a 5-legged caterpillar T with n vertices admit a 1-planar packing if and only if $n \geq 6$ and each vertex has degree at most $n - 3$ in T.*

The technique behind the previous theorem constructs 1-planar drawings with $O(n)$ crossings. A natural question is whether it is possible to compute a 1-planar packing with $O(1)$ crossings. We can prove that seven (resp. fourteen) crossings suffice for three paths (resp. cycles). It is worth remarking that a 1-planar packing of three paths (resp. cycles) has at least three (resp. six) crossings because it has $3n - 3$ edges (resp. $3n$ edges).

Theorem 4. *Three paths with $n \geq 6$ vertices can be packed into a 1-plane graph with at most 7 edge crossings. Three cycles with $n \geq 20$ vertices can be packed into a 1-plane graph with at most 14 edge crossings.*

We finally extend the study of 1-planar packings from triples of graphs to quadruples of graphs. A 1-planar packing of four graphs does not exist if all graphs are connected, because the number of edges of the four graphs is higher than the number of edges allowed in a 1-planar graph. We consider therefore a quadruple consisting of three paths and a perfect matching. Notice that, in this case the number of vertices n has to be even.

Theorem 5. *Three paths and a perfect matching with $n \geq 12$ vertices admit a 1-planar packing. If $n \leq 10$, the quadruple does not admit a 1-planar packing.*

References

1. Aichholzer, O., Hackl, T., Korman, M., van Kreveld, M., Löffler, M., Pilz, A., Speckmann, B., Welzl, E.: Packing plane spanning trees and paths in complete geometric graphs. Inf. Process. Lett. **124**, 35–41 (2017). https://doi.org/10.1016/j.ipl.2017.04.006
2. Frati, F.: Planar packing of diameter-four trees. In: Proceedings of the 21st Annual Canadian Conference on Computational Geometry, pp. 95–98 (2009). http://cccg.ca/proceedings/2009/cccg09_25.pdf
3. Frati, F., Geyer, M., Kaufmann, M.: Planar packing of trees and spider trees. Inf. Process. Lett. **109**(6), 301–307 (2009). https://doi.org/10.1016/j.ipl.2008.11.002
4. García Olaverri, A., Hernando, M.C., Hurtado, F., Noy, M., Tejel, J.: Packing trees into planar graphs. J. Graph Theory **40**(3), 172–181 (2002). https://doi.org/10.1002/jgt.10042
5. Geyer, M., Hoffmann, M., Kaufmann, M., Kusters, V., Tóth, C.D.: Planar packing of binary trees. In: Dehne, F., Solis-Oba, R., Sack, J.-R. (eds.) WADS 2013. LNCS, vol. 8037, pp. 353–364. Springer, Heidelberg (2013). https://doi.org/10.1007/978-3-642-40104-6_31
6. Geyer, M., Hoffmann, M., Kaufmann, M., Kusters, V., Tóth, C.D.: The planar tree packing theorem. JoCG **8**(2), 109–177 (2017). https://doi.org/10.20382/jocg.v8i2a6
7. Hedetniemi, S., Hedetniemi, S., Slater, P.: A note on packing two trees into K_n. Ars Combinatoria **11**, January 1981
8. Mahéo, M., Saclé, J., Wozniak, M.: Edge-disjoint placement of three trees. Eur. J. Comb. **17**(6), 543–563 (1996). https://doi.org/10.1006/eujc.1996.0047
9. Oda, Y., Ota, K.: Tight planar packings of two trees. In: 22nd European Workshop on Computational Geometry (2006)
10. Pach, J., Tóth, G.: Graphs drawn with few crossings per edge. Combinatorica **17**(3), 427–439 (1997). https://doi.org/10.1007/BF01215922
11. Sauer, N., Spencer, J.: Edge disjoint placement of graphs. J. Comb. Theory, Ser. B **25**(3), 295–302 (1978). https://doi.org/10.1016/0095-8956(78)90005-9
12. Wang, H., Sauer, N.: Packing three copies of a tree into a complete graph. Eur. J. Comb. **14**(2), 137–142 (1993). https://doi.org/10.1006/eujc.1993.1018
13. Wozniak, M., Wojda, A.P.: Triple placement of graphs. Graphs Comb. **9**(1), 85–91 (1993). https://doi.org/10.1007/BF01195330

Geographic Network Visualization Techniques: A Work-In-Progress Taxonomy

Sarah Schöttler$^{(\boxtimes)}$, Tobias Kauer, and Benjamin Bach

University of Edinburgh, Edinburgh, UK
`sarah@schoettler.email, bbach@inf.ed.ac.uk`

Abstract. This poster presents a survey of visualization techniques for geographic networks. Based on 60 techniques, we provide an initial taxonomy based on categorizing each technique across four facets: how the geographic aspect is represented, how the network aspect is represented, how these two visual representations are integrated, and whether the technique relies on user interaction. The current collection can be found online: https://geographic-networks.github.io.

Keywords: Geographic networks · Survey · Taxonomy

1 Scope and Methodology

Geographic network data describes the relationships between geolocated entities. Examples include airports connected by commercial flights, trading networks, migration, geographic social networks or public transport networks in cities. Yet, visualizing these networks remains challenging: overlap and clutter frequently make visualizations difficult to read or even misleading. Often, there is a trade-off between computational complexity, visual quality, and the specific task at hand (analyzing geographic locations, analyzing network topology, correlating both, etc). *No taxonomy specific to these techniques exists.*

To qualify for inclusion into our survey, a paper has to either be focused entirely on geographic networks, or, at a minimum, demonstrate its applicability to geographic networks with a case study. Techniques that can theoretically be applied to geographic networks, but do not visualize the geographic aspect of the network, were not considered. Papers come from different venues: IEEE VIS, ACM CHI, EuroVis, PacificVis, and Graph Drawing. Our search resulted in 191 papers which we manually narrowed down to 40. Through additional manual search, the number increased back to 60 papers/techniques.

2 Taxonomy

A—Geographic Representation. This facet describes how the geographic aspect of the network is represented visually. We found visualizations to differ in the way they distort and abstract that geographic representation: **Map** is the

© Springer Nature Switzerland AG 2019
D. Archambault and C. D. Tóth (Eds.): GD 2019, LNCS 11904, pp. 608–611, 2019.
https://doi.org/10.1007/978-3-030-35802-0

least distorted technique [4, 12, 14, 15, 23]. **Distorted map** includes any visualization that is still recognizable as a map, but distorted beyond the distortion introduced by the map projection [1, 5, 15, 19]. **Abstract** techniques represent geography in some non-geographic (abstract) form such as grouping nodes in a circular layout [11].

B—**Network Representation.** Initially, we thought to categorize according to the type of visualization. However, we quickly found that approx. 90% of all techniques use node-link diagrams, some matrices. Thus, we decided to again look for 'abstraction' in the network representation. Since a network consists of nodes *and* edges, we classify techniques along both axes: node abstraction and edge abstraction. The node representation is *explicit* when nodes are shown as points in a node-link diagram and *abstract* if not; the edge representation is *abstract* when edges are shown different than links in a node-link diagram. Another way of looking at this is whether it is theoretically possible to extract the precise network data from the visualization—independent from clutter due to potential overlap and occlusion. **Explicit nodes & explicit edges:** Includes all techniques that explicitly visualize nodes and edges: edge bundling, edge routing, 3D globes etc. [12, 14] **Explicit nodes & abstract edges:** Techniques in this category explicitly show the nodes of the network, but use abstract means of showing the connections between them. Examples include omitting edges [1] or using alternative representations [4]. **Abstract nodes & explicit edges:** Abstracting the nodes but not the edges, e.g. aggregating nodes [7, 8]. **Abstract nodes & abstract edges:** Both nodes and edges are abstracted, e.g. OD maps or aggregating both nodes and edges [3, 21].

C—**Integration** describes how geography and topology are integrated in the visualization, simplifying the approach in [10]. **Geography-as-basis:** The majority (44) of the surveyed visualization techniques use the geography representation as their basis and overlay a network visualization [1–3, 8, 9, 21]. A **balanced** integration is one where neither geography nor network are clearly dominant [13, 23]. **Network-as-basis:** Only one technique uses the network representation as its basis [11].

D—**Interaction:** classifies techniques into *none* [13, 18, 21], *optional* [4, 22], *required* [1, 6, 23], and *technique-is-interaction*; meaning that a technique is a pure interaction technique such as a fisheye lens [5], *EdgeLens* [19], link bundling [17], link plucking [20] or *Bring & Go* and *Link Sliding* [16].

3   Open Challenges

We are currently working to extend our collection and refine our taxonomy. However, many techniques remain to be explored; e.g., not taking interaction into account, there are 36 possible combinations of the different categories across facets of the taxonomy. Besides the groups discussed in the paper, we could identify the following open challenges for which we could find few or no techniques: **uncertainty** visualization of geographic positions and areas, **dynamic** geographic networks, **network-focused techniques** that preserve geography well, and precise **task and data taxonomies** that can inform future techniques, design spaces and interaction techniques.

References

1. Alper, B., Sümengen, S., Balcisoy, S.: Dynamic visualization of geographic networks using surface deformations with constraints. In: Proceedings of the Computer Graphics International Conference (CGI). Computer Graphics Society, Petrópolis, Brazil (2007)
2. Andrienko, G., Andrienko, N., Fuchs, G., Wood, J.: Revealing patterns and trends of mass mobility through spatial and temporal abstraction of origin-destination movement data. IEEE Trans. Vis. Comput. Graph. **23**(9), 2120–2136 (2017). https://doi.org/10.1109/TVCG.2016.2616404
3. Andrienko, N., Andrienko, G.: Spatial Generalization and aggregation of massive movement data. IEEE Trans. Vis. Comput. Graph. **17**(2), 205–219 (2011). https://doi.org/10.1109/TVCG.2010.44
4. Boyandin, I., Bertini, E., Bak, P., Lalanne, D.: Flowstrates: an approach for visual exploration of temporal origin-destination data. Comput. Graph. Forum **30**(3), 971–980 (2011). https://doi.org/10.1111/j.1467-8659.2011.01946.x
5. Brown, M.H., Meehan, J.R., Sarkar, M.: Browsing graphs using a fisheye view (Abstract). In: Proceedings of the INTERACT 1993 and CHI 1993 Conference on Human Factors in Computing Systems, CHI 1993, Amsterdam, The Netherlands, pp. 516. ACM, New York, NY, USA (1993). https://doi.org/10.1145/169059.169474
6. Cox, K.C., Eick, S.G., He, T.: 3D geographic network displays. SIGMOD Rec. **25**(4), 50–54 (1996). https://doi.org/10.1145/245882.245901
7. Elzen, S.V.D., Wijk, J.J.V.: Multivariate network exploration and presentation: from detail to overview via selections and aggregations. IEEE Trans. Vis. Comput. Graph. **20**(12), 2310–2319 (2014). https://doi.org/10.1109/TVCG.2014.2346441
8. Guo, D.: Flow mapping and multivariate visualization of large spatial interaction data. IEEE Trans. Vis. Comput. Graph. **15**(6), 1041–1048 (2009). https://doi.org/10.1109/TVCG.2009.143
9. Guo, D., Zhu, X.: Origin-destination flow data smoothing and mapping. IEEE Trans. Vis. Comput. Graph. **20**(12), 2043–2052 (2014). https://doi.org/10.1109/TVCG.2014.2346271
10. Hadlak, S., Schumann, H., Schulz, H.J.: A survey of multi-faceted graph visualization. In: Eurographics Conference on Visualization (EuroVis), vol. 33, pp. 1–20. The Eurographics Association Cagliary, Italy (2015)
11. Hennemann, S.: Information-rich visualisation of dense geographical networks. J. Maps **9**(1), 68–75 (2013). https://doi.org/10.1080/17445647.2012.753850
12. Holten, D., Wijk, J.J.V.: Force-directed edge bundling for graph visualization. Comput. Graph. Forum **28**(3), 983–990 (2009). https://doi.org/10.1111/j.1467-8659.2009.01450.x
13. Hong, S.H., Merrick, D., do Nascimento, H.A.D.: Automatic visualisation of metro maps. J. Vis. Lang. Comput. **17**(3), 203–224 (2006). https://doi.org/10.1016/j.jvlc.2005.09.001
14. Lambert, A., Bourqui, R., Auber, D.: 3D edge bundling for geographical data visualization. In: 2010 14th International Conference Information Visualisation, pp. 329–335, July 2010. https://doi.org/10.1109/IV.2010.53

15. Merrick, D., Gudmundsson, J.: Increasing the readability of graph drawings with centrality-based scaling. In: Proceedings of the 2006 Asia-Pacific Symposium on Information Visualisation, APVis 2006, Tokyo, Japan, vol. 60, pp. 67–76. Australian Computer Society Inc., Darlinghurst, Australia (2006). https://dl.acm.org/citation.cfm?id=1151914

16. Moscovich, T., Chevalier, F., Henry, N., Pietriga, E., Fekete, J.D.: Topology-aware navigation in large networks. In: Proceedings of the SIGCHI Conference on Human Factors in Computing Systems, CHI 2009, Boston, MA, USA, pp. 2319–2328. ACM, New York, NY, USA (2009). https://doi.org/10.1145/1518701.1519056

17. Riche, N.H., Dwyer, T., Lee, B., Carpendale, S.: Exploring the design space of interactive link curvature in network diagrams. In: Proceedings of the International Working Conference on Advanced Visual Interfaces, AVI 2012, Capri Island, Italy, pp. 506–513. ACM, New York, NY, USA (2012). https://doi.org/10.1145/2254556.2254652

18. Romat, H., Appert, C., Bach, B., Henry-Riche, N., Pietriga, E.: Animated edge textures in node-link diagrams: a design space and initial evaluation. In: Proceedings of the 2018 CHI Conference on Human Factors in Computing Systems, CHI 2018, Montreal QC, Canada, pp. 187:1–187:13. ACM, New York, NY, USA (2018). https://doi.org/10.1145/3173574.3173761

19. Wong, N., Carpendale, S., Greenberg, S.: Edgelens: an interactive method for managing edge congestion in graphs. In: IEEE Symposium on Information Visualization 2003 (IEEE Cat. No. 03TH8714), pp. 51–58, October 2003. https://doi.org/10.1109/INFVIS.2003.1249008

20. Wong, N., Carpendale, S.: Supporting interactive graph explorationusing edge plucking. In: Proceedings of IS&T/SPIE 19th Annual Symposium on Electronic Imaging: Visualization and Data Analysis (2007). https://doi.org/10.1.1.230.7985

21. Wood, J., Dykes, J., Slingsby, A.: Visualisation of origins, destinations and flows with OD maps. Cartogr. J. 47(2), 117–129 (2010). https://doi.org/10.1179/000870410X12658023467367

22. Yang, Y., Dwyer, T., Goodwin, S., Marriott, K.: Many-to-many geographically-embedded flow visualisation: an evaluation. IEEE Trans. Vis. Comput. Graph. 23(1), 411–420 (2017). https://doi.org/10.1109/TVCG.2016.2598885

23. Yang, Y., Dwyer, T., Jenny, B., Marriott, K., Cordeil, M., Chen, H.: Origin-destination flow maps in immersive environments. IEEE Trans. Vis. Comput. Graph. 25(1), 693–703 (2019). https://doi.org/10.1109/TVCG.2018.2865192

On the Simple Quasi Crossing Number of K_{11}

Arjun Pitchanathan[1(✉)] and Saswata Shannigrahi[2]

[1] International Institute of Information Technology, Hyderabad 500032, India
arjun.p@research.iiit.ac.in
[2] Saint Petersburg State University, St Petersburg 199034, Russia
saswata.shannigrahi@gmail.com

Abstract. We show that the simple quasi crossing number of K_{11} is 4.

A *quasi-planar* graph [2] is one that can be drawn in the plane without any triples of pairwise crossing edges. A drawing of a graph in the plane is *simple* if every pair of edges meets at most once, either at an intersection point or at a common endpoint. Accordingly, a graph is *simple quasi-planar* if it has a simple drawing in the plane without any triples of pairwise crossing edges. We define the *simple quasi crossing number* of a graph G, denoted by $cr_3(G)$, to be the minimum number of such triples in a simple drawing of G in the plane.

It has been shown [1] that a simple quasi-planar graph with $n \geq 4$ vertices has at most $6.5n - 20$ edges and this bound is tight up to an additive constant. We use this bound and follow the proof of the crossing number inequality [4, 6] to obtain a lower bound on the simple quasi crossing number of a graph as follows.

Let G be a graph with $n \geq 4$ vertices and e edges. Consider a simple drawing of G with $cr_3(G)$ triples of pairwise crossing edges. We can remove each such triple by removing an edge. In this way, we can obtain a simple quasi-planar graph with at least $e - cr_3(G)$ edges and n vertices. By the above-mentioned bound, we have

$$cr_3(G) \geq e - 6.5n + 20. \tag{1}$$

We improve this bound by the probabilistic method, as in the proof of the crossing number inequality. Let p be a parameter between 0 and 1, to be chosen later. Consider a random subgraph of H obtained by including each vertex of G independently with a probability p, and including the edges for which both vertices are included. Let n_H, e_H and $cr_3(H)$ be the random variables denoting the number of vertices, number of edges and the simple quasi crossing number of H, respectively. By applying the inequality (1) and taking expectations, we obtain $\mathbb{E}[cr_3(H)] \geq \mathbb{E}[e_H] - 6.5 \, \mathbb{E}[n_H] + 20$. By the independence of the choices, we have $E[e_H] = p^2 e$ and $\mathbb{E}[n_H] = pn$. In any triple of pairwise crossing edges, there are exactly six distinct vertices involved. Therefore, we have $\mathbb{E}[cr_3(H)] \leq p^6 cr_3(G)$. Setting $p = \alpha n/e$ and simplifying, we obtain

$$cr_3(G) \geq \left(\frac{\alpha - 6.5}{\alpha^5} \right) \frac{e^5}{n^4} + \frac{20e^6}{\alpha^6 n^6} \tag{2}$$

© Springer Nature Switzerland AG 2019
D. Archambault and C. D. Tóth (Eds.): GD 2019, LNCS 11904, pp. 612–614, 2019.
https://doi.org/10.1007/978-3-030-35802-0

for graphs satisfying $e \geq \alpha n$ since the probability p must be at most 1. The value of α maximizing $(\alpha - 6.5)/\alpha^5$ is 8.125, which implies that $cr_3(G) \geq (1.625/8.125^5)e^5/n^4 + (20/8.125^6)e^6/n^6$ for graphs satisfying $e \geq 8.125n$.

In this paper, we are particularly interested in the values of $cr_3(K_n)$ where K_n denotes the complete graph on n vertices. For $n \leq 10$, K_n is known to be simple quasi-planar [3, 5] and therefore we have $cr_3(K_n) = 0$. For $n = 11$, we obtain $cr_3(K_{11}) \geq 3.5$ from (1). Therefore, there must be at least four triples of pairwise crossing edges in any simple drawing of K_{11}. In the following, we present a drawing (Fig. 1d) which shows that $cr_3(K_{11}) = 4$. In each of the Figures a-d, the triples of pairwise crossing edges introduced in the figure are marked with red circles.

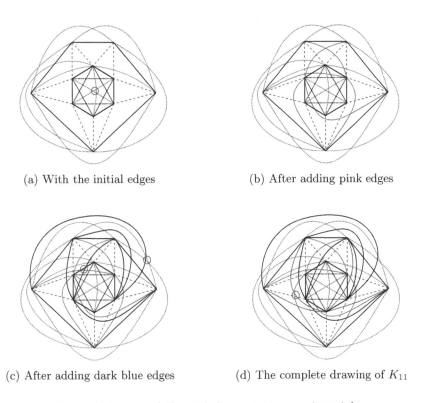

(a) With the initial edges   (b) After adding pink edges

(c) After adding dark blue edges   (d) The complete drawing of K_{11}

Fig. 1. A drawing of K_{11} with four pairwise crossing triples

For $n \geq 12$, we are not aware of the exact values of $cr_3(K_n)$. For each such n, the best known lower bound can be obtained from one of the inequalities (1) and (2). It is an open problem to obtain non-trivial upper bounds on $cr_3(K_n)$ for $n \geq 12$. Another open problem is to find a general drawing that provides a good upper bound on $cr_3(K_n)$ for large n.

References

1. Ackerman, E., Tardos, G.: On the maximum number of edges in quasi-planar graphs. J. Comb. Theory Ser. A. **114**(3), 563–571 (2007)
2. Agarwal, P.K., Aronov, B., Pach, J., Pollack, R., Sharir, M.: Quasi-planar graphs have a linear number of edges. Combinatorica **17**(1), 1–9 (1997)
3. Aichholzer, O., Krasser, H.: The point set order type data base: a collection of applications and results. In: Proceedings of 13th CCCG, Waterloo, Ontario, Canada, pp. 17–20 (2001)
4. Ajtai, M., Chvátal, V., Newborn, M.M., Szemerédi, E.: Crossing-free subgraphs. Theory and practice of combinatorics, North-Holland Mathematics Studies, 60, North-Holland, Amsterdam, MR 0806962, pp. 9–12 (1982)
5. Brandenburg, F.J.: A simple quasi-planar drawing of K_{10}. In: Proceedings of 24th GD, Athens, Greece, pp. 603–604 (2016)
6. Leighton, T.: Complexity Issues in VLSI. Foundations of Computing Series. Cambridge, MIT Press (1983)

Minimising Crossings in a Tree-Based Network

Jonathan Klawitter[1]([✉]) [iD] and Peter Stumpf[2] [iD]

[1] School of Computer Science, University of Auckland, Auckland, New Zealand
jo.klawitter@gmail.com
[2] Faculty of Computer Science and Mathematics,
University of Passau, Passau, Germany
Peter.Stumpf@uni-passau.de

Abstract. A tree-based network N is a rooted graph that has a spanning tree T which is the subdivision of a rooted binary tree. We show that crossing minimisation for drawings of N is either NP-hard or polynomial time solvable depending on the drawing style used for non-T edges.

1 Introduction

Phylogenetic trees and *networks* are rooted, leaf-labelled graphs used to model and visualise the evolutionary history of a set of taxa, e.g. species or languages [3]. A *tree-based network* N is a phylogenetic network with a spanning tree T that is the subdivision of a binary phylogenetic tree [1]. For fixed T, we refer to edges not covered by T in N as *cross edges*. We assume that vertices of N are assigned a height such that leaves have height zero and otherwise only two endpoints of a cross edge may have the same height. We only consider drawings of N where T is planar. Motivated by examples in the literature [4, 5] we identified the following drawing styles for cross edges. First, cross edges may be drawn x-monotone such that they are *horizontal, snakes* (i.e. have to bends), *curves*, or *straight lines*. Second, we consider the *ear* and *ear** drawing styles where a cross edge (u, v) is drawn with two bends such that the vertical segment is to the right of the subtree containing u and v or of T, respectively. See Fig. 1(a)–(d) for examples.

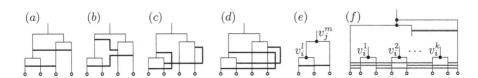

Fig. 1. (a)–(d): The horizontal, snake, ear, and ear* drawing styles, respectively; (e), (f): edge and vertex gadgets.

D. Archambault and C. D. Tóth (Eds.): GD 2019, LNCS 11904, pp. 615–617, 2019.
https://doi.org/10.1007/978-3-030-35802-0

2   Horizontal Drawing Style

Theorem 1. *Crossing minimisation for a tree-based network with the horizontal drawing style is NP-complete.*

We sketch a polynomial-time reduction of an instance G of MAX-CUT to an instance of crossing minimisation of a tree-based network N. Assume without loss of generality that G contains no vertex of degree 1. Sort $V(G)$ in an arbitrary order, say $v_1, v_2, \ldots, v_n$.

We build the base tree T of N as follows. A gadget for each edge $\{v_i, v_j\}$, $i < j$, as shown in Fig. 1(e), where v_i^l (v_j^m) is a representative of v_i (resp. v_j), induces zero crossings if v_i, v_j are rotated differently and one crossing otherwise. Thus it represents whether the edge is in the cut or not. We connect the edge gadgets with a binary tree. For $i \in \{1, \ldots, n\}$, let $v_i^1, \ldots, v_i^k$ be all vertices corresponding to $v_i \in V(G)$. We give v_i^l the height $i + l\epsilon$ for sufficiently small ϵ. Furthermore, to force them to all have the "same" rotation we add two vertices above the so far build tree and bundles of horizontal edges as shown in Fig. 1(f). Further bundles of horizontal edges fix the part of T that connects the edge gadgets and the upper part of vertex gadgets.

Note that several of the drawing styles can be reduced to the horizontal drawing style, which gives us the following corollary.

Corollary 1. *Crossing minimisation for a tree-based network with the snake, monotone curve, or straight line drawing style is NP-complete.*

3   Ears Drawing Style

Theorem 2. *Crossing minimisation for a tree-based network N on n vertices and k cross edges drawn with the ears drawing style can be solved in $\mathcal{O}(nk)$ time.*

Let v be a vertex of T. The key observation is that the rotation of v only determines crossings of edges from its left subtree through its right subtree or vice versa. Thus the best rotation can be determined independently for each vertex of T. The following algorithm runs in $\mathcal{O}(nk)$ time. First we determine the lowest common ancestor (lca) for each pair of endpoints of a cross edge in $\mathcal{O}(n)$ time [2]. Then we sweep from the leaves of T towards the root. At every endpoint v of a cross edge we determine for every vertex u of T the width of its left and right subtree at the height of v in $\mathcal{O}(n)$ time. Then from v up to its cross edge lca, we add up for each vertex of T the width of the subtree not containing v to a counter. When reaching a vertex u of T, we can decide its best rotation based on this counter. Lastly, we extend a partial order of groups of nested vertical segments to a total order to minimise crossings between cross edges.

Adjusting this algorithm to propagate the height up to the root instead of only to the lca, we get the following corollary.

Corollary 2. *Crossing minimisation for a tree-based network N on n vertices and k cross edges drawn with the ears* drawing style can be solved in $\mathcal{O}(nk)$ time.*

Acknowledgements. The first author thanks the New Zealand Marsden Fund for their financial support. We thank the anonymous referees.

References

1. Francis, A.R., Steel, M.: Which phylogenetic networks are merely trees with additional arcs? Syst. Biol. **64**(5), 768–777 (2015). https://doi.org/10.1093/sysbio/syv037
2. Gabow, H.N., Tarjan, R.E.: A linear-time algorithm for a special case of disjoint set union. J. Comput. Syst. Sci. **30**(2), 209–221 (1985). https://doi.org/10.1016/0022-0000(85)90014-5
3. Huson, D.H., Rupp, R., Scornavacca, C.: Phylogenetic Networks: Concepts, Algorithms and Applications. Cambridge University Press (2010). https://doi.org/10.1093/sysbio/syr055
4. Kumar, V., Lammers, F., Bidon, T., Pfenninger, M., Kolter, L., Nilsson, M.A., Janke, A.: The evolutionary history of bears is characterized by gene flow across species. Sci. Rep. **7**(1) (2017). https://doi.org/10.1038/srep46487
5. Vaughan, T.G., Welch, D., Drummond, A.J., Biggs, P.J., George, T., French, N.P.: Inferring ancestral recombination graphs from bacterial genomic data. Genetics **205**(2), 857–870 (2017). https://doi.org/10.1534/genetics.116.193425

Crossing Families and Their Generalizations

William Evans and Noushin Saeedi[✉]

University of British Columbia, Vancouver, Canada
{will, noushins}@cs.ubc.ca

Introduction. Let P be a set of n points in general position in the plane. A collection of line segments, each joining two of the points, is called a *crossing family* if every two segments intersect internally. Let $\mathrm{crf}(P)$ denote the size of the maximum crossing family in P, and let $\mathrm{crf}(n) = \min_{|P|=n} \mathrm{crf}(P)$, where the minimum is taken over all n-point sets P in general position in the plane. Aronov et al. [1] studied the size of $\mathrm{crf}(n)$. They noted that a set of n points chosen at random in a unit disc "almost surely" has a linear-sized crossing family and that there are point sets whose maximum crossing family uses at most $\frac{n}{2}$ of the points. They proved that any set of n points contains a crossing family of size at least $\Omega(\sqrt{n})$. Very recently, Pach et al. [4] improved the bound to $\Omega(n/2^{O(\sqrt{\log n})})$. It is conjectured that $\mathrm{crf}(P) = \Theta(n)$.

Our Results. We improve the upper bound on $\mathrm{crf}(n)$. We also study combinatorial and geometric generalizations of crossing families, and give several lower and upper bounds on their sizes.

A point set A *separates* point set B from C if A and $B \cup C$ are separable by a line and every line through two points in A has all of B on one side and all of C on the other side. We show that all segments of any crossing family in a point set $A \cup B \cup C$, where A separates B from C, emanate from one set (A or B or C). We exploit this property among subsets of points to design a template for constructing n-point sets with maximum crossing family of size[1] at most $5\lceil \frac{n}{24} \rceil$.

We study a relaxation of crossing families with bi-coloured segments in two-coloured point sets. Two bi-coloured segments are *side compatible* if each pair of same-coloured endpoints are on the same side of the line joining the other two endpoints—the segments with same-coloured endpoints are "parallel" (cannot be crossing or "stabbing"). A *side compatible family* is a set of bi-coloured segments that are pairwise side compatible. Any pair of crossing bi-coloured segments are side compatible but a side compatible pair may be crossing or parallel (but not stabbing). We consider sets of red and blue points such that all red points are on the same side of any line through two blue points, and call them 1-*avoiding* point sets. We study the characteristics of side compatible families in 1-avoiding

---

An extended version of this work appears in arXiv:1906.00191 [3]. This work was funded by an NSERC Discovery grant and in part by the Institute for Computing, Information and Cognitive Systems (ICICS) at UBC.

[1] Pach et al. [4] cite an $n/5$ upper bound by Aichholzer by personal communication in their recent work.

© Springer Nature Switzerland AG 2019
D. Archambault and C. D. Tóth (Eds.): GD 2019, LNCS 11904, pp. 618–620, 2019.
https://doi.org/10.1007/978-3-030-35802-0

point sets in combinatorial terms and in the dual plane. Note that in the dual plane, all intersections of blue lines are on the same side of all red lines.

We abstract the combinatorial information of the dual line arrangement using a "bar stack", which represents intersections between red lines, and "wires", which are horizontal lines that represent blue lines. A *bar stack* $\mathcal{B}_{l,n}$ is a set of l bars (horizontal segments) $B_1 \ldots B_l$ such that $B_i = [(a_i, i), (b_i, i)]$, where $1 \leq a_i < b_i \leq n$, and no two bars have the same horizontal interval. Bar B_i can be used to represent the intersection of red lines numbered a_i and b_i and represents the i-th intersection from below among red lines. The pair of a bar stack and wires encode where the intersections of red lines are with respect to the blue lines. While a bar stack can represent the intersections of a line arrangement, it is more general. We refer to the vertical lines through the endpoints of bars as *pillars* (which represent red lines). A *marble* (point) at the intersection of a wire and a pillar represents a bi-coloured segment. A side compatible family is visualized, roughly speaking, as a set of marbles such that for every bar both of whose pillars contain a marble, both marbles are above or both below the bar. Using counting arguments, we show that any bar stack $\mathcal{B}_{n,n}$ together with a set of $n+1$ wires such that there is a bar between every two consecutive wires has a side compatible family of size n. In particular, we show that there exists a "side compatible marbling" such that the marble associated with an endpoint of a bar is below the bar if and only if the bar is above some fixed wire. Our proof can be generalized to get a linear-sized side compatible family for any set of n wires.

We also study geometric generalizations of crossing families. A *spoke set* [2, 5] for a point set P is a set L of pairwise non-parallel lines such that each open unbounded region defined by L has at least one point of P. If $\text{crf}(P) = k$, the size of the largest spoke set for P is at least k (infinitesimal clockwise rotation of the supporting lines of segments in a crossing family yields a spoke set). Schnider [5] studied spoke sets in the dual plane. We introduce a generalized notion for the dual of a spoke set. A pseudoline ℓ is *monotonically semialternating* or *M-semialternating* in a bi-coloured line arrangement if the level of every other cell it visits is a non-decreasing sequence, and between these cells it crosses a distinct line of each color. If a subarrangement $\mathcal{A} \subseteq \mathcal{L}$ admits an M-semialternating pseudoline, $\mathcal{L}$ has an M-semialternating path of size $|\mathcal{A}|$. The dual of a spoke set defines a special M-semialternating path for which the level sequence is constant.

A bi-coloured line arrangement $\mathcal{L}$ is *color-separable* if the vertical line to the left of all intersection points of $\mathcal{L}$ intersects all of one then all of the other color. We show (by rotation) that the minimum sizes—over all color-separable arrangements of n lines—of the largest M-semialternating paths with constant and strictly increasing level sequences are the same (for a fixed arrangement they may be different). M-semialternating paths with different level sequences in a line arrangement correspond to sets of lines with different properties with respect to the dual point set. We exploit this correspondence and the connection between the sizes of different M-semialternating paths to improve the upper bound on the size of spoke sets from $\frac{9n}{20}$ to $\frac{n}{4} + 1$. The upper bound is constant if we consider M-semialternating lines rather than pseudolines. Our proof technique (duality and rotation) together with recursive use of hamsandwich cuts, implies that the largest set of pairwise stabbing or crossing segments in any set of n points is $\frac{n}{2}$.

References

1. Aronov, B., Erdős, P., Goddard, W., Kleitman, D.J., Klugerman, M., Pach, J., Schulman, L.J.: Crossing families. Combinatorica **14**(2), 127–134 (1994)
2. Bose, P., Hurtado, F., Rivera-Campo, E., Wood, D.R.: Partitions of complete geometric graphs into plane trees. In: Pach, J. (ed.) GD 2004. LNCS, vol. 3383, pp. 71–81. Springer, Heidelberg (2005). https://doi.org/10.1007/978-3-540-31843-9_9
3. Evans, W., Saeedi, N.: On problems related to crossing families. arXiv:1906.00191, June 2019
4. Pach, J., Rubin, N., Tardos, G.: Planar point sets determine many pairwise crossing segments. arXiv:1904.08845v1, April 2019
5. Schnider, P.: A generalization of crossing families. arXiv:1702.07555, February 2017

Which Sets of Strings Are Pseudospherical?

R. Bruce Richter and Xinyu L. Wang$^{(\boxtimes)}$

Department of Combinatorics and Optimization, University of Waterloo,
Waterloo, Canada
{brichter, x654wang}@uwaterloo.ca

The idea of pseudospherical drawings comes from a natural generalization of pseudolinear drawings, where each edge extends to a pseudoline and every pair of pseudolines crosses exactly once. The geometry of these drawings can be applied to techniques for crossing numbers, for example, to prove special cases for the Harary–Hill conjecture about the crossing number of K_n. Pseudospherical arrangements can be thought of as drawings where each edge extends to a simple closed curve, every pair of which intersect exactly twice, and no edge crosses any of the simple closed curves more than once.

A **string** is a simple bounded arc. We consider collections of strings embedded in $\mathbb{S}^2$. Such a collection is denoted by Σ, which will be extended at each string from their endpoints to **pseudocircles**, simple closed curves. A **pseudospherical arrangement** is an extension of Σ to pseudocircles where every string e is extended to a pseudocircle γ_e so that:

PS1: For each string e, no vertex except an endpoint of e is contained in γ_e.
PS2: For distinct e, f, $|\gamma_e \cap \gamma_f| = 2$, and all intersections are crossings.
PS3: For any edge e, if its endpoints u and v are contained in the closure Δ of one of the components of $\mathbb{S}^2 \setminus \gamma_e$, then $e \subset \Delta$.

For any cycle, vertices (where strings intersect) on the cycle are categorized as either rainbow or reflecting: v is **reflecting** in C if two edges incident to v from inside C or on C belong to the same string, and v is **rainbow** in C if all edges incident to v from inside or on C belong to distinct strings [1].

When drawing K_n as a pseudospherical arrangement, each side of a great circle induces a pseudolinear drawing of a smaller complete graph [2]. There is a known classification of obstruction cycles where a set of strings fails to be pseudolinear by Arroyo et al. if and only if the obstruction cycle has at most two rainbow vertices [1]. These cycles turn out to be related to pseudocircular obstructions except for one type of cycle, which we will detail below.

Every collection of strings with a pseudolinear extension also has a pseudospherical extension which can be obtained from the former by contracting the ends of the pseudolines into a point and perturbing the edges.

There exist four types of pseudolinear obstruction cycles. We will refer to cycles with no rainbow vertices as **clouds**, to those with one rainbow vertex as **fish**, to those with two adjacent rainbow vertices as **shrubs**, and to those with two non-adjacent rainbow vertices as **croissants**. The last type turns out to be different than the previous ones in the context of pseudocircles.

© Springer Nature Switzerland AG 2019
D. Archambault and C. D. Tóth (Eds.): GD 2019, LNCS 11904, pp. 621–623, 2019.
https://doi.org/10.1007/978-3-030-35802-0

Theorem 1. *Every cloud, fish, and shrub type cycle has some pseudocircle extended from one of its strings contained entirely in the cycle's closure in any pseudospherical extension. However, croissant type cycles can have all their strings extended into a pseudospherical arrangement so that each pseudocircle has some portion outside the cycle.*

We will label cloud, fish, and shrub type cycles as **(pseudospherical) obstruction cycles**. From the above result we deduce that if we have two disjoint obstruction cycles in some set of strings Σ, then Σ has no pseudospherical extension.

This condition is not necessary for collections of strings without pseudospherical extensions. If every string on an obstruction cycle is intersected by some string with both ends outside the cycle, then we cannot find a pseudospherical extension for the set of strings, since some pseudocircle is in the cycle and contains some part a the string with both ends outside the cycle, which violates **PS3** (Fig. 1).

Theorem 2. *Every cloud type cycle has at least 4 pseudocircles contained entirely in its interior in any pseudospherical extension (Fig. 1(left)).*

Theorem 3. *Every fish or shrub type cycle has at least 2 pseudocircles contained entirely in its interior in any pseudospherical extension (Fig. 1(right)).*

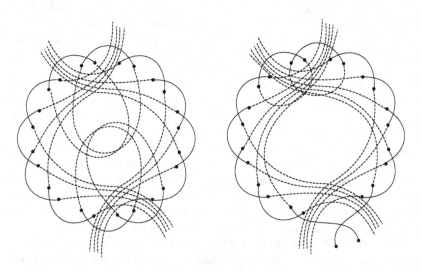

Fig. 1. Examples of extending cloud (left) and fish (right) cycles pseudospherically containing a minimum number of pseudocircles inside the cycle.

Question: Is there an excluded minor characterization of pseudospherical drawings along the lines of the pseudolinear characterization?

References

1. Arroyo, A., Richter, R.B., Sunohara, M.: Extending drawings of complete graphs into arrangements of pseudocircles (Submitted)
2. Arroyo, A., Bensmail, J., Richter, R.B.: Extending Drawings of Graphs to Arrangements of Pseudolines (Submitted)

Author Index